M. Nicolle

Kate-

It has been a pleasure working with you. May this book help you on your new path. I am always here as a mentor & friend.

Michelle

HISTOPATHOLOGIC METHODS AND COLOR ATLAS OF SPECIAL STAINS AND TISSUE ARTIFACTS

LEE G. LUNA

Library of Congress Catalog Card Number 92-070140

Printed by Johnson Printers, Downers Grove, IL

Preface

This book on histopathologic methods is unique in several ways. First, the book deals strictly with histopathologic methods which are directed at the bench technologist. The book does not contain chapters for performing high resolution microscopy (plastic) sections, immunochemistry, electron microscopy, histochemistry, equipment, safety or computers in the laboratory, although it is recognized that these are all parts of the field of histotechnology armamentarium. The reason for eliminating these particular subjects is justified on the following basis. It is felt that to include chapters on these subjects one would provide insufficient information for the novice or the experienced individual. More likely it would simply be a wasted effort. A fair number of pages would be occupied providing information that would provide little value. At the present time there are many excellent books on these subjects which would better serve the novice and experienced person. In addition, a great deal of information is constantly being disseminated through various newsletters, lectures and other means of communication on equipment and safety. The absence of any information related to computers in the histopathology laboratory is simply due to the fact that not many people in histopathology currently use or are familiar with computer operations. It is my feeling that the field of histotechnology has a lot to offer and is in fact an emerging science. I think, therefore, that this specialty deserves its own book. I'm sure the number of pages and the information contained in this book will justify this decision. It is strongly recommended, to those that are interested in the subjects not contained in this book, that efforts be made to seek the information through various means available to us. I want to express my appreciation to the following individuals for their dedication, hard work, time and effort in helping to produce this book: Darryl Luna, Nancy Luna and Alita Luna who typed the book, produced the graphics and assisted me in photographing many of the color photographs which are contained in this book. My appreciation is also extended to Jean Verfuerth for editing the book and Charles West, Roberta Mosedale and Lamar Jones for their assistance in performing the final review and suggestions made during the development and writing of this book.

The use of xylene in this book and other chemical hazards.
The use of xylene as a decerating and clearing agent in this book in no way suggests strong endorsement of this agent. It is used in this book because I have not found other clearing agents as effective as xylene. This is primarily because not one of the many commercially available replacements have been universally accepted by laboratories, including a good replacement for xylene. Because of this, every laboratory has to decide on their preference for a clearing agent.

Dedication

This book is dedicated to: my wife Iris; son Randall, his wife Beverly; daughter Alita, her husband Michael; son Darryl, his wife Nancy; my two grandchildren, Natalie and Adrian. This finished product was made possible through their unselfishness, invasion of time, patience in a very difficult situation of illness, emotional strength and support.

In addition, it is my desire to dedicate this book to Ricky R. Goss, born in Dallas, Texas on June 28, 1954. It was my pleasure to meet Ricky at the San Antonio National Society for Histotechnology Symposium/Convention in 1990. At the time of our meeting I was unaware that he had been diagnosed as a carrier of the HIV virus. Ricky exhibited an abundance of enthusiasm toward his career in histology. I was very impressed with his attitude and feel this was particularly significant as he no doubt was under a great deal of mental and emotional stress as the results of the HIV infection. His career as a histotechnologist has spanned 20 years, most of which have been in the Dallas area hospitals.

The dedication of this book to him is also a means of indicating my strong support to others who may be facing similar adversities.

Lee G. Luna

Lee G. Luna

Lee G. Luna, D.Lit., (Hon.) H.T. (ASCP)

July 9, 1931 - February 27, 1992

In memory of a loving husband, a devoted, caring father and grandfather, we acknowledge the tremendous dedication, hours of research, channeled discipline and perseverance which enabled Lee to meet the challenge of reaching his ultimate goal of completing this book. He was not one to waver and even in the midst of adversity he weathered the storm of a devastating illness with an abundance of inner strength.

He was very special and with humility thrived on being a teacher, listener, mentor and a friend to the histotechnology community. His philosophy of life was one of integrity and sharing. He desired for all to be intently involved and to excel. As a result of his countless contributions he was honored with many tributes and awards which he gratefully accepted and treasured.

Lee's moral, spiritual commitment and love for his family held first priority in his life. He was a perfect role model in every way and our time together was "quality time". He truly possessed the ability to live life to its fullest.

Lee was a crusader who trusted his instincts, had intuition and courage, was brave and became a giant filled with wisdom. He looked not to the right nor to the left, but straight ahead, keeping his eyes fixed on his goals. Many dreams came true.

Now in this "Decade of Progress" as we look into the scope of the future of histotechnology it is hoped that this book be a reservoir of knowledge for the advancement of the profession he dearly loved.

"Lee carved his name on many hearts"

The Luna Family

Chapter 1

Agents and Methods for Specimen Fixation

Definition of Fixation

The word fixation is derived from the Latin word fixatio (to fix) and means the act or operation of holding, suturing, or fastening in a fixed position; also, the condition of being held in a fixed position. In histotechnology, fixation is the treatment of material so that its structure may be examined in greatest detail with minimal alteration of the normal state. In some cases the purpose of fixation is to provide information concerning the chemical properties of protoplasmic substances by interpretation of fixation reactions.

A fixative is an agent (e.g., formaldehyde) employed in the preparation of histologic or pathologic specimens for the purpose of maintaining the existing form and structure of all their constituent elements[1]. A good fixative is one that changes the cell chemistry the least and, at the same time, preserves the cell structure the best.

A fixative solution is a homogeneous mixture of one or more agents (solute) dispersed molecularly in a sufficient quantity of dissolving medium (solvent). The agent may be a gas (formaldehyde), liquid (alcohol), or solid (mercuric chloride). The solvent for most fixatives is usually a liquid, most often water or alcohol.

Functions of Fixation

The functions of fixation are many and, to a great extent, the anticipated specific results depend on the action of a given fixative agent on specific protoplasmic constituents. Thus, we can consider the actions of all fixing agents in only general terms when we discuss the functions of fixation. Those functions are:

To Preserve

- tissue as near to lifelike as possible, thereby retaining prefixation microanatomical tissue constituent characteristics.
- protoplasmic products (nucleic acids, enzymes, etc.) which can be demonstrated by subsequent staining procedures.
- stainability of the different parts of the cell.

To Prevent

- disruption of tissue constituents, (Fig. 1).
- distortion or dissolving of cell parts.
- tissue chemical alterations.
- tissue from drying out.
- tissue from swelling after death.
- tissue from overhardening (which inhibits good processing and microtomy).
- growth of mold.

- invasion of saprophytes (organisms such as bacterium which live on decaying or dead tissue), (Fig. 2).
- hardening of the naturally soft tissue, thereby allowing easy non disturbant manipulation during subsequent steps.

To Reduce

- potential shrinking of tissue when exposed to subsequent steps in tissue processing.
- autolysis (self digestion of tissue by enzyme activity), (Fig. 3).
- putrefaction (enzyme decomposition, especially of proteins, with the production of foul-smelling compounds such as hydrogen sulfide and ammonia).

To satisfactorily carry out the process of fixation and achieve the above objectives, one should take into account several important details before, during and subsequent to fixation.

Before fixation it is important to determine the purpose for which the tissue is to be preserved, the pathologic and/or normal entities one wishes to demonstrate, and the processing for staining techniques to be used. Also, one must consider the thickness and size of the specimen, the volume of fixative needed, the rate of fixative penetration, and the duration of tissue exposure to the chosen fixative. All of these factors will influence the choice of fixative to be used.

Likewise, during fixation it is important to consider the length of fixation, the frequency of solution changes needed to insure proper results, the replacement of any fixative solution which does not look correct, and the addition of a fixative accelerator (heat, agitation, aeration, vacuum, etc.) if the need dictates. (The use of accelerators and changing of the fixative solution are very important if one is dealing with large or thick specimens.) The location and temperature of the formalin fixative solution must also be considered, as ultra violet light (sunlight) will accelerate oxidation of formaldehyde to formic acid in non-buffered fixative solutions and the pH of some fixative solutions changes with the temperature of the solution. (A formalin fixative solution which is neutral at 20°C will be pH 6.8 at 38°C.) Furthermore, while fixatives proceed more rapidly at higher temperatures, so do postmortem changes. Ultra violet light and/or temperature may have other deleterious effects on other fixatives; thus, the above remarks should apply to all fixative mixtures.

Subsequent to fixation it is important to determine whether the tissue was fixed properly (additional fixation exposure time may be in order) and if the fixative used or specimen fixed requires additional special treatment.

Osmolarity and Fixation[2]

The osmolarity (osmotic pressure) of a fixative solution plays an important part in the expansion/contraction action of cellular constituents. It is for this reason that the following information on osmotic pressure of solutions is provided.

It is instructive to define some terms used commonly in discussing the subject of osmolarity. Molarity (mol/l): Amount of substance in moles divided by the volume of the mixture (solution). Molality (mol/kg): Amount of substance in moles divided by the mass of the solvent. Osmolarity (Osmol/l): The molarity that an ideal solution of a nondissociating substance must possess in order to exert the same osmotic pressure as the test solution. Osmolality (Osmol/kg): The molality that an ideal solution of a nondissociating substance must possess in order to exert the same osmotic pressure as the test solution.

The term osmolarity is used here strictly with regard to the response of cells immersed in a solution. Solutions should be defined in osmolarity measurements because they are independent of temperature.

Explaining Osmolarity

The sum of the atomic weights in the formula of a substance is its formula weight. Formula weight of sulfuric acid (H_2SO_4) is (2x1.0) + (1x32.1) + (4x16.0) = 98.1. The formula weight of a substance expressed in grams is a gram-formula weight, or a mole of that substance. Thus, the gram-formula weight of sulfuric acid is 98.1 grams, whereas its formula weight is simply 98.1. When the true formula for a molecule of a substance is known, the molecular weight can be calculated by adding the atomic weights of the atoms given in the formula of the substance. The use of the term "mole" means a gram-formula weight of ionic compounds and a gram-molecular weight of molecular substances (either elements or compounds). It is often convenient in quantitative chemical work to use millimoles in place of moles of a substance. A millimole of a substance is its formula weight expressed in milligrams. For example, the formula weight of H_2SO_4 is 98.1, one mole is 98.1 gm, and one millimole is 98.1 mg. In other words, one millimole is one thousandth of a mole.

Osmotic Pressure of Solutions

When a solution and its pure solvent are separated by a semipermeable membrane (one through which the solvent, but not the solute, passes), the pure solvent will diffuse through the membrane and dilute the solution. This process is known as osmosis. Actually, the solvent diffuses through the membrane in both directions simultaneously; however, the rate of flow is greater from the pure solvent to the solution. The force causing diffusion of liquids is called osmotic pressure. (Osmotic pressure is defined as the amount of pressure that must be applied to prevent the flow of water, or osmosis, through the membrane). Osmotic pressure depends only on the number of solute particles in the solution. The greater the concentration of solute, the greater the pressure that must be applied to prevent osmosis. Osmotic pressure may be measured in the following way. A piece of differentially permeable membrane is placed in a container and one side is filled with a concentrated sugar solution, the other side with water. As water flows through the membrane into the solution, a slow rise of liquid in the container will be observed. If the membrane is sufficiently strong to withstand the pressure and if the container is long enough, the liquid will rise until its hydrostatic pressure is equal to the osmotic pressure. When this condition is achieved, the vapor pressure of the water in the solution under pressure equals the vapor pressure of the pure water outside the container; a condition of equilibrium. The height to which the liquid will rise depends upon the concentration of the solution; thus, the osmotic pressure is proportional to the number of solute particles in a definite volume of liquid (Fig. 4a, b and c).

The osmolarity (tonicity) of a fixative has a direct effect on the appearance of the fixed specimen. Cell size and shape may be affected by a change in buffer osmolarity. It appears, though, that various tissue types differ in the degree of their response to the difference in osmolarity between the fixative solution and the cell's normal external environment. The effect produced is probably partly correlated with "density" or compactness of the tissue. The effect of osmolarity generally appears to be minimal on relatively compact or hard tissues, whereas osmolarity exerts a marked effect on "less compact" tissues. It should be noted that any tissue specimen having more than one cell type may require different osmolalities of the fixative for each cell type. For most mammalian tissues, the recommended total osmolality ranges from 500 to 700 mosmols, whereas the vehicle osmolality is 300 mosmols. Lower total osmolalities (300-350 mosmols) are desirable for delicate specimens.

Isotonic, Hypertonic and Hypotonic Fixative Solutions

Isotonic fixative solutions: A term denoting a solution in which body cells can be bathed without a net flow of water across the semipermeable cell membrane; also denoting a solution having the same tonicity (isosmotic pressure) as some other solution with which it is compared, such as physiological salt solution and blood serum. An isotonic solution exerts an osmotic pressure equal to that exerted by cell cytoplasm. A fixative solution is isotonic with a cell (or tissue specimen) if the cell neither shrinks nor swells when immersed in it. The two solutions are isosmotic with each other if both exert the same osmotic pressure; in other words, both solutions possess the same solute particle concentration. An isosmotic fixative solution may or may not be isotonic (Fig. 5a).

Hypertonic fixative solution: A term denoting a fixative solution which, when bathing body cells, causes a net flow of water across the semipermeable cell membrane out of the cell. The term also denotes a fixative solution having greater tonicity (osmotic pressure) than another solution, e.g., the blood, with which it is compared (Fig. 5b).

Hypotonic fixative solution: A term denoting a solution which, when bathing body cells, causes a net flow of water across the semipermeable cell membrane into the cell. Also, denoting a solution having less tonicity (osmotic pressure) than another solution, e.g., the blood, with which it is compared (Fig. 5c).

Characteristics of Agents Used as Fixatives

There are good and bad fixative agents for the demonstration of any given tissue constituent. For example, neither the suggestion that alcohol is an "ideal fixative" for chemical studies, nor the notion that formaldehyde, osmic acid, and chromates hardly affect the valence and the molecular interrelationship of tissue constituents, nor even the notion that freeze-drying is the best fixation method for use in cytochemistry, can be regarded as accurate. Most agents cause chemical changes in the protein, and the choice of any one of them as a fixative should depend on the specific aim of the investigator[1].

The process of fixation is further complicated by the fact that fixing agents need time to reach the different tissue structures. Many studies have revealed that rapid fixation of the external layers interfered with the penetration of fixatives to internal structures. Likewise, theoretical considerations make plausible the possibility of a barrier created by the fixed external layer just as penetration through a finely interlinked mesh would be expected to be slower than that through a looser structure. Nevertheless, it has been found that the penetration of liquid fixatives into a plasma clot obeys the physical laws of diffusion through a homogeneous system. Because of this diversity in the actions (or inactions) of various fixing agents, the characteristics of some agents used as fixatives are provided below.

Formaldehyde Gas

At ordinary temperatures formaldehyde is a flammable, colorless gas, having a pungent, suffocating odor. It is very soluble in water (up to 55%). A very reactive reagent, it combines readily with many substances and polymerizes easily in aqueous solutions.

Formaldehyde Solution (Formalin)

Formalin is a solution of formaldehyde gas in water, of about 37% formaldehyde by weight. This solution is the same strength as that known as formalin 40%, which signifies that it contains 40 gms formaldehyde in 100 ml of solution and is equivalent to 37% by weight. This formalin solution is treated as a 100% solution in making other "percent formalin" solutions (10 ml formalin + 90 ml water = 10% formalin). In the truest sense, this is approximately a 4% solution.

Formalin is a fairly rapidly penetrating fixative. It penetrates 2.5 mm in 4 hours. Because it oxidizes into formic acid, formalin should be buffered when used for fixation. Buffering prevents tissue staining properties from being altered.

Reactions of the formalin with tissue proteins are numerous and complex, but it is an accepted fact that formalin does not precipitate proteins and only lightly precipitates the other constituents of the cell. Formalin reacts readily with various functional groups of biological macromolecules in a cross-linking fashion such as with proteins, glycoproteins, nucleic acids and polysaccharides.

Formalin neither preserves nor destroys fat. It is good for complex lipids, but has no effect on neutral fats. However, long-term storage in formalin solutions has a marked influence on the composition of brain lipids.

Formalin forms methylene compounds with amino, amide, and hydroxy groups and therefore affects the solubility and reactivity of proteins. There is some suggestion that formalin blocks 1:2 glycol groups.

Formalin is not a fixative for carbohydrates. It preserves the proteins which, in turn, traps glycogen so that it is not easily dissolved.

Luna's Method for Detecting the Presence of Formalin (Aldehyde Groups)[3]

The following tests for the presence of formalin solutions is provided because of the extensive use of this agent in the histopathology laboratory. These methods may prevent potential health hazards or destruction of specimens. During the course of daily activities in a busy laboratory, one may inadvertently make a mistake in preparing 10% Formalin. Consequently, there often arises a need to check formalin specimen bottles to see if formalin has actually been added. Frequently, although a known health hazard, we check for the presence of formalin by smelling the solution to detect formalin fumes. To assist in detecting formalin and eliminating the associated health hazard of smelling, the following procedure may be used to determine the presence of formalin in a questionable solution.

Mix equal parts (2 to 5 mls) of Coleman's leucofuchsin with the solution being tested in a small glass or clear plastic container. The presence of formalin (aldehyde groups) will produce a purple color in the combined solution within a few seconds (Fig. 6a, b, c and d). The leucofuchsin solution should be kept refrigerated and used directly from the refrigerator. It is also suggested that a negative control be employed to ensure the reaction obtained is taking place due to the presence of formalin. Deionized water may be used for the negative control. It should be understood that any solution containing aldehyde groups such as formaldehyde (formalin), glutaraldehyde, paraformaldehyde, etc., will produce a positive reaction. It should also be noted that this is a qualitative and not a quantitative procedure.

Color Coding of Formalin Solutions[4]

Food coloring can be used to color code formalin solutions in the laboratory by utilizing the following procedure. When 10% neutral buffered formalin is made or whenever a container of commercial 10% neutral buffered formalin is opened in the histology laboratory, red food coloring (available at a grocery) is added to the formalin before use. One milliliter of the food color is added to each five gallons of formalin and mixed by shaking the container. The formalin then becomes slightly pink and is now "color coded". This color code exists so no one in the laboratory has to wonder if the solution on the specimen or in a given container is formalin. Specimen surfaces pick up a slight pinkish tint but this has not been a hindrance to the pathologists during gross description of the specimens. Staining artifacts are not produced as the slight tinting of the tissue surfaces washes out during the subsequent paraffin processing.

Glacial Acetic Acid

Glacial acetic acid causes the cell constituents to swell. It is therefore, never used alone but is combined with some other reagent having a shrinking effect. It penetrates rapidly, does not affect fats, and precipitates nucleoproteins, but dissolves most cytoplasmic proteins. It is most useful because of its good fixation of nuclei. However, acetic acid destroys mitochondria and the Golgi apparatus. It leaves tissues soft and inhibits hardening from the subsequent use of alcohols. Acetic acid is used in numerous fixative mixtures, including formalin, to eliminate a form of tissue vacuolization artifacts (see Artifact chapter).

Zinc Chloride

The following remarks refer to the use of zinc chloride as a substitute for mercuric chloride in Zenker's fixative. This salt is somewhat similar to mercuric chloride in that it penetrates uniformly and therefore the shrinkage of cell constituents is uniform. That is, all cell constituents are affected to a similar extent. The effect on tissue hardness are similar to mercuric chloride. The same can be said of the staining of nuclear chromatin. Excessive zinc chloride is soluble in the tissue processing agents and therefore is completely removed from tissue. This negates the need for further treatment of tissue for removal of artifactual metal, which is one of the problems with the use of mercuric chloride.

According to Russell[5], the staining qualities of all tissue constituents are similar to those of Zenker's fixative when mercuric chloride is used. Zinc chloride precipitates protein as well as mercuric chloride. Because of the similarities in tissue reactions, zinc chloride is recommended as a good substitute for mercuric chloride in Zenker's fixative solution.

Zinc Sulfate

Zinc sulfate used with formalin has been found very useful for preservation of a number of antigens[6].

Chromic Acid (Chromium Trioxide)

Chromic acid solutions are made by dissolving its anhydride, chromium trioxide, in water. Chromium trioxide absorbs moisture quite rapidly and therefore should be kept tightly sealed at all times. Specimens fixed in solutions containing chromic acid turn green because of chromium oxide which develops as the result of an oxidation process. Chromic acid coagulates proteins. It makes fat insoluble in fat solvents. Chromic acid is generally used in mixtures designated to fix nuclei, but it penetrates relatively slowly and fixes tissue unevenly at different levels. It has a tendency to shrink cytoplasm, especially when the fixative mixture is at a low pH. Specimens fixed in mixtures containing chromic acid undergo moderate hardening. An insoluble precipitate, chromium suboxide, may be produced when tissues are exposed to alcohols during tissue processing. This problem can be prevented by washing chromic acid-fixed specimens in running water for 2 hours prior to processing.

Ethyl Alcohol (Ethanol)

Ethyl alcohol is often used as a preservative for glycogen. In the process ethyl alcohol also precipitates nucleic acids and other proteins by denaturization (changing the nature of the protein without forming a new compound or adding other atoms). Ethyl alcohol causes collapse of protein structures; therefore, exposure of tissue must be limited to 8 hours or less[7]. When exposure is limited, nucleic acid is not precipitated and remains soluble in water. Ethyl alcohol makes nucleic staining somewhat difficult, however, it does not interfere with cytoplasmic staining. Ethyl alcohol extracts some lipids from tissue and causes considerable shrinkage and hardening.

Methyl Alcohol (Methanol)

This alcohol is probably one of the best fixatives for fixing films and impression smears on slides. It is also used for frozen cut sections. Like ethyl alcohol, methyl alcohol extracts some lipids from tissue; however, carbohydrates are not dissolved. This alcohol does not harden tissue as much as ethyl alcohol.

Acetone

This agent is occasionally used for fixation in special circumstances. Like ethyl alcohol, acetone can denature tissue if exposure is more than a few hours. Extraction of lipids is extensive and, like ethyl and methyl alcohols, acetone preserves carbohydrates. Acetone is extremely useful for fixation of blood smears for the preservation of leucocyte surface antigens.

Mercuric Chloride (Bichloride of Mercury, Corrosive Sublimate)

A simple solution of mercuric chloride produces a rapid and usually rather fine coagulation of most cytoplasmic and nucleic proteins, although nucleoproteins are coagulated less effectively than cytoplasmic ones. In solutions which are more acid than the isoelectric point of the proteins (i.e., fixatives containing acetic acid) the mercury is added to the proteins as an ion. This reactivity is facilitated by the presence of sodium chloride. However, when the solution of mercuric chloride is close to the isoelectric point of the proteins, sodium chloride interferes with the reactivity of mercuric chloride with the proteins. This action may even encourage the dissolution of the coagulum. On this account, mixtures of mercuric chloride and acetic acid are generally preferable to mercuric chloride alone. An exception occurs when one wishes to study some substance, such as mucin, which would be affected adversely by the acid.

Mercuric chloride has little effect on triglyceride lipids, although it is suspected that in cells treated with mercuric chloride these become less stainable with dyes of the Sudan group, than is the case with comparable cells fixed in formalin. Phospholipids form compounds with mercuric chloride, but the subsequent fate of these in solvents of lipids has not been studied.

Mercuric chloride penetrates with moderate rapidity, and its hardening effect is also moderate. By itself it induces little shrinkage, but tissue fixed in it will generally shrink considerably after passage through various processing solvents and melted paraffin. Nonetheless, material fixed in mercuric chloride has a fairly good cutting consistency.

Prolonged immersion will cause excessive mercury precipitates to form in the tissues, (Fig. 7a, b, c, d, e, f, g and h). The time required for fixation varies according to the strength of the solution, the temperature, and the size and permeability of the specimen. (Some suggestions as to the time necessary for fixing various types of specimens will be given in connection with each fixative mixture described later.)

Picric Acid (Trinitrophenol)

The yellow crystals of picric acid are sold in a slightly moist condition in order to assure safety in handling. (Picric acid is highly explosive in the dry state). Its solubility in water at room temperature approaches 1.5%.

Picric acid acts by coagulating proteins, and probably becomes combined with them. It generally causes formation of a network in both nucleus and cytoplasm. By itself, picric acid induces only moderate shrinkage, but tissue fixed in it becomes badly shrunken after passage through processing solvents and melted paraffin. It causes only very slight hardening, and this is one of its assets. It penetrates rather slowly. Picric acid seems to have little or no effect upon lipids. Mitochondria are preserved by picric acid, but the best-known picric acid mixtures (such as Bouin's solution) contain acetic acid or other acids which tend to destroy mitochondria.

Although the coagulates seem not to be dissolved by water, washing in alcohol rather than water is recommended. It is certainly a safe procedure, and there is some evidence that, in the case of cells from testis, ovary, bone marrow, and certain other delicate tissues, material washed in water yields inferior preparations.

Potassium Dichromate (Potassium Bichromate)

Potassium dichromate is valuable in fixative mixtures used for the fixation of lipids, particularly phospholipids. It preserves phosphotides and is useful for preserving mitochondria. Potassium dichromate, like formalin, is one of the major noncoagulant fixing agents. It does not fix chromatin unless the fixative solution has an acid pH, and for this reason, it is unsuitable for work involving nuclei and chromosomes. Acidified (acetic acid) potassium dichromate reacts in much the same manner as chromic acid solutions, in that it coagulates chromatin and cytoplasmic proteins as networks. Potassium dichromate will oxidize lipids if the tissue is exposed for an extended period (i.e., longer than 12 hrs). This oxidation prevents lipids from dissolving in alcohol and fat solvents during tissue processing. It hardens tissues slowly and confers upon them moderately good cutting consistency. Tissues fixed in potassium dichromate solutions shrink very little; however, a fair amount of shrinkage will occur subsequently during processing and paraffin impregnation. Tissues fixed in potassium dichromate solutions should be washed well after fixation and before processing. Generally, 2 to 3 hours of washing in running water is sufficient.

Fixative Mixtures

Some of the more commonly used fixative mixtures are provided below. There are some instances in which tissues should be washed in running water after the use of a given fixative before further tissue processing. This is particularly true if a formalin-based fixative is used. The pH of the various fixative mixtures has been included since the pH of a fixative plays an important part in the tinctorial effects produced by various staining procedures. (For example, unbuffered formalin may have a pH as low as 3.0 versus a pH of 6.9-7.1 for a buffered formalin solution). The use of formalin with a pH of 3.0 may result in poor nuclear staining. Some significant morphologic and chromatic alterations of tissue and/or staining reactions have been reported by Luna and Cawley[8].

10% Buffered Neutral Formalin Solution

Concentrated formalin solution (37%)	100.0	ml
Distilled water	900.0	ml
Sodium phosphate, monobasic	4.0	gm
Sodium phosphate, dibasic (anhydrous)	6.5	gm

This is the most useful fixative; therefore, strongly recommended for routine use. Formalin pigment is not produced when this fixative is used. Although it leaves tissues softer than most fixatives, tissue sectioning will not be affected if the tissue is properly dehydrated, cleared, and impregnated with paraffin. The buffering of formalin often produces artifactual vacuolization of tissue structures (Fig. 219 and 220). This is the result of the coalescence of nuclear and cytoplasmic materials. This problem can be easily overcome by treating the tissue with formalin alcohol-acetic acid solution (see page 11).

Formaldehyde gas when combined with water forms methylene glycol. In this aqueous solution the fixative reactions of methylene glycol are the same as those of formalin; therefore, one speaks of these reactions as if they contained formalin exclusively. Methylene glycol forms rather slowly (takes 2 to 3 days) in unbuffered neutral formalin solutions. For this reason unbuffered solutions should be made several days before used as a fixative. Conversely, depolymerization occurs very rapidly in buffered neutral formalin solutions and thus these solutions can be used immediately. Formalin's solid polymer is paraformaldehyde, which occurs with the continued addition of the spontaneously occurring methylene hydrate molecules. The paraformaldehyde can be seen as a white powder at the bottom of the glass container. Methyl alcohol is added to formaldehyde solutions (formalin) to prevent polymerization to paraformaldehyde.

Formalin (Unbuffered)

Concentrated formalin solution (37%)	100.0	ml
Distilled water	900.0	ml

This 10% formalin solution is widely used as a fixative but it must be recognized that some staining reactions and tissue elements may be altered, and others removed, by the deleterious and insidious effects of unbuffered formalin. Formalin is oxidized to formic acid as the result of Cannizzaro reaction, (a reaction which aldehydes may undergo when brought in contact with tissue specimens). In this reaction one molecule of the aldehyde formalin is reduced to the corresponding alcohol, methyl alcohol, and the other molecule is simultaneously oxidized to the corresponding acid, formic acid. The pH of the solution gradually falls as more formic acid is produced. Generally, one does not know how much formic acid is present in a concentrated (37%) formalin solution when a fixative solution is prepared. Therefore, one knows very little regarding the final pH of the fixative solution. The formation of formic acid may produce a number of ruinous effects on tissue if fixation exposure is extended beyond several hours.

Formalin-Sodium Acetate Solution

Concentrated formalin solution (37%)	100.0	ml
Sodium acetate	20.0	gm
Distilled water	900.0	ml

This is an excellent fixative in which to store gross blocks of tissue. I would choose this second only to buffered formalin fixative solution for routine tissue fixation. There is a tendency for the pH to change as tissue specimens are introduced into this solution. The staining qualities are very good after use of this fixative.

Formalin-Saline Solution

Concentrated formalin solution (37%)	100.0	ml
Sodium chloride	9.0	gm
Tap water	900.0	ml

This fixative is most useful for the preservation of carbohydrates. Because it fixes nuclear chromatin in a diffused homogeneous manner, it is impossible to visualize chromatin distinctly. Long storage does not create excessive hardening or damage to tissue, but may cause formation of formalin pigment.

Formalin-Ammonium Bromide Solution

Concentrated formalin solution (37%)	150.0	ml
Ammonium bromide	20.0	gm
Distilled water	850.0	ml

This is an excellent fixative for brain tissue on which the silver and gold techniques are to be performed.

Formalin-Alcohol-Acetic Acid Solution

Concentrated formalin solution (37%)	100.0	ml
Alcohol (80%)	900.0	ml
Glacial acetic acid	50.0	ml

This is a very rapid fixative which can be used for routine fixation. Tissue should not measure more than 3 mm in thickness and exposure time should be limited to 5 hours. Acetic acid insures fixation of nucleoprotein, providing an improved histological picture.

Baker's Formol-Calcium Solution

Concentrated formalin solution (37%)	100.0	ml
Calcium chloride, anhydrous	10.0	gm
Distilled water	900.0	ml

This fixative is especially useful when studies on lipids, phospholipids, and cholesterols are to be performed.

Zenker's Solution

Distilled water	1000.0	ml
Mercuric chloride	50.0	gm
Potassium dichromate	25.0	gm
Sodium sulfate	10.0	gm
Glacial acetic acid	50.0	ml

This mixture has been used since 1894, when Zenker suggested the addition of mercuric chloride to Muller's fluid to improve the fixation of nuclei. Tissues preserved by this method (or any of its modifications) stain well with many staining methods. Small pieces (3 mm in thickness) are usually fixed in 2 hours. After fixation wash tissue in running tap water for 2 hours.

It is no longer necessary to wait to add the acetic acid to Zenker's solution just before use. Acetic acid now may be added to the fixative mixture when it is compounded. Modern manufacturing methods have eliminated the manufacturing impurities which previously precluded its addition to Zenker's fixative. The mixture has a shelf life of three to five months.

Tissue fixed in Zenker's undergo three distinct changes during the process of fixation: (Fig. 8a, b, c and d).

In the first change, nuclei shrink drastically during the first several hours of tissue exposure to the fixative (1 to 3 hours). Microscopically the chromatin will appear shrunken and condensed.

Secondly, nuclei undergo some swelling (usually between 4 to 6 hours exposure to the fixative). At this point the nuclear chromatin is distinct. It is at this point of nuclear swelling that tissue should be removed from the fixative solution.

Thirdly, if tissue is allowed to remain beyond the point suggested above, the result will be monochromic staining of tissue, a condition known as "overzenkerization". That is, the nuclear chromatin and cytoplasm stain similarly. Microscopically the section will appear purple with no tinctorial differentiation between the nuclei and cytoplasm.

This fixative solution preserves nuclei, cytoplasm, and connective tissue elements quite well; but, owing to the presence of acetic acid, it destroys mucin, mitochondria, zymogen granules, and some other elements of the cytoplasm. In embryology it serves very well for the preservation of mammalian embryos.

The mercuric chloride contained in this fixative (or any mercuric chloride-containing fixative) deposits a precipitate (crystals) throughout the tissue specimens. This so-called Zenker's Crystals is deposited in a wide variety of colors and shapes, (Fig. 7a, b, c, d, e, f, g and h). These crystals should be removed before the slides are stained. See page 360 Luna's method for the specific identification of Zenker's crystals, and page 361 for mercuric chloride precipitate (Zenker's Crystals) removal method.

Longer exposure to hematoxylin may be required for Zenker-fixed tissues to insure a good nuclear chromatin stain, while the counterstain may have to be diluted or the staining time decreased to prevent excessive staining. Progressive, Mayer's-type hematoxylins may not produce the proper stain intensity and, therefore, one may have to use regressive Harris-type hematoxylins.

Helly's Solution

Distilled water	1000.0	ml
Mercuric chloride	50.0	gm
Potassium dichromate	25.0	gm
Sodium sulfate	10.0	gm
Concentrated formalin solution (37%)	50.0	ml

Zenker-formalin, or Helly's solution, is a modification of Zenker's solution. Formalin is used in place of glacial acetic acid. It is especially good for demonstrating mitochondria. It can be used routinely, especially for surgical specimens. Specimens measuring 4 mm in thickness fix well in 6 hours. Overexposure of tissue should be avoided since staining properties will be adversely affected. Further, overexposure will result in tissue becoming excessively hard which, in turn, creates problems in microtomy.

Russell's Solution

Distilled water	1000.0	ml
Zinc chloride	50.0	gm
Potassium dichromate	25.0	gm
Sodium sulfate	10.0	gm
Glacial acetic acid	50.0	ml

The Russell's[5] modified-Zenker solution utilizes zinc chloride in place of mercuric chloride. The change is advisable in order to prevent the environmental hazards associated with the use of mercury. The staining results produced by this fixative are not as good as those produced by Zenker's solution. Consequently, this fixative should be used sparingly.

Cadmium chloride may be substituted for zinc chloride. Zinc and cadmium chloride are removed from sections by the usual dehydrating procedures. Therefore, no extra steps are required to remove crystals of the salts, such as those required after fixation in solutions containing mercuric chloride.

Zinc Formalin

Concentrated formalin (37%)	370.0	ml
Distilled water	730.0	ml
Zinc sulfate	10.0	ml

This fixative mixture is excellent for the demonstration of a number of antigens. Special stains can also be performed satisfactorily after the use of this fixative.

B-5 Fixative Solution

Mercuric chloride	60.0	gm
Sodium acetate, anhydrous	12.5	gm
Distilled water	900.0	ml
Add immediately before use:		
Concentrated formalin solution (37%)	100.0	ml

This fixative has come into extensive use for bone marrow and lymph node specimens. It is extremely useful for nuclear chromatin and provides a nice, even-staining of cytoplasm. The removal of mercuric chloride crystals is necessary after the use of this fixative.

The unusual name for this fixative derives from the fact that it was the fifth fixative of series "B" on which its inventor, Dr. R. D. Lillie, was working. Others of the series were never put into use since they were not as effective.

Bouin's Solution

Picric acid, saturated aqueous solution	750.0	ml
Concentrated formalin solution (37%)	250.0	ml
Glacial acetic acid	50.0	ml

Bouin's solution is a dependable and convenient fixative for most routine work in zoology, histology, embryology, and parasitology. It penetrates rapidly, fixes evenly, and preserves many cell constituents with a minimum of alteration. Tissues fixed in this solution for a limited time usually have a favorable consistency for sectioning. However, extended periods of fixation will cause some tissue constituents, such connective tissue, to harden excessively. This hardness will result in cutting difficulties. A specimen measuring 4 mm will fix well in 10 hours.

Bouin's solution is a good fixative for nuclei and chromosomes. It is very useful in cytological studies on mitosis, ogenesis, and spermatogenesis. Bouin's solution is also good for delicate cytoplasmic structures such as cilia, mitotic spindles, and astral radiations. In embryology, it works well for eggs, embryos, and larvae of coelenterates, flatworms, annelids, echinoderms, tunicates, hemichordates, and Branchiostoma[9]. It is also an excellent fixative for vertebrate embryos. Embryos of considerable size (72-hour chick embryos, 5 to 8 mm mammalian embryos) or thin slices of vertebrate tissue should remain in the solution for at least 5 to 8 hours. Fix larger embryos or pieces of tissue over 5 mm thick for 24 hours.

Bouin's solution fails to fix lipids, leaving them soluble in alcohol as well as in toluene, xylene, and related solvents. It does not preserve mitochondria, Golgi, neurofibrils, or zymogen granules. In addition, it causes mucin to swell and does not fix glycogen. Bouin's solution should not be used for salivary glands or other organs in which it is desired to preserve mucin. Staining of striations in skeletal or cardiac muscle is generally poor after this fixative.

After fixation, wash specimens in several changes of 50% alcohol for 1 hour, agitating constantly to insure proper removal of picric acid. The removal of picric acid from tissues is essential to insure proper staining. It has also been determined that tissue morphology is adversely affected when the picric acid is not removed from the specimen. The harmful effects to the tissue continue within the paraffin-impregnated specimen for a number of years (Fig. 9a and b). A good section can be produced up to several months after fixation, but subsequent recuttings will produce increasingly poorer sections as more time passes. Thus, the need to remove most, if not all, the picric acid from the specimen cannot be overemphasized.

Carnoy's Solution

Absolute alcohol	600.0	ml
Chloroform	300.0	ml
Glacial acetic acid	100.0	ml

This is one of the fastest-penetrating fixatives known; generally, 1 1/2 to 3 hours exposure is adequate for 3 mm thick specimens. No washing is necessary subsequent to fixation; tissue may be transferred directly to absolute alcohol for processing. Carnoy's solution gives excellent nuclear fixation when preserving Nissl substances, plasma cells, and glycogen, but it hemolyzes red blood cells. It is the fixative of choice for the demonstration of ribonucleic acid (RNA) and deoxyribonucleic acid (DNA) when the methyl green-pyronin stain is to be used.

Clarke's Solution

Absolute alcohol	750.0	ml
Glacial acetic acid	250.0	ml

This fixative penetrates rapidly, resulting in good nuclear fixation and reasonably good preservation of cytoplasmic elements. It is an excellent fixative for smears or coverslip preparation of cell cultures. Small tissue specimens (2 mm in thickness) are fixed completely in 3 hours. Coverslip preparations of cell cultures are fixed in 5 minutes. Tissue specimens should be placed in 70% alcohol for 1 hour after fixation to remove acetic acid. A 5 minute exposure to 70% alcohol is necessary for smears and coverslip preparations.

Newcomer's Solution

Isopropyl alcohol	600.0	ml
Propionic acid	300.0	ml
Petroleum-ether	100.0	ml
Acetone	100.0	ml
Dioxane	100.0	ml

Primarily used for chromosome studies, this fixative preserves the nuclear chromatin better than Carnoy's solution and gives an improved Feulgen reaction. Fix 4 mm thick specimens for 8 hours; small pieces (2 mm in thickness) for 3 hours. Following the use of Newcomer's solution, place tissue in 60% alcohol for 2 hours.

Karpechenko's Solution

Chromic acid	5.0	gm
Distilled water	545.5	ml
Glacial acetic acid	50.0	ml
Concentrated formalin solution (37%)	400.0	ml

The chromic acid may be dissolved in the distilled water and kept as a stock solution to which the other ingredients are added immediately before each use.

Karpechenko's solution was designed for work in plant cytology, particularly for the fixation of nuclei and chromosomes. It is an excellent fixative for this purpose. It penetrates rapidly and produces a most delicate and beautiful fixation of both interphase and dividing nuclei. Karpechenko's fixative has been used very little for animal material, but has been found to be quite good for stages in the transformation of testicular spermatids. Small specimens or thin slices of tissue fix well in 8 hours. Keep the fixative container in darkness or subdued light. After fixation, wash specimens in running water for about 2 hours.

Perenyi's Solution

Chromic acid, 1% aqueous solution	150.0	ml
Nitric acid, 10% aqueous solution	400.0	ml
Alcohol, 95%	300.0	ml
Distilled water	150.0	ml

Mix immediately before use.

This is a special solution for specimens which ordinarily become extremely hard and difficult to section if an ordinary histological fixative is used. Among these specimens is the entire vertebrate eye. Perenyi's fluid fixes the retina with very little separation of its layers, and it does not harden the lens excessively. Fix the entire eye for 12 (human) to 24 (cow) hours, depending on its size, keeping the container in darkness or subdued light. Place specimen in 70% alcohol for 2 hours or longer, changing the alcohol once or twice. Then proceed with desired processing method.

This mixture is used also for the yolky eggs of arthropods, molluscs, fishes and other animals. This fixative leaves the yolk soft enough to be sectioned. Fix such tissues for 10 to 12 hours. Subsequent to fixation transfer specimen to 70% alcohol for a short time before processing.

Following this fixative, nuclear material stains rather poorly. Use alum hematoxylin and eosin for general structure, iron hematoxylins for details.

Orth's Solution

Potassium dichromate	25.0	gm
Sodium sulfate (may be omitted)	10.0	gm
Distilled water	1000.0	ml
Add immediately before use:		
Concentrated formalin solution (37%)	100.0	ml

This fixative can be employed with success for rather large pieces of tissue. It fixes the majority of tissues fairly well, preserving chromaffin cell granules and producing, in most cases, very little alteration in the shape of cells. Orth's fixative solution is quite useful for bulky pieces from the central nervous system. Excellent preservation of human and other mammalian kidneys has also been achieved, but results with testis and lung tissue have been poor. A serious disadvantage of this mixture is that it often causes nuclei to assume an unnatural, hollow appearance. The fixation of nuclei can be improved by adding 50 ml of glacial acetic acid to the above formula.

During fixation, it is best to keep the containers in darkness or subdued light. Wash fixed specimens in running water for 2 hours. If sections show a brownish color, bleach them with potassium permanganate and oxalic acid. Material fixed in Orth's fluid stains well, though quite slowly, with alum hematoxylin or iron hematoxylin.

Maximow's Solution

Zenker's fixative solution	1000.0	ml
Add immediately before use:		
Formalin, solution (10%)	100.0	ml

In order to preserve lipids, 100 ml of 2% osmium tetroxide solution may also be added.

The effects of this fixative are similar to those produced by Zenker's fixative, except that Maximow's solution preserves mucin, zymogen granules, and other cytoplasmic granules. For study of cytoplasmic granules, as in the islet cells of the pancreas, use 10% neutral buffered formalin in preparing the solution.

Entire eyes of monkeys, fixed without being punctured, will exhibit the structural details of the retina well. However, Maximow's solution is a poor fixative for testis.

The time required to fix specimens in this solution is similar to that stated for Zenker's solution. Wash fixed material in running water for 2 hours before processing.

Schaudinn's Solution

Mercuric chloride, saturated aqueous solution	666.0	ml
Alcohol, 95%	333.0	ml
Add immediately before use:		
Glacial acetic acid	50.0	ml

Schaudinn's solution has been widely used as a fixative for protozoa, especially intestinal organisms which can be fixed in smears of fecal material on coverglasses. The smears are dropped face down on the fixative for 15 to 20 minutes. The fixative may be warmed to 60 or 70°C, which reduces the time of fixation by 5 or 10 minutes. Intestinal protozoa may also be fixed in bulk by diluting the fecal material with physiological saline solution, straining it to remove large objects, and adding several times its volume of hot Schaudinn's solution[9].

Gilson's Solution

Distilled water	880.0	ml
Alcohol, 70%	100.0	ml
Nitric acid, 80%	16.0	ml
Glacial acetic acid	4.0	ml
Mercuric chloride	20.0	gm

Gilson's solution is one of the better fixatives for entire preparations of several types of invertebrates. It penetrates well, acts rapidly, and does not cause the muscles of some worms to contract as much as other fixing solutions do. In addition, it leaves less mercury precipitates in thc tissues than other mercurial-type fixative solutions.

This solution has its greatest use in the fixation of both free-living and parasitic flatworms. Fix small species for 3 hours or longer. Large worms should be left in the fixative for 12 to 24 hours. Material may remain in this solution for several days without suffering any harm, though prolonged exposure is not necessary. After fixation, wash specimens in 2 changes of 50% alcohol.

Gilson's solution is not recommended for routine surgical specimens. It coagulates the cytoplasm in a coarse reticulum and induces the formation of unusual vacuoles. On the other hand, it confers an unusually good cutting consistency upon some difficult-to-section specimens, including pieces of large nematodes, and preserves them well enough for ordinary work in morphology or histology.

Heidenhain's "Susa" Solution

Mercuric chloride, saturated solution in physiological saline	500.0	ml
Trichloroacetic acid	20.0	gm
Concentrated formalin solution (37%)	200.0	ml
Glacial acetic acid	40.0	ml
Distilled water	300.0	ml

This is a good general fixative solution for specimens of vertebrates and higher invertebrates. It penetrates rapidly and fixes nuclei, cytoplasm, and connective tissue constituents well enough for critical microscopic work. It preserves delicate epithelia with a minimum of distortion and is perhaps better than any other fixative for kidney specimens. The acids present cause mucin to swell and zymogen granules are not preserved.

Tissues fixed in this solution cut somewhat more easily than those fixed in formalin, Zenker's, or Bouin's solutions. It is especially good for the ovaries of fishes, amphibia, and birds, which contain yolky eggs. The uterus and other organs which contain large masses of connective tissue and muscle fibers also cut very well when this fixative is used.

The time required for fixation ranges from 3 to 4 hours for small embryos or thin slices, to 24 hours for large embryos or slices of tissue more than 5 mm thick. Transfer the specimens from the fixing solution to 50% alcohol for 2 hours before processing.

The staining properties of specimens fixed in this solution mixture are exceptionally good. Heidenhain's "Azan" stain and other trichrome stains yield preparations in which the structures are differentiated clearly and beautifully. This fixative solution, followed by the proper staining procedure, is very useful for demonstrating myocardial infarction. Alum hematoxylin and eosin give excellent results, as does iron hematoxylin[9].

Kleinenberg's Solution

Distilled water	1000.0	ml
Sulfuric acid, concentrated	20.0	ml

Picric acid sufficient to saturate the above mixture.

This solution causes little shrinkage of delicate tissues and is good for arthropod materials because it penetrates the exoskeleton rather quickly. Another important property of this solution is that it hardens yolk less than many other fixatives, making it useful for yolky eggs of molluscs and teleost fishes. Likewise, it is especially good for crustaceans and other arthropods which are covered by impermeable membranes, including lobsters, crabs, shrimp, wood lice, water fleas, and barnacles. Place specimens in 70% alcohol until ready to process. Impermeable membranes, thick chorions, or surrounding jelly should be pricked with a sharp needle to allow penetration of fixative solution. Carmine and hematoxylin stains are often quite effective after this fixative.

Duboscq and Brasil's Solution

Alcohol, 80%	750.0	ml
Picric acid	5.0	gm
Concentrated formalin solution (37%)	300.0	ml
Glacial acetic acid	75.0	ml

This modification of Bouin's solution is especially good at penetrating refractory objects, such as small crustacea, insects, and nematodes. It will penetrate the cuticle and preserve the internal organs well enough for detailed study. Specimens may remain safely in the fluid for 2 or 3 days. Transfer material to 50% alcohol, and change this several times to remove the picric acid.

Allen's B-15 Solution

Bouin's solution	1050.0	ml
Heat to 37°C and add:		
Chromic acid	15.0	gm
Stir until dissolved, then add:		
Urea	20.0	gm
Continue to stir until this is dissolved. Use at once.		

This fixative solution is used for critical work on mammalian chromosomes, especially in the cells of the testis. Fix small pieces of tissue immediately after the death of the animal. A better alternative is to inject the warm (37°C) fixing solution into the genital arteries or aorta and, after 15 minutes, remove the testis, cut sandwich-type slices, and place them in an adequate volume of the fixative. Approximately 2 hours are required to fix pieces of testis 4 mm or less in thickness. Larger or denser objects require

4 hours, or slightly longer. Replace the fixative with 70% alcohol, adding a little at a time. Alum hematoxylin and eosin will stain fairly well but with less striking color differentiation than is obtained after Bouin's or Zenker's solution.

Hollande's Solution

Picric acid	40.0	gm
Cupric acetate	25.0	gm
Distilled water	1000.0	ml
Formalin concentrated solution (37%)	100.0	ml
*Glacial acetic acid	15.0 ml to 50.0 ml	

Dissolve the cupric acetate in the distilled water; then add the picric acid; then, the formalin and acetic acid.

*The amount of acetic acid specified in the original formula is 15.0 ml for each 1100 ml of the other liquid ingredients. However, increasing the amount of acetic acid to 50.0 ml does not seem to affect the utility of Hollande's solution for silver impregnation or staining, and may lessen the extent of tissue shrinkage during fixation.

This solution is one of the best fixatives for certain kinds of protozoa. Hollande's solution is also one of the better fixatives for flagellates and ciliates which are to be demonstrated with silver stains. It is also a good fixative for routine surgical specimens.

After fixation, transfer the specimens to 70% alcohol and leave them in this until the picric acid and cupric acetate have been thoroughly washed out (approximately 2 hours).

General Comments on Fixation

The following information is presented in the form of comments rather than in narrative form in order to provide considerably more information on fixation in limited space. It is also felt that the reader may gain knowledge more quickly by reading general comments on fixation.

- A crude model of protoplasm can be made and utilized to perform simple fixation studies. The model is made by combining gelatin and albumin which results in a gel.

  Gelatin-Albumin Gel

Egg white	100.0	ml
Distilled water	200.0	ml
Gelatin	45.0	gm

  It is important to warm the water and egg white to allow the gelatin to dissolve.

  The gelatin-albumin gel is then poured into a flat mold and placed in the refrigerator to set until hard. The gelatin-albumin gel results in a refractive index of 1.365, close to protoplasm's index of 1.353. The protein content is 19.4 and is about the same as protoplasm. The model of protoplasm contains globular protein (albumin) and a fibrous protein (gelatin), a derivative of callogen. Simple or globular proteins include albumins, globulins, histones and protamines and are generally soluble in water or salt solutions. Fibrous or fibrillar protein (structural proteins) are generally insoluble; major types are collagen, elastin, keratin, and actin-myosin. Conjugated or compound proteins are those in which the protein molecule is united with a nonprotein molecule or molecules. They include nucleo proteins, mucoproteins lipoproteins, chromoproteins, and phosphoproteins.

- The foundation of all good histological preparations is rapid and complete fixation.

- Tissue should be placed in the fixative immediately upon removal from the body. One must remember that autolytic processes begin within blood vessels as soon as blood stops flowing in them.

- A saprophyte is a microorganism which thrives on decaying matter. It is observed after autolysis commences (Fig. 2). The organism is noted for its invasive characteristics. It is impossible to restore nuclear staining properties of autolyzed specimens.

- A very common artifact is starch from surgical dusting powder. The classic appearance is a hexagonal crystal which is grey after staining with H&E, green with the Geimsa stains, and PAS positive. When polarization is performed, a Maltese cross is observed. Starch is most frequently introduced by the use of powdered rubber gloves during tissue grossing. It can also be deposited by the surgeon during the surgical procedure.

- Specimens should be cut thin enough so that the fixative penetrates the tissue within a reasonably short time. To this end most tissue specimens should not exceed 5 mm in thickness.

- The amount of fixative used should be at least 10 times the volume of the tissue specimens. The fixative should be changed after the first hour of tissue exposure.

- The choice of fixing solution should be based upon the purpose for which the tissue is to be stained. The effects of various fixatives on staining reactions was studied by Luna[8].

- Formalin solution penetrates 2 mm of most tissue in the first 4 hours of exposure; continued exposure results in fixation penetration of 10 mm in 24 hours. It should be recognized that penetration of formalin fixative solution can be improved by the use of mechanical stimulants, such as agitation, heat, and aeration. More significant, however, is the speed of fixation produced by the new, closed processing systems which are now in widespread use in histology. For example, one such system is capable of completely fixing a 5 mm thick piece of tissue in one hour[10].

- Formaldehyde forms methylene compounds with amino, amide and hydroxy groups, and thus affects the solubility and reactivity of proteins.

- Tissue-bound formalin cannot be removed by water washing and/or dehydration procedures used in histology and histochemistry. Only adsorbed and loosely bound formalin is displaced by washing in running water. However, such formalin is not likely to play an important role in fixation. Bonds linking formalin to tissue differ in stability. Some formalin can be dislodged by prolonged treatment with water, but residual bound formalin cannot be removed even by weeks of washing at room temperature.

- The pH of the fixing solutions is a controlling factor in the reaction of formalin with proteins in aqueous media. In unbuffered formalin solutions (approximately pH 3) many amino and guanidyl groups are protonated, hence formalin can react only with relatively few uncharged groups. As the reaction proceeds, other groups give off their protons and combine with formalin. Consequently, fixation of tissue proceeds slowly. In buffered neutral formalin (pH 7) amino groups are discharged and react avidly with formalin. For example, the amount of bound formalin increases from 0.05 mol/gm collagen at pH 4, to 0.4 mol/gm at pH 7 to 8. Maximum tissue fixation, i.e. cross-linking (Fig. 10a and b), occurs in the pH range 4 to 5.5; no increase in tissue stabilization is observed above pH 5.5. The increased amount of formaldehyde bound at higher pH levels only blocks numerous reactive groups. Thus, fixation in buffered neutral formalin inactivates more tissue groups than unbuffered formalin without improving fixation, i.e. cross-linking.

- Studies have shown that formalin forms primary additive compounds with amines or other groups containing an active hydrogen atom. The methylol (CH_2OH) compounds formed may then react with amines or other groupings containing active hydrogen situated on the same or another protein molecule, thus forming a methylene bridge ($-CH_2-$). It is believed that the second reaction, involving the cross-linking, is responsible for the hardening and fixing effects of formalin.

- Concentrated formalin will undergo virtually spontaneous condensation reactions with itself when stored in a concentrated form. This reaction results in the formation of paraformaldehyde, a solid polymer which is sometimes visible at the bottom of formalin solutions as a white precipitate. This reaction can be inhibited at room temperature by including modest amounts of alcohols in the solution as preservatives or inhibitors. About 10% methanol is added by the manufacturer to concentrated formalin as a preservative.

- Formalin fixation does not produce "over fixation"; that is, tissues do not become hardened unpredictably.

- When tissues are immersed in formalin solutions, they are penetrated rapidly by methylene glycol and the fraction of formaldehyde present. Equilibrium between the components of formalin, carbonyl formaldehyde, and methylene glycol explains most of the mystery as to why formaldehyde penetrates rapidly (as methylene glycol) and fixes slowly (as carbonyl formaldehyde).

- Because unbuffered formaldehyde solutions are acidic, part of the degradation of brain lipids can be ascribed to slow hydrolysis (the splitting of a compound into fragments by the addition of water; the hydroxyl group being incorporated into one fragment and the hydrogen atom into the other). After a very prolonged fixation a more or less stationary situation is reached.

- It is known that in the concentrations used for fixation (approximately 4% formaldehyde in water), formaldehyde exists mainly as a monomer, most of it in the hydrated form, as methylene glycol. Concentrated formalin contains mostly the polymerized form, and its dilution therefore, involves a depolymerization process which is not instantaneous. Although the final equilibrium does not seem to depend on the pH of the solution, the speed of depolymerization increases with higher pH values, being very rapid in the neutral and slightly alkaline ranges.

- The pH of unbuffered concentrated formalin (reagent grade) ranges from 3.0 to 4.0. If tissue is fixed in pH 4 concentrated formalin the hydrogen-ion concentration remains constant. If, however, the original pH is 3.0 it reaches 4.0 in 48 hours with larger amounts of tissue. The pH rise is most rapid in the first hours; with a 1/125 ratio of tissue to fluid the pH at one hour is 3.4 and the final pH after 48 hours is 4.0.

- Calcium carbonate is frequently used as a neutralizer in some formalin fixative solutions. Calcium carbonate crystallization occurs within the specimen resulting in cellular displacement (Fig. 11a and b). The appearance is misleading because it resembles an in vivo reaction.

- Unbuffered 10% formalin has a pH of 3.8 to 4.1 depending on the acidity of the concentrated formalin from which it is made. The pH changes from 3.8 to 4.6 during the first hour of fixation, and at the end of 8 hours is 5.1. After 8 hours there is no further change; smaller amounts of tissue causes a rise to pH 4.5.

- Ten percent formalin neutralized with calcium carbonate and filtered becomes acid within three hours after the start of tissue exposure to fixation. Conversely, neutralized, filtered 10% formalin containing no tissue maintains a pH of 7.

- The pH of formol-calcium in Baker's solution is constant if no tissue is added to the solution. It changes from 9 to 6.4 if a large amount of tissue is added, and to 7.3 with a smaller amount. Conversely, formalin with a phosphate buffer maintains a constant pH at all times.

- Tissue shrinkage is not caused by the formalin fixative, as claimed by some, but by the dehydration, clearing and impregnation of tissue which follows.

- Concentrated formalin diluted to 10% with distilled water and then treated with an excess of Magnesium carbonate shows a pH of about 9.6, owing to the appreciable solubility of the latter. Concentrated formalin first "neutralized" with Magnesium carbonate and then diluted to 10% with distilled water has a pH of 7.6. The pH of fixing fluids containing formalin consequently varies with the pH of the formalin used. For example, Regaud's fluid freshly prepared with "neutralized" formalin has a pH of 4.7, whereas when prepared with unneutralized formalin, it has a pH of 3.6.

- Coagulant fixatives include alcohols, picric acid, mercuric chloride, and chromium trioxide. The major noncoagulant fixatives are formalin, potassium dichromate, osmium tetroxide, and acetic acid (which fixes nucleoproteins only).

- The capacity of buffers occurring naturally in cells to maintain a constant pH in the presence of weakly buffered or unbuffered fixatives is rather limited. Also, it is recognized that tissues fixed in unbuffered fixative solutions form artifacts, since acidification of tissues precedes their fixation. By buffering the fixing solution the acidification can be neutralized, and thus damage to the tissue is minimized. Buffer solutions contain a weak acid and its salt or a weak base and its salt and they resist changes in hydrogen ion concentration when small amounts of a strong acid or base are added.

- The quality of fixation is influenced not only by the pH but also by the type of ions present in the buffer, since buffers are active reagents in the fixation process.

- Under normal conditions the pH of cells is usually on the alkaline side of the isoelectric pH of proteins. For instance, the pH of cytoplasm in various types of animal cells has been found to be in the neighborhood of 7.0, whereas nuclei have a pH range of 7.6 - 7.8.

- The pH of the fixative must remain close to the pH of the tissue because a change in the tissue pH brought about by the fixative is bound to radically alter the structure and behavior of tissue proteins in solution.

- An ideal fixative should kill tissue specimens quickly, causing minimum shrinkage or swelling. The speed of killing is dependent primarily on the rate of penetration of a fixative into the tissue. Chemicals of low molecular weight, such as formaldehyde, are usually most effective; however, a direct relationship between molecular weight and rate of penetration of a fixture does not always exist. (Mercuric chloride penetrates rapidly, although it has a relatively high molecular weight.) The rate of penetration is also affected by the number of reacting radicals in the molecule of a fixative independent of its molecular weight. Two of the intrinsic properties of a fixative that influence the rate of penetration are its solubility in lipids and the polarity of its molecule.

- Factors that influence the rate and depth of penetration are tissue type, temperature, duration and mode of fixation, concentration of the fixative, osmolality of the fixative vehicle, and the addition of electrolytes or nonelectrolytes to the vehicle.

- For a specific tissue, duration of fixation is controlled by the type of fixative and buffer used, specimen size and type, temperature, staining method to be used, and the objective of the study.

- Very little is known of the effects of overfixation except that it results in the extraction of tissue constituents. Since the extraction caused by chemical fixation is progressive in time, the need to determine the optimal duration is obvious.

- In general, low concentrations of fixative agent require longer durations of fixation. Longer duration of fixation tends to cause diffusion of enzymes, extraction of cellular materials, and shrinkage or swelling of the tissue.

- Zenker's solution, which contains no reducing agent, does not precipitate, change color, or undergo any significant pH change in a 48 hour period.

- Zenker-formol, which causes the greatest tissue shrinkage of all formalin-based fixatives, is widely used because of its excellent cytoplasmic preservation.

- The chemical alterations caused by Bouin's solution are not yet understood. It is absolutely imperative that there be no residual picric acid in Bouin's fixed tissue. Residual picric acid will alter the tissue and affect the staining properties. Eventually, almost all cellular structures will be distorted, unrecognizable, or obliterated. Remove picric acid by placing specimen (prior to processing) in saturated lithium carbonate in 70% ethyl alcohol until all yellow color disappears (approximately 4 hours for specimens 4 mm in thickness).

- Bouin's fixative has an adverse effect on hyaluronic acid and it cannot be demonstrated by staining. It is not known if other mucosubstances are adversely affected. Buffered neutral formalin, however, does not produce any gross alterations affecting the staining properties of hyaluronic acid.

- Bouin's fixative contains formalin, glacial acetic acid, and trinitrophenol (picric acid). It produces crystals (precipitates) in tissue sections similar to those produced by unbuffered formalin.

- Bouin's solution does not convert the cell cytoplasm into a coagulum. What appears in optical microscopy to be a coagulum is, in fact, endoplasmic reticulum, the cisternae of which are identifiable but the ribosomes are not. Mitochondria and Golgi apparatus are destroyed, though many pancreatic zymogen granules survive. The nuclear and cell membrane are preserved but little is seen of their fine structure. The nucleolus is not seen clearly. Most viral particles do not survive fixation in Bouin's solution.

- When use of fixatives containing acetic acid is desired, the addition of an equimolar amount of sodium chloride or a similar salt will serve to protect erythrocytes and soluble cell granules from lysis during fixation.

- Mitochondria and the Golgi apparatus are not preserved by fixatives containing acetic acid.

- Flemming's solution without acetic acid is one of the best fixatives for limiting shrinkage of muscle specimens.

- Helly's, Regaud's, Orth's, and Karpechenko's fluids are unstable solutions in which the pH rises gradually when standing without tissue. In Helly's or Regaud's the final pH is almost the same whether no tissue, a small amount, or a large amount is fixed.

- Regaud's solution gives a satisfactory fixation from the cytological standpoint. Unfortunately, Regard's solution has serious faults. It is unstable and begins to decompose within a few hours. The rate of penetration is slow and only very small pieces of tissue can be fixed. It has the tendency to harden tissues excessively. Notwithstanding such drawbacks, it can be used successfully for the study of mitochondria, which it preserves well.

- The process of most types of fixation is complicated by the fact that fixing agents need time to reach the different tissue structures. Most workers conclude that fixation of the external specimen layers interferes with the penetration of fixative ingredients (just as penetration through a finely interlinked mesh would be expected to be slower than through a more loosely woven one).

- Fixation involves the unfolding of globular proteins and an increased availability of reactive groups of the protein molecules. This enhances stainability by acid or basic dyes and increases the reducing activity which can be visualized by silver impregnation methods. The various fixatives affect the binding capacity of acid and basic dyes differently. Fixatives which affect the basic group of protein molecules (e.g., formalin) will shift the isoelectric point to the acid range, thus decreasing acidophilia. On the other hand, fixatives which mainly bind the acid groups (e.g., potassium dichromate, picric acid) will render the tissue less basophilic[1].

- A change in tissue density, much as a concentration of a structure by loss of bound water, will cause a stronger and more distinct staining even when the total number of the dye-binding groups remain unchanged. Thus, the fixative used in the preparation of a tissue specimen for microscopic study affects its stainability.

- One of the most important factors affecting tissue proteins is the hydrogen ion concentration (pH) of the fixative solution. The combination of proteins with anions or cations to form insoluble compounds is closely related to the pH of the fixing fluid.

- The acidity of the fixative determines the physical characteristics and chemical combining power of tissue proteins and, in turn, chemical reactions may be reflected by pH changes in an unbuffered fixing solution.

- As would be expected, a large amount of tissue produces greater pH changes than a small amount in the same quantity of fixative.

- In general, when tissue is added to a fixative more acid than pH 5, the pH will rise; it will fall if the fixative is more alkaline. This probably occurs because the isoelectric point of most tissue proteins is in the vicinity of pH 5.

- The special fixative properties of Cajal's formol-ammonium bromide and formol-uranium nitrate solutions are due to a pH of 1.5 and 2.7 respectively. Fifteen percent formalin, made acid to pH 1.5, reproduces the results given by formol-ammonium bromide in Cajal's gold sublimate method. Similarly, after fixation either in formol-uranium nitrate or in formol at pH 2.7, astroglia and Golgi apparatus stained by Cajal's methods yield identical results. Thus, the acidity of the fixative is very important in determining its usefulness. There is a small pH rise in these fixatives, whether or not tissue is present, during the first 12 hours followed by a return to approximately the original value. The total change is not more than 0.3 pH unit.

- The pH of solutions such as Zenker's, Bouin's, and those containing mercuric chloride change less than 0.5 pH unit when tissue is added.

- Isotonic saline is superior to distilled water as a formalin diluent, reducing tissue swelling during formalin fixation.

- Shrinkage of tissue during dehydration and infiltration with hot paraffin exceed that produced by fixation alone.

- Studies of specific fixing agents show considerable variation in results. This is explained in part by the use of different tissues, various sized blocks, different schedules, and reagents that do not have exactly the same composition.

- Tissue shrinkage and density caused by fixation may explain in part the increased tissue hardness encountered on subsequent microtomy.

- Tissues which shrink greatly during fixation tend to show less change under subsequent dehydration, clearing, and paraffin impregnation.

- It is not unusual to expect that proteins of apparently the same type and character are precipitated differently by the same fixative solution.

- The type of fixation used has a marked influence on the weight, volume, and lineal dimensions of tissues.

- There is an important correlation between the pH of a dichromate and its fixation image. If it is too acid, it will fix the chromatin but not the mitochondria; if it is too basic, it will fix the mitochondria but not the chromatin.

- Fixing agents such as potassium dichromate increase the weight of tissue specimens during fixation by as much as 15% over a period of 5 days. The increase in weight is in direct proportion to the dilution of the salt and freshness of the tissue.

- The degree and type of tissue shrinkage during fixation depends upon the penetrative ability of a salt. Salts can be classified roughly into two groups based on their penetration ability. Some examples of uniformly penetrating salts are aluminum acetate, aluminum sulfate, mercuric chloride, and zinc chloride. Examples of non-uniformly penetrating salts are aluminum chloride, cadmium chloride, uranyl nitrate and ferric sulfate. Whatever the type of fixative penetration, tissue shrinkage may be either generalized, affecting all structures to a similar extent, or cellular, primarily affecting the cell constituents.

- Tissue constituents subjected to identical staining methods show different affinity to the same stains when different fixing solutions have been used. These changes in staining reaction are considered to be as a consequence of shifting of the isoelectric point of proteins under the influence of different fixatives.

- It seems that in tissue fixation the first toxic substance to reach the tissue is apt to determine the fixation image.

- The nucleolus is fixed by both acid and basic fixatives.

- Nuclear chromatin is fixed only by acid fixatives.

- Mitochondria are fixed only by solutions of dichromates more basic than pH 4.2.

- Some fixatives (e.g., mercuric chloride, picric acid) promote the precipitation or clumping of proteins. Others (e.g., formalin, glutaraldehyde) promote cross-linking but not coarse precipitation of proteins. All fixatives have both desirable and undesirable effects.

Chapter 2

Agents and Methods for Specimen Processing

All specimens being prepared for processing and ultimate embedding in a given medium must be identified well. The adequate processing of tissue specimens is often subject to the inconsistencies inherent with histological technique. Namely, variables in fixation and/or fixatives used, dehydrating and clearing agents used, paraffins and/or paraffin additives used, thickness of the tissue specimen, and adequacy of processing schedules. One or all of the above factors can produce a number of problems which will result in the production of poorly processed tissue specimens. In addition, inadequate fixation will produce shrinkage of tissue constituents when exposed to dehydrating and clearing agents during processing.

In histological technique, tissue fixation is closely followed by tissue processing. Tissue processing is defined as a series of operations, steps, or events leading to the achievement of a specific result; in this case, the formation of a matrix by the aggregation (in tissue) of a specific medium which ties together tissue structure. This tying in with an impregnating medium prevents disruption of tissue structures during microtomy.

The steps generally involved in processing are dehydration, clearing, impregnation, and embedding. These events will be described in turn.

Dehydration

Dehydration is derived from Latin, de = away and Greek, hydor = water, and is defined as the removal of water from a substance. In histological technique, dehydration is the removal of unbound water from tissue constituents during tissue processing. Bound water is not affected by normal tissue dehydration.

In order to infiltrate tissue with an embedding medium, it is necessary first to displace the unreactive aqueous or alcoholic fixatives from the tissue with a dehydrant before introducing a clearing agent. Some dehydrating agents are hydrophilic, i.e., having strong polar groups that readily interact with water. In others, dehydration is accomplished by diffusion of the contained fluid out of the tissue specimen into a surrounding fluid which, in turn, diffuses into and replaces the unbound fixing solution.

Followed closely by isopropyl and denatured alcohol, ethyl alcohol has been the tissue dehydrant of choice for many years. Perhaps this is so, to some extent, because it has been used extensively as a fixative or fixative agent. Ethyl alcohol also is an excellent dehydrant but has the disadvantage that it absorbs water from air more rapidly than isopropyl. It is miscible with water; methyl, ethyl, and denatured alcohols; ether; chloroform; xylene; acetone; benzene; and dioxane.

My preference for dehydrating tissue specimens destined for paraffin embedding is isopropyl alcohol. This decision was reached after many years of experience with the use of ethyl, isopropyl, methyl and denatured alcohols, as well as many commercially available dehydrating agents. Isopropyl alcohol penetrates as well as ethyl and better than most other alcohols. It denatures tissue less than other alcohols; that is, it removes extractable water without affecting molecules of bound water which are so essential for subsequent good staining. Isopropyl alcohol also provides better overall dehydration without affecting tissue constituents at the submicroscopic level. Tissue can remain submerged in this alcohol for extended periods without tissue constituents being adversely affected. Isopropyl alcohol is miscible with water; methyl, ethyl and denatured alcohols; ether; chloroform; xylene; acetone; benzene; dioxane; gums; shellacs; and many other agents. Mixed with some common paraffin solvents, its tolerance for small amounts of water, on the average, equals that of ethyl alcohol (Fig. 12).

Other Dehydrating Agents

A large number of other dehydrating solutions have been made available by different manufacturers in the last few years. I think it is safe to say that many of these products are nothing more than alcohols which have been denatured by the addition of a small amount of one or several kinds of other alcohols. Regardless of the source of the denatured product, one should not use a dehydrant unless the agent used to denature it is known. It is important to remember that adulterated dehydrants may affect tissue constituents badly, resulting in poor microtomy and/or staining results. Because of the proven values attributed to isopropyl alcohol as a dehydrant, all procedures which follow will call for the use of isopropyl as the dehydrant of choice.

Specimen Clearing

The word clear is derived from the Latin word Claru = clear, bright, and means free from cloudiness, muddiness, blemishes; transparent, not turbid. The word also means to free a thing or person from something. In the case of histological techniques, clearing is the removal of a dehydrating agent from tissue constituents in order to prepare the tissue for an impregnating medium. This is accomplished by passing the tissue specimen through clearing agents. (The term "clearing agent" probably came from the fact that tissue is, in fact, cleared to varying degrees by clearing agents such as xylene, toluene, benzene acetone, methyl salicylate, cedarwood oil and many other clearing agents found in the literature.)

The sum effects of clearing is the changing of the refractive index of the tissue. The refractive index is a measure of the relative velocity of light passing through a particular medium. For example, a piece of clear glass is more easily seen in air than if it is immersed in an oil. This is because the refractive index of glass (1.52) is very different from air (1.00) but close to that of oil. Similarly, if a certain constituent of a cell has a distinctly higher refractive index than the cytoplasm which surrounds it, it should be distinguishable with the microscope. In histopathological technique, the natural refractibility of a tissue is usually deliberately suppressed by the use of a mounting medium with a refractive index close to that of the anhydrous material constituting the tissue section (approximately 1.53). Almost all contrast is produced artificially by staining. Clearing agents having a high refractive index render pieces of tissue transparent. This means that clearing is literally complete and is responsible for the commonly used term "clear" in histopathological technique.

Following are the refractive indices of some clearing agents. These clearing agents are miscible with dehydrating agents, melted paraffin wax, and resinous mounting media. Tissue becomes transparent only when the refractive index of the clearing agent is higher than 1.47.

Agent	Index	Agent	Index
Methyl salicylate	1.54	Chloroform	1.45
Cedarwood Oil	1.51	*Dioxane	1.42
Xylene	1.50	*Cellosolve	1.41
Toluene	1.50	*Tetrahydrofuron	1.41
Benzene	1.50	N-butanol	1.40
Terpineol	1.48	*Tertiary Butyl Alcohol	1.39
Carbon Tetrachloride	1.46		

*These agents are miscible with both water and melted paraffin and should not be used to dissolve resinous mounting media. All other agents above are not miscible with water.

Paraffin Wax Impregnation

The word impregnate is derived from the Latin word Impraegnalus and means to make pregnant, to fill or saturate, or cause to be permeated. In histological technique, the permeation and subsequent saturation of tissue by specific media results in the tying together of tissue constituents, forming a matrix which prevents tissue structure disruption during microtomy.

Paraffin wax is a compound known as a hydrocarbon which is made from carbon and hydrogen atoms. As a matter of fact, many different compounds can be made by combining these two elements. The simplest hydrocarbon is methane, followed by ethane (Fig. 13a and b), both of which are the first of a long series of compounds called paraffins. By stringing carbon atoms together one can make bigger and bigger molecules of the same kind. All one has to do to complete the molecule is to fill up the spare bonds of the carbon by linking them with hydrogen atoms. The paraffin molecule is a long chain hydrocarbon such as that illustrated in Fig. 13c. These long chain hydrocarbon molecules come together to form a liquid mixture when paraffin is in the molten state but form a mixture of solid hydrocarbons when cooled to room temperature. These solid hydrocarbons appear as crystals when viewed microscopically.

The amount of hydrocarbon molecules which aggregate in a given tissue interstice determines the density of the paraffin wax matrix formed. This, in turn, determines how well tissue has been impregnated. Tissue interstices can be infiltrated with paraffin by various means. Namely, submersion in a hot paraffin bath with no agitation, use of an automatic tissue processor with circular tissue agitation, or use of a closed system processor which provides one or more of the following: pressure, heat, vacuum, and agitation. In my opinion, those systems which provide vacuum and pressure are to be considered best for impregnation of paraffin, since the use of vacuum removes (pulls) air from tissue interstices which allows optimum impregnation of these spaces by the paraffin wax molecules (Fig. 14 a, b and c).

Specimen Embedding

The word embed means to set or fix firmly in a surrounding mass. In histological technique this refers to the operation which results in the "fixation" or setting of a tissue specimen in a firm medium. This firm medium (paraffin, plastic, water, or other substances) keeps the tissue intact during microtomy. Contrary to general belief, this step is very important in determining the quality of microscopic slides. For example, the transfer of tissue from the impregnating container to the embedding mold should be done quickly to prevent the paraffin from solidifying on the surface of the specimen. This simple action will eliminate the "hairline cracking" often experienced at the surface of the specimen during cooling of the paraffin block. The hairline cracking often results in poor microtomy since there is a lack of firmness at the knife/paraffin/tissue interface.

The introduction of plastic processing/embedding cassettes and embedding centers has had an enormous impact on both the quality and quantity of specimen embedding. Each laboratory must decide for itself which type of cassette and embedding center to use. However, it is recommended that several types be tried and the following points be considered when making this important decision.

Plastic Embedding Cassettes

Some of the factors which should be considered in purchasing embedding cassettes are: 1) How well does the writing surface trap pencil graphite or pen ink? (This is important since graphite or ink should not run off or be smeared by hand rubbing.) 2) The cassette lid closing element (latch) should insure closure during processing and yet be easy to open. 3) The cassettes should be made of sturdy material to prevent buckling during exposure to various processing solutions and the possible loss of specimens. 4) The cassette should be large enough to allow a 4 mm thick piece of tissue to move freely inside, thus insuring better transfer of solution through the tissue specimen. 5) The holes in the plastic cassette should be large enough to allow good, uninhibited transfer of processing fluids and prevent surface tension activity which could result in poor exchange of fluids. The surface tension activity which may occur would take place with those specimens at the top of processing containers.

Embedding Centers

In purchasing embedding centers one should look for the following: 1) The center should be comfortable for the operator. In particular, the system should not be too high, which would result in the operator working in an unbalanced position or with elbows elevated. 2) All functioning elements responsible for heating and cooling should maintain temperatures for extended periods of time at full load capacity. 3) The center should be easy to clean of paraffin. That is, much of the system's exposed surfaces should be smooth to allow fast and easy scraping with a plastic scrapper. 4) Both cold and hot plates should have large working surfaces to permit embedding of many specimens.

Embedding Paraffin Wax Medium

As in the case of plastic embedding cassettes and embedding centers, the purchase of impregnating and embedding medium is important. One should select that medium which best suits the purpose. Most paraffin waxes, regardless of the source, will suffice as a tissue structural matrix. This includes those used for candles, shellwax, essowax, etc. This is because all are straight chain hydrocarbons and, therefore, can be expected to react somewhat similarly to all the forces produced by the advancing knife blade facet, (see chapter on Microtomy). The use of paraffins as an embedding matrix for special purposes is addressed in the chapter on special techniques.

Paraffin Waxes' Contraction/Expansion Factor

For conventional tissue processing one should look for a paraffin wax with some elasticity (expansion factor). This can be determined by embedding a cross section of a circular piece of tissue (i.e., aorta, large vein, eye globe). Paraffin with some elasticity will allow the section to expand to almost its normal size on the floatation bath at 45°C in 2 minutes. This type of paraffin is essential for those laboratories involved in sectioning eye globes or other circular structures. Paraffin with a good expansion factor is also recommended for embedding ordinary material, be it human or animal. Paraffin should not shrink (contraction factor) more than 15% when cooling from its molten state (60°C) to room temperature (25°C). It is best to use a paraffin that shrinks not more than 12%. In shrinking more than the tissue, the paraffin will exert a compressive force on the surface of the tissue. This inward force will produce problems during subsequent exposure to the floatation water bath. The tissue section will not be able to expand to its normal size, resulting in the development of wrinkles. The paraffin shrinking (contraction factor) will also have an effect on the quality of a section during microtomy. Namely, if the paraffin shrinks more than 12% it will result in the creation of spaces (paraffin voids) within the tissue (Fig. 15a and b). The overall effect of these spaces is similar to, and produces the same results as, inadequate paraffin wax impregnation; specifically the cutting of poor sections because the tissue specimen lacks a good solid and complete paraffin wax matrix.

Method for Determining Shrinking (Contraction Factor) of Paraffin Waxes

The shrinking factor of paraffin wax can be determined by the following simple procedure: Fill a beaker with 500 ml of molten paraffin as indicated by the meniscus line of the paraffin. Allow the paraffin to solidify quickly by placing in the refrigerator. Remove beaker from refrigerator and allow it to come to room temperature. Pour sufficient water into the beaker containing the solidified paraffin until the added water reaches the 500 ml mark. Measure the water added to the paraffin wax container by pouring into a graduated flask or beaker. The amount of water used to reach the 500 ml mark reflects how much the paraffin shrank.

Some embedding medium contain additives, such as resins or plastic polymers, to provide better support of tissue structures during microtomy. These modified paraffins are especially good for cutting thin sections.

Method for Determining Quality of Paraffin Wax Impregnation

As indicated earlier, proper processing is extremely important to the ultimate goal of achieving good tissue sections from paraffin embedded specimens. The following method can be used to determine if the impregnation of tissue has been performed properly. For this purpose, a piece of kidney will serve as the control. Be sure the kidney specimen contains glomeruli. Process the specimen using the processing schedule to be tested. Be particularly mindful that the kidney specimen should not be overexposed to the impregnating medium (paraffin). Section the specimen at 6 micrometers and place on the floatation bath. Pick it up on a slide and allow it to dry on a slide warmer (37°C) overnight or in a 37°C oven for 2 hours. It is important that the paraffin on the slide does not dissolve. The non-deparaffinized section is then viewed under polarized microscopy. Polarized microscopy will reveal that paraffin has some anisotropic properties. This property can be used to determine the quality of paraffin wax impregnation (Fig. 16a and b).

Procedure for Evaluating Quality of Tissue Impregnation

There is a certain percentage of tissue specimens which will always fall into the under or overprocessed category. This is to be expected because of the many variables in histotechnology; that is, specimens grossed too thick, poor agitation of agents used in the processing systems, use of processing solutions which have a tendency to over-harden tissue, overpacked processing system, etc.

These problems notwithstanding, there is a way of determining the expected norm of improperly processed tissue specimens. Using the "evaluating quality of tissue impregnation" test presented below, one can determine the likely percentage of poorly processed specimens. Quality control of tissue processing can be performed on a daily basis by submitting a sampling of processed and embedded specimens to this test.

Incomplete impregnation of tissue can be readily identified by the application of moisture (piece of lukewarm water-soaked cotton) to the surface of a faced paraffin block. Tissue that has not been properly impregnated with paraffin will turn white and swell somewhat. The degree of impregnation can be estimated by the speed at which the exposed specimen turns white (Fig. 18).

The last two categories, good (20%) and excellent (5%), are the ranges within which most surgical services process tissue. In the 5% range all tissues should be impregnated well enough to allow good microtomy of all specimens. However, in some instances it may be necessary to freeze the surface of the block more in order to obtain good sections. This extra freezing will produce a firmer paraffin matrix, which will result in less tissue distortion as the cutting blade advances. Improperly impregnated tissue specimens can be expected to "explode" or disintegrate when placed on the floatation bath. The degree of disintegration will depend on how poorly the specimen was impregnated with paraffin and/or the temperature of the floatation bath.

Proper Specimen Alignment in Embedding Matrix

Ansman[11] in her paper "Blocking Procedures in Histopathological Techniques" points out that proper alignment of tissue specimens in an embedding mold is a much neglected procedure. The modified illustrations of blocking specimens demonstrate both right and wrong ways to orient tissue in the embedding cassette or mold (Fig. 17a - t). The recommended methods should not be taken lightly since they are essential to good microtomy, resulting in improved tissue sections. There are many other ways of blocking tissue specimens. The method of choice is dependent upon the laboratory's requirements.

Methods for Processing Specimens

The advent of closed-system processing, using vacuum/pressure and heat has much improved tissue processing technology, since it allows a great deal of processing latitude. Namely, it offers the ability to program a processing system for short, quick schedules to accommodate small tissues, or lengthy schedules for extra large specimens. Presented below are a wide variety of schedules which should suit most requirements. It should be understood that the schedules provided can, and should be, altered to each laboratory's needs. Alterations to processing schedules should be based on the quality of tissue dehydration, clearing, and impregnation achieved. Any needed alteration in the processing schedule can be determined from the tissue impregnation information found on page 33 (Procedure for Evaluating Quality of Tissue Impregnation). It is particularly important to consider the system employed for processing before selecting a schedule. Processing will be quicker and more effective with the use of so-called enclosed systems which incorporate vacuum, pressure and heat. Conversely, the same schedule will require longer exposure periods in each solution if an older processing system (which exposes tissue to air as a result of its up and down movement) is used.

Overnight Surgical Processing Schedule

(To be used with non-enclosed processing systems)

In order to insure complete fixation and processing in conventional tissue processors, tissue must not be thicker than 5 mm.

Station	Reagents	Time
1.	10% Buffered neutral formalin	3 hours
2.	10% Buffered neutral formalin	3 hours
3.	80% Isopropyl alcohol	1 hour
4.	95% Isopropyl alcohol	30 min.
5.	95% Isopropyl alcohol	30 min.
6.	100% Isopropyl alcohol	1 hour
7.	100% Isopropyl alcohol	1 hour
8.	Xylene or clearing agent of choice	1 hour
9.	Xylene or clearing agent of choice	1 hour
10.	Xylene or clearing agent of choice	30 min.
11.	Paraffin	2 hours
12.	Paraffin	1 hour

Embed and section.

Note: A 30-minute exposure of tissue to a third paraffin bath is recommended prior to embedding. Rotate solutions every other day and change the entire setup every week or after processing 500 tissue cassettes, whichever comes first. It is assumed that the liquid content of each of the above stations is approximately 1000 ml. Isopropanol is a very good alcohol for dehydrating tissues prior to exposure to the clearing agent. However, any alcohol can be used in above dehydrating stations.

Semi-Rapid Surgical Processing Schedule

(To be used with enclosed processing systems)

Tissue must not be thicker than 5 mm. As with all processing schedules, the one presented below is to be used as a guide which should be altered to suit individual laboratory requirements. This is necessary since there are some enclosed systems which may not process tissue as effectively as others and therefore the schedule may have to be altered. The reagents and time schedule provided below requires that the enclosed processing system being utilized, possess at least two of the three operational capabilities; vacuum, pressure and heat.

Station	Reagents	Time
1.	10% Buffered neutral formalin	2 hours
2.	10% Buffered neutral formalin	2 hours
3.	70% Isopropyl alcohol	30 min.
4.	80% Isopropyl alcohol	30 min.
5.	95% Isopropyl alcohol	30 min.
6.	95% Isopropyl alcohol	30 min.
7.	100% Isopropyl alcohol	45 min.
8.	100% Isopropyl alcohol	45 min.
9.	Xylene or clearing agent of choice	45 min.
10.	Xylene or clearing agent of choice	45 min.
11.	Paraffin	1 hour
12.	Paraffin	1 hour
Embed and section.		

Note: Some processing systems may not require the 4 hours indicated above for fixation. For example, those systems which utilize heat, vacuum and pressure can fix a 5 mm piece of tissue well in 2 hours. Solutions used in the enclosed systems do not have to be changed as often as in conventional processing systems. In this case, rotate solutions every third day and change the complete setup every 10 days or after processing 800 tissue cassettes, whichever comes first. It is assumed that the liquid content of each of the above stations is approximately 1000 ml. Isopropyl alcohol is recommended for dehydrating tissue prior to exposure to the clearing agent, but any alcohol may be used.

Biopsy Processing Schedules[12]

Provided below are several schedules for rapid processing of small surgical specimens (biopsies). The schedules should be used with enclosed systems.

Method No. 1

Four (4) hour processing schedule for tissue specimens which do not exceed 3 mm in thickness.

Station	Reagents	Time
1.	Formalin-alcohol solution	55 min.
2.	95% Isopropyl alcohol	10 min.
3.	100% Isopropyl alcohol	10 min.
4.	100% Isopropyl alcohol	10 min.
5.	100% Isopropyl alcohol	15 min.
6.	100% Isopropyl alcohol	20 min.
7.	100% Isopropyl alcohol	20 min.
8.	Xylene or clearing agent of choice	15 min.
9.	Xylene or clearing agent of choice	15 min.
10.	Xylene or clearing agent of choice	15 min.
11.	Paraffin	20 min.
12.	Paraffin	40 min.

Embed and section.

Method No. 2

Two (2) hour processing schedule for tissues which do not exceed 2 mm in thickness. Since this schedule accommodates very small pieces of tissue, it is suggested that all specimens be placed in lens paper, teabags, or between biopsy pads for processing, to prevent loss.

Station	Reagents	Time
1.	Formalin-alcohol solution	25 min.
2.	95% Isopropyl alcohol	5 min.
3.	100% Isopropyl alcohol	5 min.
4.	100% Isopropyl alcohol	5 min.
5.	100% Isopropyl alcohol	5 min.
6.	100% Isopropyl alcohol	10 min.
7.	100% Isopropyl alcohol	10 min.
8.	Xylene or clearing agent of choice	15 min.
9.	Xylene or clearing agent of choice	10 min.
10.	Paraffin	15 min.
11.	Paraffin	15 min.

Embed and section.

Method No. 3

The following two (2) hour processing schedule differs from that above primarily in its use of Bouin's fixative solution instead of formalin-alcohol. The tissue specimens should not exceed 2 mm in thickness. As with the method above, the specimens should be placed in lens paper, teabags, or between biopsy pads for processing.

Station	Reagents	Time
1.	Bouin's solution	25 min.
2.	95% Isopropyl alcohol	5 min.
3.	100% Isopropyl alcohol	5 min.
4.	100% Isopropyl alcohol	5 min.
5.	100% Isopropyl alcohol	5 min.
6.	100% Isopropyl alcohol	10 min.
7.	100% Isopropyl alcohol	10 min.
8.	Xylene or clearing agent of choice	15 min.
9.	Xylene or clearing agent of choice	10 min.
10.	Paraffin	15 min.
11.	Paraffin	15 min.
Embed and section.		

Processing Schedules for Scrapings, Washings, and Needle Biopsies[13]

The following multipurpose schedules are designed for the processing of a wide variety of tissue specimens. Specimens included in this group are endometrial scrapings, washings, muscle and nerve needle biopsies. It should be pointed out that other small specimens which do not measure more than 2 mm in thickness can also be processed with these methods.

Tissue Scrapings

1. Fix scrapings in 10% buffered neutral formalin for 1 hour. (Carnoy's or Bouin's solutions are also excellent fixatives since they provide very good cellular detail.)
2. Place lens paper over a small sieve and pour scrapings on lens paper, allowing liquid to filter through. Tissue scrapings will collect on the lens paper. Fold the paper gently around collected material and place in tissue processing cassette. Process according to the schedule provided below.

Station	Reagents	Time
1.	95% Isopropyl alcohol	10 min.
2.	95% Isopropyl alcohol	10 min.
3.	95% Isopropyl alcohol	10 min.
4.	100% Isopropyl alcohol	10 min.
5.	100% Isopropyl alcohol	10 min.
6.	100% Isopropyl alcohol	10 min.
7.	Xylene or clearing agent of choice	15 min.
8.	Xylene or clearing agent of choice	15 min.
9.	Xylene or clearing agent of choice	15 min.
10.	Paraffin	20 min.
11.	Paraffin	20 min.
12.	Paraffin	20 min.
Embed and section.		

Remove cassettes from paraffin container and quickly, but gently, remove scrapings from lens paper with a dull knife or spatula. Place scrapings in molten paraffin and arrange as desired.

Muscle, Nerve Biopsies and Small Tissue

Fix in 10% buffered neutral formalin for 4 hours (1 hour if using enclosed processing system with vacuum, pressure and/or heat). The processing schedule provided above for scrapings should be used for processing these specimens. The only significant change involves the embedding of muscle or nerve specimens. These tissues should be embedded to demonstrate both cross and longitudinal sections, so as to make available the greatest amount of information to the investigator.

Endocardial Biopsies

Presented below is a method for the rapid processing of endocardial biopsies from transplanted hearts. The technique allows prompt assessment of the post-transplant status of the patient.

1. Immediately after removal from the patient, the biopsy specimen is placed in a cassette and fixed in the fixative of choice for a minimum of 30 minutes.
2. Rinse specimen with 95% isopropanol and place in approximately 100 ml of 95% isopropanol for 30 minutes on a shaker.
3. Rinse with 100% isopropanol and place in approximately 100 ml of 100% isopropanol for 30 minutes on a shaker.
4. Place the cassette containing the specimen in a container of xylene for 20 minutes on a shaker.
5. Place the cassette in 2 changes of paraffin for 15 minutes each.
6. Embed and section.

Methods for Preparing Cell Blocks

There are many methods available for producing cell blocks and the choice of method must be left to the individual laboratory. However, the method provided below has proven itself for the production of good cell blocks.

Cell Block (Routine Method)

1. Centrifuge specimen at 2000 rpm for 20 minutes or at 5000 rpm for 10 minutes.
2. Decant supernatant and transfer remaining sediment to a centrifuge tube.
3. Add 10% buffered neutral formalin to sediment and re-spin as in step 1.
4. Carefully remove tube from centrifuge.
5. Remove and discard supernatant.
6. Place concentrated sediment on a filter paper, wrap and secure with tie.

7. Process cell block using any of the processing schedules provided in this chapter depending upon cell block size.
8. Embed and section.

Cell Block - Agar-Agar or Gelatin Method

1. Pour a small amount of 10% agar-agar or 10% gelatin onto a glass surface and allow to gel.
2. Place concentrated sediment on the solidified gel.
3. Pour warm agar-agar or gelatin over sediment and allow to solidify.
4. Cut solidified gel containing specimen into 2-3 mm blocks if desired.
5. Process cell block using any of the processing schedules provided in this chapter, depending upon cell block size.
6. Embed and section.

Cell Block - Thrombin-Plasma Method

1. Centrifuge specimen at 2000 rpm for 20 minutes or at 5000 rpm for 10 minutes.
2. Decant supernatant.
3. Place one drop of thrombin into a clean centrifuge tube.
4. Add concentrated sediment to the thrombin and agitate.
5. Add one drop of plasma to centrifuge tube containing specimen.
6. After clotting occurs, fix specimen in 10% buffered neutral formalin for 2 hours.
7. Process cell block using any of the processing schedules provided in this chapter, depending upon cell block size.

Cell Block - Normal Plasma-Thromboplastin Method

1. Centrifuge specimen at 2000 rpm for 10 minutes.
2. Decant supernatant.
3. Add to the sediment:

a. Three drops of fresh, normal plasma. Mix gently with wooden applicator stick.
b. Three drops of thromboplastin (e.g., Simplastin).
c. Three drops .025% calcium chloride. Mix sediment gently with wooden applicator stick.
d. Let stand for five minutes.

4. Remove clot from centrifuge tube by inverting tube over lens paper.
5. Wrap clot in lens paper; place in tissue processing cassette.
6. Fix in 10% neutral buffered formalin for 2 hours or more.
7. Process cell block using any of the processing schedules provided in this chapter, depending upon cell block size.

Remarks:
Normal plasma, thromboplastin and .025% calcium chloride can be obtained from the hematology or coagulation department of the hospital. Refrigerate all reagents and observe expiration dates. This method eliminates the inherent difficulties encountered in removing centrifuge sediment intact from centrifuge tube.

Cell Block Methods Using Bouin's Solution[14]

Bouin's solution has been found to be an excellent fixative for cytological diagnosis of cell blocks. The method offers several advantages: a) Bouin's fixed material stains brilliantly and exhibits excellent distinct nuclear detail. b) It affords the advantage of allowing for the preparation of cell blocks from material that ordinarily would be too "soft" when prepared in the usual manner. c) Unlike material fixed in picric acid alone, material fixed in Bouin's solution does not require prolonged fixing before processing. In most cases, the material will be sufficiently hardened so that wrapping in lens paper is not necessary.

Sputum and Bronchial Washings

1. Collect specimen by method of choice.
2. Add Bouin's solution directly to a small amount of the specimen.
3. Mix well, but gently, with an applicator stick.
4. Place specimen in a tissue cassette and process according to any of the processing schedules provided in this chapter, depending upon cell block size.

Effusions, Gastrics, Urines, Etc.

1. Place sample in four centrifuge tubes, spin down, and prepare smears.
2 Combine the remaining sediments in one centrifuge tube and add approximately 10 ml of Bouin's solution. Mix well and allow to fix for 30 minutes.
3. Centrifuge specimen at 2000 rpm for 10 minutes.
4. Pour off Bouin's solution and invert tube on a paper towel so that all of the excess Bouin's solution drains off.
5. Loosen the "button" gently with the point of a sharp knife or the end of an applicator stick which has been shaved to a point on one side.
6. Transfer the button to a cassette and process according to any of the processing schedules provided in this chapter, depending upon cell block size.

Cell Blocks Containing No Visible Button[15]

The consistency of embedded cell blocks often varies. Some are firm and easily processed with other surgical specimens, while others are loosely formed or not visible at all. These are often impossible to process. The following procedure facilitates handling and processing of all types of cell blocks. This procedure may be used alone or in conjunction with membrane filtration techniques.

The following solutions should be prepared just before use:

Citrated Plasma (Ci-Trol*)

Add 1.0 ml of distilled water to pre-measured citrated plasma vial, reinsert stopper and invert gently to mix.

Thromboplastin-Calcium Chloride

Thoroughly mix equal amounts of activated thromboplastin** and calcium chloride, 0.02m***.

Procedure:

1. Centrifuge specimen for 10 minutes at 1000 or 1500 rpm.
2. Decant as much supernatant as possible. Leave drop in tip of centrifuge tube if no button is visible.
3. Depending on size of button, add 1-3 drops of freshly prepared Ci-Trol solution. (If there is no visible button, add 1 drop of the Ci-Trol solution.) Agitate gently.
4. Add twice as many drops (2-6) of freshly prepared thromboplastin-calcium chloride. Note: Agitate gently while adding thromboplastin-calcium chloride in order to trap entire button in clot.
5. Let stand for 1 minute; then add 10% buffered formalin or Bouin's solution. Invert test tube gently to insure that clot is completely immersed in fixative solution.
6. After fixation, place clot in cassette and process according to any of the processing schedules provided in this chapter, depending upon cell block size.

Remarks:
Clot may be wrapped in tissue paper to prevent loss of specimen. Centrifuge tubes must be rinsed several times in distilled water before use.

*Ci-Trol (Dade, cat. no. B4224-10)
**Activated Thromboplastin (Dade, cat. no. B4210-2)
***Calcium Chloride 0.02m (Dade, cat. no. B4211-2)

Cell Block Using Polyvinyl Alcohol Foam[16]

Body fluids to be embedded in cell block preparations often present a problem to both the histotechnologist and the pathologist. To obtain a satisfactory representation of the cells suspended in a fluid for diagnostic purposes, a sponge-like material, polyvinyl alcohol foam*, can be used. The polyvinyl alcohol foam is purchased in

12"x9" sheets, 5/32" thick. The sheets arrive in a moist state and first can be cut into squares 0.5x0.5 cm and allowed to air dry.

Procedure:

1. Collected, fixed fluid submitted to the laboratory is first examined for its gross appearance. (If the specimen arrives unfixed, fixative is added immediately and allowed to set for 30 minutes.)
2. The specimen is poured into one or more clear glass test tubes then centrifuged for 15 minutes at 1800 rpm.
3. The tubes are removed from the centrifuge and examined for the formation of concentrated cell buttons. If a button has formed, the supernatant is decanted and the button is resuspended (step 2). Then a square of the foam is dropped into the bottom of each tube to absorb the cells. If no visible button has formed, the supernatant is carefully decanted, leaving a few drops in the bottom of the tube. The sponge is then dropped in the test tube and the tube is agitated.
4. After absorbing the concentrated cells, the sponge is carefully removed from the test tube, either using forceps or by inverting the tube over a small piece of lens paper (approximately 5x5 cm).
5. Wrap the sponge block in lens paper and place into a cassette and process according to any of the processing schedules provided in this chapter, depending upon cell block size.

Remarks:
The polyvinyl alcohol foam has no detrimental effect on the cutting edge of the microtome blade and does not interfere with routine or special staining methods. The block can be wrapped in lens paper and processed along with any other routine tissues.

* Unipoint Laboratories, High Point, NC 27260.

Aqueous Taps And Vitrectomies[17]

Recent advances in ophthalmic surgery have included the vitrectomy procedure, and much earlier, the aqueous tap. Because of the minute amount of material obtained from these procedures, new methods are necessary to produce adequate slides.

With the use of the Shandon Cytospin SCA-0030* good, artifact clear slides can be obtained from as little as 0.5 ml of material. The specimen can be processed fresh or fixed, and the method also allows procedures for determining the presence of fats, enzymes, or other substances which are usually lost by paraffin embedded methods.

The Cytospin produces a monolayer of well-separated cells, since they spin out according to weight of the cells. There is almost no distortion, so the cellular detail is excellent and the staining is good.

Procedure:

1. If more than 10 ml of fluid are present with the specimen, spin down in a regular centrifuge and combine the resulting residues into a 10 ml sample or less.

2. If the amount of sample submitted is extremely small and four or more slides are desired, dilute the sample with up to 10 ml of 10% buffered neutral formalin.
3. Fill the Cytospin head with clean blotters and slides which have been cleaned with alcohol.
4. Using a 1 ml pipette, insert a maximum of 0.7 ml of specimen fluid into each well. (In order to get a representative sample, be sure to shake bottle before each insertion.)
5. Place cover on Cytospin head.
6. Set speed at 1200 rpm.
7. After a 10 minute spin, remove slides, add accession number and place on slide warming plate to dry until ready to stain, but for at least one hour.

Remarks:
This procedure may be used for any body fluid, including bone marrow specimens. Most stains can be performed with no further fixation of slides. However, if a Giemsa stain is to be done, the slide first must be fixed with methanol after drying.

* Shandon-Lipshaw, Inc., Pittsburgh, PA 15143.

Brain Processing Methods

The fixation, processing, embedding, sectioning and staining of brain specimens requires special handling and special procedures. It would be a difficult and mammoth undertaking to expound on all facets of brain technology, since many factors would have to be considered. For example, size of the original brain, size of the specimen selected for processing, whether the brain is from a human baby or from an animal. All these factors determine the procedure one uses. The following comments are limited to certain aspects of brain technology, but it is felt they will be extremely helpful in the production of quality microscopic slides of this organ, the processing of which presents an interesting challenge to histotechnologists.

The procedures below are intended for handling entire cross sections of brain specimens measuring not more than 10 mm in thickness. The procedures are intended for use with all conventional non-enclosed processing systems. (This would also include processing by hand.) Again, as with all processing schedules, these schedules may require some modification to achieve desired results. If an enclosed processing system is used, one can reduce the processing time of the schedule provided below by one fourth and achieve the desired results.

Human Adult Brain Method

1. The entire brain should be suspended in 20% buffered neutral formalin within a large container for 48 to 72 hours. No incisions should be made on the brain prior to suspending in fixative solution.

20% Buffered Neutral Formalin

Formaldehyde, concentrated (37-40%)	200.0	ml
Distilled water	900.0	ml
Sodium phosphate monobasic	4.0	gm
Sodium phosphate dibasic (anhydrous)	6.5	gm

The use of 20% formalin speeds the penetration of the fixative.

2. Place the entire brain in fresh 10% buffered neutral formalin for 4 weeks. Change fixative solution every two days.
3. At the end of 4 weeks the brain can be grossed (macrosectioned) either into entire slices or blocked into conventional-sized specimens ready for processing. If the brain specimens show signs of improper fixation, place specimens in a fresh change of fixative solution until they are completely fixed.
4. Wash brain specimens in cold running tap water for 4 hours.
5. Transfer brain specimens to 80% alcohol for 3 hours.
6. Place brain specimens on tissue processor utilizing the following 24 hour schedule:

Brain Method (24-Hour Schedule)

(This schedule is for cross sections of entire brain)

Station	Reagents	Time
1.	95% Alcohol	2 hours
2.	95% Alcohol	2 hours
3.	95% Alcohol	2 hours
4.	100% Alcohol	2 hours
5.	100% Alcohol	2 hours
6.	100% Alcohol	2 hours
7.	100% Alcohol	2 hours
8.	Xylene-alcohol (equal parts)	2 hours
9.	Xylene or clearing agent of choice	2 hours
10.	Xylene or clearing agent of choice	2 hours
11.	Paraffin	2 hours
12.	Paraffin	2 hours
13.	Paraffin	2 hours

Embed and section with selected method.

Human Baby Brain Method

(This method is for cross sections of entire brain)

The consistency of baby brains differs from that of adults and therefore will require a modification in the manner in which they are processed. Follow steps 1-3 of the schedule above for processing adult human brain specimens and then proceed with the following steps:

4. Wash in slowly running tap water for 1 hour.

5. Transfer specimens to 70% alcohol for 24 hours.
6. Place specimens in 80% alcohol for 24 hours.
7. Place specimens on tissue processor utilizing the entire 24-hour processing schedule provided for human adult brains.

Monkey Brain Method

(This method is for cross sections of entire brain)

Monkey brains are fixed and processed according to the same schedule, steps 1-3, provided above for processing adult human brain specimens. Then proceed with the following steps:

4. Wash specimens in tap water for 4 hours.
5. Transfer specimens to 80% alcohol for 24 hours.
6. Place in 80% alcohol for 24 hours.
7. Place in 80% alcohol for 24 hours.
8. Place specimens on tissue processor utilizing the entire 24 hour processing schedule provided for human adult brains.

Embedding Brain Specimens

1. Entire cross sections of brain specimens should be placed as flat as possible in a suitable embedding pan. Do not use too much force when flattening, since brain tissue may be brittle and crack with the force, introducing an artifact.
2. Cool paraffin to room temperature. Then place pan or embedding mold in cold running water to complete paraffin hardening.
3. It is important to prevent movement of the brain specimen while the paraffin is hardening. Movement at this point may produce bubbles in the paraffin, which, in turn, create lines across the tissue section during microtomy.
4. Entire brain specimens should be embedded individually since multiple embedding may cause sectioning problems.

Sectioning Brain Specimens

1. When sectioning brain specimens, never apply coolant to the tissue. It is preferable to cool the microtome knife only. (This will produce wrinkle-free sections.)
2. Floatation water bath temperature should be 43°-45°C.
3. Slide warming table temperature should be 40°C. If slides are dried in an oven, maintain the oven temperature at 40°C.
4. Trim excess paraffin from each side of the paraffin block. This enables the sections to stretch out easily as they come off the microtome knife.
5. If one experiences difficulty in obtaining a compression-free section, try one or all of the following suggestions:

a. Turn the microtome wheel slowly 50-60 revolutions per minute, but with an even motion.
b. Soak the block in cold water. Be careful not to oversoak the brain specimen as this may produce thick sections (more than 6 micrometers) which in turn, may fall off the slide during the staining process.

There are many good microtomes for sectioning entire cross sections of brain specimens. One should try a number of them before purchase to insure they meet the needs of the department.

Processing Of Large Specimens

The need for processing large tissue specimens is an infrequent occurrence. However, when the need arises, it can be very frustrating not to have a method for processing and embedding readily available. It is for this reason that a method for processing and embedding large specimens (up to 1.5 cm thick) is included here. The processing of these specimens can be accomplished by the use of the human adult brain method (24-hour schedule) found on page 44. After processing with that method, specimens are embedded in the manner outlined below.

Embedding Of Large Specimens In Paraffin

As indicated before, specimen embedding is the orientation of tissue in melted paraffin, which, when solidified, provides a firm medium for keeping intact all parts of the tissue when sections are cut. Each laboratory has a preference as to the method used for embedding; the most efficient for small specimens are plastic embedding cassettes. Embedding of large specimens can be accomplished by the use of a shallow pan which can be purchased commercially or made by a machinist. This method of embedding is used exclusively for embedding large specimens. Pans 8 by 10 inches in size and 1 to 2 inches deep with sloping sides are satisfactory. Sloping sides on the pan prevent it from sinking when placed in cold water.

Placed on a convenient stand, the pan is held about 6 inches above the desk top and is warmed gently with a Bunsen burner or a suspended 250-watt (115-120 volt) reflector infrared heat lamp. (This heat lamp may be purchased at most hardware or drug stores.) To aid in removal of the paraffin block the pan may be pre-coated lightly with glycerin. Next, fill the pan with melted paraffin and place each piece of tissue in position with an appropriate identifying tag. In order to achieve a flat surface for cutting, gently press specimen down with forceps while rubbing an ice cube across the bottom of the pan. Be sure that all parts of the tissue are flat. When the paraffin has cooled to the point that a heavy film has formed across the top, the pan should be carefully floated on cold water. The paraffin, when hardened throughout, will contract from the sides of the pan. Then the solid block can be lifted out and individual blocks cut to the appropriate sectioning size.

Whole Organ Processing[18]
Materials and Methods

These schedules are very flexible. Smaller specimens or thinner specimens would allow the use of fewer changes of each concentration of solutions and/or less time in each program step.

The processed tissue from each of these schedules should be embedded in paraffin that has been brought to the same temperature as the infiltration paraffin and the tissue specimen.

Supplies:
Gauze sponges
Thin piece of wire or string
Plastic food container
Chrome-alum gelatin adhesive
Large glass slides (6"x8", 4"x5", 3"x4")
Large coverglasses, (to accommodate the above slides)

Carefully wrap the specimen in a single layer of gauze and place it in the processor's specimen basket. Use a very thin, flexible piece of wire or string to anchor the specimen to the inside bottom of the basket. The purpose of this is to keep the solution that is being drawn into the retort from dislocating the specimen. The single layer of gauze does not interfere with the processing but does protect the specimen from trauma.

Schedules:
Whole Organ Brain - Cross section of both lobes, approximately 1 1/2 inches thick by 5 1/2 inches by 6 inches.

Station	Reagents	Time
1.	95% Alcohol	5 hours
2.	95% Alcohol	5 hours
3.	95% Alcohol	5 hours
4.	100% Isopropyl alcohol	5 hours
5.	100% Isopropyl alcohol	5 hours
6.	100% Isopropyl alcohol	6 hours
7.	100% Isopropyl alcohol	6 hours
8.	Xylene	5 hours
9.	Xylene	5 hours
10.	Xylene	8 hours
11.	Paraffin	3 hours
12.	Paraffin	3 hours

Whole Organ Heart - Cross section of both ventricles, approximately 3/4-inch thick.

OR

Whole Organ Lung - Cross section of lobe, approximately 1 1/2 inches thick.

Station	Reagents	Time
1.	95% Ethyl alcohol	1 hour
2.	95% Ethyl alcohol	2 hours
3.	95% Ethyl alcohol	2 hours
4.	100% Isopropyl alcohol	1 hour
5.	100% Isopropyl alcohol	2 hours
6.	100% Isopropyl alcohol	2 hours
7.	100% Isopropyl alcohol	2 hours
8.	Xylene	2 hours
9.	Xylene	3 hours
10.	Xylene	3 hours
11.	Paraffin	3 hours
12.	Paraffin	3 hours

Processing Methods Using Different Dehydrating or Clearing Agents

Methyl Salicylate Processing Method

Methyl salicylate has been used successfully for clearing tissue specimens, particularly large bone specimens. It is a colorless, yellow, or reddish, oily liquid with the aroma of oil of wintergreen. It is slightly soluble in water and is miscible with alcohol. It is used in the fragrance industry and for flavoring candy. Its toxicity to the histotechnologist is difficult to assess. Although the method provided below was designed for large bone specimens, any specimen up to 5 cm in thickness can be processed successfully with it. After fixation, process the specimen according to the following schedule.

Station	Reagents	Time
1.	Running water	3 hours
2.	70% Isopropyl alcohol	8-11 hours
3.	95% Isopropyl alcohol	2 hours
4.	100% Isopropyl alcohol	6 hours
5.	Methyl salicylate	overnight
6.	Methyl salicylate	3 hours
7.	Paraffin	8 hours
8.	Paraffin	8 hours
9.	Paraffin	8 hours
10.	Embed in large pan.	

Remarks:
As with all processing schedules, the suggested processing time can be adjusted to suit the specimen's size and density. For example, processing of large eyes (cow or horse) is possible with this method but the processing time is reduced considerably. In this case, the entire schedule can be reduced by half with the exception of the paraffins, which are reduced to 3 hours each and two changes instead of three.

Cedarwood Oil Processing Method

(Specimens 5mm thick or less)

This method is useful for very delicate as well as hard or dense fibrous tissues, such as skin and uterus, because it has little hardening effect. In fact, tissues may be left in cedarwood oil for months without any excessive hardening.

Station	Reagents	Time
1.	80% Alcohol	1 hour
2.	80% Alcohol	1 hour
3.	95% Alcohol	1 hour
4.	95% Alcohol	1 hour
5.	100% Isopropyl alcohol	1 hour
6.	100% Isopropyl alcohol	1 hour
7.	100% Isopropyl alcohol	1 hour
8.	100% Alcohol and xylene (equal parts)	1 hour
9.	100% Alcohol and xylene (equal parts)	1 hour
10.	Oil of cedarwood	2 hours
11.	Oil of cedarwood	2 hours
12.	Oil of cedarwood	1 hour
13.	Paraffin	2 hours
14.	Paraffin	1 hour
15.	Paraffin	1 hour
16.	*Paraffin	1 hour

Embed and cool quickly. Section as desired.
*Vacuum, if possible, in this station.

Dioxane Processing Method

(Specimens 5mm thick or less)

This method has several advantages. Since dioxane will act as both a dehydrant and a clearing agent, it can be used for rapid processing. Specimens may be left in dioxane for long periods without excessive hardening. Thus, tissue specimens can be left in this reagent overnight and routine processing continued in the morning. Dioxane is toxic and should be used under a hood and with extreme care.

Station	Reagents	Time
1.	Dioxane and water (equal parts)	1 hour
2.	Dioxane	1 hour
3.	Dioxane	2 hours
4.	Paraffin	1 hour
5.	Paraffin	1 hour
6.	Paraffin	1 hour
7.	*Paraffin	1 hour

Embed and cool quickly. Section as desired.
*Vacuum, if possible, in this station.

Tetrahydrofuran Processing Method

(Specimens 5mm thick or less)

Tetrahydrofuran can also serve as both a dehydrant and clearing agent, accelerating processing time considerably. It is possible to process thin tissue specimens that do not exceed 3 mm in thickness in 5 hours. Tetrahydrofuran is less toxic than dioxane but should be used under a hood or in a well-ventilated area.

Station	Reagents	Time
1.	Tetrahydrofuran and water (equal parts)	1 hour
2.	Tetrahydrofuran	1 hour
3.	Tetrahydrofuran	1 hour
4.	Paraffin	1 hour
5.	Paraffin	1 hour
6.	*Paraffin	1 hour

Embed and cool quickly. Section as desired.

*Vacuum, if possible, in this station.

Celloidin-Paraffin Double Impregnation Method

This method of embedding in a combination of celloidin and paraffin may prove advantageous in instances where the preservation of the morphological relationships is paramount. It is for specimens not in excess of 10 mm in thickness.

Clearing Oil Solution

Chloroform	20.0	ml
Oil of origanum	20.0	ml
Oil of cedarwood	40.0	ml
100% Alcohol	10.0	ml
Phenol (carbolic acid)	10.0	ml

Station	Reagents	Time
1.	60% Alcohol	3 hours
2.	80% Alcohol	3 hours
3.	80% Alcohol	3 hours
4.	95% Alcohol	2 hours
5.	95% Alcohol	2 hours
	(For convenience, the specimen may be left overnight in this solution.)	
6.	100% Alcohol	2 hours
7.	100% Alcohol	2 hours
8.	100% Alcohol	2 hours
9.	Ether - alcohol (equal parts)	2 hours
10.	Ether - alcohol (equal parts)	2 hours
11.	Celloidin, 2%	24 hours
12.	Clearing oil solution	8 hours
13.	Benzene	4 hours

continued on next page

Station	Reagents	Time
14.	Benzene	4 hours
	(For convenience, the specimen may be left overnight in this solution.)	
15.	Paraffin	1 hour
16.	Paraffin	1 hour
17.	Paraffin	1 hour
18.	Paraffin	1 hour
Embed in paraffin and cool quickly. Section specimen as desired.		

Special precautions should be taken when using chloroform and benzene as they are carcinogenic; refer to the appropriate safety information before using these agents. Also, take special precautions with the use of both ether and celloidin; they are extremely flammable. Cutting double-embedded specimens may present some problems since the normal characteristic of paraffin is to expand on the floatation bath, while at the same time celloidin has a tendency to shrink. To prevent difficulties, float section in 95% alcohol for 20 minutes to soften the celloidin before placement on floatation bath.

Carbowax Method

Polyethylene glycols (Carbowax) are water-soluble waxes. Thus, their use eliminates the need for dehydration and clearing, consequently saving valuable processing time. Most stains can be applied to the tissue section after processing in carbowax. Very little shrinkage and/or tissue distortion is produced with the use of carbowax.

1. Fix small sections of tissue in the desired fixative.
2. Wash well in running water.
3. Impregnate in 56°C oven for 3 hours in a mixture of nine parts carbowax 4000 to one part carbowax 1500. (This mixture should be prepared in advance and kept in an incubator.) The quantity of each component of the mixture may be varied as required by the individual laboratory and dictated by local conditions in order to control the hardness of the resulting embedding medium.
4. After 2 to 3 hours in incubated carbowax mixture, place sample in a fresh mixture of melted carbowax in a small embedding pan, paper boat, or plastic cassette. Chill in refrigerator for 30 minutes. Remove block from embedding mold.
5. Cut on rotary microtome at room temperature. Place sections directly on the slide and press down with a gently rolling motion of the index finger. If this technique is not satisfactory, the sections may be floated and fixed to the slide using the following solution in the floatation waterbath:

Potassium dichromate	0.2	gm
Gelatin	0.2	gm
Distilled water	1000.0	ml

6. Place slide on warming table at a temperature not exceeding 40°C.
7. Stain as desired.

Remarks:
Carbowax will become brittle if overheated, and crumbling may occur during microtomy. In addition, the block must not be allowed to come in contact with ice orwater, as carbowax is very hygroscopic. If hardening of the block is necessary to eliminate section compression during cutting, it can be hardened with a commercially available skin refrigerant or dry ice. A convenient and useful method is the placement of a 32-oz funnel filled with dry ice in a ring stand with the nipple of the funnel 3-4 inches above the tissue specimen (which is oriented on the microtome ready to be cut). The cold fumes from the dry ice, being heavier than room-temperature atmosphere, shoot downward, enveloping the specimen. This method keeps the immediate area of the block knife in a considerably colder state, facilitating carbowax cutting. Store carbowax-embedded blocks in a cool, low-moisture area.

Method for Reprocessing/Processing Dried Tissue Specimens[19]

There are numerous ways in which tissue specimens can become "dried out" and thus require "reprocessing". Drying may be the result of a malfunction in older model tissue processors, the tissue having dried before or after fixation and/or improperly processed paraffin blocks. In any case, the specimens may be salvaged by using part or all of the reprocessing schedule below.

Tissue Processor Malfunction

Processor malfunction or breakdown may leave the tissue basket hanging in the air or drop it on the lip of the beaker preventing proper insertion into solution. This causes the tissue to dry before being exposed to the paraffin bath. When this happens, follow steps 6 and 7 of the reprocessing schedule below to correct the problem.

Dried Tissue Before or After Fixation

Drying of tissue can also occur when tissue is inadvertently left out of fixative overnight or the fixative evaporates from the container. Specimens obtained from a cadaver that has been exposed to the elements for several days would likewise be dry, hard and brittle. In these circumstances, when tissue is dried either before or after fixation but before processing, follow steps 6 and 7 of the reprocessing schedule below.

Improperly Processed Paraffin Blocks

Finally, if at microtomy paraffin blocks do not section well because of improper impregnation with paraffin, reprocessing can be accomplished by following all steps of the reprocessing schedule.

Tissue Softening Solution

The following solution is used for softening dried tissue specimens prior to reprocessing or processing.

Formol-Sodium Acetate (Stock) Solution		
Concentrated formalin (37%) solution	10.0	ml
Sodium acetate	2.0	gm
Tap water	90.0	ml

Formol-Glycerol (Working) Solution		
Formalin-sodium acetate (stock)	90.0	ml
Glycerol (glycerin)	10.0	ml

Reprocessing Schedule

1. Place paraffin block in molten paraffin (60°C) for 2 hours.
2. Remove tissue sample and place in 3 changes of xylene for 1 hour each.
3. Place in 100% absolute alcohol, 2 changes for 1 hour each.
4. Place in 95% ethyl alcohol, 2 changes for 1 hour each.
5. Place under running tap water for 20 minutes
6. Place in formol-glycerol working solution until tissue becomes soft. (In most instances this will occur within 5-8 hours; extended exposure to the formol-glycerol will not harm tissue specimens.)
7. Place specimen in tissue processor and proceed with routine processing schedule. However, if your processing schedule includes several stations of formalin fixation, it is not necessary for tissue to be exposed to those steps.

Remarks:
The author has used this formol-glycerol tissue softening method successfully on many occasions. Remember, however, that the procedure may require modification of the exposure time (step 6) to achieve desired results.

Alternate Method for Rehydration of Desiccated Tissue[20]

The following method is useful for tissues that have undergone desiccation due to tissue processor malfunction or breakage of specimen container during mailing. This procedure involves boiling the tissue in a wetting agent for 20 minutes, allowing removal of any paraffin and complete rehydration. Processing, cutting, and staining can be accomplished in the same day while excellent microscopic histology is retained. Many common wetting agents, including laboratory detergents, can be used; however, *Leconal has been found to be the most effective.

1. Leave tissue in processing cassette and immerse cassette in a 10% aqueous solution of Leconal.
2. Bring solution to a boil and allow to boil for 20 minutes.
3. Place container in a pan of ice water. As the solution cools, the paraffin will rise to the top and solidify. Remove paraffin from water surface.
4. Wash tissue in warm water for 10 minutes.
5. Reprocess tissues and embed in conventional manner.

Remarks:
Only one disadvantage to this procedure has been encountered, and that is in the processing of lungs which have been fixed with formaldehyde gas and air dried. Although the cellular histology is excellent, the lung tissue shrinks by approximately one-half. This is a disadvantage, however, only when morphometry is critical.

*Leconal, Scientific Products, Evanston, IL 60201.

General Comments

- Tissue processing is the formation of a matrix by the aggregation of specific medium which ties together tissue structure. This tying in with an impregnating medium prevents disruption of tissue structure during microtomy.

- Dehydration is the removal of all extractable water by a dehydrant diffusing through the tissue, and in the process, diluting itself 2 - 4%.

- Alcohol is the most commonly used dehydrant. This includes denatured and isopropyl alcohol.

- Isopropyl alcohol is an excellent dehydrant which can be used in tissue processing as well as dehydrating hematoxylin and eosin stained slides.

- Bound water is water molecules in tissues of the body bound to macromolecules or organelles. Bound water must remain attached to macromolecules to allow for reabsorption of moisture during microtomy and to provide even, consistent staining throughout the tissue section.

- Dehydration must be gradual and thorough. There is a great deal of difference in density between water (specific gravity of 1.00) and absolute alcohol (specific gravity of 0.79). Transferring tissue from water or weak alcohol directly to strong alcohol results in violent diffusion, possibly causing shrinkage and distortion of tissue structures.
- Isopropyl alcohol (99%) is the best all-around substitute for ethyl alcohol for dehydrating tissue. This is particularly beneficial since one does not have to maintain a narcotics' log for isopropyl alcohol. In mixture with various paraffin solvents it tolerates small amounts of water at least as well, on the average, as ethyl alcohol mixtures.

- Since most water-miscible fluids are not paraffin solvents, it is usually necessary to dehydrate first with a water miscible fluid, and then to replace the dehydrating agent with a paraffin solvent which is also miscible with the dehydrating agent used.

- Silica gel can be used to remove moisture from dehydrating agents, thereby prolonging their usefulness.

- Tertiary butyl alcohol can be used both as a dehydrant and clearing agent. Tertiary butyl alcohol produces 10% less shrinkage than denatured, ethyl, or isopropyl alcohol, when the tissue is exposed to subsequent treatment with clearing agents and/or paraffin.

- Dioxane is a rapid dehydrant, but the fumes are toxic and its use requires careful control in well-ventilated areas.

- Dioxane can be used as both a dehydrant and clearant. Specimens remain fairly soft after dehydration and clearing in dioxane.

- Dioxane should not be used for routine processing. It tends to harden tissue excessively.

- The use of dioxane is recommended when it becomes necessary to expose tissue to a dehydrant or clearing agent for extended periods of time (up to 48 hours) as with large specimens.

- In general, tissues fixed in formalin and dehydrated/cleared in dioxane stain more poorly than those dehydrated with alcohol and cleared with xylene.

- Tetrahydrofuran possesses dehydrating and clearing qualities similar to dioxane; namely, tissue can remain in it for extended periods of time without becoming overly hardened.

- Xylene is the most commonly used clearing agent in histotechnology. It is one of the most difficult clearing agents to remove from tissue specimens, generally requiring longer tissue exposure to paraffins. Xylene carryover will not evaporate from paraffin pots as rapidly as other clearing agents; therefore the paraffin baths must be changed more frequently. Xylene has a slightly higher viscosity than most other clearing agents - 0.7 compared to 0.4 for acetone and 0.6 for chloroform. Worse yet is dioxane at 1.2. The viscosity of clearing and dehydrating fluids affects the rate at which they penetrate tissues. This is because the larger the molecule the higher the viscosity and the slower the rate of penetration. This is particularly important in the impregnating stages of tissue processing. Heat reduces the viscosity of impregnating substances, and therefore it is an essential part of the paraffin wax impregnation process.

- One of the most frequent problems encountered in processing small biopsies and/or body fluids for cell blocks is the lack of sufficient material. The following simple procedure makes finding minute specimens much easier. When a specimen is received in the laboratory, immediately fix it in 50 ml of 50% ethyl alcohol to which 3 to 5 ml of 0.4% eosin has been added. Allow the specimen to remain in this solution for approximately 30 minutes before placing on lens paper and then in the tissue processor. The specimen, stained by eosin, will retain the dye during processing, making it much easier to find for embedding.

- Vas deferens are easily identified by the following method. Fix vas deferens in 10% formalin. After fixation, dip the right vas in Mercurochrome and blot dry with a paper towel. The left vas is not dipped in the Mercurochrome. Both can now be processed together. Label slide "L & R" for identification of specimens. After microtomy, pick up section so that red stained vas is immediately below "R" marked on slide.[21]

- Cleaning paraffin off the counter and from around embedding centers can be quite a time-consuming effort. Tissue-Tek Mold Release compound (Miles, Cat No. 4141) can be used successfully for the purpose of removing paraffin from embedding centers. First, rub the compound on the embedding center and other areas on to which paraffin usually drips, before beginning work. Upon completion of the embedding process, simply take a gauze pad and wipe paraffin from the counter and embedding center. The paraffin will wipe off with ease and does not require scraping.[22]

- When transferring tissue from paraffin impregnating container to embedding mold, do not allow a thin layer of paraffin to solidify around the specimen. If it does, it must be completely remelted to avoid a hairline separation between the tissue and the embedding medium which will cause problems during microtomy.

- Paraffin must be cooled rapidly after tissue embedding since slow cooling often produces minute air bubbles that result in the cutting of poor quality sections.

- The poor embedding practice of placing tissues in the embedding mold in any direction produces potential cutting problems. Extra care must be used when embedding multiple pieces of tissue in the same block in matching types and densities of tissue. It is also very important to orient tissues so that the final mass forms a square or rectangle.

- Tissue specimens shrink approximately 20% during processing through alcohol clearing agents and impregnation in paraffin.

- Tissue specimens that shrink greatly during fixation will normally shrink less during processing; conversely, tissues showing little shrinkage during fixation will shrink considerably during processing. The 60°C heat used in paraffin processing will cause increased shrinkage of tissue specimens. For example, a formalin-fixed specimen infiltrated with a 34°C paraffin will shrink less than a similar specimen exposed to a 60°C paraffin. Also, the specimen exposed to the 34°C paraffin will remain quite a bit softer for sectioning. It is interesting to note that the microscopic picture of both are similar. It is believed that hardening of tissue excessively exposed to hot paraffin occurs only if the specimen has not been completely dehydrated. In my opinion, all stages of section preparation, from fixation to impregnation, play a part in tissue shrinkage, either singularly or collectively.

- Incomplete dehydration will result in poor clearing and subsequent impregnation with paraffin.

Chapter 3

Microtomy of Specimens

The word microtomy is derived from the Greek words mirkos (micro) which is a combining form meaning small, and temnein, which means cut. Mirkos + temnein defines the word microtomy (histotomy), which is the process of cutting thin sections. Many factors must come together for the cutting of thin sections, and even more for the cutting of high quality ones. The degree and quality of patient care could easily be compromised by the quality of the section produced in the laboratory. Therefore, every effort should be made to ensure that all facets of histotechnology leading up to and including microtomy be performed with diligence and care. Microtomy can only be as good as the quality of the tissue fixation, dehydration, clearing, and impregnation which preceded it. Equally important to achieving good microtomy is the histologist's expertise. Since one must handle extremely small and delicate structures and/or instruments, excellent manual dexterity is a must. Likewise, the histologist's training and practical experience should provide sufficient knowledge upon which to base decisions concerning potential problems, so that poor sectioning is unlikely (Fig. 19 a-j). This knowledge should include familiarity with all equipment being used, a thorough understanding of the theory of sectioning and the amount of moisture or coolant needed, and the ability to recognize when the best possible sectioning has been obtained in a given circumstance. One should be able to recognize, for example, that some tissues (such as keratin, decalcified bone, chitin and hemorrhage) require a good deal of moisture for proper sectioning, while others (such as kidney tubules, sebaceous glands, fat cells, and intestinal villi) require very little. One should also recognize when the use of a coagulant fixative may necessitate the application of additional moisture, because of excessive tissue hardening. Finally, the histologist must be able to determine the quality of a microtome blade's facet surface and understand at the molecular level the action of the blade on tissue and embedding media. Only then can one determine whether bad sectioning is due to a poor knife edge or to poor processing/embedding of the specimen.

Theory of Cutting

The mechanism of microtomy can best be explained by briefly stating what constitutes a sharp knife and, secondly, what constitutes the cutting process. The sharpness of a knife may be visualized as the convergence of the beveled facets to a mathematically exact bevel angle (intersection of two planes). This convergence may be defined in terms of the average radius of edge curvature. A small average radius of edge curvature implies relative sharpness to a mathematically exact bevel. It has been estimated that the radius of curvature of a maximally sharp microtome knife is 0.3 micrometers (Fig. 29). The second aspect has to do with the cutting process which was proposed by Wachtel et al[23]. These authors claimed that cutting is the rupturing of molecular bonds which lay in the path of the advancing knife edge (Fig. 20a, b, c and d).

The scheme illustrated above does not help in defining the best possible edge for the rupture of molecular bonds, nor in explaining empirical findings concerning knife effectiveness. Numerous reports exist in the literature in which various methods or schemes have been tried to improve knife effectiveness: Nitric Acid Etchings[24], Tungsten Coating[25], and Reduced Facet Angel[26]. These reports, and many others, suggest improved cutting efficiency with the use of selected treatments to the knife facet.

The Knife's Amorphous Bevel Surface

In my opinion, the cutting effectiveness (efficiency) of a microtome blade is determined not by sharpness alone (the convergence of the beveled facet to a mathematically exact angle), but also by the molecular status of the knife's facets. Specifically, the surface of the facets should be in an amorphous state, not a crystalline state (Fig. 21, 22 and 23). That being the case, it is essential to produce an amorphous surface on the microtome blade's facets, or to purchase disposable blades which have been coated with an amorphous substance. Regardless what kind of blade one uses, it is impossible to produce high quality tissue sections without an amorphous blade surface.

An amorphous surface can be produced by polishing the knife's facets. This can be performed by hand or by mechanical means. Polishing the blade's facets alters the crystalline structure of the metal surface and causes a molecular flow whereby the surface becomes covered with an amorphous layer, which may be likened to a highly viscous liquid. The development of an amorphous surface on a knife's facet at the molecular level can best be explained by the following analogy: With continual polishing the surface of a crystalline structure is seen to fade slowly as if one were observing through a powerful microscope the dissolving of crystal supernatant on a solvent. This principle is explained in Fig. 21, 22 and 23. The schematic shows the surface (at the molecular level) of a metal microtome blade with the free end of the crystalline structure projecting at the facet surface.

Manual Method For Producing An Amorphous Surface

The proper polishing of a microtome blade's facets can be accomplished by proper stropping on a linen or leather strop for 15 to 30 seconds. This degree of polishing is sufficient to develop the amorphous layer on the knife facets. During the polishing process it is important that the blade facet be completely flat on the stropping system used. Too much facet angle will result in a dull knife, because excessive polishing takes place on or close to the intersection of the two facet planes. Similarly, insufficient facet angle will result in the polishing of the heel of the facet, again making polishing ineffective. Figure 25 illustrates the frame work which can be constructed for producing a stropping (polishing) system for use with linen cloth. Figure 26 provides a diagrammatical representation of proper stropping. The first requirement is that the microtome blade must have a back and handle. The stropping system is held firmly with the operator's left or right hand as illustrated. (For safety's sake, it is best to develop a portable, but sturdy, clamp holder to secure the stropping system in a fixed position during stroppings.) The mechanics of learning proper stropping can be confusing, and therefore, diligence and perseverance are required during the stropping process. It is suggested that the stropping process be practiced with the use of a completely blunted microtome blade. This would eliminate the possibility of nicking the stropping system or causing injury. As illustrated in Fig. 26, the knife handle is held from the side so that the blade's facet faces toward the technician's other hand.

At this point it is well to practice rotating the knife with your hand and fingers; the blade should move in a fanning motion from left to right. The knife is placed on the linear strop with the facet facing the left side of the operator (Fig. 26a). The stropping process is started by placing the toe of the blade on the lower edge of the strop and pushing the blade at a 45° angle across the strop. When the blade reaches the other end of the strop, the blade is rotated toward the right (the blade facet will now face to your right). At this point the knife is pulled toward you and the facet allowed to rest again on the strop surface. The blade is now pushed/pulled toward the left of the strop at a 45° angle as illustrated in Fig. 26a. This process is continued for six to eight stropping motions. After this is completed, the blade sharpening pattern is repeated but starts with the heel end of the knife on the strop (Fig. 26b). The blade is pulled toward the right and down at a 45° angle. It is then rotated with the fingers and hand (counter to what was done before) and the knife pushed away across the strop. The knife is now pulled down and toward you at the 45° angle discussed earlier. Stropping of a microtome blade on a linen or leather strap can be accomplished in 15 to 30 seconds. This degree of polishing is sufficient to develop the amorphous layer on the knife facet. It should also be mentioned that the stropping of the blade toe-to-heel and heel-to-toe as illustrated in Fig. 26a and b, provides crisscrossing of the stropping patterns. This insures that most, if not all, of each facet's surface is converted from the normal metal crystalline state to an amorphous one at the molecular level. Also, I have found that a linen strop is preferable. A linen strop will "give" during the stropping process, allowing the linen to bend itself around the knife's facets.

Automatic Methods for Producing an Amorphous Surface

An automatic or mechanical knife sharpener must also produce an amorphous bevel surface. Therefore, the purchase of such a system, to some degree, must be based on how well the final polishing of the facet is performed. Is the final action of the knife sharpening system sufficient to provide a good polish of the knife's facet? Various polishing systems have been incorporated into knife sharpness. In some, the polishing is performed by a very fine abrasive, glass plates, or a semi-viscous liquid. Other systems incorporate special materials, such as leather or psuedo-leather, while still others may incorporate specially treated paper for final polishing. Regardless of the polishing system used, the most important thing to remember is that metal microtome blade's facets must possess an amorphous surface in order to achieve improved coefficiency as the knife travels through the paraffin/tissue block.

Disposable Microtome Blades and the Amorphous Surface

Most, if not all, disposable blades are sharpened (and polished) by the manufacturer in the same way. Namely, the blades are sharpened and polished while still in long bands. The bands are exposed to rotating wheels which are made of some amorphous material, such as leather, plastic or glass. This method of sharpening and polishing includes blades which are used for shaving. One must remember that regardless of which blades one considers, the coefficiency of the blade facet surface is determined by whether the facet contains an amorphous surface or not. I'm certain that every histologist has experienced varied cutting efficiency with different blades from the same distributor.

The reason that some disposable blades, even from the same container, cut better than others is due to the effectiveness and quality of polishing of that specific run of blades by the manufacturer. There are, however, some manufacturers of disposable blades who do not use polishing to develop an amorphous knife surface. Instead of being polished, the facet surface of the blade is coated with a non-crystalline material. This bevel surface coating can replace the formation of an amorphous layer by polishing, because at the molecular level it possesses the same qualities.

The Coefficiency of an Amorphous Facet Surface

The effects of an amorphous surface can be equated to a highly viscous layer or super cooled liquid. This results in the facet's surface exhibiting the properties of plastic flow; that is, being conformable, capable of being molded or malleable at the molecular level. This capability allows the surface of the blade's facet to be fashioned (shaped) to the needs and dictates of the specimen being cut. This results in reduction of friction between the knife's bevel and the specimen being sectioned. Consequently, an amorphous surface cuts 20 to 30% better than a crystalline surface. Many cutting artifacts and other microtomy problems are the direct result of the lack of an amorphous surface. The reason the glass knives used for plastic sections cut very thin and work so well, is that glass is an amorphous material. This is also true of diamond knives used for electron microscopy.

Even though the amorphous knife facet is a critical component of microtomy, other factors are also involved in the production of good sections. For example, the proper knife rake angle, facet clearance angle, and the quality of the paraffin. In addition to these factors is the need to moisturize tissue constituents. The preparation of sections can best be performed after one understands the theory of the cutting process, the important aspects of an amorphous surface on the microtome knife's facets and the effects of moisture in tissue on the cutting process. The requirements of a good technician were explained in the introduction to this chapter, and will not be repeated here.

Microtome Blades with Different Profiles

In the author's opinion there is no need for the use of microtome blades with different cutting profiles to section tissues with different densities, as suggested in many books on histological techniques. Most modern manufactured microtome blades (including disposables) will cut material of varying hardness and composition, as long as all the requirements on microtomy, cited earlier in this chapter, have been met.

Microtome Knife Blade's Facet Angle

The microtome blade's facet angle is important, to some extent, in reducing resistance between the knife and material being sectioned. Figure 27 provides useful information on the appropriate blade facet angle used during microtomy. As indicated in Fig. 27, the critical clearance angle, that is, the angle that is the dividing line between efficient and inefficient cutting of a section, is 5-10° A microtome blade's facet with the proper angle will cut a section well (Fig. 24a), whereas an angle which is more negative will result in a plowing effect through the tissue and production of microvibrations (Fig.

24b). It should be recognized that pressure points (small arrows) result in the production of microvibrations. The effect here is competing pressures points which results in tissue compressing at that point and bending on contact. The thick and thin effects (microvibrations) are obvious when the section is layed flat (unfolded) on the surface of the floatation bath. This particular type of microvibration artifact can be reduced considerably by a more negative blade rake angle and the moisturizing process mentioned later in this chapter.

The Tissue Sectioning Process

Orientation of Specimen Blocks

Before one begins the sectioning process of a specimen block, it is important to achieve proper orientation of the block in the microtome so that one obtains the most representative section possible. The long axis of the paraffin block should be parallel to the blade facet's free end. Tissue capsules, dense and hard to cut tissue (keratin, fibrous tissue, etc.) should be at the upper portion of the specimen block. Sufficient paraffin should be left around the specimen to allow easy section separation on the floatation water bath. Paraffin should be trimmed from the sides of difficult-to-section specimens to allow improved sectioning and better spreading of the section on the floatation water bath. Note: Early sections should be evaluated while on the floatation water bath to insure that the best possible sections have been obtained. If the sections are unsatisfactory, one should take whatever corrective actions are necessary before attempting to cut more sections.

Macrosectioning (Rough Cutting) of Specimen Blocks

Set the microtome micrometer gauge at desired micron setting. Macrosectioning (rough cutting/macromicrotomy) of sections should be done carefully since the moisture in the specimen has been reduced considerably during processing. Careless macrosectioning results in cracking and gouging (by the knife) of the specimen's surface, which then results in section artifacts. Secondly, there is always the possibility of the specimen, or a portion of it, popping out of the embedding matrix.

Moisturizing the Specimen

The introduction of water (moisturizing) into tissue constituents is best performed by first soaking a piece of cotton in the floatation water bath. The cotton is then gently squeezed to rid it of excess water. Place the moist cotton on the surface of the previously faced (macrosectioned) paraffin block for a few seconds. The length of time for moisture impregnation is determined by a number of factors. Namely, the hardness of the specimen as a result of fixation and/or processing, the type and density of the specimen, etc. Only through experience can one learn how to determine the duration of moisturizing needed for various tissue densities. However, five seconds of moisturizing with the piece of cotton is generally sufficient for most adequately processed specimens.

The quality of moisturization will be determined by the quality of cotton used. A good absorbent cotton ball provides the best moisturizing, because it presents a homogeneous wetting surface to the specimen block. In addition, it retains more water than paper towels, gauze, or Kimwipe type products, whose use is not recommended.

Moisturization also can be accomplished by using one's thumb. Merely dip the thumb in the floatation bath and rub the surface of the specimen block. The warmth of the thumb will assist in the impregnation of water into the specimen. Additional methods for introducing moisture include blowing on the faced specimen (not recommended), placing the paraffin block in the refrigerator or face down on ice or cold water and placing ice against the faced surface of the block. Shapiro[27] suggests other methods of tissue softening prior to microtomy. It is interesting to note that most histotechnologists apply moisture in some manner prior to microtomy, but few realize the importance of its effects on the quality of the section. The depth of moisture penetration into the specimen will determine the quality of the section. Figures 30 and 31 demonstrate that good sections will be obtained when moisture has penetrated the tissue specimen well. Conversely, microvibration will be produced in areas of the specimen where moisture has not penetrated the specimen sufficiently. Regardless of the method used for introduction of moisture into the tissue, this step is essential for the elimination of microvibrations and, consequently, production of high quality sections. Figure 28a and b dramatically illustrate the difference between a tissue which has been moisturized and one which has not.

Microvibrations in Tissue Sections

As stated earlier, microvibrations occur when tissue has been overprocessed, leaving the specimen dried and hard (similar to a dried sponge). It is the brittleness of the tissue which causes microvibration as the microtome knife travels through it. This, in a simplified way, can be compared to taking a kitchen knife and cutting a handful of uncooked spaghetti. In this case the strands of spaghetti are difficult to cut and break at different lengths. Conversely, cooked (water containing) spaghetti can be cut with ease.

The following experiment clearly demonstrates the profound effect moisturization of tissue specimens has during microtomy. First, process a piece of conventional sized kidney in a routine manner, making sure specimen is not underprocessed. Then cut a section at 6 micrometers without exposing the specimen to ice, water, skin refrigerant, or other procedures generally used to improve sectioning. (This and all subsequent sections are picked up from the floatation water bath and dried in an oven in the usual manner.) For the second section, dip your thumb in the warm floatation water bath and rub it on the surface of the specimen one time for two seconds. Without icing, cut the section. Cut the third section after the surface of the specimen has been exposed to the water-soaked thumb three times for two seconds each. Cut four or five more sections in a similar manner, except increase the number and duration of moisturizing treatments by a factor of one for each subsequent section. Stain all slides as usual. Microscopic examination of the slides will reveal a dramatic gradual improvement in each subsequent section as the time of water treatment and seconds of exposure increase. The improvement in quality encompasses all artifacts often seen after microtomy, including microvibration.

Cooling the Specimen

The specimen is cooled prior to microtomy by using an ice cube, spraying with skin refrigerant, or precooling in a refrigerator. Regardless of the cooling method employed, it must be effective. Different tissues require different cooling exposure periods for effective hardening of the specimen and embedding matrix. For example, tissues that are well impregnated will not require as much cooling as those which are underprocessed, particularly if they lack sufficient paraffin impregnation. Extended periods of cooling may be required in order to overcome the difficulties created by the reduced embedding matrix of underprocessed tissue. For the same reason, a well processed liver specimen normally requires less cooling than a piece of well processed bone specimen. The major point to remember is that different paraffin embedded specimens will require different cooling exposure periods in order to obtain the best possible section.

Cooling of the specimens has a tendency to do two things. First, as the specimen and embedding media cool, they contract and the density of the block increases. This results in the block becoming harder. This hardness provides support to the entire block as the knife travels through it. Secondly, as the coldness continues beyond the freezing point, the specimen and paraffin molecules expand, creating an open lattice structured effect. This lattice structure of molecules provides a pathway for the water from the melting surface ice to penetrate the tissue, thereby introducing additional moisture into the tissue. This is the reason that paraffin embedded blocks cut better after being placed upside down in ice water or ice cubes. The cooling of specimen blocks with skin refrigerant prior to microtomy is not recommended because of the potential hazards to technician and the environment. Skin refrigerants (which are sold under many names) are primarily Dichlorodifluoromethane (CCL_2F_2), commonly known as freon. These fluorocarbons are used as refrigerants and as aerosol propellents. Excessive and/or continued use of dichlorodifluoromethane could possibly have an adverse effect on one's health. In addition, these chemicals are harmful to the earth's ozone layer.

The Sectioning Process (Microtomy)

After correctly orienting the blocks, retract the specimen from the microtome knife to prevent the tissue from being gouged or popping out of the paraffin block. Failure to retract the specimen block before commencing to cut could result in a ruined block. To obtain the best possible ribbon and tissue section at the designated micrometer setting, the microtome hand wheel should be turned at approximately 120 turns per minute. This speed is the most efficient for performing paraffin embedded tissue sectioning. Nevertheless, extremely slow microtome hand wheel speed (15-20 turns per minute) is often necessary in order to obtain sections from extremely difficult-to-cut specimens. On the other hand, a very fast turning speed (160-180 turns per minute) is never advised, as it results in the cutting of sections thicker than where the microtome's micrometer has been set. It is possible to have the micrometer set at 3 micrometers and produce cut sections as thick as 5 micrometers by turning the hand wheel too fast. In addition, turning the microtome hand wheel in a jerky manner or at variable speeds will result in the cutting of uneven sections. All these poor practices of microtomy should be corrected, since failure to do so will result in the production of poor quality slides. Good microtomy is as much an art as a science.

Floating Ribbon on Floatation Water Bath

Touch your wet thumb to the free end of the paraffin ribbon and remove from the microtome blade surface. Pull gently to stretch the ribbon away from the blade. Detach the other end of the ribbon from the blade with the aid of a camel's hair brush, laying this end of the ribbon on the floatation bath first. The remainder of the ribbon is lowered gradually (i.e., ribbon descending at an even, moderate inclination) with a slight pull. If done properly, bubbles should not form underneath the sections and none, or few, wrinkles will form in the ribbon.

Section Separation

The separation of the ribbon into individual sections can be performed in many ways. These include using a heated tissue separator; forcing sections apart with curved forceps; pricking at the paraffin attachment line with a teasing needle; or partially picking up the ribbon from the floatation bath with a glass slide and scoring the paraffin between sections on the edge of the slide with the wooden portion of a camel's hair brush.

Picking Up Sections from Floatation Bath

Insert a clean, dustless slide into the floatation bath at a 30-40° angle. Tease the section onto the glass slide with the aid of a camel's hair brush. Orientate the section so that it is centrally placed, allowing margins for the coverslip and label. Then angle the slide to an almost vertical position (70-80°) and pull from the water bath. Air bubbles can be removed easily if too much water is not allowed to drain from underneath the section. This is done by gently brushing the bubble toward the edge of the slide with a fine camel's hair brush. Place slide in a vertical position on a draining board for approximately one minute before final drying in an oven at 60°C for 20-30 minutes, or on a slide warming table at 37°C overnight. Proper drying of slides at this point is extremely important since it is possible to over dry the slide, which will cause tissue staining problems. I believe that exposure to 60°C drying ovens results in evaporation of additional water molecules from tissue, producing a unique form of denaturation.

Wrinkle Formation in Tissue Sections

Wrinkles in tissue sections are extremely disturbing to both pathologists and histotechnologists. Wrinkles are produced by different factors related to tissue processing and/or microtomy. Very often the pattern of wrinkle formation provides clues to its origin. Wrinkles which run in various directions are generally produced by underprocessed tissue specimens. This wrinkle pattern occurs because the tissue specimen lacks sufficient paraffin matrix within its interstices and therefore is squeezed in all directions by the paraffin contraction/expansion factor (see page 32). When solidifying, the paraffin around the tissue contracts 12-15%, causing the tissue to shrink. The brief exposure to the 45°C water bath may not be sufficient to counteract this compression, resulting in the section being contained within the paraffin framework with little possibility of expanding to its original size. Hence, it appears wrinkled when subsequently viewed through the microscope.

Wrinkles which appear to be parallel to each other may be caused by several factors. These include the paraffin being too soft or excessively dull microtome blade, blade facets without an amorphous surface, or a blade having too much of a positive angle (i.e., angled too sharply toward the microtome).

Method for Removal of Wrinkles in Tissue Sections

Regardless of the reason for wrinkle formation in paraffin sections, there are simple ways to reduce or remove most of the wrinkles from tissue sections.

The most effective method is:

1. Pick up sections from floatation water bath on a section adhesive-coated slide. (Use adhesive of choice.)
2. Allow slide to drain in a vertical position for approximately 30 to 60 seconds. (Do not allow excessive water to drain from the section or become dry since wrinkles will "set fast" and will not come out.)
3. Place slide in a staining rack in a horizontal position.
4. Place rack in a 60°C oven for approximately 20 minutes. (Do not place slides flat on a slide warmer or an oven at 60°C or hotter, because both will produce artifacts.)
5. Remove slide from oven and allow to come to room temperature before staining with desired stain.

This simple procedure will insure wrinkle-free sections.

Chapter 4

Methods for Staining Cellular Entities

Introduction to Hematoxylin

The generic name of hematoxylin is Haematoxylon campechianum; "Haematoxylon" meaning "bloodwood", and "campechianum" because both the tree and its discovers, the Nahuatl Indians, are native to the part of Mexico bordered by the Bay of Compeche. Haematoxylon campechianum comes from a family of many useful plants, the legumes, which is the third largest in the plant kingdom. Related to it are the peanut, soybean, and clover, which harbor nitrogen-fixing bacteria; other dye-yielding plants, including indigo; the balsam, used in present day shampoos; and the licorice plant.

The compound extracted from Haematoxylon campechianum is hematoxylin, whose major oxidation product is hematein. The ortho-hydroxyquinone moiety is responsible for the color properties and is called the chromophore. All dyes contain chromophores. These are structures with alternating or conjugated double bonds. The double bonds allow the dye to absorb light of discrete wavelengths. When one color is absorbed, the unbalanced complementary color is reflected and this is the color that we see. The hematoxylin chromophore absorbs green to yellow light in the 560 to 600 nm range; hence, we see the color complements violet and blue.

Hematoxylin is a stain whose fastness and richness are predicated on use with metal salts. Dyeing with metals takes three forms: (1) mordant dyeing, in which the fabric or tissue is first treated with a metal salt followed by treatment with dye; (2) afterchrome dyeing, in which the tissue is treated first with dye and then with metal; and (3) metachrome dyeing in which dye and metal salts are premixed before the tissue is added for staining. The mechanism of metal dyeing is a matter of speculation. The metal may bind covalently to functional moieties of biomolecules of the tissue and then attract the dye. The dye may chelate the metal with the resulting complex having new absorbance characteristics and tissue binding capacity. Additionally, the metal may absorb light itself. With hematoxylin, it is known that copper mordants yield a blue dye, alum hematoxylin produces a violet stain, and iron mordants result in a black color. By increasing the combining ratio of metal to dye, deeper stains can be obtained.

Poor hematoxylin staining often can be attributed to improper mixing of the aluminum ammonium sulfate or aluminum potassium sulfate. These chemicals serve as mordants for the hematoxylin and, therefore, must be in complete solution with the other ingredients. The directions given for mixing hematoxylin should be followed closely. When mixing hematoxylins, be certain that all ingredients are in solution.

Hematoxylin vs. Hematein[28]

The term "hematoxylin" is used to refer to a number of stains, including Harris hemalum and similar stains. This is misleading. Hematoxylin is the unoxidized, almost colorless dye that reacts readily with various tissue structures but does not impart color. Hematein, the oxidation product of hematoxylin, forms a blue chelate with aluminum, called "hemalum". Hemalum then reacts with nuclei and, depending on staining conditions, with other tissue structures, coloring them blue. Despite the misnomer, the term "hematoxylin" will be used in this book because of its familiarity to the histotechnology community.

Oxidation of Hematoxylin to Hematein

During oxidation to hematein, two hydrogens are removed from hematoxylin; one from a phenolic -OH group and one from the central carbon atom (Fig. 32a). Usually, hematein is shown with a quinonoid structure that lacks a negative charge. Removal of a hydrogen ion from a phenolic -OH group leaves a negatively charged $-O^-$ group, i.e., a phenolate ion (Fig. 32b). This negative charge of hematein readily explains the formation of monopotassium salts and the interaction with cationic sites in tissues. Loss of the hydrogen from the central carbon atom (indicated by large dot), leaves the carbon atom coordinatively unsaturated.

Hematein resembles triphenylmethyl dyes, except that the third valency is occupied by an aliphatic group. In the colorless hematoxylin, all four valencies are occupied; it corresponds to leuco Pararosanilin (leucofuchsin). Removal of the hydrogen from the central carbon atom permits passage of π-electrons from ring to ring. Consequently, leuco Pararosanilin becomes deep red Pararonsanilin. By the same mechanism, almost colorless hematoxylin becomes brownish hematein.

Parenthetically, the negatively charged hematein anion is not an "acid" dye, but an electron-donating Brönsted-Lowry and Lewis anionic base. The positively charged Al^{+++} ion is not a base, but an electron-accepting cationic Lewis acid. The classification of "acid" and "basic" dyes in biological staining is chemically incorrect.

Formation of Hematoxylin

Aluminum: Al^{+++} has the coordination number 6; it can accommodate six electron pairs of ligands in its unoccupied orbitals. When alum is dissolved in aqueous solutions, Al^{+++} forms the hexaquo complex $[Al(H_2O)_6]^{+++}$. Each water molecule shares an electron pair of its oxygen atom with the metal, forming a coordinate bond. Similar complexes are obtained with methyl alcohol and ethyl alcohol. Alcohols are more easily displaced from metals than water and, therefore, favor chelate formation with dyes. During the reaction of metals with hematein, the hydrogen ion of the -OH group ortho to the $-O^-$ group is removed and hematein acquires a second negative charge. The two $-O^-$ groups displace two water molecules from the hexaquo complex and share an electron pair each with Al^{+++} (Fig. 33). The other electrons (small dots) shown at the oxygen atoms are the two unshared electron pairs. The reaction can be written:

$$hem^{--} + [Al(H_2O)_6]^{+++} \longrightarrow [Al(hem)(H_2O)4]^{+} + 2H_2O$$

The two negative charges of hematein cancel two positive charges of the trivalent Al cation. This 1:1 hematein:metal chelate is red. The 1:1 hematein:metal chelate promptly reacts with a second hematein ion, which displaces two water molecules from the complex and donates an electron pair of each -O$^-$ group to form coordinate bonds with the metal; i.e. $[Al(hem)(H_2O)_4]^+$ + hem^{--} $[Al(hem)_2H_2O)2]^-$ + 2 H_2O. This 2:1 hematein chelate is blue and carries a negative charge. In other words, it is an anionic dye. The misleading hypothesis that metal:hematein chelates are cationic ("basic") dyes was based on their ability to stain nuclei. Extensive chemical studies demonstrated a negative charge of 1:2 chelates of trivalent metals and dihydroxy dyes lacking anionic groups. Fe^{+++}, like ferric chloride or alum, and Cr^{+++} form similar chelates with hematein. However, these transition metals hold their ligands much more firmly than Al^{+++}. Furthermore, Fe^{+++} and Cr^{+++} can remove protons from hematoxylin; prior oxidation of hematoxylin to hematein is not necessary.

Color Changes of Hematein:Metal Chelates

The color changes of brownish hematein to red 1:1 and blue 2:1 hematein: metal chelates, can be explained by molecular orbital theory. As already discussed, each -O- substituent of hematein shares an electron pair with the metal. This attachment of the metal facilitates the release of unshared electrons of the oxygen atoms into the π-electron system of the aromatic ring. The metal also donates electrons from its own orbitals to form a π-bond with the dye; electrons of the metal are distributed over ligands. These alterations of the π-electron cloud shift the color to red. When the 1:1 chelate reacts with a second hematein ion, π-electrons are delocalized over both hematein ligands and the chelate becomes blue. See page 77 for the effects of decolorizing and blueing of tissue sections.

Binding of Hematoxylin

During staining with hematoxylin, the metal and the remaining two phenolic -OH groups of each hematein ligand contribute to dye binding. The aluminum atom reacts with -O$^-$ and -OH groups in tissues. Each group replaces a water molecule in the hematoxylin complex and shares an electron pair with the metal to form a coordinate bond. This reaction is similar to the complex formation with dyes containing -O$^-$ or -OH groups, like hematein. Without acid in the dye bath, most tissue structures are colored. However, as aluminum is not a transition metal, these bonds are easily broken. The two phenolic-OH groups of each hematein ligand form hydrogen bonds with suitably spaced sites in tissues. To form hydrogen bonds, the reactive groups of dye and tissue must approach to within 2 to 3 angstroms. Furthermore, as the hematoxylin ion is rigid and hydrogen bonds are directional, (i.e., they can only be formed when atomic orbitals are correctly aligned) not all tissue components possess a distribution of hydrogen bonding groups that matches the spatial requirements of metal hemateins. Therefore, various structures differ in their reactivity with hematoxylin. These variations in hydrogen bonding, as with muscle fibers and collagen, are very evident in Heidenhain's iron hematoxylin stain and the phosphotungstic acid-hematoxylin (PTAH) reaction.

Binding of Hematoxylin by Nuclei

It has been suggested that aluminum in hematoxylin forms a complex with diphosphate groups of nucleic acids. However, this cannot be reconciled with chemical data. In deoxyribonucleic acid (DNA) the monophosphate groups are separated from each other by deoxyribose rings. In molecular models the distances between phosphate groups do not permit complex formation of a metal atom with two phosphate groups. Such bond formation is sterically impossible. Staining of nuclei with metal hemateins after extraction of DNA indicates dye binding by cationic nuclear proteins. Because hematoxylin, iron and chromium hemateins are anionic chelates, reactions with cationic components of nuclei seem probable. Chemical data indicate dye binding by nucleic acids and chromatin via van der Waals and hydrophobic forces. The large molecular orbital of 2:1 hematein:metal chelates and the four phenolic _OH groups per hematoxylin anion can be expected to contribute significantly to the non-ionic finding of hematoxylin by nuclei.

Colors of Hematoxylin Staining Solutions

The color of a hematoxylin solution often determines the quality of its stain reaction. The following information may be useful in quality controlling hematoxylin staining solutions.

a. Purple solution is generally most vigorous (Harris hematoxylin).
b. A reddish colored solution indicates the stain is less vigorous (Mayer's hematoxylin).
c. The appearance of an orange-brown color marks the end of the hematoxylin's useful life.

Hematoxylin's Chromatin Staining Results

Optimal hematoxylin alum staining should result in the chromatin staining blue to blue-black. However, this is not always achieved. Possible causes for various staining results are given below.

a. No color in nuclear chromatin - Hematoxylin solution is probably exhausted, or there may be too much chlorine in the tap water used for washing or blueing slides.
b. Red to purple color in nuclear chromatin - Blueing done in tap water which is too acid; blueing too briefly; blueing solution is exhausted.
c. Gray color in nuclear chromatin - Hematoxylin solution overoxidized (this may occur during or after preparation of the solution).
d. Brown color in nuclear chromatin - Hematoxylin badly overoxidized when prepared or is the result of long-term exposure to air or sunlight.

Commercially Available Hematoxylin Solutions

There are a number of ready-made hematoxylins available from various commercial firms. An evaluation by the author of these hematoxylins yielded the following conclusions:

a. All hematoxylins tested stained nuclear chromatin (from a variety of organs) very well, including nuclear chromatin from tissue specimens which had been altered by inadequate fixation or overexposure to an acid decalcifying solution.
b. Most hematoxylins stained tissue sections rapidly and would therefore, meet the "rush" demands of a busy laboratory.
c. All hematoxylins were capable of staining a minimum of 300 slides (using 50 slide racks) with the same solution.
d. All hematoxylins stained nuclear chromatin very well after six months storage in the laboratory.

The shelf life of most commercial hematoxylins could easily be a year or more if stored under proper conditions; that is, away from light, heat, and fumes from other chemicals. In addition, hematoxylins should be kept in a brown bottle or opaque plastic containers. The report concluded that all commercially available hematoxylin solutions designated for histological use are very good for staining tissue specimens. In addition, problems related to poor quality H&E slide staining could be attributed directly to the person performing the stain. (I estimate that 80% of the hematoxylin and eosin [H&E] stained slides worldwide are not stained to optimum.)

Guide to Optimum H&E Staining

Assuming all related factors (i.e. fixation, processing and embedding) have been performed properly and well, a good hematoxylin and eosin stained slide should demonstrate the following characteristics:

a. The nuclear chromatin should be blue to blue-black and very distinct.
b. The nucleoli in most cells should exhibit a reddish or reddish-purple color.
c. If phloxine eosin has been used as the counterstain, the cytoplasm should exhibit two tinctorial qualities, reddish yellow (collagen) and reddish pink (muscle).
d. If only eosin has been used as the counterstain, the cytoplasm should exhibit a reddish yellow color.
e. Eosinophilic granules should be well defined and appear orange-red. Lack of well-defined eosinophilic granules suggests overstaining by either hematoxylin or eosin, or both.

The reactions between tissue entity and stain solution is often dictated by the fixative and staining solution used. Therefore, above suggestions can be used as a guide only. It should also be recognized that personal preference plays a roll in the interpretation of what constitutes a good H&E. Regardless of this reality it should be recognized that there is such a thing as an optimum stained H&E slide. A tissue section that appears purple and monochromatic under the microscope has not been blued sufficiently or the eosin was not differentiated sufficiently. Proper blueing of the H&E stained slide is essential. The biggest problem with blueing is generally related to cold running water used to blue slides, or used after slides have been exposed to a blueing agent. This problem can be corrected by the use of lukewarm (37-40°) water for blueing and/or for any steps that require working in running tap water.

These suggestions for quality controlling H&E staining are not all-inclusive but should be sufficient to help remedy staining problems. A further suggestion is the use of a good microscope for daily review of H&E stained slides. Slides are reviewed to determine what may be a staining problem or tendency. A mental note of the problem or tendency is made so necessary staining changes can be instituted the next day.

Sub-Optimum H&E Staining

As stated earlier I estimate that 80% of the H&E slides produced worldwide are not stained to optimum. This conclusion was reached after studying this problem for over 30 years. The lack of optimum staining is due to many factors, including:

a. Poor technician insight into what constitutes a good H&E.
b. Inability to recognize poor staining solutions.
c. Inability to recognize selected tissue entities which reveal whether the hematoxylin and/or eosin stains have been decolorized properly, resulting in precise differentiation of nucleic and cytoplasm entities.
d. Overexposure of specimens to acids during the decalcification process.
e. Delay in placing specimens into a fixative solution, resulting in autolytic changes.
f. Overexposure of specimens to some fixatives (e.g., Zenker's or Bouin's).

It is important that the technician be able to recognize which of the above factors is impacting on the staining results. The following guidelines should be useful.

Guidelines for Determining Quality of H&E Staining

To determine if the hematoxylin and eosin have been decolorized and/or applied properly in the tissue sections, microscopically review the stained nucleoli. They (nucleoli) should be stained red to reddish-purple in most cells. Conversely, a purple or blue nucleoli often suggests lack of good hematoxylin decolorization from the cellular entities.

To determine if poor staining is due to events prior to staining or simply poor staining technique, microscopically review the nuclear chromatin. In the case of poor staining, the nuclear chromatin of a selected cell will reveal a three-dimensional effect at high power (x400) as one subjects the section to the up and down movement of the microscope's focal plane. (In other words, one should see a specific chromatin structure clearly at three focal levels, although it may not be colored strongly by the hematoxylin.) On the other hand, nuclear chromatin in tissue that has been poorly fixed or overexposed to acid during the decalcification process will present a flat, non-three-dimensional focal plane microscopically. Proficiency in this procedure will require practice, of course. However, it is felt that every histotechnologist should learn to employ this system since it does provide answers to some staining dilemmas.

Proper Decolorization of Eosin

Many technicians pay little attention to proper decolorization and/or application of the eosin counterstain when performing an H&E. This is unfortunate, since counterstains more often than not provide some diagnostic information. Further, proper eosin differentiation assures blue nuclear chromatin staining. Decolorization of the eosin counterstain should be carried out as outlined below, since only this approach assures good and proper differentiation of the many tissue entities. Successful counterstaining will selectively demonstrate inclusion bodies rich in ribonucleic acid; tinctorial differentiation between muscle and collagen; eosinophilic hyaline bodies; cell granules, and many other cell entities. Slides should be dipped several times in the first 95% alcohol after the eosin, before they are allowed to set for the designated time. This will insure proper action between the alcohol and tissue entities being decolorized.

1. Thirty seconds of eosin staining followed by 2 changes of 95% alcohol for 30 seconds each.
2. One minute of eosin staining followed by 2 changes of 95% alcohol for 1 minute each.
3. Two minutes of eosin staining followed by 3 changes of 95% alcohol for 1 minute each.

Dipping slides in alcohol to achieve proper decolorization is not recommended. The practice of dipping slides in the alcohol to remove excess eosin results in inadequate decolorization of the eosin. This causes some eosin overshadowing of nuclear chromatin producing a purple-stained chromatin.

Although this schedule may seem troublesome and time-consuming, it will insure proper differentiation of tissue entities.

Sulfate Contaminated Eosin: A Staining Problem[29]

Eosin should have a dye content of 88% and should not contain sodium sulfate. Sodium sulfate is used as a filler in some eosin dyes. When eosin solutions containing sodium sulfate are prepared, a fine granular precipitate forms. Staining with the use of this eosin preparation results in poor-to-absent cytoplasmic staining. The precise chemical nature of the precipitate which forms and of the deterioration of cytoplasmic staining in the presence of sodium sulfate has not been determined.

Acids and Bases[30]

Acids and basic (alkaline) solutions are used in many staining methods in the histotechnology laboratory, particularly in the performance of hematoxylin and eosin staining. Consequently, an understanding of their chemical properties and interactions is essential. It is for this reason that the following information on acids and bases is included in this text.

A critical balance between acids and bases must be maintained in all parts of the living organism to permit normal physiological reactions to occur. For example, cells in human stomach lining produce hydrochloric acid (HCl), which is very corrosive. In fact, at the concentrations secreted by the stomach (0.1 moles/liter), hydrochloric acid is able to dissolve a piece of zinc and is deadly to cells. In the stomach, this acid works to kill bacteria in foods, to soften foods, and to convert the inactive enzyme pepsinogen into its active form, pepsin, to begin the digestion of protein. Hydrochloric acid reacts with water in the stomach fluid to form hydronium ions (H_30^+, or hydrated protons) and chloride ions:

$$HCI + H_20 \longrightarrow H_30^+ + CI^-$$

It is the presence of the hydronium ion that makes the stomach contents acidic. Sometimes the stomach produces excess acid, resulting in indigestion and heartburn. The various products sold as antacids all contain compounds that can react with hydronium ions to neutralize the excess acid. The "active ingredient" in antacid tablets is calcium carbonate ($CaC0_3$). Dissolved in the water mixture in the stomach, calcium carbonate breaks apart to form the calcium ion and the carbonate ion:

$$CaC0_{3(s)} \longleftarrow H_20 \;\; Ca^{2}+(aq) + C0_3^{2-}(aq)$$

The carbonate ions then react with and neutralize hydronium ions in the stomach, relieving the indigestion and heartburn.

$$C0_3^{2-}(aq) + 2H_30^+ \longrightarrow 3H_20 + C0_{2(g)}$$

Calcium carbonate is an excellent antacid for many reasons. It acts rapidly, can neutralize a large amount of acid, and has a long-lasting effect. However, it can also have serious side effects when used over long periods of time. Recent research has shown that calcium carbonate can stimulate increased secretion of stomach acid and, in effect, become self-defeating. When the balance between acid and base is upset, normal physiological reactions are also disturbed. Just as the balance between acid and base affects the living organism, it also plays a critical role when staining reactions of any kind are taking place in tissue entities.

What are Acids and Bases

Strong acids and bases can be quite harmful; they will destroy tissue by dissolving protein material and drawing out water. For example, concentrated sulfuric acid is a strong dehydrating agent and will injure tissues on contact. Likewise, concentrated bases will react with the fats that make up the protective membranes of cells, destroying such membranes and causing even more widespread destruction to tissue than acids. This explains why strong laundry soaps and detergents, which contain bases, should not be used to clean clothes containing wool and silk (which are animal proteins). The base in the soap will cause the fibers of these materials to shrink and partially dissolve.

In general, acids are compounds which, when dissolved in water, produce solutions that conduct electricity; react with metals, such as zinc or magnesium, to produce hydrogen gas; taste sour; and turn paper containing litmus dye from blue to red. Bases also form solutions that conduct electricity. However, they taste bitter, feel slippery to the touch, and turn litmus paper from red to blue. Bases will react with acids to neutralize each other's properties. As cited earlier, medicines used today for the relief of indigestion often contain the base bicarbonate ($HC0_3$-), which will neutralize the hydrochloric acid (HCl) in the stomach.

Ionization of Water

Water has the ability to dissolve ionic compounds and to ionize some highly polar molecular compounds. To a very small extent, water molecules are also able to ionize other water molecules. One water molecule out of every 550 million will be pulled apart to form a hydronium ion, H_30^+, and hydroxide ion, OH^-. In a liter of water the concentration of H_30^+ and OH^- is $1x10^{-7}$ moles.

$$H_20 + H_20 \rightleftharpoons H_30^+ + OH^-$$

For convenience, the equation for the ionization of water is often written as follows:

H_20	$\rightleftharpoons$	H^+ Hydrogen ion (proton)*	+	OH^- Hydroxide ion

*A hydrogen atom has just one proton and one electron in it's nucleus; therefore, the hydrogen ion (H+, a hydrogen atom that has lost one electron) is simply a proton.

However, it is important to remember that although one often uses the terms "protons" or "hydrogen ions" when talking about aqueous solutions, single hydrogen ions never exist in water. They are always associated with a water molecule in the form of a hydronium ion (hydrated proton).

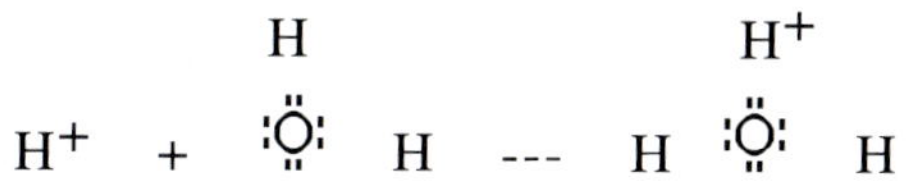

In the covalent bond that is formed between the oxygen and the hydrogen ion, the oxygen molecule donates both of the electrons. This type of bond, in which one atom donates both of the electrons, is known as a coordinate covalent bond.

The Brönsted-Lowry Definition of Acid and Bases

What makes a compound an acid or a base? Although there are several definitions that can be used, it is convenient to use the definitions proposed in 1923 by the Danish chemist J.N. Brönsted and the English chemist T.M. Lowry:

An acid is a substance that can donate a proton, H^+
A base is a substance that can accept a proton, H^+

Let's look at some examples. Consider the following reaction:

HNO_2	+	H_20	$\rightleftharpoons$	$N0_2^-$	+	H_30^+
Nitrous Acid				Nitrite Ion		

In this reaction, nitrous acid is donating a proton to water. Therefore, nitrous acid is the acid (the proton donor) and water acts as the base (the proton acceptor).

$:NH_3$	+	H_20	$\rightleftharpoons$	NH_4^+	+	OH^-
Ammonia				Ammonium Ion		Hydroxide Ion

In this reaction, ammonia (NH_3) accepts a proton from a water molecule. Therefore, ammonia is the base, and water (as the proton donor) is the acid. You can see that water may function as either an acid or a base depending upon the substance with which it is reacting.

Consider for a moment the reverse reaction in equation (1). Here, the hydronium ion (H_30^+) is the acid and the nitrate ion ($N0^{2-}$) is the base. Similarly, in the reverse of equation (2), the ammonium ion (NH_4^+) is the acid and the hydroxide ion (OH^-) is the base.

The Strength of Acids and Bases

A strong acid is one that completely, or almost completely, ionizes to donate all of its protons. The result of adding a strong acid to water is a large increase in the concentration of hydronium ions. For example, nitric acid (HNO_3) is a strong acid. In 0.1M HN03, 92% of the nitric acid molecules are ionized to hydronium ions and nitrate ions, and only 8% remain as whole molecules.

$$HN0_3 \quad + \quad H_20 \quad \rightleftharpoons \quad H_30^+ \quad + \quad N0_3^-$$

Other strong acids are hydrochloric acid (HCl), hydrobromic acid (HBr), hydroiodic acid (HI), and sulfuric acid (H_2SO_4).

A weak acid only partly ionizes in water to donate protons. Therefore, the addition of a weak acid to water results in only a small increase in the concentration of hydronium ions. Acetic acid is an example of a weak acid. In a 0.1M CH_3COOH solution, only 1.3% of the molecules will be ionized.

$$CH_3COOH \quad + \quad H_20 \quad \rightleftharpoons \quad H_30^+ \quad + \quad CH_3COO^-$$

Other weak acids are nitrous acid ($HN0_2$), carbonic acid (H_2CO_3), and boric acid (H_3BO_3).

Similarly, a strong base will have a very large attraction for protons. A weak base will have a weak attraction and only a small percentage of its molecules will accept protons. The hydroxide ion is a strong base, whereas ammonia is a weak base.

Decolorization and Blueing of H&E Stained Tissue Sections

When hematoxylin stained sections are decolorized* in acid solutions, protons attack the metal-hematein bonds. This disturbance of the bonding symmetry causes a hypsochromic** shift; the blue 2:1 hematein:metal chelate becomes reddish. This same shift to a more reddish color occurs in acidified hematoxylin solutions, such as Mayer's. Conversely, during washing of sections in running water or blueing in alkaline solutions, the protons are removed, the bonding symmetry of the hematoxylin is restored, and the stained entity becomes blue. Also read the original article by S.N. Meloan and H. Puchtler[1] for more excellent information on this subject.

*The words decolorize and differentiate are often used synonymously but in reality, sections are decolorized to allow microscopic differentiation of different entities.

**Refers to an atom or group whose introduction into a compound shifts the compound's absorption maximum to a shorter wavelength.

Hematoxylin Decolorizing Solutions

Hematoxylin staining solutions are referred to as progressive (passing successively from one station [solution] of a series to the next) or regressive (going back to an earlier station [solution] of a series). Progressive hematoxylin staining solutions are those which do not require decolorization of the section to differentiate the nuclear chromatin from other tissue entities. In other words, staining solutions which are not decolorized in an acid, due to hematoxylin overstaining. Regressive hematoxylins are those solutions which, due to overstaining in strong hematoxylins, require decolorization of tissue sections in a weak acid solution in order to differentiate nuclear chromatin from cytoplasmic detail. Provided below are useful solutions for decolorizing hematoxylin from tissue sections.

1% Acid Alcohol

Hydrochloric acid, concentrated	1.0	ml
70% Alcohol	100.0	ml

This solution is used for decolorizing strong hematoxylins such as Harris' or Bullard's. It should be recognized that some entities (such as calcium, copper and iron) contained in tissue will be removed or diminished by exposure to this solution. The extent of removal will be determined by the time of exposure to the decolorizing solutions.

0.25% Citric Acid

Citric acid	0.25	gm
Distilled water	100.0	ml

This weak decolorizing solution is not used on strong deep-staining hematoxylins. It is intended for use in the following instances: a) When using a progressive-type hematoxylin and one wishes a more delicate differentiation of nuclear chromatin from the cell cytoplasm, exposure of section to this solution results in "clearing" of the background. b) To remove excessive hematoxylin stain caused by section adhesives (e.g., gelatin).

2% Phosphotungstic Acid

Phosphotungstic acid	2.0	gm
Normal saline	100.0	ml

The normal saline is 0.85% sodium chloride in distilled water (8.5 grams per liter).

This decolorizing solution is used in Ziegler's hematoxylin and eosin stain but can also be used for other hematoxylins.

Hematoxylin Blueing Solutions

The ability to achieve blue instead of purple nuclear chromatin staining depends on several factors: a) the quality of the hematoxylin solution, b) the mordants and/or oxidizers used and c) the type of blueing solution used. Provided below are some of the more useful solutions for blueing nuclear chromatin in hematoxylin and eosin stained sections. Microscopically, nuclear chromatin should appear blue to blue-purple in well-stained hematoxylin and eosin stained slides. This result is achieved only by two means: a) proper blueing of chromatin and b) proper decolorization of the counterstain from the nuclear chromatin.

Scott's Tap Water

Magnesium sulfate	20.0	gm
Potassium bicarbonate	2.0	gm
Distilled water	1000.0	ml

Dissolve each chemical separately in part of the distilled water and then combine. A few crystals of thymol or 5-10 ml of concentrated (37%) formaldehyde may be added to prevent the growth of mold.

0.5% Lithium Carbonate

Lithium carbonate	0.5	gm
Tap water	100.0	ml

0.5% Ammonium Hydroxide

Ammonium hydroxide	0.5	ml
Tap water	100.0	ml

Running Tap Water

Placing slides in lukewarm (approximately 37°C) running tap water can serve as an excellent way of blueing tissue sections stained with hematoxylin solutions. Simply place staining rack holding slides in a pan under hot water faucet. Blueing of sections in lukewarm water provides two distinct effects: a) Sharper blue tinctorial quality of nuclear chromatin is developed and, b) the hematoxylin has less tendency to fade during long storage. Blueing in running tap water requires a minimum of ten minutes. However, normal exposure time is usually fifteen minutes.

Methods for Restoration of Basophilic Staining Properties

One of the most important aspects of histopathological technique is staining. Properly fixed, washed, decalcified, and processed specimens should pose no problem in producing optimum staining. Frequently, however, one encounters specimens which have been overexposed to unbuffered formalin (acid formalin), Zenker's or Bouin's solution and/or an acid decalcifying solution. Improper washing of specimens after fixation (residual fixative) will present staining difficulties similar to those noted in fixative overexposure. All these factors often prevent nuclei from staining properly. Fortunately one of the following methods can be employed to restore the nuclear chromatin staining properties of most sections.

Overexposure to Zenker's Solution

1. Place decerated sections in 5% aqueous sodium bicarbonate for 4 hours.
2. Wash slides in running water for 5 minutes.
3. Stain with H&E or desired stain.

Unbuffered Formalin and Over-Decalcification

The best method for staining nuclear chromatin in specimens which have been overexposed to unbuffered (acid) formalin and/or an acid decalcifying agent is the staining of sections with Weigert's iron hematoxylin.

SOLUTIONS

Weigert's Iron Hematoxylin Solution

Solution A

Hematoxylin crystals	1.0	gm
95% Alcohol	100.0	ml

Solution B

29% Aqueous ferric chloride	4.0	ml
Distilled water	95.0	ml
Concentrated hydrochloric acid	1.0	ml

Working Solution

Mix equal parts of Solutions A and B and stir vigorously.

1. Stain decerated sections in Weigert's hematoxylin for 30 minutes.
2. Rinse sections in running tap water. (Note: If staining results are too dark, decolorize in 1% acid alcohol.)
3. Wash slides in running tap water for 10 minutes.
4. Counterstain with eosin or desired counterstain.

The staining properties of some extensively over-decalcified bone specimens cannot be restored, regardless of the method used.

Overexposure to Bouin's Solution

1. Place decerated slides in 5% aqueous lithium carbonate solution for 1 hour.
2. Wash in running tap water for 10 minutes.
3. Stain with H&E or desired stain.

The following methods can also be used for restoring basophilic staining properties in any of the situations cited. They should be tried on any section which does not demonstrate optimum staining qualities.

5% Sodium Bicarbonate Method

1. Decerate and hydrate slides to water.
2. Place slides in a 5% aqueous sodium bicarbonate solution for 3 hours.
3. Wash in tap water for 5 minutes.
4. Stain with H&E or desired stain.

5% Ammonium Sulfide Method

1. Decerate and hydrate slides to water.
2. Place slides in a 5% aqueous solution of ammonium sulfide for 3 hours.
3. Wash in tap water for 5 minutes.
4. Stain with H&E or desired stain.

5% Periodic Acid Method

1. Decerate and hydrate slides to water.
2. Place in 5% aqueous periodic acid for 30 minutes.
3. Rinse slides in distilled water, 3 changes.
4. Stain with H&E or desired stain.

Although the methods provided above are very useful, poor nuclear staining properties can be avoided by insuring good fixation and prompt removal of specimens from decalcifying fluid.

Staining Solutions for Nuclear Chromatin

Provided below are a wide variety of nuclear chromatin (DNA) staining solutions. Most of the solutions presented are hematoxylins and produce blue stained chromatin. The remaining solutions will provide a spectrum of nuclear chromatin stained colors. The latter solutions are presented since the need for non-hematoxylin nuclear staining exists in some staining reactions.

Mayer's Hematoxylin

Hematoxylin	1.0	gm
Distilled water	1000.0	ml
Sodium iodate	0.2	gm
Ammonium or potassium alum*	50.0	gm
Citric acid	1.0	gm
Chloral hydrate	50.0	gm

Dissolve the alum in water, without heat; add and dissolve the hematoxylin in this solution. Then add the sodium iodate, citric acid, and the chloral hydrate. Shake until all components are in complete solution. Omission of chloral hydrate, a controlled substance, from Mayer's hematoxylin does not noticeably affect the staining properties and useful life of the solution.[2] Alternatively, glycerin can be used (in the same proportions) to replace chloral hydrate. The final color of the stain is reddish violet. Stain keeps well for months.

*ammonium alum = aluminum ammonium sulfate
potassium alum = aluminum potassium sulfate

Harris' Hematoxylin

Hematoxylin	5.0	gm
100% Alcohol	50.0	ml
Ammonium or potassium alum	100.0	gm
Distilled water	1000.0	ml
Mercuric oxide (red)	2.5	gm

Dissolve the hematoxylin in the alcohol and the alum in the water with the aid of heat. Remove from heat and mix the two solutions together. Bring to a boil as rapidly as possible. (Limit this heating to less than one minute and stir often.) Remove from heat and add the mercuric oxide slowly. Reheat and simmer until solution becomes dark purple. Remove from heat immediately and set the vessel in a basin of cold water to cool. The stain is ready for use as soon as it cools. The addition of 2 to 4 ml of glacial acetic acid* per 100 ml of solution increases the precision of the nuclear stain. Filter before use.

*Glacial acetic acid contains not less than 99.4% of acetic acid by weight. Acetic acid generally implies an acid which is obtainable in diluted form.

Clark-Dodd Hematoxylin

Lugol's Iodine Solution (Stock)

Iodine	2.0	gm
Potassium iodide	4.0	gm
Distilled water	100.0	ml

Hematoxylin Solution (Stock)

Hematoxylin	10.0	gm
95% Ethyl alcohol	100.0	ml

Potassium Aluminum Sulfate Solution (Stock)

Potassium aluminum sulfate	6.0	gm
Distilled water	500.0	ml

Hematoxylin Solution (Working)

Hematoxylin (stock)	1.5	ml
Lugol's (stock)	1.5	ml
Potassium aluminum sulfate (stock)	100.0	ml

Mix and let solution stand for 15 minutes before staining slides.

This hematoxylin results in excellent staining and is a progressive stain. Slides stained for 30 minutes stain identical to slides stained for 5 minutes. The staining capability of the solution remains constant for a week to 10 days (It is possible to stain 500 slides in 500 ml of the solution without noticeable changes in staining quality.)

Gill #1 Hematoxylin

Distilled water	730.0	ml
Ethylene glycol	250.0	ml
Hematoxylin	2.0	gm
Sodium iodate	0.2	gm
Aluminum sulfate	17.6	gm
Glacial acetic acid	20.0	ml

Combine the above reagents in the order given and stir for 1 hour at room temperature using a magnetic stirrer.

REMARKS
Excellent for staining exfoliated cytology preparations and surgical or autopsy specimens.

Gill #2 Hematoxylin

Distilled water	710.0	ml
Ethylene glycol	250.0	ml
Hematoxylin	4.0	gm
Sodium iodate	0.4	gm
Aluminum sulfate	35.2	gm
Glacial acetic acid	40.0	ml

REMARKS
Excellent for routine tissue specimens and exfoliated cytology preparations and plastic sections 0.5 to 2 micrometers thick.

Gill #3 Hematoxylin

Distilled water	690.0	ml
Ethylene glycol	250.0	ml
Hematoxylin	6.0	gm
Sodium iodate	0.6	gm
Aluminum sulfate	52.8	gm
Glacial acetic acid	60.0	ml

REMARKS

This hematoxylin stains tissue structures intensely and therefore will require decolorization in an acid solution for proper differentiation of nuclear entities. This is a regressive stain and is excellent for routine tissue specimens and plastic sections measuring 0.5 to 2 micrometers thick.

The following applies to all Gill hematoxylin solutions:

a. If crystalline hematoxylin is used as opposed to powdered, substitute the following amounts:

Gill #1	2.36	gm
Gill #2	4.12	gm
Gill #3	7.98	gm

b. One gram of citric acid can be substituted for each 20 ml of glacial acetic acid.
c. Hematoxylin stains may be used immediately; however, a slightly longer staining time may be required initially.
d. Filter stains before use even though there will be no visible precipitate.

Delafield's Hematoxylin

Hematoxylin	8.0	gm
95% Alcohol	50.0	ml
Ammonium or potassium alum, saturated aqueous solution (approx. 15 gm/100 ml)	800.0	ml

Add the hematoxylin, which has been dissolved in the alcohol, to the alum solution and expose to light and air in an unstoppered bottle for 5 days. Filter and add:

Glycerin	200.0	ml
95% Alcohol	200.0	ml

In a tightly stoppered bottle expose the solution to daylight for 3 days, then filter and store tightly closed in an amber bottle.

Bohmer's Hematoxylin

Hematoxylin	5.0	gm
Ammonium or potassium alum	100.0	gm
95% Alcohol	60.0	ml
Distilled water	1000.0	ml

Dissolve the hematoxylin in the alcohol and the alum in distilled water. Ripen by exposure to light and air for 2 weeks.

Bullard's Hematoxylin

Hematoxylin crystals	8.0	gm
80% Alcohol	144.0	ml
Glacial acetic acid	16.0	ml
Dissolve using gentle heat and add mixture of:		
Ammonium or potassium alum	20.0	gm
Distilled water	250.0	ml
Heat to boiling, remove from flame and add slowly:		
Mercuric oxide	8.0	gm
Cool quickly, filter and add mixture of:		
95% Alcohol	275.0	ml
Glycerin	330.0	ml
Glacial acetic acid	18.0	ml
Ammonium or potassium alum	40.0	gm

Ehrlich's Hematoxylin

Hematoxylin	4.0	gm
95% Alcohol	200.0	ml
Ammonium or potassium alum	120.0	gm
Distilled water	200.0	ml
Glycerin	200.0	ml
Glacial acetic acid	20.0	ml

Dissolve the hematoxylin in the alcohol and the alum in distilled water; then mix. After these are in complete solution, add the glycerin and glacial acetic acid.

Weigert's Iron Hematoxylin

(See Page 79)

Not recommended for routine use. A very good and useful nuclear stain for special stains requiring hematoxylin or for sections requiring restoration of nuclear properties.

Lithium-Carmine

This nuclear stain may be used with most fixatives. Staining in lithium-carmine is accomplished in 5 minutes. It is important that slides go directly into 1% acid alcohol following staining. Exposure to 1% acid alcohol for 3 minutes fixes the lithium-carmine dye in the nuclear chromatin.

Add 2.5 to 5 gm carmine to following solution:

Lithium carbonate	1.25	gm
Distilled water	100.0	ml

Boil solution for 15 minutes. When cool, add 1 gm thymol to prevent mold growth. Filter before use.

Kernechtrot (Nuclear Fast Red)

This is a well-recognized nuclear chromatin stain which has been used extensively as a counterstain for many methods, particularly silver reactions.

Kernechtrot	0.1	gm
5% Aluminum sulfate	100.0	ml

Mix and dissolve with the aid of heat. Cool, filter and add a grain of thymol to prevent mold growth.

The staining time in nuclear fast red is 5 to 10 minutes.

Iron Alum-Celestine Blue B

This stain provides good nuclear staining and can be used to replace hematoxylin in the Papanicolaou and most hematoxylin and eosin stains. Additionally, it can be used as a nuclear stain for frozen oil red O stained sections. It may also prove useful in other procedures requiring nuclear staining. The staining time varies, depending upon the intensity desired, but in most cases 2 to 3 minutes will provide good results.

Celestine Blue B (Stock)

Celestine blue B	1.0	gm
Distilled water	100.0	ml

Iron (Ferric) Alum (Stock)

Iron alum	4.0	gm
Distilled water	100.0	ml

Iron Alum-Celestine Blue B (Working)

Mix equal parts celestine blue B (stock) and iron alum (stock) solutions. This stains nuclear chromatin a blue-black color.

Celestine Blue B

The following solution has been used successfully as a substitute for hematoxylin in the H&E staining procedure. It provides very good staining of nuclear chromatin.

Celestine Blue B Solution

Celestine blue B	0.5	gm
Ferric ammonium sulfate	5.0	gm
Concentrated hydrochloric acid	2.0	ml
Glycerin	15.0	ml
Distilled water	85.0	ml

Dissolve the ferric ammonium sulfate in the distilled water, add the celestine blue B and dissolve by gently boiling solution for 3 minutes. Cool solution and add the glycerin and hydrochloric acid solution is now ready for use.

Debidin's Modified Harris' Hematoxylin

This hematoxylin differs from Harris' in that no alcohol is used for dissolving the hematoxylin. The advantages of preparing this hematoxylin are: It is easier to prepare, it is less time-consuming; there is less likelihood of "spills" and is less of a fire hazard.

Hematoxylin	2.5	gm
Potassium or ammonium alum	50.0	gm
Distilled water	500.0	ml
Mercuric oxide (red)	1.25	gm
2% Glacial acetic acid	10.0	ml

Dissolve the potassium or ammonium alum in distilled water with the aid of a magnetic stirrer. Add the hematoxylin. Cover the flask and leave mixture in 60°C oven overnight. Remove from oven, slowly add mercuric oxide and stir until solution is cold. Filter and add the glacial acetic acid. The solution is ready for use.

REMARKS
It is advisable to prepare small quantities of this solution each month. Nuclear staining with this hematoxylin is sharp and reproducible.

Cytoplasmic Stain

Many cytoplasmic stains (counterstains) exist and give almost identical results. It is for this reason that only those stains which give different staining results will be listed below.

The ideal counterstain adds color contrast in order to give additional information and orientation without obscuring the variations in intensity of the principal stain. Pink or red counterstains contrast slightly better with the bluish color of ammonium alum hematoxylin than do other colored counterstains. Greenish and bluish counterstains work well with potassium alum hematoxylin. In any case, the counterstain should be relatively pale as a rule, although individual preferences may vary.

Eosin Y Stain

1% Alcoholic Eosin (Stock)

Eosin Y, water soluble	1.0	gm
Distilled water	20.0	ml
Dissolve and add:		
95% Alcohol	80.0	ml

Eosin Y (Working)

1% Alcoholic eosin (stock)	1 part
80% Alcohol	3 parts

Just before use, add 0.5 ml of glacial acetic acid to each 100 ml of stain and stir. This cytoplasmic stain produces only slight tinctorial differentiation of various tissue entities, but is very useful nonetheless.

A Light-Resistant Permanent Eosin Stain

Many eosin dyes stain cytoplasm in varying shades, often masking the tissue component under study. This eosin Y solution provides a counterstain for consistent, permanent contrast among tissue components, even in the presence of light. This counterstain fulfills these objectives through the use of a stock preparation of precipitated acidified eosin Y.

Eosin Y	2.5	gm
Distilled water	500.0	ml
Concentrated hydrochloric acid	4.0	ml

Dissolve eosin Y in distilled water in a beaker. Add concentrated hydrochloric acid with a dropper or pipet. Allow the solution to stand for 30 minutes or until the supernatant is clear. Decant the supernatant solution through filter paper, taking care not to lose any of the precipitate. (Vacuum filtration is helpful.) Repeat this step three times, using an additional 500 ml of distilled water for each washing. Filter with suction and dry the final precipitate in an oven at 56°C for 24 hours. Scrape the dry precipitate from the filter paper with a metal spatula. Dissolve each 0.5 gm precipitate in 100 ml of 95% alcohol. Check pH, which should be about 5.6. The resultant solution constitutes the counterstain for use with hematoxylin staining solutions.

REMARKS

Paraffin embedded sections are uniformly colored a bright orange-yellow, which contrasts well with the bluish-purple nuclear staining of hematoxylin. The stability of acid-washed eosin is excellent, but take care to prevent excessive evaporation of the alcohol in which the acid-washed eosin is dissolved.

Eosin-Phloxine B Stain

1% Eosin Y (Stock)

Eosin Y	1.0	gm
Distilled water	100.0	ml

1% Phloxine B (Stock)

Phloxine B	1.0	gm
Distilled water	100.0	ml

Eosin Phloxine B (Working)

1% Eosin Y (stock)	100.0	ml
1% Phloxine B (stock)	10.0	ml
95% Alcohol	780.0	ml
Glacial acetic acid	4.0	ml

Make up working solution as needed. Working solution should be changed at least once a week.

REMARKS

The combination of eosin and phloxine B produces an excellent cytoplasmic stain which tinctorially differentiates many tissue entities. For example, muscle stains slightly darker red than collagen and inclusion bodies are demonstrated better than by the use of an eosin solution alone.

Eosin B Stain

Eosin B	0.5	gm
Distilled water	100.0	ml

The substitution of eosin B for eosin Y gives a slightly pinker cytoplasmic counterstain lacking the yellowish tinge of eosin Y.

Congo Red Stain

Congo red	0.5	gm
Distilled water	100.0	ml

This counterstain provides a pink cytoplasmic stain but does not stain collagen as well as eosin.

Sirius Red Stain

Sirius red F3BA	0.05	gm
Distilled water	100.0	ml

This counterstain provides good staining of connective tissue. It not only stains collagen well but also reticulum and basement membrane.

Aniline-Chromotrope 2R Stain

Chromotrope 2R	1.0	gm
80% Alcohol	100.0	ml
Aniline*	3.0	ml
Glacial acetic acid	1.0	ml

Note: Some of the tinctorial effects of this counterstain are, erythrocytes stain salmon and darker than with other counterstains, it stains some granules, such as zymogen, eosinophil, mast cell, etc., more distinctly than does eosin. Cytoplasmic staining will require approximately 10 minutes. Slides are washed in water before dehydration in alcohol and clearing in xylene prior to mounting coverglass.

*Aniline should be handled with care. See chemical safety data sheets before using this chemical.

Phloxine B-Safran-Picric Acid Stain

These counterstains are used at different steps in the H&E staining procedure.

2% Picric Acid

Picric acid	2.0	gm
Distilled water	100.0	ml

1% Phloxine B

Phloxine B	1.0	gm
Distilled water	100.0	ml

2% Safran

Safran du Gatinais*	2.0	gm
100% Ethyl alcohol	100.0	ml

Dissolve the safran in the alcohol by placing the solution in a 56-60°C oven. It is recommended the safran-alcohol solution be kept in the oven at all times. The bottle must be well stoppered to prevent excessive alcohol evaporation. This is an excellent counterstain for routine H&E's, since tinctorial differentiation results in several distinct colors. For example, collagen stains yellow while muscle stains red. The value of this counterstain can only be appreciated from one's own experience. The procedure is longer and more complicated than most counterstains, but the distinctive tinctorial differentiation of various tissue entities may justify its use.

*Safran du Gatinais can be replaced by the use of safran, the dried, orange-colored stigmas used in flavoring and coloring foods. Safran can be found in the spice section of grocery stores.

Mayer's Hematoxylin and Eosin Staining

Mayer's hematoxylin is used in this method because it is not necessary to decolorize the section in acid and blue in an alkaline solution. It is a progressive stain which produces a stained section with clearly defined nuclear chromatin and clear background. In the past the biggest objection to the use of Mayer's hematoxylin was that stained slides often faded after two or three years. This problem can be eliminated if the slides are washed in running tap water for a minimum of 10 minutes after hematoxylin staining.

This method gives consistent results even when more than one person stains sections from the same block. Also, slides may be left in hematoxylin for hours without overstaining. Because of the simplicity of the technique, it is possible to teach others how to use it in a very short time.

FIXATION 10% Buffered neutral formalin.
PROCESS Paraffin, frozen, or celloidin.
MICROTOMY Cut sections at 5 or 6 micrometers.
CONTROL Use control number 27.
SOLUTIONS

Mayer's Hematoxylin
(See Page 81)

Eosing Y Stain
(See Page 86)

Gram's or Lugol's Iodine
(See Page 187 or 81)

STAINING PROCEDURE

1. Decerate slides in xylene, absolute alcohol and 95% alcohol, 2 changes, 2 minutes each and rinse in distilled water.
2. If sections are Zenker or B-5 fixed, remove the mercuric chloride crystals with Gram's or Lugol's iodine and sodium thiosulphate (hypo).
3. Place in Mayer's hematoxylin for 15 minutes.
4. Wash in lukewarm running tap water for 10 minutes.
5. Counterstain with eosin from 30 seconds to 2 minutes depending on the age of the eosin and the intensity of the counterstain desired. For even staining results dip slides several times before allowing them to sit in the eosin for the desired time.
6. Dehydrate slides in 95% alcohol and absolute alcohol 3 changes each. Clear in xylene, 3 changes.
7. Mount coverglass with appropriate media (refractive index 1.48-1.56).

RESULTS

Nuclear chromatin ....blue with some metachromasia
Cytoplasm ....various shades of pink identifying different tissue components

REMARKS

Some adhesives used to attach sections onto the slides (gelatin, egg albumen, etc.) sometimes stain around the edge of the section. This will give the slides a slightly dark appearance but in no way affects the nuclear staining. To remedy this, use 1% aqueous citric acid to "clean" the slides after step 3. If followed with a few dips in saturated aqueous lithium carbonate, the nuclei will blue immediately. This is optional since the 10 minute wash in running water is sufficient to blue the nuclei. This step will in no way alter or minimize the staining of the nuclei.

Mayer's Hematoxylin Eosin-Phloxine B Stain

FIXATION Any fixative (10% buffered neutral formalin preferred).
PROCESS Paraffin, frozen, or celloidin.
MICROTOMY Cut sections at 6 micrometers.

CONTROL SOLUTIONS — Use control number 27.

Mayer's Hematoxylin
(See Page 81)

Eosin-Phloxine B Stain
(See Page 87)

STAINING PROCEDURE

1. Decerate slides in xylene, absolute alcohol and 95% alcohol, 2 changes, 2 minutes each and rinse in distilled water.
2. Stain slides in Mayer's hematoxylin for 15 minutes.
3. Blue slides in lukewarm running tap water for 15 minutes.
4. Counterstain slides in working eosin-phloxine B solution for 2 minutes.
5. Decolorize eosin-phloxine in 3 changes of 95% alcohol for 2 minutes each.
6. Place slides in 3 changes of absolute alcohol for 1 minute each.
7. Clear slides in 3 changes of xylene for 2 minutes each.
8. Mount coverglass with appropriate media (refractive index 1.48-1.56).

RESULTS

Nuclear chromatin....................blue
Cytoplasm....................pink to red

REMARKS

Use of ethyl alcohol in the above schedule is preferred, although isopropyl alcohol may be substituted. The use of eosin-phloxine B results in cytoplasmic staining which varies tinctorially from light pink to red. For example, muscle stains red and collagen stains pink. To insure even staining, dip slides several times in eosin-phloxine B before allowing them to remain in the solution for the prescribed time. The alcohol rinses after eosin are very important. The time specified in each alcohol is essential for proper differentiation of eosin. A blueing solution, such as ammonia water or lithium carbonate, can be used to blue sections. Rinsing in distilled water before staining in hematoxylin (step 2) insures longer shelf life of the hematoxylin solution.

Hematoxylin Aniline-Chromotrope 2R Stain

FIXATION — Any fixative (10% buffered neutral formalin preferred).
PROCESS — Paraffin, frozen, or celloidin.
MICROTOMY — Cut sections at 5 or 6 micrometers.
CONTROL — Use control number 23.
SOLUTIONS

Aniline Chromotrope 2R
(See Page 87)

Mayer's Hematoxylin
(See Page 81)

STAINING PROCEDURE

1. Decerate slides in xylene, absolute alcohol and 95% alcohol, 2 changes, 2 minutes each and rinse in distilled water.
2. Proceed with desired hematoxylin stain (Mayer's preferred).
3. After final water wash, stain in aniline-chromotrope 2R solution for 10 minutes.
4. Wash in running tap water for 2 minutes.
5. Dehydrate in 95% and absolute alcohols, 2 changes of 2 minutes each.
6. Clear in xylene, 2 changes of 2 minutes each.
7. Mount coverglass with appropriate media (refractive index 1.48-1.56).

RESULTS

Nuclear chromatin	blue
Cytoplasm	salmon
Zymogen granule, eosinophile granule	salmon
Mast cells	salmon
Erythrocytes	red

Harris' Hematoxylin and Eosin Stain

FIXATION — Any fixative.
PROCESS — Paraffin, frozen, or celloidin.
MICROTOMY — Cut sections at 5 or 6 micrometers.
CONTROL — Use control number 27.
SOLUTIONS

Harris' Hematoxylin
(See Page 81)

1% Acid Alcohol
(See Page 77)

Ammonia Water

Ammonium hydroxide	2-3	ml
Tap water	1000.0	ml

1% Lithium Carbonate

Lithium carbonate	1.0	gm
Distilled water	100.0	ml

Eosin Y Stain
(See Page 86)

STAINING PROCEDURE

1. Decerate slides in xylene, absolute alcohol and 95% alcohol, 2 changes, 2 minutes each and rinse in distilled water.
2. If sections are Zenker or B-5 fixed, remove the mercuric chloride crystals with Gram's or Lugol's iodine and sodium thiosulphate (hypo).
3. Place in Harris' hematoxylin for 15 minutes (shorter staining times may be used, if desired).

4. Rinse in tap water.
5. Decolorize in 1% acid alcohol, 3 to 10 quick dips. Check the differentiation with a microscope. Nuclei should be distinct and the background very light to colorless.
6. Rinse well in tap water.
7. Dip in ammonia water or lithium carbonate water until sections are bright blue (3 to 5 slow dips).
8. Wash in running tap water for 5 minutes. (If washing is inadequate, eosin will not stain evenly.)
9. Stain with eosin for 30 seconds to 2 minutes, depending on the age of the eosin and the degree of counterstaining desired. For even staining results dip slides several times before allowing them to sit in the eosin for the desired time.
10. Dehydrate in 95% and absolute alcohols, 3 changes for 2 minutes each.
11. Clear in xylene, 3 changes, 2 minutes each.
12. Mount coverglass with appropriate media (refractive index 1.48-1.56).

Hematoxylin-Phloxine B-Safran-Picric Acid Stain

This method is often used for color differentiation of various tissue entities (see results) and may be valuable for that reason.

FIXATION Any fixative.
PROCESS Paraffin or frozen.
MICROTOMY Cut sections at 5 or 6 micrometers.
CONTROL Use control number 27.
SOLUTIONS

Mayer's Hematoxylin
(See Page 81)

2% Picric Acid
(See Page 88)

1% Phloxine B
(See Page 88)

2% Safran
(See Page 88)

STAINING PROCEDURE
1. Decerate slides in xylene, absolute alcohol and 95% alcohol, 2 changes, 2 minutes each and rinse in distilled water.
2. Place in 2% picric acid for 5 minutes.
3. Rinse in tap water until all the picric acid is removed.
4. Stain in Mayer's hematoxylin for 15 minutes.
5. Wash in running tap water for 15 minutes.
6. Stain in 1% phloxine B solution for 2 minutes.
7. Wash in tap water for 3 minutes.
8. Dehydrate in 3 changes of absolute alcohol.

9. Stain in 2% safran for 5 minutes.
10. Rinse in 2 changes of absolute alcohol.
11. Clear in xylene, 2 changes, 2 minutes each.
12. Mount coverglass with appropriate media (refractive index 1.48-1.56).

RESULTS

Nuclear chromatin..blue
Cytoplasm...reddish pink to yellow
Collagen..yellow
Muscle ....red

REMARKS

This method is often used for color differentiation of various tissue entities. The method is longer and more complicated but the distinct tinctorial effects on various tissue structures may justify its use.

Ziegler's Hematoxylin-Eosin Stain

The most essential feature of this staining technique is the introduction of phosphotungstic acid as a self-limiting differentiating agent. The nuclear chromatin staining results are generally much sharper when this method of differentiation is used. The average time for the performance of this stain is only about ten to twelve minutes.

SOLUTIONS

Delafield's or Harris' Hematoxylin
(See Page 83 or 81)

Normal Saline

Sodium chloride	0.85	gm
Distilled water...	100.0	ml

(8.5 grams per liter)

2% Phosphotungstic Acid
(See Page 78)

2% Sodium Citrate

Sodium citrate	2.0	gm
Normal saline	100.0	ml

Eosin Y Stain
(See Page 86)

STAINING PROCEDURE

1. Decerate slides in xylene, absolute alcohol and 95% alcohol, 2 changes, 2 minutes each and rinse in distilled water.
2. Stain slides in hematoxylin for 3 minutes.
3. Dip slides in normal saline 10 times.
4. Dip slides in phosphotungstic acid 10 times.

5. Dip slides in normal saline 10 times.
6. Dip slides in sodium citrate 10 times.
7. Dip slides in normal saline 10 times.
8. Counterstain in eosin for 1 minute.
9. Dehydrate in absolute ethyl alcohol, 3 changes, 2 minutes each.
10. Clear in xylene, 2 changes, 2 minutes each.
11. Mount coverglass with appropriate medium (refractive index 1.48-1.56).

REMARKS
This H&E stain is an excellent differential stain. Practically all cells and intercellular substances will stain with different tinctorial qualities and densities of pink or blue. Isopropyl alcohol cannot be used in this procedure, since it decolorizes the tissue section.

Linder-Hopps-Johnson Method for Inflammatory Cells[31]

FIXATION	10% Formal-saline or formal-calcium acetate (see remarks).
PROCESS	Paraffin or frozen (see remarks).
MICROTOMY	Cut section at 4 to 6 micrometers.
CONTROL	Use control number 24.
SOLUTIONS	

Pyronin Y (Stock)

Pyronin Y (see remarks)	0.08	gm
100% Ethyl alcohol	75.0	ml

2% Methyl Green (Stock)

Methyl green	2.0	gm
Distilled water	100.0	ml

Dissolve the dye in water. Place in a separatory funnel. Extract the methyl violet impurity by repeatedly shaking the solution with equal volumes of chloroform and draining until no more methyl violet is extracted.

5% Phenol (Stock)

Phenol (carbolic acid) crystals	5.0	gm
Distilled water	100.0	ml

0.2 M Sodium Acetate (Stock)

Sodium acetate (trihydrate)	27.2	gm
Distilled water	1000.0	ml

0.25% Orange G (Stock)

Orange G	0.25	gm
Distilled water	100.0	ml

Pyronin Y-Methyl Green (Working)

Pyronin Y (stock)	15.0	ml
2% Methyl green (stock)	12.5	ml
5% Phenol (stock)	5.0	ml

Adjust pH to 4.6 with 0.2 M sodium acetate and add sufficient water to bring the solution volume to 100 ml. Add:

0.25% Orange G (stock) .... 3.0 ml

Ehrlich's Hematoxylin
(See Page 84)

STAINING PROCEDURE

1. Decerate slides in xylene, absolute alcohol and 95% alcohol, 2 changes, 2 minutes each and rinse in distilled water.
2. Stain with Ehrlich's hematoxylin for 1 minute.
3. Rinse in distilled water.
4. Blue by washing in tap water. Check microscopically to see that the nuclei are faintly stained.
5. Rinse in distilled water for 30 seconds.
6. Place in pyronin Y-methyl green working solution for 15 minutes. Verify that the nuclei are blue, cytoplasm red, and connective tissue pink-yellow.
7. Dehydrate slides in absolute alcohol, 3 changes.
8. Clear in xylene, 3 changes each.
9. Mount coverglass with appropriate media (refractive index 1.48-1.56).

RESULTS

Nuclear chromatin (DNA) ....blue
Basophilic cytoplasm ....red
Connective tissue ....orange
Epithelial cells ....pale pink
Erythrocytes ....orange red

REMARKS

Although the procedure described above is reliable, several aspects are critical. The most important is fixation, which should not be continued for more than three days. With small pieces, of tissue overnight fixation is followed by transfer to 70% alcohol. This preserves nucleic acids with minimal depolymerization*. To avoid depolymerization tissues must not be exposed to acids. High temperatures, likewise, cause disruption of the nuclear material with an ensuing loss of methyl green staining. Take care to ensure that impregnating and embedding temperatures are kept as close to 60°C as possible. Dry sections at not more than 40°C.

Dye concentrations are important and vary greatly between different batches of pyronin Y. Purchase pyronin Y with a 20% dye content, 0.4 gm of which yields the 0.08 gm of pure dye needed for the stock solutions. The pH of the stain affects its staining behavior. At low pH pyronin Y predominates over methyl green, whereas the reverse is true at a higher pH.

*The conversion of a compound into one of the smaller molecular weight and different physical properties without changing the percentage relationships of the elements composing it.

Gill's #1 Hematoxylin-Eosin-Phloxine Stain

FIXATION 10% Buffered neutral formalin.
PROCESS Paraffin, frozen, or celloidin.

MICROTOMY Cut sections at 5 or 6 micrometers.
CONTROL Use control number 27.

This staining procedure is often used for exfoliated cytological preparations but can also be used for staining routine tissue preparations, if one desires a progressive stain. Nuclear chromatin staining is clear and distinct with this staining procedure.

SOLUTIONS

Gill's #1 Hematoxylin
(See Page 82)

Scott's Tap Water
(See Page 78)

Eosin-Phloxine B Stain
(See Page 87)

STAINING PROCEDURE

1. Fix cytologic samples in 2 changes of 95% ethyl alcohol, 5 minutes each.
2. Rinse in 2 changes of tap water, 5 dips each.
3. Stain in Gill's #1 hematoxylin solution for 5 minutes.
4. Rinse in 2 changes of tap water, 5 dips each.
5. Blue in Scott's tap water or desired blueing agent for 15 seconds.
6. Rinse in 2 changes of tap water, 10 dips each.
7. Dehydrate in 2 changes of 95% ethyl alcohol, 10 dips each.
8. Counterstain with eosin-phloxine B for 15 seconds.
9. Rinse in 2 changes of 95% ethyl alcohol, 10 dips each.
10. Rinse in 2 changes of absolute alcohol, 10 dips each.
11. Rinse in 3 changes of xylene, 10 dips each.
12. Mount coverglass with appropriate media (refractive index 1.48-1.56).

RESULTS

Nuclear chromatin....................blue
Cytoplasm....................reddish pink

Gill's #2 Hematoxylin-Eosin-Phloxine Stain

FIXATION 10% Buffered neutral formalin.
PROCESS Paraffin, frozen, or celloidin.
MICROTOMY Cut sections at 5 or 6 micrometers.
CONTROL Use control number 27.
SOLUTIONS

Gill's #2 Hematoxylin
(See Page 82)

1% Acid Alcohol
(See Page 77)

Scott's Tap Water
(See Page 78)

Eosin-Phloxine B Stain
(See Page 87)

STAINING PROCEDURE

1. Decerate slides in xylene, absolute alcohol and 95% alcohol, 2 changes, 2 minutes each and rinse in distilled water.
2. Stain in Gill's #2 hematoxylin solution for 5 minutes.
3. Rinse slides in running tap water for 2 minutes.
4. Decolorize in 1% acid alcohol for 10 seconds or until desired cellular components have been differentiated properly.
5. Wash in running tap water for 2 minutes.
6. Blue in Scott's tap water, 4 or 5 dips.
7. Wash in running tap water for 2 minutes.
8. Counterstain in eosin-phloxine B, 15 seconds to 2 minutes depending on intensity of stain desired.
9. Dehydrate in 95% alcohol, 2 changes, 2 minutes each.
10. Absolute alcohol, 2 changes, 2 minutes each.
11. Clear in xylene, 2 changes, 2 minutes each.
12. Mount coverglass with appropriate medium (refractive index 1.48-1.56).

RESULTS

Nuclear chromatin ....bluish-gray
Cytoplasm ....varying shades of pink to red

Gill's #3 Hematoxylin-Eosin Phloxine Stain

FIXATION 10% Buffered neutral formalin.
PROCESS Paraffin, frozen, or celloidin.
MICROTOMY Cut sections at 5 or 6 micrometers.
CONTROL Use control number 27.
SOLUTIONS

Gill's #3 Hematoxylin
(See Page 82)

1 % Acid Alcohol
(See Page 77)

Scott's Tap Water
(See Page 78)

Eosin-Phloxine B Stain
(See Page 87)

STAINING PROCEDURES

1. Decerate slides in xylene, absolute alcohol and 95% alcohol, 2 changes, 2 minutes each and rinse in distilled water.

2. Stain in Gill's #3 hematoxylin solution for 5 minutes.
3. Wash slides in running tap water for 2 minutes.
4. Decolorize in 1% acid alcohol for 20 seconds.
5. Wash in running tap water for 2 minutes.
6. Blue in Scott's tap water, or desired blueing agent, for 1 minute.
7. Wash in running tap water for 2 minutes.
8. Counterstain in eosin-phloxine B for 1 minute.
9. Dehydrate slides in 95% alcohol, absolute, 3 changes.
10. Clear in xylene, 3 changes, 1 minute each.
11. Mount coverglass with appropriate media (refractive index 1.48-1.56).

RESULTS

Nuclear chromatin ....bluish-gray
Cytoplasm ....varying shades of pink to red

Celestine Blue B-Eosin Stain

The following staining procedure is an acceptable substitute for the routine hematoxylin and eosin staining procedure. Microscopically the sections appear similar to those stained with hematoxylin.

FIXATION	10% Buffered neutral formalin, Zenker's or Bouin's solutions.
PROCESS	Paraffin, frozen, or celloidin.
MICROTOMY	Cut sections at 6 micrometers.
CONTROL	Use control number 27.
SOLUTIONS	

Celestine Blue B (Stock)
(See Page 85)

Eosin Y Stain
(See Page 86)

STAINING PROCEDURE

1. Decerate slides in xylene, absolute alcohol and 95% alcohol, 2 changes, 2 minutes each and rinse in distilled water.
2. Stain slides in celestine blue B solution for 15 minutes.
3. Dip slides in distilled water until excess celestine blue B solution is washed off.
4. Counterstain slide with eosin solution (working) for 1 minute, or until desired eosinophilia has been achieved.
5. Dehydrate slides in 95% alcohol, absolute, 3 changes.
6. Clear in xylene, 3 changes, 1 minute each.
7. Mount coverglass with appropriate media (refractive index 1.48-1.56).

RESULTS

Nuclear chromatin...blue
Cytoplasm ....red

Luna's "Stat" Method

This is a good method for staining "stat" frozen-cut sections for rapid diagnosis.

FIXATION — Tissue may be fixed in 10% formalin with the aid of mild heat or cut unfixed on a cryostat.

PROCESS — Frozen.

CRYOTOMY — Cut sections at desired thickness and pick up section on a clean glass slide containing selected section adhesive (see chapter 17).

SOLUTIONS

Harris' Hematoxylin
(See Page 81)

1% Acid Alcohol
(See Page 77)

Sodium Carbonate Saturated Aqueous Solution

Sodium carbonate	1.0	gm
Distilled water	100.0	ml

This solution is somewhat harsh, may also use Lithium Carbonate Saturated Aqueous solution.

Eosin Y Stain
(See Page 86)

STAINING PROCEDURE
(See fixation above before staining.)

1. 100% Alcohol .... 1 dip
2. 1% Acid alcohol .... 1 dip
3. Tap water .... 1 dip
4. Hematoxylin .... 1 minute
   (Place hematoxylin on slide with dropper; hold slide near slide warmer)
5. Warm (37°C) tap water .... 1 dip
6. 1% Acid alcohol .... 1 dip
7. Tap water .... 1 dip
8. Sodium carbonate, saturated aqueous .... 1 dip
9. Warm (37°C) tap water .... 1 dip
10. 80% Alcohol .... 1 dip
11. Eosin Y solution .... 1 dip
12. 95% Alcohol .... 1 dip
13. 100% Alcohol .... 1 dip
14. 100% Alcohol .... 1 dip
15. Xylene .... 1 dip
16. Xylene .... 1 dip
17. Mount coverglass with appropriate media (refractive index 1.48-1.56).

RESULTS

Nuclear chromatin .... blue
Cytoplasm .... pink

Terry's Polychrome Methylene Blue "Stat" Method[32]

(For Fresh Unfixed Tissue)

FIXATION Fresh tissue (unfixed).
PROCESS Frozen.
CRYOTOMY Cut sections at 10 micrometers and pick up section on a clean glass slide.
SOLUTIONS

1% Methylene Blue (Stock)

Methylene blue	1.0	gm
Distilled water	100.0	ml

1% Potassium Carbonate (Stock)

Potassium carbonate	1.0	gm
Distilled water	100.0	ml

Terry's Methylene Blue (Working)

1% Methylene Blue (stock)	6.0	ml
1% Potassium carbonate (stock)	6.0	ml
Distilled water	18.0	ml

Combine above ingredients and bring to a boil for exactly 2 1/2 minutes. Cool solution under running tap water. After the solution has cooled, q.s. up to 30 ml with distilled water to replace any water lost during boiling. Filter solution before use.

STAINING PROCEDURE

1. Cover the section with a small amount of Terry's working methylene blue solution for 5 seconds.
2. Drain the stain off the slide and rinse in distilled water.
3. Section is examined while still wet (see remarks).

RESULTS

Nuclei and epithelial cells	blue
Connective tissue fibrils	red to colorless
Smooth muscle	purple

REMARKS
This stain gives its best results when used on frozen sections of living, unfixed tissue. As soon as the cells die, the staining is unsatisfactory. The section may be coverslipped by employing a drop of water on the undersurface of the coverslip. The time the solution is boiled is very important in developing polychrome. If the alkaline stain is boiled longer that 2 1/2 minutes or if it is slowly cooled, more red is present, more crystalline precipitate forms, and the stain is less intensely active.

Terry's Acid Polychrome Methylene Blue "Stat" Method[32]

(For Fixed Tissue)

FIXATION 10% Buffered neutral formalin.
PROCESS Frozen.

CRYOTOMY Cut sections at 10 micrometers and pick up section on a clean glass slide containing selected section adhesive (see chapter 17).

SOLUTIONS

1% Methylene Blue (Stock)
(See Page 100)

1% Potassium Carbonate (Stock)
(See Page 100)

10% Acetic Acid (Stock)

Glacial acetic acid	10.0	ml
Distilled water	90.0	ml

Terry's Acid Polychrome Methylene Blue (Working)

1% Methylene blue (stock)	20.0	ml
1% Potassium carbonate (stock)	20.0	ml
Distilled water	60.0	ml

Combine and heat to boiling. Boil gently for 2 1/2 minutes. Cool rapidly under running tap water. Add 10 ml 10% acetic acid to cooled solution. Shake the solution for 1 minute and q.s. up to 100 ml with distilled water to replace any water lost during boiling. Filter solution before use.

STAINING PROCEDURE

1. Cover the section with a small amount of Terry's acid polychrome methylene blue working solution for 3 to 5 seconds.
2. Drain the stain off the slide and rinse in distilled water.
3. Section is examined while still wet (see remarks).

RESULTS

Nuclei and epithelial cells	blue
Connective tissue	bright red to colorless
Smooth muscle	purple

REMARKS

If less glacial acetic acid is added to the stain, connective tissue staining will be a brighter red. Conversely, more acetic acid results in pale to colorless staining of connective tissue. The time the solution is boiled is very important in developing polychrome. Boiling for 2 to 2 1/2 minutes seems to be best, although good acid stains may be made from solutions boiled only 30 seconds or as long as 5 minutes.

Luna-Verfuerth "Stat" H&E Method

FIXATION 10% Buffered neutral formalin.

PROCESS Frozen.

CRYOTOMY Cut sections at 10 micrometers and pick up section on a clean glass slide containing selected section adhesive (see chapter 17).

SOLUTIONS

10% Buffered Neutral Formalin

Concentrated formaldehyde	100.0	ml
Distilled water	900.0	ml
Sodium phosphate, monobasic	4.0	gm
Sodium phosphate, dibasic	6.5	gm

Bullard's Hematoxylin
(See Page 83)

Eosin-Phloxine B Stain
(See Page 87)

1% Acid Alcohol
(See Page 77)

1% Lithium Carbonate

Lithium carbonate	1.0	gm
Distilled water	100.0	ml

STAINING PROCEDURE

1.	10% Buffered neutral formalin	30	seconds
2.	Tap water (change frequently)	5	dips
3.	Bullard's hematoxylin	1	minute
4.	Warm (37-40°C) tap water (change frequently)	5	dips
5.	Differentiate in 1% acid alcohol	1	dip
6.	Tap water (change frequently)	5	dips
7.	Saturated lithium carbonate	1	dip
8.	Tap water (change frequently)	5	dips
9.	80% Alcohol	3	dips
10.	Eosin-phloxine B	30	seconds
11.	95% Alcohol	3	dips
12.	100% Alcohol	3	dips
13.	100% Alcohol	3	dips
14.	Xylene	3	dips
15.	Xylene	3	dips
16.	Mount coverglass with appropriate media (refractive index 1.48-1.56).		

RESULTS

Nuclear chromatin	blue
Cytoplasm	pink to red

Chapter 5

Methods for Softening Specimens

Methods for Decalcification of Specimens

Introduction

Bone calcified tissue consists of specialized cells performing the functions of structural support, protection for underlying soft tissues, and a reservoir of calcium and phosphate for the body. Bone is a meshwork of fibers and cells impregnated with calcium salts.

Calcified tissues consist of:

1. An inorganic fraction.
2. An organic fraction, in general, comprised of:
   a. cells
   b. intercellular substance

Bone, dentine, cementum and enamel are composed of needle-shaped crystals of inorganic salts associated with an organic matrix. The composition of the crystals in each of the tissues is very similar, consisting of hydroxyapatite, a calcium phosphate which is analogous to the mineral fluorapatite. However, there are important differences between the size of the crystalites in bone, dentine and cementum. In bone they are very small (300x25x25 Å), and larger in enamel (10,000x500x500 Å), although still small in relation to the limit of resolution of the optical microscope. The calcium salts present in bone tissue prevent the technician from sectioning this tissue in the routine manner. Therefore, decalcification (the process of removing the calcium salts) is necessary. This extra step is needed so that the bone tissue may be sectioned in a routine manner and the need for special equipment and procedures is eliminated.

Failure to completely decalcify a tissue specimen will result in torn, ragged, incomplete sections and could possibly damage the cutting edge of the microtome knife.

Note: It should be pointed out that undecalcified bone can be sectioned, but one must use special equipment, procedures and embedding media. For this purpose one should review the literature dealing with plastic embedding methods. The selection of a suitable reagent for decalcification, from the large number that are available, can be difficult for those who are unfamiliar with the process. High concentrations of inorganic acid decalcifying solutions, although rapid in their action, may produce artifacts and also degrade the staining of the cellular elements. Conversely, the use of weak or dilute reagents merely prolongs the period necessary to complete the dissolution of mineralized components and may not improve the quality of the staining. The choice is therefore critical.

In my opinion, there is no such thing as the "best" or "worst" acid decalcifying solution. There are only decalcifying solutions which decalcify at various speeds–from relatively slow, as in the case of 5% formic acid with sodium citrate to those which decalcify fairly rapidly, as in the case of many commercially available solutions. The above comments are significant because many histotechnologists often ask to be directed to the best decalcifying solution. (Often the "best" is defined as a solution which will not adversely affect subsequent staining properties.) In fact, what they are really looking for is a decalcifying fluid which will decalcify rapidly and will allow for good H&E staining results. What is missed in all of this is the fact that poor staining properties are not the result of how fast a solution decalcifies, but rather of unexpeditious removal of the bone specimens from the decalcifying solution, once the process of decalcification ceases. The degree of poor H&E staining is directly related to the length of time tissue remains in the acid decalcifying solution after the end point of decalcification. Overexposure of specimens to acid decalcifying solutions will result in the reduction of staining properties by approximately 10% for every 2 hours tissue remains in the decalcifying solution, after the end point of decalcification. Soft tissue staining qualities will not be affected as long as the process of decalcification is taking place. This is because the calcium carbonate and phosphate serve as a buffer in and around the tissue as it is being liberated (Fig. 34), regardless of the time tissue is exposed to the acid decalcifying fluids. For example, it normally takes weeks to decalcify a tooth and up to six months to decalcify a temporal bone, yet soft tissues are not adversely affected and therefore stain well. It goes without saying that any decalcifying fluid which contains an agent (e.g., sodium citrate) that can serve as a buffer to the tissue, after the end point of decalcification, will replace the protective effects of calcium carbonate and phosphate. It is for this reason that organic chelating agents used as decalcifying fluid seldom produce poor staining results after long tissue exposure.

Naturally, before decalcification is begun, the tissue specimen must be cut and trimmed to an appropriate size to insure that the processing and microtomy equipment is adequate to handle the specimen. The specimen must be fixed properly and should be washed in running water to remove excess fixative when fixation is completed.

Decalcification Using Acid Solutions

The use of acid solutions is the most common method of tissue decalcification. The process is based on the solubilities of various salts. Calcium in the bone occurs chiefly as carbonate and phosphate salts. The acid releases the calcium from its bond and forms a new calcium salt which is soluble in water. (The original calcium salts are only slightly soluble in water.)

HCl	+	Ca^{++}	$\longrightarrow$	$CaCL_2$	and	CO_2
Hydrochloric Acid		Calcium Salts Yield		Calcium Chloride		Carbon Dioxide

Acids can be used in a wide variety of dilutions and formulas. They are fast-acting and preserve cellular details well, when used at room temperature. Acid decalcifying solutions will cause swelling of the tissue if used with heat (anything hotter than room temperature).

There are several qualities of a decalcifying solution which should be considered when selecting the one to be used in your laboratory. These include:

1. Complete removal of the calcium salts within a limited time.

2. The absence of swelling or shrinkage of cells during tissue exposure to the decalcifying solution.
3. The absence of harmful effects on subsequent staining, especially on the staining of the nuclear chromatin.

The procedure for acid decalcification is relatively simple. After proper trimming and fixation, place the specimen in a large amount of acid solution (approximately ten times the volume of the specimen). Change the solution daily until decalcification is complete. Stirring, agitation, aeration and the use of vacuum will hasten the process of decalcification. To remove all traces of the acid after complete decalcification is achieved, the specimen should be washed in running tap water for 20 minutes to 1 hour depending on the size of the specimen. Prolonged washing could cause tissue swelling. Alternatively, the specimen may be placed in several changes of 5% lithium carbonate followed by a short water wash. Once the specimen has been decalcified and neutralized, either with water or lithium carbonate, it can be processed in a conventional manner. (See chapter 2 for a wide variety of processing schedules.)

Chemical Aspects[33]

Decalcification or demineralization consists of the solution of the inorganic components of mineralized tissue in a suitable reagent, which results in the tissue becoming soft enough to be cut with a microtome knife.

The inorganic components of normal calcified tissues of vertebrates consist of crystals with an apatite type of structure and, for the present purpose, they may be considered as being composed of pure hydroxyapatite, $Ca_{10}\ (PO_4)_6(OH)_2$. The minute apatite crystals of bone, dentine and cementum are completely embedded in, and therefore surrounded by, the organic matrix, which is largely composed of collagen. Any reagent which brings about demineralization, must diffuse through this matrix before it reaches the surface of the crystal to effect its solution.

The apatite crystals of mature enamel are much larger and, by comparison, lie much closer to one another than those of conventional bone. However, narrow spaces do exist between the crystals; these are probably partially occupied by the much less abundant organic matrix and serve for the diffusion of the decalcifying agent.

The two types of reagents used for decalcification are acids and complexing agents. Acids and acidic buffer solutions have been used more widely than complexing agents for histological purposes. Acids are readily available in considerable variety. Moreover, judicious selection of the particular acid used, together with careful monitoring of other conditions, especially concentation and temperature, provides control of the demineralizing process to suit specific needs. Several features of the chemistry and ionization of acids must be understood for these possibilities to be realized.

For our purposes, an acid may be considered as a substance which ionizes or dissociates to produce hydrogen ions (protons) when it is dissolved in a liquid, such as water; that is to say, an acid is a donor of hydrogen ions.

$$HA \rightleftharpoons H^+ + A^-$$

Acid	Hydrogen ion	Anion

Strong acids, such as the mineral acids (hydrochloric, sulfuric and nitric acid) dissociate almost completely, so that the equilibrium represented by the chemical equation favors the right-hand side; consequently

$$HCl \rightleftharpoons H^+ + CL^-$$

Hydrochloric acid (little)	Hydrogen ion (much)	Chloride ion (much)

The concentration of hydrochloric acid in solution will be quite low, whereas the hydrogen ion concentration will be relatively high. Since the hydrogen ions represent the central idea of acidity, acids which furnish a high concentration of these are strong acids.

Weak acids (such as most organic acids, including formic and acetic acids), dissociate only slightly so that the equilibrium favors the left-hand side.

$$CH_3COOH \qquad H^+ + CH_3COO^-$$

Acetic Acid (much)	Hydrogen ion (much)	Acetic ion (little)

The concentration of hydrogen ions produced by a weak acid will therefore be much less than would be produced by an equivalent amount of a strong acid. If, however, hydrogen ions are removed from a solution of a weak acid by a chemical reaction (for example, with hydroxyapatite) some of the larger reserve of undissociated acid will immediately dissociate and produce more hydrogen ions, so as to restore the ionic equilibrium represented by the equation. A weak acid, therefore, maintains a kind of reserve supply of acidity which tends to keep the hydrogen ion concentration constant if there is any tendency to disturb it. A weak acid constitutes the simplest kind of buffer solution.

A more effective buffer solution would be produced by adding, to the solution of weak acid, a high concentration of a salt having the same anion as the weak acid; for example, sodium acetate would be a suitable salt for acetic acid solutions. This has the effect of maintaining the concentration of anion on the right-hand side of the equation, so that the hydrogen ion concentration is kept constant within very narrow limits indeed.

It is important to distinguish between the strength of an acid in the chemical sense and its concentration. The strength is a fixed property for each particular acid, whereas the concentration of any acid may be varied as required between zero and an upper limit set by its solubility.

Hydroxyapatite is unstable and tends to pass into solution in the range below pH 7. It is only very slightly soluble in the region pH 6 to 7, but the solubility increases progressively as the pH is further reduced.

The dissolution of hydroxyapatite can be represented in terms of a simple reaction with hydrogen ions with the appearance of calcium and phosphate ions in solution:

$$(1)\quad Ca_{10}(PO_4)_6(OH)_2 + 8H^+ \rightleftharpoons 10Ca^{++} + 6HPO_4{}^{=} + 2\,H_2O$$

Phosphoric acid is tribasic and has three ionizable hydrogens, each with its own characteristic pK value:

$$(2)\quad H_3PO_4 \rightleftharpoons \underset{pK\ 1.96}{H_2PO_4{}^- + H^+} \rightleftharpoons \underset{pK\ 7.12}{HPO_4{}^{=} + 2H^+} \rightleftharpoons \underset{pK\ 12.32}{PO_4{}^{\equiv} + 3\,H^+}$$

Thus, the doubly negatively charged monohydrogen phosphate ions produced according to equation (1) would predominate only above pH 7. In acid solution, further hydrogen ions would be required to convert them to the slightly charged di-hydrogen phosphate ions, which are the most abundant species in the range pH 2 to 7, by the mechanism of equation (2) acting in the reverse direction.

Methods For Decalcifying Specimens

Some of the most useful decalcifying methods are presented below.

Formic Acid-Sodium Citrate Method

This technique gives better staining results than the nitric acid methods cited below, since sodium citrate is used. Sodium citrate continues the buffering process of the specimen after decalcification is complete. This solution prevents or retards the destructive action of acids on soft tissues. This has become my method of choice for all forms of decalcification, including eye globes.

SOLUTIONS

Formic Acid-Sodium Citrate Solution

Solution A

Sodium citrate	50.0	gm
Distilled water	250.0	ml

Solution B

Formic acid, 90%	125.0	ml
Distilled water	125.0	ml

Mix solutions A and B in equal portions for use.

PROCEDURE

1. Place calcified specimen in large quantities of formic acid-sodium citrate solution until decalcification is complete. (Change solutions daily for best results.)
2. Wash in running water for 30 minutes to 1 hour.
3. Dehydrate, clear, and impregnate with paraffin or process as desired.

Ippolito-Lavelle-Pedrini Method for Decalcification and Preservation of Cartilage Proteoglycans

Histological and histochemical studies of paraffin embedded cartilage often require decalcification of mineralized cartilage or surrounding bone. The growth plate is a typical example of a tissue requiring such treatment. Decalcification extracts organic components from the tissues and the amount of material extracted is related to the decalcifying solution employed.

Proteoglycans are one of the main macromolecular components of the cartilage matrix. Five percent formic acid decalcifies quickly and causes only minimal loss of uronic acid from calcified cartilage, which cannot be appreciated by histochemical means. Formic acid should be considered as an excellent decalcifying agent for proteoglycan histochemistry in paraffin embedded cartilage.

SOLUTION

5% Formic Acid

Formic acid	5.0	ml
Distilled water	100.0	ml

PROCEDURE

1. Place calcified specimen in large quantity of 5% formic acid until decalcification is complete. (Change solution daily and agitate.)
2. Wash in running tap water for 2 hours or more depending on size of the specimens.
3. Dehydrate, clear and impregnate with paraffin or process as desired.

Kristensen's Method

This method of decalcification of bone is excellent for the preservation of nissl substance when Einarson's gallocyanin method is used. In addition, shrinkage is minimal, speed is moderate, staining properties are excellent and tissue damage is minimal.

SOLUTION

Formic Acid-Sodium Formate (Stock)

8N formic acid	50.0	ml
1N sodium formate	50.0	ml

Mix equal parts just prior to use.

PROCEDURE

1. Place tissue in formic acid-sodium formate solution until decalcification is complete.
2. Wash in running tap water for 30 minutes.
3. Dehydrate, clear and impregnate with paraffin as desired.

Perenyi's Method

Perenyi's solution is slow for decalcifying hard bone but is excellent for small deposits of calcium; e.g., calcified arteries, coin lesions, thyroid colloid and calcified glands. It is also an excellent decalcifying fluid for avian eye globes, which normally contain bone at the corneal-scleral junction, and for human globes that contain calcium due to pathologic conditions. Little hardening of the tissue takes place, and excellent morphologic detail is preserved. This solution is also very useful for softening chitin.

SOLUTION

Perenyi's Solution		
10% Aqueous nitric acid	40.0	ml
100% Alcohol	30.0	ml
0.5% Aqueous chromic acid	30.0	ml

The component solutions can be kept in stock, but the decalcifying fluid should be mixed immediately before use. This solution may acquire a blue-violet tinge after a short period, but this will have no affect on its decalcifying properties.

PROCEDURE

1. Place calcified specimens in large quantities of Perenyi's solution until decalcification is complete.
2. Wash in running water for 30 minutes. Store in 70% alcohol if processing is not to be performed immediately.
3. Dehydrate, clear, and impregnate with paraffin or process as desired.

Fenyves' Method for Decalcifying Teeth[34]

This procedure calls for decalcification of teeth with nitric acid and the aid of heat. Decalcification of teeth can be hastened by increasing the temperature of the decalcifying solution to 40°C.

SOLUTION

5% Nitric Acid		
Concentrated nitric acid	5.0	ml
Distilled water	95.0	ml

PROCEDURE

1. Fix tooth in 10% buffered formalin for 48 hours or longer.
2. Place tooth in 5% nitric acid solution at room temperature for 8 hours.
3. Place tooth in a fresh change of nitric acid solution and incubate at 40°C overnight. NOTE: After this step it is possible to gross (macrosection) the tooth without difficulty.
4. Place tooth in a fresh solution of nitric acid at room temperature for 8-12 hours.
5. Wash tooth in running tap water for 4 hours.
6. Place tooth in 70% alcohol. The specimen is now ready for processing.

Nitric Acid Method

This is a good and rapid decalcifying method. Tissue will not swell in a 5-10% acid solution. They will swell if acid concentrations exceed 15%. Bone specimens must be removed from the decalcifying solution shortly after decalcification is complete, since overexposure to nitric acid impairs or destroys nuclear chromatin staining. Nitric acid is the acid of choice for decalcifying temporal bones.

SOLUTION

5% Nitric Acid
(See Page 109)

PROCEDURE

1. Place calcified specimen in large quantities of 5% nitric acid solution until decalcification is complete. (Change solution daily for best results.)
2. Wash in running water for 30 minutes or neutralize for a minimum of 30 minutes in 10% formalin to which an excess of calcium or magnesium carbonate has been added.
3. Wash in running water for 30 minutes.
4. Dehydrate, clear, and impregnate with paraffin or process as desired.

Alcoholic-Nitric Acid Method

This decalcifying solution is suggested as a fairly rapid decalcifying fluid, due to the action of alcohol on some soluble fats.

SOLUTION

Alcoholic-Nitric Acid

80% Alcohol	95.0	ml
Concentrated nitric acid	5.0	ml

PROCEDURE

1. Place calcified specimens in large quantities of nitric acid-alcohol solution until decalcification is complete. (Change solution daily for best results.)
2. After specimens are completely decalcified, transfer directly to a solution of 4% aqueous sodium sulfate for 30 minutes.
3. Wash specimens in running water for 30 minutes to 1 hour to remove sodium sulfate.
4. Dehydrate, clear and impregnate with paraffin or process as desired.

Method for Decalcifying Bone, Dentine and Cementum

For general diagnostic purposes, only two reagents need be considered; namely, a normal solution of nitric acid and 4N formic acid. The nitric acid may be used when the time factor is important. However, treatment with this reagent should not exceed 15 hours and some impairment of nuclear staining should be anticipated.

SOLUTIONS

1N Nitric Acid		
Concentrated nitric acid (sp. gr. 1.42)	63.0	ml
Distilled water	937.0	ml

4N Formic Acid		
Formic acid (sp. gr. 1.21)	170.0	ml
Distilled water	830.0	ml

PROCEDURE

1. Reduce the thickness of the specimen and wherever possible insure that it does not exceed 3 mm.
2. Select a glass container with a wide neck; cover the base with several pieces of gauze.
3. Place the specimen on top of the gauze. Add 200 ml of one of the decalcifying solutions. Seal the flask with a piece of cotton. The solution should be maintained at room temperature.
4. Do not change the fluid, but gently agitate it by hand for a few seconds, two or three times each day.

Chelates as Decalcifying Agents

Organic chelating agents have been recommended in special instances for decalcification of bone specimens. The staining quality is excellent and specific entities are preserved well after the use of these agents. The slow speed at which these agents decalcify demands that the size of the specimen not exceed 4 mm in thickness.

The method most useful in routine histology is presented below. For other procedures, see the following references:

1. Freiman, D.G.: Organic chelating agent in demineralization of bone for histochemical study of alkaline phosphates. *Amer. J. Clin. Path.* 24: 227-231, 1954.
2. Trott, J.R.: The presence of glycogen in the rat liver following in vitro processing in decalcifying agents. *J. Histochem. Cytochem.* 9: 699-702, 1961.
3. Balogh, K: Decalcification with versene for histochemical study of oxidative enzyme systems. *J. Histochem. Cytochem.* 10: 232-233, 1962.
4. Carlquist, S.: The use of ethylenediamine in softening hard plant structure for paraffin sectioning. *St. Techn.* 57:5, 1982.

Birge-Imhoff Ethylenediaminetetra Acetic Acid (EDTA) Method

SOLUTION

EDTA Solution

EDTA	10.0	gm
Distilled water	100.0	ml

PROCEDURE

1. Fix specimen in 10% buffered neutral formalin or fixative of choice.
2. Place calcified specimen in a saturated solution of EDTA until decalcification is complete.
3. Place in 70% alcohol holding station.
4. Dehydrate, clear, and impregnate with paraffin or process as desired.

REMARKS

Most bone specimens (4 mm thick) are decalcified in two to four days, but specimens may be left in EDTA solution for as long as 14 days without noticeable adverse effects on staining quality. Decalcification in this solution prevents removal of metals in tissue, such as iron and copper. It is possible to remove these metals from tissue if acids such as nitric or hydrochloric are used for decalcification. The only disadvantage is that it takes considerably longer to decalcify with the EDTA method than those using acids.

Method for Decalcifying Enamel In Situ

SOLUTION

4N Sodium Acetate-Hydrochloric Acid

Add 217 ml of concentrated hydrochloric acid (sp. gr. 1.18) to 500 ml of water and stir. To this solution slowly add 246 gm of anhydrous sodium acetate; stir until dissolved. Measure the pH and adjust to pH 3.55 by the addition of hydrochloric acid or sodium hydroxide. Dilute the solution with water to 750 ml.

PROCEDURE

1. Cut the specimen into slices no more than 3 mm in thickness.
2. Cover the base of a wide-mouthed glass container with a layer of gauze.
3. Add 400 ml of the sodium acetate-hydrochloric acid solution to the container.
4. Place the specimen on the surface of the gauze.
5. Leave the specimen in the fluid until demineralized.
6. Rinse specimen in running tap water for 1 hour.
7. Dehydrate, clear and impregnate with paraffin or process as desired.

Method for Decalcifying Bone, Dentine, and Cementum with Ethylenediaminetetra Acetic Acid (EDTA)

SOLUTION

Ethylenediaminetetra Acetic Acid (EDTA) Solution pH 7.4

EDTA, disodium salt (versene, sequestric acid)	100.0	gm
1N sodium hydroxide	280.0	ml
Distilled water	600.0	ml

Dilute to 1 liter.

PROCEDURE

1. Reduce the thickness of the specimen and wherever possible insure that it does not exceed 3 mm.
2. Select a glass container with a wide neck. Cover the base with several pieces of gauze.
3. Place the specimen on top of the gauze. Add 200 ml of the decalcifying solution. Seal the flask with a plug of cotton. Decalcification may be carried out at a temperature within a range of 5-60°C.
4. Do not change the fluid but gently agitate it by hand for a few seconds, several times each day.
5. Check the progress of demineralization with x-ray or other method.
6. Wash the EDTA from the tissues by immersing in running water for 1 hour.
7. Dehydrate, clear and impregnate with paraffin or process as desired.

REMARKS

The temperature used to decalcify (step 3) is determined by the entity being demonstrated. For example, 5°C temperature should be used for preservation of enzymes and antigens.

Freiman's Method for Decalcification and Preservation of Alkaline Phosphatase

This method permits demineralization of calcified material without destruction of alkaline phosphatase activity by means of the organic chelating agent, ethylenediaminetetra acetic acid (EDTA).

SOLUTIONS

5% Ethylenediaminetetra Acetic Acid (EDTA)

EDTA	5.0	gm
Distilled water	100.0	ml

Adjust pH to 6-6.5 with a solution of sodium hydroxide (see step 3).

A 5% solution is usually adequate, but the concentration is unimportant, provided that a sufficient excess of the chelating agent is available to bind all the calcium present in the bone.

1% Magnesium Chloride

Magnesium chloride	1.0	gm
Distilled water	100.0	ml

PROCEDURE
1. Fix calcified specimens in cold 80% alcohol for 24 hours.
2. Wash in running tap water for 1 hour.
3. Decalcify in a solution of 5% EDTA adjusted with sodium hydroxide to a pH of 6.0 to 6.5 until specimens are decalcified. (This step should be carried out in the refrigerator for the first 24 hours. Then the calcified specimen may be left in the chelating agent at room temperature until decalcification is complete.)
4. Wash in running water for 1 hour.
5. Dehydrate, clear and impregnate with paraffin. (Dehydration, clearing and impregnation at refrigerated temperatures, followed by rapid paraffin embedding below 56°C, is the method of choice.)
6. Paraffin blocks should be refrigerated, if cutting is delayed.
7. Sections are cut in the usual manner, but heating above 37°C is omitted.
8. Reactivate enzyme by placing decerated sections in a 1% solution of magnesium chloride for 6 hours.
9. Demonstrate enzyme activity by either the Gomori or Diazo techniques.

REMARKS
The exact pH level of the EDTA is not critical, since satisfactory results can be obtained at values ranging from 5.0 to 7.2. A slightly acid pH is preferable, however, since alkaline phosphatase is much more stable in acid than in alkaline solution. This method is simple and convenient to perform and it permits excellent preservation of enzyme activity.

Wakita-Kobayashi-Shiai Method for Decalcification and Preservation of Sections for Electron Microscopy[35]

This method is very useful for electron microscopy of mammalian tooth germs and bone, after fixation in glutaraldehyde paraformaldehyde mixture. The method has the advantage that decalcification is complete within a quarter of the time required for ethylenediaminetetra acetic acid (EDTA) methods. The fine structure of ameloblasts and hard tissue is preserved as well as with EDTA.

SOLUTIONS

L-ascorbic Acid Sodium Chloride

L-ascorbic acid	2.0	gm
Sodium chloride	0.9	gm
Distilled water	100.0	ml

Add a small amount of thymol to prevent bacterial contamination.

PROCEDURE
1. Suspend specimens in upper layer of the solution.
2. Stir the solution constantly and keep in darkness to avoid oxidation of ascorbic acid.

3. After decalcification, specimens may be processed in the usual manner for electron microscopy.

REMARKS
L-ascorbic acid (vitamin C) is a readily available nontoxic material. Decalcification with ascorbic acid, as with other acid decalcifying solutions, may cause some enzyme activity to be lost. EDTA should be used as the decalcifying agent for preservation of enzyme and electron microscopy, since it provides good preservation of cell components and enzyme activity.

Decalcification Using Commercial Decalcifying Solutions

The commercially available decalcifying solutions have proven useful in many laboratories, particularly since they are ready for use and therefore can be time savers. One should recognize, however, that the faster the decalcifying action, the more rapid the decrease of nuclear chromatin once the staining with hematoxylin end point of decalcification is passed. Therefore, one should purchase a solution which results in a medium speed decalcifying action. Determining the decalcifying end point becomes critical when using these decalcifying fluids. It is my personal belief that the ingredients of the decalcifying solution should be known to the user. This is necessary to determine what effects, if any, are produced by the decalcifying fluids in tissue structures and constituents.

There are numerous commercially available decalcifying solutions, each of which can be used satisfactorily by following these general directions:

1. Place calcified specimen in selected solution until decalcification is complete. Change solution daily for best results and follow manufacturer's directions.
2. Wash in running water for 30 minutes or follow manufacturer's directions.
3. Dehydrate, clear, and impregnate with paraffin or process as desired.

Method for Determining End Point of Decalcification

The biggest problem encountered in decalcification is the failure of the histotechnologist to understand that the selection of a decalcifying fluid is not the most important consideration. More important is the removal of specimens from decalcifying fluids soon after decalcification is complete. If this is not done, the chances of subsequent good staining reactions are reduced 10% for every two hours the tissue remains in the decalcifying solution. It must be recognized that 90% of the poor staining qualities demonstrated in decalcified tissues is due to this one factor. The loss of staining properties is not difficult to understand if one recalls the chemistry involved. Normal calcified tissue of vertebrates consists of crystals with an apatite type of structure which we may consider as being composed of pure calcium hydroxyapatite, $Ca_{10} (PO_4)_6 (OH)_2$. Calcium hydroxyapatite reacts with hydrogen ions (H+) provided by the decalcifying acid to form phosphoric acid ($6HP0_4$), calcium ions (10Ca), and water ($2H_20$). The important end product here is phosphoric acid, since it releases cations (H+) at a high pH and captures them at a low pH. The reaction attempts to maintain the tissue at pH

6.0. A pH of 6.0 prevents the decalcifying acid from altering the cellular structures and the potential loss of staining properties during decalcification. When decalcification is complete, phosphoric acid is no longer produced and the acid ions are free to attack and thereby alter the cellular structures and affect staining properties. The methods provided below are for determining the end point of decalcification.

It is difficult to determine when testing with these methods should begin, since the time required for decalcification of any specimen is determined by the size of the specimen or the decalcifying solution used. Care should be taken not to contaminate any of the test solutions. All glassware should be cleaned carefully to prevent false results.

Arnim's Method

SOLUTIONS

5% Ammonium Hydroxide

Concentrated ammonium hydroxide (58%)	5.0	ml
Distilled water	100.0	ml

5% Ammonium Oxalate

Ammonium oxalate	5.0	ml
Distilled water	100.0	ml

PROCEDURE

1. Draw approximately 5 ml of the decalcifying solution from the bottom of the container in which decalcification is taking place.
2. Add 5 ml each of 5% ammonium hydroxide and 5% ammonium oxalate.
3. Mix and let stand 15 minutes.

RESULTS

A cloudy solution (caused by calcium oxalate) indicates that the specimen is not thoroughly decalcified. In this case, place the specimen in a fresh change of decalcifying solution. Repeat the test at a later time. The specimen is completely decalcified when a cloudy solution is no longer produced from the mixture. The test can be performed as frequently as necessary.

X-Ray Method

Undoubtedly the best method for determining complete decalcification is by x-raying the specimen. However, the expense involved in this method may limit its use.

Specimen Flexibility Method

It is possible to determine the end point of decalcification by the way a specimen bends, or how it "feels" to the fingers of the histotechnologist. However, this technique requires a great deal of experience. To test for complete decalcification, bend the specimen very gently. If it bends with little resistance, it should be well-decalcified. Next, press thumbnail gently into specimen; areas containing calcium will feel slightly harder than areas which have been fully decalcified.

Eastoe's Method

SOLUTIONS

5% Citric Acid

Citric acid	5.0	gm
Distilled water	100.0	ml

0.25N Potassium Hydroxide (Stock)

Potassium hydroxide	1.4	gm
Distilled water	100.0	ml

3N Potassium Hydroxide

Potassium hydroxide	16.8	gm
Distilled water	100.0	ml

0.004% Calcein*

Calcein	0.004	gm
0.25N Potassium hydroxide (stock)	1.00	ml

1. Take 0.1 ml of the decalcifying solution and place in a test tube.
2. Shake the decalcifying fluid containing the specimen and take 0.1 ml and place in a second test tube.
3. To each of the tubes add the following solutions:

5% Citric acid	1.5	ml
0.004% Calcein solution	0.4	ml
3N Potassium hydroxide	1.0	ml

Mix well and titrate with 0.005M solution of ethylenediaminetetra acetate until the green fluorescence disappears.

4. Observe the reaction in strong daylight against a black background.

*Fluorescein bismethylene imino diacetic acid.

RESULTS

A green fluorescence appears within a few seconds if calcium is present. Decalcification is complete when the fluorescence fails to appear. Repeat the titration in step 3 daily until the amount of EDTA required remains the same for three consecutive days. The specimen is then assumed to be decalcified.

Method for Testing Acid Content in Specimens

A pH indicator can be used to determine if all acid has been removed from bone specimens after the decalcification process.

SOLUTIONS

Universal pH Indicator Solution (Stock)

This solution can be obtained from a number of laboratory supply houses.

2% Indicator Solution (Working)

Universal pH indicator solution	2.0	ml
Distilled water	100.0	ml

PROCEDURE

1. Remove specimen from decalcifying solution and wash in running tap water for 30 minutes.
2. Rinse in distilled water.
3. Place the specimen in an appropriate size container.
4. Add sufficient 2% indicator solution to submerge the specimen and agitate the container occasionally.
5. Compare the pH of the solution with the color chart provided, which will indicate the pH of the solution. This, in turn, indicates whether all acid has been removed from the specimen. A pH reading of 6.5 suggests all acid has been removed from the specimen. The test can be repeated at hourly intervals until the pH readings are near 6.5.

Methods For Softening Keratin

There are often instances when one must obtain microscopic tissue sections of specimens containing hard and difficult-to-section keratin. Examples of this material are toe or thumb nails and animal hoofs. The following methods have proven useful for the softening of keratin, thereby allowing one to obtain excellent microscopic sections. One should remember that the hardness of keratinized specimens is due to disulfide bonds. These bonds are not tightly held by fixation in formalin, and therefore the clearing of cystine is possible. This action results in sufficient softening of keratin to allow good microtomy. Conversely, it is important that the crosslinking produced by fixation takes place, in order to prevent the entire specimen from dissolving completely in the keratin-softening solutions.

Shapiro's Method[36]

This method uses permanent wave or depilatory lotion and works well since the former usually contains thioglycolate and the latter either thioglycolate or inorganic sulfide. It is a useful procedure for softening specimens containing keratin which have been embedded in paraffin wax.

FIXATION 10% Phosphate buffered formalin.
PROCESS Paraffin.
MICROTOMY Cut sections at 6 micrometers.
SOLUTIONS Commercially available permanent wave or depilatory lotion.

PROCEDURE
1. Soak paraffin block face down in a shallow pan containing either of the two types of commercial cosmetic preparations, for 1 hour or until specimen is sufficiently soft to section.
2. Wash off excess preparation in water for a few seconds.
3. Cut sections at desired thickness.

Luna's Method[37]

A depilatory lotion for softening keratin has been used for many years. Although Neet* is used in this procedure, it is felt that any hair removing lotion will perform satisfactorily if the following method is used.

FIXATION 10% Phosphate buffered formalin (see remarks).
PROCESS Paraffin.
MICROTOMY Cut sections at 6 micrometers.
SOLUTIONS Neet or any good depilatory lotion.

PROCEDURE
1. After fixation place specimen in Neet or other depilatory cream for several hours.
2. Check to determine if specimen has softened sufficiently to section. (Specimen should bend easily.)
3. Wash specimen in running tap water for 10 minutes.
4. Dehydrate, clear and impregnate with paraffin or process as desired.
5. Cut sections at desired thickness.

REMARKS
Fixation in 10% buffered formalin is necessary to produce crosslinking and thereby prevent keratin from dissolving completely in Neet or other hair removing lotion. (Formalin fixation of keratin introduces methylene bridges between amino groups in keratin and thus helps to preserve the structural integrity of keratin.) Excellent results have been obtained with the use of this method.

*A commercial cosmetic hair removal product found in most drug stores.

Dotson's Method[38]

This method is used for treating (softening) keratinized tissue after tissue processing.

FIXATION 10% Buffered neutral formalin.
PROCESS Paraffin.
MICROTOMY Cut sections at 6 micrometers.

SOLUTIONS Ammonium thioglycolate, 50% solution. (Available commercially in aqueous solution.)

Softening Solution

Aerosol OT	0.1	ml
Distilled water	100.0	ml
Ammonium thioglycolate	5-10	drops

The amount of ammonium thioglycolate used is determined by the size of the specimen and degree of keratinization desired.

PROCEDURE

1. Macrosection (rough cut) specimen blocks to expose tissue.
2. Soak a small piece of cotton with the softening solution.
3. The cotton containing the solution is than applied to the face of the block for 10-15 minutes.
4. Cut sections at desired thickness.

REMARKS

It is always better to dekeratinize tissue before processing as suggested by Luna's method. However, for laboratories with heavy workloads and deadlines, little time is available for handling individual biopsies differently. In the author's hands this method has provided a quick, efficient and inexpensive means of softening keratin.

Turner-Holtom Method for Softening Mummified Tissue[39]

In recent years there has been a reawakening of interest in the histological investigation of mummified tissues. A variety of methods and fluids have been employed in the past to reconstitute this tissue. Presented here is a procedure which utilizes a diluted solution of fabric softener in normal saline to soften mummified tissue. There are many fabric softeners on the market which may make it difficult to select a proper softener. The principle ingredient of the proper fabric softener is dimethyl ammonium chloride and should be the major ingredient of the softener used below.

SOLUTION

Fabric Softener

Fabric softener (dimethyl ammonium chloride)	0.2	ml
Normal saline	100.0	ml

PROCEDURE

1. Place tissue in fabric softening solution at room temperature overnight. The solution should be approximately 20 times the volume of the tissue. The solution should be agitated gently (20 rpm) if possible.
2. Fix tissue in 10% formal-saline or 10% formalin. (Formalin fixative should not be buffered or neutralized.)
3. Process in conventional paraffin method and embed in paraffin of choice.
4. Perform microtomy in the conventional manner by cutting 5-6 micrometer sections.
5. Stain with preferred methods.

Method for Softening Chitin

The best procedure for softening chitin is Perenyi's solution which can be found on page 109 of this chapter.

PROCEDURE
1. The specimen is fixed in desired fixative, but formalin based fixatives work best.
2. Wash specimen in running tap water for 20 minutes.
3. Place in Perenyi's solution until chitin feels soft.
4. Wash specimens in running tap water for 30 minutes.
5. Process as desired.

General Comments

- The risk of tissue swelling and maceration can be reduced if heat is avoided when decalcifying in acid reagents. However, the precise effect that temperature has on the tissues is unknown. It is the increased reactivity at high temperatures of some substances used in the processing solutions (for example, protein hydrolysis by acids) which is largely responsible for deleterious changes. A high temperature accelerates the penetration of the reagents and does not appear to produce adverse changes in the tissues, in and of itself.

- The effect on the tissues produced by the sudden increase in temperature during infiltration with paraffin wax has not been fully investigated. However, this factor probably plays an important part in the hardening and shrinkage observed after the routine preparation of sections in this medium. When fixation and decalcification are performed at 20°C, it is preferable to dehydrate and clear the tissues at a temperature of 37°C before transferring them to molten wax at 60°C. It is believed that two increases in temperature of approximately 20°C produces less hardening and shrinkage than a single increase of 40°C.

- Simultaneous fixation and decalcification in acid reagents produces considerable swelling of the tissue and, in the case of acetic and nitric acids, partial maceration occurs.

- Overexposure of tissue specimens to acid based decalcifying solutions, results in hydrolysis. This in turn results in poor staining results.

Chapter 6

Special Procedures and Uncommon Methods

There is often the need for special methods to accomplish unusual goals in the demonstration of pathologic and/or normal entities or structures. Provided in this chapter are a group of methods which, though uncommon, never the less can be very useful to the practice of histologic technique.

Methods for Eye Specimens

The methods outlined here are for human and animal eye specimens. In general, there are more requirements for the proper sectioning of human eye specimens than animal. In the case of human eyes, it is generally a requirement that the sections obtained demonstrate optic nerve, both cross-sectional and longitudinal; pupil, lens, fovea, and suspected abnormal features. In the case of animals, it is usually required that sections be obtained that demonstrate the optic nerve, both longitudinal and cross-sectional, the lens and the middle of the pupil.

It is also important to realize that retinal detachment is a common occurrence in paraffin processed eye sections. This is due primarily to the loss of the vitreous body when the eye specimen is grossed. In this case, as it oozes out of the eye specimen, the viscous vitreous body pulls the retina away from the sclera, creating a detached retina. The retina will shrink even further during exposure to the various processing solutions.

Eye specimens can be processed in a variety of ways including wet and dry celloidin methods. The latter methods may be used to prevent some of the tissue structure disorientation which takes place with the paraffin method.

Although there are some problems (in the form of artifacts) with the use of the paraffin methods, the advantages outweigh the disadvantages. The problems of tissue shrinkage, retinal displacement and difficulty in lens sectioning encountered in paraffin processing are generally accepted to gain the following advantages: faster processing, microtomy and staining; ability to perform more special stains (including immunochemical stains) and to cut thinner sections than with celloidin methods; and the elimination of hazardous agents such as ether and celloidin.

Fixation of Eye Specimens

The enucleated eye should be fixed immediately in 10% buffered neutral formalin in a volume 10 to 15 times that of the eye. Fixation is usually complete in 24 hours and the specimen ready for processing. Cutting windows or injecting the fixative into the eye globe is neither necessary nor desirable. The technique after receipt of the eye specimens in the laboratory is as follows:

1. Wash eye specimens in running water for 2 hours to remove excess fixative.

2. Place in 60% alcohol until ready to cut. This restores much of the specimen's normal color, which has been dulled by the fixative. It also is the first step in dehydration and it can be used as a good storage solution.
3. Write a complete description of the specimen, including the external appearance, measurements, general aspects of the globe, and note any scars from injury and/or operation.
4. Open (macrosection) the globe with a flat razor blade, using a sawing motion starting several millimeters from either side of the optic nerve and passing through the cornea just outside the limbus. (The center block containing the pupil and optic nerve is the portion used for processing, while the two sides, calottes [caps] are held in formalin for possible future use.)

Macrosectioning of the Eyeball

Figure 35a, b, c and d. After fixation, the posterior ciliary blood vessels (Fig. 35a) are more prominent on the nasal side. The horizontal diameter of the cornea is greater than the vertical. The distance between the optic nerve and the limbus is less on the nasal side (Fig. 35b). The superior oblique muscle (Fig. 35c) inserts temporally above the optic nerve. The inferior oblique muscle (Fig. 35d) inserts on the temporal side of the optic nerve, extending interiorly from the posterior ciliary vessels. Describe the eye in mm, through the following planes: Anteroposterior, horizontal and vertical. Transillumination of the eye should be performed to detect abnormal shadowing which may suggest a lesion. This can be performed by placing the eye in front of a bright light, preferably in the dark. Sections are usually made along the horizontal plane in order to get the optic disc and macula lutea in a single line, but when necessary, the eye may be cut in any plane. The first section must be made through the peripheral cornea and adjacent to the optic nerve (Fig. 35e and f). The second section is made parallel to the first one, on the other side of the cornea and optic nerve (Fig. 35g and h). At this point describe internal structures of the center portion indicating known or suspected lesions as well as abnormal appearance. Also incorporate any special instructions to enable the technician to demonstrate any abnormal lesions or conditions in relation to the normal structures. A thin cross section of the optic nerve (Fig. 35i) should be cut and processed. Remaining portions of the optic nerve and caps or calottes should be placed in 60% alcohol or 10% buffered neutral formalin. The specimen is now ready for processing or embedding in paraffin or celloidin.

5. Orientation and embedding instructions should be written on an identification tag or label which accompanies the specimen through the processing. Place specimen in 80% alcohol to await processing.
6. If calcified material is observed on macrosectioning, the specimen should be placed in a decalcifying solution for several days. The solution of choice for eye tissue is sodium citrate and formic acid. Following the use of the decalcifying solution, the specimen must be washed in running water for 2 hours before further processing.
7. Processing on an automatic tissue processor may be accomplished using the following schedule:

Station	Human Eye Processing Schedule Reagents	Time
1.	95% Isopropyl alcohol	1 hour
2.	95% Isopropyl alcohol	1 hour
3.	100% Isopropyl alcohol	1 hour
4.	100% Isopropyl alcohol	1 hour
5.	100% Isopropyl alcohol	1 hour
6.	Xylene	1 hour
7.	Xylene	1 hour
8.	Xylene	1 hour
9.	Paraffin, Peel-A-Way*	1 hour
10.	Paraffin, Peel-A-Way*	1 hour
11.	Paraffin, Peel-A-Way*	1 hour
Embed in Peel-A-Way paraffin wax.		

* Peel-A-Way, Polyscience, Inc., Warrington, PA, 18976

NOTE: It is important that Peel-A-Way paraffin wax be used for embedding eye specimens since this particular paraffin allows the sectioned eye specimen to expand to its original size in the floatation water bath. This prevents the formation of wrinkles and other artifacts commonly seen in tissues which do not expand to their fullest in the floatation bath. It is my feeling that Peel-A-Way paraffin performs well when used on round specimens because it does not shrink the approximate 14% that other paraffins shrink when they go from molten to a solidified state. It is reasonable to assume that the paraffins shrinking more than the tissue will exert a compressing force upon it, causing the tissue to also shrink. This shrinkage is difficult to overcome when tissue sections are placed on a 45-50°C floatation water bath. The compressive effect produced by the paraffin can be surmounted if the temperature of the floatation bath is raised to 60°C. Peel-A-Way is sold in 3 melting points; 53-55°, 56-58° and 60-62°C. The type which melts at 56-58°C is preferred for eye specimens.

Embedding Human Eye Specimens

Embedding of eye specimens can be accomplished by several ways, the size of the specimen usually determining the method chosen. Most human eye specimens can be embedded in plastic embedding cassette systems. Of course, it is important to select a deep embedding base mold which will accommodate the entire specimen. This is necessary to prevent squeezing and distortion of the eye specimen, retinal detachment, and/or lens displacement. Careful evaluation of the numerous embedding systems available is needed to determine which best suits one's purpose.

Embedding Large Eye Specimens

Metal embedding pans can be used quite successfully for multiple embedding of large eye specimens. This procedure is especially useful for embedding eyes with attached tumors, or from large animals, such as dogs, cows, baboons, etc. Regardless of the method employed, it is important to transfer the eye quickly from the impregnating paraffin to the embedding mold. This eliminates the possible development of hairline separations between the sclera and surrounding paraffin.

It is also very important to prevent the formation of air bubbles within the eye specimens, as these bubbles will adversely affect the quality of the sections. In the case of metal pan embedding, one must hasten the paraffin's solidification by placing a piece of ice underneath the pan while carefully orienting the eye specimens with a pair of forceps. The paraffin containing the embedded specimens should be cooled quickly, by placing the pan in cold water, to prevent minute air bubbles forming within the paraffin. The paraffin block is removed from the pan and carefully scored between eye specimens with a sharp paring knife. The blocks are then broken apart by hand and excess paraffin trimmed, allowing approximately 2 mm from the specimen on each side for proper production of a paraffin ribbon during microtomy. These larger specimens will have to be mounted on microtomes with a special holder. One of the most useful is the Hopkins clamp. There are some microtomes which will allow the insertion of the paraffin block into their "vise type" block holder.

Small Eye Processing Schedule

This schedule deals with eye specimens of smaller animals, such as rats and mice. This particular eye size is so small that attempts to macrosection them properly requires some patience. The results are often not worth the effort, so these often are processed in total, i.e., without macrosectioning. To achieve high quality sections with limited effort the following procedure is suggested.

Fixation

The fixation of these specimens is identical to that employed above for human eye specimens. The processing schedule for human eye specimens can also be used for small eyes. However, if the eye specimens appear brittle or hard to section during microtomy, the processing schedule can be reduced by 15 minutes at each station. This should eliminate most problems related to overprocessing. As stated earlier, macrosectioning of small animal eyes is difficult; nevertheless, final results can be improved if small windows are cut at opposite poles of the horizontal plane. This can best be performed by the use of a single edge razor blade; use a sawing motion for best results. Alternatively, the windows can be cut out with a sharp pair of scissors. In this case, care must be exercised to prevent retinal detachment which may be caused by the squeezing action of the scissors. The eye specimen is now ready for processing with the use of plastic cassettes found on page 31.

Microtomy of Eye Specimens

The following steps should be followed closely for the production of high quality microscopic tissue sections of eye specimens.

1. Place the eye on the microtome block holder in the horizontal plane; i.e., scleral sides parallel to knife, cornea and optic nerve on either side.
2. Rough cut into the eye specimen slowly to prevent dislodging of the hard lens.
3. Place a small piece of water-soaked cotton on the surface of the specimen. (It is best to soak the cotton in the floatation bath since this provides warm water which, in turn, will soften the eye structures faster. It is important to realize that water soaking of eye specimens is essential if one is to obtain good sections. This simple

step minimizes dislodging, tearing and compression of the lens, gelatinous exudate and blood.)

4. Chill the block containing the specimen and the microtome blade with a flat piece of ice for a few seconds.
5. Cut sections at desired thickness (generally 6-8 microns) by using a smooth, slow turn of the microtome wheel (resoak paraffin block's surface with water-soaked cotton if lens does not cut well or shows signs of shattering). Take required sections through the optic nerve and pupil as well as suspected abnormal features. At this point it is essential to have the following items: a large rectangular glass dish filled with cold distilled water and a floatation water bath filled with water at a temperature 10°C above the paraffin's melting point. This water bath should contain 3 teaspoons of 5% gelatin solution as a section adhesive. Alternatively, one may use any section adhesive which will ensure good section adhesion, (see chapter 17).
6. Place the section ribbon on the surface of the cold distilled water dish and separate with a teasing needle. This is best done by needle pricking along the section ribbon contact lines.
7. Pick up a separated section on a glass slide and transfer to the hot floatation water bath. Slowly dip slide into the floatation bath, allowing a small portion of the section to float on the surface of the water bath. Much of the paraffin melts away at this point, allowing the section to expand to its normal size. Pull the slide containing the section from the water and turn in order to allow the remaining half of the eye specimen to be dipped in the floatation bath for stretching. This procedure is repeated as long as necessary to insure stretching of the specimen to its original form and size.
8. Dry sections overnight on a slide warmer, in a 60°C oven for 40 minutes or by preferred method. The sections are now ready for staining with desired stain.

Kuwabara-Kogan Method for Eye Retinal Vessels[40]

The structure of the eye allows a unique opportunity to study structures such as retinal vessels, corneal and trabecular endothelium, lens epithelium, and retinal pigment epithelium in flat mounts. Improved visualization of retinal vascular architecture is made possible by preliminary enzyme digestion using the following method (Fig. 36). Material usually used is taken from the two sides of the globe, or the calottes, after the center portion containing pupil and optic nerve has been processed for routine pathological evaluation.

SOLUTIONS

0.15 M Tris Buffer (pH 7.8)

Tris Buffer (M.W. 121-136)*	18.17	gm
Distilled water	1000.0	ml

Hydrochloric acid, concentrated, (to adjust pH if necessary)

*Sigma Chemical, St. Louis, MO 63118

Trypsin Digestion Solution

Trypsin	3.0	gm
Tris Buffer	100.0	ml
Merthiolate (as a preservative)	1	drop

Make fresh for immediate use.

Dissection of the Retina

Work under the dissecting microscope with the eye calotte immersed in distilled water in a deep clear plastic or glass bowl. Starting at the posterior pole, with the "heel" of the dissecting needle separate the retina proper from the choroid-sclera (which is held firm with jeweler's forceps). Separate at the pars plana. When the choroid adheres to the retina, it is often easier to partially digest both before attempting to separate them. (An initial digestion in trypsin of the entire calotte sector for 10 to 20 minutes may be helpful.)

Digestion Procedure

Suspend the retina in the trypsin digestion solution at 37°C for half to 1-1/2 hours. Incubation time varies with each eye. Terminate digestion when the medium becomes cloudy and the tissue shows signs of disintegration. Transfer to distilled water by picking up the retina on the heel of the dissecting needle.

Preparation of the Retinal Vessels

Work with the dissecting microscope with the retina submerged in a clear glass or plastic bowl of distilled water. The internal limiting membrane can be peeled off in one sheet by careful manipulation. Hold the internal limiting membrane with jeweler's forceps, and disengage the retinal vessels from it with the heel of the dissecting needle. (The point of the needle may adhere to and tear the vessels.) Grasp the internal limiting membrane at a point close to the line of separation. The network of vessels can be freed of retinal tissue by gentle shaking. If undigested tissue remains, it may be necessary to incubate longer. The final retinal vessel preparation will appear as an open meshwork. The veins and arteries branch and fan out and are interconnected by a capillary plexus (Fig. 36).

To mount the retinal specimen on a slide, replace the water in the bowl with clean distilled water. Skim dust from the surface. Float the vessels (retinal specimen) on a clean slide and with the specimen/slide partly immersed, use the heal of the dissecting needle and the surface tension of the water to arrange the vessels in their original configuration. Air dry slide in a vertical position overnight.

Staining Method for Retinal Vessels

The periodic acid Schiff method with Mayer's hematoxylin as a counterstain is used for demonstrating blood vessel walls and cellular entities. In this case, vessel walls stain magenta and nuclei blue. As a general rule, Schiff staining will demonstrate arteries more intensely than veins. Hematoxylin and eosin may be used for staining vessels routinely, resulting in blue nuclei, pink cytoplasm and reddish-yellow red blood cells. It is often necessary to increase the time in the eosin for better staining results of the red blood cells.

SOLUTIONS

M5 Sodium Acetate

Sodium acetate	13.6	gm
Distilled water	500.0	ml

Periodic Acid Solution

Periodic acid	0.8	gm
Distilled water	20.0	ml
Dissolve and add:		
M5 sodium acetate	10.0	ml
95% Alcohol	70.0	ml

Prepare fresh each time.

Reducing Rinse Solution

Potassium iodide	2.0	gm
Sodium thiosulfate	2.0	gm
Distilled water	40.0	ml
Dissolve and add:		
95% Alcohol	60.0	ml
2N Hydrochloric acid	1.0	ml

Prepare fresh each time.

Schiff's Solution

Dissolve 0.1 gm basic fuchsin in 200 ml hot distilled water. Bring to boiling point. Cool to 50°C. Filter and add 20 ml normal hydrochloric acid. Cool further and add 1 gm anhydrous sodium bisulfite, or sodium metabisulfite. Keep in the dark for 48 hours until solution becomes straw colored. Store in a refrigerator.

Mayer's Hematoxylin
(See Page 81)

STAINING PROCEDURE

1. Place slides in 70% alcohol for 5 minutes.
2. Oxidize in periodic acid solution for 5 minutes.
3. Place in 70% alcohol, 2 changes, 2 minutes each.
4. Place in reducing rinse solution for 5 minutes.
5. Place in 70% alcohol, 2 changes, 2 minutes each.
6. Rinse in distilled water.
7. Stain in Schiff's leucofuchsin solution for 35 minutes.
8. Rinse under running tap water to develop color for 10 minutes.
9. Stain in Mayer's hematoxylin for 10 minutes.
10. Wash in running water for 15 minutes.
11. Dehydrate in 95% alcohol, absolute alcohol, and clear in xylene, 2 changes each.
12. Mount coverglass with appropriate medium (refractive index 1.48-1.56).

RESULTS

Vessel walls	magenta
Nuclei	blue
Aneurysms	magenta

Luna's Method for Processing and Sectioning Eye Lenses

The following procedure has been developed for processing eye lenses.

FIXATION 10% Buffered neutral formalin.

After fixation of not less than 8 hours, place lens in tissue cassette and wash specimen in running water for 30 minutes. Place specimen in 80% alcohol until ready to place on tissue processor.

Processing Schedule

Station	Reagents	Time
1.	95% Alcohol	1 hour
2.	95% Alcohol	1 hour
3.	95% Alcohol	1 hour
4.	Absolute alcohol	1 hour
5.	Absolute alcohol	1 hour
6.	Absolute alcohol	1 hour
7.	Xylene	1 hour
8.	Xylene	1 hour
9.	Paraffin (see remarks)	1 1/2 hours
10.	Paraffin	1 1/2 hours
11.	Paraffin, under vacuum	1/2 hour

Embed specimen in embedding paraffin wax medium of choice.

MICROTOMY

1. Rough cut (macrosection) paraffin block carefully and slowly until the entire width of the lens is exposed.
2. Soak specimen surface with a piece of cotton which has been saturated with room temperature water. The specimen should be soaked for several minutes.
3. Chill the surface of the block and knife with an ice cube.
4. Section specimen with a slow even turn of the microtome wheel.
5. Float sections on albuminized slides (or adhesive of choice) and dry on slide warmer overnight or in a 60°C oven for 45 minutes.
6. Stain lenses as desired.

REMARKS

Do not use paraffins which contain plastic polymers since the sections will not spread out adequately on the floatation bath. The success of the procedure is dependent on two factors: (1) a sharp knife and (2) adequate soaking of the specimen. The specimen has not been soaked adequately if it shatters during microtomy.

Roger's Method for Small Biopsies and Tissue Fragments

Because of the small size and nature of some tissues sent to the laboratory, it is very important to orient the specimen properly to prevent loss of tissue. Traditionally, small biopsies and friable tissue are wrapped in some form of paper and oriented at the embedding stage. A method using agar and the Tissue-Tek Cassette System* for handling small biopsies and tissue fragments is presented below.

For this method a 3% solution of oxoid DST agar prepared in 10% buffered neutral formalin (pH 7.0) is suitable. A stock solution should be prepared each week, then divided into 5x30 ml aliquots and stored at 4°C. An aliquot is sufficient for each day's work. The aliquot is melted at the beginning of the grossing session and stored in a 56°C oven. The tissues are subsequently handled as follows.

*Miles Inc., Elkhart, IN 46515.

Method for Bone Marrow Aspirates

1. Fix marrow aspirates for 1 hour in Zenker's solution or fixative of choice (fresh) in a conical container.
2. Wash in distilled water for 5 minutes.
3. Discard the red blood cells that are left floating in the supernatant. (The heavier particles will sink to the bottom rapidly.) In fatty specimens, skim the marrow globules off the surface into a clean glass vial.
4. Pour the marrow aspirates thus harvested into a 20 ml plain glass tube and spin at 2000 rpm for 30 seconds.
5. Discard supernatant.
6. Remove agar from the oven and allow to cool to between 40° and 45°C.
7. Pour 1 to 2 mls of agar into the glass vial containing the aspirate.
8. Shake sample to mix well, then spin at 2000 rpm for 1 minute.
9. Allow the aspirate, which will form a button at the bottom of the vial, to cool for 5 minutes.
10. Wrap the vial in a piece of Kimwipe, then sharply tap the end of the vial furthest from the aspirate. The agar button will slide out. The specimen can then be grossed and placed in a Tissue-Tek Biopsy Uni-Cassette (Miles No. 4172) for processing in the normal manner.

Method for Biopsies and Friable Tissues

1. Fill a 15x15 mm Tissue-Tek II metal base mold (Miles No. 4162) with agar solution that has been cooled to between 40-45°C (as above). For large specimens, use 24x24 mm molds (Miles No. 4163).
2. Number the embedding cassette with the appropriate number.
3. Place biopsy in the molten agar-filled mold and orientate as desired.
4. Allow the agar to stand for 5 minutes to solidify.
5. Once the block is solid, place a scalpel down one side of the mold to detach the specimen. It can then be trimmed and the agar can be notched to show the surface to be embedded, if required.
6. The agar block is placed in a cassette and fixed for 1 to 2 hours. The tissue processor can be used for this purpose.

REMARKS

The problems encountered with the use of agar fall into two areas. First, not all tissues are suitable for the double embedding techniques. Skin and all needle biopsies (i.e., liver, kidney and bone marrow) do not cut easily with agar surrounding the tissue. It is very difficult to get a thin section of renal tissue with agar embedding.

Secondly, if the agar blocks are not adequately fixed, the agar acts as a barrier to the processing fluids. The size of the agar block must be one that allows adequate fixation of both block and tissue. Rapid processing of agar embedded tissue can be detrimental; a 3-hour processing schedule using a vacuum and heat should not be used. A 16-hour overnight processing schedule is recommended.

In addition, there is a background halo around the tissue which corresponds to the shape of the agar block shape. This is also noticeable in other stains, such as periodic acid Schiff.

The shelf life of molten agar stored at 56°C is very limited and varies from batch to batch. Prepare stock solution at the beginning of each week, test it for solidity, divide it into 5 aliquots of approximately 30 mls each and then store them at 4°C. Melt a fresh bottle each day and discard the one used on the previous day.

If the agar is too hot when tissue is placed in it, the exterior layer of cells may be damaged. It is therefore advisable to let biopsy specimens fix for 1 to 2 hours before embedding them in agar, and to keep the agar temperature just above the melting point when embedding the tissue.

Film Method for Mounting Sections

FIXATION	Any fixative.
MICROTOMY	Cut paraffin embedded sections at 6 micrometers.
SOLUTIONS	Clear plastic spray enamel*.
EQUIPMENT	Film: 35mm Cronar, P 40 B Leader Film.
	Reel: Nikor auto load, from any photographic supply house.

Film holders, to hold film in position while viewing film strips on a microscope, are not commercially available. They can be fabricated from a 3-1/4x4" microscopic glass plate, two 3x1" slides and four 22x40 mm coverglasses and a tube of Duco cement. A strip of paper approximately 1 mm wider than the film is placed in the center of the long axis of the glass slide. (The paper can be kept in place with masking tape.) The four coverglasses are butted against the edge of the paper with Duco cement, the two 3x1" slides are then cemented on top of the coverglasses and must extend inwardly (toward the middle) 2 mm over the edge of the coverglasses.

*Manufactured by E.I. du Pont de Nemours and Co., Wilmington, DE 19803. Available from du Pont de Nemours photographic products dealers.

PROCEDURE

1. Cut paraffin sections in the usual manner and float ribbon on floatation water bath.
2. The leader film is maneuvered under the ribbon and picked up as on a slide until the entire film is filled. Label film with a diamond pencil or suitable ink and place in a 56°C oven for 30 minutes to dry.

3. Load film into a 35 mm developing reel (three strips of 18 inches in length can be accommodated on one reel). Stain as desired.
4. After staining, remove one strip of film at a time from the clearing xylene and place face up on a paper blotter. It must not be allowed to dry. Note: The film must lie flat because buckling will result in an uneven coat of plastic. Metal weights may be placed at each end of the plastic to keep it flat.
5. Hold can of plastic spray approximately 6 inches above the film and apply the spray quickly until the surface appears clear and smooth. (This step should be done under a hood.)
6. Allow the plastic to harden for 10 minutes. Repeat spraying the plastic twice and let dry. The strips should be ready for microscopic examination after 2 hours.
7. If necessary, the plastic coat may be removed by placing in xylene for 10 minutes. (This step may be necessary when spray is applied unevenly.)
8. Label dry film with suitable ink or paper labels.
9. Use the fabricated film holder to view sections under the microscope. It will keep the sections flat and facilitate advancement. The sections may be filed in special cabinets, stapled to the record, suspended from racks, or rolled and stored in 35 mm film cans. Before permanent filing, strips should be dried for 2 or more days.

Methods for Removal of Coverglasses from Slides

When the need for restaining slides arises, it becomes necessary to remove the coverglass. This can be done in one of several ways. When employing any one of these methods, remember that the older the slide, the longer it will take to remove the coverglass. Also, remember to take all proper precautions when using heated xylene or popping of coverglasses. It would be very easy for a piece of "pop off" glass to hit an eye during this process.

Method One (Xylene)

1. Place slide in xylene until coverglass detaches of its own accord. (Leave slide in xylene until all of the mounting medium is dissolved.)
2. Hydrate to water.
3. Stain as desired.

Method Two (Heat)

1. Heat underside of slide around coverglass edges, by passing over a flame (Bunsen burner, cigarette lighter, etc.) about 2 times a second. Do not hold slide directly on the flame as the section may be burned and/or the slide will crack if the flame is directed to one spot for any length of time.
2. Ease coverglass off the slide with forceps when the first air bubble is noted beneath the coverslip. If coverslip does not move easily, flame slide until it does. Do not force the coverslip as this may be injurious to the section.

3. Cool to room temperature.
4. Place slide in xylene until all the mounting medium is dissolved.
5. Hydrate to water.
6. Perform desired procedure.

Method Three (Heat and Xylene)

1. Place slide in coplin jar filled with preheated (60°C) xylene.
2. Close coplin jar with lid and place on 40° or 45°C slide warming table.
3. Check slide every 30 minutes until coverglass is off.
4. Hydrate to water.
5. Perform desired procedure.

Method Four (Cooling with Skin Refrigerant)

This method is extremely effective and very rapid if performed properly.

1. Spray the coverglass gently with skin refrigerant. (It is best to spray in short rapid spurts to allow the coverglass and slide to cool gradually.)
2. Place the edge of a razor blade at the corner of the coverglass and pry gently. The coverglass should pop off from the slide. Re-cool the coverglass if it does not pop-off immediately.
3. Allow slide to warm to room temperature.
4. Dip in xylene to remove excess mounting medium from the slide.
5. Perform desired procedure.

Method Five (Cooling)

1. Place slide in freezer for 15 minutes.
2. Place the edge of a razor blade at the corner of the coverglass and pry gently. The coverglass should pop off from the slide.
3. Allow slide to warm to room temperature.
4. Dip in xylene to remove excess mounting medium from the slide.
5. Perform desired procedure.

Yoshiki-Nikaido-Yamazaki-Tachikawa Method for Osteoid and Unmineralized Matrices in Bone[41]

FIXATION	10% Buffered neutral formalin. After fixation wash bone specimen for 3-5 hours in running water (see step 1).
PROCESS	Paraffin (see steps 14 and 15).
MICROTOMY	Cut sections at 6 to 8 micrometers (see steps 16 and 17).

SOLUTIONS

2% Silver Nitrate

Silver nitrate	2.0	gm
Distilled water	100.0	ml

Reducing Solution

Sodium hypophosphite(NaH_2PO_2)	5.0	gm
Distilled water	100.0	ml
0.1N Sodium hydroxide	0.2	ml

5% Sodium Thiosulfate

Sodium thiosulfate	5.0	gm
Distilled water	100.0	ml

Cyanuric Chloride

Cyanuric chloride	0.5	gm
100% Methyl alcohol	100.0	ml
N-Methyl morpholine	1.0	ml

10% Formic Acid

Formic acid	10.0	ml
Distilled water	100.0	ml

Mayer's Hematoxylin
(See Page 81)

0.1% Eosin

Eosin Y	0.1	gm
Distilled water	100.0	ml

Do not use acetic acid since it will cause some decalcification of the specimen.

STAINING PROCEDURE

1. Cut bone specimen into segments measuring not more than 5 mm thick.
2. Wash the bone specimens in several changes of distilled water for 3 hours.
3. Place specimens in 2% silver nitrate solution for 48 hours in complete darkness.
4. Rinse in 3 changes of distilled water 20 seconds each. Wash well in running tap water for 4 hours.
5. Place specimens in reducing solution for 48 hours.
6. Wash in running tap water for 1 hour.
7. Place in 5% aqueous sodium thiosulphate for 24 hours.
8. Wash in running tap water for 1 hour.
9. Dehydrate in 80%, 95%, and 100% alcohol, 2 changes each for 30 minutes each.
10. Place in cyanuric chloride solution for 48 hours.
11. Rinse specimens briefly in 100% alcohol and hydrate in 95% and 80% alcohol, 2 changes each for 20 minutes each.
12. Decalcify specimens in 10% aqueous formic acid until decalcification is complete.
13. Wash in running tap water for 1 hour to remove formic acid.
14. Process bone specimens with preferred processing system.
15. Embed in paraffin.
16. Section specimens in the conventional way, making sure that enough moisture is added to the block during microtomy to ensure good sections.

17. Pick up sections on an adhesive coated glass slide from the floatation water bath.
18. Stain sections with Mayer's hematoxylin for 10 minutes.
19. Wash slides in tap water for 15 minutes to develop the hematoxylin blue.
20. Counterstain in 0.1% aqueous eosin for 1 minute.
21. Rinse briefly in water, decolorize in 70% alcohol for 2 minutes.
22. Dehydrate slides in 3 changes each of 95% alcohol, absolute, and clear in xylene.
23. Mount coverglass with appropriate medium (refractive index 1.48-1.56).

RESULTS

Osteoid matrix	vivid pink
Center regions of calcified trabeculae surface	light yellow
Diffusely calcified matrix	black
Osteoid seams	red
Unmineralized matrix	red
Cartilaginous matrix	light red
Nuclei	purple

REMARKS

This method works well only if formic or trichloroacetic acids are used to decalcify specimens. Nitric or hydrochloric acid should not be used to decalcify specimens when using this method. Poor differential staining of osteoid matrix will result if these strong inorganic acids are used.

Mizgala-Schmitz-Hollinger Method for Bone-Metal Interfaces[42]

Stainless steel and titanium alloys are being used with increasing frequency for reconstructive surgery. The presence of metal in osseous tissues, however, makes it difficult to prepare microscopic specimens for routine histologic examination, because the metal implants require a contiguous osseous interface. Conventional microtomy and manual grinding techniques are unsatisfactory, due to the impairment of the critical assessment zones and the traces of metallic filings left behind by hand grinding that often obscure essential histologic elements.

The method provided below produces 100 to 150 micrometer thick sections of undecalcified bone-metal sections for microscopic examination with transmitted light. The procedure produces bone-metal sections that are especially useful for routine histologic examination and histomorphometry.

FIXATION Fix bone implant specimens in 70% ethyl alcohol for 24 hours or longer.
MICROTOMY Cut sections 100 to 150 micrometers (see step 1 below).
SOLUTIONS

70% Ethyl Alcohol

95% Ethyl alcohol	737.0	ml
Distilled water	263.0	ml

80% Ethyl Alcohol (Stock)

95% Ethyl alcohol	842.0	ml
Distilled water	1580.0	ml

1% Eosin (Stock)

Eosin Y	1.0	gm
Distilled water	100.0	ml

Alcoholic-Eosin (Stock)

1% Eosin	50.0	ml
80% Ethyl alcohol	150.0	ml

Xylene-Eosin (Working)

Alcoholic eosin (stock)	200.0	ml
Xylene	800.0	ml

STAINING PROCEDURE

1. After fixation cut bone implant into 0.5 to 1.0 cm pieces with a Buehler* Isomet low speed saw. Keep specimens at 0-4°C until they are embedded.
2. Place sections in 70% ethyl alcohol for 2 hours at 0-4°C.
3. Place in 80% ethyl alcohol 2 changes, 2 hours each at 0-4°C.
4. Place in 95% ethyl alcohol 2 changes, 2 hours each at 0-4°C.
5. Place in 100% ethyl alcohol 3 changes, 2 hours each at 0-4°C.
6. Infiltrate and embed in polymethyl methacrylate (PMM).
7. Section at 100 to 150 micrometers using the Buehler low speed saw and a diamond wafering blade.
8. Place section on a clean glass slide and cover with polyethylene methacrylate strips.
9. Apply lead weights to ensure that the section is flat and wrinkle free. Heat on a 50°C warming table for 12 hours.
10. Carefully remove section with a #10 scalpel blade.
11. Mount on a clean glass slide using 2 to 3 drops of coverglass mounting medium. Let set for 12 hours.
12. Immerse slide in xylene eosin working solution in a shallow, covered container. This solution should remove the plastic in 24 to 48 hours and stain the section.
13. Mount coverglass with resinous mounting medium and let dry under lead weights for 24 hours.

*Buehler Inc, Lake Bluff, IL 60044.

Method for Processing Temporal Bones

The petrous portion of the temporal bone houses the inner and middle ear structures as well as other important tissues, such as part of the carotid artery and the facial nerve. Specimens taken for the study of the middle ear and inner ear structures should include bone anterior to the internal auditory meatus and posterior into the mastoid region. The desired structures will be found between these two landmarks.

FIXATION

1. Place fresh bone specimens in 20% phosphate buffered formalin for 1 hour, then

change to fresh 20% formalin. Tip slowly, several times, using glass rods or forceps to handle the specimen. Tipping allows trapped air to leave the mastoid and other air spaces.

2. Fix in 20% formalin for 24 hours.

Use degassed water to dilute stock formalin. Degassed water is used in order to prevent bubbles from forming on the surfaces of the bone or within the labyrinth. (Water may be degassed by subjecting it to a vacuum from a faucet filter or a pump for a few hours. Use a thick-walled flask which will withstand the vacuum. After degassing, disconnect slowly from the vacuum; a rapid disconnection will agitate the water causing bubble formation. When using the water, transfer by siphon so that it mixes with air as little as possible.)

A 400 ml quantity of fixative is adequate for a pair of bones. In a pint jar or similar container, add 80 ml of formalin (a 40% solution of formaldehyde in water) to 320 ml of degassed water. Use a tall pint glass jar with glass lid sealed by a rubber ring for each pair of bones. The pair should be kept together.

3. Place bones in fresh 10% formalin for 24 hours. (Make this by mixing 40 ml of formalin with 360 ml of degassed water.) Tip the bone with each change of solution to remove the trapped air from mastoid and other air spaces.
4. Place in fresh 10% formalin for 24 hours. Bones may be stored in this solution.
5. Trim the bone specimen with a circular saw or a band saw. The specimen should be trimmed with vertical cuts made anterior and posterior to the inner and middle ear structures. These cuts are perpendicular to the tentorial ridge. The anterior cut should be as far anterior to the internal auditory meatus as is permitted by the size of block.
6. To remove bone dust, carefully grasp the bone with stout toothed forceps and shake rapidly with a back-and-forth motion in 10% formalin in a tall jar. Turn the specimen 90° and repeat. Turn once again and repeat. Each shaking should be vigorous and should last 15 seconds. (Use of a tall jar minimizes splashing.)
7. Remove specimen and examine it. Gross pathology notes may be dictated at this time.

DECALCIFICATION

8. Suspend each bone by a string which passes through the tentorial margin, through soft tissue near the glenoid fossa, or around the bone. Either nylon suture material or white carpet thread (preferably linen) that has been soaked in melted paraffin can be used. Suspend the bones so that the tops are at least one inch below the fluid surface when 400 ml of fluid is in the jar. The bottom of the bone should be at least one inch above the bottom of the jar.
9. Decalcify the bones at room temperature in 1% nitric acid in 10% formalin. Use a separate glass jar for each pair of bones. For decalcification, each jar should contain 356 ml of degassed water, 40 ml of formalin, and 4 ml of nitric acid. Mix carefully to avoid forming bubbles. Transfer each pair of bones to a fresh decalcifying solution daily. When transferring bones to fresh solution, have the fresh solution in a thoroughly clean jar standing near the jar containing the bones. Transfer rapidly to minimize air-trapping. Tip the bones with glass rods to release any trapped air.

Test the decalcifying solution periodically after two or three weeks have passed, to determine when decalcification is sufficient. The test should be performed just before the bones are transferred to fresh decalcifying solution. Test the fluid that is directly beneath the suspended bone specimen; this has the greatest concentration of dissolved calcium salts, provided the fluid has not been agitated. The amount needed for the test is 5 to 10 ml. Add this to a fresh solution of testing fluid consisting of approximately 5 ml of 5% ammonium hydroxide solution added to about 5 ml of 5% aqueous ammonium oxalate solution in a test tube. If no precipitate forms, allow test tube to stand overnight before reading the test as clear. For each pair of bones, a record should be kept of the tests, using terms such as "distinct", "faint", "very faint", etc., to describe the precipitate. Decalcification is complete for histological purposes when no precipitate is visible for two days in succession, when the tube is closely examined by swirling near a good light source, such as a window.

This is an empirical test to determine an endpoint for histological sectioning; a more delicate chemical test will still indicate the presence of calcium at a stage when the bone is able to be sectioned. Human temporal bones usually require 20 to 30 days for decalcification.

10. When decalcification is complete, specimens may be held in 10% formalin until all specimens are ready for the next step.
11. Place in 5% sodium sulfate for 24 hours. For this, dissolve 20 gm of sodium sulfate in 380 ml of water for each pair of bones. Degas just before using.
12. Change to fresh degassed 5% sodium sulfate for 24 hours.
13. Wash in either (a) running degassed water for 48 hours, or (b) hourly changes of degassed water during working hours for three consecutive days.

PROCESSING

14. Place specimens in 35% ethyl alcohol for 24 hours. Make this and the following dilutions of alcohol with degassed water.
15. Place in 50% ethyl alcohol for 24 hours.
16. Place in 80% ethyl alcohol for 24 hours.
17. Place in 90% ethyl alcohol for 24 hours.
18. Place in 95% ethyl alcohol for 24 hours.
19. Place in 95% ethyl alcohol for 24 hours.
20. Place in 100% ethyl alcohol for 24 hours.
21. Place in 100% ethyl alcohol for 24 hours.
22. Place in a fresh mixture of equal parts 100% ethyl alcohol and ethyl ether for 24 hours.
23. Transfer specimens to nitrocellulose or parlodion solution to begin the infiltration. Nitrocellulose solutions used in these steps should be mixed as follows from a stock solution of 30% nitrocellulose.

6% Nitrocellulose

30% Nitrocellulose (stock)	50.0	ml
Absolute alcohol..	100.0	ml
Ethyl ether	100.0	ml

12% Nitrocellulose

30% Nitrocellulose (stock)	100.0	ml
Absolute alcohol..	75.0	ml
Ethyl ether	75.0	ml

20% Nitrocellulose

30% Nitrocellulose (stock)	200.0	ml
Absolute alcohol..	50.0	ml
Ethyl ether	50.0	ml

30% Nitrocellulose

Use stock solution which may be purchased as tissue embedding solution.

35% Nitrocellulose

Add 5 grams of dry* nitrocellulose to each 100 ml of 30% nitrocellulose stock solution.

*To dry, expose to air until dry to the touch. Use ethyl ether precautions.

Thick Nitrocellulose

Old, used 35% nitrocellulose is used for mounting the nitrocellulose block on a fiber or wooden block; its concentration is not critical.

24. Place bone specimens in 6% nitrocellulose (or 4% parlodion) for 3 weeks, in a sealed glass jar. This may be re-used later if clear. Tip specimens to release trapped air bubbles.
25. Transfer to 12% nitrocellulose (or 8% parlodion) for 3 weeks. Tip to release trapped air bubbles.
26. Transfer to 20% nitrocellulose (or 12% parlodion) for 3 weeks. Tip to release trapped air bubbles.
27. Transfer to 30% nitrocellulose (or 16% parlodion) for 3 weeks. Tip to release trapped air bubbles.

EMBEDDING

28. Embed in 35% nitrocellulose (or 16% parlodion), in 100 mm diameter Stender dishes of 50 mm depth, with lid sealed with petroleum jelly and held with weights. (Stender dishes for this purpose should have individually ground covers so that the cover fits exactly. After purchasing dishes, mark the top of the cover and the side of the dish with an identifying number so that the two are identified as belonging together.)

    To embed, pour 35% nitrocellulose into the Stender dish and pour a few drops of ether-alcohol onto the surface; put petroleum jelly rimmed cover on, add weights, and wait for the bubbles to come out (4 to 5 hours). For weights use two full boxes of glass microslides (50x75 mm size). Then remove the weights and lid, cut string from the bone, and transfer the bone to the 35% nitrocellulose. Tip to release trapped air bubbles. Add a few drops of ether-alcohol to the surface to remove surface bubbles. Reseal the lid with petroleum jelly and replace weights on the top.

    Four bones (two pairs) can be embedded in one dish. Note the position of the bones in the dish on a sketch, using the number on the side of the dish as guide to the positions of the specimens.

29. After the bones have been in the sealed dish for 3 weeks, remove the weights from the top during the day and replace them at night. Do this each day for 3 weeks, provided that the relative humidity is 50% or less. If it is greater than this, do not remove the weights. The nitrocellulose should have an even consistency

throughout. If a surface crust forms, reseal and replace weights; solvent will ascend in the sealed dish and will soften the crust.

30. When the surface of the nitrocellulose is no longer "tacky" but yields with gentle finger pressure, flood the surface with chloroform. Replace the top, and add the weights. Allow 2 to 3 days for the block to harden.

31. Pour off the chloroform. Pour on 80% alcohol to fill the dish. Replace the cover and let it stand for 1 day.

32. Cut block out of dish. Cut blocks by scoring V-shaped grooves on the top of the hardened nitrocellulose between the bones in the dish. Continue deepening these grooves, using a thin-bladed scalpel to separate individual blocks.

33. Nitrocellulose blocks should be kept moist at all times with 80% alcohol.

34. Mount nitrocellulose block on fiber, wood or appropriate specimen mounting system. Prepare the block by wrapping a strip of paper around the sides of the fiber or wooden block, so that the paper extends above the top of the block by about one centimeter. Secure the paper with a rubber band. This makes a paper-sided container around the top of the block, which will hold the thick nitrocellulose used to mount the nitrocellulose block on the fiber or wooden block. Place the block with its paper cuff in equal parts of absolute alcohol and diethyl-ether; store in this until ready to use.

    Soften the bottom of the nitrocellulose block with alcohol-ether in a shallow watch glass while keeping the top of the specimen moist with 80% alcohol by use of a brush. Remove the mounting block from the alcohol-ether container. Pour thick nitrocellulose into the paper container on the top of the block.

    Push the nitrocellulose block into the thick nitrocellulose on top of the mounting block quickly, before the mounting medium begins to harden, and push it firmly down onto the grooved surface. Hold it firmly for 1 minute. Allow it to set for 5 to 10 minutes. Then immerse it in 80% alcohol to harden, handling it by the fiber or wood block, not the specimen. On the following day, the paper can be removed. Store the mounted blocks in 80% alcohol.

MICROTOMY

35. Cut sections on a sliding microtome at 24 micrometers, keeping the block wet with 80% alcohol. Alcohol should drip onto the block; the sections on the knife should also be kept moist with alcohol. Select every tenth section for staining. Place the other nine, in sets of nine, in order on Bantam #9 manifold paper strips labeled with the section number. The paper strips are wrapped in unglazed paper and then in gauze, tied with a string, and stored in 80% alcohol.

STAINING PROCEDURE

36. Stain the sections overnight in Ehrlich's hematoxylin. Arrange the sections in order, in the stain, in a large glass pie plate. Because of differences in the stain, it is best to stain a few sections for a trial before staining an entire series.

Ehrlich's Stock Solution

Hematoxylin	5.0	gm
Absolute alcohol	250.0	ml
Distilled water	250.0	ml
Glycerin	250.0	ml
Glacial acetic acid	25.0	ml
Potassium alum, to excess	50.0	gm

To prepare, dissolve the hematoxylin crystals in the absolute alcohol, add the glacial acetic acid, then add the glycerin. Dissolve the alum in the water, and add the remaining ingredients. Loosely cover flask with filter paper. The paper cap should be lifted off each day and the flask swirled to mix in the precipitated alum.

The flask should be large enough to be only partially full. Allow to stand at room temperature on an open shelf, not in direct sunlight, until ripened (approximately three months). When ripened, the mixture will give off a fruity smell, like burgundy wine.

Ehrlich's Hematoxylin Stain

Add 10 to 20 drops of Ehrlich's stock solution to 300 ml of water. The number of drops will depend on the degree of ripening of the stock. Seldom are more than 20 drops needed. Use this fresh diluted solution for the stain.

37. Blue in (a) dilute ammonium hydroxide (3 drops of ammonium hydroxide in 200 ml of water) or (b) cold tap water or (c) dilute lithium carbonate (a few drops of lithium carbonate to 200 ml water). This will make sections paler as well as bluer. If too blue, remove stain with one drop of a dilute hydrochloric acid in 100 ml of water; then re-blue.
38. Wash twice in distilled water if ammonium hydroxide is used. Wash four or more times if lithium carbonate is used to ensure removal of all lithium carbonate.
39. Stain in eosin stain for 5 minutes.

Eosin (Stock)

Eosin Y	2.0	gm
95% Alcohol	50.0	ml
Distilled water	150.0	ml

Dissolve eosin in liquids.

Eosin Stain

Eosin Y (stock)	6.0	ml
Distilled water	150.0	ml

Dilute eosin in water.

40. Place in 95% alcohol for 3 minutes, 2 changes.
41. Place in steam-distilled pine oil for at least 15 minutes.
42. Take sections from pine oil and place on slide. Blot the sections dry with lint-free Whatman #1 chromatography grade filter paper. Press hard to remove the oil.

43. Use approximately 2 drops of appropriate mounting medium (refractive index 1.48 to 1.56) and place coverglass on the slide. Press gently to remove bubbles. With #1 coverglass, the margins of the coverglass will bend down and adhere closely to the microslide, decreasing the chance of air coming under the coverglass and forming bubbles. Use this thickness, rather than #2 coverglass.
44. Label slides.

Nitrocellulose is Pyroxylin, a product of the action of a mixture of nitric acid and sulphuric acid on cotton, consisting chiefly of cellulose tetranitrate. Celloidin is a concentrated preparation of Pyroxylin. Parlodion is celloidin in shredded form.

Lund's Method for Preliminary Embedding in Agar

Preliminary embedding in agar-agar is helpful in safely carrying small biopsies through the process of dehydration and embedding, in orientating small specimens correctly, and in combining multiple small biopsies into a single block for individual identification.

Agar-Agar Solution

Agar-agar	4.0	gm
Distilled water	96.0	ml

Dissolve by heating on a boiling water bath. (Do not use direct heat.) Add 2 crystals of thymol to prevent fungal and bacterial growths.

The agar is dispensed in test tubes kept at 56° or 60°C. Tubes of agar are kept where gross tissue is examined and cut.

PRELIMINARY EMBEDDING

Small specimens are properly orientated on a flat, clean surface. It is helpful to have small solidified particles of agar on hand to prop up thin or limp pieces of tissue and maintain them on edge. Using an applicator stick, hot agar is dropped upon the tissue to form a small pool (Fig. 37). This is allowed to solidify and is trimmed into a small block. The block of agar is processed as a biopsy specimen, and embedded in paraffin in the desired position. (Agar tends to harden during dehydration and embedding.)

Combining several small biopsies into one block is accomplished by arranging them in a pattern which makes the first of the series distinct. The first of the series is placed at right angle to the rest. Subsequent biopsies are orientated on a vertical plane (Fig. 37 lower). The agar block is trimmed to a point on the side where the right angled specimen is embedded.

REMARKS

Agar stains faintly with hematoxylin, and may stain intensely if a slide is "heavy on the hematoxylin". Agar-agar boiled on a direct flame becomes more basophilic than agar dissolved carefully over a hot water bath. Agar also stains with orcein, methenamine silver and other special stains, making an unattractive slide. However, the agar surrounds the tissue and does not permeate it, so the staining of the tissue itself is not altered.

Acrylic Spray Enamel for Coating (Covering) Sections

Krylon* or other similar clear acrylic spray coatings, can be used for spray coating tissue sections mounted on microscopic slides. These acrylics can be used in place of a coverglass in certain instances, such as smears, blood films, tissue sections which are destined for quick review and discard, and many other applications.

PROCEDURE

1. Slides containing material to be coated are placed in xylene for 5 minutes.
2. Place slides on filter paper, preferably under a hood.
3. Spray the slide with plastic spray enamel.
4. Allow sections to dry thoroughly.
5. Sections are ready for microscopic viewing.

The refractive index is lower than that of resin coverglass mounting medium. Therefore, the clarity of the section will be reduced somewhat.

*Krylon is a Sherwin-Williams paint store product.

Method for Lifting Sections from Glass Slides

Commercially available plastics, Krylon or similar acrylic spray coating, can be utilized easily and advantageously as medium for lifting stained microscopic sections in the form of plastic films, for the preparation of multiple sections, study sets, large sections, removal of multiple sections from one slide and conversion to several slides and smears, as well as repair of broken slides.

FIXATION	Any fixative.
PROCESS	Paraffin sections, blood smears and finger prints.
SOLUTIONS	Commercially available clear plastics.

PROCEDURE

1. Cut sections and place on the floatation bath in the usual manner.
2. Pick up sections on slides and place on the slide warmer overnight or in 60°C oven for 30 minutes.
3. Decerate and hydrate slide in usual manner.
4. Leave in xylene until ready for plastic.
5. Coat with acrylic by spraying surface of the slide at least 7 times. The spray must be allowed to dry completely after each spray.
6. Dry thoroughly for 2 to 4 hours.
7. Incise film near the edges of the slide with a sharp instrument.
8. Soak slides in cold water for 30 minutes.
9. Peel film from slide gently and firmly. (If film does not come off easily, return slides to the cold water for additional soaking).
10. Dry film between sheets of filter paper.
11. Number film with India ink and store between sheets of tracing paper, cut to appropriate sizes. Sections can be kept in this form for an indefinite period. Sections can also be mounted on slides as indicated below.

Remounting Procedures

1. Cut sections free from the plastic film with scissors or with a razor blade.
2. Place a small drop of mounting medium on the slide.
3. A second drop is placed upon the film and the coverslip applied immediately. Avoid air bubbles.

Microscopy

For microscopy, the film is placed on a glass slide slightly larger than the film. Good optical quality and minimal thickness are required for good results.

For visual observation, regular dry systems can be used, or oil immersion objective can be used directly on the film. For optimal results a drop of oil should be placed underneath the section as well as on the top. Do not use xylene or other solvents to remove oil immersion from film, since they will dissolve the plastic. Oil can be wiped off with a clean cloth.

Repairing Broken Slides

Slides sent through the mail frequently arrive broken. The use of plastic to repair broken slides has greatly aided the pathologist since no time or expense is lost in corresponding with the contributor to request more material. By utilization of the plastic technique, broken slides are repaired as follows:

1. Recover all or most of the pieces from the broken slide, especially those pieces which contain the tissue section.
2. As in a jigsaw puzzle, assemble all the pieces together on top of another slide, attempting to preserve the full section if possible.
3. Soak the slides with xylene in a shallow container (petri dish, etc.) until all pieces of the coverslip are removed. Allow section to dry, spray with Krylon acrylic spray, completely flooding slide.
4. Dry thoroughly for 2 to 4 hours.
5. Proceed with Method used for Lifting Sections from Glass Slide, steps 6-11, provided above.
6. Employ slide remounting procedure provided above.

Removing Multiple Sections

Frequently only one slide with several sections of tissue is available. In such instances, one or more sections may be removed from the original slide and transferred to individual slides.

PROCEDURE

1. Place original slide with multiple sections in xylene to remove coverslip.
2. When coverslip is removed, let slide dry thoroughly.
3. Proceed with Method used for Lifting Sections from Glass Slides, steps 6-11, provided above.
4. Employ slide remounting procedure provided above.

Multiple Section Staining on One Slide

The following technique is useful when one desires multiple special stained sections on one slide.

PROCEDURE

1. Decerate slides in 3 changes of xylene.
2. Cover sections not to be stained with Dow Corning 7 Compound* (Dow Corning compound is applied while slide is xylene damp.)
3. Dip in 80% alcohol and distilled water.
4. Perform stain desired. Only the uncovered section will stain.
5. Place in xylene, several changes, until covering medium is completely dissolved.
6. Repeat steps 2 through 5 for as many stains as required or number of sections available.

*Dow Corning 7 Compound, Dow Corning Company, Midland, MI 07605. Available also from Fisher Scientific, Pittsburgh, PA, 15219.

Repassy and Lopis' Method for Plastic Implants[43]

It is important to know what reactive processes are, or have taken place, in plastic or other types of prosthetic tissue transplants. Provided below is a simple procedure which can be used for sectioning tissue specimens containing prosthetic implants. The procedure is for specimens which measure 5 mm in thickness or less. Processing schedules should be adjusted for thicker specimens.

PROCEDURE

1. Fix tissue containing prosthetic implant in 10% buffered neutral formalin or fixative of choice.
2. Process specimen through 2 changes of 80%, 95%, and 100% ethyl or isopropyl alcohols, 1 hour each.
3. Clear in 2 changes of xylene for 2 hours each.
4. Place in the following paraffin-resin mixture for impregnation from 4 hours to overnight, depending on specimen's thickness. (5 mm specimens require 4 hours; larger thicker specimens may require overnight exposure.)

Paraffin-Resin Embedding Medium

Pine resin (Rosin)	50.0	gm
Yellow beeswax	15.0	gm
Paraffin, without additives	5.0	gm

Dissolve each ingredient with heat and mix in proportions given.

5. Embed specimen in paraffin-resin embedding medium.
6. Carefully face (rough cut) paraffin-resin block to expose tissue.
7. Apply clear cellophane tape on the surface of the paraffin-resin block.
8. Section specimen with a slow careful rotation of the microtome wheel.
9. Remove the section from the block together with the tape.
10. Press the section bearing tape onto a slide containing egg albumen or other suitable section adhesive.
11. Place slide in 56-60°C oven for 30 minutes.
12. Place slide in xylene for 1 to 2 hours.
13. Remove plastic adhesive tape from slide. (The section remains on the slide and the embedding medium has dissolved out.)
14. Hydrate section to water in the usual manner.
15. Stain with desired procedure.

REMARKS
Pine resin is quite hard and brittle and, therefore, unsuitable for cutting by itself. The addition of beeswax and paraffin in the amounts suggested above, results in a proper hardness to softness ratio, which allows cutting good sections. The paraffin-resin embedding medium impregnates tissues well. The resin used on violin strings can be substituted in this procedure in the same proportions given above.

Hinshaw-Pearse Methods for the Differential Staining of Burned and Normal Tissues[44]

The methods presented below, to differentiate burned from normal tissue, yield a striking contrast between the two tissues. The methods are relatively simple to do, and the results can be interpreted without difficulty. Method One is my method of choice.

Method One

FIXATION 10% Buffered neutral formalin.
PROCESS Paraffin.
MICROTOMY Cut sections at 6 micrometers.
SOLUTIONS

Iodine Solution (Stock)

Iodine	2.0	gm
Potassium iodide	4.0	gm
Distilled water	100.0	ml

10% Ferric Chloride (Stock)

Ferric chloride	10.0	gm
Distilled water	100.0	ml

2% Ferric Chloride

Ferric chloride	2.0	gm
Distilled water	100.0	ml

Hematoxylin-Ferric Chloride

Hematoxylin	1.0	gm
100% Alcohol	20.0	ml
Filter and add:		
10% Ferric chloride	8.0	ml
Iodine solution	8.0	ml

STAINING PROCEDURE

1. Decerate slides in xylene, 2 changes, 2 minutes each. Place in absolute alcohol, 2 changes, 2 minutes each; 95% alcohol, 2 changes, 2 minutes each; then rinse in distilled water.
2. Immerse sections in hematoxylin-ferric chloride solution for 20 to 30 minutes.
3. Decolorize in the 2% aqueous solution of ferric chloride. A rack containing about 25 slides can be dipped up and down in this solution 20 times or so, in 20 to 30 seconds.
4. Wash immediately and thoroughly in running tap water.
5. Place in 95% alcohol for 1 minute.
6. Wash in running tap water thoroughly for 5 minutes.
7. Stain for 15 seconds in Van Gieson's stain.
8. Dip slides once in tap water.
9. Dehydrate slides in 3 changes each of 95% alcohol and absolute, then clear in xylene
10. Mount coverglass with appropriate medium (refractive index of 1.48-1.56).

RESULTS

Normal tissue	gray
Burned tissue	pale yellow
Normal collagen	yellow-brown
Burned collagen	black or purple

REMARKS

The staining solution should be used within 24 hours. Fresh staining solution should be used after each 75 slides.

Method Two

FIXATIVE 10% Buffered neutral formalin.
PROCESS Paraffin.
MICROTOMY Cut sections at 6 micrometers.
SOLUTIONS

0.1M Citric Acid (Stock)

Citric acid (anhydrous)	19.2	gm
Distilled water	1000.0	ml

0.2M Disodium Phosphate (Stock)

Disodium phosphate (anhydrous)	28.3	gm
Distilled water	1000.0	ml

Buffer

0.1M Citric acid (stock)	12.29	ml
0.2M Disodium phosphate (stock)	7.70	ml

0.1% Azure A (Stock)

Azure A	0.1	gm
Distilled water	100.0	ml

0.1% Eosin B (Stock)

Eosin B	0.1	gm
Distilled water	100.0	ml

Buffer-Azure Eosinate

0.1% Azure A	3.5	ml
0.1% Eosin B	3.5	ml
Add:		
Buffer	2.0	ml
Acetone	5.0	ml
Distilled water	26.0	ml

STAINING PROCEDURE

1. Decerate slides in xylene, 2 changes, 2 minutes each. Place in absolute alcohol, 2 changes, 2 minutes each; 95% alcohol, 2 changes, 2 minutes each; then rinse in distilled water.
2. Stain 1 hour in the buffer-azure eosinate.
3. Dehydrate in 2 or 3 changes of acetone.
4. Clear in 1 change of equal parts acetone and xylene and 2 changes of xylene.
5. Mount coverglass with appropriate resinous medium (refractive index 1.48-1.56).

RESULTS

Normal tissue surface	blue
Burned tissue surface	pink

Trombetta-Forbes Method for Sectioning Heavily Sclerotized Insects[45]

A double infiltration method using n-butyl alcohol in conjunction with celloidin and paraffin wax is provided below. Sections of heavily sclerotized insect cuticle fixed with alcoholic Bouin's solution can be cut easily at 4 mm. The architecture of the cuticle and the morphology of the underlying tissue show little distortion and excellent staining. If the insect is to be embedded and sectioned entirely, openings should be made in the cuticle to permit the flow of fixative as well as dehydrating and infiltration fluids.

FIXATION Bouin's or any fixative that will not harden tissue excessively can be used. After fixation process the insect according to the schedule which follows.

DEHYDRATION

1. Place in 10% n-butyl alcohol equal parts in 45% ethyl alcohol for 2 hours.
2. Place in 20% n-butyl alcohol equal parts in 62% ethyl alcohol for 2 hours.
3. Place in 35% n-butyl alcohol equal parts in 77% ethyl alcohol for 4 to 6 hours.
4. Place in 55% n-butyl alcohol equal parts in 90% ethyl alcohol for a minimum of 6 hours (stopping stage).

5. Place in 75% n-butyl alcohol equal parts in 90% ethyl alcohol for 2 hours with 2 changes.
6. Place in pure n-butyl alcohol for 6 to 12 hours, 4 changes.

CELLOIDIN INFILTRATION AND EMBEDDING

1. Place in 1% celloidin in methyl benzoate for 3 to 5 hours. Shake solution occasionally.
2. Place in fresh 1% celloidin in methyl benzoate for 3 hours.
3. Place in fresh 1% celloidin in methyl benzoate for a minimum of 24 hours (stopping stage).
4. Place in benzene for 30 minutes with 1 change.
5. Place in benzene that has been saturated with paraffin wax at 30°C for 30 minutes.
6. Place in paraffin wax at 57°C for minimum of 3 hours and maximum of 4 hours.
7. Embed in pure paraffin wax. Trim block to within 1 mm of tissue and place in ice water overnight.

REMARKS

Sectioning of insect cuticle on a standard microtome can be easily accomplished by this method. Although most tissue can be cut at a clearance angle of 7 degrees, cuticle which is very heavily sclerotized may require an increased clearance angle up to 15 degrees. The microtome knife and paraffin block should be kept cold at all times.

Method for Removing Mercuric Chloride from Waste Solutions

Zenker's and B-5 type fixative solutions contain mercuric chloride. The toxicity of mercury is such that the element and its compounds should not be allowed to contaminate air or water. The following procedure is recommended for removing mercuric chloride from solutions prior to disposal.

SOLUTION

13% Thioacetamide

Thioacetamide	13.0	gm
Distilled water	100.0	ml

Mix thoroughly and store solution in a tightly capped bottle. This solution is stable for approximately one year.

PROCEDURE

1. (USE HOOD). To each liter of solution which contains mercuric chloride add 20 ml of 13% thioacetamide solution.
2. Mix the combined solution thoroughly and cap tightly.
3. Place the solution under a hood and leave undisturbed for 24 hours or until precipitate of mercuric sulfide is formed.
4. Filter the solution through fluted filter paper or a Buchner filter. The clear filtrate can be discarded into a sink. The remaining residue, which contains the mercuric salt, can be stored indefinitely in a capped container or turned over to the safety officer for disposal.

Method for Demonstrating Carbonate and Phosphate in Unstained Sections

This method provides a specific test for the microscopic identification of calcium carbonate and calcium phosphate with the use of glacial acetic acid.

SOLUTION

Concentrated Glacial Acetic Acid
(This solution is sold commercially.)

PROCEDURE

1. Decerate slides in xylene, 2 changes, 2 minutes each. Place in absolute alcohol, 2 changes, 2 minutes each; 95% alcohol, 2 changes, 2 minutes each; then rinse in distilled water.
2. Apply coverglass to the section with water.
3. Place section under hood to prevent exposure to acids.
4. Place slides under microscope. Run glacial acetic acid under one edge of the coverglass while drawing water out of the opposite end with a piece of filter paper. Note: Special care should be taken to insure minimum exposure to acid fumes.

RESULTS

Microscopically, deposits will be seen to dissolve. If deposit is calcium carbonate, formation of gas bubbles will occur. If phosphate, no gas bubbles will appear.

REMARKS

If sulfuric acid is used in place of glacial acetic acid, the deposits dissolve but monoclinic gypsum crystals are formed. If 10% oxalic acid is used the characteristic cubic calcium oxalate crystals appear.

Information for Making up Solutions[46]

Over the years, more and more ready made and pre-diluted reagents have become available to histotechnologists. In most laboratory situations, convenience and/or economics play a very important part in the way technologists order their reagents. There is generally nothing wrong with utilizing commercially prepared reagents in the laboratory. However, many histotechnologists still prefer to prepare reagents from scratch. For them, the following information may be useful.

The calculations involved in preparing solutions for the histology laboratory are easy and straightforward, providing the technologist has some basic understanding of chemistry and mathematics. In the following text, percent concentration, dilutions, hydrated chemicals, total dye content, molarity and normality will be discussed.

First, however, several chemistry terms need to be defined. A SOLUTE is a substance that is dissolved in a solvent. Generally, the solute is a lesser amount than the solvent. A SOLVENT is a substance which contains the dissolved solute. A SOLUTION is a mixture of two or more substances. There are several types of solutions:

1. Unsaturated Solution - contains a solute dissolved in a given quantity of solvent.
   (a) dilute - contains a small amount of solute in relation to solvent.
   (b) concentrated - contains a large amount of solute in relation to solvent.
2. Saturated Solution - contains all the solute that will dissolve in the solvent under given conditions.
3. Supersaturated Solution - contains more solute than will dissolve in the solvent (in normal situations).

PERCENT CONCENTRATION:

The percent concentration of a solution is a simple proportion given in terms of weight, weight and volume, or volume.

1. Weight Percent - Generally used to express the concentration of commercial aqueous reagents.

   $$\text{weight percent (w/w)} = \left(\frac{\text{weight of solute}}{\text{weight of solution}}\right) \times 100$$

   Example: Nitric acid is sold as a 70% (w/w) solution; therefore, the nitric acid reagent contains 70 grams of nitric acid per 100 grams of solution.

2. Volume Percent - Used to express the concentration of the volume of a liquid solute that is added to a liquid solvent.

   $$\text{volume percent (v/v)} = \left(\frac{\text{weight of solute}}{\text{weight of solution}}\right) \times 100$$

   Example: Prepare 250 mls of 30% (v/v) aqueous ethyl alcohol.

   $$30\%\text{(v/v)} = \left(\frac{\text{"?" ml(s)}}{\text{250 mls}}\right) \times 100$$

   75 mls = “?” mls
   Add 75 mls of ethanol to 175 mls of distilled water.

3. Weight-Volume Percent - Used to express the concentration of solid reagents dissolved in an appropriate solvent.

   $$\text{weight-volume percent (w/v)} = \left(\frac{\text{weight of solute-gm(s)}}{\text{volume of solution-ml(s)}}\right) \times 100$$

   Example: Prepare 300 ml of 5% (w/v) aqueous silver nitrate.

   $$5\%\ \text{(w/v)} = \left(\frac{\text{"?" gm(s)}}{\text{300 mls}}\right) \times 100$$

   15 gm = “?” gm
   Dissolve 15 gm of silver nitrate in 300 ml of distilled water.

DILUTIONS:

A DILUTION is a solution of a known concentration which needs to be diluted to a weaker concentration. There are two ways of calculating how to make a dilution.

1. Diluting a percent, molar, or normal solution.
   The product of the volume (V_1) and concentration (C_1) of the known solution is equal to the product of the volume (V_2) and concentration (C_2) of the weaker solution.

$V_1 \times C_1 = V_2 \times C_2$

Example: Prepare 50 mls of 2% ferric chloride from 29% ferric chloride.
$V_1 \times C_1 = V_2 \times C_2$
("?" ml(s)) (29%) = (50 mls) (2%)
"?" ml(s) = 3.45 mls
Add 3.45 mls of 29% ferric chloride to 46.55 mls of distilled water.

2. Diluting a ratio.
A ratio is a proportion that is made between the solute and solvent. One part of the original solution (or solute) is diluted with a specified part of the solvent to give the total volume. A 1:4 dilution means that one part of the original solution (solute) is diluted with three parts of the solvent to give a total volume of four.

Example: Prepare normal goat serum diluted 1:100 in phosphate buffered saline. Combine one part of normal goat serum and 99 parts of phosphate buffered saline for a total volume of 100 parts.

HYDRATED AND ANHYDROUS CHEMICALS:
A hydrated chemical has attached water molecules while an anhydrous chemical has no water molecules. These factors need to be taken into account when preparing solutions. This is particularly important since one may have the hydrated version of a given chemical when the procedure being performed calls for the anhydrous form or vice versa. This problem can be solved easily. Simply find the formula weight of both the anhydrous and hydrated forms of the chemical. Then, a ratio/equation is developed:

$$\left(\frac{\text{anhydrous formula weight}}{\text{hydrated formula weight}}\right) = \left(\frac{\text{weight of anhydrous reagent needed}}{\text{weight of hydrated reagent needed}}\right)$$

Example: Prepare 500 mls of a 1.5% solution of magnesium chloride. The stain procedure you are following states to use 7.5 gm of $MgCl_2$ (FW = 95.22) in 500 mls of water. You find only the hydrated magnesium chloride, $MgCl_2 \times 6H_20$ (FW = 203.31), on your shelf.

$$\left(\frac{\text{FW of } MgCl_2}{\text{FW of } MgCl_2 \times 6H_2O}\right) = \left(\frac{\text{weight of } MgCl_2 \text{ needed}}{\text{weight of } MgCl_2 \times 6H_2O \text{ needed}}\right)$$

$$\left(\frac{95.22}{203.31}\right) = \left(\frac{7.5 \text{ gm}}{\text{"?" gm(s)}}\right)$$

Add 16.01 gms of $MgCl_2 \times 6H_2O$ to 500 mls of distilled water.

TOTAL DYE CONTENT:
On rare occasions the total dye content of a particular dye may need to be known. A ratio is also set up in order to obtain the correct amount of dye needed to prepare the solution.

$$\left(\frac{\text{Total Dye Content}}{100}\right) = \left(\frac{\text{amount "pure" dye needed}}{\text{amount to weight}}\right)$$

Example: Prepare 400 mls of 1% toluidine blue solution (The total dye content of the toluidine blue is 60%.)

$$1\% = \left(\frac{\text{"?" gm(s)}}{\text{400 ml}}\right) \times 100$$

Need 4 gm of toluidine blue. Therefore,

$$\left(\frac{60\%}{100}\right) = \left(\frac{4\text{ gm}}{\text{"?"}}\right)$$

Add 6.67 gm of toluidine blue (60% dye content) to 400 mls of distilled water.

MOLARITY:

The molar concentration of a solution defines the number of moles of solute per liter of solution.

$$\text{Molarity} = \left(\frac{\text{moles of solute}}{\text{liters of solution}}\right)$$

In order to work with molarity, several terms and equations need to be discussed. The formula weight of a solute is defined as the grams of solute per mole of solute. The number, which represents the formula weight, is the sum of the atomic weights of all atoms shown by the formula of the solute. This is generally found on the chemical label.

Example: Sodium Hydroxide (NaOH)

Na	=	22.990
O	=	15.999
H	=	1.008
		39.997

Formula weight of NaOH - 40.0 gm/mole.

$$\text{Formula weight} = \left(\frac{\text{grams of solute}}{\text{moles of solute}}\right)$$

Now, rearrange the equation:

$$\text{Moles of solute} = \left(\frac{\text{grams of solute}}{\text{formula weight of solute}}\right)$$

Combine this equation with the molarity equation:

$$\text{Molarity} = \left(\frac{\left(\frac{\text{grams of solute}}{\text{formula weight of solute}}\right)}{\text{liters of solution}}\right)$$

Example: Prepare 0.5 liters of 0.4M solution of sodium hydroxide (NaOH). (FW = 40.0 gm/mole.)

$$\text{Molarity} = \left(\frac{\left(\frac{\text{grams of solute}}{\text{formula weight of solute}}\right)}{\text{liters of solution}}\right)$$

$$0.4\text{M} = \left(\frac{\left(\frac{\text{"?" gm}}{\text{40 gm/mole}}\right)}{\text{0.5 liters}}\right)$$

8.0 gm = “?”

Dissolve 8.0 gm of NaOH in enough distilled water to make 0.5 liters. (Use volumetric flask.)

Example: Prepare 0.5 liters of a 0.1M solution of hydrochloric acid. (FW = 36.5 gm/mole, specific gravity = 1.18 gm/mole, assay = 37.3%.)

Specific gravity is defined as the ratio of the weight of the solute to the weight of an equal volume of water. The assay states the weight percent of the solute.

The specific gravity tells us that one milliliter of "pure" HCl weighs 1.18 gm (or one liter contains 1180.0 gm). Since the assay of HC1 is 37.3%, then

(1180.0 gm/1) x (37.3%) = 440.14 gm/liter.

This states that there are 440.14 gm of HCl in one liter of solution.

$$\text{Molarity} = \left(\frac{\left(\frac{\text{grams of solute}}{\text{formula weight of solute}}\right)}{\text{liters of solution}}\right)$$

$$\text{Molarity} = \left(\frac{\left(\frac{440.14\text{ gm}}{36.5\text{ gm/mole}}\right)}{1\text{ liter}}\right)$$

Molarity = 12.06 moles/liter

This means that the stock bottle of the hydrochloric acid has a concentration of 12.06 molar. Since we know the molarity of the stock HCl, we just need to dilute it down to what we need.

$V_1 \times C_1 = V_2 \times C_2$
("?" liters) (12.06M) = (0.5 liters) (0.1M)
"?" liters = 0.004 liters

The solution: Add 4 mls of concentrated HCl to 496 mls of distilled water.

NORMALITY:
The normal concentration or normality of a solution defines the gram-equivalents of solute per liter of solution.

$$\text{gram-equivalent weight of solute} = \left(\frac{\text{formula weight of solute}}{\text{number of reactive ions in solute}}\right)$$

$$\text{gram-equivalents of solute} = \left(\frac{\text{grams of solute}}{\text{gram-equivalent weight of solute}}\right)$$

$$\text{Normality} = \left(\frac{\text{gram-equivalents of solute}}{\text{liters of solution}}\right)$$

Combine the equations:

$$\text{Normality} = \left(\frac{\left(\frac{\text{gram of solute}}{\left(\frac{\text{formula weight of solute}}{\text{\# of reactive ions in solute}}\right)}\right)}{\text{liters of solution}}\right)$$

Example: Prepare 0.250 liters of a 0.1N solution of sodium carbonate. (FW = 106.0 gm/mole, Na_2CO_3).

$$0.1N = \left(\frac{\left(\frac{\text{"?" grams}}{\left(\frac{106.0\ \text{gm/mole}}{2}\right)}\right)}{0.25\ \text{liters}}\right)$$

1.325 = "?" grams

Dissolve 1.33 gm of sodium carbonate in enough distilled water to make 0.250 liters.

Example: Prepare 1.5 liters of a 4.0N solution of sulfuric acid (H_2SO_4). (FW = 98.08 gm/mole; specific gravity = 1.8337, Assay = 95%)

The specific gravity tells us that 1 ml of pure H_2SO_4 weighs 1.8337 gm; therefore, the weight of 1 liter of pure sulfuric acid is 1833.7 gm. Since the assay of H_2SO_4 is 95% then,

(1833.7 gm/liter) x (95%) = 1742.0 gm/1iter

Therefore, $$\left(\frac{1742.0\ \text{gm/liter}}{\left(\frac{98.08\ \text{gm/mole}}{2}\right)}\right) = \text{Normality}$$

$$\left(\frac{35.2\ \text{grams equivalents}}{\text{liters}}\right) = \text{Normality}$$

Now,

$V_1 \times C_1 = V_2 \times C_2$

("?" liters) (35.52N) = (1.5 liters) (4.ON)

"?" = 0.16928 liters

Add 169.3 mls of concentrated H_2SO_4 to 1330.7 mls of distilled water.

Rapid Method for Frozen Sections

FIXATION 100% Methyl alcohol.

CRYOTOMY Cut frozen sections at 6 micrometers. Mount section on slide with adhesive of choice, see chapter 17.

SOLUTIONS

Toluidine Blue

Toluidine blue	2.0	gm
100% Methyl alcohol	100.0	ml

STAINING PROCEDURE

1. Fix slide in 100% methyl alcohol for 1/2 to 2 minutes.
2. Stain in toluidine blue for 6-10 seconds.
3. Place in distilled water, 2 dips.

4. Dip in acetone, 3 dips each of 2 changes.
5. Xylene
6. Mount coverglass with appropriate medium (refractive index 1.48-1.56).

RESULTS

Nuclear chromatin.. deep purple to black
Cytoplasm.. varies
Elastic tissue... refractile green-blue
Smooth muscle... pale blue
Collagen... pale red, blue or unstained
Epithelium .... pale blue to purple
White blood cells .... pale green or unstained
Plasma cells .... dark lavender
Mast cells... red-purple
Myelin .... light to dark red (enhanced by polarized light)
Fibrin.. pale yellow-green or unstained
Mucus (epithelial) .... red, purple or unstained

Reasons for Stains Fading[47, 48]

The useful lifetimes of dyes employed in histotechnology vary widely. Some fluorochromes fade within minutes when exposed to ultraviolet light; other stains remain unchanged for 20 years or longer. Furthermore, the durability of a dye can depend on the procedure used. Aniline blue WS, for example, fades fairly rapidly in the allochrome procedure, but is almost indestructible in Gomori-type one-step trichrome methods. Such differences can be predicted from the fastness properties of dyes. The resistance, or sensitivity of dyes, to light, acids, alkalis, and other reagents has been studied extensively. This information is readily available in the technical and chemical literature. Data on fastness properties of dyes can be invaluable for improvement of histological techniques, for which reason the following information is provided.

Fastness: In the field of dyes and dyeing, fastness indicates the resistance to fading (loss of color) under the influence of light, chemicals, and other factors.

Fugitive: Dyes fading rapidly when exposed to light or other conditions are termed fugitive. Certain fastness data for various dyes, (e.g., against acids, alkalis) cannot not be readily found in the available literature.

Light Fastness Standards

Two light fastness standards are currently in use, the ISO and the AATCC standards. In the ISO system of the International Organization for Standardization, pieces of wool or woolen fabrics are dyed with eight blue dyes of different light fastness (see below). The standards are chosen so that the next higher standard will take twice as long to fade than the one below it; standard 7 is about four times as fast as standard 6. The sample to be tested is exposed together with these standards until it fades. This specimen is then assigned the light fastness number of the standard showing the same degree offading. Dyes with light fastness 1 are the most fugitive, fading noticeably after one hour's exposure to bright summer sunlight; dyes with light fastness 8 show little fading after exposure for several years.

ISO Standards for Light Fastness

Light Fastness

	Rating	Dye	Color Index (CI) Number
1	very poor	Alcian Brilliant Blue FFR	42735
2	poor	Alcian Brilliant Blue FFB	42740
3	moderate	Supranol Cyanine 6B	42660
4	fair	Supramine Blue EG	50310
5	good	Alcian Fast Blue RX	62085
6	very good	Alizarin Light Blue 4GL	61125
7	excellent	Anthrasol 06B	73066
8	outstanding	Anthrasol Blue AGG	73801

The trade names of dyes used as standards for light fastness vary from country to country, but their CI numbers are identical.

The AATCC system of the American Association of Textile Chemists and Colorists differs from the international ISO standards. Various amounts of two batches of wool dyed with a fugitive and light fast dye, respectively, are blended; the ratio of the fugitively dyed wool varies from 100% for light fastness 2 to 5% for light fastness 1. This system has not been accepted outside the U.S. Light fastness ratings obtained with the ISO and the AATCC tests are not directly comparable.

The AATCC system has one convenience for histotechnology; correlations of the light fastness ratings and the number of hours of exposure required to produce onset of fading. The lists below provide light fastness values and the corresponding exposure times to sunlight and to artificial light in a fade-o-meter. These values were obtained on dry textile material tested in air. Fastness properties can be unfavorably affected by moisture in sections, reactive sites in tissues, and by reagents used in histotechnology. The durations of exposure given for various light fastness ratings refer to daylight or its equivalent in the fade-o-meter and are usually not applicable to fluorescence microscopy with strong UV exciting light. Fluorescent dyes with light fastness 1-2 tend to fade within minutes.

AATCC Tests for Light Fastness

Light fastness	1	2	3	4	5	6	7	8
Standard Sun Test Exposure hours	1.5 to 3	3 6	6 12	12 24	24 48	48 96	96 192	192 384
Fade-O-Meter Test Exposure hours	0 to 2.5	2.5 5	5 10	10 20	20 40	40 80	80 160	160 320

Effects of Dye Concentration on Light Fastness

Numerous studies have demonstrated the relationship between the amount of various dyes bound by substrates and their light fastness; light fastness tends to improve as the depth of dyeing increases. This effect has been ascribed to aggregation of dyes in fibers and other substrates. The Color Index usually lists three light fastness ratings; material dyed at half the normal (customary) strength, normal, and twice normal dye concentrations in substrates. This range is listed in the tables below. Thus, strong coloration of sections can be expected to prolong the useful life of stains. However, some dyes show little effect of the concentration in substrates. For instance, the light fastness of rhodamine 3GO increases only from 1 to 2; rhodamine B and 6GO have light fastness 2 at all three concentrations.

Other Fastness Tests

Fastness of dyed substrates to acid, alkali, chlorine, and other reagents is rated from 1 to 5 (see below). The dyed samples to be tested are wrapped with various white materials; polyamides, silk, cellulosic fibers. Two effects of reagents are determined: alteration of the color or loss of dye from dyed sample, and staining of adjacent white material. In the textile industry, tests for fastness to acids and alkalis are carried out under conditions more stringent than those usually encountered in histotechnology. However, these data allow estimates of the behavior of dyes in various staining techniques. In our experience, for example, dyes with poor to fair acid fastness (acid fuchsin, aniline blue WS) are prone to fade when used in picro-dye methods, whereas dyes with acid fastness 4 to 5 and good light fastness (sirius red F3BA) show no alterations in sections stored for 25 years. Therefore, dyes with low alkali fastness may be affected by acids, or alkalis tend to be sensitive to counterstains containing such reagents.

Fastness of Dyes to Acids, Alkali, Peroxide, Chlorine, etc.

Fastness Rating	Alterations of Shade	Staining of Adjacent Materials
1 = poor	great loss or alteration of color	adjacent white deeply dyed
2 = moderate	distinct loss or alteration of color	adjacent white deeply dyed
3 = fair	appreciable loss or alteration of color	appreciable staining of adjacent white
4 = good	very slight loss or alteration of color	very slight staining of adjacent white
5 = very good	shade unaltered	no staining of adjacent white

Light Absorption by Dyes and Fastness

Dyes do not react with the whole UV and visible spectrum emitted by light sources, but absorb only in certain regions. Only light at wavelengths absorbed by the dye can cause fading. Whether or not fading occurs depends also on the energy of the light (photons) and on the structure of the dye. The energy of photons is expressed in kcal/mole and decreases with increasing wavelength of the light. UV and violet light are most detrimental to many dyes. For example, exposures of rhodamine B to light of wavelength 546 nm produce no detectable fading, but UV light cause distinct changes. Light of wavelengths around 550 nm has much less energy than UV light.

Differences Among Light Sources

Xenon lamps used in light fastness testing have an emission spectrum similar to daylight. In contrast, carbon arcs emit much energy-rich UV and violet light in the 330-410 nm band, greatly accelerating fading of many dyes. The peak emissions of high pressure mercury (Hg) lamps are concentrated in a few bands between 300 and 450 nm and 540-600 nm; therefore, light fastness ratings do not necessarily indicate the behavior of dyes under Hg lamps. For example, if a main absorption band of a dye coincides with an emission peak of the Hg lamp in the UV region, fading can be rapid. If the major absorption band matches a "hole" in the spectrum of the Hg lamp, a dye may be more light fast than expected from its rating. Tungsten lamps emit mainly in the visible and near infrared region, and their light does not have enough energy to cause noticeable fading of dyes during usual procedures of microscopy.

Fluorochromes

Many dyes commonly used in fluorescence microscopy have low light fastness, even when tested under daylight conditions. Some dyes have absorption maxima in the UV and/or blue region (primulin at 350 nm, thioflavin S at 365 nm) and tend to fade rapidly when exposed to UV light. Several fluorochromes (acridine red 3 B, acridine yellow G and T, acriflavine) are considered obsolete by professional dyers; hence fastness ratings are not listed in the Color Index. Xanthene dyes have very poor to fair light fastness; fluorescein, erythrosin Y and rose Bengal are not even assigned light fastness 1. Parenthetically, according to some sources rose bengal and the erythrosins are non-fluorescent in aqueous solutions. When bound to tissues, these dyes emit strong, though short-lived, fluorescence. New intensely fluorescent compounds with good fastness to light and various reagents are widely used in the textile industry (phorwhites, ultraphors, blancophors, calcofluors, tinopal, etc.) but have received little attention in histotechnology. Fluorescent brightening agents with high selectivity for different substrates (cellulosic material, proteins, polyamides) have been synthesized. When used in aqueous salt solutions, phorwhite BBU, a fluorescent brightening agent for cellulosic fibers, imparts intense selective fluorescence to amyloid. Fluorescent brightening agents for other substrates, like proteins and polyamides, show different reactive patterns. Because of the good light fastness of these reagents, sections can be studied and photographed leisurely without fading of fluorescence. There is normally no noticeable decrease in fluorescence, of sections stored for ten years in the dark, as in a drawer.

REMARKS
The history, fastness rating, light absorption, fluorochromes, dye structure, molecular orbital system, metals, substrates, moisture, and chemical reactions of dye fading can be found in the following reference: Puchtler, H., McGowan, A.L. and Meloan, S.N.: Why Do Some Stains Fade? Part I. History, Fastness Ratings, Light Absorption, and Fluorochromes, *J of Histotechnology,* 12:1, p-57, 1989. Why Do Some Stains Fade? Part II. Effects of Dye Structure, Molecular Orbital Systems, Metals, Substrates, Moisture, and Chemical Reactions, *J of Histotechnology,* 12:2, p-129, 1989.

Greenberg Method for Use of Phorwite Fabric Brighteners[49]

Certain members of the phorwite group of fabric brighteners can render a useful service in histology. In fluorescent microscopy, they sharply delineate various cell and tissue components. Their most notable qualities are the sharpness with which they delineate parenchymal vascular structures, congestion, and petechial hemorrhage; and the ability to resist quenching, a significant asset in photographically recording observations.

The Phorwite fabric brighteners should be prepared as 0.1% aqueous solutions with a pH of about 6. Deparaffinized 5 micrometer sections can be placed in these solutions for 10 minutes, washed in distilled water, dehydrated, cleared, and mounted in low-fluorescent mounting medium. The epithelial cells will be green to red-brown; muscle, yellow; bone, dark green; red blood cells, bright yellow; and white blood cells, deep green. In general, the cells and histologic features are better displayed with these dyes than with acridine orange.

Bosch Method for Paraffin Sectioning of Minute Specimens[50]

This method has been devised for placing minute specimens in viscous aqueous sodium alginate solution, which is polymerized in 0.25 molar calcium chloride. The resulting matrix can then be dehydrated, embedded, and sectioned, maintaining the orientation of the specimens. The sectioned alginate can be dissolved in 0.1 molar EDTA if desired. Tissue sections respond to normal staining procedures.

FIXATION 10% Buffered neutral formalin, Bouin's solution or alcohol-formalin-acetic acid. (Following fixation, the specimens are brought to water.)

MICROTOMY Cut sections at 6 micrometers.

SOLUTIONS

1% Sodium Alginate

Algenic acid, sodium salt	0.1	gm
Distilled water	10.0	ml

Boil briefly and filter through paper using vacuum suction.

PROCESSING PROCEDURE

1. On the stage of a dissecting microscope, place a flexible 22 mm square plastic coverglass horizontally onto a 20 mm diameter low glass vial, leaving the edges of the coverslip free for grasping.
2. Place a small plastic "O" ring (approximate inside diameter 15 mm, outside diameter 17 mm, and 1.5 mm high) in the center of the coverglass. Adjust the substage mirror to illuminate the ring.
3. Fill the inside of the plastic ring with 4 drops of the sodium alginate solution.
4. Transfer the specimens to the solution using fine forceps, and place in desired position.
5. Lift the coverglass, without jarring, and lower horizontally under the surface of a 0.25 M aqueous calcium chloride solution in a small dish. Then flex the edges of the coverslip slightly so that the solution will touch the bottom surface of the alginate within the ring.
6. After a few seconds, tilt the coverglass slightly to allow the ring to slide off into the solution. Allow the ring to remain in the solution for 5 minutes, turning it over twice.
7. Wash the disc of polymerized calcium alginate, still inside the ring, in gently running water for 3 minutes and dehydrate through alcohol. In 90% or 95% alcohol, remove the plastic ring by cutting around its edge with a needle. The "O" ring can be washed and reused.
8. After thorough dehydration in 100% alcohol, clear in 2 changes of xylene, infiltrate with 2 changes of molten paraffin, embed and section as desired.
9. After slides have been cut and rehydrated, they can be briefly immersed in 0.1M EDTA to remove the polymerized alginate, which will stain pink with eosin.

Riley-Slocum Method for Sectioning Fresh Frozen Muscle[51]

Longitudinally oriented muscle sections are advantageous for investigating properties that vary along the length of the muscle fibers. However, longitudinal sections of quick-frozen living muscles undergo a destructive Ca^{++}- dependent contraction when thawed on glass slides. The fume-fixation method provided below prevents contraction of thick longitudinal frozen sections and is compatible with histochemical staining of enzymes.

PROCEDURES

1. Place muscles on thin cards and quick-freeze by plunging them into Freon 12 cooled to near freezing by liquid nitrogen.
2. Mount portions of the frozen muscles on cryostat chucks with Tissue-Tek OCT and section at thicknesses of 8 to 100 micrometers.
3. Cut sections in the cryostat and transfer with a cold wooden spatula onto precooled glass slides.
4. Fume fix sections in the cryostat at -25°C by covering them with fixation chambers containing 37% formaldehyde. (Prepare the fixation chambers by lining the plastic caps of 3 ml disposable syringes* with 2 disks of Whatman #2 filter paper. The paper in the cap should be moistened at room temperature with 4 to 5 drops of 37% formaldehyde.) After applying petroleum jelly as a seal to the rim of the cap, the cap is pressed onto the cold slide in the cryostat to cover and fume-fix the sections. The chamber is removed after 30 seconds to 20 minutes, depending upon the thickness of the sections.

5. Thaw sections by bringing the slide to room temperature.
6. Dry slides for 15 minutes and stain as desired.

REMARKS
While clean glass slides work well for most applications, section adherence is improved by coating slides with 2% gelatin.

*Monoject, St. Louis, MO 63103.

Liem-Van Willigen Method for Staining Peripheral Nervous Tissue[52]

This is a simple quantitative method for staining in toto peripheral nerves.

SOLUTIONS

3% Potassium Hydroxide

Potassium hydroxide	3.0	gm
Distilled water	100.0	ml

Bleaching Solution

3% Hydrogen peroxide	3	drops
3% Potassium hydroxide	100.0	ml

Sihler's Solution I

Glacial acetic acid	1	part
Glycerin	1	part
1% Aqueous chloral hydrate	6	parts

Sihler's Solution II

Ehrlich's hematoxylin stock solution	1	part
Glycerin	1	part
1% Aqueous chloral hydrate	6	parts

Ehrlich's Hematoxylin Solution
(See Page 84)

PROCEDURE
1. Anesthetize animals by ether inhalation or procedure of choice.
2. Perfuse transcardially through the left ventricle with a brief initial wash of 0.9% saline followed by 10% unneutralized formalin.
3. Explant the hard palate, or other tissue, and post-fix in 10% unneutralized formalin.
4. Wash in water, then macerate in bleaching solution. The bleaching solution should be changed each week. Maceration is complete when fine nerve branches become visible as white filaments under the dissecting microscope (about 21 days).
5. Decalcify at room temperature in Sihler's Solution I. Change the solution every 2 days for about 14 days. Progress of decalcification should be checked twice a week by means of radiographs. Decalcification is adequate when structures are uniformly radiolucent.

6. After decalcification, transfer specimens to Sihler's Solution II for staining. Leave specimens in this solution until they are purplish-blue (at least 20 days).
7. Destain tissues in Sihler's Solution I with frequent agitation. Check progress under a dissecting microscope and stop just before the finest branches of the nerves begin to fade.
8. Wash specimens in a weak (0.25%) aqueous solution of lithium carbonate to blue.
9. Clear specimens with changes of 40% and 50% glycerin, 6 hours each, then 75% glycerin for at least 24 hours until the excess stain has washed out. Store in pure glycerin with a few thymol crystals added.

RESULTS

Nerves .... deep blue
Connective tissue and muscle .... light pink

REMARKS

If the specimens are stained longer than 20 days, growth of molds can be prevented by adding a few thymol crystals.

Optimal maceration is important, but difficult to determine. It is therefore advisable to macerate a number of specimens simultaneously and stop the process at intervals of 3 hours, until the fine nerve branches can be seen as thin white filaments under the dissecting microscope. Accurate determination of decalcifying endpoint is important. If incomplete, the transparency of the specimen will be impaired. If prolonged, staining will suffer.

Rubin-Baur Method for Multiple Specimen Embedments[53]

This technique allows for large numbers of hollow viscera, blood vessels, or nerve trunk cross sections to be arranged in small groups for identification purposes.

PROCEDURE

1. Use a segment of larger diameter tissue, such as a 2 cm long piece of murine gut, as a casing in which to pack parallel tissue samples (Fig. 38a, b, c and d).
2. Surgical needles prethreaded with braided black 6-0 silk, can be used to string the specimens (Fig. 38a). With the end of the silk knotted, pass the needle through the diameter of each sample in the groups, at about 1 mm from one of the cut ends. (a) Then pass the needle through the lumen of the casing (intestinal segment) and gently pull the sample within (Fig. 38b). To identify each group, markers can be threaded along side of the samples (Fig. 38c). Human hair and plant fiber strands make useful markers. After samples and markers are encased, the ends are closed with small threads of surgical silk and each casing is cut in half perpendicular to its long axis (Fig. 38d). The casings are then processed and embedded in paraffin with cut ends facing down. Several casings can be embedded in the same block.

Shelf Life of Solutions, Dyes and Chemicals

Histology laboratories are frequently faced with the dilemma of shelf life (expiration date) for staining solutions, dyes and chemicals. To my knowledge, no satisfactory system exists for determining the shelf life of these items, yet each laboratory is expected to have a verification program. These requirements are expected from various agencies or professional societies dealing with laboratory inspections and quality assurance programs related to drug or chemical studies. A search of the literature indicates that there are no written programs or information dealing with the shelf life of these items. The one exception is a stability study of certified biological stains performed by Drs. Victor M. Emmel and Elmer H. Stotz for the Biological Stain Commission in 1974 (Biological Stain Commission Laboratories, University of Rochester Medical Center, Rochester, NY 14642). That report concluded with the following statement: "These results affirm and support the widely held view that under ordinary conditions of laboratory storage, biological stains in dry powder form have a useful shelf-life of many years." Laboratories are familiar with expiration dates provided by manufacturers, but this information does not indicate the actual shelf life of dyes, solutions and chemicals. The following method is suggested for quality controlling dyes, solutions and chemicals, and therefore, provides a system for determining the shelf life of these diversified items in the laboratory.

Method for Quality Control of Staining Solutions, Dyes and Chemicals

1. Upon receipt of dyes, chemicals or solutions in the laboratory, document the following:

   Date of receipt:
   Date opened:

   For a staining solution or chemical solution made in the laboratory, indicate:

   Date made:
   Date first used:

   This documentation establishes a starting point to quality control these items.

2. Date the staining solution every time the solution is used and provides good staining results. The same applies to the chemicals, dyes, or compounded chemicals used in the laboratory. Continue to date a solution each time it is used as long as it works well. This process should continue until one notices that the staining solution no longer works, or the chemical solution prepared with certain chemicals, is not good. When a combination of chemicals is used to make up a dye or solution and the results are not satisfactory, all chemicals and dyes used should be discarded. The ideal method of dating is to apply a small label to the container when the item is received or prepared in the laboratory. Labels can be 1x3" self adhesive, or those used as slide foot labels (3/8x1").

Swan-Davis' Method for Tissue Orientation[54]
"The Cucumber Unit"

This method was developed primarily for the accurate orientation of cervical biopsies, but there is little doubt that it can be useful for orientation of specimens of all kind.

PREPARATION OF CUCUMBER SLICES
A fresh cucumber is skinned and cut into quarters. After the seeds are scraped from the surface, each quarter is divided into horizontal sections approximately 3 mm thick. Depending upon the biopsy size desired, each horizontal section is then cut into rectangular pieces of various dimensions.

The cucumber sections are treated as follows:

1. Suspend cucumber slices in absolute alcohol (in airtight container) for 24 hours.
2. Place cucumber slices in a fresh change of absolute alcohol for 24 hours.
3. Place cucumber slices in the third change of absolute alcohol for 24 hours.
4. Store slices in a fresh change of absolute alcohol until used.
5. Place a drop of egg albumen on a cucumber slice and spread evenly over its surface.
6. Press the biopsy gently onto the albuminized surface of the cucumber.
7. Wait three minutes before placing cucumber biopsy unit into the fixative of choice for required period.
8. Dehydrate, clear and impregnate cucumber biopsy unit with paraffin, or process as desired.

REMARKS
Before a biopsy is taken, a slice of cucumber is removed and placed on an absorbent towel. As much fluid as possible is pressed out of the slice. This is done to prevent the cucumber biopsy unit from floating in liquid paraffin.

It is important to place the biopsy along one longitudinal edge and trim the opposite side; this prevents preparing slides with only cucumber on them.

The cucumber biopsy unit is processed in toto. There is no interference with H&E staining. A fine mesh network may be seen outside the tissue.

Defatting Procedure for Breast Tissue

The purpose of this short procedure is to present a method for the extraction of lipids to facilitate the infiltration of paraffin into the fat cell (adipocyte). Adipocytes occur in two major forms; unilocular, with a single large inclusion of lipid, and multilocular, with smaller lipid inclusions. The adipocyte of concern here is unilocular (white adipose). Between 60-85% of the weight of white adipose is lipid. Simple lipids are esters of fatty acids with alcohols and include fats, oils and waxes. Fats are neutral esters of glycerol with saturated fatty acids. Compound lipids consist of a fatty acid, an alcohol (usually glycerol) and one or more additional groups such as phosphorous or nitrogen. Different lipids are removed by different extraction processes. Cold acetone, for example, removes glycerides; hot acetone removes cerebrosides; hot ether removes lecithins and cephalins; a mixture of hot chloroform and methanol will remove all lipids.

Defatting Procedure:

1. Fix tissue in 10% neutral buffered formalin for a required period.
2. Place tissue in metal or nonacetone reacting cassettes.
3. Place cassettes containing tissue in acetone at a volume of at least 20 times that of the tissue to be defatted.
4. After 30-45 minutes, check solution for a cloudy or yellowish appearance. If fluid appears cloudy or yellow, change acetone; if not, change after 1 hour.
5. One hour later, change solution again.
6. Two hours later, check the tissue in one of the cassettes for oily appearance of the fat. If the fat does not appear oily, discontinue procedure. Use good judgement not to discontinue procedure too soon.
7. If procedure is to be continued, change solution for the final time. The defatting procedure should be complete after 6 hours with proper changes of acetone.
8. Wash tissue containing cassettes in cool running tap water for 3-5 minutes to remove residual acetone.
9. Process on Tissue-Tek V.I.P.* as usual.

*Miles, Inc., Elkhart, IN 46515.

Method for Neutralizing Free Aldehyde Groups After Glutaraldehyde or Paraformaldehyde Fixative

Tissue specimens fixed in gluteraldehyde produce two distinct problems when mucopolysaccharide procedures are performed. 1) The mucopolysaccharides are less vividly stained with the alcian blue and colloidal iron staining procedures and 2) The periodic acid Schiff reaction produces a more intense, generalized, diffuse staining of all tissue structures. Application of Schiff's reagent without oxidation with periodic acid yields similar intense staining, indicating that the gluteraldehyde treated specimens provide free aldehyde groups to react with the Schiff's reagent. Gluteraldehyde produced reactivity may be neutralized by the use of the following procedure.

SOLUTION

Aniline

Aniline	88.0	ml
Glacial acetic acid	12.0	ml

NEUTRALIZING PROCEDURE

1. Decerate slides in xylene, 2 changes, 2 minutes each. Place in absolute alcohol, 2 changes, 2 minutes each; 95% alcohol, 2 changes, 2 minutes each; then rinse in distilled water.
2. Place slides in aniline solution for 1 hour.
3. Rinse slides quickly in 2 changes of 95% alcohol.
4. Wash slides in running tap water for 10 minutes.
5. Perform periodic acid Schiff procedure in the usual manner.

REMARKS

Sections should be mounted on salinated or Poly-L-Lysine coated slides to prevent them from falling off the slide.

Molecular Weights of Reagents Commonly used in Buffer Solutions

Reagent	Molecular Weight
Acetic acid - $CH^3.COOH$	60.03
Borax (sodium tetraborate) $Na^2B^7O^7.10\ H^20$	381.43
Boric acid - $B(OH)^3$	61.84
Citric acid (anhydrous) - $C^3H^4(OH)(COOH)^3$	192.12
Citric acid crystals - $C^3H^4(OH)(COOH)^3.H^2O$	210.14
Glycine - NH^2CH^2COOH	75.07
Hydrochloric acid - HCl	36.465
Potassium acid phosphate (potassium dihydrogen phosphate) - KH^2PO^4	136.09
Potassium hydroxide - KOH	56.104
Sodium acetate (anhydrous) - CH^3COONa	82.04
Sodium acetate crystals - $CH^3COONa.^3\ H^2O$	136.09
Sodium acid phosphate (sodium dihydrogen phosphate) - $NaH^2PO^4.H^2O$	138.01
Sodium citrate crystals - $C^3H^4OH(COONa)^3.5\ 1/2\ H^2O$	357.18
Sodium citrate, granular - $C^3H^4OH(COONa)^3.2\ H^2O$	294.12
Sodium chloride - NaCl	58.46
Sodium hydroxide - NaOH	40.0
Sodium phosphate, dibasic - Na^3HPO^4	141.98
Sulphuric acid - H^3SO^4	98.082

Walpole's Sodium Acetate-Hydrochloric Acid Buffer (pH range 0.65-5.2)

m/1 Sodium acetate, prepared by dissolving 82.04 gm (anhydrous), or 136.09 gm crystals in distilled water, and making up to 1 liter in a measuring flask.

m/1 Hydrochloric acid, prepared by diluting concentrated hydrochloric acid which is approximately 10 M, and titrating against a normal alkali.

Working Solution

To prepare buffer solution, take the appropriate amount of each solution and make up with distilled water to 500 ml.

pH	m/1 Sodium Acetate (ml)	m/1 Hydrochloric Acid (ml)
0.65	100	200
0.91	100	160
1.09	100	140
1.24	100	130
1.42	100	120
1.71	100	110
1.99	100	105
2.32	100	102
2.72	100	99.5

continued on next page

pH	m/1 Sodium Acetate (ml)	m/1 Hydrochloric Acid (ml)
3.09	100	97
3.29	100	95
3.5	100	92.5
3.79	100	85
3.95	100	80
4.19	100	70
4.58	100	50
4.76	100	40
4.92	100	30
5.2	100	20

Sodium Citrate-Hydrochloric Acid Buffer (pH range 1.1-7.5)

0.1 M Sodium citrate, prepared by dissolving 35.718 gm (crystals) or 29.412 gm (granular) in distilled water, and making up to 1 liter.

0.1M Hydrochloric acid, prepared by diluting concentrated acid (approximately 10 M), and titrating against a normal alkali.

pH	0.1 M Hydrochloric Acid (ml)	0.1 M Sodium Citrate (ml)
1.1	100	0
1.35	90	10
1.85	80	20
2.5	76	24
3.6	70	30
4.51	60	40
5.1	50	50
5.49	40	60
5.75	30	70
6.0	20	80
6.4	10	90
6.73	4	96
6.99	2	98
7.5	0	100

Walpole's Acetic Acid-Sodium Acetate Buffer (pH range 3.6-6.8)

0.1 N Acetic acid, prepared by diluting glacial acetic acid (mol. wt 60.03) and titrating against 0.1 N alkali.

0.1 N Sodium acetate, prepared by dissolving 8.204 gm (or 13.6 gm of crystals) in distilled water, and making up to 1 liter in a measuring flask.

Working Solution

To prepare buffer solution take the appropriate amount of acetic acid and sodium acetate solutions.

pH	0.1 M Hydrochloric Acid (ml)	0.1 M Sodium Citrate (ml)
3.6	185	15
3.8	176	24
4.0	164	36
4.2	147	53
4.4	126	74
4.6	102	98
4.8	80	120
5.0	59	141
5.2	42	158
5.4	29	171
5.6	19	181
6.0	10	190
6.3	5	195
6.8	0	200

Sorensen's Phosphate Buffer (pH range 5.3-8)

M/15 Sodium phosphate dibasic (Na_2HPO_4), prepared by dissolving 9.465 gm of the salt in distilled water and making up to 1 liter in a measuring flask.

M/15 Potassium acid phosphate (KH_2PO_4), prepared by dissolving 9.07 mg in distilled water, and making up to 1 liter.

Working Solution

To prepare M/15 buffer solution, take the appropriate amount of each solution in milliliters, as given below. The resulting 100 ml of buffer solution may be diluted with distilled water to 1000 ml for certain staining techniques.

pH	M/15 Sodium Phosphate Dibasic (ml)	M/15 Potassium Acid Phosphate (ml)
5.3	2.5	97.5
5.6	5.0	95
5.91	10	90
6.24	20	80
6.47	30	70
6.64	40	60
6.8	49.6	50.4
6.98	60	40
7.2	72	28
7.38	80	20
7.73	90	10
8.04	95	5

Michaelis' Veronal-Hydrochloric Acid Buffer (pH range 4.5-9.2)

Veronal Acetate Solution

Sodium acetate	1.943	gm
Veronal (sodium barbituate)	2.943	gm
Distilled Water	to 100.0	ml

Working Solution

To prepare a buffer solution add to 5 ml of the veronal acetate solution the amount of M/10 hydrochloric acid and distilled water given below.

pH	Veronal Acetate Solution (ml)	M/10 Hydrochloric Acid (ml)	Distilled Water (ml)
4.5	5	11	9
4.95	5	9	11
5.3	5	8	12
6.1	5	7	13
6.75	5	6.5	13.5
7	5	6	14
7.25	5	5.5	14.5
7.4	5	5	15
7.66	5	4	16
7.9	5	3	17
8.2	5	2	18
8.6	5	1	19
8.7	5	0.75	19.25
8.9	5	0.5	19.5
9.2	5	0.25	19.75

Tris (hydroxymethyl) Aminomethane-Maleic Acid (Gomori's Buffer) (pH range 5.08-8.45)

M/1 Maleic acid ($C_4H_4O_4$). Dissolve 116 gm in distilled water and make up to 1 liter in a measuring flask.

M/1 Tris (hydroxymethyl) aminomethane $(CH_2OH)_3C\text{-}NH_2$. Dissolve 121 gm in distilled water and make up to 1 liter in a measuring flask.

M/2 Sodium hydroxide (NaOH). Dissolve a little over 20 gm in 1 liter of distilled water and titrate against a normal acid.

Working Solution

To prepare the buffer solution take the appropriate amounts of each solution in milliliters as given below. The resulting 50 ml of buffer solution may be diluted up to 250 or 500 ml for certain techniques.

pH	M/1 Maleic Acid (ml)	M/1 Tris (hydroxy methyl) Aminomethane (ml)	M/2 Sodium Hydroxide (ml)	Distilled Water (ml)
5.08	5	5	1	39
5.30	5	5	2	38
5.52	5	5	3	37
5.70	5	5	4	36
5.88	5	5	5	35
6.05	5	5	6	34
6.27	5	5	7	33
6.5	5	5	8	32
6.86	5	5	9	31
7.20	5	5	10	30
7.5	5	5	11	29
7.75	5	5	12	28
7.97	5	5	13	27
8.15	5	5	14	26
8.30	5	5	15	25
8.45	5	5	16	24

Gomori's 0.2 M "Tris" Buffer (pH range 7.2-9.1)

0.2 M Tris (hydroxymethyl) aminomethane (mol. wt. 121.14)
0.1 N Hydrochloric acid

Working Solution

To prepare buffer solution take the appropriate amount of each solution and make up with distilled water to 100 ml.

pH	02. M Tris (ml)	0.1 N Hydrochloric Acid (ml)
7.19	25	45.0
7.36	25	42.5
7.54	25	40.0
7.66	25	37.5
7.77	25	35.0
7.87	25	32.5
7.96	25	30.0
8.05	25	27.5
8.14	25	25.0
8.23	25	22.5
8.32	25	20.0

continued on next page

pH	02. M Tris (ml)	0.1 N Hydrochloric Acid (ml)
8.41	25	17.5
8.51	25	15.0
8.62	25	12.5
8.74	25	10.0
8.92	25	7.5
9.10	25	5.0

<u>Holmes' Boric Acid-Borax Buffer (pH range 7.4-9.1)</u>

M/5 Boric acid, prepared by dissolving 12.368 gm of boric acid in distilled water in a 1 liter measuring flask, and making up to 1 liter with distilled water.

M/20 Borax (sodium tetraborate), prepared by dissolving 19.071 gm of borax in distilled water in a 1 liter measuring flask, and making up to 1 liter with distilled water.

Working Solution

To prepare buffer solution mix the appropriate amounts of each solution, as given below.

pH	M/5 Boric Acid (ml)	M/20 Borax (ml)
7.4	90	10
7.6	85	15
7.8	80	20
8.0	70	30
8.2	65	35
8.4	55	45
8.7	40	60
9.0	20	80

<u>Sorensen and Walbaum's Glycine-Sodium Chloride-Sodium Hydroxide Buffer (pH range 8.45-12.77)</u>

0.1 M Glycine in 0.1 M NaCl, prepared by dissolving 5.846 gm of NaCl in a quantity of distilled water in 1 liter measuring flask. Add 7.507 gm of glycine and more distilled water to dissolve it; make the resultant solution up to 1 liter with distilled water.

0.1 M NaOH, prepared by dissolving 4 gm of NaOH in distilled water, and making up to 1 liter with distilled water. This solution should be checked by titrating against a decimolar acid.

Working Solution

To prepare a 0.1 M buffer solution mix the appropriate amounts of each solution as given below.

pH	0.1 M Glycine in 0.1 M NaCl (ml)	0.1 M NaOH (ml)
8.45	95	5
8.79	90	10
9.22	80	20
9.56	70	30
9.98	70	40
10.32	60	45
10.9	55	49
11.14	50	50
11.39	49	51
11.92	45	55
12.21	40	60
12.48	30	70
12.66	20	80
12.77	10	90

Use of Polymer Film Carrier in Cryostat Sectioning of Large Fragile Samples

This problem dealt with obtaining histological sections of a subcutaneous gel implant. The gel implant was very hydrophilic, which prohibited the use of standard aqueous fixation procedures, - i.e., 10% formalin solution. A decision was made to quick-freeze the biopsy on CO_2 and proceed with standard cryosectioning techniques.

Initial attempts to obtain a section were frustrating. The high water content and size of the gel implant caused the sample to shatter during sectioning. As the gel shattered, the capsular structure lost support. The lower portion of the capsule curled during sectioning despite attempts to increase support of the block with additional matrix material. Following trimming of the block, a swatch of polyvinylidene chloride was gently placed over the planed face of the block. As the film was smoothed to the block face with a finger, the frozen matrix material melted and refroze, rapidly securing the film to the block. The film-section complex was then slowly cut with the section sandwiched between the film and the blade. The film-section complex was retrieved and carefully placed on an albumin-coated microscope slide, section down. Again, contact with a warm finger on the underside of the slide secures the section to the slide. Placing the slide immediately on a block of dry ice releases the polyvinylidene chloride film from the section-slice complex. The section was now ready for normal H&E staining procedures.

Luna's STAT Method

CRYOTOMY PROCEDURE

1. Select tissue to be sectioned and place on a mound of cryostat mounting medium (OCT etc.) previously frozen on a specimen holder and kept in the refrigerated cabinet. The medium gives the metal specimen holder greater clearance from the knife edge.
2. Freeze from the top surface of the specimen down with Freon 22 aerosol spray (-40°C). This decreases the freezing artifact on the surface of the specimen, which is where the specimen will be cut from, and most likely used for interpretation. (Freon 22 is recommended over Freon 12).
3. Mount specimen holder with sufficient knife clearance.
4. Use sharp knife to section.
5. Cut using anti-roll device (wire "trapeze type" works better for larger sections). If anti-roll device does not work, use a camel hair brush.
6. Pick up desired section on chrome gelatin coated slide (kept at room temperature).
7. FIX IMMEDIATELY in absolute methanol for 1 minute and then stain with desired procedure. The nuclear and cytoplasmic detail is lost if there is any delay in fixation. Fixed sections are permanent.

SOLUTIONS

Harris' Hematoxylin
(See Page 81)

Lithium Carbonate
(See Page 91)

Eosin Y Stain
(See Page 86)

2% Toluidine Blue

Toluidine blue	2.0	gm
Methyl alcohol	100.0	ml

STAINING PROCEDURE

Harris' Hematoxylin and Eosin Method:

1. Place in Harris' hematoxylin for 1 minute.
2. Rinse in distilled water, 2 changes.
3. Place in lithium carbonate, 2 dips.
4. Rinse in distilled water for 30 seconds, 2 changes.
5. Place in eosin Y stain, 5 dips.
6. Place in 80% alcohol, 5 dips.
7. Place in 95% alcohol, 5 dips.
8. Place in absolute alcohol, 5 dips, 2 changes.
9. Place in acetone, 5 dips.
10. Place in xylol, 5 dips.
11. Mount coverglass with appropriate medium (refractive index 1.48-1.56).

RESULTS

Nuclear chromatin .... blue
Ctyoplasm .... pink

Toluidine Blue Method:

1. Place in toluidine blue, 5 dips.
2. Rinse in distilled water for 30 seconds.
3. Place in acetone, 5 dips.
4. Place in xylol, 5 dips.
5. Mount coverglass with appropriate medium (refractive index 1.48-1.56).

RESULTS

Nuclear chromatin .... blue
Cytoplasm .... light blue

Polyvinyl Pyrrolidone as a Water-Soluble Mounting Medium (Refractive Index 1.43)

Fifty grams of polyvinyl pyrrolidone (PVP) (Plasdone C) are dissolved in 50 ml of distilled water and permitted to stand overnight. Two milliliters of glycerol are then added and the mixture stirred. (A crystal of thymol may be added as a preservative.) The medium may be conveniently applied from a 2-ounce dropper bottle. The cap may be lubricated with silicone grease in order to prevent sticking. The refractive index (1.43) increases after the evaporation of some of the water.

Preservation of the following stains or reactions is possible, with the use of this medium.

- Oil Red O stain for fat
- Sudan III stain for fat
- Esterase-alpha-naphthyl acetate; naphthol AS acetate and complex naphthol AS derivatives, AS-D, AS-MX, AS-LT, and AS-LC esters
- Alkaline phosphatase-alpha-naphthyl phosphate (red RC salt); and glycerophosphate calcium-cobalt
- Periodic acid-Schiff method
- Hematoxylin, chromalum
- Celestin blue
- Aldehyde fuchsin
- Eosin

Coverglasses became immovable within an hour after mounting, and the formation of bubbles is minimal. The clarity of the preparations was not altered during several months of observation. This PVP water-soluble medium is a useful substitute for glycerol-gelatin and should simplify the mounting of frozen sections.

Permanent Coverglass Mounting Methods

Canada Balsam (Refractive Index 1.52)

Canada balsam is an oleoresin collected from blisters in the bark of balsam fir trees which are native to Canada. Canada balsam adheres firmly to glass and sets to a hard consistency without granulation. The only disadvantage is that this mounting medium darkens slightly but does not affect the microscope observation or photography of microscopic slides. The biggest advantage is that it does not draw back from the edge of the coverglasses. Properly diluted, slides that were coverslipped in 1922 are still in perfect condition. Xylene oxidizes Canada balsam to tuluic and phthalic acids, which may cause some stains to fade.

Clarite or Clarite X (Refractive Index 1.54)

This is a good coverglass mounting medium. It is a synthetic resin which is used as a 60% solution in xylene. It should be noted that there are many good resinous and plasticized mounting mediums on the market. The choice of which to use has to be decided by the individual laboratory.

Distrene - Tricresylphosphate (D.P.X.) (Refractive Index 1.52)

This is made of the resins distrene plasticized by the addition of tricresylphosphate or dibutylphthalate. The disadvantage is that D.P.X. shrinks somewhat on drying. This mounting medium is not fluorescent, therefore, it is suitable for fluorescence microscopy.

Temporary Semipermanent Coverglass Mounting Medium

There are some tissue preparations which are adversely affected by mounting coverglasses on tissue samples. Some mounting medium with hydration base xylene, benzene, coluene, etc., dissolves some essential staining; e.g., fat stains. These preparations can be mounted with some of the following mounts.

Water

Low refractive index, is moderately transparent, and afford reasonably good visibility. However, it is the least permanent and does not permit an oil objective to be used.

Glycerin (Refractive Index 1.46)

Glycerin is a very stable temporary mounting medium. This mounting medium will last for months if sealed with clear finger nail polish. It has a good preservation action. Slight dilution with water aids in visibility.

Liquid Paraffin (Mineral Oil)

Liquid paraffin is a useful temporary coverglass mounting medium. This is especially true for Romanowsky stained smears. These stained slides should be air dried well before mounting the coverglass.

Glycerin Jelly (Refractive Index 1.4-1.47)

This mounting medium sets quite hard and will keep for years.

SOLUTIONS

Gelatin (Stock)

Gelatin	5.0	gm
Distilled water	25.0	ml

Place in 60°C oven until gelatin is dissolved.

Glycerin (Stock)

Glycerin	35.0	ml
Distilled water	40.0	ml
Cresol (or 10 mg of thimerosal-merthiolate)	0.25	ml

Glycerin Jelly (Working)

Mix gelatin (stock) solution with the glycerin (stock) solution. Small bottles may be used for storing this mounting medium, before mounting sections. Place in 55-60°C oven to melt medium when ready to use.

Farrant's Gum Acacia (Refractive Index 1.43)

Gum acacia	50.0	gm
Distilled water	50.0	ml

Dissolve gum acacia by placing container in a 55°C water bath. Shake solution frequently. When solution is dissolved add 25 ml glycerin, mix well and filter. This mounting medium is not as good as Apathy's gum syrup.

Levulose Fructose Syrup (Refractive Index 1.5)

This is a good temporary mounting medium because of its high refractive index and also because it does not leach metachromatic stains.

SOLUTION

Levulose	30.0	gm
Distilled water	20.0	ml

Place in stoppered bottle in a 37°C oven for 24 hours.

Apathy's Gum Syrup (Refractive Index 1.52)

This is an excellent mounting medium since it sets as hard as resins.

SOLUTION

Gum Arabic (Gum Acacia)

Gum arabic	50.0	gm
Cane sugar	50.0	gm
Distilled water	50.0	ml

The viscosity of this solution can be increased by addition of distilled water. Dissolve ingredients by placing containers in 60°C water bath with frequent shaking. Add 50 mg of thymol as a preservative to prevent bleeding of metachromatic staining of amyloid by methyl or crystal violet. Add 30 gms of potassium acetate or 10 gms of sodium chloride.

For Histochemical and Immunofluorescent Stained Slides

1. 20 gm polyvinyl alcohol (Gelvatol 20-30).
2. 80 ml PBS*. Mix 1 and 2 and agitate on magnetic stirrer for 16 hours or until completely dissolved. To facilitate dissolving Gelvatol, heat mixture to 50-60°C.
3. Add 50 ml glycerol and mix well until solution is uniform. DO NOT SHAKE.
4. Add several small crystals of thymol for preservative.
5. Filter warm mixture 50-60°C under vacuum. Filter through coarse fritted glass filter funnel.
6. Store stock in airtight container in refrigerator, but warm to room temperature before using. Keep small container at room temperature for daily use.
7. If bubbles appear, place in vacuum oven for 30 minutes to 1 hour or until dissolved.
8. Dispense like routine mounting medium.

*pH may be altered by changing buffer if necessary.

PBS Solution

Sodium chloride	8.0	gm
Potassium chloride	0.2	gm
Potassium dihydrogen	0.2	gm
Phosphate disodium phosphate	1.15	gm
Distilled water to 1 liter.		

Mounting medium possesses the following attributes:

1. Non-fluorescent.
2. Sets to semi-solid gel.
3. Evaporates without trapping air bubbles.
4. pH between 6 and 7.
5. Slides remain unaltered for at least 1 year.
6. Always use fluid at room temperature.

Quality Control of Special Stains

Special Stains

The first thing to consider is the use of a good microscope to review finished slides. Luke warm (37-40°C) water should be used whenever a given method calls for a running water wash. Tap water washes at this temperature, in my opinion, are essential in any staining procedure where washing is required, including silver reactions. The movement of staining solution molecules is extremely important if not essential. Many poor staining results are directly related to the lack of good molecular movement. Inadequate molecular movement may be the result of a laminar flow effect (a flow of various fluid in which neighboring "layers" are not mixed). In staining solutions, a laminar flow type effect is produced by the rough surface of the staining containers. In other words, molecules move slower the closer they are to the surface of the inner wall of the staining containers (Fig. 39a). This slower movement of molecules eventually affects the staining results. It is for this reason that one often obtains different staining results on identical slides which have been stained in the same container. One must also remember that in microwave staining, results can vary considerably between the top and lower portion of a slide. This is because solutions which are heated with microwave radiation can vary up to a 15°C difference in temperature between the top and bottom levels of the solution (Fig. 39b). Therefore, in order to equalize the temperature of the solution, the slides must be dipped up and down in the staining step(s) of the staining procedure. Staining slides in a convection slide drying type oven can also pose some staining problems. Slides situated near the outer walls of the stain container will stain darker than those in the middle. This is because there will be more heat on the outer walls than in the middle of the staining container (Fig. 39c). One should be aware of this possibility when using a convection oven for staining. The second potential problem has to do with a solution reaching a stated temperature. For example, if a stain requires 30 minutes exposure in a 60°C oven, one should place the staining solution in the oven prior to the introduction of the slides. This will ensure exposure to 60°C heat for the entire 30 minutes, thus eliminating uneven staining.

Silver Reactions

The key to silver staining revolves around the fact that a reducing site must be available to reduce silver ions. Provided below is an example of what may go wrong in one form of silver staining.

Oxidizers

The various methods of silver impregnation of basement membranes and reticular fibers include an oxidative step following fixation and prior to the action of the silver salts themselves. The aim of this oxidative step is to increase the number of argyrophilic sites so significantly, that a selective blackening of these structures can be obtained. Concomitantly, a considerable diminution in the number of argyrophilic sites takes place in neighboring structures, leading to a colorless background. There are many reasons for assuming that, following oxidation, some potentially unstainable groups are transformed to stainable groups, and numerous argyrophilic radicals are converted into nonargyrophilic ones (Fig. 39d).

Chapter 7

Bacteria, Fungi, and Spirochetes

Fite's Method for Leprae Bacilli and Nocardia[55a]

(See Fig. 40a and b)

FIXATION 10% Buffered neutral formalin.
PROCESS Paraffin or celloidin.
MICROTOMY Cut sections at 6 micrometers.
CONTROL Use control number 30.
SOLUTIONS

Xylene-Peanut Oil

Peanut oil	1	part
Xylene	2	part

Ziehl-Neelsen Carbol Fuchsin

Phenol crystals, melted	2.5	ml
100% Alcohol	5.0	ml
Basic fuchsin	0.5	gm
Distilled water	50.0	ml

Filter before use.

1% Sulfuric Acid

Concentrated sulfuric acid	1.0	ml
Distilled water	100.0	ml

Methylene Blue (Stock)

Methylene blue	1.4	gm
95% Alcohol	100.0	ml

Methylene Blue (Working)

Methylene blue (stock)	10.0	ml
Tap water	90.0	ml

STAINING PROCEDURE

1. Decerate slides in 2 changes of xylene-peanut oil solution for 12 minutes each.
2. Drain, wipe off excess oil and blot to opacity. The residual oil helps prevent shrinkage and injury of section.
3. Place in Ziehl-Neelsen carbol fuchsin solution for 30 minutes.
4. Wash in running tap water for 3 minutes.
5. Decolorize slides individually with 1% sulfuric acid solution until sections are faint pink - about 1 minute.
6. Wash in running tap water for 3 minutes.
7. Counterstain lightly with working methylene blue solution.
8. Rinse off excess methylene blue in running tap water.
9. Blot and let stand for a few minutes to air-dry thoroughly.

10. Dip slides in xylene before mounting.
11. Mount coverglass with appropriate medium (refractive index 1.48-1.56).

RESULTS

Acid-fast bacilli	red
Nocardia filaments	red
Leprae bacilli	red
Background	pale blue

REMARKS

This technique is one of the most useful of the acid-fast procedures for the demonstration of Leprae bacilli and filaments of nocardia, for the following reasons: (1) Peanut oil coats the microorganism; (2) the oil coating, applied around the filaments of nocardia and leprae bacilli, is not dissolved because of the elimination of alcohol from both the hydration and dehydration processes; and (3) the decolorization of the background with a weak solution of aqueous sulfuric acid, instead of the usual 1% hydrochloric acid alcohol, decolorizes the pathologic entities slowly and allows for better control of the differentiation of the tissue entity which is an important step.

Truant's Fluorescent Method for Tubercle and Leprae Bacilli[55b]

(See Fig. 41)

FIXATION	10% Buffered neutral formalin.
PROCESS	Paraffin, frozen, or celloidin.
MICROTOMY	Cut sections at 6 micrometers.
CONTROL	Use control number 30.
SOLUTIONS	

Weigert's Iron-Hematoxylin
(See Page 79)

Auramine-Rhodamine

Auramine	1.5	gm
Rhodamine B	0.75	gm
Glycerin, U.S.P.	75.0	ml
Phenol crystals, melted	10.0	ml
Distilled water	50.0	ml

1% Acid Alcohol
(See Page 77)

STAINING PROCEDURE

1. Decerate slides in xylene, 2 changes, 2 minutes each. Place in absolute alcohol, 2 changes, 2 minutes each; 95% alcohol, 2 changes, 2 minutes each; then rinse in distilled water.
2. Place in Weigert's iron-hematoxylin solution for 10 minutes.
3. Wash in running tap water for 10 minutes.
4. Place in auramine-rhodamine solution at 60°C for 10 minutes.

5. Wash in running tap water for 1 minute.
6. Decolorize in 1% acid alcohol solution by dipping slides 10 times.
7. Wash in running tap water for 5 minutes.
8. Rinse in distilled water.
9. Dehydrate slides in 95% alcohol and absolute alcohol 3 changes each. Clear in xylene, 3 changes.
10. Mount coverglass with appropriate medium (refractive index 1.48-1.56).

RESULTS
Bacilli will fluoresce reddish-yellow or other color depending on filters used. Artifacts will not have the reddish color and will remain yellow.

REMARKS
The Weigert's hematoxylin is used as a fluorescence inhibitor but also gives a good cytoplasmic stain. The use of 2 to 3 micrometer sections will reduce background staining.

Kinyoun's Method for Tubercle Bacilli and Lipofuscin[55c]

(See Fig. 42 and 43)

FIXATION 10% Buffered neutral formalin.
PROCESS Paraffin, frozen, or celloidin.
MICROTOMY Cut sections at 6 micrometers.
CONTROL Use control number 41 for tubercle bacilli or 27A for lipofuscin.
SOLUTIONS

Kinyoun's Carbolfuchsin

Basic fuchsin	4.0	gm
Phenol crystals, melted	8.0	ml
95% Alcohol	20.0	ml
Distilled water	100.0	ml

1% Acid Alcohol
(See Page 77)

Methylene Blue (Stock)

Methylene blue	1.4	gm
95% alcohol	100.0	ml

Methylene Blue (Working)

Methylene blue solution (stock)	10.0	ml
Tap water	90.0	ml

STAINING PROCEDURE
Use control slide.
1. Decerate slides in xylene, 2 changes, 2 minutes each. Place in absolute alcohol, 2 changes, 2 minutes each; 95% alcohol, 2 changes, 2 minutes each; then rinse in distilled water.
2. Place in Kinyoun's carbol fuchsin solution at 56°C for 30 minutes.
3. Wash in running tap water for 5 minutes.

4. Decolorize in 2 changes of acid alcohol solution until tissue is pale pink.
5. Wash in running tap water for 2 minutes.
6. Counterstain in working methylene blue solution for 2 seconds. (See remarks for counterstaining lipofuscin.)
7. Dehydrate slides in 95% alcohol and absolute alcohol 3 changes each. Clear in xylene, 3 changes.
8. Mount coverglass with appropriate medium (refractive index 1.48-1.56).

RESULTS

Tubercle bacilli	bright red
Lipofuscin	yellowish-red
Background	light blue

REMARKS

This is one of the best methods for the demonstration of tubercle bacilli and lipofuscin. The demonstration of lipofuscin can be enhanced by the substitution of the working methylene blue solution (step 6) with a saturated aqueous solution of picric acid. The counterstain can usually be accomplished by 5 to 10 dips in the picric acid.

Ziehl-Neelsen Method for Tubercle Bacilli and Lipofuscin[55d]

FIXATION 10% Buffered neutral formalin.
PROCESS Paraffin, frozen, or celloidin.
MICROTOMY Cut sections at 6 micrometers.
CONTROL Use control number 41 for tubercle bacilli and 27A for lipofuscin.
SOLUTION

Ziehl-Neelsen Carbol Fuchsin

Phenol crystals, melted	2.5	ml
100% Alcohol	5.0	ml
Basic fuchsin	0.5	gm
Distilled water	50.0	ml

Filter before use.

1% Acid Alcohol
(See Page 77)

1% Sulfuric Acid
(See Page 180)

Methylene Blue (Working)
(See Page 180)

STAINING PROCEDURE
1. Decerate slides in xylene, 2 changes, 2 minutes each. Place in absolute alcohol, 2 changes, 2 minutes each; 95% alcohol, 2 changes, 2 minutes each; then rinse in distilled water.
2. Place in carbol fuchsin solution for 30 minutes.

3. Wash in running tap water for 5 minutes.
4. Decolorize with acid alcohol solution or sulfuric acid solution until sections are pale pink.
5. Wash in running tap water for 5 minutes.
6. Counterstain by dipping one slide at a time in working methylene blue solution. Sections should be pale blue.
7. Wash with running tap water, then rinse in distilled water.
8. Dehydrate slides in 95% alcohol and absolute alcohol 3 changes each. Clear in xylene, 3 changes.
9. Mount coverglass with appropriate medium (refractive index 1.48-1.56).

RESULTS

Tubercle bacilli and lipofuscin .... bright red
Erythrocytes .... yellowish-orange
Background .... pale blue
Lipofuscin .... yellow

Rojas-Darby-Nochomovitz Method for Nocardia, Tubercle and Leprae Bacilli[56]

The Fite procedure is a universally accepted staining method for the demonstration of microorganisms like Nocardia species and leprae bacilli. The technique has assumed an almost "routine" status over the last few years in the identification of mycobacterial species in the tissues of patients with Acquired Immunodeficiency Syndrome (AIDS). This modification takes about 40 minutes and results in cleaner staining and easier interpretation. This method has been used successfully on tissue sections and imprints with excellent results.

FIXATION 10% Buffered neutral formalin, B-5, or Bouin's solution.
PROCESS Paraffin, frozen, or celloidin.
MICROTOMY Cut section at 3 micrometers.
CONTROL Use control number 41.
SOLUTIONS

Ziehl-Neelsen Carbol Fuchsin Solution
(See Page 180)

Methylene Blue (Working)
(See Page 180)

STAINING PROCEDURE

1. Decerate slides in xylene, 2 changes, 2 minutes each. Place in absolute alcohol, 2 changes, 2 minutes each; 95% alcohol, 2 changes, 2 minutes each; then rinse in distilled water.
2. Air dry.
3. Dip in xylene 3 times.
4. Place in mineral oil at room temperature for 5 minutes.
5. Drain slides on a piece of gauze or paper towel.
6. Wipe off excess mineral oil from the back and sides of the slides.

7. Dip slides individually in xylene, 6 dips.
8. Drain excess xylene and dip slides twice in absolute alcohol.
9. Place in Ziehl-Neelsen carbol fuchsin for 10 minutes.
10. Rinse in running tap water for 2 minutes.
11. Decolorize one slide at a time with 2 quick dips in 2% sulfuric acid and rinse in running tap water. Repeat 3 times and check microscopically.
12. Rinse in running tap water for 2 minutes.
13. Wipe dry back of slides and around tissue.
14. Counterstain with working methylene blue, 3 dips.
15. Rinse in running tap water for 2 minutes.
16. Air dry.
17. Dip quickly in absolute alcohol, 2 dips.
18. Clear in 3 changes of xylene for 2 minutes each.
19. Mount coverglass with appropriate medium (refractive index 1.48-1.56).

RESULTS

Tubercle and leprae bacilli	red
Nocardia	red
Background	pale blue

Wade's Method for Tubercle and Leprae Bacilli[55e]

FIXATION	Zenker's solution or 10% buffered neutral formalin.
PROCESS	Paraffin.
MICROTOMY	Cut sections at 6 micrometers.
CONTROL	Use control number 41 for tubercle bacilli and 30 for lepra bacilli.

SOLUTIONS

Decerating Solution

Turpentine	2	parts
Paraffin oil (liquid petrolatum)	1	part

Carbol-New Fuchsin Solution

New fuchsin, magenta III	0.5	gm
Phenol crystals, melted	5.0	ml
100% Alcohol	10.0	ml
Distilled water	85.0	ml

1% Potassium Permanganate Solution

Potassium permanganate	1.0	ml
Distilled water	100.0	ml

2% Oxalic Acid Solution

Oxalic acid	2.0	gm
Distilled water	100.0	ml

5% Sulfuric Acid Solution

Concentrated sulfuric acid	5.0	ml
Distilled water	95.0	ml

Modified Van Gieson Solution

Acid fuchsin	0.01	gm
Picric acid	1.0	gm
Distilled water	100.0	ml

STAINING PROCEDURE

1. Decerate slides in xylene, 2 changes, 2 minutes each. Place in absolute alcohol, 2 changes, 2 minutes each; 95% alcohol, 2 changes, 2 minutes each; then rinse in distilled water.
2. Drain, wipe off excess fluid, blot to opacity, and place in water for 5 minutes.
3. Place in carbol-new fuchsin solution, overnight.
4. Wash in running tap water for 2 minutes.
5. Place in full strength formalin for 5 minutes (see note). The bacilli will become blue. The sections may turn blue or remain predominantly reddish.
6. Wash in running tap water for 2 minutes.
7. Place in sulfuric acid for 1 minute; then wash well in running tap water for 5 minutes.
8. Place in 1% potassium permanganate 3 minutes. Rinse in running tap water.
9. Bleach in 2% oxalic acid solution for 30 to 60 seconds; then wash thoroughly in running tap water. (Use 5% oxalic acid for specimens that do not bleach readily.)
10. Place in modified Van Gieson solution for 3 minutes.
11. Dehydrate slides in 95% alcohol and absolute alcohol 3 changes each. Clear in xylene, 3 changes.
12. Mount coverglass with appropriate medium (refractive index 1.48-1.56).

RESULTS

Tubercle bacilli	deep ultramarine blue, or blue-black
Connective tissue elements	red
Background	yellowish

Garvey-Fathi-Bigelow-Carpenter-Gimenez Method for Gram-Positive and Gram-Negative Organisms[57]

FIXATION	10% Buffered neutral formalin.
PROCESS	Paraffin, frozen, or celloidin.
MICROTOMY	Cut section at 4 micrometers.
CONTROL	Use control number 20.

SOLUTIONS

1% Crystal Violet (Stock)

Crystal violet	1.0	gm
Distilled water	100.0	ml

2.5% Sodium Bicarbonate (Stock)

Sodium bicarbonate	2.5	gm
Distilled water	100.0	ml

Gram's Iodine

Iodine	1.0	gm
Potassium iodide	2.0	gm
Distilled water	300.0	ml

Cellosolve (ethylene glycol monoethyl ether)

Carbol Fuchsin (Stock)
Solution A

New fuchsin	1.0	gm
Ethyl alcohol	5.0	ml

Solution B

Phenol	5.0	ml
Distilled water	90.0	ml

Combine solutions A and B and filter before use.

Carbol Fuchsin (Working)

Carbol fuchsin (stock)	25.0	ml
Distilled water	75.0	ml

0.01% Tartrazine

Tartrazine	0.01	gm
Distilled water	100.0	ml
Glacial acetic acid	2.0	ml

Crystal Violet (Working)

1% Crystal violet (stock)	1.0	ml
2.5% Sodium bicarbonate (stock)	0.2	ml

Make just before use.

STAINING PROCEDURE

1. Decerate slides in xylene, 2 changes, 2 minutes each. Place in absolute alcohol, 2 changes, 2 minutes each; 95% alcohol, 2 changes, 2 minutes each; then rinse in distilled water.
2. Place in working crystal violet for 1 minute.
3. Wash in running tap water briefly to remove excess crystal violet.
4. Place in Gram's iodine for 1 minute.
5. Wash in running tap water to remove excess iodine.
6. Using a wash bottle, decolorize with cellosolve for 2 minutes or until extraction of blue color stops.
7. Rinse in absolute alcohol.
8. Quickly rinse in running tap water.
9. Place in working carbol fuchsin solution for 90 seconds.
10. Rinse briefly in running tap water to remove excess dye.
11. Counterstain in 0.01% tartrazine for 15 seconds.
12. Within 15 seconds dehydrate slides in 95% alcohol and absolute alcohol 3 changes each. Clear in xylene, 3 changes.
13. Mount coverglass with appropriate medium (refractive index 1.48-1.56).

RESULTS

Gram-positive microorganisms .... blue
Gram-negative microorganisms .... red
Nuclei .... red
Background .... yellow

Brown-Brenn Method for Gram-Positive and Gram-Negative Organisms[55f]

FIXATION 10% Buffered neutral formalin.
PROCESS Paraffin.
MICROTOMY Cut sections at 6 micrometers.
CONTROL Use control number 20.
SOLUTIONS

1% Crystal Violet

Crystal violet	1.0	gm
Distilled water	100.0	ml

5% Sodium Bicarbonate

Sodium bicarbonate	5.0	gm
Distilled water	100.0	ml

Gram's Iodine
(See Page 187)

Ethyl Ether-Acetone

Ethyl ether	50.0	ml
Acetone	50.0	ml

0.25% Basic Fuchsin (Stock)

Basic fuchsin	0.25	gm
Distilled water	100.0	ml

Basic Fuchsin (Working)

Basic fuchsin solution (stock)	10.0	ml
Distilled water	100.0	ml

0.1% Picric Acid-Acetone

Picric acid	0.1	gm
Acetone	100.0	ml

Acetone-Xylene

Acetone	50.0	ml
Xylene	50.0	ml

STAINING PROCEDURE

1. Decerate slides in xylene, 2 changes, 2 minutes each. Place in absolute alcohol, 2 changes, 2 minutes each; 95% alcohol, 2 changes, 2 minutes each; then rinse in distilled water.

2. Place slides in a staining rack. Pour on approximately 1 ml (or 20 drops) of working crystal violet solution and add 5 drops of sodium bicarbonate. Agitate gently for 1 minute. (If preferred, the two solutions may be mixed just before use.)
3. Rinse in running tap water.
4. Flood slides with Gram's iodine solution for 1 minute.
5. Rinse in water and blot with filter paper to dry completely.
6. Decolorize with ethyl ether-acetone solution dropped on slides until no more color runs off.
7. Place in working basic fuchsin solution for 1 minute. Wash in water. Blot gently but do not allow sections to dry completely.
8. Dip in acetone to start decoloring reaction.
9. Decolorize immediately with picric acid-acetone solution until sections are yellowish-pink.
10. Rinse quickly in acetone, then in acetone-xylene solution.
11. Dehydrate slides in 95% alcohol and absolute alcohol 3 changes each. Clear in xylene, 3 changes.
12. Mount coverglass with appropriate medium (refractive index 1.48-1.56).

RESULTS

Gram-positive organisms	blue
Gram-negative organisms	red
Filaments of nocardia and actinomyces	blue
Nuclei	red
Other tissue elements	yellow

REMARKS

This method is invaluable for the demonstration of the filaments of nocardia and actinomyces. It must be realized, however, that these filaments are not strongly Gram-positive. It is possible to obtain either Gram-positive or Gram-negative stain results depending on the degree of decolorization. It is suggested, therefore, that at least two slides be run with different degrees of decolorization in order to lessen the possibility of over-differentiation of entities at step 8 in the staining procedure.

MacCallum-Goodpasture Method for Gram-Positive and Gram-Negative Organisms[55g]

FIXATION 10% Buffered neutral formalin.
PROCESS Paraffin.
MICROTOMY Cut sections at 6 micrometers.
CONTROL Use control number 20.
SOLUTIONS

Goodpasture's Solution

Basic fuchsin	0.59	gm
Aniline*	1.0	ml
Phenol crystals, melted	1.0	ml
30% Alcohol	100.0	ml

*Care should be exercised when using aniline since it has been declared a potential carcinogen (see material safety data sheets).

Gram's Iodine
(See Page 187)

5% Crystal Violet

Crystal violet	5.0	gm
100% Alcohol	10.0	ml
Aniline	2.0	ml
Distilled water	88.0	ml

1.25% Picric Acid

Picric acid	1.25	gm
Distilled water	100.0	ml

This is a saturated solution.

STAINING PROCEDURE

1. Decerate slides in xylene, 2 changes, 2 minutes each. Place in absolute alcohol, 2 changes, 2 minutes each; 95% alcohol, 2 changes, 2 minutes each; then rinse in distilled water.
2. Place in Goodpasture's solution for 10 minutes.
3. Rinse in distilled water.
4. Decolorize in concentrated full strength formalin for a few minutes. This also fixes Goodpasture's stain.
5. Wash in running tap water for 3 minutes.
6. Place in 1.25% picric acid for 3 minutes.
7. Rinse in distilled water.
8. Decolorize in 95% alcohol for 30 seconds.
9. Rinse in distilled water.
10. Place in 5% crystal violet for 3 minutes.
11. Rinse in distilled water.
12. Place in Gram's iodine for 1 minute.
13. Rinse in distilled water. Blot, but leave moist.
14. Decolorize in a solution of equal parts of aniline and xylene, several changes, until section appears light purplish-red.
15. Dehydrate slides in 95% alcohol and absolute alcohol 3 changes each. Clear in xylene, 3 changes.
16. Mount coverglass with appropriate medium (refractive index 1.48-1.56).

RESULTS

Gram-positive microorganisms	blue
Gram-negative microorganisms	red
Other elements	various shades of red to purple

Taylor's Method for Gram-Positive and Gram-Negative Organisms[58]

FIXATION 10% Buffered neutral formalin.
PROCESS Paraffin.
MICROTOMY Cut sections at 6 micrometers.
CONTROL Use control number 20.

SOLUTIONS

Harris' Hematoxylin
(See Page 81)

1% Acid Alcohol
(See Page 77)

Hucker's Solution

Crystal violet, 10% alcoholic	2.0	ml
Distilled water	18.0	ml
Ammonium oxalate, 1% aqueous	80.0	ml

Mix and filter before use.

Basic Fuchsin (Stock)

Basic fuchsin	0.1	gm
Methyl alcohol	95.0	ml
Distilled water	5.0	ml

Basic Fuchsin (Working)

Basic fuchsin (stock)	5.0	ml
Distilled water	60.0	ml

Saturated Lithium Carbonate

Lithium carbonate, approx.	0.80	gm
Distilled water	100.0	ml

Ethyl Ether-Acetone

Ethyl ether	50.0	ml
Acetone	50.0	ml

0.1% Picric Acid-Acetone

Picric acid	0.1	gm
Acetone	100.0	ml

Acetone-Xylene #1

Acetone	1	part
Xylene	2	parts

Acetone-Xylene #2

Acetone	1	part
Xylene	3	parts

STAINING PROCEDURE

1. Decerate slides in xylene, 2 changes, 2 minutes each. Place in absolute alcohol, 2 changes, 2 minutes each; 95% alcohol, 2 changes, 2 minutes each; then rinse in distilled water.
2. Stain in Harris' hematoxylin for 5 minutes.
3. Wash in running tap water for 1 minute.
4. Decolorize in 1% hydrochloric acid alcohol.
5. Wash in running tap water for 3 minutes.
6. Blue in saturated lithium carbonate. From this point on, stain only one slide at a time.

7. Wash in running tap water for 5 minutes.
8. Stain in Hucker's solution for 2 minutes.
9. Wash quickly in water.
10. Mordant in Gram's iodine for 1 minute.
11. Wash in water. Blot, but do not allow to dry, using filter paper moistened with water.
12. Decolorize in ethyl ether-acetone until blue color no longer comes off. Blot, but do not allow to dry, using filter paper moistened with ethyl ether-acetone.
13. Stain in working basic fuchsin for 3 minutes.
14. Wash in water. Blot, but do not allow to dry, using filter paper moistened with water.
15. Dip in acetone until sections begin to decolorize.
16. Pass quickly to picric acid-acetone to decolorize until section becomes reddish-brown to yellow, approximately 15 seconds.
17. Pass quickly through acetone and xylene solution #1, then acetone and xylene solution #2.
18. Dehydrate slides in 95% alcohol and absolute alcohol 3 changes each. Clear in xylene, 3 changes.
19. Mount coverglass with appropriate medium (refractive index 1.48-1.56).

RESULTS

Gram-positive microorganisms	blue to blue-black
Gram-negative microorganisms	bright red
Nuclear chromatin	brownish-red
Erythrocytes	red to greenish-yellow
Necrotic tissue	yellowish-green
Cytoplasm	yellow
Connective tissue	red

Churukian-Schenk Method for Gram-Positive and Gram-Negative Organisms[59]

FIXATION 10% Buffered neutral formalin.
PROCESS Paraffin.
MICROTOMY Cut section at 5 micrometers.
CONTROL Use control number 20.
SOLUTIONS

10% Crystal Violet (Stock)

Crystal violet	10.0	gm
Absolute alcohol	100.0	ml

1% Ammonium Oxalate (Stock)

Ammonium oxalate	1.0	gm
Distilled water	100.0	ml

Hucker-Conn Crystal Violet (Working)

10% Crystal violet (stock)	2.0	ml
1% Ammonium oxalate (stock)	98.0	ml

Modified Gram's Iodine

Iodine	2.0	gm
Potassium iodide	4.0	gm
Distilled water	20.0	ml
Dissolve the above completely and add:		
Distilled water	400.0	ml

Alcohol-Acetone

100% Alcohol	100.0	ml
Acetone	100.0	ml

0.5% Basic Fuchsin (Stock)

Basic fuchsin	0.5	gm
Distilled water	100.0	ml

Place on a magnetic stirrer. Agitate and heat until completely dissolved.

Basic Fuchsin (Working)

Basic fuchsin (stock)	5.0	ml
Distilled water	45.0	ml

0.1% Picric Acid-Acetone

Picric acid	0.1	gm
Acetone	100.0	ml

Acetone-Xylene

Acetone	100.0	ml
Xylene	100.0	ml

STAINING PROCEDURE

1. Decerate slides in xylene, 2 changes, 2 minutes each. Place in absolute alcohol, 2 changes, 2 minutes each; 95% alcohol, 2 changes, 2 minutes each; then rinse in distilled water.
2. Stain in modified Hucker-Conn crystal violet for 2 minutes.
3. Wash briefly in running tap water.
4. Place in modified Gram's iodine for 1 minute.
5. Wash briefly in running tap water.
6. Blot the slide but not the tissue section (see remarks).
7. Decolorize with alcohol-acetone until blue color no longer comes off. This usually requires 8-10 dips.
8. Stain in working basic fuchsin for 1 minute. Dip several times before timing.
9. Rinse in distilled water.
10. Blot the slide but not the tissue (see remarks).
11. Place in acetone for 3 seconds.
12. Differentiate in picric acid-acetone for 10 seconds.
13. Dip a few quick times in acetone-xylene.
14. Dehydrate slides in 95% alcohol and absolute alcohol 3 changes each. Clear in xylene, 3 changes.
15. Mount coverglass with appropriate medium (refractive index 1.48-1.56).

RESULTS

Gram-positive microorganisms	blue to blue-black
Gram-negative microorganisms	red

Filaments of nocardia and actinomyces....................blue or blue and red
Nuclei, elastic fibers, paneth cells .... red
Other tissue elements .... yellow

REMARKS
The only problem which is occasionally encountered is an absence of, or inadequate differentiation of, the staining achieved by basic fuchsin. This problem occurs if the tissue either partially or completely dries out after being stained with basic fuchsin. The part of the tissue section which has dried out cannot be decolorized and remains red. This procedure will not demonstrate Legionella pneumophila.

Brown-Hopps Method for Gram-Positive and Gram-Negative Microorganisms[60]

(See Fig. 44, 45, 46, 47 and 48)

FIXATION	10% Buffered neutral formalin, glutaraldehyde, paraform-aldehyde, Zenker's or Bouin's solution.
PROCESS	Paraffin, frozen, or celloidin.
MICROTOMY	Cut sections at 6 micrometers.
CONTROL	Use control number 20.
SOLUTIONS	

1% Crystal Violet

Crystal violet	1.0	gm
Distilled water	100.0	ml

0.25% Basic Fuchsin (Stock)

Basic fuchsin	0.25	ml
Distilled water	100.0	ml

Filter solution through #4D Whitman paper.

Basic Fuchsin (Working) A
(For tissues fixed in glutaraldehyde and Bouin's solution)

Basic fuchsin (stock)	1.0	ml
Distilled water	25.0	ml

Basic Fuchsin (Working) B
(For tissues fixed in formalin, Zenker's or formal-Zenker's solution)

Basic fuchsin	5.0	ml
Distilled water	25.0	ml

Gram's Iodine
(See Page 187)

Gallego's Differentiating

Distilled water	50.0	ml
Concentrated formalin	1.0	ml
Glacial acetic acid	0.5	ml

0.1% Picric Acid-Acetone

Picric acid	0.5	gm
Acetone	1000.0	ml

Acetone-Xylene

Acetone	50.0	ml
Xylene	50.0	ml

STAINING PROCEDURE

1. Decerate slides in xylene, 2 changes, 2 minutes each. Place in absolute alcohol, 2 changes, 2 minutes each; 95% alcohol, 2 changes, 2 minutes each; then rinse in distilled water.
2. Place slides in crystal violet solution for 2 minutes.
3. Rinse in distilled water to remove excess stain.
4. Mordant in Gram's iodine solution for 5 minutes.
5. Rinse in running tap water to remove excess iodine.
6. Blot, but do not allow to dry.
7. Decolorize in acetone until blue color no longer streams away from the section (approximately 5 to 10 seconds).
8. Quickly rinse in running tap water to remove acetone.
9. Place in selected basic fuchsin solution (A or B) for 5 minutes.
10. Rinse in running tap water.
11. Place in Gallego's differentiating solution for 5 minutes (differentiates and "fixes" the basic fuchsin).
12. Rinse thoroughly in running tap water.
13. Dip in acetone, 3 quick dips.
14. Dip in picric acid-acetone, 3 quick dips.
15. Dip in acetone, 3 quick dips.
16. Dip in acetone-xylene, 5 quick dips.
17. Place in xylene, 3 changes, 10 dips each. Slides may remain in xylene until ready for mounting.
18. Mount coverglass with appropriate medium (refractive index 1.48-1.56).

RESULTS

Gram-positive microorganisms	blue
Gram-negative microorganisms	red
Background	yellow

REMARKS

This technique has proven consistently reliable in identifying and differentiating both Gram-positive and Gram-negative microorganisms in sections of tissue from a variety of lesions in man and animals. Other species of Gram-positive and Gram-negative bacteria that have been demonstrated with this method include Pasteurella pestis, aerobacter-klebsiella, bacillus anthracis, staphylococcus aureus, streptococcus, pseudomonas aeruginosa, mycobacterium tuberculosis, mycobacterium leprae, hemophilus influenza, and Legionella pneumophila.

This stain is also very useful as a screening method for the following infectious agents: those causing actinomycosis, nocardiosis, coccidiomycoses, blastomycosis, cryptococcosis, aspergillosis, rhinoporidiosis, condidiasis, amebiasis, balantidiasis, pneumocytosis, enterobiasis, and paragonimiasis (ova).

Greer-Chandler-Hicklin Method for Legionella Pneumophila-Rickettsia, Chlamydia and Bacteria[61]

(See Fig. 49)

The Giménez stain, originally developed for the demonstration of rickettsiae, readily stains the Legionnaires' disease bacterium (Legionella pneumophila) in frozen sections and smears of fresh or formalin fixed tissue.

FIXATION	10% Buffered neutral formalin.
PROCESS	Frozen.
CRYOTOMY	Cut sections at 4 to 6 micrometers. Pick up sections on gelatinized slides, or other good section adhesive. Air dry. Place sections in 60°C oven for 15 to 30 minutes before staining.
CONTROL	Use control number 29.
SOLUTIONS	

Phenol-Gelatin Adhesive

Gelatin	2.0	gm
Distilled water	400.0	ml
Heat until gelatin is dissolved and solution is clear; then add		
Phenol (melted)	2.24	ml

Store at room temperature. If a precipitate forms during storage, heat and redissolve. To coat slides, heat this solution until melted. Place in a warm coplin jar, dip slides to coat and air dry. Slides may be pretreated and stored in a refrigerator.

10% Alcoholic Basic Fuchsin (Stock)

Basic fuchsin	10.0	gm
95% Ethyl alcohol	100.0	ml

4% Phenol (Stock)

Phenol crystals	4.0	gm
Distilled water	100.0	ml

Carbol Fuchsin (Stock)

10% Alcoholic basic fuchsin	100.0	ml
4% Phenol	250.0	ml
Distilled water	650.0	ml

Stir until the basic fuchsin dissolves and then hold at 37°C for 48 hours.

0.1M Sodium Phosphate Monobasic (Stock)

Sodium phosphate monobasic (anhydrous)	3.8	gm
Distilled water	100.0	ml

0.1M Sodium Phosphate Dibasic (Stock)

Sodium phosphate dibasic (anhydrous)	14.2	gm
Distilled water	1000.0	ml

0.1M Phosphate Buffer pH 7.5 (Stock)

0.1M Sodium phosphate, monobasic (stock)	35.0	ml
0.1M Sodium phosphate, dibasic (stock)	155.0	ml

Carbol Fuchsin (Working)

Carbol fuchsin (stock)	10.0	ml
0.1M Phosphate buffer pH 7.5 (stock)	25.0	ml

Filter; use within 48 hours.

0.8% Malachite Green

Malachite green	0.8	gm
Distilled water	100.0	ml

STAINING PROCEDURE

1. Stain on a staining rack with freshly filtered working carbol fuchsin for 2 minutes.
2. Wash well in running tap water.
3. Stain with malachite green for 15 to 20 seconds.
4. Wash well in running tap water. Sections should appear blue-green; if they are red, restain with Malachite green until they are blue-green.
5. Blot sections gently with filter paper; allow to air dry.
6. Dehydrate slides quickly in absolute alcohol and clear in xylene, 2 changes each.
7. Mount coverglass with appropriate medium (refractive index 1.48-1.56).

RESULTS

Legionella pneumophila	bright red
Rickettsiae	bright red
Chlamydia	bright red
Nuclei	greenish-blue
Background	light green

Roy's Method for Legionella Pneumophila[62]

FIXATION 10% Buffered neutral formalin.
PROCESS Paraffin, frozen, or celloidin.
MICROTOMY Cut section at 5 micrometers.
CONTROL Use control number 29.
SOLUTIONS

5% Silver Nitrate

Silver nitrate	5.0	gm
Distilled water	100.0	ml

3% Methenamine

Methenamine*	3.0	gm
Distilled water	100.0	ml

*Hexamethylenatetramine

10% Potassium Dichromate

Potassium dichromate	10.0	gm
Distilled water	100.0	ml

3% Sodium Metabisulfite

Sodium metabisulfite	3.0	gm
Distilled water	100.0	ml

2% Gold Chloride

Gold chloride	2.0	gm
Distilled water	100.0	ml

5% Sodium Thiosulfate

Sodium thiosulfate	5.0	gm
Distilled water	100.0	ml

2% Neutral Red

Neutral red	2.0	gm
Distilled water	100.0	ml

5M Boric Acid (Stock)

Boric acid	12.4	gm
Distilled water	1000.0	ml

20M Borax (Stock)

Borax	19.1	gm
Distilled water	1000.0	ml

Gomori-Burtner Methenamine Silver (Stock)

5% Silver nitrate	5.0	ml
3% Methenamine	100.0	ml

Gomori-Burtner Methenamine Silver (Working)

5M Boric acid (stock)	8.0	ml
20M Borax	2.0	ml
Gomori-Burtner methenamine silver	30.0	ml

STAINING PROCEDURE

1. Decerate slides in xylene, 2 changes, 2 minutes each. Place in absolute alcohol, 2 changes, 2 minutes each; 95% alcohol, 2 changes, 2 minutes each; then rinse in distilled water.
2. Place in 10% potassium dichromate for 1 hour at room temperature.
3. Wash in running tap water for 5 minutes.
4. Place in 3% sodium metabisulfite for 2 minutes.
5. Wash in running tap water for 2 minutes.
6. Place in working Gomori-Butner methenamine silver in a coplin jar and place in hot-air oven at 56°C until pale golden-yellow; approximately 3 to 4 hours.
8. Wash 3 times in distilled water.
9. Tone in 2% gold chloride for 2 minutes.
10. Place in 3 changes distilled water, for 10 seconds each.
11. Fix in 5% sodium thiosulfate for 2 minutes.
12. Wash in running tap water for 3 minutes.
13. Counterstain in 1% neutral red for 2 minutes.
14. Dehydrate slides in 95% alcohol and absolute alcohol 3 changes each. Clear in xylene, 3 changes.
15. Mount coverglass with appropriate medium (refractive index 1.48-1.56).

RESULTS

Legionella pneumophila	black
Nucleoli	red-black
Collagen	silver

Gridley's Method for Fungi[55h]

(See Fig. 50 and 51)

FIXATION	10% Buffered neutral formalin.
PROCESS	Paraffin, frozen, or celloidin.
MICROTOMY	Cut sections at 6 micrometers.
CONTROL	Use control number 18.

SOLUTIONS

Normal Hydrochloric Acid

Hydrochloric acid, (sp. gr. 1.19)	83.5	ml
Distilled water	916.5	ml

4% Chromic Acid

Chromic acid (chromium trioxide)	4.0	gm
Distilled water	100.0	ml

Coleman's Feulgen Reagent

Dissolve 1.0 gm basic fuchsin in 200.0 ml hot distilled water. Bring to a boil. Cool and add 2.0 gm potassium metabisulfite, 10.0 ml normal hydrochloric acid. Let bleach for 24 hours, then add 0.5 gm activated carbon. Shake for 1 minute and filter through coarse filter paper. Repeat filtration until solution is colorless. Store in refrigerator.

Aldehyde-Fuchsin

Basic fuchsin*	1.0	gm
70% Alcohol	200.0	ml
Concentrated hydrochloric acid	2.0	ml
Paraldehyde	2.0	ml

Let stand at room temperature for 2 to 3 days or until stain is deep purple in color.

*Use pararosanilin in place of basic fuchsin to make this solution (see remarks).

0.25% Metanil Yellow

Metanil yellow	0.25	gm
Distilled water	100.0	ml
Glacial acetic acid	0.25	ml

STAINING PROCEDURE

1. Decerate slides in xylene, 2 changes, 2 minutes each. Place in absolute alcohol, 2 changes, 2 minutes each; 95% alcohol, 2 changes, 2 minutes each; then rinse in distilled water.
2. Oxidize in 4% chromic acid for 1 hour.
3. Wash in running tap water for 5 minutes.
4. Place in Coleman's leucofuchsin for 15 minutes.
5. Wash in running tap water for 15 minutes.

6. Rinse in 70% alcohol, several changes.
7. Place in aldehyde-fuchsin solution for 30 minutes.
8. Rinse off excess stain with 95% alcohol.
9. Rinse in distilled water.
10. Counterstain lightly with metanil yellow solution for 1 minute.
11. Rinse in distilled water.
12. Dehydrate slides in 95% alcohol and absolute alcohol 3 changes each. Clear in xylene, 3 changes.
13. Mount coverglass with appropriate medium (refractive index 1.48-1.56).

RESULTS

Mycelia	deep rose-purple
Candida	deep rose to purple
Background	yellow
Elastic tissue and mucin	deep purple

REMARKS

Basic fuchsin is not a pure dye. It can be a mixture of pararosanilin, rosanilin, magenta II and new fuchsin. Pararosanilin can be substituted for basic fuchsin in all staining methods. It must be used to prepare this aldehyde fuchsin.

The morphologic detail of yeast forms and hyphae are usually visible, although very old fungi (which were probably nonviable at the time of fixation) are not as well stained as by the Grocott's methenamine silver technique. The filaments of nocardia and actinomyces are not stained by this method.

Grocott's Method for Fungi (GMS)[55i]

(See Fig. 52, 53, 54, 55, 56, 57 and 58)

PROCESS	Paraffin, frozen, or celloidin.
MICROTOMY	Cut sections at 6 micrometers.
CONTROL	Use control number 18.
SOLUTIONS	

4% Chromic Acid
(See Page 199)

5% Silver Nitrate
(See Page 197)

3% Methenamine
(See Page 197)

5% Borax

Borax (sodium borate dehydrate)	5.0	gm
Distilled water	100.0	ml

Methenamine-Silver Nitrate (Stock)

5% Silver nitrate	5.0	ml
3% Methenamine	100.0	ml

A white precipitate forms but immediately dissolves upon shaking. Clear solution remains usable for months. Store in refrigerator.

Methenamine-Silver Nitrate (Working)

Methenamine-silver nitrate (stock)	25.0	ml
Distilled water	25.0	ml
5% Borax	2.0	ml

Make fresh. Mix borax with distilled water, then add the methenamine-silver nitrate solution.

1% Sodium Bisulfite

Sodium bisulfite	1.0	gm
Distilled water	100.0	ml

0.1% Gold Chloride

1% Gold chloride	10.0	ml
Distilled water	90.0	ml

This solution may be used repeatedly.

2% Sodium Thiosulfate (Hypo)

Sodium thiosulfate	2.0	gm
Distilled water	100.0	ml

0.2% Light Green (Stock)

Light green, SF yellowish	0.2	gm
Distilled water	100.0	ml
Glacial acetic acid	0.2	ml

Light Green (Working)

Light green solution (stock)	10.0	ml
Distilled water	50.0	ml

STAINING PROCEDURE

1. Decerate slides in xylene, 2 changes, 2 minutes each. Place in absolute alcohol, 2 changes, 2 minutes each; 95% alcohol, 2 changes, 2 minutes each; then rinse in distilled water.
2. Oxidize in 4% chromic acid solution for 1 hour.
3. Wash in running tap water for 2 minutes.
4. Place in 1% sodium bisulfite solution for 1 minute to remove any residual chromic acid.
5. Wash in running tap water for 5 minutes.
6. Rinse in 4 changes of distilled water.
7. Place in freshly mixed working methenamine-silver nitrate solution then in oven at 58-60°C for 60 minutes or until section turns yellowish-brown.
8. Rinse in 6 changes of distilled water.
9. Tone in gold chloride solution for 2 minutes.
10. Rinse in distilled water.
11. Remove unreduced silver with 2% sodium thiosulfate (hypo) for 2 minutes.

12. Wash in running tap water for 3 minutes.
13. Counterstain with working light green solution for 30 seconds.
14. Dehydrate slides in 95% alcohol and absolute alcohol 3 changes each. Clear in xylene, 3 changes.
15. Mount coverglass with appropriate medium (refractive index 1.48-1.56).

RESULTS

Fungi	sharply delineated in black
Nocardia and actinomyces	black filaments
Mucin	taupe to dark grey
Inner parts of mycelia and hypha	rose
Background	pale green

REMARKS

This technique gives all forms of fungi a black-brown coloration. In staining fungi, especially filaments of nocardia, it has been found that the time of exposure to methenamine-silver nitrate solution, for complete and proper development, may have to be varied according to type and/or strains suspected. Therefore, if nocardia is suspected, two slides should be stained (one for 60 minutes and one for 90 minutes), thereby assuring a positive demonstration of any filaments of nocardia that may exist.

Slides previously stained with most other stains may be used by removing coverglasses and staining as above. The chromic acid treatment will remove any existing stain.

This technique has proven useful as a scanning method for identifying entamoeba histolytica in tissue sections, since the methanamine-silver nitrate stains the glycogen present in amoeba. To stain ingested red blood cells in ameoba, place sections in alcoholic eosin for 2 minutes after step 1.

Pickett's Fluorescence Method for Fungi[55j]

FIXATION	10% Buffered neutral formalin.
PROCESS	Paraffin, frozen, or celloidin.
MICROTOMY	Cut sections at 6 micrometers.
CONTROL	Use control number 18.
SOLUTIONS	

Weigert's Iron-Hematoxylin
(See Page 79)

0.1% Acridine Orange

Acridine orange	0.1	gm
Distilled water	100.0	ml

STAINING PROCEDURE
1. Decerate slides in xylene, 2 changes, 2 minutes each. Place in absolute alcohol, 2 changes, 2 minutes each; 95% alcohol, 2 changes, 2 minutes each; then rinse in distilled water.
2. Place in Weigert's iron-hematoxylin solution for 5 minutes.

3. Wash in running tap water for 4 minutes.
4. Immerse in acridine orange solution for 2 minutes.
5. Rinse in distilled water.
6. Dehydrate slides in 95% alcohol and absolute alcohol 3 changes each. Clear in xylene, 3 changes.
7. Mount coverglass with non-fluorescing media.

RESULTS

Fungi .... brightly fluorescent

Mahan-Sale's Method for Pneumocystis Carinii and Fungi[63]

FIXATION 10% Buffered neutral formalin.
PROCESS Paraffin, frozen, celloidin, aspirates and touch preparations.
MICROTOMY Cut sections at 6 micrometers.
CONTROL Use control number 37A.
SOLUTIONS

10% Chromic Acid

Chromic acid (chromium trioxide)	10.0	gm
Distilled water	100.0	ml

5% Silver Nitrate (Stock)
(See Page 197)

3% Methenamine (Stock)
(See Page 197)

5% Borax (Stock)
(See Page 200)

Methenamine Silver (Working)

3% Methenamine (stock)	40.0	ml
5% Silver nitrate (stock)	2.0	ml
Shake and add:		
5% Borax (stock)	3.0	ml
Distilled water	35.0	ml

Heat solution to approximately 95°C until it becomes golden brown just before use. Do not overheat.

1% Gold Chloride

Place a 1 gm glass vial of gold chloride (from which paper and other extraneous materials have been removed) in a clean, heavy glass container with a glass stopper. Shake the container vigorously until the vial is broken. Add 100 ml distilled water.

1% Sodium Metabisulfite

Sodium metabisulfite	1.0	gm
Distilled water	100.0	ml

5% Sodium Thiosulfate

Sodium thiosulfate	5.0	gm
Distilled water	100.0	ml

Mayer's Hematoxylin
(See Page 81)

Eosin-Phloxine B
(See Page 87)

Light Green Solution (Stock)

Light green, SF yellowish	0.2	gm
Distilled water	100.0	ml

Light Green Solution (Working)

Stock light green	10.0	ml
Distilled water	100.0	ml

STAINING PROCEDURE

1. Decerate slides in 2 changes of xylene, 2 minutes each (air dry smear at this point); absolute alcohol, 2 changes, 2 minutes each; and 95% alcohol, 2 changes, 2 minutes each. Rinse in distilled water 3 changes, 2 minutes each.
2. Place in 10% chromic acid for 10 minutes. (Discard after use.)
3. Wash slides in running tap water for a few seconds to remove excess chromic acid.
4. Place in 1% sodium metabisulfite for 1 minute.
5. Rinse in hot running tap water (50°C) for 1 minute or until coplin jar is hot.
6. Pour hot (50°C) working methenamine silver solution on slides. Leave until sections are golden brown. This may take 2 minutes or longer.
7. Rinse in hot tap water and cool gradually with tap water to avoid cracking the coplin jar.
8. Rinse in distilled water.
9. Tone in 1% gold chloride for 10 seconds. (Save solution.)
10. Rinse in distilled water.
11. Place in 5% sodium thiosulfate for 3 minutes.
12. Wash well in running tap water.
13. Counterstain with routine hematoxylin-eosin staining or briefly in light green.
14. Dehydrate slides in 95% alcohol and absolute alcohol 3 changes each. Clear in xylene, 3 changes.
15. Mount coverglass with appropriate medium (refractive index 1.48-1.56).

RESULTS

Pneumocystis carinii	black
Fungi	black
Glycogen	black
Mucin	black
Melanin	black
Lipofuscin	black
Other Elements	dependant upon counterstain(s) used

Churukian-Schenk-Clark Method for Fungi and Pneumocystis Carinii[64]

FIXATION	10% Buffered neutral formalin. Smears and touch preparations should be fixed in 95% alcohol for 3 minutes.
PROCESS	Paraffin.
MICROTOMY	Cut sections at 5 micrometers.
CONTROL	Use control number 18 for fungi, 37A for Pneumocystis carinii.
SOLUTIONS	

5% Chromic Acid

Chromic acid (chromium trioxide)	5.0	gm
Distilled water	100.0	ml

1% Sodium Metabisulfite

Sodium metabisulfite	1.0	gm
Distilled water	100.0	ml

10% Silver Nitrate (Stock)

Silver nitrate	10.0	gm
Distilled water	100.0	ml

10% Potassium Hydroxide (Stock)

Potassium hydroxide	10.0	gm
Distilled water	100.0	ml

Gomori's Ammoniacal Silver

10% Silver nitrate	10.0	ml
10% Potassium hydroxide	2.0	ml

Add concentrated ammonia hydroxide drop by drop with constant shaking until the precipitate which forms dissolves. Add sufficient distilled water to bring the volume to 1000 ml. Store in a refrigerator.

0.2% Gold Chloride

Place a 1 gm vial of gold chloride (from which paper and other extraneous materials have been removed) in a clean, heavy 500 ml glass container with a glass stopper; shake the container vigorously until the vial is broken. Add 500 ml distilled water.

2% Sodium Thiosulfate

Sodium thiosulfate	2.0	gm
Distilled water	100.0	ml

Fast Green, (Stock)

Fast green, FCF	0.2	gm
Distilled water	100.0	ml
Glacial acetic acid	0.2	ml

Fast Green (Working)

Fast green (stock)	10.0	ml
Distilled water	100.0	ml

STAINING PROCEDURE

1. Decerate slides in xylene, 2 changes, 2 minutes each. Place in absolute alcohol, 2 changes, 2 minutes each; 95% alcohol, 2 changes, 2 minutes each; then rinse in distilled water.
2. Place in 5% chromic acid that has been preheated to 43°C in a water bath. Transfer solution and slides immediately to a 58°C water bath for 5 minutes. (Discard solution after use.)
3. Rinse in 3 changes of distilled water.
4. Treat with 1% sodium metabisulfite for 30 seconds.
5. Wash in running tap water for 30 seconds.
6. Rinse in 3 changes of distilled water.
7. Microwave 45 ml of refrigerated, cold ammoniacal silver at power level 7 for 1 minute. Immediately place the slides in this hot solution, return to the microwave oven and microwave at power level 1 (10% of 1450 watts or about 150 watts) for 1 minute. Allow the slides to remain in the solution for 1 more minute or until the sections become yellowish-brown. Check control under a wet scope to see that organisms are brown-black (see remarks).
8. Rinse in 4 changes of distilled water.
9. Tone in 0.2% gold chloride for 30 seconds.
10. Rinse in 2 changes of distilled water.
11. Place in 2% sodium thiosulfate for 30 seconds.
12. Rinse in 4 changes of distilled water.
13. Counterstain in working fast green solution for 30 seconds.
14. Rinse in 3 changes of distilled water.
15. Dehydrate slides in 95% alcohol and absolute alcohol 3 changes each. Clear in xylene, 3 changes.
16. Mount coverglass with appropriate medium (refractive index 1.48-1.56).

RESULTS

Pneumocystis carinii .... black
Fungi .... black
Actinomyces filaments .... black
Glycogen .... black
Mucin .... black
Lipofuscin .... black
Melanin .... black
Background .... pale green

REMARKS

If a microwave oven is not available, slides can be stained in 45 ml of ammoniacal silver in a glass coplin jar placed in a 43°C water bath for 2 minutes. Transfer immediately to a 58°C water bath for 15 minutes or until the sections become yellowish-brown. (To avoid cracking the coplin jar, allow it to warm up to near room temperature before placing in initial water bath. Alternatively, plastic coplin jars may be used. However, time of exposure to the warm solution may have to be extended by 10 minutes.)

Swartz-Medrek-Robboy Method for Fungi[65]

FIXATION	10% Buffered neutral formalin.
PROCESS	Paraffin.
MICROTOMY	Cut section at 6 micrometers.
CONTROL	Use control number 18.
SOLUTIONS	

0.1M Citric Acid

Citric acid, monohydrate	21.0	gm
Distilled water, q.s. to	1000.0	ml

0.2M Disodium Phosphate

Sodium phosphate, dibasic, anhydrous	28.4	gm
Distilled water, q.s. to	1000.0	ml

0.2M McIlvaine Buffer, pH 8.0

0.1M Citric acid	5.5	ml
0.2M Disodium phosphate	194.5	ml

This solution should be adjusted to pH 8.0 by adding the appropriate solution drop by drop and mixing well between pH measurements. If it is above pH 8.0, add more 0.1M citric acid; if below, add more 0.2M disodium phosphate.

Shear's Mounting Solution, pH 8.0

Potassium acetate	0.4	gm
0.2M McIlvaine buffer, pH 8.0	20.0	ml
Dissolve and add:		
Glycerin	120.0	ml
95% Ethyl alcohol	180.0	ml

Solution A

Shear's mounting solution, pH 8.0	50.0	ml
Rose bengal	0.25	gm
Dissolve and then add:		
Clove oil, USP	50.0	ml

Allow to stand at room temperature for 20 minutes and then pour equal amounts into two centrifuge tubes (50 ml). Use the 2 filled tubes to counterbalance each other and centrifuge at 3000 rpm for 10 minutes. Decant and save the supernatant. Discard the sediment.

Solution B

Evan's blue	0.25	gm
Physiologic saline	50.0	ml
Dissolve and add:		
Clove oil, USP	50.0	ml

Allow to stand at room temperature for 20 minutes and then centrifuge as described above at 3000 rpm for 10 minutes. Decant and save the supernatant. Discard the sediment.

STAINING PROCEDURE

1. Decerate slides in xylene, 2 changes, 2 minutes each. Place in absolute alcohol,

2 changes, 2 minutes each; 95% alcohol, 2 changes, 2 minutes each; then rinse in distilled water.
2. Rinse in running tap water.
3. Stain in solution A for 15 minutes.
4. Rinse slides in a gentle stream of running tap water.
5. Dry and allow slides to stand at room temperature for about 5 minutes.
6. Stain with solution B for 40 minutes.
7. Dehydrate slides in 95% alcohol and absolute alcohol 3 changes each. Clear in xylene, 3 changes.
8. Mount coverglass with appropriate medium (refractive index 1.48-1.56).

RESULTS

Fungi capsule .... dark blue
Fungal cytoplasm................................ usually light blue (see remarks)
Background .... rose-red

REMARKS

Occasionally, dark blue bodies are found in the cytoplasm of some hyphae and blasto spores of Candida albicans.

This stain is easy to prepare and use and clearly differentiates the fungi from the invaded tissue. This stain is also of value because, unlike most other stains for fungi, the cytoplasm is stained and toned differently from the cell wall. The reasons for the reproducible variations in the staining procedure (the dark blue wall, light blue and occasionally, rose-red cytoplasm) are unknown.

Actinomyces israelii and nocardia asteroides cannot be identified with this stain. This may be additional evidence that the organisms in these two genera are not true fungi but bacteria.

Churukian-Schenk Method for Pneumocystis Carinii and Fungi[64]

Cryostat sections of tissue may be used, but should be fixed briefly with 10% buffered neutral formalin. Satisfactory results can be obtained on smears and touch preparations. These should be fixed with 95% alcohol for 3 minutes. The method is also useful with paraffin-embedded tissue sections.

FIXATION 10% Buffered neutral formalin. Smears and touch preparations should be fixed in 95% ethyl alcohol for 3 minutes.
PROCESS Paraffin, frozen smears, or touch preparations.
MICROTOMY Cut section at 5 to 6 micrometers.
CONTROL Use control number 37A.
SOLUTIONS

5% Chromic Acid

Chromic acid (chromium trioxide)	5.0	gm
Distilled water	100.0	ml

1% Sodium Metabisulfite (Stock)

Sodium metabisulfite	1.0	gm
Distilled water	100.0	ml

3% Methenamine (Stock)

Methenamine (hexamethylenetetramine)	3.0	gm
Distilled water	100.0	ml

Silver nitrate	5.0	gm
Distilled water	100.0	ml

5% Borax (Stock)

Borax (sodium borate, dehydrate)	5.0	gm
Distilled water	100.0	ml

Methenamine Silver (Stock)

3% Methenamine	100.0	ml
5% Silver nitrate	5.0	ml

Shake to dissolve precipitate. Store in refrigerator.

Methenamine Silver (Working)

Methenamine silver (stock)	25.0	ml
Distilled water	25.0	ml
5% Borax	2.0	ml

Fast Green (Stock)

Fast green, FCF	0.2	gm
Distilled water	100.0	ml
Glacial acetic acid	0.2	ml

Fast Green (Working)

Fast green (stock)	10.0	ml
Distilled water	40.0	ml

0.2% Gold Chloride
(See Page 205)

2% Sodium Thiosulfate

Sodium thiosulfate	2.0	gm
Distilled water	100.0	ml

STAINING PROCEDURE

1. Use fixed smears or cryostat sections brought to water, or decerate slides in 2 changes of xylene, 2 minutes each; absolute alcohol 2 changes, 2 minutes each; and 95% alcohol 2 changes, 2 minutes each. Rinse in distilled water, 2 changes.
2. Preheat 5% chromic acid and working methenamine in a 43°C water bath for 15 minutes in coplin jars.
3. Transfer 5% chromic acid to a 58°C water bath and add slides.
4. Oxidize slides in 5% chromic acid for 5 minutes.
5. Wash in running tap water for 2 minutes.
6. Place in 1% sodium metabisulfate for 30 seconds.
7. Wash in running tap water for 15 seconds.

8. Rinse in 4 changes of distilled water.
9. Place slide in container of freshly made working methenamine silver at 43°C and set container in a 58°C water bath for 20 minutes. Check controls microscopically to be sure organisms are demonstrated.
10. Rinse in 4 changes of distilled water.
11. Tone in 0.2% gold chloride for 30 seconds.
12. Rinse in 2 changes of distilled water.
13. Place in 2% sodium thiosulfate for 30 seconds.
14. Wash in running tap water for 15 seconds.
15. Counterstain with working fast green for 30 seconds.
16. Wash briefly in running tap water and rinse in 2 changes of distilled water.
17. Dehydrate slides in 95% alcohol and absolute alcohol 3 changes each. Clear in xylene, 3 changes.
18. Mount coverglass with appropriate medium (refractive index 1.48-1.56).

RESULTS

Fungi and Pneumocystis carinii .... black
Glycogen, mucin .... rose taupe to grey
Background .... pale green

REMARKS

Two water baths, one at 43°C and the other at 58°C, are used to prevent coplin jars from cracking, which tends to occur when they are placed directly in a 58°C water bath. All solutions in the procedure keep well at room temperature except the methenamine-silver nitrate, which keeps for about a month when stored in the refrigerator.

Hummer's Method for Pneumocystis Carinii[66]

FIXATION Saccomanno's fixative (2% v/v polyethylene glycol 1450: 50% ethanol).
PROCESS Cytospin preps, or sputum smears.
CONTROL Use control number 37A.
SOLUTIONS

Toluidine blue O

Toluidine blue O	0.17	gm
Distilled water	60.0	ml
Concentrated hydrochloric acid	2.0	ml
100% Ethyl alcohol	140.0	ml

Dissolve dye in water, add hydrochloric acid, then the alcohol. Solution is stable for 1 year. Filter solution before use.

Sulfation Solution

Glacial acetic acid (chilled to 10°C)	45.0	ml
Concentrated sulfuric acid	15.0	ml

Slowly and carefully add concentrated sulfuric acid to chilled glacial acetic acid and mix. Perform this function under a hood. Discard after 1 week.

STAINING PROCEDURE

1. Place slides in sulfation solution for 5 minutes.
2. Wash in running tap water for 1 minute.

3. Place in toluidine blue O for 3 minutes.
4. Dehydrate slides in 95% alcohol and absolute alcohol 3 changes each. Clear in xylene, 3 changes.
5. Mount coverglass with appropriate medium (refractive index 1.48-1.56).

RESULTS

Pneumocystis carinii cysts	purple
Pulmonary macrophages	pale yellow
Background	green and pale blue

Genaw's Method for Pneumocystis Carinii[67]

(See Fig. 59)

FIXATION	10% Buffered neutral formalin. Fix air dried smears in absolute alcohol.
PROCESS	Paraffin, smears.
MICROTOMY	Cut sections at 4 micrometers.
CONTROL	Use control number 37A.
SOLUTIONS	

10% Cresyl Violet Acetate (Stock)

Cresyl violet acetate	10.0	gm
Distilled water	100.0	ml

Buffer Solution

Sodium phosphate, monobasic	0.2	gm
Distilled water	76.0	ml
Concentrated hydrochloric acid	0.7	ml

Make fresh. Mix sodium phosphate in water until dissolved. Slowly and carefully add the hydrochloric acid. Discard after use.

Cresyl Violet Acetate, pH 1.5 (Working)

Buffer solution	50.0	ml
10% Cresyl violet acetate	0.5	ml

Make fresh, filter, check pH and discard after use.

1% Acetic Acid

Glacial acetic acid	2.0	ml
Distilled water	198.0	ml

Martius Yellow

1% Acetic acid solution	100.0	ml
Martius yellow	0.01	gm

Sulfation Reagent

Make fresh. Place a coplin jar containing 30.0 ml of concentrated glacial acetic acid in a beaker of cold water. Add 10.0 ml of concentrated sulfuric acid very slowly and mix well. Perform this function under a hood. Use within an hour.

STAINING PROCEDURE

1. Decerate slides in xylene, 2 changes, 2 minutes each. Place in absolute alcohol, 2 changes, 2 minutes each; 95% alcohol, 2 changes, 2 minutes each; then rinse in distilled water.
2. Place in sulfation solution for 10 minutes. Agitate occasionally to remove air bubbles.
3. Rinse in running tap water for 5 minutes.
4. Place in working cresyl violet acetate for 30 minutes (check microscopically for metachromatic color).
5. Rinse in running tap water for 3 minutes.
6. Place in martius yellow for 1 minute.
7. Dehydrate slides in 95% alcohol and absolute alcohol 3 changes each. Clear in xylene, 3 changes.
8. Mount coverglass with appropriate medium (refractive index 1.48-1.56).

RESULTS

Pneumocystis carinii cysts, mucin, cartilage .... rose-purple
Other fungi .... light rose-purple
Connective tissue .... blue-purple to green
Cytoplasm, erythrocytes .... yellow

REMARKS

The slides must not be allowed to go past absolute alcohol before exposure to the fresh sulfation solution. Staining in cresyl violet may take longer than 30 minutes in some cases. The slides should be checked microscopically.

Other Staining Methods for Pneumocystis Carinii

Pneumocystis carinii is an organism that appears to be widespread in the environment and is probably acquired by the airborne route; i.e., the respiratory route from carriers or individuals with infections. Pneumocystis carinii causes interstitial pneumonia in infants, children and individuals with congenital or acquired immunodeficiency diseases. It is for the latter reason that Pneumocystis carinii has assumed increasing importance with the emergence of AIDS in the last few years. Organisms appear to multiply both as trophozoites and within cysts. Both trophozoites and cysts are found in alveoli of the lung. Trophozoites are 2 to 5 µm in diameter and are round or cup shaped.

The diagnosis of Pneumocystis carinii depends on its demonstration in lung tissue. It is for this reason the results of the following stains are presented here. Also presented is a schematic of an alveolar space containing Pneumocystis organisms to provide a better understanding of the general overall morphology of this disease process.

Hematoxylin and Eosin (H&E)

Hematoxylin and eosin stained sections reveal numerous Pneumocystis organisms within the alveolar spaces, and the intracystic bodies are slightly visible within the cysts. The staining of the intracystic bodies by hematoxylin-eosin fails to give a good contrast for diagnosis.

Periodic Acid-Schiff Reaction (PAS)

The PAS procedure stains the walls of the cysts. Some of the cysts appear smaller than others and are approximately the size of Klebsiella (2 to 3 micra in diameter). With this stain, occasional small bluish-stained dots, which might represent a nucleus, can be seen in some organisms. Inside some of the larger cysts, intracystic bodies are stained slightly grayish-blue. Macrophages may be present near groups of cysts containing PAS positive material, some in the form of rounded bodies which may represent phagocytosed cysts or elementary bodies of the Pneumocystis. The PAS stain with diastase digestion shows no change from the PAS without digestion.

Gomori Methenamine Silver (GMS)

The GMS silver stain is very useful for demonstrating the cyst walls. It also stains bodies within the cysts, which take the form of rounded circles or parentheses-like structures. In addition, portions of the cyst wall stain in a denser fashion in weak GMS preparations. In deeper staining GMS preparations, the cyst wall shows no characteristic denser zone of staining. In these stronger preparations, the intracystic body is densely stained and sometimes appears to be double. Within the alveolar exudate some erythrocytes are present which stain faintly with the GMS and have a faint body in the central portions, therefore simulating the cysts of Pneumocystis carinii.

Giemsa

In a well-stained Giemsa preparation, the cyst walls stain faintly blue, and the intracystic bodies are well delineated and stain deep blue to purple.

Gallocyanin (Einarson's Method)

A Gallocyanin stain is of extreme interest because it stains the intracystic bodies without staining the cyst wall. These bodies are seen within the foamy material in the alveolar spaces. Occasionally some of these intracystic bodies, can be seen in the cytoplasms of macrophages. The Gallocyanin stain is of special interest because close observation suggests that the alveolar lining cells are entering the alveolar spaces and phagocytosing the organisms within the alveolar spaces.

Weigert Method for Fibrin

In many of the organisms, the Weigert stain reveals dense bluish purple stained foci in the rim of some of the cyst walls. In some areas it appears that there is an eosinophilic cytoplasmic mass extending from this dense rim into the cyst. In some of the cyst, the densely stained body appears in the more central position within the cyst and occasionally has the configuration and staining properties of a small nucleus. The crystal violet stains the walls of the cyst and also the intracystic bodies, faintly. The densely stained portions of the cyst walls appear to represent nuclear material of the microorganisms. Within the alveolar spaces there are areas containing filamentous purplish-stained fragments that appear to be fibrin.

Warthin-Starry

The Warthin-Starry stains the walls of the cyst and intracystic bodies. The Warthin-Starry also demonstrates the small forms which are about the size of Klebsiella (2 to 3 micra) in cross sections and may represent elementary bodies, or possibly very early small cysts.

Bodian-Van Gieson

A Bodian-Van Gieson stains the walls of the cysts brown to black, but the intracystic bodies are not stained well.

Bodian-Gallocyanin

A Bodian-Gallocyanin stains the walls of the cyst black; it also stains the small extracystic bodies. In focal areas, one can see that these small elementary bodies are quite numerous outside the cysts, but may also occur inside the cysts.

Bodian-Alcian Blue

A Bodian-alcian blue stains the walls of the cyst gray to black, but the intracystic bodies are not well stained.

PAS-Gallocyanin

A combination PAS-Gallocyanin stains the cyst walls well, but does not stain the intracystic bodies well.

PAS-Giemsa

A combination PAS-Giemsa stains the cyst walls well, but does not stain the intracystic bodies well.

Brown-Brenn

A Brown and Brenn Gram stain fails to stain either the cyst walls or the intracystic bodies.

Masson

The Masson stains the cyst walls bluish-purple and the intracystic bodies a dark red, but microscopic details of the intracystic bodies are not very good.

Elliott's Method for Leptospira[68]

FIXATION 10% Buffered neutral formalin.
PROCESS Paraffin.
MICROTOMY Cut section at 5 micrometers.
CONTROL Use control number 40.
SOLUTIONS

0.2M Acetate Buffer, pH 3.7

1.2% Acetic acid	62.0	ml
2.7% Sodium acetate	38.0	ml

1% Silver Nitrate

Silver nitrate	1.0	gm
0.2M Acetate buffer, pH 3.7	100.0	ml

2% Silver Nitrate

Silver nitrate	2.0	gm
0.2M Acetate buffer, pH 3.7	100.0	ml

6% Gelatin-Acetate Buffer

Gelatin	3.0	gm
0.2M Acetate buffer, pH 3.7	45.0	ml

Hydroquinone-Acetate Buffer

Hydroquinone	0.3	gm
0.2M Acetate buffer, pH 3.7	6.0	ml

0.5% Iodine

Iodine	0.5	gm
70% Alcohol	100.0	ml

5% Sodium Thiosulfate

Sodium thiosulfate	5.0	gm
Distilled water	100.0	ml

STAINING PROCEDURE

1. Decerate slides in xylene, 2 changes, 2 minutes each. Place in absolute alcohol, 2 changes, 2 minutes each; 95% alcohol, 2 changes, 2 minutes each; then rinse in distilled water.
2. Place in 0.5% iodine for 5 minutes.
3. Place in 5% sodium thiosulfate for 20 seconds.
4. Wash in running tap water for 5 minutes.
5. Rinse thoroughly in distilled water.
6. Place slides in a coplin jar containing 1% silver nitrate which has been preheated to 60°C.
7. Add 9 ml of 2% silver nitrate to 45 ml of 6% gelatin solution. Incubate the slides in the silver-gelatin for 1 hour at 60°C. At the same time, warm the hydroquinone solution in a 60°C oven.
8. Pour off silver-gelatin solution. Add hydroquinone solution to gelatin-silver solution. Mix briefly, and pour developer onto the slides in the coplin jar.
9. Agitate the slides to ensure even development.
10. When sections are golden brown, stop the development by rinsing in hot tap water for 5 minutes.
11. Dehydrate slides in 95% alcohol and absolute alcohol 3 changes each. Clear in xylene, 3 changes.
12. Mount coverglass with appropriate medium (refractive index 1.48-1.56).

RESULTS

Leptospira.. black
Nuclei... brown
Background.. yellow

REMARKS

The treatment of the slides with the 0.5% iodine solution, in step 3, dramatically enhances the staining of Leptospira.

Methods for Staining Helicobacter* Pylori[69]

In recent years laboratories have received requests to perform special procedures on gastric biopsies to aid in the identification of helicobacter pylori, a bacteria associated with active chronic gastritis and peptic ulcer disease.

In 1983, Warren[70] posed the question of unidentified curved bacilli on gastric epithelium, in active cases of chronic gastritis. He observed a small, curved S-shaped bacilli in numerous gastric biopsy specimens that were stained with a Warthin-Starry silver method. In 1984, Jones, et al.[71], reported that a campylobacter-like organism had been isolated from patients with gastritis. Using various special stains, including a Gram stain, periodic acid Schiff, Giemsa, and Warthin-Starry, he was able to demonstrate the organism histologically. The organisms are located on the surface of the mucosa in the layer of mucin, and in the gastric pits. Goodwin, et al[72] further classified the organism as campylobacter pyloridis, a true spiral bacteria possessing large amounts of urease enzyme. (True spiral bacteria are distinguished from spirochetes by the absence of endoflagella and periplasmic fibers.) The organism was shown to be strongly associated with active chronic gastritis.

The histotechnologist has a choice of several special staining procedures which will adequately demonstrate the organism in gastric biopsy sections. Control material for the procedure can usually be obtained by reviewing past cases of gastric resections for chronic gastritis or peptic ulcer. Although helicobacter pylori can be demonstrated with the H&E under close examination, they are best demonstrated with the following special stains, or most silver stains that also demonstrate spirochetes.

The size of the organisms in the stained section is dependent upon the stain employed. Although they can be visualized at 400x, the organisms are best distinguished at 1000x. Under oil immersion with silver impregnation techniques, the organism appears fat or stumpy. On the other hand, the Giemsa procedure stains the inner portion of the organism, making the organism appear thinner and more difficult to distinguish in the stained section.

*Formerly Campylobacter.

Warthin-Starry for Spirochetes

(See Fig. 60a and b, 61, 62, 63, 64 and 65)

Fixation	10% Buffered neutral formalin. Do not use chromate fixatives.
Microtomy	Cut paraffin sections at 6 micrometers.
Solutions	Use chemically clean glassware.

Acidulated Water

Triple distilled water	1000.0	ml

Add enough 1% aqueous citric acid to bring water to pH 4.0.

1% Silver Nitrate (For Impregnation)

Silver nitrate, ACS crystals	1.0	gm
Acidulated water	100.0	ml

2% Silver Nitrate (For Developer)

Silver nitrate, ACS crystals	2.0	gm
Acidulated water	100.0	ml

5% Gelatin

Gelatin, high grade	5.0	gm
Acidulated water	100.0	ml

0.15% Hydroquinone Solution

Hydroquinone crystals, photographic grade	0.15	gm
Acidulated water	100.0	ml

Place the 2% silver nitrate, 5% gelatin and the hydroquinone in separate flasks in a 54°C floatation bath, during staining procedure.

Developer Solution

Silver nitrate solution, 2%	1.5	ml
Gelatin solution, 5%	3.75	ml
Hydroquinone solution, 0.15%	2.0	ml

Combine in order given in a small beaker, making certain solutions are mixed well after each addition. Prepare immediately before use.

Staining Procedure

1. Deparaffinize and hydrate to triple distilled water.
2. Place slides in the 1% silver nitrate impregnating solution in a water bath or oven at 43°C for 30 minutes.
3. Prepare the developer solution making sure to mix well.
4. Flood sections, that have been placed on a glass rod slide rack, with the developer solution. Allow sections to develop until they are light brown.
5. Wash quickly and thoroughly in hot tap water, approximately 56°C.
6. Rinse in distilled water.
7. Dehydrate in 95% alcohol, absolute alcohol and clear in xylene, 2 changes each.
8. Mount coverglass with resinous media.

Results

Spirochetes, Donovan bodies ....black
Melanin..black
Background..pale yellow to light brown

Remarks

It may be necessary to prolong development of sections for the demonstration of Donovan bodies. Use paraffin coated forceps, particularly at steps 2 and 3.

Modified Steiner Technique for Spirochetes and Helicobacter Pylori[73]

FIXATION	10% Buffered neutral formalin.
PROCESS	Paraffin, frozen, or celloidin.
MICROTOMY	Cut paraffin sections at 4 micrometers.
CONTROL	Use control number 22A.
SOLUTIONS	

1% Uranyl Nitrate

Uranyl nitrate	1.0	gm
Distilled water	100.0	ml

1% Silver Nitrate

Silver nitrate crystals	1.0	gm
Distilled water	100.0	ml

2.5% Gum Mastic

Gum mastic	2.5	gm
Absolute Alcohol	100.0	ml

0.04% Silver Nitrate

Silver nitrate crystals	0.04	gm
Distilled water	100.0	ml

2% Hydroquinone

Hydroquinone crystals	1.0	gm
Distilled water	50.0	ml

Reducing Solution
(Prepare just before use.)

2.5% Gum mastic	20.0	ml
2% Hydroquinone	50.0	ml
100% Ethyl alcohol	10.0	ml

Combine the gum mastic solution and the hydroquinone solution. Add the alcohol and mix well. Filter through Whatman #4 filter paper. Immediately before placing slides in solution, add 5 ml of 0.04% silver nitrate.

STAINING PROCEDURE

1. Decerate slides in xylene, 2 changes, 2 minutes each. Place in absolute alcohol, 2 changes, 2 minutes each; 95% alcohol, 2 changes, 2 minutes each; then rinse in distilled water.
2. Place in 1% uranyl nitrate, preheated to 60°C, for 15 minutes.
3. Rinse thoroughly in distilled water.
4. Place in 1% silver nitrate for 1 hour at 60°C.
5. Rinse in 2 changes of distilled water.
6. Dehydrate in 2 changes of 95% and 2 changes of 100% alcohol.
7. Place in 2.5% gum mastic for 5 minutes.
8. Allow sections to air dry for 1 minute.
9. Rinse in 2 changes of distilled water.
10. Place sections in reducing solution that has been preheated to 45°C. Allow sections to develop for 15 to 20 minutes, occasionally agitating sections.
11. Rinse in distilled water to stop reduction.
12. Dehydrate slides in 95% alcohol and absolute alcohol 3 changes each. Clear in xylene, 3 changes.
13. Mount coverglass with appropriate medium (refractive index 1.48-1.56).

RESULTS

Spirochetes, fungi, helicobacter and most bacteria	dark brown to black
Background	yellow to brown

Gimenez Method for Helicobacter Pylori[74]

FIXATION 10% Buffered neutral formalin.
PROCESS Paraffin, frozen, or celloidin.
MICROTOMY Cut paraffin sections at 3 to 5 micrometers.
CONTROL Use control number 22A.
SOLUTIONS

Carbol Fuchsin (Stock)

Basic fuchsin	1.0	gm
95% Alcohol	10.0	ml
4% Phenol, aqueous	25.0	ml
Distilled water	65.0	ml

Stock solution is stable for several months.

0.1M Phosphate Buffer, pH 7.5

0.1M sodium phosphate monobasic	7.0	ml
0.1M sodium phosphate dibasic	31.0	ml

Carbol Fuchsin (Working)

Stock carbol fuchsin	15.2	ml
0.1M phosphate buffer, pH 7.5	38.0	ml

A precipitate will form immediately. Mix well and filter. Filter again immediately before use. Working solution is stable for 48 hours.

0.8% Malachite Green

Malachite green	0.8	gm
Distilled water	100.0	ml

STAINING PROCEDURE

1. Decerate slides in xylene, 2 changes, 2 minutes each. Place in absolute alcohol, 2 changes, 2 minutes each; 95% alcohol, 2 changes, 2 minutes each; then rinse in distilled water.
2. Place in working carbol fuchsin solution for 3 minutes.
3. Wash thoroughly in running tap water for 2 minutes.
4. Place in 0.8% malachite green for 15 seconds.
5. Wash thoroughly in running tap water for 1 minute.
6. Repeat steps 4 and 5 until sections become blue-green.
7. Blot and allow sections to air dry.
8. Mount coverglass with appropriate medium (refractive index 1.48-1.56).

RESULTS

Rickettsia and bacteria	bright red
Helicobacter pylori	red
Nuclei	blue-green
Background	light green

Modified One Hour Giemsa Stain

FIXATION 10% Buffered neutral formalin.
MICROTOMY Cut paraffin sections at 4 to 5 micrometers.
SOLUTIONS

Azure II Eosin

Azure II eosin (Giemsa)	1.3	gm
Glycerin	80.0	ml
Incubate at 56°-60°C for 2 hours. Mix and Cool.		
Add:		
Methyl alcohol	170.0	ml
Acetone	290.0	ml

May-Grunwald Stain

May-Grunwald stain (Jenner)	0.15	gm
Methyl alcohol	290.0	ml
Acetone	290.0	ml

Giemsa Solution (Stock)

Combine Azure II eosin and May-Grunwald solutions. This stock solution can be used immediately but improves with aging.

Acetic Water Solution

Glacial acetic acid	0.01	ml
Distilled water	1000.0	ml

Giemsa Solution (Working)

Giemsa stock solution	10.0	ml
Acetic water	50.0	ml

STAINING PROCEDURE

1. Decerate slides in xylene, 2 changes, 2 minutes each. Place in absolute alcohol, 2 changes, 2 minutes each; 95% alcohol, 2 changes, 2 minutes each; then rinse in distilled water.
2. Immerse sections in working Giemsa solution for 1 hour.
3. Dehydrate sections in 3 changes of absolute alcohol.
4. Clear in xylene, 3 changes for 2 minutes each.
5. Mount coverglass with appropriate medium (refractive index 1.48-1.56).

RESULTS

Helicobacter and other bacteria	deep blue
Nuclei	blue
Mast cell granules	purple
Cytoplasm	pink

A Warthin-Starry Silver Method for the Demonstration of Bacillus Piliformis[75]

(See Fig. 66)

The Warthin-Starry silver technique is modified for use in demonstrating bacillus piliformis in paraffin sections. Microwave heating is used to reduce staining time, but more importantly, when used in conjunction with Walpole's buffer, it dramatically improves the reliability of the technique.

Bacillus piliformis, the pathogenic bacterium that causes Tyzzer's disease, is a major concern for all rodent breeding laboratories as well as research facilities that use rodents. An outbreak of this disease often means that an entire colony must be broken down, equipment disinfected and the animals destroyed. The devastating effect of Tyzzer's disease both financially and on long term research is immeasurable. Thus, the Warthin-Starry silver technique is a valuable tool in prevention, detection and eradication of this pathogen.

The modification of this procedure further refines an excellent diagnostic tool for veterinary histologists.

SOLUTIONS

5M Acetic Acid (Stock)

Glacial acetic acid	11.8	ml
Deionized water	1000.0	ml

5M Sodium Acetate (Stock)

Sodium acetate	16.4	gm
Deionized water	1000.0	ml

Walpole's Buffer

5M Acetic acid (stock)	17.6	ml
5M Sodium acetate (stock)	2.4	ml
Deionized water	480.0	ml

2% Silver Nitrate

Silver nitrate	2.0	gm
Walpole's buffer	100.0	ml

1% Silver Nitrate

Silver nitrate	1.0	gm
Walpole's buffer	100.0	ml

5% Gelatin

Gelatin	5.0	gm
Walpole's buffer	100.0	ml

3% Hydroquinone

Hydroquinone	3.0	gm
Walpole's buffer	100.0	ml

Developer

2% Silver nitrate	1.5	ml
Gelatin	7.5	ml
3% Hydroquinone	0.5	ml

STAINING PROCEDURE

1. Deparaffinize and hydrate to deionized water.
2. Rinse in Walpole's buffer.
3. Place slides in a plastic coplin jar, cover with 1% silver nitrate, and microwave for 1.5 minutes at a medium setting (see #4 in remarks). Let slides stand for 5 minutes in this solution.
4. Rinse in Walpole's buffer.
5. Flood slides with developer which has been freshly prepared during the preceding 5 minute standing time.
6. Wash in warm tap water.
7. Dehydrate slides in 95% alcohol and absolute alcohol 3 changes each. Clear in xylene, 3 changes.
8. Mount with Permount.

RESULTS

Bacillus piliformis	black
Background	light brown to golden-brown

REMARKS

1. In our laboratory we prepare a large volume of the gelatin solution in advance and then aliquot 7.5 ml into 15 ml screw top vials, which are then stored until needed in the refrigerator. We can then take one vial of gelatin, heat it in warm (56°C) tap water, add the solutions and then discard the vial after the procedure has been finished. Using a disposable vial greatly decreases the chance of contamination.
2. Most Warthin-Starry variations used in the past, have required triple distilled water; this modification works fine with deionized water.
3. Plastic coplin jars used in this procedure should be acid cleaned.
4. The microwave used in this procedure was a 600 watt Sharp Carousel-II.
5. As with most silver stains, the slides should never be touched with metal instruments.

Modification of the Dieterle Spirochete Stain

FIXATION 10% Buffered neutral formalin.
MICROTOMY Cut paraffin sections at 6 micrometers.
SOLUTIONS

5% Alcoholic Uranyl Nitrate

Uranyl nitrate	5.0	gm
70% Ethyl alcohol	100.0	ml

10% Alcoholic Gum Mastic

Gum mastic	10.0	gm
Absolute ethyl alcohol	100.0	ml

1% Silver Nitrate

Silver nitrate	1.0	gm
Distilled water	100.0	ml

Developing Solution

Mix in order:

Hydroquinone	15.0	gm
Sodium sulfite	2.5	gm
Distilled water	600.0	ml
Acetone	100.0	ml
Formaldehyde (37%-40%)	100.0	ml
Pyridine	100.0	ml
10% Alcoholic gum mastic	100.0	ml

Swirl gently as each solution is added. Solution turns milky yellow as the gum mastic is added and medium brown if standing in a well lighted area. As the brown streaks appear, swirl the flask gently. Developer is ready to use in about 6 hours and may be used until color becomes a dark brown, usually in 2 to 3 days.

STAINING PROCEDURE

1. Preheat 5% alcoholic uranyl nitrate and the 1% silver nitrate in 56-58°C oven for at least 30 minutes.

2. Decerate and hydrate sections to distilled water.
3. Place slides in preheated 5% alcoholic uranyl nitrate in 55-58°C oven for 1 hour.
4. Dip slides 1 time in distilled water.
5. Dip slides 1 time in 95% alcohol.
6. Place slides in 10% alcoholic gum mastic for 3 minutes.
7. Dip slides 1 time in 95% alcohol.
8. Dip slides 1 time in distilled water.
9. Place slides in preheated 1% silver nitrate solution in 55-58°C oven for 4 hours.
10. Dip slides 2 times in distilled water.
11. Dip slides in developing solution until sections are pale yellow to light tan.
12. Dip slides 2 times in distilled water.
13. Dehydrate slides in 2 changes each of 95% alcohol and acetone, dip slides 2 times in each change.
14. Clear slides in xylene.
15. Mount coverglass with resinous media.

RESULTS
Spirochetes and some bacteria stain dark brown to black. Background stains yellow to light tan. Other structures that stain are melanin granules, chromatin, formalin pigment and some foreign material in macrophages.

Luna's Method for Bacteria

FIXATION 10% buffered neutral formalin.
PROCESS Paraffin, frozen smears, or imprints.
MICROTOMY Cut paraffin sections at 6 micrometers.
CONTROL Use control number 20.
SOLUTIONS

0.5% Thionin

Thionin	0.5	gm
20% Ethyl alcohol	100.0	ml

STAINING PROCEDURE
1. Decerate slides in xylene, 2 changes, 2 minutes each. Place in absolute alcohol, 2 changes, 2 minutes each; 95% alcohol, 2 changes, 2 minutes each; then rinse in distilled water.
2. Stain slides in 0.5% thionin for 15 minutes.
3. Dehydrate slides in 95% alcohol and absolute alcohol 3 changes each. Clear in xylene, 3 changes.

RESULTS

Bacteria	purple
Mast cells	purple
Nuclei	blue
Cytoplasm	various shades of blue

Luna-Parker Method for Nocardia Asteroides

(See Fig. 67)

FIXATION 10% Buffered neutral formalin.
PROCESS Paraffin.
MICROTOMY Cut paraffin sections at 6 micrometers.
CONTROL Use control of a known case of nocardia.
SOLUTIONS

Giemsa Solution (Stock)

Giemsa powder	1.0	gm
Glycerin	66.0	ml
Methyl alcohol	66.0	ml

Mix glycerin and Giemsa powder. Place in a 60°C oven for 2 hours. Let cool, then add the 66 ml of methyl alcohol.

Buffer Solution pH 6.0

Potassium phosphate, monobasic	5.0	gm
Sodium phosphate, dibasic	1.0	gm
Distilled water	1000.0	ml

Giemsa Solution (Working)

Giemsa solution (stock)	5.0	ml
Methyl alcohol	3.0	ml
Buffer solution	100.0	ml

10% Rosin Alcoholic Solution (Stock)

Rosin, white	10.0	gm
100% Alcohol	100.0	ml

Rosin Alcohol Solution (Working)

Rosin solution (stock)	2.5	ml
95% Alcohol	100.0	ml

Acetate Rinse Solution

Sodium acetate, trihydrate	2.72	gm
Distilled water	1000.0	ml

The acetate rinse is used to remove the working buffer salt solution. The buffer has a tendency to precipitate when the slides are placed in rosin alcohol.

STAINING PROCEDURE

1. Decerate slides in xylene, 2 changes, 2 minutes each. Place in absolute alcohol, 2 changes, 2 minutes each; 95% alcohol, 2 changes, 2 minutes each; then rinse in distilled water.
2. Stain in working Giemsa solution overnight (8 hours).
3. Place slides in acetate rinse for 5 minutes.
4. Differentiate by agitating gently in working rosin alcohol solution for 4 minutes. If background is too dark, agitate for an additional 2 minutes. Check differentation microscopically, one slide at a time.
5. Dehydrate slides in 95% alcohol and absolute alcohol 3 changes each. Clear in xylene, 3 changes.
6. Mount coverglass with appropriate medium (refractive index 1.48-1.56).

RESULTS

Nocardia filaments .... purplish-red
Nuclei .... blue
Collagen .... light pink
Erythrocytes .... red to copper
Bacteria .... blue
Background .... light blue

REMARKS

Attempts to demonstrate other fungi with the modified staining procedure were unsuccessful with one exception. The filaments of actinomycosis were demonstrated well.

Allen-Luna Method for Spirochetes, Donovan Bodies and Other Microorganisms

This method is very useful for demonstrating bacilli from cat scratch disease, Legionella pneumophila, helicobacter pylori and other similar type organisms.

FIXATION	10% Buffered neutral formalin. Avoid chromate fixatives.
PROCESS	Paraffin, frozen, or celloidin.
MICROTOMY	Cut sections at 6 micrometers.
CONTROL	See page 564 to select control based on entity being stained. Use chemically clean glassware.

SOLUTIONS

Acidulated Water

Distilled water	1000.0	ml

Add enough 1% aqueous citric acid to bring water to pH 4.0.

1% Silver Nitrate (For Impregnation)

Silver nitrate	1.0	gm
Acidulated water	100.0	ml

2% Silver Nitrate (For Developer)

Silver nitrate	2.0	gm
Acidulated water	100.0	ml

5% Gelatin

Sheet gelatin, high-grade	10.0	gm
Acidulated water	200.0	ml

0.15% Pyrocatechol

Pyrocatechol crystals	0.15	gm
Acidulated water	100.0	ml

Keep 2% silver nitrate, 5% gelatin, and 0.15% pyrocatechol in 50 ml flasks or tri-pour beakers, in a floatation bath at 54°C until developer is made.

Developer
(Prepare immediately before use.)

2% Silver nitrate solution	1.5	ml
5% Gelatin solution	3.75	ml
0.15% Pyrocatechol solution	2.0	ml

Combine in the order given in small flask or beaker, making certain solutions are mixed well.

STAINING PROCEDURE

1. Decerate slides in xylene, 2 changes, 2 minutes each. Place in absolute alcohol, 2 changes, 2 minutes each; 95% alcohol, 2 changes, 2 minutes each; then rinse in distilled water.
2. Impregnate with silver nitrate solution heated in an oven or floatation water bath at 43°C for 30 minutes. Prepare the developer solution at this point.
3. Flood sections (which have been laid across glass rods) with the developer solution, which must be used as soon it is mixed. Allow sections to develop until they are light brown or yellow.
4. Wash quickly and thoroughly in hot running tap water, approximately 56°C. This step is important since it is essential that all reducible silver which may be attached to the gelatin, be removed.
5. Rinse in distilled water.
6. Dehydrate slides in 95% alcohol and absolute alcohol 3 changes each. Clear in xylene, 3 changes.
7. Mount coverglass with appropriate medium (refractive index 1.48-1.56).

RESULTS

Spirochetes, Donovan bodies, helicobacter pylori, Legionella pneumophila and cat scratch fever	black
Background	pale yellow to light brown

REMARKS

It may be necessary to prolong development of sections for the demonstration of Donovan bodies. Certain hematogenous pigments, nuclei and melanin have a greater attraction for silver than do spirochetes, and therefore it is difficult to stain the spriochetes in close proximity to these elements. By lowering the pH of the acidulated solution to 3.6 and prolonging the development, you may demonstrate the spirochetes in the areas of competition; however, the part of the section not containing competitive elements may be overstained and useless. Sections can be restained to increase the amount of development, if microscopic observation of the known positive tissue shows pale spirochetes or none at all. Use paraffin or plastic coated forceps, particularly at step 2 and step 3. While in the impregnating and reducing solutions, slides should be agitated as often as possible for complete and even staining.

Price's Method for Protozoa and Bacteria[55k]

FIXATION	10% Buffered neutral formalin.
PROCESS	Paraffin.
TECHNIQUE	Cut paraffin sections at 4 micrometers.
CONTROL	Use control of entity being stained.

SOLUTIONS

Giemsa (Stock)

Giemsa powder	1.0	gm
Methyl alcohol	65.0	ml
Glycerin	40.0	ml

Pour glycerin into an acid cleaned brown bottle. Use the measured methyl alcohol to rinse the glycerin from container and into the bottle. Add sterile glass beads to glycerin and methyl alcohol mixture. Combine the methyl alcohol and glycerin by shaking slightly. Add Giemsa powder, tighten cap, and shake well. SHAKE WELL about once an hour for 8 hours. It is best to make a large volume (100 ml) of this stock and store in this bottle. When ready to use, filter through coarse filter paper into small 50 ml dark dropper bottles for use as working solutions.

Buffer (Stock)

Disodium monohydrogen phosphate, anhydrous	6.0	gm
Monopotassium di-hydrogen phosphate	5.0	gm

Mix buffer salts and weigh out 1 gm units and place them in small well stoppered vials or test tubes. One of these units is dissolved in one liter of distilled water to give a buffer solution of approximately pH 7.0.

Giemsa (Working)

Giemsa (stock)	3.0	ml
Buffer (stock)	97.0	ml

To avoid contamination of the stock solution, never introduce a pipette into the stock container. If dropper bottles are not used, pour a small amount of solution from the stock bottle into another container and pipette required amount from this second container. Discard the unused portion of Giemsa stock solution.

Glacial Acetic Acid-Water Solution

Glacial acetic acid	1.0	ml
Distilled water	499.0	ml

STAINING PROCEDURE

1. Decerate slides in xylene, 2 changes, 2 minutes each. Place in absolute alcohol, 2 changes, 2 minutes each; 95% alcohol, 2 changes, 2 minutes each; then rinse in distilled water.

   If necessary, remove mercuric chloride crystals with alcoholic iodine. (Iodine is removed with exposure to subsequent alcohols.)
2. Place in 80% alcohol for 2 changes, 1 minute each.
3. Rinse in distilled water for 15 seconds.
4. Place in buffer solution for 30 minutes.
5. Stain in working Giemsa solution overnight.
6. Quick rinse in stock buffer solution.
7. Place in glacial acetic acid-water solution for 1 minute. (See remarks.)
8. Place in absolute alcohol for 2 changes, 15 seconds each.
9. Clear in xylene for 2 changes, 2 minutes each.
10. Mount coverglass with appropriate medium (refractive index 1.48-1.56).

RESULTS

Malarial parasites .... blue
Malarial pigment.. black
Nuclear chromatin .... blue
Collagen and other elements .... pale pink
Bacteria.. dark blue
Egg shells of schistosomes .... blue
Erythrocytes .... pink to rose

REMARKS

Total time for steps 7 and 8 should be exactly 90 seconds.

Chapter 8

Rickettsia, Inclusion Bodies and Protozoa

Lendrum's Method for Viral Inclusion Bodies[551]

(See Fig. 68a, b, c and 69)

FIXATION 10% Buffered neutral formalin.
PROCESS Paraffin, frozen, or celloidin.
MICROTOMY Cut sections at 6 micrometers.
CONTROL Use control number 43.
SOLUTIONS

Mayer's Hematoxylin
(See Page 81)

0.5% Phloxine

Phloxine	1.0	gm
70% Alcohol	200.0	ml
Calcium chloride	1.0	gm

2.5% Tartrazine

Tartrazine	2.5	gm
Cellosolve (ethylene glycol monoethyl ether)	100.0	ml

STAINING PROCEDURE

1. Decerate slides in xylene, 2 changes, 2 minutes each. Place in absolute alcohol, 2 changes, 2 minutes each; 95% alcohol, 2 changes, 2 minutes each; then rinse in distilled water.
2. Place in Mayer's hematoxylin solution for 5 minutes.
3. Blue in running tap water for 5 minutes.
4. Place in phloxine solution for 15 minutes.
5. Rinse briefly in distilled water and drain slides well.
6. Place in tartrazine solution until inclusion bodies stand out a bright red on a yellow background.
7. Dehydrate slides in 95% alcohol and absolute alcohol 3 changes each. Clear in xylene, 3 changes.
8. Mount coverglass with appropriate medium (refractive index 1.48-1.56).

RESULTS

Viral inclusion bodies	red
Nuclei	blue
Background	yellow

Bosch's Methods for Negri Bodies[55m]

FIXATION Bouin's solution or 10% buffered neutral formalin.
PROCESS Paraffin, frozen, or celloidin.
MICROTOMY Cut sections at 6 micrometers.
CONTROL Use control number 34A.
SOLUTIONS

Mayer's Hematoxylin
(See Page 81)

Phenol-Phloxine B

Phloxine B	1.5	gm
Distilled water	5.0	ml

Mix and add, with vigorous shaking:
5% Phenol to make 100.0 ml

Add 20% glacial acetic acid drop by drop with vigorous shaking until solution becomes cloudy. Place in a closed vessel and heat in a 56°C oven for 24 hours. Use the supernatant as the staining solution.

0.2% Lithium Carbonate

Lithium carbonate	0.2	gm
Distilled water	100.0	ml

STAINING PROCEDURE

1. Decerate slides in xylene, 2 changes, 2 minutes each. Place in absolute alcohol, 2 changes, 2 minutes each; 95% alcohol, 2 changes, 2 minutes each; then rinse in distilled water.
2. Mordant in Bouin's solution for 1 hour at 60°C if tissue has been fixed in formalin.
3. Wash in running tap water for 5 minutes, then rinse in distilled water. (There is no need to perform steps 2 and 3 if Bouin's was the initial fixative.)
4. Place in phenol-phloxine B solution for 5 minutes.
5. Rinse in distilled water for 1 minute.
6. Place in Mayer's hematoxylin solution for 5 minutes.
7. Wash in running tap water for 15 seconds.
8. Differentiate and blue in 0.2% lithium carbonate solution for several seconds until a pale pink cloud comes from the sections and the bluish tone of hematoxylin appears.
9. Wash in running water for 5 minutes.
10. Dehydrate slides in 95% alcohol and absolute alcohol 3 changes each. Clear in xylene, 3 changes.
11. Mount coverglass with appropriate medium (refractive index 1.48-1.56).

RESULTS

Negri bodies	deep red
Nuclei, nucleoli and Nissl substance	deep blue
Background	light pink
Erythrocytes	bright red

REMARKS
The best results are obtained after initial fixation in Bouin's fluid.

Schleifstein's Method for Negri Bodies[55n]

(See Fig. 70 and 71)

FIXATION	Zenker's solution.
PROCESS	Paraffin, frozen, or celloidin.
MICROTOMY	Cut sections at 6 micrometers.
CONTROL	Use control number 34A.
SOLUTIONS	

Schleifstein's (Stock)

Basic fuchsin	1.8	gm
Methylene blue	1.0	gm
Glycerin	100.0	ml
Methyl alcohol	100.0	ml

This solution will keep indefinitely.

Potassium Hydroxide

Potassium hydroxide	0.01	gm
Tap water	4000.0	ml

It is important to achieve a slightly alkaline solution. It may be necessary to vary the amount of potassium hydroxide to accomplish this.

Schleifstein's (Working)

Schleifstein's solution (stock)	10	drops
Potassium hydroxide solution	20.0	ml

Mix the working solution immediately before use.

STAINING PROCEDURE

1. Decerate slides in xylene, 2 changes, 2 minutes each. Place in absolute alcohol, 2 changes, 2 minutes each; 95% alcohol, 2 changes, 2 minutes each; then rinse in distilled water.
2. Remove mercuric chloride crystals with iodine and clear with sodium thiosulfate.
3. Place sections on ring stands, then flood with freshly prepared working Schleifstein's solution. Gently heat the bottom of each slide until vapor is produced. Allow slide to cool to room temperature.
4. Wash quickly in tap running water.
5. Differentiate each slide individually by gently agitating in 90% alcohol until sections are a faint violet color.
6. Dehydrate slides in 95% alcohol and absolute alcohol 3 changes each. Clear in xylene, 3 changes.
7. Mount coverglass with appropriate medium (refractive index 1.48-1.56).

RESULTS

Negri bodies	deep magenta
Cytoplasm	bluish-violet
Erythrocytes	copper

REMARKS

This technique is excellent for demonstrating Negri bodies in Zenker-fixed specimens only. Attempts to post fix sections in Zenker's solution prior to staining are not successful.

Luna's Method for Negri Bodies

(See Fig. 72 and 73)

FIXATION 10% Buffered neutral formalin.
PROCESS Paraffin, frozen, or celloidin.
MICROTOMY Cut sections at 6 micrometers.
CONTROL Use control number 34A.
SOLUTIONS

Weigert's Iron Hematoxylin
(See Page 79)

1% Biebrich Scarlet (Stock)

Biebrich scarlet	1.0	gm
Distilled water	100.0	ml

Hematoxylin-Biebrich Scarlet (Working)

Weigert's iron hematoxylin solution	40.0	ml
1% Biebrich scarlet (stock)	10.0	ml

1% Acid Alcohol
(See Page 77)

0.5% Lithium Carbonate

Lithium carbonate	0.5	gm
Distilled water	100.0	ml

STAINING PROCEDURE

1. Decerate slides in xylene, 2 changes, 2 minutes each. Place in absolute alcohol, 2 changes, 2 minutes each; 95% alcohol, 2 changes, 2 minutes each; then rinse in distilled water.
2. Stain in working hematoxylin-Biebrich scarlet solution for 5 minutes.
3. Decolorize in 1% acid alcohol until desired nuclear detail is achieved (usually 8 dips).
4. Rinse in running tap water to remove acid alcohol.
5. Dip in lithium carbonate solution until section turns blue and Negri bodies bright red (usually 5 dips).
6. Wash in running tap water for 2 minutes.
7. Dehydrate slides in 95% alcohol and absolute alcohol 3 changes each. Clear in xylene, 3 changes.
8. Mount coverglass with appropriate medium (refractive index 1.48-1.56).

RESULTS

Negri bodies	red
Erythrocytes	red
Background	blue

Luna's Giemsa Method for Rickettsia[55o]

(See Fig. 74 and 75a, b, c, d)

FIXATION	10% Buffered neutral formalin or Zenker's.
PROCESS	Paraffin, frozen, or celloidin.
MICROTOMY	Cut paraffin sections at 6 micrometers.
CONTROL	Use control number 38A.
SOLUTIONS	

Buffered Water, pH 6.8

Buffer salts (pH 6.8)*	1.0	gm
Distilled water	1000.0	ml

*Commercially available as buffer salt dry, pH 6.86 at 25°C.

Giemsa (Stock)

Giemsa powder	1.0	gm
Glycerin, reagent grade	66.0	ml
Methyl alcohol, acetone-free, reagent grade	66.0	ml

Giemsa (Working)

For paraffin processed tissue

Giemsa (stock)	1.0	ml
Buffered water, pH 6.8	50.0	ml

Giemsa (Working)

For plastic processed tissue

Giemsa (stock)	10.0	ml
Buffered water, pH 6.8	50.0	ml

0.2% Glacial Acetic Acid

Glacial acetic acid	0.2	ml
Distilled water	100.0	ml

Rosin-Alcohol

Rosin*	5.0	gm
100% Alcohol or absolute	100.0	ml

*The rosin used for treating violin strings can be used to make this solution.

STAINING PROCEDURE

1. Decerate slides in xylene, 2 changes, 2 minutes each. Place in absolute alcohol, 2 changes, 2 minutes each; 95% alcohol, 2 changes, 2 minutes each; then rinse in distilled water.
2. Mordant in buffered water solution for 30 minutes.
3. Place in working Giemsa solution overnight.
4. Rinse in buffered water solution.
5. Place in 0.2% glacial acetic acid solution for 1 minute.
6. Rinse in buffered water solution.
7. Decolorize sections individually in working rosin-alcohol solution. Check frequently with a microscope, until rickettsia appear as violet colored granules (may take up to 3 minutes).

8. Dehydrate slides in 95% alcohol and absolute alcohol 3 changes each. Clear in xylene, 3 changes.
9. Mount coverglass with appropriate medium (refractive index 1.48-1.56).

RESULTS

Rickettsia	violet
Nuclei	blue
Cytoplasm, connective tissue	pink
Erythrocytes	salmon

REMARKS

The differentiation in step 7 is most critical; only if this is properly done can one expect good results. Use the microscope as suggested above for this purpose.

Luna's Basic Fuchsin Method for Rickettsia

(See Fig. 76)

The rickettsia of Rocky Mountain Spotted Fever are more easily demonstrated in paraffin sections than are those of typhus. After fixation in Zenker's or Regaud's solution, sections may be stained by the Giemsa method or by the following method.

FIXATION Zenker's, Regaud's or 10% buffered neutral formalin.
PROCESS Paraffin, frozen, or celloidin.
MICROTOMY Cut sections at 6 micrometers.
CONTROL Use control number 38A.
SOLUTIONS

1% Methylene Blue

Methylene blue	1.0	gm
Distilled water	100.0	ml

0.25% Basic Fuchsin

Basic fuchsin	0.25	gm
Distilled water	100.0	ml

0.5% Citric Acid

Citric acid	0.5	gm
Distilled water	100.0	ml

STAINING PROCEDURE

1. Decerate slides in xylene, 2 changes, 2 minutes each. Place in absolute alcohol, 2 changes, 2 minutes each; 95% alcohol, 2 changes, 2 minutes each; then rinse in distilled water.
2. Remove mercuric chloride crystals with iodine and clear with sodium thiosulfate.
3. Place in methylene blue solution overnight.
4. Rinse in 95% alcohol for 5 seconds or blue color will be lost.
5. Rinse quickly in distilled water for 2 to 3 seconds.
6. Place in 0.25% basic fuchsin solution for 30 minutes.
7. Decolorize rapidly in 0.5% citric acid solution for 1 to 2 seconds, never more than 3 seconds.

8. Continue differentiation in absolute alcohol until nuclei stand out blue and the rickettsia red.
9. Dehydrate slides in 95% alcohol and absolute alcohol 3 changes each. Clear in xylene, 3 changes.
10. Mount coverglass with appropriate medium (refractive index 1.48-1.56).

RESULTS

Rickettsia .... bright red
Nuclei .... reddish-blue

REMARKS

This is by far the most reliable method available for demonstrating rickettsia in paraffin sections. The preferred fixative is Regaud's or Zenker's solution.

The following changes to the above method stains the rickettsia deep blue against a background which is partly red and partly blue.

1-5. Use procedures as above.
6. Wash lightly in distilled water.
7. Counterstain again in 1% aqueous methylene blue for 5 seconds.
8. Differentiate in 95% alcohol.
9. Dehydrate in absolute alcohol, 2 changes, 2 minutes each.
10. Clear in xylene 2 changes, 2 minutes each.
11. Mount coverglass with appropriate medium (refractive index 1.48-1.56).

Pinkerton's Adaptation of Machiavello's Stain for Rickettsia

(See Fig. 77 and 78)

FIXATION Zenker's, Regaud's, or 10% buffered neutral formalin.
PROCESS Paraffin, frozen, or celloidin.
MICROTOMY Cut section at 4 to 5 micrometers.
CONTROL Use control number 38A.
SOLUTIONS

1% Methylene Blue

Methylene blue	1.0	gm
Distilled water	100.0	ml

0.25% Basic Fuchsin

Basic fuchsin	0.25	gm
Distilled water	100.0	ml

0.5% Citric Acid

Citric acid	0.5	gm
Distilled water	100.0	ml

STAINING PROCEDURE

1. Decerate slides in xylene, 2 changes, 2 minutes each. Place in absolute alcohol, 2 changes, 2 minutes each; 95% alcohol, 2 changes, 2 minutes each; then rinse in distilled water.
2. Stain in 1% methylene blue solution overnight.
3. Decolorize in 95% alcohol.
4. Wash quickly in distilled water.
5. Counterstain in 0.25% basic fuchsin solution for 30 minutes.
6. Decolorize rapidly in 0.5% citric acid solution for 1 or 2 seconds, never more than 3 seconds.
7. Differentiate rapidly in absolute alcohol.
8. Dehydrate slides in 95% alcohol and absolute alcohol 3 changes each. Clear in xylene, 3 changes.
9. Mount coverglass with appropriate medium (refractive index 1.48-1.56).

RESULTS

Rickettsia .... bright red
Nuclei .... blue

Tomlinson-Grocott Method for Staining Malaria Protozoa and Other Parasites in Paraffin Sections[76]

FIXATION 10% Buffered neutral formalin or formalin-alcohol.
PROCESS Paraffin or celloidin.
MICROTOMY Cut sections at 5 micrometers.
CONTROL Use control number 37B.
SOLUTIONS

Polychrome Toluidine Blue

Toluidine blue	1.0	gm
Lithium carbonate	0.5	gm
Distilled water to make	75.0	ml

Allow solution to ripen for 3 to 4 days or until a red-purple deposit appears on beaker above the evaporation line. The ripening may be accelerated by incubating at 37°C for 12 to 24 hours. Return volume to 75.0 ml with distilled water. Then add:

Glycerin	20.0	ml
Ethyl alcohol	5.0	ml

1% Calcium Chloride (Stock)

Calcium chloride	1.0	gm
Distilled water	100.0	ml

1% Phloxine B

Phloxine B	1.0	gm
Distilled water	98.0	ml
When dissolved, add:		
1% Calcium chloride (stock)	1.0	ml

1% Orange G Solution (Stock)

Orange G	1.0	gm
Distilled water	100.0	ml

5% Rosin (Stock)

Rosin*	5.0	gm
Acetone to make	100.0	ml

*Rosin used to treat violin strings may be used for this purpose.

Differentiating Solution
(Make fresh)

5% Rosin (stock)	75.0	ml
100% Ethyl alcohol	25.0	ml
1% Orange G (stock)	0.5	ml

Mix well.

Glacial Acetic Acid Solution
(Make fresh)

Glacial acetic acid	0.2	ml
Distilled water to make	100.0	ml

Ammonium Sulfide Solution

The ammonium sulfide solution for removing malaria pigment from the sections is prepared by diluting ammonium sulfide (light) solution to half strength, or approximately 5%.

STAINING PROCEDURE

1. Decerate slides in xylene, 2 changes, 2 minutes each. Place in absolute alcohol, 2 changes, 2 minutes each; 95% alcohol, 2 changes, 2 minutes each; then rinse in distilled water.
2. Stain for 5 minutes in the 1% phloxine B solution.
3. Rinse in tap water, 1 minute.
4. Stain for 45 seconds to 1 minute in the polychrome toluidine blue solution. Agitate gently.
5. Rinse in running tap water for 1 minute and drain.
6. Place in acetone for 2 minutes.
7. Transfer to the glacial acetic acid solution and agitate gently until clouds of blue no longer appear, about 1 minute.
8. Place in acetone for 2 minutes.
9. Transfer to differentiating solution for 1 to 2 minutes.
10. Transfer to acetone for 3 minutes.
11. Run through 2 changes of acetone-xylol, equal parts, for 3 minutes each.
12. Dehydrate slides in 95% alcohol and absolute alcohol 3 changes each. Clear in xylene, 3 changes.
13. Mount coverglass with appropriate medium (refractive index 1.48-1.56).

RESULTS

Cytoplasm granules	pale rose-red
Trypanosoma cruzi	deep red
Entamoeba histolytica	deep blue

Support stroma .... pale orange-red
Erythrocytes .... brilliant red to orange-red
Malaria parasites .... pale blue
Nuclear chromatin .... deep purple

Richard's Method for Malaria Protozoa[77]

FIXATION Equal parts ethyl ether and ethyl alcohol.
PROCESS Smears.
CONTROL Use control number 37B.
SOLUTIONS

M/15 Phosphate Buffer

Potassium phosphate, monobasic	6.76	gm
Sodium phosphate, dibasic	2.37	gm
Distilled water	100.0	ml

Adjust to pH 6.4 by adding minute quantities of potassium phosphate monobasic if the pH is too high or minute quantities of sodium phosphate dibasic if the pH is too low.

Acridine Orange (Stock)

Acridine orange	0.1	gm
Distilled water	99.0	ml

Acridine Orange (Working)

Acridine orange (stock)	1.0	ml
M/15 Phosphate buffer	99.0	ml

There is some tendency for the pH of acridine orange working solution to shift; if the pH is below 6.4 the organisms will stain a dull orange rather than a vivid red. At higher pH fluorescence is less intense and specific.

M/15 Calcium Chloride

Calcium chloride	16.5	gm
Distilled water	1000.0	ml

STAINING PROCEDURE

1. Let air dry.
2. Fix in equal parts ethyl ether and ethyl alcohol for 15 minutes.
3. Hydrate slides in alcohol dilutions of 80%, 70% and 50% ethyl alcohol.
4. Rinse in 2 changes of M/15 phosphate buffer solution.
5. Stain in acridine orange for 45 seconds.
6. Rinse in M/15 phosphate buffer for 3 minutes.
7. Differentiate in M/15 calcium chloride for 2 1/2 minutes.
8. Mount coverglass with buffer and scan immediately.

RESULTS

Malaria parasite .... red
Leukocytes and platelet .... green fluorescence
Erythrocytes .... faint green
RNA .... red
DNA .... green

Thomas' Method for Malaria Protozoa[55p]

(See Fig. 79 and 80)

FIXATION	10% Buffered formalin, B-5, or Zenker's solution.
PROCESS	Paraffin, frozen, or celloidin.
MICROTOMY	Cut sections at 6 micrometers.
CONTROL	Use control number 37B.
SOLUTIONS	

0.5% Phloxine B

Phloxine B	0.5	gm
Distilled water	100.0	ml
Glacial acetic acid	0.2	ml

A slight precipitate forms but settles on the bottom. Filter before use.

Methylene Blue-Azure B

Methylene blue	0.25	gm
Azure B	0.25	gm
Borax	0.25	gm
Distilled water	100.0	ml

0.2% Glacial Acetic Acid

Glacial acetic acid	0.2	ml
Distilled water	100.0	ml

STAINING PROCEDURES

1. Decerate slides in xylene, 2 changes, 2 minutes each. Place in absolute alcohol, 2 changes, 2 minutes each; 95% alcohol, 2 changes, 2 minutes each; then rinse in distilled water.
2. Place in phloxine B solution for 2 minutes.
3. Rinse well in distilled water.
4. Place in methylene blue-azure B solution for 1 minute.
5. Decolorize with glacial acetic acid solution.
6. Complete differentiation in 95% alcohol, 3 changes.
7. Dehydrate slides in 95% alcohol and absolute alcohol 3 changes each. Clear in xylene, 3 changes.
8. Mount coverglass with appropriate medium (refractive index 1.48-1.56).

RESULTS

Nuclei	blue
Plasma cell cytoplasm	blue
Malaria protozoa	blue
Erythrocytes	pink
Other tissue elements	shades of rose or red

Note: This procedure is very good for the demonstration of malaria parasites in paraffin sections.

Gaffney's Method for Bacteria, Rickettsia and Protozoa[55at]

The following modification of Bell's method has proven to be a simple, consistent and reproducible stain for all types of tissue.

FIXATION	10% Buffered neutral formalin.
PROCESS	Paraffin, frozen, or celloidin.
MICROTOMY	Cut paraffin sections at 6 micrometers.
CONTROL	Use selected control.
SOLUTIONS	

Azure II Eosin Solution

Place a few sterile glass beads into a 500 ml amber bottle.

Azure II eosin	1.3	gm
Glycerin	80.0	ml

Incubate at 56-60°C for 2 hours. Mix and cool.

Add:

Methyl alcohol	170.0	ml
Acetone	170.0	ml

May-Grunwald Stain

Combine in 1L Erlenmeyer flask.

May-Grunwald stain	0.15	gm
Methanol	290.0	ml
Acetone	290.0	ml

Combine Azure II eosin and May-Grunwald solutions. This stock solution can be used immediately but improves with aging.

Acetic Water Solution

Glacial acetic acid	0.01	ml
Distilled water	1000.0	ml

Giemsa Solution (Working)

Giemsa solution (stock)	10.0	ml
Acetic water	50.0	ml

STAINING PROCEDURE

1. Decerate slides in xylene, 2 changes, 2 minutes each. Place in absolute alcohol, 2 changes, 2 minutes each; 95% alcohol, 2 changes, 2 minutes each; then rinse in distilled water.
2. If tissue has been fixed in Zenker's solution, remove mercuric chloride crystals with iodine and sodium thiosulfate (hypo) in the usual manner.
3. Stain in working Giemsa solution for 1 hour.
4. Dehydrate in absolute alcohol, 3 changes.
5. Dehydrate slides in 95% alcohol and absolute alcohol 3 changes each. Clear in xylene, 3 changes.
6. Mount coverglass with appropriate medium (refractive index 1.48-1.56).

RESULTS

Nuclei	blue
Erythrocytes	red
Mast cell granules	purple
Cytoplasm	pink
Eosinophilic granules	red
Bacteria	dark blue
Fungus (cryptococcus)	blue
Rickettsia	violet
Entamoeba histolytica	blue purple
Malarial pigment	black

REMARKS

The only problem encountered with this procedure has been in staining under-decalcified bone specimens, or tissue improperly fixed in an acid solution (Bouin's/Zenker's).

Attwood's Method for Viral Inclusion Bodies and Mucin

FIXATION 10% Buffered neutral formalin.
PROCESS Paraffin, frozen, or celloidin.
MICROTOMY Cut sections at 6 micrometers.
CONTROL Use control number 43.
SOLUTIONS

1% Alcian Green

Alcian green	1.0	gm
Distilled water	100.0	ml
Glacial acetic acid	2.0	ml

Phloxine-Calcium Chloride

Phloxine	0.5	gm
Calcium chloride	0.5	gm
Distilled water	100.0	ml

Tartrazine-Cellosolve

Tartrazine	2.5	gm
Cellosolve	100.0	ml

Scott's Tap Water
(See Page 78)

0.5% Oxalic Acid

Oxalic acid	0.5	gm
Distilled water	100.0	ml

0.5% Potassium Permanganate

Potassium permanganate	0.5	gm
Distilled water	100.0	ml

STAINING PROCEDURE

1. Decerate slides in xylene, 2 changes, 2 minutes each. Place in absolute alcohol, 2 changes, 2 minutes each; 95% alcohol, 2 changes, 2 minutes each; then rinse in distilled water.
2. Place in alcian green solution for 2 minutes.
3. Rinse in distilled water.
4. Place in potassium permanganate solution for 2 minutes.
5. Rinse in distilled water.
6. Place in oxalic acid solution for 2 minutes.
7. Rinse in distilled water.
8. Place in Mayer's hematoxylin solution for 3 minutes.
9. Rinse and blue in Scott's tap water for 1 minute.
10. Place in phloxine solution for 10 minutes.
11. Rinse in distilled water, then in cellosolve.
12. Differentiate each slide with tartrazine-cellosolve solution; check microscopically until the phloxine is removed from erythrocytes.
13. Dehydrate in cellosolve and clear in xylene, 2 changes each.
14. Mount coverglass with appropriate medium (refractive index 1.48-1.56).

RESULTS

Nuclei .... blue-black
Inclusion bodies .... red
Mucin, ground substance of cartilage, granules of
certain mast cells .... green
Squamous epithelium, fibrin .... red
Erythrocytes, plasma, collagen .... yellow

REMARKS

This method has been used very successfully for the demonstration of viral inclusion bodies. For this purpose, however, differentiation at step 12 must be rigidly controlled. NOTE: The permanganate-oxalic acid bleach assists in subsequent differentiation of the phloxine and tends to remove any background staining by alcian green. Any excess staining with tartrazine can be removed by rinsing slides in water until the desired intensity is obtained.

Wilson-Fairchild Method for Malaria Protozoa[78]

FIXATION — 10% Buffered neutral formalin.
PROCESS — Paraffin, frozen, or celloidin.
MICROTOMY — Cut section at 5 micrometers.
CONTROL — Use control number 37B.
SOLUTIONS

Picric Acid-Alcohol

Picric acid	6.0	gm
95% Ethyl alcohol	100.0	ml

Sulfuric Acid

Distilled water	40.0	ml
1N Sulfuric acid	60.0	ml

Always add the acid to the water.

1.5% Ferric ammonium sulfate (stock)	0.75	gm
Sulfuric acid	50.0	ml

Use freshly crushed lavender crystals of ferric ammonium sulfate; dissolve into the acid with gentle heat.

1% Alcoholic Hematoxylin (Stock)

Hematoxylin, certified	0.5	gm
95% Ethyl alcohol	50.0	ml

Dissolve completely before continuing.

Modified Whipf's Iron Hematoxylin

1% Alcoholic hematoxylin (stock)	50.0	ml
1.5% Acetic ferric ammonium sulfate (stock)	50.0	ml

Let stand overnight. Stir and filter. Let ripen for 3 to 5 days. Discontinue use when overripened. Overripening is indicated by inadequate or muddy brown staining of nuclei.

1% Acetic Acid

Glacial acetic acid	1.0	ml
Distilled water	99.0	ml

Biebrich Scarlet-Orange G (Stock)

Biebrich scarlet	0.25	gm
Orange G	0.25	gm
1% Acetic acid	50.0	ml

Phosphotungstic-Phosphomolybdic Acid (Stock)

Phosphotungstic acid	0.5	gm
Phosphomolybdic acid	0.5	gm
1% Acetic acid	50.0	ml

Biebrich Scarlet-Orange G (Working)

Biebrich scarlet-orange G (stock)	50.0	ml
Phosphotungstic-phosphomolybdic acid (stock)	50.0	ml

1.5% Picric Acid (Stock)

Picric acid	1.5	gm
Distilled water	100.0	ml

Let sit overnight. When using to make other solutions, be sure to transfer a few undissolved granules of picric acid to the new solution.

Picro-Aniline Blue Solution

Aniline blue WS	0.1	gm
1.5% Picric acid (stock)	100.0	ml

75% Ethyl Alcohol

95% Ethyl alcohol	79.0	ml
Distilled water	21.0	ml

50% Ethyl Alcohol

95% Ethyl alcohol	52.6	ml
Distilled water	47.4	ml

Alcohol-Acetic Acid

100% Ethyl alcohol	200.0	ml
Glacial acetic acid	0.1	ml

STAINING PROCEDURE

1. Decerate slides in xylene, 2 changes, 2 minutes each. Place in absolute alcohol, 2 changes, 2 minutes each; 95% alcohol, 2 changes, 2 minutes each; then rinse in distilled water.
2. Wash in running tap water until yellow color disappears.
3. Dehydrate in 75% ethyl alcohol.
4. Stain in modified Whipf's hematoxylin for 5 to 10 minutes. Staining time will depend upon degree of ripening. Check slides microscopically.
5. Rinse in distilled water for 2 minutes.
6. Stain with working Biebrich scarlet-orange G solution for 10 minutes.
7. Differentiate Biebrich scarlet in 50% ethyl alcohol for 10 seconds. Remove rapidly to distilled water; rinse for 1 minute.
8. Stain with picro-aniline blue for 1 to 3 minute; rinse in distilled water for 1 minute.
9. Place in alcoholic-acetic acid for 1 minute.
10. Dehydrate slides in 95% alcohol and absolute alcohol 3 changes each. Clear in xylene, 3 changes.
11. Mount coverglass with appropriate medium (refractive index 1.48-1.56).

RESULTS

Malaria protozoa	blue
Nuclei	black
Skeletal and cardiac muscle	salmon to brick red
Erythrocytes	golden yellow
Connective tissue	bright blue

REMARKS

All unused staining solutions are stable. It is necessary to replace the hematoxylin every one or two weeks, depending on the number of slides stained. Other tissue protozoa are also demonstrated easily. Significantly improved staining of connective tissue is obtained compared to other trichrome-type stains.

This polychrome stain is unusual in two respects: (1) the use of a Van Gieson picric acid connective tissue counterstain, and (2) the use of a staining solution containing cytoplasmic dyes combined with both phosphomolybdic and phosphotungstic acid. The clarity and detail achieved by this stain can be attributed to the use of this combined solution and the mordanting action of the acids in that solution.

Luna's Method for Malaria Protozoa in Blood Smears and Tissue Sections

The following technique was developed for the demonstration of malaria protozoa in blood smears and tissue sections.

SOLUTIONS

Giemsa (Stock)

Giemsa powder	1.0	gm
Glycerin	66.0	ml
Methyl alcohol	66.0	ml

Mix glycerin and giemsa powder. Place in a 60°C oven for 2 hours. Let cool, then add 66.0 ml of methyl alcohol.

Buffer Solution, pH 6.0

Potassium phosphate, monobasic	5.0	gm
Sodium phosphate, dibasic	1.0	gm
Distilled water	1000.0	ml

Giemsa (Working)

Giemsa solution (stock)	1.0	ml
Buffer solution pH 6.0	50.0	ml

STAINING PROCEDURE

For paraffin embedded sections, decerate slides in the usual manner and rinse in distilled water. Continue with steps 1-5 as outlined below.

For blood smears, go directly to step 1.

1. Fix thin blood films by immersion in methyl alcohol for 3 to 4 minutes.
2. Place slides, without drying, in working giemsa solution for 45 minutes.
3. Rinse in distilled water.
4. Dehydrate rapidly in 95% alcohol, absolute alcohol and clear in xylene, 2 changes each.
5. Mount coverglass with appropriate medium (refractive index 1.48-1.56).

RESULTS

Erythrocytes	salmon-pink
Cytoplasma	blue
Chromatin	red
Malarial protozoa	greenish black

Massignani-Malferrari Method for Negri Bodies[79]

FIXATION	10% Buffered neutral formalin.
PROCESS	Paraffin.
MICROTOMY	Cut sections at 6 micrometers.
CONTROL	Use control number 34A.

SOLUTIONS

Harris' Hematoxylin (Without Acetic Acid)
(See Page 81)

0.5% Hydrochloric Acid

Concentrated hydrochloric acid	0.5	ml
Distilled water	100.0	ml

Lithium Carbonate

Lithium carbonate, saturated aqueous	1.0	ml
Distilled water	200.0	ml

Eosin-Phosphotungstic Acid Solution

Grind together 1 gm of eosin and 0.7 gm of phosphotungstic acid. Add 10 ml of distilled water and make up to 200 ml with absolute alcohol. Add 2 drops of saturated lithium carbonate and stir for 10 minutes. Filter. (For best results, it is necessary that the eosin and phosphotungstic acid be ground together in the solid state.)

STAINING PROCEDURE

1. Decerate slides in xylene, 2 changes, 2 minutes each. Place in absolute alcohol, 2 changes, 2 minutes each; 95% alcohol, 2 changes, 2 minutes each; then rinse in distilled water.
2. Place in Harris' hematoxylin for 2 minutes.
3. Wash for 5 minutes in running tap water.
4. Dip 8 times in hydrochloric acid solution and wash 5 minutes in running tap water.
5. Place in diluted lithium carbonate solution for 1 minute.
6. Wash in running tap water 10 minutes; then dehydrate in 50% alcohol, 70% alcohol, 80% alcohol, 90% alcohol, and 95% absolute alcohol, 10 dips each.
7. Stain 8 minutes in eosin solution. Wash thoroughly in water and dehydrate by short dips in 50% alcohol, 70% alcohol, 80% alcohol, and 90% alcohol and longer dips in 95% alcohol. Then use 2 or 3 changes of absolute alcohol, 4 minutes each.
8. Blot with filter paper.
9. Clear in 2 changes of xylene, 2 minutes each.
10. Mount coverglass with appropriate medium (refractive index 1.48-1.56).

RESULTS

Negri bodies	deep red
Nuclei	light blue

REMARKS

To obtain a good stain it is important not to overstain in Harris' hematoxylin, to differentiate very well in dilute HCl and follow the instructions exactly regarding the dips needed after staining in eosin.

Gridley's Method for Ingested Red Blood Cells and Entamoeba Histolytica

FIXATION	10% Buffered neutral formalin.
PROCESS	Paraffin, frozen, or celloidin.
MICROTOMY	Cut sections at 6 micrometers.
CONTROL	Use control number 37B.
SOLUTIONS	

Harris' Hematoxylin Solution
(See Page 81)

Weigert's Iron Hematoxylin Solution
(See Page 79)

Aniline-Eosin Solution

Eosin Y	1.5	gm
80% Alcohol	100.0	ml
Aniline (see remarks)	3.0	ml
Glacial acetic acid	1.0	ml

Naphthol Green B Solution

Naphthol green B	1.0	gm
Distilled water	100.0	ml
Glacial acetic acid	1.0	ml

STAINING PROCEDURE

1. Decerate slides in xylene, 2 changes, 2 minutes each. Place in absolute alcohol, 2 changes, 2 minutes each; 95% alcohol, 2 changes, 2 minutes each; then rinse in distilled water.
2. Place in Harris' hematoxylin solution for 10 minutes or Weigert's iron hematoxylin solution for 3 minutes.
3. Wash in running tap water for 2 minutes.
4. Decolorize in 1% acid alcohol, 5 dips.
5. Wash in running tap water for 5 minutes.
6. Blue in ammonia water.
7. Wash in running tap water for 5 minutes.
8. Place in aniline-eosin solution for 5 minutes.
9. Rinse well in distilled water. Sections should be a deep rose color.
10. Place in naphthol green B solution for 5 minutes.
11. Dehydrate slides in 95% alcohol and absolute alcohol 3 changes each. Clear in xylene, 3 changes.
12. Mount coverglass with appropriate medium (refractive index 1.48-1.56).

RESULTS

Ingested red blood cells	red
Entamoeba histolytica	green
Background	green

REMARKS
Use aniline oil under hood and follow instructions indicated on the material safety sheets.

Shorr's Method for Inclusion Bodies

FIXATION Any well-fixed tissue.
PROCESS Paraffin or celloidin.
MICROTOMY Cut sections at 6 micrometers.
CONTROL Use control number 43.
SOLUTIONS

1% Acid Alcohol

Concentrated hydrochloric acid	1.0	ml
70% Alcohol	99.0	ml

Shorr's Stain

50% Alcohol	100.0	ml
Biebrich scarlet	0.5	gm
Orange G	0.25	gm
Fast green, FCF	0.075	gm
Phosphotungstic acid	0.5	gm
Phosphomolybdic acid	0.5	gm
Glacial acetic acid	1.0	ml

Harris' Hematoxylin
(See Page 81)

Ammonia Water

Ammonium hydroxide	2-3	ml
Tap water	1000.0	ml

STAINING PROCEDURE

1. Decerate slides in xylene, 2 changes, 2 minutes each. Place in absolute alcohol, 2 changes, 2 minutes each; 95% alcohol, 2 changes, 2 minutes each; then rinse in distilled water.
2. Remove mercury precipitates by running slides through iodine and sodium thiosulfate solutions, if necessary.
3. Stain in Harris' hematoxylin for 5 minutes.
4. Rinse in running tap water.
5. Decolorize in 1% acid alcohol until there is no hematoxylin in the cytoplasm of the cells. Check under microscope.
6. Wash in running tap water.
7. Place sections in ammonia water to blue. Check under microscope.
8. Wash for 10 minutes in running tap water.
9. Place sections in Shorr's staining solution for 1 minute.
10. Dip 4 times in 70% alcohol.
11. Dip 5 times in 95% alcohol.
12. Rinse in 2 changes of 100% alcohol.
13. Clear in 2 changes of xylene.
14. Mount coverglass with appropriate medium (refractive index 1.48-1.56).

RESULTS

Inclusion bodies	brilliant red
Connective tissue	light green
Elastic tissue	purplish-red
Muscle	red
Keratin	orange
Erythrocytes	orange-red
Nuclei	blue

Chapter 9

Hematological and Nuclear Elements

Maximow's Method for Bone Marrow Cells[55q]

FIXATION Zenker's, B-5 or Carnoy's solution.
PROCESS Paraffin, frozen, or celloidin.
MICROTOMY Cut sections at 6 micrometers.
CONTROL Use bone marrow specimen.
SOLUTIONS

Mayer's Hematoxylin
(See Page 81)

0.1% Eosin (Stock)

Eosin Y, yellowish	0.1	gm
Distilled water	100.0	ml

0.1% Azure II (Stock)

Azure II	0.1	gm
Distilled water	100.0	ml

Eosin-Azure II (Working)

0.1% Eosin (stock)	10.0	ml
Distilled water	100.0	ml
0.1% Azure II (stock)	10.0	ml
Glacial acetic acid	2	drops

Must be prepared fresh for each use with distilled water or buffer solution, pH 6.8-7.0.

STAINING PROCEDURE

1. Decerate slides in xylene, 2 changes, 2 minutes each. Place in absolute alcohol, 2 changes, 2 minutes each; 95% alcohol, 2 changes, 2 minutes each; then rinse in distilled water.
2. Place in Mayer's hematoxylin solution for 5 minutes.
3. Wash in tepid running tap water for 10 minutes.
4. Place in working eosin-azure II solution for 3 hours.
5. Decolorize in 95% alcohol until blue ceases to flow into the alcohol and erythrocytes and collagen are pink.
6. Dehydrate slides in 95% alcohol and absolute alcohol 3 changes each. Clear in xylene, 3 changes.
7. Mount coverglass with appropriate medium (refractive index 1.48-1.56).

RESULTS

Nuclei	blue
Basophil leucocyte	purple to violet
Mast cell granules	purple to violet

Cartilage	purple
Erythrocytes	pink
Eosinophil granules	pink
Cytoplasm	blue to pink

Verfuerth-Luna Method for Eosinophil Granules

(See Fig. 86)

FIXATON 10% Neutral formalin, phosphate buffered
PROCESS Paraffin, frozen, or celloidin.
MICROTOMY Cut paraffin sections at 6 micrometers.
SOLUTIONS

1% Lissamine-Rhodamine B

Lissamine-rhodamine B	1.0	gm
Distilled water	100.0	ml

Borax Methylene Blue

Methylene blue	0.1	gm
Borax, dehydrate	0.1	gm
Distilled water	100.0	ml

STAINING PROCEDURE

1. Decerate slides in xylene, 2 changes, 2 minutes each. Place in absolute alcohol, 2 changes, 2 minutes each; 95% alcohol, 2 changes, 2 minutes each; then rinse in distilled water.
2. Stain in 1% lissamine-rhodamine B for 15 minutes.
3. Wash in running tap water for 2 minutes to remove excess stain.
4. Dip slides, individually, 8 times in borax methylene blue.
5. Wash in running tap water 1 minute or less (take care not to wash out borax methylene blue counterstain).
6. Dehydrate slides in 95% alcohol and absolute alcohol 3 changes each. Clear in xylene, 3 changes.
7. Mount coverglass with appropriate medium (refractive index 1.48-1.56).

RESULTS

Eosinophil granules	red
Mast cells	bluish-purple
Background	blue

Kass-Cooper-Choi Method for Eosinophil Cell Granules[80]

FIXATION B-5 or formalin.
PROCESS Paraffin, smears (see remarks).
MICROTOMY Cut sections at 6 micrometers.

SOLUTIONS

B-5 - Formalin

37% Formalin	10.0	ml
B-5 fixative	90.0	ml

Lugol's Solution

Iodine	1.0	gm
Potassium iodide	2.0	gm
Distilled water	100.0	ml

5% Sodium Thiosulfate

Sodium thiosulfate	5.0	gm
Distilled water	100.0	ml

10% Acid Red 137*

Acid red 137	10.0	gm
Distilled water	100.0	ml

*Powdered form obtained as Alcian Crocein FL (Miles, Inc., Mobay Road, Pittsburgh, PA 15205).

STAINING PROCEDURE

1. Decerate slides in xylene, 2 changes, 2 minutes each. Place in absolute alcohol, 2 changes, 2 minutes each; 95% alcohol, 2 changes, 2 minutes each; then rinse in distilled water.
2. Place in Lugol's solution for 3 minutes.
3. Rinse in distilled water.
4. Place in 5% sodium thiosulfate for 3 minutes.
5. Wash in running tap water and rinse in distilled water.
6. Stain in unfiltered 10% acid red 137 for 10 minutes.
7. Dehydrate slides in 95% alcohol and absolute alcohol 3 changes each. Clear in xylene, 3 changes.
8. Mount coverglass with appropriate medium (refractive index 1.48-1.56).

RESULTS

Eosinophils	bright red-orange
Nuclei	light tan
Background	pale cream

REMARKS

For smears and coverglass of blood and bone marrow, nasal secretions (patients with allergic rhinitis) the slides are fixed in B-5 - Formalin. Sections are then treated according to steps 2-6, rinsed in distilled water for 30 seconds and air dried. Mount coverglass as outlined in step 8.

Methods for Staining Charcot-Leyden Crystals

Charcot-Leyden crystals derive their name from the two German physicians, Jean Martin Charcot and Ernst Victor von Leyden, who first described them. These crystalline structures, which are protein in nature, are found wherever eosinophilic leukocytes are undergoing fragmentation. The development of the crystal represents an uncommon fundamental intracellular process, with rapid protein crystal formation.

Charcot-Leyden crystals have been found in the tissue of patients with eosinophilic granuloma of bone, in the sputum of patients with bronchial asthma, in the stools of patients with amebic dysentery, in ulcerative diseases of the colon and a variety of other diseases.

The crystals are hexagonal or elongated dipyramidal shapes, depending on the plane in which they are sectioned. Some crystals exhibit coalescence of eosinophilic granules from which three or more sides of a crystal emerge. Some large crystals incorporate the granules into the crystal proper, while still other crystals appear perfect in shape without evidence of included granules. Charcot-Leyden crystals do not stain with sudan black although eosinophil granules do. The crystals are negatively birefringent, insoluble in lipid solvents and do not stain with fat soluble dyes.

The following procedures have been used successfully for the demonstration of these Charcot-Leyden crystals in paraffin sections.

Luna's Modified Brown-Brenn Method for Charcot-Leyden Crystals

(See Fig. 87, 88 and 89)

FIXATION	10% Buffered neutral formalin.
PROCESSING	Paraffin, frozen, or celloidin.
MICROTOMY	Cut sections at 4 to 6 micrometers.
CONTROL	Use control number 9A.
SOLUTIONS	

1% Crystal Violet

Crystal violet	1.0	gm
Distilled water	100.0	ml

5% Sodium Bicarbonate

Sodium bicarbonate	5.0	gm
Distilled water	100.0	ml

Gram's Iodine

Iodine	1.0	gm
Potassium iodide	2.0	gm
Distilled water	300.0	ml

Basic Fuchsin (Stock)

Basic fuchsin, to saturation, approx.	0.25	gm
Distilled water	100.0	ml

Basic Fuchsin (Working)

Basic fuchsin (stock)	0.1	ml
Distilled water	100.0	ml

Picric Acid-Acetone Solution

Picric acid	0.1	gm
Acetone	100.0	ml

Ether-Acetone Solution

Ether	50.0	ml
Acetone	50.0	ml

STAINING PROCEDURE

1. Decerate slides in xylene, 2 changes, 2 minutes each. Place in absolute alcohol, 2 changes, 2 minutes each; 95% alcohol, 2 changes, 2 minutes each; then rinse in distilled water.
2. Remove mercury precipitates with iodine and sodium thiosulfate solutions if necessary.
3. Place slides on staining rack. Pour on approximately 1.0 ml (or 20 drops) of 1% crystal violet solution, and add 5 drops of 5% sodium bicarbonate solution. Agitate gently for 1 minute. If preferred, the two solutions may be mixed just before use.
4. Wash in running tap water to remove excess stain.
5. Flood slides with Gram's iodine solution for 1 minute.
6. Rinse in running tap water and blot with filter paper to dry completely.
7. Decolorize with ether-acetone dropped on slides until wash is clear.
8. Stain with working basic fuchsin solution for 1 minute.
9. Dip slide in water; blot gently, but do not dry completely.
10. Dip slide in acetone to start reaction.
11. Differentiate immediately with 0.1% picric acid-acetone until sections are yellowish pink.
12. Rinse quickly in acetone and clear tissue in a mixture of equal parts of acetone and xylene.
13. Mount coverglass with appropriate medium (refractive index 1.48-1.56).

RESULTS

Charcot-Leyden crystals	red
Other tissue elements	yellow

Luna's Method for Charcot-Leyden Crystals and Eosinophil Granules[81]

(See Fig. 88, 89 and 90)

FIXATION 10% Buffered neutral formalin.
PROCESSING Paraffin embedded specimens.
MICROTOMY Cut sections at 6 micrometers.
CONTROL Use control number 9A.
SOLUTIONS

Hematoxylin (Stock)

Hematoxylin	1.0	gm
100% Ethyl alcohol	100.0	ml

Ferric Chloride (Stock)

Ferric chloride, 29% aqueous	4.0	ml
Distilled water	95.0	ml
Concentrated hydrochloric acid	1.0	ml

Hematoxylin Solution (Working)

Hematoxylin (stock)	50.0	ml
Ferric chloride (stock)	50.0	ml

Biebrich Scarlet-Acid Fuchsin

Biebrich scarlet, 1% aqueous	80.0	ml
Acid fuchsin, 1% aqueous	20.0	ml
Glacial acetic acid	2.0	ml

1% Acid Alcohol Solution

95% Ethyl alcohol	100.0	ml
Concentrated hydrochloric acid	1.0	ml

Biebrich Scarlet-Hematoxylin Stain

Biebrich scarlet-acid fuchsin	5.0	ml
Hematoxylin solution (working)	45.0	ml

Lithium Carbonate

Lithium carbonate	0.5	gm
Distilled water	100.0	ml

STAINING PROCEDURE

1. Decerate slides in xylene, 2 changes, 2 minutes each. Place in absolute alcohol, 2 changes, 2 minutes each; 95% alcohol, 2 changes, 2 minutes each; then rinse in distilled water.
2. Place slides in Biebrich scarlet-hematoxylin staining solution for 5 minutes.
3. Dip slides in distilled water to remove excess hematoxylin.
4. Differentiate slides in 1% acid alcohol by dipping slide 8 times.
5. Rinse slides in distilled water to remove acid alcohol.
6. Dip slides in lithium carbonate solution until sections turn blue, usually 8 dips.
7. Wash sections in running tap water for 1 minute.
8. Dehydrate slides in 95% alcohol and absolute alcohol 3 changes each. Clear in xylene, 3 changes.
9. Mount coverglass with appropriate medium (refractive index 1.48-1.56).

RESULTS

Charcot-Leyden crystals	red
Eosinophil granules	reddish-orange
Erythrocytes	reddish-orange
Nuclear elements	blue

Mallory's Method for Hemofuscin[55r]

FIXATION	Zenker's solution, absolute alcohol, or 10% buffered neutral formalin.
PROCESS	Paraffin, frozen, or celloidin.
MICROTOMY	Cut sections at 6 micrometers.
CONTROL	Use control number 22.
SOLUTIONS	

Alum-Hematoxylin

Hematoxylin	1.0	gm
Aluminum ammonium or potassium sulfate	20.0	gm
Distilled water	400.0	ml
Thymol	1.0	gm

0.5% Basic Fuchsin

Basic fuchsin	0.5	gm
95% Alcohol	50.0	ml
Distilled water	50.0	ml

STAINING PROCEDURE

1. Decerate slides in xylene, 2 changes, 2 minutes each. Place in absolute alcohol, 2 changes, 2 minutes each; 95% alcohol, 2 changes, 2 minutes each; then rinse in distilled water.
2. Place in hematoxylin for 5 minutes.
3. Wash thoroughly in running tap water.
4. Place in basic fuchsin solution for 30 minutes.
5. Wash in running tap water.
6. Differentiate in 95% alcohol until hemofucsin granules stand out sharply against a gray background.
7. Dehydrate slides in 95% alcohol and absolute alcohol 3 changes each. Clear in xylene, 3 changes.
8. Mount coverglass with appropriate medium (refractive index 1.48-1.56).

RESULTS

Nuclear chromatin	blue
Hemofuscin granules	bright red

Puchtler-Sweat Method for Hemoglobin and Hemosiderin[82]

FIXATION	Zenker-formol.
PROCESS	Paraffin, frozen, or celloidin.
MICROTOMY	Cut sections at 6 micrometers.
CONTROL	Use control number 22.
SOLUTIONS	

2% Potassium Ferrocyanide (Stock)

Potassium ferrocyanide	2.0	gm
Distilled water	100.0	ml

2% Hydrochloric Acid (Stock)

Concentrated hydrochloric acid	2.0	ml
Distilled water	100.0	ml

Potassium Ferrocyanide-Hydrochloric Acid (Working)

Potassium ferrocyanide (stock)	50.0	ml
Hydrochloric acid (stock)	50.0	ml

Mix just before use.

5% Tannic Acid

Tannic acid	5.0	gm
Distilled water	100.0	ml

1% Phosphomolybdic Acid

Phosphomolybdic acid	1.0	gm
Distilled water	100.0	ml

Phloxine B

Phloxine B	5.0	gm
Glacial acetic acid	30.0	ml
Methyl alcohol	80.0	ml

Adjust pH to 3.7.

Methyl Alcohol-Glacial Acetic Acid

Glacial acetic acid	10.0	ml
Methyl alcohol	90.0	ml

STAINING PROCEDURE

1. Decerate slides in xylene, 2 changes, 2 minutes each. Place in absolute alcohol, 2 changes, 2 minutes each; 95% alcohol, 2 changes, 2 minutes each; then rinse in distilled water.
2. Mordant in Zenker-formol for 3 hours if originally fixed in 10% formalin.
3. Remove mercuric chloride crystals with iodine and clear with sodium thiosulfate.
4. Rinse in distilled water, several changes. Avoid any contact with metal.
5. Place in potassium ferrocyanide-hydrochloric acid solution for 30 minutes.
6. Rinse in distilled water, 3 changes.
7. Place in tannic acid for 10 minutes.
8. Rinse in distilled water, 3 changes.
9. Place in phosphomolybdic acid solution for 10 minutes.
10. Rinse in distilled water, 3 changes.
11. Place in phloxine B for 5 minutes. Do not filter phloxine B solution.
12. Rinse in methyl alcohol-glacial acetic acid, 2 changes.
13. Dehydrate slides in 95% alcohol and absolute alcohol 3 changes each. Clear in xylene, 3 changes.
14. Mount coverglass with appropriate medium (refractive index 1.48-1.56).

RESULTS

Hemoglobin	red
Hemosiderin..	dark blue to blue
Lipofucsin	green

Lillie's Method for Deoxyribonucleic Acid (DNA)[55s]

(See Fig. 91)

FIXATION 10% Buffered neutral formalin.
PROCESS Paraffin, frozen, or celloidin.
MICROTOMY Cut sections at 6 micrometers.
CONTROL Use control number 15.
SOLUTIONS

Schiff's Solution

Dissolve 1.0 gm basic fuchsin in 200.0 ml hot distilled water. Bring to boiling point. Cool to 50°C. Filter and add 20.0 ml normal hydrochloric acid. Cool further and add 1.0 gm anhydrous sodium bisulfite, or sodium metabisulfite. Keep in the dark for 48 hours until solution becomes straw-colored. Store in refrigerator.

Normal Hydrochloric Acid

Concentrated hydrochloric acid	83.5	ml
Distilled water	916.5	ml

0.05% Sodium Bisulfite

Sodium bisulfite	0.05	gm
Distilled water	100.0	ml

0.01% Fast Green

Fast green, FCF	0.01	gm
95% Alcohol	100.0	ml

STAINING PROCEDURE

1. Decerate slides in xylene, 2 changes, 2 minutes each. Place in absolute alcohol, 2 changes, 2 minutes each; 95% alcohol, 2 changes, 2 minutes each; then rinse in distilled water.
2. Place slides in normal hydrochloric acid solution at 60°C for 15 minutes.
3. Place in Schiff's for 10 minutes.
4. Immerse in 0.05% sodium bisulfite solution, 3 successive baths for 2 minutes each.
5. Wash in running tap water for 5 minutes.
6. Immerse in 0.01% fast green solution for 5 seconds.
7. Dehydrate slides in 95% alcohol and absolute alcohol 3 changes each. Clear in xylene, 3 changes.
8. Mount coverglass with appropriate medium (refractive index 1.48-1.56).

RESULTS

Deoxyribonucleic acid (DNA)	deep reddish-purple
Cytoplasm	light green

Note: The fast green solution does not wash out in alcohol, but if it is too intense, it may be removed promptly in water.

Menzie's Method for Deoxyribonucleic (DNA) and Ribonucleic (RNA) Acids[55t]

FIXATION 10% Buffered neutral formalin.
PROCESS Paraffin, frozen, or celloidin.
MICROTOMY Cut sections at 6 micrometers.
CONTROL Use control number 15.
SOLUTIONS

Hydrochloric Acid-Tetrahydrofuran

Concentrated hydrochloric acid	10.0	ml
Tetrahydrofuran	90.0	ml

1% Azure B (Stock)

Azure B	1.0	gm
Distilled water	100.0	ml

0.1% Basic Fuchsin (Stock)

Basic fuchsin	0.1	gm
Distilled water	100.0	ml

Azure B-Basic Fuchsin (Working)

1% Azure B (stock)	30.0	ml
0.1% Basic fuchsin (stock)	8.0	ml
Glacial acetic acid	2.0	ml

STAINING PROCEDURE

1. Decerate slides in xylene, 2 changes, 2 minutes each. Place in absolute alcohol, 2 changes, 2 minutes each; 95% alcohol, 2 changes, 2 minutes each; then rinse in distilled water.
2. Place in hydrochloric acid-tetrahydrofuran solution at 37°C for 3 minutes.
3. Place in working azure B-basic fuchsin solution for 30 minutes.
4. Rinse in acetone, clear in xylene, 2 to 3 changes each.
5. Mount coverglass with appropriate medium (refractive index 1.48-1.56).

RESULTS

Deoxyribonucleic acid (DNA)	red
Ribonucleic acid (RNA)	blue

REMARKS

This method is excellent for rat DNA and RNA. May be used on human tissue but the results are not as good.

Emanuele's Method for Deoxyribonucleic and Ribonucleic Acid

(See Fig. 92, 93 and 94)

FIXATION Carnoy's is the preferred fixative; however, 10% buffered neutral formalin may be used if tissue is not more than 3 mm thick and fixation limited to 7 or 8 hours maximum.

PROCESS Paraffin, frozen, or celloidin.

MICROTOMY Cut paraffin sections at 4 micrometers.

CONTROL Use control number 39.

SOLUTIONS

Methyl Green (Stock)

Methyl green	2.1	gm
Distilled water	75.0	ml

Pyronine GS (Stock)

Pyronine GS	0.9	gm
Distilled water	75.0	ml

Add a few glass beads to aid in dissolving the pyronine.

Sodium Acetate (Stock)

Sodium acetate	1.64	gm
Distilled water	100.0	ml

1.2% Glacial Acetic Acid (Stock)

Glacial acetic acid	1.2	ml
Distilled water	100.0	ml

Prepare the above solutions separately and agitate occasionally for 4 days following preparation.

Methyl Green-Pyronine GS (Working)

Methyl green (stock)	75.0	ml
Pyronine GS (stock)	75.0	ml

Cap this solution tightly and place in a 60°C oven for 2 hours.
Cool and add:

Sodium acetate (stock)	90.0	ml
Glacial acetic acid (stock)	60.0	ml

Allow solution to stand overnight to achieve a pH of 4.8.

STAINING PROCEDURE

1. Decerate slides in xylene, 2 changes, 2 minutes each. Place in absolute alcohol, 2 changes, 2 minutes each; 95% alcohol, 2 changes, 2 minutes each; then rinse in distilled water.
2. Place slides in working methyl green-pyronine solution for 15 dips.
3. Blot slides dry with bibulous paper.
4. Differentiate slides in distilled water, 1 or 2 rapid dips.
5. Dip slides rapidly 3 times each in 2 changes of cold (4°C) 80% alcohol.
6. Dip slides 3 times each in 2 changes of acetone.
7. Dip slides 2 times in a mixture of acetone-xylene, equal parts.
8. Clear slides in 3 changes of xylene for 1 minute each.
9. Mount coverglass with resinous medium.

RESULTS

Deoxyribonucleic acid	blue to blue-green
Ribonucleic acid	red

REMARKS

Check results microscopically. If nuclear chromatin appear too bluish-green, repeat the 2 rapid dips in distilled water in step 4. If nuclear chromatin appear too red, dip slides in the cold 80% alcohol as in step 5. This alcohol should be kept in the refrigerator until ready to use.

Method for Digestion of Deoxyribonucleic Acid (DNA)

FIXATION 10% Buffered neutral formalin, Carnoy's or Bouin's solution. (Fixatives containing mercuric, chromic, or other heavy metal ions cannot be used.)

PROCESS Paraffin, frozen, or celloidin.

MICROTOMY Cut sections at 6 micrometers. Cut 2 sections, one to be used for control (see step 2).

CONTROL Use control number 15.

SOLUTIONS

Tris Buffer (Stock)

Solution A

Tris(hydroxymethyl)aminomethane	24.228	gm
Distilled water	1000.0	ml

Solution B

Concentrated hydrochloric acid	8.35	ml
Distilled water	991.7	ml

Tris Buffer, pH 7.3 (Working)

Solution A	10.0	ml
Solution B	17.0	ml
Distilled water	13.0	ml

Deoxyribonuclease

Deoxyribonuclease	0.001	gm
0.2M Magnesium chloride	22.5	ml
0.2M Calcium chloride	2.5	ml
0.1M Tris buffer, pH 7.3 (working)	20.0	ml
Distilled water	55.0	ml

Deoxyribonuclease activity should be 400 to 600 kunit3 per mg protein.

DIGESTION PROCEDURE

1. Decerate slides in xylene, 2 changes, 2 minutes each. Place in absolute alcohol, 2 changes, 2 minutes each; 95% alcohol, 2 changes, 2 minutes each; then rinse in distilled water.

2. (a) Treat one section with deoxyribonuclease solution for 2 hours at 37°C.
   (b) Treat the second section in working tris buffer solution for 2 hours at 37°C.
3. Remove both sections and wash together in running tap water for 5 minutes.
4. Stain as desired.

RESULTS
The material that stains in the section incubated in working tris buffer but does not stain in the section treated with deoxyribonuclease is presumably deoxyribonucleic acid.

Method for Digestion of Ribonucleic Acid (RNA)

FIXATION — 10% Buffered neutral formalin, Carnoy's or Bouin's solution. (Fixatives containing mercuric, chromic, or other heavy metal ions cannot be used.)
PROCESS — Paraffin, frozen, or celloidin.
MICROTOMY — Cut sections at 6 micrometers. Cut 2 sections, one to be used for control (see step 2).
CONTROL — Use control number 15.
SOLUTIONS

Ribonuclease

Ribonuclease	0.5	gm
Distilled water, pH 6.8	100.0	ml

Adjust pH with 0.1N sodium hydroxide or hydrochloric acid. Ribonuclease activity should be 50 to 100 kunit3 per mg protein.

DIGESTION PROCEDURE
1. Decerate slides in xylene, 2 changes, 2 minutes each. Place in absolute alcohol, 2 changes, 2 minutes each; 95% alcohol, 2 changes, 2 minutes each; then rinse in distilled water.
2. (a) Treat one section in ribonouclease solution for 3 hours at 37°C.
   (b) Treat the second section in distilled water for 3 hours at 37°C.
3. Remove both sections and wash together in running tap water for 10 minutes.
4. Stain as desired.

RESULTS
The material that stains in the section incubated in distilled water but does not stain in the section treated with ribonuclease solution is presumably ribonucleic acid.

Guard's Method for Sex Chromatin (Barr Bodies)[83]

These two methods can be employed satisfactorily for chromatin as outlined below.

FIXATION — Fix immediately in 95% alcohol.
PROCESS — Smears (buccal mucosal cells).
CONTROL — Use control smear obtained from the buccal mucosa of a female.

SOLUTIONS

1% Biebrich Scarlet

Biebrich scarlet	1.0	gm
Phosphotungstic acid	0.3	gm
Glacial acetic acid	5.0	ml
50% Alcohol	100.0	ml

0.5% Fast Green

Fast green FCF	0.5	gm
Phosphomolybdic acid	0.3	gm
Phosphotungstic acid	0.3	gm
Glacial acetic acid	5.0	ml
50% Alcohol	100.0	ml

Hematoxylin

Harris' hematoxylin (modified)	0.5	ml
50% Alcohol	100.0	ml

STAINING PROCEDURE - Method 1

1. From 95% alcohol fixative transfer specimen to 70% alcohol for 2 minutes.
2. Place in 1% Biebrich scarlet for 2 minutes.
3. Rinse in 50% alcohol.
4. Differentiate in 0.5% fast green solution for 1 to 4 hours (see remarks).
5. Place in 50% alcohol for 5 minutes.
6. Dehydrate slides in 95% alcohol and absolute alcohol 3 changes each. Clear in xylene, 3 changes.
7. Mount coverglass with appropriate medium (refractive index 1.48-1.56)

RESULTS

Sex chromatin (Barr bodies) .... red
Background .... green

STAINING PROCEDURE - Method 2

1. From 95% alcohol fixative transfer specimen to 70% alcohol for 2 minutes.
2. Immerse in hematoxylin solution for 15 seconds.
3. Immerse in 1% Biebrich scarlet solution for 2 minutes.
4. Place in 50% alcohol for 5 minutes.
5. Differentiate in 0.5% fast green solution for 1 to 4 hours (see remarks).
6. Dehydrate slides in 95% alcohol and absolute alcohol 3 changes each. Clear in xylene, 3 changes.
7. Mount coverglass with appropriate medium (refractive index 1.48-1.56).

RESULTS

Sex chromatin (Barr bodies) .... red
Nuclear chromatin .... blue
Cytoplasm .... green

REMARKS

Check the differentiation (step 4, method 1, step 5, method 2) under a microscope at hourly intervals. When all the cells reveal green cytoplasm and the vesicular nuclei are also green, the reaction is complete, usually in about 4 hours. The pyknotic nuclei, however, will not differentiate and will be bright red in color.

Klinger-Ludwig Method for Sex Chromatin (Barr Bodies)[84]

FIXATION Equal parts of 95% alcohol and ethyl ether.
PROCESS Smears (buccal mucosal cells).
SOLUTIONS

Normal Hydrochloric Acid
(See Page 199)

1% Thionin (Stock)

Thionin	1.0	gm
50% Alcohol	100.0	ml

Buffer (Stock)

Sodium acetate	9.714	gm
Sodium barbituate	14.714	gm
Distilled water to make	550.0	ml

Thionin (Working)

1N Hydrochloric acid	32.0	ml
Buffer solution (stock)	28.0	ml
1% Thionin solution (stock)	40.0	ml

STAINING PROCEDURE

1. Obtain relatively thick smears of buccal mucosal cells on slides.
2. Fix smears promptly in equal parts of 95% alcohol and ethyl ether for 15 minutes.
3. Place in absolute alcohol for 3 minutes.
4. Immerse for 2 minutes in 0.2% parlodion made in equal parts absolute alcohol and ethyl ether.
5. Air-dry for 15 minutes.
6. Immerse in 70% alcohol for 5 minutes.
7. Place in distilled water, 2 changes, for 5 minutes each.
8. Hydrolyze in normal hydrochloric acid solution at 56°C for 5 minutes.
9. Place in distilled water, 2 changes, 5 minutes each.
10. Place in working thionin solution for 5 minutes.
11. Dehydrate slides in 95% alcohol and absolute alcohol 3 changes each. Clear in xylene, 3 changes.
12. Mount coverglass with appropriate medium (refractive index 1.48-1.56).

RESULTS

Sex chromatin (Barr bodies)	purplish-red
Other cell structures	light blue

Leder's Method for Neutrophilic Myeloid and Tissue Mast Cells[85]

FIXATION 10% Buffered neutral formalin. Do not use methyl alcohol.
PROCESS Paraffin, frozen, or celloidin.
MICROTOMY Cut sections at 6 micrometers. (Will work on unfixed blood films.)
CONTROL Use control number 31.
SOLUTIONS

4% Pararosaniline (Stock)

Pararosaniline hydrochloride	1.0	gm
Distilled water	20.0	ml
Concentrated hydrochloric acid	5.0	ml

Gently warm the solution after mixing. Cool, then filter and refrigerate. Solution is stable for several months.

4% Sodium Nitrite (Stock)

Sodium nitrite	0.2	gm
Distilled water	5.0	ml

Prepare fresh each time.

Michaelis Veronal Acetate Buffer (Stock)

Sodium acetate, trihydrate	9.714	gm
Sodium barbiturate	14.714	gm
Distilled water, carbon dioxide free, to make	500.0	ml

Refrigerate. Solution stable for several months. If sodium acetate anhydrous salt is used, use 5.85 gm.

Michaelis Veronal Acetate Buffer (Working)

Michaelis veronal acetate buffer (stock)	42.4	ml
0.1 Normal hydrochloric acid	37.6	ml

Make fresh each time.

Esterase Substrate (Stock)

Naphthol AS-D chloroacetate esterase	20.0	mg
N-N Dimethyl formamide	2.0	ml

Prepare just before use.

Leder's Stain

4% Paraosaniline (stock)	2	drops
4% Sodium nitrite (stock)	2	drops
Michaelis veronal acetate buffer, working	60.0	ml
Esterase substrate	2.0	ml

To mix, add 2 drops of the 4% pararosaniline solution and 2 drops of the 4% sodium nitrite solution to a 100 ml beaker using a 1 ml serological pipette. Mix and wait 60 seconds. Then add 60 ml of the Michaelis veronal acetate buffer (working). At this point adjust the pH of the solution to 6.3 by adding 1N hydrochloric acid, drop by drop. The premixed esterase substrate is then added to the beaker and stirred vigorously, forming a red milky looking solution. After stirring, filter with a fine grade paper. The solution then becomes clear and slightly reddish or hot pink and is ready to use. Prepare fresh each time.

STAINING PROCEDURE

1. Decerate slides in xylene, 2 changes, 2 minutes each. Place in absolute alcohol, 2 changes, 2 minutes each; 95% alcohol, 2 changes, 2 minutes each; then rinse in distilled water.
2. Place Leder's staining solution at room temperature for 30 minutes to 1-1/2 hours. (Tissue sections usually take longer than films.)
3. Wash in running tap water for 2 minutes.
4. Counterstain in Mayer's hematoxylin for 5 minutes.
5. Wash in running tap water for 10 minutes, then distilled water.
6. Mount coverglass with glycerol gelatin.

RESULTS

Neutrophilic myeloid cells .... scarlet
Tissue mast cells.. scarlet
Chediak-Higashi inclusions .... scarlet
Nuclei .... blue

REMARKS

The time in the Leder's staining solution may be extended up to three hours. Do not use methyl alcohol unless it is used to provide a negative control. This method is unique, in that it is an enzyme stain that will work on paraffin sections.

Kass' Method for Human Megakaryocytes[86]

FIXATION See step 1.
PROCESS Paraffin, frozen, or celloidin.
MICROTOMY Cut sections at 5 micrometers.
CONTROL Use bone marrow smear for control.
SOLUTIONS

Lycramine Brilliant Blue JL

Liquid dye solution*..	3	drops
Deionized water..	10	ml

*Lycramine brilliant blue JL is available in liquid form from Ugine Kuhlmann Division, ICI Americas, Charlotte, NC.

STAINING PROCEDURE

1. Film of bone marrow particles is placed on clean coverglass (use methyl alcohol to clean) and fixed in absolute methyl alcohol for 15 minutes. Allow to air-dry and stain for 10 minutes in a solution of lycramine brilliant blue JL. Wash in distilled water, air-dry, and mount face down with appropriate medium (refractive index 1.48-1.56).

RESULTS

Mature and immature megakaryocytes .... metachromatic-lavender

Kass' Method for Cells of Granulocytic Origin[87]

FIXATION	Biopsy cores of bone marrow are fixed in B-5 or Bouin's solution.
PROCESS	Paraffin, frozen, or celloidin.
MICROTOMY	Cut sections at 5 micrometers.

Note: Coverglass preparations of peripheral blood buffy coat and bone marrow are made on methyl alcohol cleaned glass slide and air-dried. Fix in Bouin's solution for 5 minutes.

SOLUTIONS

0.5% Acid Red 52*

Acid red 52	0.5	gm
Distilled water	100.0	ml

0.1% Acid Red 52

Acid red 52	0.1	gm
Distilled water	100.0	ml

*Purchased as Intracid Rhodamine B from Crompton and Knowles Co., Mahwah, NJ 07430.

STAINING PROCEDURE - Biopsy Core

1. Decerate slides in xylene, 2 changes, 2 minutes each. Place in absolute alcohol, 2 changes, 2 minutes each; 95% alcohol, 2 changes, 2 minutes each; then rinse in distilled water.
2. Wash in running distilled water for 30 seconds.
3. Stain in 0.1% acid red 52 solution for 15 seconds.
4. Wash with distilled water, air-dry and mount with appropriate medium (refractive index 1.48-1.56).

STAINING PROCEDURE - Blood and Bone Marrow Smears

1. After fixation, wash specimen in running distilled water for 30 seconds.
2. Dry by blotting with bibulous paper.
3. Stain in 0.5% acid red 52 for 5 minutes.
4. Wash with distilled water, air-dry and mount coverglass with appropriate medium (refractive index 1.48-1.56).

RESULTS

Normal peripheral blood

Granules in neutrophils	brilliant pink
Granules in eosinophils and basophils	dark red
Granules in lymphocytes, monocytes and platelets	unstained
Erythrocytes	red-orange

Normal bone marrow

Granules in immature cells of neutrophilic granulocytic series	bright pink

Doe-Gryzbac-Schumacher Method for Myeloperoxidase[88]

FIXATION — 10% Formaldehyde-ethyl alcohol (see step 1 and 2).

PROCESS — Blood smears, bone marrow smears and touch preparations (see steps 1 and 2).

SOLUTIONS

10% Formaldehyde Ethyl Alcohol

37% Formalin	5.0	ml
100% Ethyl alcohol	45.0	ml

4 Chloro-1-Naphthol* (Stock)

4-Chloro-1-Naphthol	0.4	gm
100% Ethyl alcohol	50.0	ml

Store in a brown bottle at 4°C and discard if it becomes discolored (deep brown).

*Sigma Diagnostics #102F-0172, St. Louis, MO 63178.

Tris-Hydrochloride (HCl) Buffer (Stock)

Trizma Base (Sigma #37C-5044)	1.39	gm
Tris HCl (Sigma Tris-HCl)	6.06	gm

Add deionized water to make 1 liter; pH should be 7.4 to 7.6.

Dilute Hydrogen Peroxide (Stock)

30% Hydrogen peroxide	0.4	ml
Deionized water	99.5	ml

Must be made fresh.

Giemsa

Giemsa stain solution (available commercially)	3.0	ml
Purified water, pH 7.0	14.0	ml

May-Grunwald Solution
(Available commercially)

Substrate Medium

4-Chlor-1-Naphthol (stock)	2.0	ml
Tris-hydrochloric (HCl) buffer (stock)	38.0	ml

Filter this mixture and then add 1.0 ml dilute hydrogen peroxide (stock).

STAINING PROCEDURE

1. Prepare the smears. Thoroughly air-dry the samples and keep them in the dark until ready to stain.
2. Fix the smears for 5 minutes in 10% formaldehyde-ethyl alcohol.
3. Prepare the substrate medium and incubate smears in the medium for 10 minutes.
4. Rinse with deionized water 15 to 30 seconds. Shake off the excess water.
5. Flood the smears with Harleco Giemsa for 4 minutes.
6. Rinse the smears with distilled water and let them dry.
7. Flood the smears with May-Grunwald stain for 1 minute.
8. Rinse the smears with deionized water and let them dry.
9. Mount coverglass with appropriate medium (refractive index 1.48-1.56).

RESULTS

Peroxidase reaction .... black
Nuclear chromatin .... blue-purple

Boon-Schleicher-Wijsman-Grootendorst-Lyon-Kok Method for Image Analysis of the Nucleolus[89]

The assessment of nucleolar size has proven to be important in evaluating the prognosis of cancer patients. The following staining methods can be used for staining nucleoli. However it is suggested that the original paper be reviewed if image analysis is to be conducted.

FIXATION 10% Buffered neutral formalin.
PROCESS Paraffin.
MICROTOMY Cut sections at 5 micrometers.
CONTROL Use control from tissue under study.
SOLUTIONS

Phloxine B

Phloxine B	0.25	gm
96% Ethyl alcohol acidified with 0.2 phosphotungstic acid to pH 7.0	100.0	ml

Aluminum Ammonium Alum

Aluminum ammonium alum	10.0	gm
Distilled water	100.0	ml

Ethyl Green

Ethyl green	0.15	gm
0.01M Phthalate buffer, pH 4.0	100.0	ml

Pyronin Y

Pyronin Y (CI 45005)	0.03	gm
0.01M Phthalate buffer, pH 4.0	100.0	ml

Ethyl Green-Pyronin Y

Ethyl green	0.15	gm
Pyronin Y	0.03	gm
0.01M Phthalate buffer, pH 4.0	100.0	ml

STAINING PROCEDURE - Protein Staining

1. Decerate slides in xylene, 2 changes, 2 minutes each. Place in absolute alcohol, 2 changes, 2 minutes each; 95% alcohol, 2 changes, 2 minutes each; then rinse in distilled water.
2. Incubate slides 5 minutes in aluminum ammonium alum solution.
3. Rinse in running tap water for 5 minutes.
4. Dip in 70% ethyl alcohol, 15 dips.
5. Dip in 96% ethyl alcohol, 15 dips.

6. Stain with phloxine B solution for 15 minutes.
7. Dip in 96% ethyl alcohol, 15 dips.
8. Clear in xylene and mount coverglass with appropriate medium (refractive index 1.48-1.56).

RESULTS

Nucleoli .... purple-red
Background .... weakly stained
Erythrocytes .... red

STAINING PROCEDURE - RNA Staining

Simultaneous Method

1. Decerate slides in xylene, 2 changes, 2 minutes each. Place in absolute alcohol, 2 changes, 2 minutes each; 95% alcohol, 2 changes, 2 minutes each; then rinse in distilled water.
2. Stain slides in ethyl green-pyronin Y for 5 minutes.
3. Rinse in distilled water, 2 changes, 5 seconds each. Shake off excess water.
4. Agitate in 3 changes of 1-butanol until clear and mount directly or after exposure to xylene.
5. Mount coverglass with appropriate medium (refractive index 1.48-1.56).

Sequential Method

1. Decerate slides in xylene, 2 changes, 2 minutes each. Place in absolute alcohol, 2 changes, 2 minutes each; 95% alcohol, 2 changes, 2 minutes each; then rinse in distilled water.
2. Stain in pyronin Y for 5 minutes.
3. Rinse in distilled water for 5 seconds.
4. Incubate in ethyl green solution for 5 minutes.
5. Rinse in distilled water, 2 changes, 5 seconds each. Shake off excess water.
6. Agitate in 3 changes of 1 butanol until clear and mount directly or after exposure to xylene.
7. Mount coverglass with appropriate medium (refractive index 1.48-1.56).

RESULTS

Nucleoli..purple-red
Background ....weakly stained

Kass' Method for Blood and Bone Marrow Cells[90]

FIXATION Unfixed, air-dried.

PROCESS Blood smears, coverglass preparations of bone marrow aspirates, and buffy coat preparations.

CONTROL Use control from tissue under study.

SOLUTIONS

1% Basic Blue 41

Basic blue	1.0	gm
100% Methyl alcohol	100.0	ml

Basic blue can be obtained as Remacryl Blue BRL from Hoerst, West Germany (see remarks).

Phosphate buffer, pH 7.2
Solution A

Potassium dihydrogen phosphate	9.0	gm
Distilled water	1000.0	ml

Solution B

Sodium phosphate dibasic heptahydrate	17.86	gm
Distilled water	1000.0	ml

To make working buffer, add 285 ml solution A to 715 ml solution B. Adjust the pH to 7.2 with drop-by-drop addition of concentrated sodium hydroxide. With the addition of several drops of phenol, the buffer is stable for one year at room temperature.

STAINING PROCEDURE

1. Place samples on a staining pan and flood with unfiltered 1% basic blue 41 for 5 minutes.
2. Add an equal amount of working phosphate buffer pH 7.2 directly to the 1% basic blue 41 solution, covering the specimens for 5 minutes.
3. Drain off excess stain by touching edge of coverslip or slide to piece of filter paper.
4. Rinse specimens in working phosphate buffer, with gentle agitation, for 5 seconds. Blot on filter paper, air-dry, and mount with appropriate medium (refractive index 1.48-1.56) on clean glass slides.

RESULTS

Eosinophil granules	turquoise
Erythrocytes	orange-green

REMARKS

Basic blue 41 is commercially known under a variety of names including: Remacryl blue BRL from Hoechst, West Germany; Astrazon blue FGGL; Diacryl blue GRLA; Basacryl blue X-3GL; Maxilon blue GRLA; Sandrocryl blue B-RLE; and Sevron blue GR. Solutions of both basic blue 41 and phosphate buffer are available in kit form under the name Panoptikon from Cytocolor, Hindsley, OH 44233.

Puchtler-Rosenthal-Sweat Method for Hemoglobin

FIXATION	Zenker's.
PROCESS	Paraffin or frozen.
MICROTOMY	Cut paraffin sections at 6 micrometers.
CONTROL	Use control number 22.
SOLUTIONS	

Zenker-Formol (Stock)

Mercuric chloride	50.0	gm
Potassium dichromate	25.0	gm
Sodium sulfate (optional)	10.0	gm
Distilled water	1000.0	ml

Zenker-Formol (Working)

Zenker-formol (stock)	95.0	ml
Formaldehyde, 37-40%	5.0	ml

Prepare immediately before use.

5% Tannic Acid

Tannic acid	5.0	gm
Distilled water	100.0	ml

Allow to stand for 48 hours before use.

1% Phosphomolybdic Acid

Phosphomolybdic acid	1.0	gm
Distilled water	100.0	ml

Buffalo Black NBR*

Buffalo Black NBR, concentrated 126%	3.15	gm
Glacial acetic acid	10.0	ml
Methyl alcohol	90.0	ml

*Also known as Amido Black 10B.

Methyl Alcohol-Glacial Acetic Acid

Glacial acetic acid	10.0	ml
Methyl alcohol	90.0	ml

STAINING PROCEDURE

1. Decerate slides in xylene, 2 changes, 2 minutes each. Place in absolute alcohol, 2 changes, 2 minutes each; 95% alcohol, 2 changes, 2 minutes each; then rinse in distilled water.
2. Mordant in Zenker-formol overnight.
3. Wash in running tap water for 15 minutes.
4. Remove mercuric chloride crystals with iodine and clear with sodium thiosulfate.
5. Wash in running tap water for 5 minutes.
6. Place in tannic acid for 10 minutes.
7. Rinse in distilled water, 3 changes.
8. Place in 1% phosphomolybdic acid for 10 minutes.
9. Rinse in methyl alcohol-glacial acetic acid solution, 2 changes.
10. Place in buffalo black NBR for 5 minutes. DO NOT FILTER.
11. Rinse in methyl alcohol-glacial acetic acid solution, 2 changes.
12. Dehydrate in absolute alcohol, then clear in xylene, 2 or 3 changes each.
13. Mount coverglass with appropriate medium (refractive index 1.48-1.56).

RESULTS

Hemoglobin casts	dark blue
Intracellular hemoglobin droplets	dark blue
Erythrocytes	dark blue
Other tissue structures	yellow

REMARKS

In my experience this method is one of the best available for the demonstration of hemoglobin, but only if tissue is fixed in Zenker's solution. Buffalo Black NBR concentrated 126% must be used to insure satisfactory results.

Taft's Method for Deoxyribonucleic and Ribonucleic Acids (DNA and RNA)[91]

FIXATION	Carnoy's solution or absolute alcohol.
PROCESS	Paraffin.
TECHNIQUE	Cut paraffin sections at 4 micrometers.
CONTROL	Use control number 15.
SOLUTIONS	

Methyl Green-Pyronin Solution

Methyl green, purified*	0.52	gm
Pyronin Y	0.1	gm

Heat 100 ml of distilled water to boiling in 250 ml flask. Remove from heat and add dyes. Mix thoroughly. Stopper flask and store for 7 days. Solution is then ready for use and should be stored in an amber glass stoppered bottle.

*Methyl green can be purified by mixing it with chloroform several times until solution is blue-green instead of lavender.

Differentiating Solution

Tertiary butyl alcohol	30.0	ml
100% Alcohol	10.0	ml

STAINING PROCEDURE

1. Decerate slides in xylene, 2 changes, 2 minutes each. Place in absolute alcohol, 2 changes, 2 minutes each; 95% alcohol, 2 changes, 2 minutes each; then rinse in distilled water.
2. Stain in methyl green-pyronin solution for 10 minutes. Filter before use.
3. Rinse briefly twice in distilled water.
4. Blot with several thicknesses of smooth filter paper.
5. Place in differentiating solution for 1 minute.
6. Dehydrate in a fresh change of the differentiating solution. (Slides may be left in this solution for 4 to 5 minutes without additional differentiation taking place.)
7. Clear in xylene, 2 changes, 2 minutes each.
8. Mount coverglass with appropriate medium (refractive index 1.48-1.56).

RESULTS

Deoxyribonucleic acid (DNA)	blue green
Ribonucleic acid (RNA)	bright red
Remainder of tissue	pale pink

Staining Methods for Blood and Bone Marrow Smears

Acidic dyes unite with the basic components of the cells (cytoplasm). Conversely, basic stains are attracted to and combine with the acidic parts of the cell (nucleic acids and nucleoproteins of the nuclei). Eosin is an acidic dye; hence "eosinophilic" and "acidophilic" are used interchangeably in describing staining qualities of cell components.

History

Ehrlich was the first to use simple aniline dyes – at first in sequence and later as premixed acidic-basic stains (neutral dyes). In this way Ehrlich evolved his historic but misnamed "triacid" stain. This is a mixture of orange G, acid fuchsin, and methyl green. The three basic groups of the methyl green were believed to combine with the two acid dyes; hence, the name "triacid". It is not used today.

Jenner found that the precipitate formed when eosin and methylene blue are mixed could be dissolved in methyl alcohol to form a useful stain combining certain properties of both parent dye-stuffs. Jenner's is very similar to the May-Grunwald stain.

Romanowsky found that when old (ripened and therefore "polychromed") methylene blue solution is mixed with eosin and the precipitate dissolved in methyl alcohol, a stain results that has a wider range in staining cell nuclei and platelet granules.

Modern Romanowsky stains, Wright's and Leishman's for example, are basically similar to Romanowsky's original method; the main difference being the method of polychroming the methylene blue.

Romanowsky Stains in Common Use

Wright's Stain

This excellent blood stain is very widely used, especially on the American continent. In its preparation the methylene blue is polychromed by heating with sodium bicarbonate. It may be purchased in solution ready for use or as powder, 1.0 gm of which is carefully dissolved in 600 ml of methyl alcohol to make the working solution.

STAINING PROCEDURE

1. Place the air-dried smear, film side up, on a staining rack (two parallel glass rods 5 cm apart).
2. Cover smear with undiluted stain and leave for 1 minute. (The methyl alcohol fixes the smear.)
3. Dilute with distilled water (approximately equal volume) until a metallic scum appears. Allow this diluted stain to act for 2 1/2 to 5 minutes.
4. Without disturbing the slide, flood with distilled water and wash until thinner parts of film are pinkish-red. Frequently distilled water does not give adequate differentiation. In that case, a phosphate buffer (pH 6.4 to 6.5) should be used, at least for washing, but may also have to be used for diluting the stain (step 3).

Leishman's Stain

This is the stain most commonly used in the British Isles. In its preparation the methylene blue is polychromed by heating a 1% solution with 0.5% sodium carbonate at 65°C for 12 hours, after which further ripening is allowed to proceed for 10 days before the mixture is mixed with an equal volume of 0.1% eosin B. After this mixture has stood for 10 hours, it is filtered and the precipitate is collected and washed with distilled water until no more color bleeds out in the washing. The precipitate is then dried and ground to a fine powder. To make up the stain, a small knife-point of powder at a time, is ground well with 10 ml of absolute methyl alcohol and allowed to settle (1 minute). Then supernatant alcoholic solution of the stain is filtered into the stock bottle. This is repeated until 0.15 gm has been dissolved in 100 ml of absolute methyl alcohol. This stain improves with age and is not satisfactory until a minimum of 3 weeks has elapsed after its preparation.

STAINING PROCEDURE

The method is similar to that used with Wright's stain except for step 3. With Leishman's stain, dilution is accomplished with approximately 2 volumes of distilled water to 1 volume of stain (the best guide is the appearance of the metallic scum). If a phosphate buffer is required for washing, one of pH 6.8 is best. This may be substituted for the distilled water when diluting the stain.

Wright's-Leishman's Stains

After staining and washing to the desired degree of differentiation (usually 1/2 to 1 minute), stand smears on end to dry. The appearance of the cellular blood elements is quite similar with the two methods. Possibly, Leishman's stain gives better staining and contrast to the nuclear structures than Wright's stain. With both, the blood cells should be a soft pink; eosinophil granules, bright reddish pink; nuclei, varying shades of lilac to purple; and neutrophil granules, lilac in color.

If the film has a bright pink appearance to the naked eye, microscopically the erythrocytes are likely to appear vivid red, the eosinophil granules a pale incandescent pink, and nuclei a faint, washed-out blue color. This defect of staining may result from acidity of the stain or buffer, or from excessive washing. Unripened Leishman stain will also give pale nuclear staining.

Conversely, if the smear has a bluish to violet appearance on naked eye inspection, then microscopically the erythrocytes are likely to appear as muddy gray to purple, eosinophil granules have a gun-metal sheen, nuclei and polymorph granules are a deep blackish-purple, nuclear and other details are obscured. The cause of this unsatisfactory appearance may lie in overstaining, underwashing, or too alkaline a stain or wash. Washing with a buffer solution (pH 6.5 or 6.8) may improve the quality. Recently stained smears, especially those that are understained or appear washed-out (overdifferentiated), may be decolorized by a rinse in 95% alcohol. Then wash with distilled water, and restain.

Generally, it must be remembered that both Wright's and Leishman's stains are somewhat empirical mixtures of polychromed dyes, so that one batch may differ from another. Also, the "ripening" or polychroming continues in the stock solution of stain. If the stock stain solution becomes contaminated with water, poor results are to be expected. To avoid such poor results, it has been suggested that all films be fixed in pure methyl alcohol before staining.

Giemsa Stain

Instead of empirically polychromed dyes, this stain employs various azure compounds (thionin and its methyl derivatives) with eosin and methylene blue. It is best purchased commercially in solution. Smears must be prefixed for 3 minutes with methyl alcohol. They are dried and immersed in dilute Giemsa stain (1 part stain to 9 to 15 parts distilled water or buffer of pH 6.8) in coplin jars for 15 minutes to 1 hour. They are then washed in distilled water, air-dried and are not mounted. This stain is not commonly used alone in hematology but is an excellent stain for inclusion bodies (e.g., inclusion conjunctivitis). However, to show the inclusions well, the smears should be allowed to stain in the dilute Giemsa stain for 16 hours (overnight).

By itself Giemsa stains red blood cells and neutrophil granules poorly, but azurophil granules (red) are well stained. In combination with Jenner and May-Grunwald stains, it constitutes "panoptic" staining. (This consists of the combination of a Romanowsky stain with another stain. Such a combination improves the staining of cytoplasmic granules and other bodies. The two most popular methods employed are the Jenner-Giemsa and May-Grunwald-Giemsa methods.)

Jenner-Giemsa

STAINING PROCEDURE

1. Fix air-dried smears 10 to 20 minutes in methyl alcohol.
2. Freshly diluted commercially purchased Jenner's stain (another Romanowsky stain, i.e., eosin-methylene blue) in the proportion of 1 volume of stain to 4 volumes of buffered distilled water, pH 6.8 (see below for formula). Stain smears in this solution for 4 minutes.
3. Do not wash smears but transfer to freshly diluted (same day) Giemsa stain (1 volume stain to 9 volumes of buffered distilled water, pH 6.8) and stain for 7 to 10 minutes.
4. Wash quickly in 3 changes of buffered (pH 6.8) distilled water and place in a fourth change of buffered water for 3 to 12 minutes for differentiation to take place. Experience, gross appearance, and low-power microscopic examination will help to control the differentiation.
5. Stand on end to dry.

May-Grunwald-Giemsa

STAINING PROCEDURE

This gives slightly better results than the Jenner-Giemsa.

1. Fix air-dried smears for 5 minutes in methyl alcohol.
2. Transfer to May-Grunwald stain (another Romanowsky dye), freshly diluted with 1 to 2 volumes of buffered (pH 6.8) distilled water. Leave 3 to 5 minutes.
3. Transfer, without washing, to fresh diluted Giemsa (see under Jenner-Giemsa) and allow to stain for 7 to 15 minutes.
4. Differentiate as for Jenner-Giemsa.

Formulas for Phosphate Buffers (for Romanowsky and other stains)

Potassium Phosphate 0.2 M KH_2PO_4 (2.722% w/v) ml	(Sodium Hydroxide) 0.2 N NaOH (0.8% w/v) ml	pH
50.0	17.8	6.6
50.0	21.0	6.7
50.0	23.7	6.8
50.0	26.5	6.9
50.0	29.6	7.0

Alternatively, prepared capsules of buffer salts may be purchased from any reputable supply house. These are available over a wide pH range. For use, the contents of one capsule are merely dissolved in the stated amount (100 ml) of distilled water.

It is important to recognize that Romanowsky-stained smears fade unless they are coverslipped. Although old smears may be decolorized and restained, the results are never satisfactory.

Cytochemistry
Although the examination of a properly prepared and stained peripheral blood film or bone marrow smear will produce an abundant amount of information, a number of special stains may be required for further elucidation. Such stains are described here although they apply mainly to the leukocytes. Control slides must always be used for specific staining reactions. It should be noted that there are a number of kits available from the following company for cytochemical staining of the leukemias. Sigma Diagnostics, St. Louis, MO 63178.

Kaplow's Method for Peroxidase Granules

The demonstration of peroxidase granules in certain leukocytes depends on their content of an iron porphyrin-containing enzyme (peroxidase) which promotes the oxidation of benzidine by hydrogen peroxide. Copper sulfate then forms a blue-green compound with the oxidized benzidine and thus renders the granules visible. The method of Kaplow is recommended because it utilizes benzine dihydochloride, which is less carcinogenic than benzidine.

FIXATIVE 95% Ethyl alcohol, 9 parts
37% Formalin, 1 part

SOLUTIONS

0.3% Benzidine Dihydrochloride

30% Ethyl alcohol	100.00	ml
Benzidine dihydrochloride	0.3	gm
3.8% Zinc sulfate	1.0	ml
Sodium acetate	1.0	gm
1N Sodium hydroxide	1.5	ml
3% Hydrogen peroxide	0.7	ml

Diluted Giemsa

Giemsa stain diluted 1 to 10 with buffered distilled water pH 6.8.

STAINING PROCEDURE

1. Fix smears in the fixative solution for 60 seconds, then wash gently in running tap water.
2. Immerse smears for 30 seconds in 0.3% benzidine dihydrochloride.
3. Wash in running tap water and counterstain with dilute Giemsa for 15 minutes.

RESULTS

Peroxidase activity is shown by the presence of green to dark blue granules in the cytoplasm. In some neutrophils and in eosinophils, rods and needle-like projections may be seen emanating from the cell margin. Basophils, lymphocytes, and erythroblasts are negative; monocytes show slight activity; myeloid cells (excepting basophils) are positive.

Hayhoe-Quaglino Method for Leukocyte Alkaline Phosphate

The reaction of this method depends on the hydrolysis of alpha-naphthyl phosphate by alkaline phosphatase to produce a colored precipitate with a diazotized amine.

FIXATIVE
Absolute methyl alcohol, 9 parts
40% Formalin, 1 part
Store in freezing compartment of refrigerator.

SOLUTIONS

Buffer (Stock)

2-amino-2-methyl propane-(1:3)-diol	10.5	gm
Distilled water	500.0	ml

Buffer (Working)

Buffer (stock)	25.0	ml
0.1 N Hydrochloric acid	5.0	ml
Distilled water	70.0	ml

Store at 4°C.

Substrate

Sodium alpha-naphthyl phosphate	20.0	mg
Brentamine-fast garnet (GBC) salt	20.0	mg
Working buffer	20.0	ml

The substrate must be prepared fresh and filtered just before use.

2% Mayer's Hematoxylin

Hematoxylin crystals	20.0	gm
Distilled water	1000.0	ml
Sodium iodate	0.2	gm
Ammonium or potassium alum	50.0	gm
Citric acid	1.0	gm
Chloral hydrate	50.0	gm

STAINING PROCEDURE

1. Using freshly prepared blood smears from finger or ear prick, fix with the fixative for 30 seconds at 4°C.
2. Wash with distilled water and allow to air-dry.
3. Immerse in, or pour on substrate and allow to remain for 10 minutes; then rinse in distilled water.
4. Stain in 2% Mayer's hematoxylin for 15 minutes.
5. Drain off excess water and mount coverglass with an aqueous mounting medium.

RESULTS

A red-brown precipitate in the cytoplasm indicates alkaline phosphatase activity. Control slides should preferably show high and low values, e.g., films from pregnant women and from patients with untreated myeloid leukemia. These may be stored in a fixed state for a considerable time.

The slides should be scanned as soon as possible after preparation since the stain tends to fade, particularly in weak reacting cells.

Naphthol AS-D Chloroacetate Esterase Method for Neutrophils

Use paraffin sections or smears fixed for 5 hours in cold formalin-methyl alcohol 1:9. Reaction solution: Mix 1 drop of 4% pararosaniline in 2N hydrochloric acid (HCl) with 1 drop of 4% sodium nitrate in distilled water. Add 30 ml of 0.1 M Michaelis buffer, pH 7.62. Adjust pH to 6.3 with 2N hydrochloric acid (HCl). Add 10 mg naphthol AS-D chloroacetate dissolved in 1 ml of N,N-dimethylformamide. Filter. Immerse slides in reaction solution for 30 minutes at room temperature. Rinse in running tap water. Counterstain with Mayer's hematoxylin. Mount coverglass with glycerin jelly.

Alpha-Naphthol Acetate Esterase for Monocytes

Use unfixed cryostat sections or smears. Reaction solution: Mix 1 drop of 4% pararosaniline in 2N hydrochloric acid (HCl) with 1 drop of 4% sodium nitrate in distilled water. Add 25 ml of 0.1M phosphate buffer, pH 7.5. Add 10 mg alpha-naphthol acetate dissolved in 1 ml acetone. Filter. Immerse slides in reaction solution for 15 to 30 minutes at room temperature. Rinse thoroughly in tap water. Counterstain with Mayer's hematoxylin. Mount coverglass with glycerin jelly.

Naphthol AS-D Acetate Esterase for Monocytes

Use unfixed cryostat sections or smears. Reaction solution: Dissolve 10 mg naphthol AS-D acetate in 1 ml acetone (solution A). Mix 0.4 ml propylene glycol with 40 ml of 0.1M phosphate buffer, pH 6.8 (solution B). Add solution A to solution B, stirring hard. Add 80 mg fast blue BB salt. Filter. Immerse slides in this solution for 30 to 60 minutes at room temperature. Rinse in running tap water. Counterstain with nuclear fast red. Mount coverglass with glycerin jelly.

Non-Specific Acid Phosphatase

Use unfixed cryostat sections or smears fixed for 30 minutes in cold acetone. Reaction solution: Mix 6 drops of 4% pararosaniline in 2N hydrochloric acid (HCl) with 6 drops of 4% sodium nitrate in distilled water. Add 30 ml of 0.1M Michaelis buffer, pH 7.62. Adjust pH to 5.0 with 2N hydrochloric acid (HCl). Add 10 mg naphthol AS-BI phosphate dissolved in 1 ml N,N-dimethylformamide. Filter. Incubate for 30 minutes to 4 hours at room temperature. Rinse in running tap water. Counterstain with Mayer's hematoxylin. Mount coverglass with glycerin jelly.

Periodic Acid-Schiff Method for Glycogen

Blood films should be fixed in methyl alcohol and brought to water before staining. The methodology can be found on Page 374. A positive reaction probably denotes the presence of glycogen.

Sudan Black Method for Lipid

This method demonstrates lipid granules in leukocytes. It is interesting to note that there is a close correlation between sudanophilia (i.e., the presence of lipid material) and a positive reaction for peroxidase granules.

SOLUTIONS

Sudan Black B

Sudan black B	0.5	gm
Absolute ethyl alcohol	100.0	ml

Shake well over a period of 1 to 2 days until all the stain is dissolved and then filter.

Buffer

Phenol	16.0	gm
Ethyl alcohol	30.0	ml
0.3% Aqueous sodium phosphate, dodecahydrate	100.0	ml

Just before use, add 60 ml of the stain to 40 ml of the buffer, mix and filter. This is the working solution.

STAINING PROCEDURE

1. Fix air-dried films in formalin vapor for 10 minutes.
2. Place films in sudan black B working solution for 30 minutes at 37°C.
3. Wash with absolute ethyl alcohol.
4. Wash with distilled water.
5. Counterstain, if desired, with 1 in 20 Giemsa's* stain for 1 hour.
6. Mount coverglass with glycerin jelly, out of water or examine directly.

RESULTS

Lipid particles .... dense black

*1 part distilled water to 20 parts Giemsa stain.

Modified Wright Method for Deoxyribonucleic Acid (DNA)

The Feulgen test is a specific method for the demonstration of deoxyribonucleic acid. Nucleoli will not stain, but bodies of nuclear origin will. This differentiation is used to distinguish certain inclusion bodies in blood cell cytoplasm. The following method is provided because it is simpler to perform than the classic technique and appears to parallel the reactions obtained by it.

STAINING PROCEDURE

1. Fix the blood film for 3 minutes in a mixture of 9 parts methyl alcohol to 1 part 40% formalin.
2. Rinse in distilled water.
3. Place in 1N hydrochloric acid (HCl) for 15 minutes at 56°C. If smears tend to detach from the slide, 0.1N hydrochloric acid (HCl) may be used.
4. Rinse in distilled water.
5. Stain with diluted Wright's stain* for 30 to 60 minutes.
6. Rinse in distilled water, allow to dry, and mount in neutral mounting medium.

RESULTS

Deoxyribonucleic acid (DNA) (including chromatin) .... blue-black
Some cytoplasmic granules .... faintly blue-black

*1 part distilled water to 10 parts Wright's stain.

Levy's Method for Heinz Bodies

Heinz bodies (also called Heinz granules) are coccoid inclusion bodies resulting from oxidative injury to and precipitation of hemoglobin, seen in the presence of certain abnormal hemaglobin and erythrocytes with enzyme defficiencies. Refractive in fresh blood smears, they are not visible when stained with Romanowsky dyes but may be stained supravitally. The method provided herein works well with both frozen and paraffin embedded specimens.

FIXATION	10% Buffered neutral formalin. Wash tissue for 1 hour. Soak specimen in gum-sucrose solution at 4°C for 1 hour.
PROCESS	Frozen and paraffin.
MICROTOMY	Cut paraffin section at 4 micrometers (frozen at 6 micrometers) and attach to slides previously cleaned in dichromate solution (see below) and coated with chrome alum Poly-L-lysine solution. Air-dry for 20 minutes.
SOLUTIONS	

Gum Sucrose

Sucrose	100.0	gm
Gum acacia	5.0	gm
Distilled water	500.0	ml
Thymol	0.1	gm

Dissolve sugar in water by heating. Cool sugar solution, add gum acacia, and allow it to dissolve. Add the thymol.

Dichromate-Sulphuric Acid

Potassium dichromate	100.0	gm
Concentrated sulphuric acid	100.0	ml
Distilled water, q.s. to	1000.0	ml

Dissolve the dichromate in distilled water, then slowly and carefully add the sulphuric acid. Wash slides in this solution until they can be completely "wetted" with tap water. Dry slides and coat them with chrome-gelatin solution.

Chrome-Gelatin

Gelatin	2.5	gm
Chrome alum	0.5	gm
Distilled water	1000.0	ml

Dips slides in chrome-gelatin solution and place in a 60°C oven for 1 hour. Slides may be stored indefinitely.

Methyl Violet 2B

Methyl violet 2B	0.1	gm
Sodium chloride	3.9	gm
Sodium sulfate	2.2	gm
Sodium phosphate, dibasic	3.0	gm
Potassium phosphate, monobasic	0.25	gm
Formaldehyde, 37%	7.5	ml
Distilled water	1000.0	ml

Dissolve all chemicals in the distilled water and let stand overnight. Filter and use.

0.5% Acetic Water

Glacial acetic acid	0.5	ml
Distilled water	100.0	ml

0.1% Eosin Y

Eosin Y	0.1	gm
Distilled water	100.0	ml

Apathy's Gum Syrup
(See Page 177)

STAINING PROCEDURE

1. Stain frozen cut section in methyl violet 2B for 10 minutes.
2. Wash in running tap water for 1 minute.
3. Differentiate in 0.5% acetic water, 4 quick dips.
4. Counterstain with 0.1% eosin Y solution, 3 minutes.
5. Rinse quickly in tap water, 3 quick dips.
6. Mount coverglass with Apathy's gum syrup.

RESULTS

Heinz bodies .... purple
Erythrocytes .... orange

REMARKS

Soaking the tissue in gum-sucrose solution is necessary to enable the cutting of thin sections and it also aids the staining process. Without proper cleaning and coating of the glass slide, the sections will not adhere. The specified mounting medium is necessary to prevent "bleeding" of the methyl violet from the Heinz bodies.

Chapter 10

Cytoplasmic Granules

Diazo's Method for Argentaffin Cell Granules[55u]

(See Fig. 95)

FIXATION 10% Buffered neutral formalin.
PROCESS Paraffin, frozen, or celloidin.
MICROTOMY Cut section at 6 micrometers.
CONTROL Use control number 3.
SOLUTIONS

1% Brentamine Fast Red B (Stock)

Brentamine fast red B	1.0	gm
Distilled water	100.0	ml

Lithium Carbonate (Stock)

Lithium carbonate	1.5	gm
Distilled water	100.0	ml

Keep at 4 to 5°C.

Brentamine Fast Red B (Working)

Brentamine fast red B (stock)	5.0	ml
Lithium carbonate (stock)	2.0	ml

Mix and let stand for 6 to 8 minutes at 4 to 5°C before use.

Mayer's Hematoxylin

(See Page 81)

STAINING PROCEDURES

1. Decerate slides in xylene, 2 changes, 2 minutes each. Place in absolute alcohol, 2 changes, 2 minutes each; 95% alcohol, 2 changes, 2 minutes each; then rinse in distilled water.
2. Immerse in Brentamine working solution at 4 to 5°C for 40 to 60 seconds.
3. Rinse in distilled water, 2 changes.
4. Place in Mayer's hematoxylin for 3 minutes.
5. Blue in warm tap water for 5 minutes.
6. Rinse in 95% alcohol, 2 changes.
7. Dehydrate slides in 95% alcohol and absolute alcohol 3 changes each. Clear in xylene, 3 changes.
8. Mount coverglass with appropriate medium (refractive index 1.48-1.56).

RESULTS

Argentaffin granules	rust red
Background	yellow
Nuclear chromatin	blue

REMARKS
Argentaffin cell granules stain yellow if Brentamine fast red B solution has broken down.

Fontana-Masson Method for Reducing Substances

(See Fig. 96 and 97)

FIXATION	10% Buffered neutral formalin.
PROCESS	Paraffin, frozen, or celloidin.
MICROTOMY	Cut sections at 6 micrometers.
CONTROL	Use control number 3.
SOLUTIONS	

Fontana Silver Nitrate

Dissolve 10 gm of silver nitrate in 100 ml of distilled water. To 95 ml of this solution, add ammonium hydroxide until a clear solution with no precipitate is obtained. Add, drop by drop, enough of the remaining 5 ml of silver nitrate solution to cause the clear solution to become slightly cloudy. When ready to use, dilute each 25 ml of silver solution with 75 ml of distilled water and filter.

Gold Chloride

1% Gold chloride, aqueous	10.0	ml
Distilled water	40.0	ml

5% Sodium Thiosulfate

(See Page 134)

Kernechtrot (Nuclear Fast Red)

(See Page 84)

STAINING PROCEDURE

1. Decerate slides in xylene, 2 changes, 2 minutes each. Place in absolute alcohol, 2 changes, 2 minutes each; 95% alcohol, 2 changes, 2 minutes each; then rinse in distilled water.
2. Place in silver nitrate solution at 56°C for 1 hour. Sections should be light brown.
3. Rinse sections in distilled water.
4. Place in gold chloride solution for 10 minutes.
5. Rinse in distilled water, 3 changes.
6. Place in sodium thiosulfate solution for 5 minutes.
7. Rinse in distilled water.
8. Place in nuclear fast red solution for 5 minutes.
9. Rinse in distilled water, 2 changes.
10. Dehydrate slides in 95% alcohol and absolute alcohol 3 changes each. Clear in xylene, 3 changes.
11. Mount coverglass with appropriate medium (refractive index 1.48-1.56).

RESULTS

Silver-reducing substances	black
Nuclei	pink

REMARKS
This method is nonspecific since many pigments and various other reducing substances can be demonstrated, including formalin pigment. It is used widely as an aid for the demonstration of melanin, but positive results should not be interpreted as melanin pigment unless other procedures confirm the results. Argentaffin cell granules are also demonstrated with this method.

Sevier-Munger Method for Argentaffin and Argyrophilic Cell Granules[55v]

(See Fig. 98)

The method was developed for the demonstration of peripheral nerve fibers. However, Sevier and Munger used the method for demonstrating argentaffin and argyrophil granules of carcinoid tumors.

FIXATION — 10% Buffered neutral formalin. Avoid chromate fixatives.
PROCESS — Paraffin, frozen, or celloidin.
MICROTOMY — Cut section at 6 micrometers.
CONTROL — Use control number 3.
SOLUTIONS

20% Silver Nitrate

Silver nitrate	20.0	gm
Distilled water	100.0	ml

10% Silver Nitrate

Silver nitrate	10.0	gm
Distilled water	100.0	ml

2% Formalin

Concentrated formalin, 37%	2.0	ml
Distilled water	100.0	ml

5% Sodium Thiosulfate
(See Page 134)

Sodium Carbonate

Sodium Carbonate, hydrated	8.0	gm
Distilled water	30.0	ml

If anhydrous sodium carbonate is used, use only 3 grams of sodium carbonate in 30 ml of distilled water.

Ammoniacal Silver Solution (Working)
Place 50 ml 10% silver nitrate solution in a 200 ml beaker. Add concentrated ammonium hydroxide, drop by drop, until the dark brown precipitate which forms has almost disappeared. Swirl vigorously between drops. Avoid complete decolorization. The end point should be a slightly cloudy solution. Add to this solution 0.5 ml of sodium carbonate solution. Shake well. Now add 25 drops of ammonium hydroxide. Shake well. Solution should now be clear.

STAINING PROCEDURE

1. Decerate slides in xylene, 2 changes, 2 minutes each. Place in absolute alcohol, 2 changes, 2 minutes each; 95% alcohol, 2 changes, 2 minutes each; then rinse in distilled water.
2. Add 4.0 ml of the 20% silver nitrate to a coplin jar and preheat the solution for 30 minutes.
3. Blot slides and place on a 40°C warming table until slides are warm and dry.
4. Add slides to the warm silver solution and let them remain in the oven for 10 to 40 minutes or until tissue is light tan in color.
5. Remove slides from oven, wash in 2 changes of distilled water, drain slightly, and place in a clean coplin jar.
6. To the working ammoniacal silver solution add 10 drops of 2% formalin solution, shaking gently while adding the formalin. Pour quickly over slides and let develop for 5 to 30 minutes until golden brown. Developing should be neither too fast nor too slow.
7. Rinse in 3 changes of distilled water.
8. Place in sodium thiosulfate solution for 2 minutes.
9. Wash in running tap water for 2 minutes.
10. Dehydrate slides in 95% alcohol and absolute alcohol 3 changes each. Clear in xylene, 3 changes.
11. Mount coverglass with appropriate medium (refractive index 1.48-1.56).

RESULTS

Carcinoid granules	black
Peripheral nerves	black
Background	yellow

REMARKS

Avoid treating slides with iodine and sodium thiosulfate since this procedure will not work after that treatment.

Staples-Grizzle Nuclear Stain for Argentaffin and Argyrophilic Cell Granules[92]

This methyl green nuclear stain is for silver reactions which demonstrate argentaffin and/or argyrophilic cell granules. It is being presented here as a separate stain entity since its use will have to be determined by the technologist or investigator. Use of the acidulated methyl green prevents masking of the neurosecretory granules used in diagnosis. The most commonly used nuclear counterstains, nuclear fast red and Mayer's hematoxylin, do not provide as much contrast.

1% Acidulated Methyl Green

Methyl green	1.0	gm
Distilled water	99.0	ml
Glacial acetic acid	1.0	ml

Filter the solution before use. The solution can be stored at room temperature for at least 2 months.

STAINING PROCEDURE

1. Decerate slides in xylene, 2 changes, 2 minutes each. Place in absolute alcohol, 2 changes, 2 minutes each; 95% alcohol, 2 changes, 2 minutes each; then rinse in distilled water.
2. Impregnate and develop sections using any of the varied silver methods for argentaffin or agyrophil cell granules.
4. Rinse in distilled water.
5. Stain the sections with 1% acidulated methyl green for 1 minute.
6. Rinse the slides in distilled water to remove excess dye.
7. Dehydrate slides in 95% alcohol and absolute alcohol 3 changes each. Clear in xylene, 3 changes.
8. Mount coverglass with appropriate medium (refractive index 1.48-1.56).

RESULTS

Argentaffin and argyrophil cell
granules .... red-brown to black
Nuclear chromatin .... green
Cytoplasm .... yellow-green
Connective tissue .... yellow-green

Staples-Grizzle Rapid Method for Argyrophilic and Argentaffin Cell Granules[93]

FIXATION	10% Buffered neutral formalin.
PROCESS	Paraffin.
MICROTOMY	Cut section at 5 micrometers.
CONTROL	Use control number 6A.
SOLUTIONS	

0.08% Silver Nitrate*
(See step 2)

Silver nitrate	0.08	gm
Distilled water	100.0	ml

Bodian's Reducer
(See step 4)

Sodium sulfite	5.0	gm
Hydroquinone	1.0	gm
Distilled water	100.0	ml

*These solutions are made with boiling distilled water and cooled to a temperature of 65-70°C.

STAINING PROCEDURE

1. Decerate slides in xylene, 2 changes, 2 minutes each. Place in absolute alcohol, 2 changes, 2 minutes each; 95% alcohol, 2 changes, 2 minutes each; then rinse in distilled water.
2. Make 0.08% silver solution with boiling water. Allow solution to cool to 65-70°C.
3. Impregnate the sections with hot silver for 3 minutes; save solution.

4. While slides are in silver, prepare Bodian's reducer using preheated distilled water cooled to 65-70°C.
5. Remove the slides from the silver and rinse off excess solution in distilled water.
6. Place in Bodian's reducer for 1 minute.
7. Rinse in distilled water.
8. Place in original silver solution for 1 minute.
9. Place the slides, after rinsing in distilled water, back into Bodian's reducer for 1 minute.
10. Rinse in distilled water.
11. Counterstain with nuclear fast red for 5 minutes.
12. Dehydrate slides in 95% alcohol and absolute alcohol 3 changes each. Clear in xylene, 3 changes.
13. Mount coverglass with appropriate medium (refractive index 1.48-1.56).

RESULTS

Argyrophilic cell granules	black
Argentaffin cell granules	black
Nuclear chromatin	pink

REMARKS

When compared to the Grimelius and Churukian-Schenk, this method stains as well, but with greater speed and less background staining.

Churukian-Schenk Method for Argentaffin and Argyrophilic Cell Granules[94]

(See Fig. 107 and 108)

FIXATION	10% Buffered neutral formalin.
PROCESS	Paraffin.
MICROTOMY	Sections cut at 6 micrometers.
CONTROL	Use control number 6A.
SOLUTIONS	

Acidified Water

Distilled water	500.0	ml

Add enough 0.3% aqueous citric acid to bring water to pH 4.2.

0.5% Silver Nitrate

Silver nitrate	0.5	gm
Acidified water	100.0	ml

Bodian's Developer

Sodium sulfite, anhydrous	5.0	gm
Hydroquinone	1.0	gm
Distilled water	100.0	ml

Kernechtrot (Nuclear Fast Red)
(See Page 84)

STAINING PROCEDURE

1. Decerate slides in xylene, 2 changes, 2 minutes each. Place in absolute alcohol, 2 changes, 2 minutes each; 95% alcohol, 2 changes, 2 minutes each; then rinse in distilled water.
2. Place in acidified water (pH 4.2) for 1 minute.
3. Place in 0.5% silver nitrate in a coplin jar and place in 43°C water bath for 2 minutes. Transfer to a 58°C water bath for 2 hours. Cracking of the coplin jar is prevented by first placing in the 43°C water bath.
4. Remove from coplin jar and rinse in 3 changes of distilled water. Save silver nitrate.
5. Transfer to Bodian's developer, previously heated to 58°C in the water bath, for 5 minutes.
6. Rinse in 3 changes of distilled water.
7. Wash in running tap water for 3 minutes.
8. Rinse in 3 changes of distilled water. Save Bodian's developer.
9. Again place in silver nitrate solution in the 58°C water bath for 10 minutes.
10. Rinse in 3 changes of distilled water.
11. Again place in the Bodian's developer in the 58°C water bath for 5 minutes.
12. Rinse in 3 changes of distilled water.
13. Place in 0.1% nuclear fast red for 3 minutes.
14. Rinse in 3 changes of distilled water.
15. Dehydrate slides in 95% alcohol and absolute alcohol 3 changes each. Clear in xylene, 3 changes.
16. Mount coverglass with appropriate medium (refractive index 1.48-1.56).

RESULTS

Argentaffin and argyrophil cell granules .... brown to black
Pancreatic islet cells (alpha and delta)............................ brown to black
Nuclear chromatin .... reddish-pink
Cytoplasm .... pale yellow

REMARKS

It should be recognized that the silver solution used at step 3 must be saved since it will be used again at step 9. The same holds true for the Bodian developer used at step 5 and again at step 11.

Luna's Method for Argentaffin Cell Granules and Goblet Cell Mucins

This method utilizes the reducing functions of potassium ferricyanide to demonstrate argentaffin cell granules and other reducing substances, like melanin granules. Additionally, the mucicarmine stain demonstrates the mucin in the goblet cells. This makes for a useful demonstration of reducing substances and mucins.

FIXATION 10% Buffered neutral formalin.
PROCESS Paraffin.
MICROTOMY Cut sections at 6 micrometers.
CONTROL Use control section of small intestine

SOLUTIONS

0.1% Potassium Ferricyanide (Stock)

Potassium ferricyanide	0.1	gm
Distilled water	100.0	ml

1% Ferric Chloride (Stock)

Ferric chloride	1.0	gm
Distilled water	100.0	ml

Ferric-Ferricyanide (Working)

1% Ferric chloride (stock)	150.0	ml
0.1% Potassium ferricyanide (stock)	50.0	ml

Adjust the pH to 2.4 with a few drops of diluted hydrochloric acid, if necessary.

Mayer's Mucicarmine

Carmine	1.0	gm
Aluminum chloride, anhydrous	0.5	gm
Distilled water	2.0	ml

Carefully mix stain in evaporating dish. Heat on electric hot plate for 2 minutes. Liquid will become almost black and syrupy. Dilute with 100 ml of 50% alcohol and let stand for 24 hours. Filter. Dilute 1 part mucicarmine with 4 parts water.

Metanil Yellow

Metanil yellow	0.25	gm
Distilled water	100.0	ml
Glacial acetic acid	0.25	ml

STAINING PROCEDURE

1. Decerate slides in xylene, 2 changes, 2 minutes each. Place in absolute alcohol, 2 changes, 2 minutes each; 95% alcohol, 2 changes, 2 minutes each; then rinse in distilled water.
2. Wash in running tap water.
3. Rinse in 3 changes of distilled water.
4. Immerse slides in 3 changes of ferric-ferricyanide solution for 15 minutes (5 minutes for each change).
5. Rinse well in distilled water.
6. Place in Mayer's mucicarmine for 30 minutes.
7. Rinse quickly in distilled water.
8. Counterstain with metanil yellow for a few seconds.
9. Rinse quickly in distilled water.
10. Dehydrate slides in 95% alcohol and absolute alcohol 3 changes each. Clear in xylene, 3 changes.
11. Mount coverglass with appropriate medium (refractive index 1.48-1.56).

RESULTS

Argentaffin cell granules	blue
Goblet cell mucin	reddish-pink
Cytoplasm	yellowish-green
Melanin	blue
Other reducing substances	blue

Grimelius' Method for Argentaffin and Argyrophilic Cell Granules[95]

(See Fig. 111)

FIXATION 10% Buffered neutral formalin.
PROCESS Paraffin.
MICROTOMY Cut sections at 6 micrometers.
CONTROL Use control number 6A.
SOLUTIONS

3% Silver Nitrate

Silver nitrate	3.0	gm
Distilled water	100.0	ml

Reducing Solution

Hydroquinone	1.0	gm
Sodium sulfite, anhydrous	5.0	gm
Distilled water	100.0	ml

0.2M Sodium Acetate Buffer, pH 5.6 (Stock)
Solution A

Glacial acetic acid	12.0	ml
Distilled water	988.0	ml

Solution B

Sodium acetate, trihydrate	27.2	gm
Distilled water	972.8	ml

0.2M Sodium Acetate Buffer pH 5.6 (Working)

Solution A (stock)	2.0	ml
Solution B (stock)	18.0	ml

Mix just before combining with silver nitrate.

Acetate Buffer-Silver Nitrate

0.2M Acetate buffer, pH 5.6 (working)	20.0	ml
3% Silver nitrate	6.0	ml
Distilled water	174.0	ml

STAINING PROCEDURE

1. Decerate slides in xylene, 2 changes, 2 minutes each. Place in absolute alcohol, 2 changes, 2 minutes each; 95% alcohol, 2 changes, 2 minutes each; then rinse in distilled water.
2. Rinse slides in 3 changes of distilled water.
3. Place in acetate buffer-silver nitrate and set in a 45°C oven or water bath for 24 hours.
4. Drain acetate buffer-silver nitrate from slides.
5. Place slides in reducer at 45°C for 7 minutes. (This solution should be preheated before use.)
6. Rinse slides in 3 changes of distilled water.
7. Dehydrate slides in 95% alcohol and absolute alcohol 3 changes each. Clear in xylene, 3 changes.
8. Mount coverglass with appropriate medium (refractive index 1.48-1.56).

RESULTS

This method is useful for demonstrating cell granules from many of the cells of the neuroendocrine system. Cell granules stain brown to black with light yellow background.

Luna's Method for Hepatitis B Surface Antigen

(See Fig. 99)

FIXATION	Any well fixed tissue (10% buffered neutral formalin preferred).
PROCESS	Paraffin.
MICROTOMY	Cut sections at 6 micrometers.
CONTROL	Use control number 21.
SOLUTIONS	

Aldehyde Fuchsin

Basic fuchsin (pararosanilin)	1.0	gm
70% Alcohol	200.0	ml
Concentrated hydrochloric acid	2.0	ml
Paraldehyde	2.0	ml

Let stand at room temperature for 3 days or until stain is deep purple in color. Filter and refrigerate for 1 day before use*. Store solution in refrigerator when not in use.

*Refrigeration before use will eliminate some background staining.

0.15% Potassium Permanganate

Potassium permanganate	0.15	gm
Distilled water	100.0	ml
Concentrated sulfuric acid	0.15	ml

1.5% Oxalic Acid

Oxalic acid	1.5	gm
Distilled water	100.0	ml

STAINING PROCEDURE

1. Decerate slides in xylene, 2 changes, 2 minutes each. Place in absolute alcohol, 2 changes, 2 minutes each; 95% alcohol, 2 changes, 2 minutes each; then rinse in distilled water.
2. Oxidize in 0.15% potassium permanganate solution for 5 minutes.
3. Rinse in 2 changes of distilled water.
4. Reduce in 1.5% oxalic acid until sections are colorless, usually 10 to 15 seconds.
5. Wash slides in running tap water for 1 minute.
6. Flood in distilled water, 2 changes, by pouring on and off.
7. Rinse slides in 3 changes of 95% alcohol.
8. Stain in aldehyde fuchsin for 1 hour.
9. Remove excessive aldehyde fuchsin by rinsing slides twice in 95% alcohol.
10. Rinse slides in absolute alcohol, 2 changes, by pouring on and off.
11. Clear slides in xylene, 3 changes.
12. Mount coverglass with appropriate medium (refractive index 1.48-1.56).

RESULTS
Liver cells which have "ground-glass" appearing cytoplasm with the hematoxylin and eosin stain, are characteristic of hepatitis B surface antigen carrier. Positively (purple) stained cytoplasm corresponds to localization of hepatitis B surface antigen.

Shikata's Method for Hepatitis B Surface Antigen and Copper Associated Protein

This method demonstrates elastic fibers, hepatitis B surface antigen, copper associated protein and gastrointestinal mucin. It allows the differentiation between collagen and fibrosis in liver disease. The demonstration of copper associated protein may be useful in the diagnosis of cholestasis where fibrosis is absent. Artifactual peripheral orceinophilia occurring in needle biopsies may mimic hepatitis B surface antigen positivity. It is important to have adequate oxidation in the potassium permanganate to obtain good contrast between the hepatic B antigen and the background.

FIXATION 10% Buffered neutral formalin or formal-saline.
PROCESS Paraffin.
MICROTOMY Cut sections at 6 micrometers.
CONTROL Use control number 21.
SOLUTIONS

0.15% Potassium Permanganate

Potassium permanganate	0.15	gm
Distilled water	100.0	ml
Before using add:		
Concentrated sulfuric acid	0.15	ml

2% Oxalic Acid

Oxalic acid	2.0	gm
Distilled water	100.0	ml

1% Orcein

Orcein	1.0	gm
70% Ethyl alcohol	100.0	ml
Concentrated hydrochloric acid	1.0	ml

1% Acid Alcohol

70% Ethyl alcohol	100.0	ml
Concentrated hydrochloric acid	1.0	ml

STAINING PROCEDURE
1. Decerate slides in xylene, 2 changes, 2 minutes each. Place in absolute alcohol, 2 changes, 2 minutes each; 95% alcohol, 2 changes, 2 minutes each; then rinse in distilled water.
2. Oxidize in 0.15% potassium permanganate for 10 minutes.
3. Rinse in distilled water.
4. Reduce in 2% oxalic acid until sections become colorless.
5. Rinse in distilled water and then wash for 2 minutes in running tap water.

6. Stain in 1% orcein for 4 hours or longer at room temperature.
7. Differentiate in 1% acid alcohol for 5 minutes.
8. Dehydrate slides in 95% alcohol and absolute alcohol 3 changes each. Clear in xylene, 3 changes.
9. Mount coverglass with appropriate medium (refractive index 1.48-1.56).

RESULTS

Hepatic B surface antigen	dark brown
Copper associated protein	brown
Gastrointestinal mucins	brown
Elastic fibers	brown

Tanaka-Mori-Suwa Method for Hepatitis B Surface Antigen[96]

FIXATION 10% Buffered neutral formalin.
PROCESS Paraffin.
MICROTOMY Cut section at 6 micrometers.
CONTROL Use control number 21.
SOLUTIONS

Victoria Blue

Gradually warm up a mixed solution of 200.0 ml distilled water, 0.5 gm dextrine, 4.0 gm Victoria blue and 4.0 gm resorcinol until it boils. Then add 25.0 ml boiling 29% ferric chloride solution and boil the mixture for another 3 minutes. After cooling, filter solution. Place the residue and filter paper in an oven at 50°C until dry. Dissolve the blue grain (slightly, brown-gold colored) material in 400.0 ml of 70% alcohol completely, then add 4.0 ml concentrated hydrochloric acid and 6.0 gm phenol. Allow solution to mature for about 2 weeks before use.

0.3% Potassium Permanganate (Stock)

Potassium permanganate	0.3	gm
Distilled water	100.0	ml

0.3% Sulfuric Acid (Stock)

Sulfuric acid	0.3	ml
Distilled water	100.0	ml

Potassium Permanganate - Sulfuric Acid (Working)

0.3% Potassium permanganate (stock)	50.0	ml
0.3% Sulfuric acid (stock)	50.0	ml

This should be freshly mixed just before use.

4% Sodium Bisulfite

Sodium bisulfite	4.0	gm
Distilled water	100.0	ml

This should be freshly mixed just before use.

Kernechtrot (Nuclear Fast Red)
(See Page 84)

STAINING PROCEDURE

1. Decerate slides in xylene, 2 changes, 2 minutes each. Place in absolute alcohol, 2 changes, 2 minutes each; 95% alcohol, 2 changes, 2 minutes each; then rinse in distilled water.
2. Oxidize in potassium permanganate-sulfuric acid working solution for 5 minutes.
3. Place in 4% sodium bisulfite solution for 1 minute.
4. Wash slides in running tap water for 2 minutes.
5. Place slides in 70% ethyl alcohol for 2 minutes.
6. Stain slides in Victoria blue solution for 8 hours to overnight.
7. Differentiate slides in 70% ethyl alcohol for 3 to 5 minutes until the background has become completely decolorized.
8. Wash slides in running tap water for 5 minutes.
9. Stain slides in nuclear fast red solution for 5 minutes.
10. Wash slides in running tap water for 5 minutes.
11. Dehydrate slides in 95% alcohol and absolute alcohol 3 changes each. Clear in xylene, 3 changes.
12. Mount coverglass with appropriate medium (refractive index 1.48-1.56).

RESULTS

Hepatitis B surface antigen	blue
Cytoplasm	red
Nuclear chromatin	red
Elastic fiber	blue
Lipofuscin	blue
Mast cell granules	blue
Bile pigment	red
Goblet cell mucin	blue

REMARKS

Proper staining (step 6) may require one to several days. Proper differentiation (step 7) may require 3 minutes to 2 hours.

Hellerstrom-Hellman Method for Pancreatic Alpha 1 and Delta Cells

FIXATION 10% Buffered neutral formalin.
PROCESS Paraffin, frozen, or celloidin.
MICROTOMY Cut paraffin sections at 4 micrometers.
CONTROL Use control number 3.
SOLUTIONS

0.1N Nitric Acid

Concentrated nitric acid	6.45	ml
95% Alcohol	993.55	ml

5% Pyrogallic Acid

Pyrogallic acid	5.0	gm
95% Alcohol	100.0	ml
Mix well and then add:		
Concentrated formalin (37%)	5.0	ml

Mix well. This is a working solution.

10% Silver Nitrate

Silver nitrate	10.0	gm
Distilled water	10.0	ml
Dissolve and add:		
0.1N nitric acid	90.0	ml

Mix well. Adjust this solution to pH 5 with concentrated (58%) ammonium hydroxide (approximately 15 drops). This is a working solution. Filter. Discard after use.

STAINING PROCEDURE

1. Decerate slides in xylene, 2 changes, 2 minutes each. Place in absolute alcohol, 2 changes, 2 minutes each; 95% alcohol, 2 changes, 2 minutes each; then rinse in distilled water.
2. Place in Bouin's solution at 37°C for 1 hour.
3. Wash in running tap water for 30 minutes.
4. Rinse once in distilled water.
5. Place in 95% alcohol for 2 minutes.
6. Place in 10% silver nitrate solution overnight in a 37°C oven.
7. Immerse for 8 seconds in 95% alcohol.
8. Develop for 60 seconds in pyrogallic acid. (Agitate)
9. Rinse in 3 changes of 95% alcohol for 1 minute each.
10. Counterstaining may be done at this point if desired.
11. Dip in 95% alcohol, 2 off-and-on changes.
12. Dip in absolute alcohol, 2 off-and-on changes.
13. Dip in xylene, 3 off-and-on changes.
14. Mount coverglass with appropriate medium (refractive index 1.48-1.56).

RESULTS

Alpha or delta cells granules .... black

REMARKS

The following conditions are critical to the success of this procedure: (1) use acid-cleaned glassware; (2) do not use metal staining apparatus; (3) adjust the pH of the nitrate silver solution to exactly 5; (4) filter the working silver nitrate solution to avoid precipitation; and (5) do not use mercurial fixation.

Lehy-Zeitoun Method for Pancreatic Alpha Cell Granules[97]

FIXATION	Bouin's solution (may not be post fixed in Bouin's) exposure to aldehydes (formalin, etc.) prevent staining reaction.
PROCESS	Paraffin.

MICROTOMY Cut section at 0.5 to 1 micrometers for plastic. Cut section at 5 micrometers for paraffin.
CONTROL Use control number 1A.
SOLUTIONS

1% Hydrochloric Acid

Concentrated hydrochloric acid	1.0	ml
Distilled water.	100.0	ml

0.1M Glacial Acetic Acid (Stock)

Glacial acetic acid	5.75	ml
Distilled water, q.s. to	1000.0	ml

0.1M Sodium Acetate (Stock)

Sodium acetate, anydrous	8.2	gm
Distilled water, q.s. to	1000.0	ml

Toluidine Blue

Toluidine	0.2	gm
0.1M Sodium acetate (stock)	140.0	ml

Adjust to pH 5.0 using 0.1M glacial acetic acid (approximately 60 ml).

Sodium Acetate Buffer

0.1M Sodium acetate (stock)	140.0	ml

Adjust to pH 5.0 using 0.1M glacial acetic acid (approximately 60 ml).

STAINING PROCEDURE

1. Decerate slides in xylene, 2 changes, 2 minutes each. Place in absolute alcohol, 2 changes, 2 minutes each; 95% alcohol, 2 changes, 2 minutes each; then rinse in distilled water.
2. Hydrolyze in 1% hydrochloric acid at 60°C for 30 minutes.
3. Rinse well in distilled water.
4. Stain in toluidine blue for 1 to 2 minutes for paraffin and 15 minutes for plastic.
5. Rinse in sodium acetate buffer.
6. Mount coverglass with neutral aqueous mounting medium.

RESULTS

Alpha cells granules	pink to red
Nuclei and background	blue

Kito-Hosoda Method for Pancreatic Islet Cells[98]

FIXATION 10% Buffered neutral formalin or Bouin's solution.
PROCESS Paraffin.
MICROTOMY Cut section at 3 micrometers.
CONTROL Use control number 1A.

SOLUTIONS

0.65% Silver Nitrate

Silver nitrate	0.65	gm
Distilled water	100.0	ml

Make fresh, just before use.

Reducing Solution

Hyroquinone	1.0	gm
Sodium sulphite	5.0	gm
Distilled water	100.0	ml

Make fresh just before use.

Aldehyde Fuchsin

Basic fuchsin	0.5	gm
95% Ethyl alcohol	63.0	ml
Distilled water	37.0	ml
Concentrated hydrochloric acid	1.5	ml
Paraldehyde	1.0	ml

Combine ingredients and allow to react at room temperature for 2 to 3 days. The color of this solution should be nearly indistinguishable from that of gentian violet. It will keep its selective staining properties for one month if stored in the refrigerator.

Saturated Ammonium Acetate (Stock)

Ammonium acetate	200.0	gm
Distilled water	150.0	ml

Let stir overnight on a magnetic stirrer plate.

5% Lead Nitrate (Stock)

Lead nitrate	5.0	gm
Distilled water	100.0	ml

Lead nitrate is a hazardous substance. Observe strict chemical precautions. Do not breathe or ingest powder. Use appropriate hood.

Lead Nitrate-Ammonium Acetate (Stock)

5% Lead nitrate (stock)	100.0	ml
Saturated ammonium acetate (stock)	100.0	ml
Filter and add:		
Concentrated formalin (37%)	4.0	ml

This stock of stabilized solution keeps at room temperature for one month.

Lead-Hematoxylin (Working)

Hematoxylin	0.2	gm
95% Ethyl alcohol	1.5	ml
Dissolve completely and add:		
Lead nitrate-ammonium acetate (stock)	10.0	ml
Distilled water	10.0	ml
Filter mixture after 30 minutes and add:		
Distilled water	75.0	ml

STAINING PROCEDURE

1. Decerate slides in xylene, 2 changes, 2 minutes each. Place in absolute alcohol, 2 changes, 2 minutes each; 95% alcohol, 2 changes, 2 minutes each; then rinse in distilled water.

2. Place slides in 0.65% silver nitrate overnight at 37°C.
3. Transfer slides to freshly prepared reducing solution (preheated to 37°C) for 1 minute.
4. Rinse in 3 changes of distilled water.
5. Stain in aldehyde-fuchsin solution for 5 minutes.
6. Rinse in 3 changes of distilled water.
7. Place in freshly prepared working lead-hematoxylin solution (preheated to 37°C) for 30 minutes.
8. Rinse in 3 changes of distilled water.
9. Dehydrate slides in 95% alcohol and absolute alcohol 3 changes each. Clear in xylene, 3 changes.
10. Mount coverglass with appropriate medium (refractive index 1.48-1.56).

RESULTS

Alpha cells....................black
Beta cells....................fine light red
Delta cells....................blue

Halmi's Method for Pancreatic Beta Cell Granules

FIXATION Modified Bouin's solution.
PROCESS Paraffin.
MICROTOMY Cut sections at 3 to 6 micrometers.
CONTROL Use control number 1A.
SOLUTIONS

Modified Bouin's Fixative

Saturated aqueous picric acid	88.0	ml
Glacial acetic acid	5.0	ml
37% Formalin	10.0	ml

Fix tissue 24 hours. Change to 70% alcohol, dehydrate and embed the same day. If possible, leave in 70% ethyl alcohol only a very short time. Long periods of time in 70% ethyl alcohol may extract some insulin from the beta granules and result in a weaker stain. The exocrine portion of mouse pancreas may section with some difficulty if stored in formalin.

Aldehyde Fuchsin (basic fuchsin)

60% Alcohol	100.0	ml
Pararosaniline hydrochloride (basic fuchsin)	0.5	gm
Paraldehyde	1.0	ml
Concentrated hydrochloric acid	1.5	ml

Ripen 2 to 4 days at room temperature; filter before use. Stain is good for 7 to 10 days.

Halmi Counterstain

Light green SF yellowish	0.3	gm
Orange G	0.5	gm
Chromotrope 2R	0.2	gm
Phosphotungstic acid	0.5	gm
Glacial acetic acid	0.5	gm
Deionized water	100.0	ml

STAINING PROCEDURE

1. Decerate slides in xylene, 2 changes, 2 minutes each. Place in absolute alcohol, 2 changes, 2 minutes each; 95% alcohol, 2 changes, 2 minutes each; then rinse in distilled water.
2. Place slides in 70% ethyl alcohol for 30 seconds.
3. Stain in aldehyde fuchsin for 15 minutes.
4. Wash twice in 95% ethyl alcohol (3 times, if many slides). First wash, 30 seconds to 1 minute; second wash, 1.5 to 3 minutes; third wash, 3 to 5 minutes.
5. Rinse slides in 70% ethyl alcohol.
6. Wash in deionized water for 30 seconds.
7. Stain in hematoxylin 5 to 15 seconds (examine microscopically for desired staining).
8. Wash in deionized water until clear, blue in Scott's tap water substitute, wash in deionized water.
9. Place slides in Halmi counterstain for 2 minutes.
10. Rinse in deionized water; check microscopically for differentiation.
11. Dehydrate quickly in 1 change 95% ethyl alcohol and 3 changes ethyl alcohol.
12. Dehydrate slides in 95% alcohol and absolute alcohol 3 changes each. Clear in xylene, 3 changes.
13. Mount coverglass with appropriate medium (refractive index 1.48-1.56).

RESULTS

Pancreatic beta cells .... fine granules, deep purple
Nuclear chromatin .... blue
Exocrine pancreas .... light green

Maldonado's Method for Pancreatic Islet Cells[99]

FIXATION Bouin's solution without glacial acetic acid.
PROCESS Paraffin.
MICROTOMY Cut section at 6 micrometers.
CONTROL Use control number 6A.

1% Phloxine
(See Page 88)

3% Phosphotungstic Acid

Phosphotungstic acid	3.0	gm
Distilled water	100.0	ml

0.05% Azure II

Azure II	0.05	gm
Distilled water	100.0	ml

Weigert's Iron Hematoxylin
(See Page 79)

STAINING PROCEDURE

1. Decerate slides in xylene, 2 changes, 2 minutes each. Place in absolute alcohol, 2 changes, 2 minutes each; 95% alcohol, 2 changes, 2 minutes each; then rinse in distilled water.

2. Stain in 1% phloxine for 10 minutes.
3. Rinse in distilled water for 10 seconds.
4. Place in 3% phosphotungstic acid for 1 minute.
5. Rinse in distilled water for 10 seconds.
6. Stain in 0.05% azure II for 30 seconds.
7. Rinse in distilled water for 5 seconds.
8. Stain in Weigert's iron hematoxylin for 1 minute.
9. Rinse in distilled water.
10. Dehydrate slides in 95% alcohol and absolute alcohol 3 changes each. Clear in xylene, 3 changes.
11. Mount coverglass with appropriate medium (refractive index 1.48-1.56).

RESULTS

Alpha cells ....purple
Beta cells..violet blue
Delta cells ....light blue with evident granules
Exocrine cells ....grayish blue with red granules

Wenger's Method for Pancreatic Beta Cell Granules[100]

The method provided below is a modification which has proven very useful and reproducible in the author's hands. The changes introduced in this method are discussed at the point where they are made, or under remarks. This method can be used to stain all other entities normally stained by aldehyde fuchsin solutions; elastica fibers, hepatitis B surface antigens, mast cells and pancreatic beta cell granules.

FIXATION 10% Buffered neutral formalin.
MICROTOMY Cut paraffin sections at 6 micrometers.
SOLUTIONS

Wenger's Aldehyde Fuchsin

This modified aldehyde fuchsin solution is offered as an alternative to purchasing paraldehyde, which is used in Gomori's aldehyde fuchsin.

Pararosanilin (basic fuchsin), 5% or more dye content...	2.5	gm
70% Ethyl alcohol	500.0	ml
Concentrated hydrochloric acid....................................	5.0	ml
Potassium dichromate	5.0	gm

Combine the above ingredients, cap the container tightly and allow to stand at room temperature for 5 days. Check solution by smelling to ensure formation of acetaldehyde. Acetaldehyde can be detected by its characteristic pungent fruity odor. If acetaldehyde is not detected within 24 hours, ethyl alcohol and/or potassium dichromate was not added, or possibly an insufficient amount of hydrochloric acid was used. At the end of the 5 day period, check the quality of the solution by staining an elastica control. If elastic tissue fibers are well demonstrated, cap the bottle tightly and refrigerate the solution until ready to use. The use of ethyl alcohol is essential to produce proper reactivity with potassium dichromate to form acetaldehyde.

0.25% Metanil Yellow

Metanil yellow	0.25	gm
Distilled water	100.0	ml
Glacial acetic acid	2	drops

Thiosulfation Solution A (Stock)

Cupric sulfate pentahydrate	1.56	gm
Tris (hydroxymethyl) aminomethane	15.14	gm
Distilled water, q.s. to	500.0	ml

Adjust to pH 9.0 if necessary, with 1% aqueous sodium hydroxide. This solution is stable at room temperature.

Thiosulfation Solution B (Stock)

Sodium metabisulfite	3.8	gm
Distilled water	100.0	ml

Make fresh.

Thiosulfation Solution (Working)

Solution A (stock)	80.0	ml
Solution B (stock)	20.0	ml

This solution is stable for up to 2 days.

Kernechtrot (Nuclear Fast Red)
(See Page 84)

STAINING PROCEDURE

1. Decerate slides in xylene, 2 changes, 2 minutes each. Place in absolute alcohol, 2 changes, 2 minutes each; 95% alcohol, 2 changes, 2 minutes each; then rinse in distilled water.
2. Place slides in fresh, working thiosulfation solution for 3 hours.
3. Wash slides in running tap water for 1 minute and rinse in distilled water.
4. Rinse slides in 95% alcohol.
5. Place slides in Wenger's aldehyde fuchsin solution for 5 minutes.
6. Rinse slides in 95% alcohol until no more stain runs off the slides (see remarks).
7. Stain slides in metanil yellow for 30 seconds.
8. Rinse slides in distilled water.
9. Stain in Kernechtrot for 3 minutes.
10. Rinse in 5 changes of distilled water.
11. Dehydrate slides in 95% alcohol and absolute alcohol 3 changes each. Clear in xylene, 3 changes.
12. Mount coverglass with appropriate medium (refractive index 1.48-1.56).

RESULTS

Pancreatic beta cell granules	purple
Background	yellow
Nuclei	red
Elastica	purple
Mast cells	purple

REMARKS

Ethyl alcohol reacts with potassium dichromate to form acetaldehyde. The major modification in this method is the substitution of tris (hydroxymethyl) aminomethane for ammonium hydroxide, which adds stability to solution A (stock), thus prolonging the shelf life of the solution. Check control slide microscopically. If beta cells are inadequately stained, repeat steps 2-6.

Glenner-Lillie Method for Pituitary Cells[55w]

FIXATION 10% Buffered neutral formalin.
PROCESS Paraffin, frozen, or celloidin.
MICROTOMY Cut sections at 6 micrometers.
CONTROL Use control number 26.
SOLUTIONS

Staining Solution

1% Aqueous eosin B	8.0	ml
1% Aqueous aniline blue	2.0	ml
0.1M Citric acid	1.1	ml
0.2M Disodium phosphate	0.9	ml
Distilled water	28.0	ml

Store at room temperature.

Acetone-Xylene

Acetone	50.0	ml
Xylene	50.0	ml

STAINING PROCEDURE

1. Decerate slides in xylene, 2 changes, 2 minutes each. Place in absolute alcohol, 2 changes, 2 minutes each; 95% alcohol, 2 changes, 2 minutes each; then rinse in distilled water.
2. Place in staining solution in oven at 60°C for 1 hour.
3. Wash in running tap water for 5 minutes.
4. Dehydrate through 50% acetone, 80% acetone, and 100% acetone, 2 changes each.
5. Clear in acetone-xylene solution, then xylene, 2 changes each.
6. Mount coverglass with appropriate medium (refractive index 1.48-1.56).

RESULTS

Basophil cell granules	blue to black
Acidophil cell granules	dark red
Chromophobe granules	slate gray to pale pink
Colloid	red to blue-violet
Erythrocytes	orange

REMARKS
This technique is also useful for duodenal enterochromaffin cells and pancreatic islet cells with the following modification. At step 2, make fresh staining solution buffered to pH 3.6 (0.1M citric acid, 1.4 ml; 0.2M disodium phosphate, 0.6 ml) and heat to 60°C. Sections are then incubated at this temperature for 3 minutes and carried through the rest of the procedure as described above.

RESULTS

Enterochromaffin cells	blue-black
Goblet cells and mucins	light blue
Paneth cells	light black
Alpha cells of the pancreatic islet	red
Beta cells of the pancreatic islet	purple
Background	varying shades ranging from blue to pale pink

Wilson-Ezrin Method for Pituitary Cells[55x]

FIXATION 10% Formol-saline solution.
PROCESS Paraffin, frozen, or celloidin.
MICROTOMY Cut sections at 6 micrometers.
CONTROL Use control number 26.
SOLUTIONS

1% Periodic Acid

Periodic acid	1.0	gm
Distilled water	100.0	ml

Schiff's Reagent
(See Page 259)

Sulfurous Acid

10% Aqueous sodium metabisulfite	60.0	ml
1N Hydrochloric acid	50.0	ml
Distilled water	1000.0	ml

1% Orange G

Orange G	1.0	gm
Distilled water	100.0	ml

5% Phosphotungstic Acid

Phosphotungstic acid	5.0	gm
Distilled water	100.0	ml

1% Methyl Blue

Methyl blue	1.0	gm
Distilled water	100.0	ml

STAINING PROCEDURE

1. Decerate slides in xylene, 2 changes, 2 minutes each. Place in absolute alcohol, 2 changes, 2 minutes each; 95% alcohol, 2 changes, 2 minutes each; then rinse in distilled water.
2. Place in 1% periodic acid for 5 minutes.
3. Rinse in distilled water.
4. Immerse in Schiff's reagent for 15 minutes.
5. Rinse in sulfurous acid, 3 changes, 3 minutes each.
6. Wash in running tap water for 10 minutes.

7. Immerse in orange G solution for 1 minute.
8. Immerse in phosphotungstic acid solution for 30 seconds.
9. Wash in running tap water for 30 seconds.
10. Place in methyl blue solution for 1 minute.
11. Wash out excess methyl blue in running tap water.
12. Dehydrate slides in 95% alcohol and absolute alcohol 3 changes each. Clear in xylene, 3 changes.
13. Mount coverglass with appropriate medium (refractive index 1.48-1.56).

RESULTS

Basophilic cell granules	red
Gamma cell granules	purple
Acidophil cell granules	yellow

Heath's Method for Pituitary Cells[101]

FIXATION	10% Buffered neutral formalin.
PROCESS	Paraffin.
MICROTOMY	Cut section at 6 micrometers.
CONTROL	Use control section of adrenal gland.
SOLUTIONS	

Peroxide-Acid Solution

30% Hydrogen peroxide	15.0	ml
Concentrated sulfuric acid	1.5	ml
Formic acid	133.5	ml

4N Sulfuric Acid

Concentrated sulfuric acid	112.6	ml
Distilled water	887.4	ml

Alcian Blue

4% Alcian blue	100.0	ml
4N Sulfuric acid	100.0	ml

0.5% Periodic Acid

Periodic acid	0.5	gm
Distilled water	100.0	ml

Schiff's Reagent
(See Page 259)

0.5% Orange G

Orange G	0.5	gm
Distilled water	100.0	ml

1% Phosphotungstic Acid

Phosphotungstic acid	1.0	gm
Distilled water	100.0	ml

1% Glacial Acetic Acid

Glacial acetic acid	1.0	ml
Distilled water	99.0	ml

STAINING PROCEDURE

1. Decerate slides in xylene, 2 changes, 2 minutes each. Place in absolute alcohol, 2 changes, 2 minutes each; 95% alcohol, 2 changes, 2 minutes each; then rinse in distilled water.
2. If Zenker fixed, remove mercuric chloride crystals with iodine and clear with sodium thiosulfate.
3. Place in peroxide-acid for 5 minutes.
4. Wash in running tap water for 10 minutes.
5. Stain in alcian blue for 1 hour.
6. Wash in running tap water for 10 minutes
7. Oxidize in periodic acid solution for 5 minutes.
8. Rinse in distilled water.
9. Place in Schiff's solution for 10 minutes.
10. Wash in tepid running tap water for 10 minutes.
11. Stain in orange G solution for 30 seconds.
12. Place in phosphotungstic acid for 30 seconds.
13. Place in glacial acetic acid for 30 seconds.
14. Dehydrate slides in 95% alcohol and absolute alcohol 3 changes each. Clear in xylene, 3 changes.
15. Mount coverglass with appropriate medium (refractive index 1.48-1.56).

RESULTS

Neurosecretory material in pars neurosa	clear blue
Basement membranes	magenta
Erythrocytes	yellow
Cytoplasms	pale magenta
Colloid	strong magenta
Basophil cell granules	blue, purple or red
Acidophil cell granules	orange to yellow

Monroe-Frommer Method for Pituitary Cells

FIXATION Zenker's solution.
PROCESS Paraffin.
MICROTOMY Cut section at 6 micrometers.
CONTROL Use control number 26.
SOLUTIONS

10% Tannic Acid

Tannic acid	10.0	gm
Distilled water	100.0	ml

1% Basic Fuchsin (Stock)

Basic fuchsin	1.0	gm
100% Alcohol	20.0	ml
Distilled water	80.0	ml

Basic Fuchsin (Working)

Basic fuchsin (stock)	50.0	ml
Distilled water	50.0	ml

Filter before use.

1% Aniline

Aniline	1.0	ml
100% Alcohol	90.0	ml
Distilled water	10.0	ml

Aniline should be used with care. It is an important cause of serious industrial poisoning associated with bone marrow depression as well as methemoglobinemia.

1% Phosphomolybdic Acid

Phosphomolybdic acid	1.0	gm
Distilled water	100.0	ml

1% Alcian Blue

Alcian blue, 8GX	1.0	gm
Distilled water	100.0	ml

Filter before use.

STAINING PROCEDURE

1. Decerate slides in xylene, 2 changes, 2 minutes each. Place in absolute alcohol, 2 changes, 2 minutes each; 95% alcohol, 2 changes, 2 minutes each; then rinse in distilled water.
2. Remove mercuric chloride crystals with iodine and clear with sodium thiosulfate.
3. Wash in running tap water for 10 minutes.
4. Place in tannic acid solution for 10 minutes.
5. Wash in running tap water for 5 minutes.
6. Stain in working basic fuchsin solution for 3 to 5 seconds. Slides should be "agitated" by dipping in and out of staining solution.
7. Wash in tap water to remove excess stain.
8. Decolorize in aniline solution until acidophils are red and basophils pink.
9. Place in phosphomolybdic acid solution for 30 seconds.
10. Rinse in distilled water.
11. Stain in alcian blue solution for 30 seconds.
12. Rinse in distilled water.
13. Dehydrate in 95% alcohol, absolute alcohol, and clear in xylene, 2 changes each.
14. Mount coverglass with Permount or Histoclad.

RESULTS

Acidophil cell granules	red
Beta basophil cell granules	green
Delta basophils	purple
Collagen fibers	bluish-green
Chromophobes	colorless
Colloid	red
Neurosecretory granules	brilliant red
Neurohypophysial fibers	blue

Luna's Aldehyde Fuchsin Method for Mast Cells[55y]

FIXATION 10% Buffered neutral formalin.
PROCESS Paraffin, frozen, or celloidin.
MICROTOMY Cut sections at 6 micrometers.
CONTROL Use control number 31.
SOLUTIONS

Aldehyde Fuchsin
(See Page 199)

Weigert's Iron-Hematoxylin
(See Page 79)

Methyl Orange

Methyl orange	0.25	gm
95% Alcohol	100.0	ml

STAINING PROCEDURE

1. Decerate slides in xylene, 2 changes, 2 minutes each. Place in absolute alcohol, 2 changes, 2 minutes each; 95% alcohol, 2 changes, 2 minutes each; then rinse in distilled water.
2. Place in aldehyde-fuchsin solution for 30 minutes.
3. Rinse in 95% alcohol.
4. Place in Weigert's hematoxylin solution for 1 minute.
5. Wash in running tap water for 10 minutes.
6. Rinse in 95% alcohol.
7. Counterstain in methyl orange solution for 5 minutes or until background is light yellow.
8. Dehydrate slides in 95% alcohol and absolute alcohol 3 changes each. Clear in xylene, 3 changes.
9. Mount coverglass with appropriate medium (refractive index 1.48-1.56).

RESULTS

Mast cell granules	purple
Elastic fibers	purple
Other cellular elements	blue
Background	yellow

REMARKS
Especially useful for demonstration of mast cells in animal tissue.

Luna's Methylene Blue Method for Mast Cells[55z]

(See Fig. 102)

FIXATION 10% Buffered neutral formalin.
PROCESS Paraffin, frozen, or celloidin.
MICROTOMY Cut sections at 6 micrometers.
CONTROL Use control number 31.

SOLUTIONS

Polychrome Methylene Blue

Methylene blue	1.0	gm
Distilled water	100.0	ml
95% Alcohol	20.0	ml
Potassium carbonate	1.0	gm

Glycerin-Calcium Chloride (Stock)

Glycerin	100.0	ml
Calcium chloride, anhydrous	20.0	ml

Heat the above mixture until the volume is reduced to 100 ml. Cool to room temperature before use.

Glycerin-Calcium Chloride (Working)

Glycerin-calcium chloride (stock)	10.0	ml
Distilled water	50.0	ml

STAINING PROCEDURE

1. Decerate slides in xylene, 2 changes, 2 minutes each. Place in absolute alcohol, 2 changes, 2 minutes each; 95% alcohol, 2 changes, 2 minutes each; then rinse in distilled water.
2. Place in polychrome methylene blue solution for 10 minutes.
3. Rinse in distilled water.
4. Decolorize in working glycerin-calcium chloride solution, for 30 seconds to 1 minute, or until section is a medium blue. Be careful not to decolorize too long.
5. Wash thoroughly in water for 2 minutes, then blot with filter paper.
6. Dehydrate rapidly in absolute alcohol.
7. Clear in xylene, 2 or 3 changes.
8. Mount coverglass with resinous medium.

RESULTS

Mast cell granules	bluish-purple
Cytoplasm	pink

REMARKS

This technique, which was originally used for the differentiation of mast cells and plasma cells, has been used for the specific demonstration of mast cells. It is especially useful for highly cellular specimens, such as lymph node and spleen.

Luna's Toluidine Blue Method for Mast Cells

FIXATION	10% Buffered neutral formalin.
PROCESS	Paraffin or frozen.
MICROTOMY	Cut sections at 6 micrometers.
CONTROL	Use control number 31.

SOLUTIONS

0.5% Toluidine Blue

Toluidine blue	0.5	gm
70% Alcohol	100.0	ml
Concentrated hydrochloric acid	0.5	ml

Filter solution before using.

0.5% Alcoholic Hydrochloric Acid

95% Alcohol	100.0	ml
Hydrochloric acid	0.5	ml

0.01% Eosin

Eosin Y	0.01	gm
95% Alcohol	100.0	ml

STAINING PROCEDURE

1. Decerate slides in xylene, 2 changes, 2 minutes each. Place in absolute alcohol, 2 changes, 2 minutes each; 95% alcohol, 2 changes, 2 minutes each; then rinse in distilled water.
2. Place in 95% alcohol for 30 seconds.
3. Stain for 1 hour at room temperature in toluidine blue.
4. Rinse briefly in 0.5% alcoholic hydrochloric acid (until the sections are grossly colorless).
5. Counterstain lightly by dipping in 0.01% eosin, 5 dips.
6. Dehydrate slides in 95% alcohol and absolute alcohol 3 changes each. Clear in xylene, 3 changes.
7. Mount coverglass with appropriate medium (refractive index 1.48-1.56).

RESULTS

Mast cell granules	blue
Background	pink

Smith's Method for Juxtaglomerular Granules[55aa]

(See Fig. 100)

FIXATION Smith's or Bouin's solution.
PROCESS Paraffin, frozen, or celloidin.
MICROTOMY Cut sections at 6 micrometers.
CONTROL Use control number 26.
SOLUTIONS

Smith's Solution

10% Formalin	100.0	ml
Potassium dichromate	5.0	gm
Glacial acetic acid	2.5	ml

Follow fixation by washing in running tap water for 2 hours. This fixative is not stable. It should be made fresh just before use.

1% Potassium Permanganate (Stock)

Potassium permanganate	1.0	gm
Distilled water	100.0	ml

1% Sulfuric Acid (Stock)

Concentrated sulfuric acid	1.0	gm
Distilled water	100.0	ml

Potassium Permanganate-Sulfuric Acid (Working)

Potassium permanganate (stock)	50.0	ml
Sulfuric acid (stock)	50.0	ml

Mix just before use.

1% Sodium Metabisulphite

Sodium metabisulphite	1.0	gm
Distilled water	100.0	ml

Ziehl's Carbolfuchsin

Basic fuchsin	1.0	gm
Phenol crystals	5.0	gm
95% Alcohol	10.0	ml
Distilled water	100.0	ml

In a mortar, grind basic fuchsin and phenol crystals. Then add 95% alcohol while still grinding. Add the distilled water in 10 ml lots to wash out the mortar. Filter the accumulated washings before use.

Acid-Alcohol

35% Alcohol	100.0	ml
Concentrated hydrochloric acid	0.5	ml

Bowie's Stain

Biebrich scarlet	1.0	gm
Distilled water	250.0	ml

Filter into a 1-liter flask while the following solution (ethyl violet) is being mixed.

Ethyl Violet

Ethyl violet	2.0	gm
Distilled water	500.0	ml

The ethyl violet solution is then filtered into the Biebrich scarlet while stirring. The end point is reached when a precipitate forms on the side of the flask. The resulting precipitate is separated by vacuum filtration, dried and added to 100 ml of absolute alcohol.

Acetone-Xylene

Acetone	50.0	ml
Xylene	50.0	ml

STAINING PROCEDURE

1. Decerate slides in xylene, 2 changes, 2 minutes each. Place in absolute alcohol, 2 changes, 2 minutes each; 95% alcohol, 2 changes, 2 minutes each; then rinse in distilled water.
2. Oxidize in freshly prepared potassium permanganate-sulfuric acid for 1 minute.
3. Rinse briefly in distilled water.
4. Decolorize in sodium metabisulphite solution for 3 minutes.
5. Wash in running tap water for 2 minutes.

6. Place in Ziehl's carbolfuchsin solution for 1 minute.
7. Wash in running tap water for 1 minute.
8. Decolorize in acid-alcohol solution until sections are pink. (Retention of red color in some nuclei is more desirable than complete extraction.)
9. Wash in running tap water for 2 minutes. Handle slides individually from this point on.
10. Place slide flat on a staining rack without draining or shaking off surface water. Drop on sufficient Bowie's stain to cover the section without flooding the slide. After 5 to 10 seconds, immerse the slide in a coplin jar of water without draining off the dye. Wait 1 to 2 minutes before dehydrating.
11. Dehydrate rapidly using 2 changes of acetone followed by equal parts acetone-xylene, 1 change.
12. Clear in xylene, 2 changes.
13. Mount coverglass with appropriate medium (refractive index 1.48-1.56).

RESULTS

Juxtaglomerular granules .... deep purple
Nuclei .... some pink to red; majority lavender
Cytoplasm .... shades of blue or purple
Elastic tissue .... deep purple
Collagen .... pink
Erythrocytes .... red

REMARKS

The greatest variable in the staining results of this method is the Bowie's stain. Trouble may arise if different dye lots are used or if the concentration in the final solution is too low. An easy way to prepare the solution is to weigh the filter paper prior to filtering the Biebrich scarlet-ethyl violet reaction mixture, then filter, dry and weigh the combination. Subtract the original paper weight to determine the amount of dye present. Place the paper and dye in a sufficient quantity of absolute alcohol to yield a 1% solution. There may still be some undissolved dye present, but this does not matter.

Mouse juxtaglomerular granules are the most prominent and easiest to stain. If good results cannot be obtained, try running the unsatisfactory stained slide back to water and apply a second application of Bowie's stain. This will sometimes yield an improvement; if not, try a different dye lot or manufacturer. It is also possible to bleach out all the stain by repeating the procedure from step 2. This sometimes yields better results and conserves research material. The thioflavine T fluorochrome method of Vassar and Culling will also stain the juxtaglomerular apparatus in mice.

Berg's Method for Spermatozoa[55ab]

FIXATION	10% Buffered neutral formalin.
PROCESS	Paraffin, frozen, or celloidin.
MICROTOMY	Cut sections at 6 micrometers.
CONTROL	Use control section of testis.

SOLUTIONS

Carbolfuchsin

New fuchsin	1.0	gm
Phenol	5.0	gm
100% Alcohol	10.0	ml
Distilled water	84.0	ml

Filter before using.

Saturated Lithium Carbonate
(See Page 191)

5% Glacial Acetic Acid-Alcohol

Glacial acetic acid	5.0	ml
100% Alcohol	95.0	ml

0.5% Methylene Blue

Methylene blue	0.5	gm
100% Alcohol	100.0	ml

STAINING PROCEDURE

1. Decerate slides in xylene, 2 changes, 2 minutes each. Place in absolute alcohol, 2 changes, 2 minutes each; 95% alcohol, 2 changes, 2 minutes each; then rinse in distilled water.
2. Stain in carbolfuchsin solution for 30 minutes.
3. Place slides directly into saturated lithium carbonate solution for 3 minutes.
4. Decolorize in 5% glacial acetic acid-alcohol solution for 5 minutes.
5. Place in absolute alcohol, 2 changes, 1 minute each.
6. Stain in methylene blue for 30 to 60 seconds.
7. Rinse rapidly in absolute alcohol, 2 changes.
8. Clear in xylene, 3 changes.
9. Mount coverglass with appropriate medium (refractive index 1.48-1.56).

RESULTS

Spermatozoa	brilliant red
Erythrocytes	pale pink
Other tissue components	blue to purple

REMARKS

Uneven staining can be avoided if slides are gently agitated throughout all stages of the procedure.

Lauren-Sorvari Method for Paneth Cell Granules[102]

FIXATION	10% Buffered neutral formalin.
PROCESS	Paraffin.
MICROTOMY	Cut section at 6 micrometers.
CONTROL	Use control from a specimen of jejunum.

SOLUTIONS

Methylation Solution

Concentrated hydrochloric acid	0.8	ml
Absolute methyl alcohol	100.0	ml

Harris' Hematoxylin
(See Page 81)

Best's Carmine Solution (Stock)

Carmine	2.0	gm
Potassium carbonate	1.0	gm
Potassium chloride	5.0	gm
Distilled water	60.0	ml
Boil gently for 3 to 5 minutes or until color darkens.		
Cool, then add: Concentrated ammonium hydroxide	20.0	ml

Allow to ripen for 24 hours at room temperature. Store in the refrigerator at 0-5°C. This stock solution is good for about 30 days.

Best's Carmine Solution (Working)

Best's carmine (stock)	20.0	ml
Concentrated ammonium hydroxide	30.0	ml
100% Methyl alcohol	30.0	ml

This solution is good for 1 day only.

Washing Mixture

100% Methyl alcohol	40.0	ml
100% Ethyl alcohol	80.0	ml
Distilled water	100.0	ml

50% Methyl Alcohol

100% Methyl alcohol	50.0	ml
Distilled water	50.0	ml

30% Methyl Alcohol

100% Methyl alcohol	30.0	ml
Distilled water	70.0	ml

STAINING PROCEDURE

1. Decerate slides in xylene, 2 changes, 2 minutes each. Place in absolute alcohol, 2 changes, 2 minutes each; 95% alcohol, 2 changes, 2 minutes each; then rinse in distilled water.
2. Place in methylation solution for 2 to 18 hours at 58° (see remarks).
3. Hydrate through two 2 minute changes each of 100%, 50% and 30% methyl alcohols into distilled water.
4. Stain in Harris' hematoxylin for 15 seconds, if nuclear staining is desired.
5. Wash in running tap water for 2 minutes.
6. Stain in freshly made working Best's carmine solution for 20 minutes.
7. Wash for 30 seconds in the washing mixture.
8. Dehydrate slides in 95% alcohol and absolute alcohol 3 changes each. Clear in xylene, 3 changes.
9. Mount coverglass with appropriate medium (refractive index 1.48-1.56)

RESULTS

Paneth cell granules	bright red
Eosinophilic leukocytes	red
Nuclei	blue

REMARKS

Methylation time can be varied to achieve the effect desired. The granules of Paneth cells stain distinctly after 2 hours of methylation, but the reaction is intensified when methylation continues for at least 18 hours. The neutral mucosaccharides and glycogen of gastric mucus still stain after about 6 hours of methylation, but not after more than 12 hours. Six hours of methylation is generally the most useful.

The collagen fibrils of connective tissue stain more strongly after long term methylation, weakening the contrast. Other structures in the intestinal wall either remain colorless, stain weakly reddish, or stain basophilically with hematoxylin.

When alcian blue is applied prior to methylation, both the core and capsule of the granules of mouse Paneth cells stain with carmine but only the capsule is alcian blue positive. The most intense carmine positivity in human intestine is in the center of the granules.

Post methylation staining with Best's carmine also gives a positive result for the serous granules of the human parotid gland and, to some extent, for the granules of the acinus cells of the pancreas. The zymogen granules of the gastric chief cells remain negative. In mouse, the granules of both the pancreatic and parotid cells react only weakly; in the rat, somewhat more strongly.

Roque's Method for Mallory Bodies, Renal Lesions and Pituitary Cells[103]

This method can be used with excellent results for demonstrating the different types of cells of the pituitary and for the study of lesions of the renal glomerulus. The basophils of the pituitary are stained blue, the acidophils bright red. If material has been fixed in formalin or Helly's solution, the chromotrope 2R stains the wire-loop lesions of renal glomerulus in lupus erythematosus bright red. The aniline blue stains diabetic lesions blue, possibly indicating a difference in the histochemical nature of the two types of lesions.

FIXATION	10% Buffered neutral formalin. Do not use Zenker's solution.
PROCESS	Paraffin, frozen, or celloidin.
MICROTOMY	Cut section at 6 micrometers.
CONTROL	Use control from a specimen containing the entity one wishes to demonstrate.

SOLUTIONS

0.02N Hydrochloric Acid (Stock)

1N Hydrochloric acid	2.0	ml
Distilled water	98.0	ml

Chromotrope-Aniline Blue Solutions

Phosphotungstic acid	1.0	gm
0.02N Hydrochloric acid	100.0	ml
Dissolve completely and add:		
Aniline blue WS	0.5	gm
Dissolve completely with the aid of gentle heat.		
Cool, then add: Chromotrope 2R	2.0	gm

Let the solution stand overnight. The pH of this solution should be 2 to 2.2 to give the best staining results. The solution keeps well for several months. Filter before use.

Dehydrant

Tertiary butyl alcohol	300.0	ml
Absolute ethyl alcohol	100.0	ml

Clearant

Terpineol	300.0	ml
Xylene	100.0	ml

STAINING PROCEDURE

1. Decerate slides in xylene, 2 changes, 2 minutes each. Place in absolute alcohol, 2 changes, 2 minutes each; 95% alcohol, 2 changes, 2 minutes each; then rinse in distilled water.
2. Stain overnight in chromotrope-aniline blue solution.
3. Rinse in distilled water.
4. Blot slides and dehydrate in the dehydrant, 3 changes, 2 minutes each.
5. Clear in the clearant, 2 changes, 5 minutes each.
6. Clear in xylene, 2 changes each.
7. Mount coverglass with appropriate medium (refractive index 1.48-1.56).

RESULTS

Mallory bodies	red
Wire-loop lesions of renal glomerulus in lupus erythematosus	red
Diabetic lesions	blue
Hyaline bodies	bright blue or red with a blue periphery
Nuclear chromatin	pink or bluish
Nucleoli	red
Cytoplasms	pink to purplish
Collagen and reticulum	blue

REMARKS

To obtain the results described, the technique should be strictly adhered to and only well-fixed tissues should be used. In sections of tissues with marked postmortem autolytic changes, only a diffuse blue staining is observed. This also occurs in sections which have been previously digested with proteolytic enzymes.

Jhung-Edelstein-Brookman-Cariglia Method for Stratum Lucidum[104]

In using this method a variety of observations may be made:

1. When the granular layer is present, the stratum lucidum can be delineated clearly by this congo red stain and can assist in differentiating the benign skin lesions, keratoacanthoma, cholesteatoma, ovarian dermoid cyst, leukoplakia and well differentiated squamous carcinoma in cervical lymph node.
2. The keratohyaline granule also has a high affinity for congo red, thus suggesting these granules may play a role in the formation of the stratum lucidum.
3. When stratum corneum is parakeratotic, the stratum lucidum is absent.
4. Malignant diseases may indirectly affect the epidermis, with resultant loss of stratum granulosum and therefore loss of the stratum lucidum. This includes melanoma and squamous carcinoma.

FIXATION	10% Buffered neutral formalin.
PROCESS	Paraffin and celloidin.
MICROTOMY	Cut section at 6 micrometers.
CONTROL	Use control number 16.
SOLUTIONS	

1% Congo Red

Congo red	1.0	gm
Distilled water	100.0	ml

Saturated Lithium Carbonate

Lithium carbonate	1.3	gm
Distilled water	100.0	ml

Harris' Hematoxylin
(See Page 81)

1% Acid Alcohol

95% Ethyl alcohol	74.0	ml
Distilled water	25.0	ml
Concentrated hydrochloric acid	1.0	ml

Diluted Ammonium Hydroxide

Tap water	1000.0	ml
Concentrated ammonium hydroxide	2.0	ml

STAINING PROCEDURE

1. Decerate slides in xylene, 2 changes, 2 minutes each. Place in absolute alcohol, 2 changes, 2 minutes each; 95% alcohol, 2 changes, 2 minutes each; then rinse in distilled water.
2. Stain sections in 1% congo red for 30 minutes.
3. Rinse off excess stain in distilled water.
4. Dip in saturated lithium carbonate solution, then rinse in distilled water and check microscopically. Repeat this step until the appropriate differentiation is achieved (see remarks).
5. Counterstain in Harris' hematoxylin for 2 minutes.

6. Wash in running tap water.
7. Differentiate in 1% acid alcohol with 2 dips, then rinse in distilled water.
8. Blue in diluted ammonium hydroxide.
9. Rinse well in distilled water.
10. Dehydrate slides in 95% alcohol and absolute alcohol 3 changes each. Clear in xylene, 3 changes.
11. Mount coverglass with appropriate medium (refractive index 1.48-1.56).

RESULTS

Stratum lucidum....................orange-red
Nuclear chromatin....................blue
Background....................clear to pale pink

REMARKS

The success of this technique is dependant upon one's ability to identify the stratum lucidum. As the name implies, the stratum lucidum is the thin homogeneous translucent layer between the stratum corneum and the stratum granulosum of the epidermis, seen best in the skin of the palm and sole. This layer consists of two to four rows of flattened closely packed cells which contain infrequent keratohyaline granule. Many cells contain droplets with a lipid-like appearance called eleidin which, however, may not be lipid but rather a derivative of keratohyaline.

Akita-Murata-Fujita-Kaneko Method for Mucous Neck Cell[105]

FIXATION 10% Buffered neutral formalin.
PROCESS Paraffin, frozen, or celloidin.
MICROTOMY Cut section at 5 micrometers.
SOLUTIONS

Aluminum Hematoxylin Staining Solution

Hematoxylin....................	1.0	gm
Distilled water....................	1000.0	ml
Sodium iodate....................	3.0	gm
Potassium aluminium....................	3.0	gm
Chloral hydrate....................	50.0	gm
Citric acid....................	1.0	gm

Prior to staining, adjust to pH 6 with saturated aqueous lithium carbonate.

STAINING PROCEDURE

1. Decerate slides in xylene, 2 changes, 2 minutes each. Place in absolute alcohol, 2 changes, 2 minutes each; 95% alcohol, 2 changes, 2 minutes each; then rinse in distilled water.
2. Stain 30 minutes in the freshly prepared aluminum hematoxylin.
3. Wash 5 minutes in running tap water.
4. Dehydrate slides in 95% alcohol and absolute alcohol 3 changes each. Clear in xylene, 3 changes.
5. Mount coverglass with appropriate medium (refractive index 1.48-1.56).

RESULTS

Mucous neck cells of gastric glands..violet
Nuclei..blue to dark blue

Gomori's Method for Chromaffin Granules[106]

(See Fig. 109 and 110)

FIXATION 10% Buffered neutral formalin or Regaud's solution.
PROCESS Paraffin.
MICROTOMY Cut section at 6 micrometers.
CONTROL Use control number 13.
SOLUTIONS

Azocarmine G

Azocarmine G	0.05	gm
Glacial acetic acid	1.0	ml
Distilled water	100.0	ml

1% Aniline-Alcohol

Aniline	1.0	ml
95% Alcohol	100.0	ml

Aniline should be used with care. It is an important cause of serious industrial poisoning associated with bone marrow depression as well as methemoglobinemia.

3% Phosphotungstic Acid

Phosphotungstic acid	3.0	gm
Distilled water	100.0	ml

Aniline Blue-Quinoline Yellow

Aniline blue	0.5	gm
Quinoline yellow or orange G	2.0	gm
Phosphotungstic acid	1.0	gm
Distilled water	100.0	ml

STAINING PROCEDURE

1. Decerate slides in xylene, 2 changes, 2 minutes each. Place in absolute alcohol, 2 changes, 2 minutes each; 95% alcohol, 2 changes, 2 minutes each; then rinse in distilled water.
2. Stain in azocarmine G solution at 58°C for 60 to 90 minutes.
3. Wash in running tap water, then blot.
4. Rinse in 95% alcohol.
5. Decolorize in aniline-alcohol. Check microscopically to the point where chromaffin cells stand out deep pink against the much paler cortical cells.
   Note: Differentiation in aniline-alcohol may take 5 to 60 minutes, depending on the type of fixation and on the thickness of the section. This is the only critical step in the entire procedure. Judging the correct degree of decolorization requires some experience.
6. Wash briefly in running tap water.
7. Place in phosphotungstic acid for 20 minutes.

8. Wash in running tap water for 1 minute.
9. Stain in aniline blue-quinoline yellow until connective tissue is deep blue; 15 to 40 minutes. (Better contrast is obtained if quinoline yellow is used in this mixture instead of orange G.)
10. Wash in running tap water.
11. Dehydrate slides in 95% alcohol and absolute alcohol 3 changes each. Clear in xylene, 3 changes.
12. Mount coverglass with appropriate medium (refractive index 1.48-1.56).

RESULTS

Chromaffin granules ....purplish-red

DeFano's Method for Golgi Complex[55ac]

FIXATION — Cobalt nitrate-formalin. Use 3 specimens, not to exceed 1 mm in thickness. (See step 1 below.)

PROCESS — Paraffin.

MICROTOMY — Cut sections at 4 micrometers.

CONTROL — Use any tissue. Golgi complex found in cytoplasm of cells.

SOLUTIONS

Cobalt Nitrate-Formalin

Cobalt nitrate	1.0	gm
Distilled water	100.0	ml
37% Formalin	15.0	ml

1.5% Silver Nitrate

Silver nitrate	1.5	gm
Distilled water	100.0	ml

Reducer

Hydroquinone	1.0	gm
10% Buffered neutral formalin	7.5	ml
Distilled water	50.0	ml
Sodium sulfite	0.25	gm

Make fresh.

STAINING PROCEDURE

1. Place the 3 specimens in the cobalt nitrate-formalin solution for 4 hours.
2. Rinse in distilled water.
3. Place in 1.5% silver nitrate solution, in the dark, for 24 hours. Use sufficient silver nitrate solution to cover all specimens.
4. Rinse well in several changes of distilled water.
5. Place the 3 specimens in reducing solution and remove the first specimen after 8 hours, the second after 16 hours, and the third after 24 hours.
6. Process specimens in the usual manner.
7. Embed in paraffin wax.
8. Cut paraffin sections at 2 to 4 micrometers.
9. Decerate in xylene, several changes.
10. Mount coverglass with appropriate medium (refractive index 1.48-1.56).

RESULTS

Golgi complex	black
Background	yellow to golden brown

Takaya's Method for Mitochondria[107]

FIXATION	Formalin-Zenker's or Regaud's solution.
PROCESS	Paraffin.
MICROTOMY	Cut sections at 2 and 5 micrometers.
CONTROL	Use control tissue from liver specimen.

SOLUTIONS

5% Glacial Acetic Acid (Stock)

Glacial acetic acid	5.0	ml
Distilled water	100.0	ml

Luxol Fast Blue

Luxol fast blue MBS	1.0	gm
95% Ethyl alcohol	100.0	ml
5% Glacial acetic acid	10.0	ml

Solution keeps well but filter before use.

0.05% Lithium Carbonate

Lithium carbonate	0.5	gm
Distilled water	1000.0	ml

70% Ethyl Alcohol

95% Ethyl alcohol	737.0	ml
Distilled water	263.0	ml

0.5% Phloxine B

Phloxine B	1.0	gm
Distilled water	200.0	ml

5% Phosphotungstic Acid

Phosphotungstic acid	5.0	gm
Distilled water	100.0	ml

STAINING PROCEDURE

1. Decerate slides in xylene, 2 changes, 2 minutes each. Place in absolute alcohol, 2 changes, 2 minutes each; 95% alcohol, 2 changes, 2 minutes each; then rinse in distilled water.
2. Stain in luxol fast blue solution for 3 to 6 hours at 58°C.
3. Rinse in 95% ethyl alcohol until no more stain comes off, then rinse well with distilled water.
4. Decolorize in 0.05% lithium carbonate for about 10 to 20 seconds. Rinse in 3 changes of 70% ethyl alcohol. Some parts of the sections will become colorless. Rinse well in distilled water.
5. Stain for 2 to 3 minutes in 0.5% phloxine B solution.

6. Differentiate in 5% phosphotungstic acid for 2 minutes. Wash in running tap water for 5 minutes.
7. Differentiate further in 95% ethyl alcohol.
8. Place in 100% alcohol and clear in xylene, 3 changes each.
9. Mount coverglass with appropriate medium (refractive index 1.48-1.56).

RESULTS

Mitochondria....................green
Cytoplasm....................pink to faintly green

Cain's Method for Mitochondria[55ad]

FIXATION Regaud's solution. Fix tissue for 6 hours and place in 3% potassium dichromate for 4 hours.
PROCESS Paraffin, frozen, or celloidin.
MICROTOMY Cut sections at 3 micrometers.
CONTROL Use control tissue from liver specimen.
SOLUTIONS

Aniline-Acid Fuchsin

Aniline	1.0	ml
Distilled water	20.0	ml
Acid fuchsin	4.0	gm

Shake aniline with distilled water for a couple of minutes. Let stand and then shake again several times at intervals over 8 hours. Filter. Add acid fuchsin, and shake at intervals over 3 hours. This solution keeps only long enough to be used the same day it is made.

0.1% Sodium Carbonate

Sodium carbonate	0.1	gm
Distilled water	100.0	ml

1% Hydrochloric Acid

Concentrated hydrochloric acid	1.0	ml
Distilled water	99.0	ml

0.5% Methyl Blue

Methyl blue	0.5	gm
Distilled water	100.0	ml

STAINING PROCEDURE

1. Decerate slides in xylene, 2 changes, 2 minutes each. Place in absolute alcohol, 2 changes, 2 minutes each; 95% alcohol, 2 changes, 2 minutes each; then rinse in distilled water.
2. Stain in hot aniline-acid fuchsin solution for 10 minutes. (Heat aniline-acid fuchsin solution to steaming, then remove solution from heat when slides are placed in it.)
3. Decolorize in 0.1% sodium carbonate solution until cytoplasm is pale pink; control microscopically.

4. Dip 2 times in 1% hydrochloric acid solution to stop decolorization and also heighten color.
5. Place in distilled water for 1 minute.
6. Stain in methyl blue solution for 30 seconds.
7. Rinse in distilled water.
8. Dip 2 times in 1% hydrochloric acid solution.
9. Rinse in distilled water.
10. Dehydrate slides in 95% alcohol and absolute alcohol 3 changes each. Clear in xylene, 3 changes.
11. Mount coverglass with appropriate medium (refractive index 1.48-1.56).

RESULTS

Mitochondria.. bright red
Nuclei.. blue to green

Laqueur's Method for Mallory Bodies

FIXATION Zenker's or 10% buffered neutral formalin.
PROCESS Paraffin, frozen, or celloidin.
MICROTOMY Cut sections at 6 micrometers.
CONTROL Use control number 49.
SOLUTIONS

Mayer's Hematoxylin
(See Page 81)

Acid Fuchsin-Aniline

Acid fuchsin	20.0	gm
Aniline	2.0	ml
Distilled water	100.0	ml

Shake the aniline in distilled water thoroughly before adding the acid fuchsin.

Alcoholic Picric Acid

Picric acid, saturated in 100% alcohol	10.0	ml
10% Alcohol	70.0	ml

1% Phosphomolybdic Acid

Phosphomolybdic acid	1.0	gm
Distilled water	100.0	ml

1% Light Green

Light green, SF yellowish	1.0	gm
Distilled water	100.0	ml
Glacial acetic acid	1.0	ml

STAINING PROCEDURE

1. Decerate slides in xylene, 2 changes, 2 minutes each. Place in absolute alcohol, 2 changes, 2 minutes each; 95% alcohol, 2 changes, 2 minutes each; then rinse in distilled water.

2. Place in Mayer's hematoxylin solution for 15 minutes.
3. Wash in running tap water for 15 minutes.
4. Flood the slides with acid fuchsin-aniline solution and heat gently with bunsen burner until fumes are visible; let stand for 5 minutes.
5. Wash in distilled water, 3 dips.
6. Differentiate in alcoholic picric acid solution until only Mallory bodies and red cells remain red, and collagen is faint gray or unstained.
7. Wash in running tap water for 3 minutes.
8. Mordant in phosphomolybdic acid solution for 2 hours.
9. Counterstain in light green solution for 1 hour.
10. Wash in running tap water for 1 minute.
11. Differentiate light green in 80% alcohol. Check under microscope. Collagen fibers are seen as distinct fibrils in well-differentiated section.
12. Dehydrate slides in 95% alcohol and absolute alcohol 3 changes each. Clear in xylene, 3 changes.
13. Mount coverglass with appropriate medium (refractive index 1.48-1.56).

RESULTS

Mallory bodies .... bright red
Erythrocytes .... red
Cytoplasm .... pale brown
Bile pigment .... green
Proteinaceous material seen occasionally in some liver cells .... red

Chapter 11

Pigments and Minerals

The success of some of the methods in this chapter depends on proper fixation and handling of the tissue during the entire procedure. It is important to remember that most minerals in tissue sections will dissolve if exposed to solutions with a pH of 5 or less. This slightly acid pH will dissolve calcium salts, iron, copper and urates. Thus, fixatives which contain acids, e.g. unbuffered formalin, Bouin's, Zenker's, will dissolve these entities and one must take this into consideration if the demonstration of these minerals is desired. Similarly, the fact that urates are soluble in water also must be taken into account when performing procedures for the demonstration of these crystals.

Dahl's Method for Calcium[108]

(See Fig. 112 and 113)

FIXATION	95% Ethyl alcohol or 10% buffered neutral formalin.
PROCESS	Paraffin.
MICROTOMY	Cut sections at 6 micrometers.
CONTROL	Use control number 9.
SOLUTIONS	

0.1% Ammonium Hydroxide Solution (Stock)

28% Ammonium hydroxide	0.1	ml
Distilled water	100.0	ml

1% Alizarin Red S

Alizarin red S	1.0	gm
Distilled water	100.0	ml
0.1% Ammonium hydroxide (stock)	10.0	ml

Stir dye into the distilled water so that only a few small grains of dye remain undissolved. Add the ammonium hydroxide slowly with constant stirring. Resulting pH should be 6.5 or above. Solution is stable for one month.

1% Light Green Solution

Light green, SF yellowish	1.0	gm
Distilled water	100.0	ml

STAINING PROCEDURE

1. Decerate slides in xylene, 2 changes, 2 minutes each. Place in absolute alcohol, 2 changes, 2 minutes each; 95% alcohol, 2 changes, 2 minutes each; then rinse in distilled water.
2. Stain in alizarin red S solution for 30 minutes.
3. Remove excess stain with distilled water, 5 or 6 rinses.
4. Counterstain in light green solution for 1 minute.
5. Rinse in distilled water for 5 to 10 seconds.

6. Dehydrate slides in 95% alcohol and absolute alcohol 3 changes each. Clear in xylene, 3 changes.
7. Mount coverglass with appropriate medium (refractive index 1.48-1.56).

RESULTS

Calcium salts ....intense reddish-orange
Background ....pale green

REMARKS

There are two important factors to consider in performing this stain. Namely, the alizarin red S solution should have a pH not lower than 6.5. Secondly, the light green solution should not contain glacial acetic acid.

Carr's Method for Calcium[55ae]

FIXATION	10% Buffered neutral formalin.
PROCESS	Paraffin, frozen, or celloidin.
MICROTOMY	Cut sections at 6 micrometers.
CONTROL	Use control number 9.
SOLUTIONS	

Chloranilic Acid

Distilled water	100.0	ml
Sodium hydroxide	0.4	gm
Chloranilic acid	1.0	gm

Agitate until all the acid is dissolved. Filter through fine filter paper.

1% Light Green
(See Page 325)

STAINING PROCEDURE

1. Decerate slides in xylene, 2 changes, 2 minutes each. Place in absolute alcohol, 2 changes, 2 minutes each; 95% alcohol, 2 changes, 2 minutes each; then rinse in distilled water.
2. Stain in chloranilic acid solution for 30 minutes.
3. Wash in running tap water for 15 minutes.
4. Counterstain in 1% light green solution for 5 minutes.
5. Rinse in distilled water.
6. Dehydrate slides in 95% alcohol and absolute alcohol 3 changes each. Clear in xylene, 3 changes.
7. Mount coverglass with appropriate medium (refractive index 1.48-1.56).

RESULTS

Calcium.. red-brown
Background .... green

Brilliantly birefringent under polarized light.

REMARKS

Some crystallization may occur in the chloranilic acid solution; however, this may be removed by filtering without affecting the usefulness of the solution.

Puchtler Method for Calcium[109]

FIXATION 10% Buffered neutral formalin.
PROCESS Paraffin, frozen, or celloidin.
MICROTOMY Cut sections at 6 micrometers.
CONTROL Use control number 9.
SOLUTIONS

0.5% Alizarin Red S

Alizarin Red S	0.5	gm
Distilled water	100.0	ml

Add 5 ml of 1% sodium hydroxide solution to 100 ml of above solution to achieve a pH of approximately 9.0.

1% Sodium Hydroxide

Sodium hydroxide	1.0	gm
Distilled water	100.0	ml

STAINING PROCEDURE

1. Decerate slides in xylene, 2 changes, 2 minutes each. Place in absolute alcohol, 2 changes, 2 minutes each; 95% alcohol, 2 changes, 2 minutes each; then rinse in distilled water.
2. Stain in 0.05% alizarin red S solution for 1 hour.
3. Rinse in 1% sodium hydroxide for 5 seconds.
4. Dehydrate slides in 95% alcohol and absolute alcohol 3 changes each. Clear in xylene, 3 changes.
5. Mount coverglass with appropriate medium (refractive index 1.48-1.56).

RESULTS

Calcium salts	intense red
Cytoplasm	unstained

REMARKS

For more detailed observations of calcium salts, counterstains are not recommended. The diffusion artifacts often seen in alizarin red S stained sections in other methods are eliminated by the pH 9 staining solution.

Von Kossa's Method for Calcium[55af]

(See Fig. 114)

FIXATION 10% Buffered neutral formalin.
PROCESS Paraffin, frozen, or celloidin.
MIRCOTOMY Cut sections at 6 micrometers.
CONTROL Use control number 9.

SOLUTIONS

3% Silver Nitrate

Silver nitrate	3.0	gm
Distilled water	100.0	ml

5% Sodium Thiosulfate
(See Page 134)

Kernechtrot (Nuclear Fast Red)
(See Page 84)

STAINING PROCEDURE

1. Decerate slides in xylene, 2 changes, 2 minutes each. Place in absolute alcohol, 2 changes, 2 minutes each; 95% alcohol, 2 changes, 2 minutes each; then rinse in distilled water.
2. Place in 3% silver nitrate solution for 60 minutes exposed to strong direct sunlight.
3. Rinse in distilled water.
4. Place in 5% sodium thiosulfate solution for 2 minutes.
5. Rinse well in distilled water.
6. Stain in nuclear fast red solution for 5 minutes.
7. Rinse in distilled water.
8. Dehydrate slides in 95% alcohol and absolute alcohol 3 changes each. Clear in xylene, 3 changes.
9. Mount coverglass with appropriate medium (refractive index 1.48-1.56).

RESULTS

Calcium deposits	black
Nuclear chromatin	red
Cytoplasm	light pink
Calcium phosphate	brown (see remarks)

REMARKS

The black deposits formed in this reaction are due to the reduction of silver by organic material. This occurs only when sections are exposed to strong light. When Von Kossa's reaction is carried out in subdued light, yellow to yellowish-brown silver phosphate is visualized selectively. Silver carbonate cannot be demonstrated with Von Kossa's technique because it dissolves in the 5% sodium thiosulfate solution. Phosphates and carbonates in calcium deposits may be associated, at least in part, with cations other than calcium. Calcium can be identified with alizarin red S or other suitable mordant dyes.

Fenton-Johnson-Zimmerman Method for Iron (Microincineration)[110]

FIXATION	10% Buffered neutral formalin.
PROCESS	Paraffin, frozen, or celloidin.
MICROTOMY	Cut sections at 6 micrometers.
CONTROL	Use control number 25.

SOLUTIONS

5% Potassium Ferrocyanide (Stock)

Potassium Ferrocyanide	5.0	gm
Distilled water	100.0	ml

0.6N Hydrochloric Acid (Stock)

Concentrated hydrochloric acid	50.0	ml
Distilled water	950.0	ml

Hydrochloric Acid-Potassium Ferrocyanide (Working)

0.6N Hydrochloric acid (stock)	50.0	ml
5% Potassium ferrocyanide (stock)	50.0	ml

Mix just before use.

STAINING PROCEDURE

1. Decerate slides in xylene, 2 changes, 2 minutes each. Place in absolute alcohol, 2 changes, 2 minutes each; 95% alcohol, 2 changes, 2 minutes each; then rinse in distilled water.
   Three sections* are stained in this technique as follows:
   a. One section stained with routine H&E.
   b. One section stained with the Prussian blue reaction.
   c. One section used for microincineration.

A. Microincineration Procedure

1. Place uncovered, unstained paraffin section near the center of a sheet of asbestos board.
2. Heat cautiously, using the outer cone from the flame of burner. (Flame which is directed down onto the section must be kept in motion to heat slide as uniformly as possible.)
3. Continue heating until specimen, which turns brown, then white, has almost completely disappeared.
4. Immediately cover slide with an inverted pie pan so slide cools gradually. (If pan is not used, cooling is too rapid and the slide may break.)
5. When the slide is cool, the presence of iron oxide in the ash is demonstrated by direct observation, with or without coverglass. Examine completely incinerated specimen under strong oblique lighting with dark field illumination.

*By comparing these three sections the location of iron deposits and the presence of "nonstainable" iron can be demonstrated.

RESULTS

Iron oxide imparts to the ash a light yellow-orange to dark red color. Specimen must be completely incinerated before interpreting colored particles as iron oxide, since in dark-field illumination, carbon particles can also give a yellowish to red color.

B. Incinerated Specimen Staining

1. Dip incinerated specimen in a thin celloidin solution (2 gm parlodion to 50 ml absolute alcohol and 50 ml ethyl ether in tightly stoppered bottle. Take 15 ml of this stock solution and make up to 2000 ml with equal parts absolute alcohol and ethyl ether).
2. Air-dry slides for a few minutes, then rapidly hydrate to distilled water.
3. Stain with working hydrochloric acid-potassium ferrocyanide for 15 minutes.

4. Rinse in distilled water, several changes.
5. Air-dry or dry in 60°C oven.
6. Mount coverglass with resinous medium.

RESULTS

Iron oxide.. blue

C. Direct Observation of Prussian Blue Reaction on Ash
1. Build a well of paraffin around the specimen.
2. Pour hydrochloric acid-potassium ferrocyanide solution in this well and observe under the microscope.

RESULTS

Iron oxide.. blue

Lillie's Method for Ferric and Ferrous Iron

(See Fig. 115 and 116)

FIXATION	10% Buffered neutral formalin.
PROCESS	Paraffin, frozen, or celloidin.
MICROTOMY	Cut sections at 6 micrometers.
CONTROL	Use control number 25.
SOLUTIONS	

Potassium Ferrocyanide - 0.06N Hydrochloric Acid

Potassium ferrocyanide	0.4	gm
0.06N Hydrochloric acid (1 ml concentrated hydrochloric acid to 199 ml distilled water)	40.0	ml

Potassium Ferricyanide - 0.06N Hydrochloric Acid

Potassium ferricyanide	0.4	gm
0.06N hydrochloric acid (see above)	40.0	ml

Make solutions fresh before use.

Basic Fuchsin

Basic fuchsin	0.5	gm
Distilled water	100.0	ml
Glacial acetic acid	1.0	ml

STAINING PROCEDURE
1. Decerate slides in xylene, 2 changes, 2 minutes each. Place in absolute alcohol, 2 changes, 2 minutes each; 95% alcohol, 2 changes, 2 minutes each; then rinse in distilled water.
2. For ferric iron place section in potassium ferrocyanide, 0.06N hydrochloric acid solution for 1 hour.
3. For ferrous iron place section in potassium ferricyanide, 0.06N hydrochloric acid solution for 1 hour.
4. Wash well in 1% aqueous glacial acetic acid.
5. Counterstain in basic fuchsin solution for 10 minutes.

6. Rinse in distilled water.
7. Dehydrate slides in 95% alcohol and absolute alcohol 3 changes each. Clear in xylene, 3 changes.
8. Mount coverglass with appropriate medium (refractive index 1.48-1.56).

RESULTS

Ferric iron	dark Prussian blue
Ferrous iron	dark Turnbull blue
Background	light red

REMARKS
For critical work omit counterstain.

Mallory's Method for Iron

FIXATION 10% Buffered neutral formalin.
PROCESS Paraffin, frozen, or celloidin.
MICROTOMY Cut sections at 6 micrometers.
CONTROL Use control number 25.
SOLUTIONS

5% Hydrochloric Acid (Stock)

Concentrated hydrochloric acid	5.0	ml
Distilled water	100.0	ml

5% Potassium Ferrocyanide (Stock)

Potassium ferrocyanide	5.0	gm
Distilled water	100.0	ml

Hydrochloric Acid-Potassium Ferrocyanide (Working)

5% Hydrochloric acid	50.0	ml
5% Potassium ferrocyanide	50.0	ml

Mix just before use.

Kernechtrot (Nuclear Fast Red)
(See Page 84)

STAINING PROCEDURE
1. Decerate slides in xylene, 2 changes, 2 minutes each. Place in absolute alcohol, 2 changes, 2 minutes each; 95% alcohol, 2 changes, 2 minutes each; then rinse in distilled water.
2. Place in hydrochloric acid-potassium ferrocyanide solution for 10 minutes.
3. Rinse thoroughly in distilled water.
4. Stain nuclear chromatin with nuclear fast red for 5 minutes.
5. Rinse well in distilled water.
6. Dehydrate slides in 95% alcohol and absolute alcohol 3 changes each. Clear in xylene, 3 changes.
7. Mount coverglass with appropriate medium (refractive index 1.48-1.56).

RESULTS

Iron pigments .... bright blue
Nuclear chromatin .... red
Cytoplasm .... light pink

Turnbull Blue Method for Hemosiderin[111]

FIXATION 10% Buffered neutral formalin or absolute alcohol.
PROCESS Paraffin.
MICROTOMY Cut sections at 6 micrometers.
CONTROL Use control number 25.
SOLUTIONS

Ammonium Sulfide, Yellow

Ammonium sulfide	25.0	ml
95% Alcohol	100.0	ml

1% Hydrochloric Acid (Stock)
(See Page 299)

20% Potassium Ferricyanide (Stock)

Potassium ferricyanide	20.0	gm
Distilled water	100.0	ml

Hydrochloric Acid-Potassium Ferricyanide (Working)

1% Hydrochloric acid (stock)	50.0	ml
20% Potassium ferricyanide (stock)	50.0	ml

Mix just before use.

Kernechtrot (Nuclear Fast Red)
(See Page 84)

STAINING PROCEDURE

1. Decerate slides in xylene, 2 changes, 2 minutes each. Place in absolute alcohol, 2 changes, 2 minutes each; 95% alcohol, 2 changes, 2 minutes each; then rinse in distilled water.
2. Place in ammonium sulfide solution for 2 hours.
3. Rinse thoroughly in distilled water.
4. Place in hydrochloric acid-potassium ferricyanide solution (working) for 20 minutes.
5. Rinse thoroughly in distilled water.
6. Place in nuclear fast red solution for 5 minutes.
7. Rinse in distilled water.
8. Dehydrate slides in 95% alcohol and absolute alcohol 3 changes each. Clear in xylene, 3 changes.
9. Mount coverglass with appropriate medium (refractive index 1.48-1.56).

RESULTS

Hemosiderin........ blue
Nuclear chromatin .... red

Luna's Method for Ferric Iron and Melanin

(See Fig. 117, 119, 120 and 121)

This procedure has been found useful for the simultaneous demonstration of melanin and ferric iron. It should be understood that the silver nitrate component of the method is that of the Fontana-Masson, which is a method for reducing substances (there are many reducing substances in tissue including melanin, Zenker crystals, formalin pigment, etc., that may react with this silver nitrate solution); therefore, any silver reactive substances should be viewed as reducing substance and not specifically melanin. For the specific demonstration of melanin, one should use other methods that provide a more definitive diagnosis.

FIXATION 10% Buffered neutral formalin.
PROCESS Paraffin, frozen, or celloidin.
MICROTOMY Cut sections at 6 micrometers.
SOLUTIONS

Silver Nitrate (Fontana)

Dissolve 10 gm of silver nitrate in 100 ml distilled water. To 95 ml of this solution, add ammonium hydroxide until a clear solution is obtained without any precipitate. Add, drop by drop, enough of the remaining 5 ml of silver nitrate solution to cause the clear solution to become slightly cloudy. For a working solution, dilute each 25 ml of silver solution with 75 ml of distilled water and filter.

Gold Chloride

1% Aqueous gold chloride	10.0	ml
Distilled water	40.0	ml

5% Sodium Thiosulfate

Sodium thiosulfate	5.0	ml
Distilled water	100.0	ml

Kernechtrot (Nuclear Fast Red)
(See Page 84)

Potassium Ferrocyanide-Hydrochloric Acid

Potassium ferrocyanide	0.4	gm
0.06 N Hydrochloric acid	40.0	ml

(1 ml concentrated HCl to 199 ml distilled water.)

STAINING PROCEDURE

1. Decerate slides in xylene, 2 changes, 2 minutes each. Place in absolute alcohol, 2 changes, 2 minutes each; 95% alcohol, 2 changes, 2 minutes each; then rinse in distilled water.
2. Place in silver nitrate solution in 56°C oven for 1 hour. Sections should turn light brown.
3. Rinse sections in distilled water.
4. Place in gold chloride solution for 10 minutes.
5. Rinse in distilled water, 3 changes.
6. Place in sodium thiosulfate solution for 5 minutes.
7. Rinse in distilled water.

8. Place section in potassium ferrocyanide-hydrochloric acid solution for 30 minutes.
9. Rinse in distilled water.
10. Place in nuclear fast red solution for 5 minutes.
11. Rinse in distilled water, 2 changes.
12. Dehydrate slides in 95% alcohol and absolute alcohol 3 changes each. Clear in xylene, 3 changes.
13. Mount coverglass with appropriate medium (refractive index 1.48-1.56).

RESULTS

Silver-reducing substances	black
Nuclei	pink
Ferric iron	dark Prussian blue
Cytoplasm	light red

REMARKS

This method demonstrates ferric iron and melanin. The method can also be used to demonstrate ferrous iron and melanin. To demonstrate ferrous iron, one must substitute the potassium ferrocyanide solution for potassium ferricyanide when combining with the 0.06N hydrochloric acid solution. The results in this case would be ferrous iron dark Turnbull blue.

Gomori's Method for Iron[112]

FIXATION	10% Buffered neutral formalin or absolute alcohol.
PROCESS	Paraffin, frozen, or celloidin.
MICROTOMY	Cut sections at 6 micrometers.
CONTROL	Use control number 25.
SOLUTIONS	

20% Hydrochloric Acid Solution (Stock)

Concentrated hydrochloric acid	20.0	ml
Distilled water	80.0	ml

10% Potassium Ferrocyanide Solution (Stock)

Potassium ferrocyanide	10.0	gm
Distilled water	100.0	ml

Hydrochloric Acid-Potassium Ferrocyanide Solution (Working)

Hydrochloric acid (stock)	50.0	ml
Potassium ferrocyanide (stock)	50.0	ml

Mix just before use.

Kernechtrot (Nuclear Fast Red)
(See Page 84)

STAINING PROCEDURE

1. Decerate slides in xylene, 2 changes, 2 minutes each. Place in absolute alcohol, 2 changes, 2 minutes each; 95% alcohol, 2 changes, 2 minutes each; then rinse in distilled water.

2. Place in hydrochloric acid-potassium ferrocyanide solution for 30 minutes.
3. Rinse thoroughly in distilled water.
4. Counterstain in nuclear fast red solution for 5 minutes.
5. Rinse in distilled water.
6. Dehydrate slides in 95% alcohol and absolute alcohol 3 changes each. Clear in xylene, 3 changes.
7. Mount coverglass with appropriate medium (refractive index 1.48-1.56).

RESULTS

Iron pigments	bright blue
Nuclear chromatin	red
Cytoplasm	light pink

REMARKS

It should be understood that some iron may be dissolved with the use of this method. This is due to the use of the high percentage of hydrochloric acid, which dissolves the iron.

Hukill-Putt Method for Ferrous Iron[113]

FIXATION 10% Buffered neutral formalin.
PROCESS Paraffin, frozen, or celloidin.
MICROTOMY Cut sections at 6 micrometers.
CONTROL Use control number 25.
SOLUTIONS

Bathophenanthroline Solution

Bathophenanthroline*	0.1	gm
3% Glacial acetic acid	100.0	ml

Shake well and place in 60°C oven overnight.

*Bathophenanthroline (4,7 – diphenyl-1,10-penanthroline).

Bathophenanthroline (Working)

Bathophenanthroline	40.0	ml
Thioglycolic acid	0.2	ml

0.5% Methylene Blue

Methylene blue	0.5	gm
Distilled water	100.0	ml

STAINING PROCEDURE.

1. Decerate slides in xylene, 2 changes, 2 minutes each. Place in absolute alcohol, 2 changes, 2 minutes each; 95% alcohol, 2 changes, 2 minutes each; then rinse in distilled water.
2. Place in working bathophenanthroline solution for 2 hours.
3. Rinse in distilled water.
4. Counterstain in 0.5% methylene blue solution for 3 minutes.
5. Rinse in distilled water, 3 changes, 1 minute each.

6. Blot and dry thoroughly in 60°C oven.
7. Dehydrate slides in 95% alcohol and absolute alcohol 3 changes each. Clear in xylene, 3 changes.
8. Mount coverglass with appropriate medium (refractive index 1.48-1.56).

RESULTS

Ferrous iron..red
Nuclear chromatin ....blue

REMARKS

This method is specific for Fe^{++} (ferrous iron) but it does not permit separate demonstration of Fe^{+++} (ferric iron) as distinct from Fe^{++} (ferrous) iron.

Perls' Method for Iron[114]

FIXATION	10% Buffered neutral formalin.
PROCESS	Paraffin, frozen, or celloidin.
MICROTOMY	Cut sections at 6 micrometers.
CONTROL	Use control number 25.
SOLUTIONS	

10% Hydrochloric Acid (Stock)

Concentrated hydrochloric acid	10.0	ml
Distilled water	90.0	ml

10% Potassium Ferrocyanide (Stock)

Potassium ferrocyanide	10.0	gm
Distilled water	100.0	ml

Potassium Ferrocyanide-Hydrochloric Acid (Working)

10% Potassium ferrocyanide (stock)	70.0	ml
Hydrochloric acid (stock)	30.0	ml

Mix just before use.

Kernechtrot (Nuclear Fast Red)
(See Page 84)

STAINING PROCEDURE

1. Decerate slides in xylene, 2 changes, 2 minutes each. Place in absolute alcohol, 2 changes, 2 minutes each; 95% alcohol, 2 changes, 2 minutes each; then rinse in distilled water.
2. Place in stock potassium ferrocyanide solution for 5 minutes.
3. Place in working potassium ferrocyanide-hydrochloric acid solution for 20 minutes.
4. Rinse well in distilled water.
5. Place in nuclear fast red solution for 5 minutes.
6. Wash well in running tap water.
7. Dehydrate slides in 95% alcohol and absolute alcohol 3 changes each. Clear in xylene, 3 changes.
8. Mount coverglass with appropriate medium (refractive index 1.48-1.56).

RESULTS

Ferric iron .... bright blue
Nuclear chromatin .... red
Cytoplasm .... light pink

REMARKS

Numerous other metals may also be demonstrated with this method. The metals demonstrated and their color reactions are as follows:

Copper .... red
Uranium .... brown
Nickel .... greenish-white*

*In some instances nickel will stain brown.

Luna's Method for Reducing Substances and Mucin

(See Fig. 131)

FIXATION	10% Buffered neutral formalin.
PROCESS	Paraffin, frozen, or celloidin.
MICROTOMY	Cut sections at 6 micrometers.
CONTROL	Use control number 3.
SOLUTIONS	

1% Ferric Chloride (Stock)

Ferric chloride	1.0	gm
Distilled water	100.0	ml

0.1% Potassium Ferricyanide (Stock)

Potassium ferricyanide	0.1	gm
Distilled water	100.0	ml

Ferric Chloride-Potassium Ferricyanide (Working)

Ferric chloride (stock)	75.0	ml
Potassium ferricyanide (stock)	25.0	ml

Adjust the pH to 2.4 with a few drops of dilute hydrochloric acid if necessary.

Southgate Mucicarmine

Carmine	1.0	gm
Aluminum hydroxide	1.0	gm
50% Alcohol	100.0	ml
Aluminum chloride (anhydrous)	0.5	gm

The carmine, aluminum hydroxide and alcohol are put into a 500 ml flask and mixed by shaking. The anhydrous aluminum chloride is then added with frequent shaking. The flask is placed on a boiling water bath for 2 1/2 minutes. The contents are then immediately cooled under running tap water and filtered when cold. Dilute 10 ml of this stock with 90 ml distilled water for use. Discard this diluted working solution after use.

0.25% Metanil Yellow Solution

Metanil yellow	0.25	gm
Distilled water	100.0	ml
Glacial acetic acid	0.25	ml

STAINING PROCEDURE

1. Decerate slides in xylene, 2 changes, 2 minutes each. Place in absolute alcohol, 2 changes, 2 minutes each; 95% alcohol, 2 changes, 2 minutes each; then rinse in distilled water.
2. Place in working ferric chloride-potassium ferricyanide solution, 2 changes, 7 minutes each.
3. Rinse in distilled water.
4. Place in Southgate mucicarmine solution for 30 minutes.
5. Rinse rapidly in distilled water.
6. Counterstain in metanil yellow solution for a few seconds.
7. Rinse rapidly in distilled water.
8. Dehydrate slides in 95% alcohol and absolute alcohol 3 changes each. Clear in xylene, 3 changes.
9. Mount coverglass with appropriate medium (refractive index 1.48-1.56).

RESULTS

Argentaffin/argyrophil granules	blue
Melanin	blue
Goblet cells, mucin	reddish-pink
Background	yellowish-green
Other reducing substances	blue

Luna's Method for Iron

FIXATION 10% Buffered neutral formalin.
PROCESS Paraffin section, air-dried smears, or touch preparations.
MICROTOMY Cut paraffin sections at 6 micrometers.
CONTROL Use control number 25.
SOLUTIONS

2% Hydrochloric Acid (Stock)

Concentrated hydrochloric acid	2.0	ml
Distilled water	100.0	ml

2% Potassium Ferrocyanide (Stock)

Potassium ferrocyanide	2.0	gm
Distilled water	100.0	ml

Hydrochloric Acid-Potassium Ferrocyanide (Working)

2% Hydrochloric acid	50.0	ml
2% Potassium ferrocyanide	50.0	ml

Kernechtrot (Nuclear Fast Red)
(See Page 84)

STAINING PROCEDURE

1. Decerate slides in xylene, 2 changes, 2 minutes each. Place in absolute alcohol, 2 changes, 2 minutes each; 95% alcohol, 2 changes, 2 minutes each; then rinse in distilled water.
2. Place slides in working hydrochloric acid-potassium ferrocyanide for 30 minutes.
3. Rinse in distilled water.
4. Counterstain in nuclear fast red for 3 minutes.
5. Rinse in distilled water.
6. Dehydrate slides in 95% alcohol and absolute alcohol 3 changes each. Clear in xylene, 3 changes.
7. Mount coverglass with appropriate medium (refractive index 1.48-1.56).

RESULTS

Iron pigment	bright blue
Nuclear chromatin	red
Cytoplasm	light pink

REMARKS

This method allows more iron to be preserved than with other methods. Therefore, none is dissolved out by the low percentage of hydrochloric acid used. (Some iron is dissolved out in other methods which use a higher percentage of hydrochloric acid.) A fine precipitate may form on the tissue section if the potassium ferrocyanide solution is more than 6 months old.

Luna's Method for Unmasking Iron in Hemal Pigments

A number of well defined iron containing compounds, such as hemoglobin malaria pigment, do not show any reaction with Prussian blue type methods; however, their iron can be revealed by destroying the organic part of the molecule ("unmasking" the iron). A good unmasking agent must preserve a good deal of tissue architecture and leave the iron at its original site. Hydrogen peroxide produces insoluble ferric oxide and allows tissue to retain its architecture. Whether all or only some of the iron-containing biological substances are unmasked, has not been determined. Nevertheless, hemoglobin gives a fairly intense Prussian blue reaction after treatment with alkalized hydrogen peroxide.

FIXATION	Methyl alcohol or 10% buffered neutral formalin.
PROCESS	Blood smears, bone marrow impression, or paraffin embedded material.
MICROTOMY	Cut paraffin embedded sections at 6 micrometers.
CONTROL	Use control number 25.
SOLUTIONS	

Alkaline-Hydrogen Peroxide

30% Hydrogen peroxide	100.0	ml
Concentrated ammonium hydroxide	5	drops

Prepare iron staining solutions for method of choice in step 4.

PROCEDURE

1. Decerate slides in xylene, 2 changes, 2 minutes each. Place in absolute alcohol, 2 changes, 2 minutes each; 95% alcohol, 2 changes, 2 minutes each; then rinse in distilled water.
2. Apply a few drops of alkalized hydrogen peroxide to the tissue section and leave for 30 minutes.
3. Wash slides under running tap water.
4. Stain slides with Perl's, Mallory's or Luna's method for iron.
5. Dehydrate slides in 95% alcohol and absolute alcohol 3 changes each. Clear in xylene, 3 changes.
6. Mount coverglass with appropriate medium (refractive index 1.48-1.56).

RESULTS

Ferric oxide .... blue
Cytoplasm .... pink
Nuclear chromatin .... red

Luna's Method for Removing Iron

Iron deposits in tissue are occasionally so extensive that they may interfere with microscopic observation of cellular detail. The following procedure may be used to remove iron deposits from tissue without adversely altering cellular components or subsequent staining.

SOLUTION

15% Hydrochloric Acid

Concentrated hydrochloric acid	15.0	ml
Distilled water	85.0	ml

PROCEDURE

1. Decerate slides in xylene, 2 changes, 2 minutes each. Place in absolute alcohol, 2 changes, 2 minutes each; 95% alcohol, 2 changes, 2 minutes each; then rinse in distilled water.
2. Place slide in 15% hydrochloric acid for 15 minutes.
3. Wash in distilled water for 5 minutes.
4. Perform desired stain.

REMARKS

Exposure time to the hydrochloric acid may have to be altered depending on the amount of iron present. In some instances, a 20% hydrochloric acid solution may be required to remove the iron present in the tissue.

Lillie-Geer-Gutierrez Method for Removing Iron[115]

There are instances when a specimen may contain massive amounts of iron which mask the cellular properties. Study of these properties can be accomplished by selective removal of the iron from the section with the following procedure.

FIXATION	10% Buffered neutral formalin.
PROCESS	Paraffin, frozen, smears, or touch preparations.
MICROTOMY	Cut sections at 6 micrometers.
CONTROL	Use control number 25.

SOLUTIONS

1.2% Acetic Acid (Stock)

Glacial acetic acid	1.2	ml
Distilled water	100.0	ml

2.7% Sodium Acetate (Stock)

Sodium acetate	2.7	gm
Distilled water	100.0	ml

0.2M Acetate Buffer (Stock)

1.2% Acetic acid (stock)	62.0	ml
2.7% Sodium acetate (stock)	38.0	ml

1% Sodium Dithionite

Sodium dithionite	1.0	gm
0.2M Acetate buffer (stock)	100.0	ml

PROCEDURE

1. Decerate slides in xylene, 2 changes, 2 minutes each. Place in absolute alcohol, 2 changes, 2 minutes each; 95% alcohol, 2 changes, 2 minutes each; then rinse in distilled water.
2. Place slides in 1% sodium dithionite for 5 minutes.
3. Rinse in 2 changes of distilled water for 2 minutes each.
4. Continue with selected staining method for iron.

RESULTS

Areas that contained iron will be seen to react negatively to the iron reaction when compared to the positive reaction seen on an untreated slide.

REMARKS

It is always advisable to run 2 slides; one treated with sodium dithionite and one untreated.

Lindquist's Rhodanine Method for Copper[116]

(See Fig. 122 and 123)

Hepatolenticular degeneration (Wilson's disease) is known to produce abnormal amounts of copper. This results in copper accumulation in the liver, kidney, brain and other organs. This method and the one following exhibiting copper in tissue sections are reliable, selective and sensitive. They are useful for the histopathologic diagnosis of this disease and for study of the relationship between copper and tissue injury.

FIXATION	10% Buffered neutral formalin.
PROCESS	Paraffin or celloidin.
MICROTOMY	Cut sections at 6 micrometers.
CONTROL	Use control number 12.
SOLUTIONS	Distilled water, preferably deionized, should be used in all solutions and slide rinses.

Rhodanine (Stock)

5 (p-dimethylaminobenzylidene) rhodanine	0.2	gm
Absolute alcohol	100.0	ml

Rhodanine (Working)

Rhodanine (stock), saturated	6.0	ml
Distilled water	94.0	ml

Shake stock solution before measuring and mixing working solutions.

Mayer's Hematoxylin
(See Page 81)

0.5% Sodium Borax

Sodium borax	0.5	gm
Distilled water	100.0	ml

STAINING PROCEDURE

1. Decerate slides in xylene, 2 changes, 2 minutes each. Place in absolute alcohol, 2 changes, 2 minutes each; 95% alcohol, 2 changes, 2 minutes each; then rinse in distilled water.
2. Incubate in working rhodanine solution at 37°C overnight.
3. Wash well in several changes of distilled water.
4. Place in Mayer's hematoxylin for 10 minutes.
5. Rinse with distilled water.
6. Quickly rinse in 0.5% sodium borax.
7. Rinse well with distilled water.
8. Dehydrate slides in 95% alcohol and absolute alcohol 3 changes each. Clear in xylene, 3 changes.
9. Mount coverglass with appropriate medium (refractive index 1.48-1.56).

RESULTS

Copper .... bright red
Nuclei .... light blue

REMARKS

It is essential that specimens destined for staining with this method be fixed in neutral fixative solutions, since acid solutions will remove copper. Copper is also diminished after a few months fixation in 10% buffered neutral formalin, in the case of fetal liver or liver with very low copper concentrations. Sections which are cut thicker than normal (8-10 micrometers) will stain darker. This fact can be used to an advantage when one is having difficulty demonstrating minute amounts of copper.

Lindquist's Alizarin Blue S Method for Copper

FIXATION 10% Buffered neutral formalin.
PROCESS Paraffin or celloidin.
MICROTOMY Cut sections at 6 micrometers.
CONTROL Use control number 12.
SOLUTIONS

Buffer Water, pH 10

Sodium borate-sodium carbonate	1	vial
Distilled water	500.0	ml

0.025% Alizarin Blue S

Alizarin blue S	0.025	gm
Buffer water, pH 10	100.0	ml

Solution should be prepared just before use.

Polyvinyl-Pyrrolidone (PVP) Mounting Medium

Polyvinyl-polrrolidinone	50.0	gm
Distilled water	50.0	ml
Glycerol	2.0	ml

Add polyvinyl-pyrrolidone to water; let stand overnight. Then add glycerol and stir the mixture. A crystal of thymol may be added as a preservative.

STAINING PROCEDURE

1. Decerate slides in xylene, 2 changes, 2 minutes each. Place in absolute alcohol, 2 changes, 2 minutes each; 95% alcohol, 2 changes, 2 minutes each; then rinse in distilled water.
2. Rinse in pH 10 buffer water.
3. Place 0.025% alizarin blue S at room temperature for 2 hours.
4. Rinse in distilled water, 3 changes.
5. Mount coverglass with PVP medium.

RESULTS

Copper	blue
Background	unstained

Uzman's Method for Copper[117]

FIXATION 10% Buffered neutral formalin.
PROCESS Paraffin.
MICROTOMY Cut sections at 6 micrometers.
CONTROL Use control number 12.

SOLUTIONS

Rubeanic Acid (Dithiooxamide) Solution

Dithiooxamide	100.0	mg
100% Ethyl alcohol	70.0	ml
Dissolve dye with gentle heat; add sufficient distilled water to make	100.0	ml

Prepare fresh before use.

Kernechtrot (Nuclear Fast Red)
(See Page 84)

6% Alcoholic Saffron

Saffron	6.0	gm
100% alcohol	100.0	ml

After preparing solution, place in a tightly corked container to prevent hydration and place in an incubator at 58°C for 48 hours before use. This is done to extract the saffron. Keep the solution in an air-tight dark bottle.

STAINING PROCEDURE

1. Decerate slides in xylene, 2 changes.
2. Place in absolute alcohol, 2 changes.
3. Stain in rubeanic acid solution for 20 minutes.
4. Add solid sodium acetate (0.2 gm/100 ml) to the staining jar and allow to settle to the bottom. Leave slides in solution overnight.
5. Place in 70% alcohol, 2 changes, 30 minutes each.
6. Rinse with distilled water.
7. Counterstain in nuclear fast red solution for 2 minutes.
8. Rinse with distilled water.
9. Place in absolute alcohol, 3 changes.
10. Counterstain in 6% alcoholic saffron solution for 15 seconds.
11. Dehydrate in absolute alcohol and clear in xylene, 3 changes each.
12. Mount coverglass with appropriate medium (refractive index 1.48-1.56).

RESULTS

Copper	greenish-black precipitate
Nuclear chromatin	light pink
Cytoplasm	light yellow

REMARKS

The final rubeanic acid solution should be clear. Otherwise, the dithiooxamide must be recrystallized from hot 100% ethyl alcohol.

Pizzolato's Method for Calcium Oxalate[55ag]

FIXATION	10% Buffered neutral formalin.
PROCESS	Paraffin, frozen, or celloidin.
MICROTOMY	Cut sections at 6 micrometers.
CONTROL	Use control number 8.

SOLUTIONS

Silver Nitrate-Hydrogen Peroxide

5% Silver nitrate	25.0	ml
30% Hydrogen peroxide	25.0	ml

Mix just before use.

Kernechtrot (Nuclear Fast Red)
(See Page 84)

STAINING PROCEDURE

1. Decerate slides in xylene, 2 changes, 2 minutes each. Place in absolute alcohol, 2 changes, 2 minutes each; 95% alcohol, 2 changes, 2 minutes each; then rinse in distilled water.
2. Pour silver nitrate-hydrogen peroxide solution onto slides in sufficient quantity to cover sections. Place 60-watt incandescent light bulb 6 inches from sections for 30 minutes.
3. Rinse thoroughly with distilled water.
4. Stain nuclear chromatin in nuclear fast red for 5 minutes.
5. Rinse in distilled water.
6. Dehydrate slides in 95% alcohol and absolute alcohol 3 changes each. Clear in xylene, 3 changes.
7. Mount coverglass with appropriate medium (refractive index 1.48-1.56).

RESULTS

Calcium oxalate	black
Nuclear chromatin	red

REMARKS

Numerous gas bubbles will develop during the light bulb treatment if a large amount of egg albumin section adhesive is present. This tends to inhibit the reaction. Should this happen, pour off the mixture and resubmerge slides in fresh silver nitrate-hydrogen peroxide solution.

Roscher's Method for Calcium Oxalate[118]

Ethylene glycol, the common agent used in antifreeze, produces severe damage to parenchymal organs following accidental or suicidal ingestion. The following technique demonstrates calcium oxalate as calcium naphthalhydroxamate following ethylene glycol poisoning.

FIXATION	10% Buffered neutral formalin.
PROCESS	Paraffin or frozen.
MICROTOMY	Cut sections at 6 micrometers.
CONTROL	Use control number 8.

SOLUTIONS

Naphthalhydroxamic Acid

Naphthalhydroxamic acid	5.0	gm
Distilled water	25.0	ml

1% Methylene Blue		
Methylene blue	1.0	gm
Distilled water	100.0	ml

STAINING PROCEDURE

1. Decerate slides in xylene, 2 changes, 2 minutes each. Place in absolute alcohol, 2 changes, 2 minutes each; 95% alcohol, 2 changes, 2 minutes each; then rinse in distilled water.
2. Place in napthalhydroxamic acid solution for 10 minutes at room temperature.
3. Wash in distilled water or in a weak solution of ammonium hydroxide.
4. Stain with methylene blue, 5 dips.
5. Dehydrate slides in 95% alcohol and absolute alcohol 3 changes each. Clear in xylene, 3 changes.
6. Mount coverglass with appropriate medium (refractive index 1.48-1.56).

RESULTS

Crystals in renal tubules .... a distinct yellow crystalline pigment

REMARKS

Although crystal precipitate with naphthalhydroxamic acid are rather distinct, sections should be examined under polarized light in order to eliminate any yellow pigment, or pigment casts, in the tubules.

Johnson-Pani Method for Calcium Oxalate[119]

FIXATION 10% Buffered neutral formalin.
PROCESS Paraffin.
MICROTOMY Cut sections at 6 micrometers.
CONTROL Use control number 8.
SOLUTIONS

2M Glacial Acetic Acid		
Glacial acetic acid	12.0	ml
Distilled water	100.0	ml

2% Alizarin Red S		
Alizarin red S	2.0	gm
Distilled water	100.0	ml

Adjust solution to pH 4.1 to 4.3 by adding ammonium hydroxide drop by drop, stirring constantly. Use glass electrode pH meter, if possible.

STAINING PROCEDURE

1. Decerate slides in xylene, 2 changes, 2 minutes each. Place in absolute alcohol, 2 changes, 2 minutes each; 95% alcohol, 2 changes, 2 minutes each; then rinse in distilled water.
2. Stain one section with H&E for general orientation. Examine under polarized light. Calcium oxalate crystals generally have irregular outlines and vary in color under polarized light.

3. Treat a second section with 2M glacial acetic acid. If bubbles evolve from the crystals or if the crystals dissolve in five minutes, do not proceed further with the test. The evolution of bubbles indicates the presence of carbonates. Dissolution of the crystals in the glacial acetic acid indicates they might be calcium salts of dicarboxylic acids other than oxalic acid.
4. Place another paraffin section near the center of a sheet of 12"x12"x1/4" asbestos board. Cautiously heat the slide by directing the flame of a high temperature Bunsen burner on it. Keep the burner in motion to heat the slide as uniformly as possible. Use the oxidizing cone of the flame. Continue heating for about 4 minutes so that the section practically disappears. At this time, cover the slide with an inverted pie pan and allow to cool gradually.
5. Examine the cooled section under incandescent light or under dark-field illumination. A white ash should be present at the sites of previous calcium oxalate deposition.
6. Place a clean coverglass over the section and attach the four corners with molten paraffin. Apply a drop of concentrated sulfuric acid to the edge of the coverglass while observing ash from crystals under the microscope. Rapid evolution of gas bubbles constitutes a positive reaction.
7. Repeat steps 4 and 5 with another section. Cover this section and apply a drop of 2% alazarin red S under the coverglass. The ash should take on a red color, indicating the presence of calcium (there is no reaction between calcium oxalate and alizarin). However, after incineration of calcium carbonate, the reaction is strong.

De Galantha's Method for Urate Crystals

(See Fig. 124 and 126)

FIXATION — Absolute alcohol (4 hours for 2 mm thick specimens).

PROCESS — Paraffin. From absolute alcohol transfer blocks to xylene-paraffin, equal parts, at 58°C for 2 hours. Place in paraffin for 1 hour. Embed.

MICROTOMY — Cut sections at 8 micrometers.

CONTROL — Use control number 42.

SOLUTIONS

10% Silver Nitrate

Silver nitrate	10.0	gm
Distilled water	100.0	ml

Developing Solution

3% Gelatin in hot water	10.0	ml
10% Silver nitrate	3.0	ml
2% Hydroquinone	2.0	ml

Prepare just before use.

STAINING PROCEDURE

1. Place in xylene, then absolute alcohol, 2 changes each.
2. Place in silver nitrate solution. Expose to strong sunlight until urates are a bright rose; approximately 3 hours.
3. Prepare developing solution and pour over slides until the urates turn black and the connective tissues are yellow.

4. Wash quickly in hot water (58°C).
5. Dehydrate slides in 95% alcohol and absolute alcohol 3 changes each. Clear in xylene, 3 changes.
6. Mount coverglass with appropriate medium (refractive index 1.48-1.56).

RESULTS

Urates .... black
Background .... yellow

REMARKS

If specimen contains bone, decalcify in absolute alcohol to which 2% nitric acid has been added. Change solution every 24 hours until bone is softened. Wash in absolute alcohol for 24 hours.

Luna-Verfuerth Fluorescent Method for Lipofuscin and Bacteria

FIXATION 10% Buffered neutral formalin.
PROCESS Paraffin.
MICROTOMY Cut sections at 6 micrometers.
CONTROL Use control number 27A.
SOLUTIONS

1% Acridine Orange

Acridine orange	1.0	gm
Distilled water	100.0	ml

0.5% Potassium Permanganate

Potassium permanganate	0.5	gm
Distilled water	100.0	ml

STAINING PROCEDURE

1. Decerate slides in xylene, 2 changes, 2 minutes each. Place in absolute alcohol, 2 changes, 2 minutes each; 95% alcohol, 2 changes, 2 minutes each; then rinse in distilled water.
2. Stain in 1% acridine orange for 30 minutes.
3. Wash in running tap water for 3 minutes, followed by distilled water.
4. Place sections in 0.5% potassium permanganate for 2 minutes.
5. Wash in running tap water for 3 minutes, followed by distilled water.
6. Dehydrate slides in 95% alcohol and absolute alcohol 3 changes each. Clear in xylene, 3 changes.
7. Mount coverglass with appropriate medium (refractive index 1.48-1.56).

RESULTS

Use fluorescent microscope.

Mast cells .... bright red
Gram positive bacteria .... yellow
Gram negative bacteria .... pink
Lipofuscin .... yellow

Carbol-Fuchsin Method for Lipofuscin

(See Fig. 127)

FIXATION	10% Buffered neutral formalin.
PROCESS	Paraffin, frozen, or celloidin.
MICROTOMY	Cut sections at 6 micrometers.
CONTROL	Use control number 27A.
SOLUTIONS	

Kinyoun's Carbol-Fuchsin
(See Page 182)

1% Acid-Alcohol
(See Page 77)

Picric Acid

Picric acid	1.18	gm
Distilled water	100.0	ml

STAINING PROCEDURE

1. Decerate slides in xylene, 2 changes, 2 minutes each. Place in absolute alcohol, 2 changes, 2 minutes each; 95% alcohol, 2 changes, 2 minutes each; then rinse in distilled water.
2. Place in Kinyoun's carbol-fuchsin solution for 1 hour.
3. Rinse in distilled water.
4. Decolorize in 1% acid alcohol, usually 5 dips or until sections are light pink.
5. Wash in running tap water for 5 minutes. Then rinse in distilled water.
6. Counterstain in picric acid solution, 8 dips or until section appears yellow.
7. Dehydrate slides in 95% alcohol and absolute alcohol 3 changes each. Clear in xylene, 3 changes.
8. Mount coverglass with appropriate medium (refractive index 1.48-1.56).

RESULTS

Lipofuscin	red
Background	yellow

REMARKS

The periodic acid Schiff and Fontana-Masson methods also stain lipofuscin. These methods should not be regarded as specific since they stain many pigments. Ceroid pigment can also be stained with these methods. Picric acid is used as a counterstain, since it avoids the masking of lipofuscin which may happen in the usual acid fast procedure in which methylene blue is used.

Oil Red O Method for Lipofuscin

(See Fig. 128)

FIXATION	10% Buffered neutral formalin.
PROCESS	Paraffin, frozen, or celloidin.
MICROTOMY	Cut sections at 6 micrometers.
CONTROL	Use control number 27.

SOLUTIONS

0.5% Oil Red O

Oil red O	0.5	gm
100% Propylene glycol	100.0	ml

Add a small amount of propylene glycol to the oil red O and mix well, crush larger pieces. Gradually add the remainder of the propylene glycol stirring periodically. Heat gently until the solution reaches 95°C. Do not allow to go over 100°C. Stir while heating. Filter through coarse filter paper while still warm. Allow to stand overnight at room temperature. Filter through Seitz filter with the aid of vacuum. If solution becomes turbid, refilter. When using the Seitz filter, put rough surface of filter up.

85% Propylene Glycol

100% Propylene glycol	85.0	ml
Distilled water	15.0	ml

STAINING PROCEDURE

1. Decerate slides in xylene, 2 changes, 2 minutes each. Place in absolute alcohol, 2 changes, 2 minutes each; 95% alcohol, 2 changes, 2 minutes each; then rinse in distilled water.
2. Place in 0.5% oil red O for 24 hours.
3. Immerse in 85% propylene glycol solution for 1 minute. Agitate slide.
4. Rinse in distilled water, 2 changes, 2 minutes each.
5. Place in Mayer's hematoxylin solution for 5 minutes.
6. Wash in running tap water for 10 minutes.
7. Mount coverglass with appropriate medium (refractive index 1.48-1.56).

RESULTS

Lipofuscin.. red
Nuclear chromatin .... blue

Warkel-Luna-Helwig Method for Melanin[120]

The Fontana-Masson silver procedure has been used extensively for the demonstration of melanin pigment. Unfortunately, other common pigments (formalin, iron, lipofuscin, etc.) also reduce silver, making the method useful but non-specific. The modified Warthin-Starry procedure presented here has proven successful for specific demonstration of melanin, since other pigments will not react with the silver at the 3.2 pH used in this method.

FIXATION 10% Buffered neutral formalin.
PROCESS Paraffin, frozen, or celloidin.
MICROTOMY Cut sections at 6 micrometers.
CONTROL Use control number 32.
SOLUTIONS

Acidulated Water

Distilled water	1000.0	ml

Add enough 1% aqueous citric acid to bring water to pH 3.2.

1% Silver Nitrate, for impregnation

Silver nitrate, ACS crystals	1.0	gm
Acidulated water	100.0	ml

2% Silver Nitrate, for developer (Stock)

Silver nitrate, ACS crystals	2.0	gm
Acidulated water	100.0	ml

5% Gelatin (Stock)

Gelatin, high grade	10.0	gm
Acidulated water	200.0	ml

0.15% Hydroquinone (Stock)

Hydroquinone	0.15	gm
Acidulated water	100.0	ml

Place the 2% silver nitrate, 5% gelatin and the hydroquinone in separate flasks in a floatation bath at 54°C until ready to use (see step 4 of staining procedure).

Silver Nitrate-Hydroquinone-Gelatin Developer

2% Silver nitrate (stock)	1.5	ml
5% Gelatin (stock)	3.75	ml
0.15% Hydroquinone (stock)	2.0	ml

When ready to use (step 4), combine in order given in a small beaker, making certain solutions are mixed well after each addition. Prepare immediately before use.

Kernechtrot (Nuclear Fast Red)
(See Page 84)

STAINING PROCEDURE - Use control slide.

1. Decerate slides in xylene, 2 changes, 2 minutes each. Place in absolute alcohol, 2 changes, 2 minutes each; 95% alcohol, 2 changes, 2 minutes each; then rinse in distilled water.
2. Place slides in the 1% silver nitrate impregnating solution in an oven at 43°C for 30 minutes.
3. Prepare the developer solution, making sure to mix well.
4. Flood sections, that have been placed on a glass rod slide rack, with the developing solution for 8 minutes.
5. Wash quickly and thoroughly in hot running tap water to remove residual gelatin, thereby preventing precipitate on slides.
6. Rinse in distilled water.
7. Stain nuclei in nuclear fast red solution for 5 minutes.
8. Rinse in distilled water, 3 changes.
9. Dehydrate slides in 95% alcohol and absolute alcohol 3 changes each. Clear in xylene, 3 changes.
10. Mount coverglass with appropriate medium (refractive index 1.48-1.56).

RESULTS

Melanin	black
Nuclear chromatin	light pink
Cytoplasm	yellow

REMARKS
Only melanin granules will reduce silver nitrate if the procedure is performed correctly. Therefore, major areas to consider are temperature and pH of the acidulated water and time of development exposure at step 4. No other pigments stain with this method.

Methods for Differentiation of Formalin Pigment, Iron and Lipofuscin from Melanin Pigment

Numerous methods can be found for the removal (bleaching) of melanin from tissue sections. The potassium permanganate and oxalic acid method (see page 356) is preferred for the following reasons:

1. The method removes melanin completely if the treatment period is adequate. Most melanins, including dense ocular melanin, will be removed with 30 minutes exposure.
2. Potassium permanganate affect formalin pigment only slightly. Some of the superficial crystals may be removed, making the remaining crystals more difficult to see microscopically, but their polarizing properties remain. The polarizing properties of formalin crystals facilitates their differentiation from melanin, iron and lipofuscin, since the latter substances do not polarize.
3. The differentiation of melanin from iron is possible since iron is affected only slightly by this treatment and can be demonstrated by using one of the conventional iron stains.
4. Lipofuscin is also only slightly affected by potassium permanganate and oxalic acid treatment and can be differentiated from melanin by subsequent staining with the acid fast method found on page 351.

   The differentiation between melanin and iron or lipofuscin is important since they often appear similar microscopically. In addition, all three pigments reduce silver nitrate. Therefore, they cannot be differentiated on the basis of silver preparations. It is also significant that on rare occasions melanin stains positive with iron stains.

Methods for Nuclear Staining after Potassium Permanganate Treatment

Since iron, lipofuscin, and formalin pigment remain after potassium permanganate treatment, it is desirable to demonstrate them (or any other pigment) clearly. The use of a specific nuclear stain following the potassium permanganate oxalic acid treatment is dictated by the results desired. Weak Mayer's hematoxylin should be used routinely. Nuclear fast red should be used when one desires to overshadow the light gray color remaining in the cytoplasm after treatment with potassium permanganate. Borax methylene blue is suggested for a better demonstration of the pigments which remain (such as iron, lipofuscin or formalin), since these pigments appear brighter microscopically after this nuclear stain.

An eosin-type stain should never be employed after melanin bleaching as most pigments mentioned above are yellowish-brown and normally found in the cytoplasm, leading to possible confusion. Some of the smaller pigment deposits may be difficult to see microscopically, since there is little color differentiation between them and the counterstain.

Mayer's Hematoxylin Method

SOLUTION

Weak Mayer's Hematoxylin		
Hematoxylin crystals	0.5	gm
Distilled water	1000.0	ml
Sodium iodate	0.1	gm
Ammonium or potassium alum	25.0	gm
Citric acid	0.5	gm
Chloral hydrate	25.0	gm

Dissolve the alum in water without heat; add and dissolve the hematoxylin in this solution. Add the iodate, citric acid and chloral hydrate. Shake until all components are in complete solution. The final color of the stain is reddish violet. Stain keeps well for months.

STAINING PROCEDURE

Melanin is removed from tissue sections by the use of Luna's method found on page 356. Sections are subsequently stained with weak hematoxylin for 15 minutes, followed by a 10 minute wash in running tap water to blue nuclei. Dehydrate, clear and mount coverglass with appropriate medium (refractive index 1.48-1.56).

Borax-Methylene Blue Method

SOLUTION

Borax-Methylene Blue		
Methylene blue	1.0	gm
Borax	1.0	gm
Distilled water	100.0	ml

STAINING PROCEDURE

After melanin removal, sections are stained as follows:

1. Rinse in distilled water.
2. Stain in borax methylene blue solution for 2 minutes or longer. Staining results will vary depending upon the length of exposure to potassium permanganate.
3. Rinse in distilled water.
4. Dehydrate, clear and mount coverglass with appropriate medium (refractive index 1.48-1.56).

Nuclear Fast Red Method

SOLUTION

Kernechtrot (Nuclear Fast Red)		
Aluminum sulfate	5.0	gm
Distilled water	100.0	ml
Nuclear fast red	0.1	gm

Dissolve aluminum sulfate in distilled water prior to adding nuclear fast red. Heat and add the nuclear fast red. Cool, filter and add a grain of thymol as a preservative.

STAINING PROCEDURE
After melanin removal, sections are stained as follows:

1. Rinse in distilled water.
2. Place slides in nuclear fast red solution for 5 minutes.
3. Rinse in distilled water, at least 3 changes.
4. Dehydrate, clear and mount coverglass with appropriate medium (refractive index 1.48-1.56).

If slides turn cloudy, they were not rinsed sufficiently in distilled water (step 3). To correct this problem, return slides to distilled water in the usual manner.

Method for Melanin Pigment Removal

There are a number of pigments referred to as melanin in both normal and abnormal conditions. These pigments may appear to be closely related, but they are probably different in chemical structure. Regardless of this possibility, we find potassium permanganate a satisfactory agent for the removal of melanin pigment.

One must keep in mind, however, that this procedure may considerably alter the staining properties of some tissues, since the oxidation that potassium permanganate causes has a very harsh affect on the tissue section. To prevent loss of tissue sections during staining, it may be necessary to coat the sections with thin celloidin (see page 582). Alternatively, one can use Poly-L-Lysine, chrome gelatin or silanated slides.

Luna's Method for of Melanin Pigment Removal

1. Decerate slides in xylene, 2 changes, 2 minutes each. Place in absolute alcohol, 2 changes, 2 minutes each; 95% alcohol, 2 changes, 2 minutes each; then rinse in distilled water.
2. Place in 0.20% aqueous potassium permanganate solution for 30 minutes to 1 hour (time to be determined by the amount of melanin pigment present).
3. Wash well in running tap water.
4. Place in 5% aqueous oxalic acid solution or hydrobromic acid solution (1 part acid, 3 parts distilled water) until sections become clear (2 minutes).
5. Wash in running tap water for 5 minutes. Rinse in distilled water.
6. Stain as desired.

Emanuele's Method for Removal of Melanin Pigment

This method has proven to be of value in removing melanin from friable or difficult-to-section tissues. It is especially useful when one experiences difficulty in retaining sections on the slides because of the oxidation effect of the potassium permanganate.

1. Cut paraffin or frozen sections at desired thickness.
2. Float sections on a flotation water bath to allow them to stretch. Pick up each section with a glass slide and float on a dish containing sufficient 0.25% aqueous potassium permanganate solution. Leave slides in solution for 1 hour, stirring often. Removal of pigment will be faster if the solution is kept as 37°C.
3. Transfer section to distilled water.
4. Float section onto a 5% aqueous oxalic or hydrobromic acid solution (similar to above) until sections become clear.
5. Transfer sections into dish containing distilled water, 3 changes for 5 minutes each.
6. Transfer sections to albuminized slide. Allow to dry on slide warmer overnight, or place in a 60°C oven for 30 minutes.
7. Dehydrate slides in 95% alcohol and absolute alcohol 3 changes each. Clear in xylene, 3 changes.
8. Mount coverglass with appropriate medium (refractive index 1.48-1.56).

Frangioni-Borgioli Method for Bleaching Melanin[121]

This method of removing (bleaching) melanin, reportedly has 3 advantages over other methods, even for highly pigmented specimens: efficiency, speed and improved maintenance of tissue integrity.

Bleaching Solution

Benzyl alcohol	20.0	ml
Acetone	10.0	ml
10% Hydrogen peroxide	5.0	ml
Concentrated ammonium hydroxide	4	drops

Each ingredient should be added in the order given.
Mix thoroughly after each addition but do not shake container.

Acetone-Distilled Water

Acetone	50.0	ml
Distilled water	50.0	ml

PROCEDURE

1. Decerate slides in 2 changes of xylene, 2 minutes each.
2. Dip slides in acetone until residual xylene is removed.
3. Place slides in bleaching solution in 37°C oven for 1-3 hours.
4. Place slides in acetone-distilled water solution for 10 minutes.
5. Place slides in distilled water for 10 minutes to remove residual pigment.
6. Stain slides as desired.

REMARKS

Tissue specimens heavily laden with melanin are generally bleached in the 3 hour period suggested. However, slides may remain in the bleaching solution up to 6 hours without appreciable tissue damage. The proportions of the ingredients of the bleaching solution are fairly critical and can be varied only slightly. Excess ammonium hydroxide solution, for example, causes the sections to detach from the slide, while insufficient ammonia slows the speed of bleaching. Conversely, excess benzyl alcohol slows the bleaching process, whereas insufficient benzyl alcohol causes detachment of sections. The bleaching of melanin is due to the action of the hydrogen peroxide, but the speed of the reaction is determined by the ammonia catalyst.

Methods for Removal of Formalin and Malarial Pigments

In performing these procedures it must be understood that no specific time can be set for the various methods given, since the results depend on the length of tissue exposure as well as the treatment of the specimens prior to processing. Thus, the tissue exposure suggested is based on experience and is given as the minimum time required for pigment removal.

Both formalin and malarial pigments can be removed quite readily (5 to 10 minutes) if the tissue specimens are washed for 6 hours in running tap water prior to exposure to the dehydrating alcohols. If washing is omitted, removal of pigments may require as much as 24 hours.

The following techniques serve to remove both formalin and malarial pigments. No satisfactory procedure has been developed to specifically differentiate these pigments in tissue sections through their resistance to, or solubility in, various chemicals. However, several of the methods presented below will remove formalin pigments more rapidly than malarial pigments. Remarks made below concerning malaria parasites relate to the staining effect produced with the H&E.

Alcoholic Picric Acid Method

SOLUTION

Alcohol-Picric Acid

Picric Acid	1.5	gm
80% Alcohol	100.0	ml

PROCEDURE

1. Decerate slides in xylene, 2 changes, 2 minutes each. Place in absolute alcohol, 2 changes, 2 minutes each; 95% alcohol, 2 changes, 2 minutes each; then rinse in distilled water.
2. Let stand in saturated alcoholic picric acid solution for 3 hours.
3. Wash well in running tap water.

REMARKS

Formalin and malarial pigments are removed equally well. Overall staining is excellent after this procedure and, due to a yet unknown cause, malaria parasites are demonstrated very well.

Alcoholic Ammonium Hydroxide Method

SOLUTIONS

Alcohol-Ammonium Hydroxide

70% Alcohol	100.0	ml
Concentrated ammonium hydroxide	3.0	ml

1% Glacial Acetic Acid

Glacial acetic acid	1.0	ml
Distilled water	100.0	ml

PROCEDURE

1. Decerate slides in xylene, 2 changes, 2 minutes each. Place in absolute alcohol, 2 changes, 2 minutes each; 95% alcohol, 2 changes, 2 minutes each; then rinse in distilled water.
2. Place slides in alcohol-ammonium hydroxide for 30 minutes.
3. Rinse in running tap water.
4. Rinse in 1% glacial acetic acid solution.
5. Wash well in distilled water and stain as desired.

REMARKS

The results after using this procedure are identical to those found in the picric acid method.

Hydrogen Peroxide-Ammonium Hydroxide Method

SOLUTION

Bleaching Solution

Acetone	50.0	ml
3% Hydrogen peroxide	50.0	ml
28% Ammonium hydroxide	1.0	ml

PROCEDURE

1. Decerate slides in 2 changes of xylene 2 minutes each; absolute alcohol 2 changes, 2 minutes each; and 95% alcohol, 2 changes, 2 minutes each.
2. Rinse well in distilled water.
3. Place in bleaching solution for 1 hour or more.
4. Wash well in running tap water and distilled water. Then stain as desired.

Alcohol-Ammonium Hydroxide Method

SOLUTION

Alcohol-Ammonium Hydroxide

95% Alcohol	50.0	ml
28% Ammonium hydroxide	15.0	ml

PROCEDURE

1. Decerate slides in xylene, 2 changes, 2 minutes each. Place in absolute alcohol, 2 changes, 2 minutes each; 95% alcohol, 2 changes, 2 minutes each; then rinse in distilled water.
2. Place slide in alcohol-ammonium hydroxide solution for 1 hour.
4. Wash well in running tap water and stain as desired.

REMARKS

This is the preferred method of choice because it removes both malarial and formalin pigments in 1 hour. In addition, the staining qualities are excellent.

Luna's Method for Demonstrating Mercuric Chloride (Zenker) Crystals

Tissues specimens exposed to Zenker's fixative or other solutions which contain mercuric chloride produce a precipitate throughout the specimens. This precipitate, often referred to as Zenker crystals, is mercuric chloride. The precipitate can be removed from tissue sections by treatment in iodine and sodium thiosulfate. In most instances mercuric chloride precipitate can be identified microscopically as they exist. There are, however, some cases in which precise identification of this material is essential; for instance, if pigment/granule staining reactions are to be performed. The microscopic identification of this precipitate is complicated by the fact that there is nothing characteristic about their color, size, position or shape. That is, the precipitate crystals will vary in color, size and configuration from case to case.

The following method is very useful for the specific demonstration of precipitates formed when tissue is exposed to mercuric chloride solutions.

SOLUTIONS

Dilute Ammonium Hydroxide

Concentrated ammonium hydroxide (58%)	10.0	ml
Distilled water	90.0	ml

0.2% Light Green Solution

Light green, SF yellowish	0.2	gm
Distilled water	100.0	ml

Do not add acetic acid, since it will dissolve out the mercury precipitate.

STAINING PROCEDURE

1. Decerate slides in xylene, 2 changes, 2 minutes each. Place in absolute alcohol, 2 changes, 2 minutes each; 95% alcohol, 2 changes, 2 minutes each; then rinse in distilled water.
2. Place slides in dilute ammonium hydroxide solution for 1 minute.
3. Wash slides in distilled water for 1 minute.
4. Dip slides in light green solution until background exhibits a faint green color.
5. Dehydrate slides in 95% alcohol and absolute alcohol 3 changes each. Clear in xylene, 3 changes.
6. Mount coverglass with appropriate medium (refractive index 1.48-1.56).

RESULTS

Mercuric chloride crystals .... black
Background .... light green

REMARKS

The black color produced by the presence of mercuric chloride crystals will fade after several weeks. Sections cannot be retreated with the ammonium hydroxide solution if the original reaction was too light.

The mercury crystals are stained by the following reaction. The aqueous ammonia reacts with mercuric chloride in an oxidation-reduction reaction to form metallic mercury and a complex salt containing bivalent mercury:

$$Hg_2Cl_2 \quad + \quad 2NH_3 \quad - \quad Hg{+}HgNH_2Cl \quad + \quad NH_4 \quad + \quad Cl-$$

Method for Removal of Mercuric Chloride Pigment

When a fixative containing mercuric chloride is employed, one can expect a random distribution of brown to black artifactual deposits throughout the tissue section. This commonly called "Zenker crystal", may be removed by the following method.

SOLUTIONS

Gram's Iodine

Iodine	1.0	gm
Potassium iodide	2.0	gm
Distilled water	300.0	ml

Lugol's Iodine

Iodine	1.0	gm
Potassium iodide	2.0	gm
Distilled water	100.0	ml

5% Sodium Thiosulfate

Sodium thiosulfate	5.0	gm
Distilled water	100.0	ml

PROCEDURE

1. Decerate slides in 2 changes of xylene, 2 minutes each; absolute alcohol 2 changes, 2 minutes each; and 95% alcohol, 2 changes, 2 minutes each.
2. Place in Gram's or Lugol's iodine solution for 15 minutes.
3. Rinse in running tap water.
4. Place in 5% aqueous sodium thiosulfate for 3 minutes.
5. Wash in running tap water for 10 minutes or longer. Stain as desired.

Hall's Method for Bilirubin (Bile Pigment)[55ah]

(See Fig. 143)

FIXATION 10% Buffered neutral formalin.
PROCESS Paraffin, frozen, or celloidin.
MICROTOMY Cut sections at 6 micrometers.
CONTROL Use control number 7.
SOLUTIONS

10% Ferric Chloride (Stock)

Ferric chloride	10.0	gm
Distilled water	100.0	ml

Fouchet's Solution

Trichloroacetic acid	25.0	gm
Distilled water	100.0	ml
Mix and then add:		
10% Ferric chloride (stock)	10.0	ml

Van Gieson's Solution

1% acid fuchsin	2.5	ml
Picric acid, saturated aqueous	97.5	ml

STAINING PROCEDURE

1. Decerate slides in xylene, 2 changes, 2 minutes each. Place in absolute alcohol, 2 changes, 2 minutes each; 95% alcohol, 2 changes, 2 minutes each; then rinse in distilled water.
2. Place in Fouchet's solution for 5 minutes.
3. Wash in running tap water for 5 minutes, then in distilled water.
4. Place in Van Gieson's solution for 5 minutes.
5. Dehydrate slides in 95% alcohol and absolute alcohol 3 changes each. Clear in xylene, 3 changes.
6. Mount coverglass with appropriate medium (refractive index 1.48-1.56).

RESULTS

Bile pigment	green
Collagen	red
Muscle	yellow

REMARKS

Bilirubin is oxidized to biliverdin, which stains olive to emerald green depending upon the concentration of bilirubin. Bile can also be well demonstrated in its normal olive to green color by the following procedure: Decerate section to water, stain with nuclear fast red (Kernechtrot) stain, dehydrate, clear and mount coverglass with appropriate medium (refractive index 1.48-1.56).

Krueger's Method for Aluminum

FIXATION	10% Buffered neutral formalin.
PROCESS	Paraffin and plastic.
MICROTOMY	Cut paraffin embedded sections at 6 micrometers. Plastic, at desired thickness - 1 or 2 micrometers.
CONTROL	Use specimen containing aluminum as the results of disease.

SOLUTIONS

Acidic Buffer

Ammonium chloride	53.49	gm
Ammonium acetate	77.08	gm
Distilled water	400.0	ml
Concentrated hydrochloric acid	17.0	ml
Deionized water	17.0	ml

Total liquid volume is 434 ml.

Aluminon

Aluminum salt (aluminon)	4.0	gm
Acidic buffer	200.0	ml

Heat this solution for 15 minutes at 75°C, stirring vigorously. Filter while hot and store in opaque bottle. Note: The 75°C temperature is very critical.

Differentiating Solution

Acidic buffer	50.0	ml
Ammonium carbonate	3.38	gm
Distilled water	22.0	ml

STAINING PROCEDURE

A. For paraffin sections:
Decerate slides in xylene, 2 changes, 2 minutes each. Place in absolute alcohol, 2 changes, 2 minutes each; 95% alcohol, 2 changes, 2 minutes each; then rinse in distilled water.

B. For plastic sections:
1. Deplasticize sections in xylene at 60°C overnight. Let slides air-dry.
2. Place acidic buffer on a slide to form a pool and leave for 10 seconds.
3. Drain excess buffer from the slide.
4. Place aluminon on slide in a pool for 15 minutes at 68-72°C. Keep slides covered and add more stain if slides begin to dry.
5. Rinse in several changes of deionized water.
6. Place differentiating solution on the slide for 10 seconds.
7. Rinse in running tap water.
8. Dehydrate slides in 95% alcohol and absolute alcohol 3 changes each. Clear in xylene, 3 changes.
9. Mount coverglass with appropriate medium (refractive index 1.48-1.56).

RESULTS

Aluminum	pink to red

Chapter 12

Polysaccharides and Mucopolysaccharides

Southgate's Method for Mucin and Cryptococcus[122]

FIXATION	10% Buffered neutral formalin.
PROCESS	Paraffin, frozen, or celloidin.
MICROTOMY	Cut sections at 6 micrometers.
CONTROL	Use control number 33.
SOLUTIONS	

Carmine (Stock)

Carmine	1.0	gm
Aluminum hydroxide	1.0	gm
50% Alcohol	100.0	ml
Aluminum chloride (anhydrous)	0.5	gm

The carmine, aluminum hydroxide, and alcohol are put into a 500 ml flask and mixed by shaking. The anhydrous aluminum chloride is then added with frequent shaking. The flask is placed in a boiling water bath for 2 1/2 minutes. The contents are then immediately cooled under running tap water and filtered when cold.

Carmine (Working)

Carmine (stock)	10.0	ml
Distilled water	90.0	ml

Discard this working solution after use.

Mayer's Hematoxylin
(See Page 81)

0.025% Metanil Yellow

Metanil yellow	0.025	gm
Distilled water	100.0	ml
Glacial acetic acid	0.025	ml

STAINING PROCEDURE

1. Decerate slides in xylene, 2 changes, 2 minutes each. Place in absolute alcohol, 2 changes, 2 minutes each; 95% alcohol, 2 changes, 2 minutes each; then rinse in distilled water.
2. Stain in Mayer's hematoxylin for 10 minutes.
3. Wash in running tap water for 10 minutes.
4. Stain in working carmine for 30 minutes.
5. Rinse quickly in distilled water.
6. Counterstain in metanil yellow for 1 minute.
7. Dehydrate slides in 95% alcohol and absolute alcohol 3 changes each. Clear in xylene, 3 changes.
8. Mount coverglass with appropriate medium (refractive index 1.48-1.56).

RESULTS

Mucin	dark rose
Capsule of cryptococcus	dark rose
Nuclei	black
Other tissue elements	yellow

Mayer's Carmine Method for Mucin and Cryptococcus

(See Fig. 149, 150 and 151)

FIXATION 10% Buffered neutral formalin.
PROCESS Paraffin.
MICROTOMY Cut sections at 6 micrometers.
CONTROL Use control number 33.
SOLUTIONS

Mayer's Hematoxylin
(See Page 81)

Carmine (Stock)

Carmine	1.0	gm
Aluminum chloride, anhydrous	0.5	gm
Distilled water	2.0	ml

Mix stain in small evaporating dish. Heat on hot plate for 2 minutes. Liquid becomes black and syrupy. Dilute with 100 ml of 50% alcohol. Filter.

Carmine (Working)

Carmine (stock)	25.0	ml
Tap water	100.0	ml

0.25% Metanil Yellow Solution
(See Page 199)

STAINING PROCEDURE

1. Decerate slides in xylene, 2 changes, 2 minutes each. Place in absolute alcohol, 2 changes, 2 minutes each; 95% alcohol, 2 changes, 2 minutes each; then rinse in distilled water.
2. Stain in Mayer's hematoxylin for 15 minutes.
3. Wash in running tap water for 10 minutes.
4. Stain in carmine solution for 60 minutes.
5. Rinse quickly in distilled water.
6. Counterstain in metanil yellow for 1 minute.
7. Rinse quickly in distilled water.
8. Dehydrate slides in 95% alcohol and absolute alcohol 3 changes each. Clear in xylene, 3 changes.
9. Mount coverglass with appropriate medium (refractive index 1.48-1.56).

RESULTS

Mucin .... deep rose to red
Capsule of cryptococcus .... deep rose to red
Nuclear chromatin .... blue
Other tissue elements .... yellow

REMARKS

This method is used for the demonstration of cryptococcus with good results. Hematoxylin and eosins or elastic counterstains may be used with this method. Note: Carminophilic properties will be obscured if sections are overstained with metanil yellow.

Bennhold's Method for Amyloid (Congo Red)

(See Fig. 146 and 147)

FIXATION 10% Buffered neutral formalin.
PROCESS Paraffin, frozen, or celloidin.
MICROTOMY Cut sections at 6 and 12 micrometers (see remarks).
CONTROL Use control number 2.
SOLUTIONS

1% Congo Red

Congo red	1.0	gm
Distilled water	100.0	ml

1% Sodium Hydroxide (Stock)

Sodium hydroxide	1.0	gm
Distilled water	100.0	ml

Alkaline Alcohol

1% Sodium hydroxide	1.0	ml
50% Alcohol	100.0	ml

Mayer's Hematoxylin
(See Page 81)

STAINING PROCEDURE

1. Decerate slides in xylene, 2 changes, 2 minutes each. Place in absolute alcohol, 2 changes, 2 minutes each; 95% alcohol, 2 changes, 2 minutes each; then rinse in distilled water.
2. Place in congo red solution for 1 hour.
3. Rinse off excess stain in water, 2 or 3 changes.
4. Differentiate in alkaline alcohol solution for 3-5 seconds. Agitate constantly until the background appears clear, and amyloid deposits stand out in a pink to red color.
5. Wash in running water for 5 minutes.
6. Stain nuclear chromatin in Mayer's hematoxylin solution for 5 minutes.
7. Wash in running water for 10 minutes.
8. Dehydrate slides in 95% alcohol and absolute alcohol 3 changes each. Clear in xylene, 3 changes.
9. Mount coverglass with appropriate medium (refractive index 1.48-1.56).

RESULTS

Amyloid .... pink to red
Nuclear chromatin .... blue

REMARKS

Sections of suspected amyloidosis may be cut at 12 micrometers if a brighter color is desired. This thick section technique is used to demonstrate minute amyloid deposits. Twelve-micrometer sections can also be used to advantage, in the crystal violet and sirius red methods for amyloid. Polarizing properties of amyloid are improved considerably in 12 micrometer sections.

Lieb's Method for Amyloid (Crystal Violet)

(See Fig. 148)

FIXATION	10% Buffered neutral formalin.
PROCESS	Paraffin, frozen, or celloidin.
MICROTOMY	Cut sections at 6 and 12 micrometers.
CONTROL	Use control number 2.
SOLUTIONS	

14% Crystal Violet (Stock)

Crystal violet	14.0	gm
95% Alcohol	100.0	ml

Crystal Violet (Working)

14% Crystal violet (stock)	10.0	ml
Distilled water	300.0	ml
Concentrated hydrochloric acid	1.0	ml

Apathy's Mounting Media

Acacia (gum arabic)	50.0	gm
Cane sugar (sucrose)	50.0	gm
Distilled water	150.0	ml
Sodium chloride	10.0	gm
Thymol	0.1	gm

Mix the acacia, cane sugar, and distilled water in a flask. Place the flask in a pan of boiling water until ingredients are dissolved. Dissolve the sodium chloride and thymol in this solution. Filter with a Seitz filter while the solution is still warm, or use coarse filter paper and place in 60°C oven, changing filter paper frequently. Refrigerate to remove air bubbles. Sodium chloride is added to prevent bleeding of the stain and thymol is added as a preservative.

STAINING PROCEDURE

1. Decerate slides in xylene, 2 changes, 2 minutes each. Place in absolute alcohol, 2 changes, 2 minutes each; 95% alcohol, 2 changes, 2 minutes each; then rinse in distilled water.
2. Stain in working crystal violet solution for 20 minutes or 5 hours (see remarks).
3. Wash well in running tap water for 5 minutes.
4. Mount coverglass with Apathy's mounting medium.

RESULTS

Amyloid.. purplish-violet
Other tissue elements.. blue

REMARKS

This stain can be performed in three ways. 1. By dipping the slides 10 times, one at a time, in the working crystal violet solution. 2. Staining in working crystal violet. 3. Staining 5 hours in working crystal violet. The 5 hour exposure should be used as the minimum staining time. (In our laboratories, 16 hours staining time demonstrated the optimum staining qualities.) The staining of amyloid for 5 to 16 hours (overnight) provides the ultimate in staining of amyloid with the crystal violet method. This can be explained on the basis that there is a reverse reaction produced by the action of the hydrochloric acid, resulting in the specific development of the tinctorial quality of the amyloid in the following manner. The entire specimen stains a dense deep purple blue after several minutes in the working crystal violet solution. The increase in color continues for several hours, at which point the section is heavily stained. At a given point, some time after 3 hours of staining, the hydrochloric acid in the staining solution begins to selectively differentiate the dye from the section. This differentiation process results in removing all of the reddish-purple color from the non-amyloid portion of the section, while the amyloid remains reddish-purple. This action results in a blue background and deep purple staining of the amyloid. (See remarks, Bennhold's method.)

Sweat-Puchtler Method for Amyloid (Sirius Red)

(See Fig. 144 and 145)

FIXATION 10% Buffered neutral formalin.
PROCESS Paraffin, frozen, or celloidin.
MICROTOMY Cut sections at 6 and 12 micrometers.
CONTROL Use control number 2.
SOLUTIONS

1% Sodium Hydroxide (Stock)
(See Page 329)

Alkaline Alcohol

80% Alcohol	100.0	ml
1% Sodium hydroxide (stock)	1.0	ml

Sirius Red

Sirius red, F3BA	1.0	gm
Distilled water	100.0	ml
Dissolve and then add:		
Sodium chloride	0.5	gm

Let stand overnight before use. Do not filter.

0.1M Borate Buffer, pH 9.0

Sodium borate	38.143	gm
Distilled water	1000.0	ml

Mayer's Hematoxylin
(See Page 81)

STAINING PROCEDURE

1. Decerate slides in xylene, 2 changes, 2 minutes each. Place in absolute alcohol, 2 changes, 2 minutes each; 95% alcohol, 2 changes, 2 minutes each; then rinse in distilled water.
2. Place in alkaline alcohol solution for 1 hour.
3. Rinse in distilled water for 10 seconds.
4. Place in preheated sirius red solution in a 60°C oven for 90 minutes.
5. Rinse 10 seconds in 2 changes of borate buffer solution.
6. Wash in running water for 5 minutes.
7. Place in Mayer's hematoxylin solution for 5 minutes.
8. Wash in running water for 15 minutes, then drain slides.
9. Dehydrate slides in 95% alcohol and absolute alcohol 3 changes each. Clear in xylene, 3 changes.
10. Mount coverglass with appropriate medium (refractive index 1.48-1.56).

RESULTS

Amyloid .... pink to red
Nuclear chromatin .... blue
Background .... unstained
Elastica .... pink to red

REMARKS

Sodium borate ($Na_2B_40_7{:}10H_20$) must be used as other borates will give a different pH.

Vassar-Culling Method for Amyloid (Thioflavine T)

The following entities are stained by this thioflavine T method: acidophil cells of the pituitary gland, Lynrogen granules in the pancreas of mice, mast cells in mice and man and juxtaglomerular apparatus in mice.

FIXATION 10% Buffered neutral formalin.
PROCESS Paraffin, frozen, or celloidin.
MICROTOMY Cut sections at 6 micrometers.
CONTROL Use control number 2.
SOLUTIONS

1% Thioflavine T

Thioflavine T	1.0	gm
Distilled water	100.0	ml

Mayer's Hematoxylin
(See Page 81)

1% Glacial Acetic Acid

Glacial acetic acid	1.0	ml
Distilled water	100.0	ml

Apathy's Mounting Medium
(See Page 367)

STAINING PROCEDURE

1. Decerate slides in xylene, 2 changes, 2 minutes each. Place in absolute alcohol, 2 changes, 2 minutes each; 95% alcohol, 2 changes, 2 minutes each; then rinse in distilled water.
2. Place in Mayer's hematoxylin solution for 5 minutes.
3. Wash in running tap water for 1 minute.
4. Place in thioflavin T solution for 3 minutes.
5. Wash in running water for 10 minutes.
6. Place in 1% glacial acetic acid for 10 minutes.
7. Wash in running tap water for 2 minutes.
8. Mount coverglass with Apathy's gum syrup.

RESULTS

Amyloid .... white fluorescence

Increased specificity of the staining results are achieved by using a shorter wave length of light (Zeiss UG5 exciter filter) and a simple (colorless) U.V. filter in the eyepiece. With this system only amyloid is yellow; all other tissue constituents are blue.

Sweat-Puchtler-Valentine (Fluorescence) Method for Amyloid[123]

(See Fig. 163)

FIXATION 10% Buffered neutral formalin, absolute alcohol, Carnoy's methacarn or Zenker's solution.

PROCESS Paraffin.

MICROTOMY Cut section at 5 micrometers.

CONTROL Use control number 3.

SOLUTIONS

Mayer's Hematoxylin

Hematoxylin	1.0	gm
Distilled water	1000.0	ml

Agitate on a magnetic stirrer until the hematoxylin is completely dissolved. Do not use heat.

Add:

Sodium iodate	0.2	gm

Dissolve completely and allow to sit until solution becomes dark reddish-brown.

Add:

Aluminum ammonium or potassium sulfate	50.0	gm

Dissolve completely and allow to sit 1-2 hours. Add:

Boric acid, monohydrate	1.0	gm
Chloral hydrate	50.0	gm

1% Sodium Hydroxide

Sodium hydroxide	1.0	gm
Distilled water	100.0	ml

Alkaline Alcohol

95% Ethyl alcohol	84.0	ml
Distilled water	16.0	ml
1% Sodium hydroxide	1.0	ml

1% Phorwhite BBU

Phorwhite BBU	1.0	gm
Distilled water	100.0	ml
Dissolve completely and add:		
Sodium chloride	0.5	gm

Do not filter. The dye will be removed by the cellulose in the filter paper.

0.4 M Boric Acid

Boric acid	12.37	gm
Distilled water	500.0	ml

0.1M Sodium Borate (Stock)

Sodium borate	19.07	gm
Distilled water	500.0	ml

0.1M Borate Buffer

0.1M Sodium borate (stock)	500.0	ml

While agitating at a moderate speed add 0.4M boric acid, drop by drop, until reaching pH 9.

STAINING PROCEDURE

1. Decerate slides in xylene, 2 changes, 2 minutes each. Place in absolute alcohol, 2 changes, 2 minutes each; 95% alcohol, 2 changes, 2 minutes each; then rinse in distilled water.
2. If necessary remove formalin pigment or mercuric precipitate using the standard method.
3. Wash in running tap water for 5 minutes. Rinse in distilled water.
4. Stain with Mayer's hematoxylin for 10 minutes.
5. Wash in warm running tap water for 10 minutes.
6. Place in alkaline alcohol solution for 20 minutes.
7. Rinse in 3 changes of distilled water.
8. Stain in fresh preheated solution of 1% Phorwhite BBU 2 hours at 60°C (see remarks).
9. Rinse briefly in 0.1M borate buffer.
10. Wash in running tap water for 5 minutes.
11. Dehydrate slides in 95% alcohol and absolute alcohol 3 changes each. Clear in xylene, 3 changes.
12. Mount coverglass with appropriate medium (refractive index 1.48-1.56).

RESULTS

Viewed with UV excitation filter Sp2 barrier filter:

Amyloid....................intense bluish fluorescence
Background....................weak moderate blue

UV-blue filter:

Amyloid....................green to yellow fluorescence

UV-blue plus tungsten light with red filter:

Amyloid....................bright yellow
Other tissue components....................red

REMARKS

Amyloid fixed in 10% buffered or unbuffered formalin solution up to 48 hours show strong to intense selective fluorescence of amyloid, similar to that seen in alcohol or Carnoy fixed sections from the same organs. Intensity of fluorescence decreases with duration of fixation or storage in 10% unbuffered formalin. Little or no fluorescence can be detected after one year. Amyloid fixed in Zenker-formal solution for approximately 16 hours and washed in running tap water for 24 hours prior to embedding shows strong greenish fluorescence. Prolonged fixation impairs fluorescence of amyloid.

The variable time allowed in step 6 allows the procedure to be adjusted to laboratory schedules. Virtually identical fluorescence occurs in 1 and 2 hour exposures.

Hinds' Method for Amyloid[124]

FIXATION 10% Buffered neutral formalin.
PROCESS Paraffin, frozen, or celloidin.
MICROTOMY Cut section at 4 micrometers.
CONTROL Use control number 2.
SOLUTIONS

1% Oxalic Acid Stock

Oxalic acid	1.0	gm
Distilled water	100.0	ml

2% Crystal Violet

Crystal violet	2.0	gm
100% Ethyl alcohol absolute	80.0	ml
1% Oxalic acid (stock)	20.0	ml

1% Glacial Acetic Acid

Glacial acetic acid	1.0	ml
Distilled water	100.0	ml

STAINING PROCEDURE

1. Decerate slides in xylene, 2 changes, 2 minutes each. Place in absolute alcohol, 2 changes, 2 minutes each; 95% alcohol, 2 changes, 2 minutes each; then rinse in distilled water.
2. Stain in 2% crystal violet for 10 minutes.
3. Wash slides in running water for 5 minutes.
4. Differentiate in 1% glacial acetic acid for 3-5 minutes. In order to control differentiation, remove the sections form the glacial acetic acid after 2 minutes and examine them under the microscope. Differentiation can be stopped in distilled water and be resumed as necessary until the amyloid is well contrasted and its purplish-red color is clearly defined against a pale, clear blue background. Wash the sections in tap water for 10 minutes and proceed to step 5. DO NOT pass sections through dehydrating alcohols.
5. Blot sections to complete dryness.
6. Mount coverglass with appropriate medium (refractive index 1.48-1.56).

RESULTS

Amyloid..purplish-red
Elastica..blue
Other tissue structures ....pale blue

Best's Method for Glycogen

(See Fig. 162)

FIXATION Carnoy's, formal-alcohol or 10% buffered neutral formalin.
PROCESS Paraffin, frozen, or celloidin.
MICROTOMY Cut sections at 6 micrometers.
CONTROL Use control number 19.
SOLUTIONS

Mayer's Hematoxylin
(See Page 81)

Carmine (Stock)

Carmine	2.0	gm
Potassium carbonate	1.0	gm
Potassium chloride	5.0	gm
Distilled water	60.0	ml

Boil in an evaporating dish, gently and cautiously, for several minutes. When cool add 20 ml of 28% ammonium hydroxide. Store in refrigerator.

Carmine (Working)

Carmine solution (stock)	10.0	ml
28% Ammonium hydroxide	15.0	ml
Methyl alcohol	15.0	ml

Differentiating Solution

100% Alcohol	20.0	ml
Methyl alcohol	10.0	ml

STAINING PROCEDURE

1. Decerate slides in xylene, 2 changes, 2 minutes each. Place in absolute alcohol, 2 changes, 2 minutes each; 95% alcohol, 2 changes, 2 minutes each; then rinse in distilled water.
2. Place in Mayer's hematoxylin solution for 15 minutes.
3. Wash in running water for 15 minutes.
4. Place in working carmine solution for 30 minutes.
5. Immerse in differentiating solution for 30 seconds.
6. Rinse quickly in 80% alcohol.
7. Dehydrate slides in 95% alcohol and absolute alcohol 3 changes each. Clear in xylene, 3 changes.
8. Mount coverglass with appropriate medium (refractive index 1.48-1.56).

RESULTS

Glycogen.. pink to red
Nuclear chromatin .... blue

REMARKS

Many workers have commented on the frequency with which the Best's carmine method precipitates on the slide. This problem only occurs when the section is stained flat on a slide rack. If staining is carried out in this manner, the ammonia present in the staining solution is driven off, leaving a methanolic aqueous solution behind, in which carminic acid is less soluble and thus precipitation occurs. This does not happen if closed coplin jars are used.

Sections should be rinsed only in methyl alcohol following staining in carmine. The use of methanolic aqueous solutions is to be avoided, because water itself removes dye non-specifically and therefore will remove dye which has stained glycogen. This is particularly important where small amounts of glycogen are to be demonstrated and could lead to false negative results. Methyl alcohol does not appear to remove any staining of the glycogen.

The staining of glycogen with ammoniacal alcohol solution of Best's carmine is due primarily to hydrogen bonding groups on the glycogen molecule. The staining of cytoplasmic elements is due to simple acid dyeing which is governed by the overall charge on the dye and the pH of the staining solution.

McManus' Method for Glycoproteins (PAS)

(See Fig. 152a, b, 153 and 207)

FIXATION	Methyl alcohol, formal-alcohol or 10% buffered neutral formalin.
PROCESS	Paraffin, frozen, or celloidin.
MICROTOMY	Cut sections at 6 micrometers.
CONTROL	Use control number 19.
SOLUTIONS	

Normal Hydrochloric Acid

Hydrochloric acid, (sp.gr. 1.19)	83.5	ml
Distilled water	916.5	ml

Coleman's Solution

Dissolve 1.0 gm basic fuchsin in 200.0 ml hot distilled water. Bring to a boiling point. Cool and add 2.0 gm potassium metabisulfite, 10.0 ml normal hydrochloric acid. Let bleach for 24 hours, then add 0.5 gm activated carbon (Norit). Shake for 1 minute and filter through coarse filter paper. Repeat filtration until solution is colorless. Store in refrigerator.

Schiff's Solution

Dissolve 1.0 gm basic fuchsin in 200.0 ml hot distilled water. Bring to boiling point. Cool to 50°C. Filter and add 20.0 ml normal hydrochloric acid. Cool further and add 1.0 gm anhydrous sodium bisulfite, or sodium metabisulfite. Keep in the dark for 48 hours until solution becomes straw colored. Store in refrigerator.

Test For Schiff's Reagent Solution

Pour a few drops of Schiff's reagent solution into 10 ml of 37-40% formaldehyde and place in a watch glass. If the solution turns reddish-purple rapidly, it is good. If the reaction is delayed and the resulting color deep blue-purple, the solution is breaking down.

0.5% Periodic Acid

Periodic acid	0.5	gm
Distilled water	100.0	ml

0.2% Light Green (Stock)

Light green, SF yellowish	0.2	gm
Distilled water	100.0	ml
Glacial acetic acid	0.2	ml

Light Green (Working)

0.2% Light green	10.0	ml
Distilled water	50.0	ml

Mayer's Hematoxylin
(See Page 81)

STAINING PROCEDURE

1. Decerate slides in xylene, 2 changes, 2 minutes each. Place in absolute alcohol, 2 changes, 2 minutes each; 95% alcohol, 2 changes, 2 minutes each; then rinse in distilled water.
2. Oxidize in 0.5% periodic acid for 5 minutes.
3. Rinse in distilled water.
4. Place in Coleman's or Schiff's solution for 15 minutes.
5. Wash in running tap water for 10 minutes for pink color to develop.
6. Place in Mayer's hematoxylin for 10 minutes, or light green counterstain for 10 seconds. (Light green is recommended for counterstaining sections in which fungi are to be demonstrated.)
7. Wash for 10 minutes in running water if Mayer's hematoxylin is used as the counterstain.
8. Dehydrate slides in 95% alcohol and absolute alcohol 3 changes each. Clear in xylene, 3 changes.
9. Mount coverglass with appropriate medium (refractive index 1.48-1.56).

RESULTS

Glycogen, mucin, reticulin, fibrin or thrombi, colloid droplets, hyalin of arteriosclerosis, hyalin deposits in glomeruli granular cells in the renal arterioles where preserved, most basement membranes, colloid of pituitary stalks and thyroid, and amyloid infiltration, may show a positive reaction of rose to purplish-red.

Nuclei	blue
Fungi	red
Background	pale green (with light green counterstaining)

REMARKS

For information on alcian blue and its combination with periodic acid-Schiff reaction see page 383.

Mammalian Polysaccharides and Mucopolysaccharides Distinguishable Histochemically

1. Polysaccharides lacking amino sugars (glycogen)
2. Neutral mucopolysaccharides (Blood group substances, fucomucins, mannose containing mucosubstances)
3. Acid mucopolysaccharides
   - (a) Sulfomucins
     - (1) strongly acidic, periodate unreactive (chondroitin and heparin sulfates)
     - (2) weakly acidic, periodate reactive
   - (b) Sialomucins
   - (c) Hyaluronic acid
   - (d) Mucins with unknown acid groups

METHODS AND RESULTS

The use of staining solutions at various pH levels will aid in the distinction between sulfated and nonsulfated mucopolysaccharides.

McManus' Method for Glycoproteins (PAS)

RESULTS

Neutral mucopolysaccharides and polysaccharides .... magenta
Acid mucopolysaccharides ....blue

Mowry's Method for Acid Mucopolysaccharides

RESULTS

Acid mucopolysaccharides .... blue to green
Hyaluronic acid ....light blue to green
Nucleic acid .... light blue to green

REMARKS

Few, if any, other acid substances remain in sufficient quantity, after formalin-fixation and paraffin-embedding, to produce a reaction.

Aldehyde Fuchsin-Alcian Blue

RESULTS

Sulfated mucopolysaccharides ....purple-blue to purple
Sialomucins and non-sulfated acid mucosaccharides .... blue

Aldehyde Fuchsin, pH 1.7 and 1.0

RESULTS
At pH 1.7 sulfated acid mucopolysaccharides as well as most mucin of epithelial origin, will give a positive purple reaction, whereas, hyaluronic acid does not react or give only a faint response. At pH values below 1.0, only strong acid compounds such as the sulfated acid mucopolysaccharides are reactive. Most mucin of epithelial origin (sialic acid containing mucin) loses its affinity for aldehyde-fuchsin, but mast cells, cartilage, and interfibrillar material of blood vessel walls remain positive.

Alcian Blue, pH 2.5 and 1.0

RESULTS
The use of solutions at various pH values will aid in the distinction between sulfated and nonsulfated acid mucopolysaccharides. At pH 2.5 most acid mucopolysaccharides give a blue color. Sialomucins and weakly acid sulfomucins stain blue; strongly acidic sulfomucins stain weakly or not at all. At pH 1.0 or lower, strongly acid substances such as the sulfated acid mucosaccharides will give a positive reaction.

Alcian Blue Method, pH 2.5

(See Fig. 154 and 155)

FIXATION — 10% Buffered neutral formalin.
PROCESS — Paraffin, frozen or celloidin.
MICROTOMY — Cut sections at 6 micrometers.
CONTROL — Use control number 10 or 23.
SOLUTIONS

3% Glacial Acetic Acid (Stock)

Glacial acetic acid	3.0	ml
Distilled water	97.0	ml

1% Alcian Blue

Alcian blue, 8GX	1.0	gm
3% Glacial acetic acid (stock)	100.0	ml

Adjust the pH to 2.5 with glacial acetic acid if necessary. Filter and add a few crystals of thymol.

Kernechtrot (Nuclear Fast Red)
(See Page 84)

STAINING PROCEDURE
1. Decerate slides in xylene, 2 changes, 2 minutes each. Place in absolute alcohol, 2 changes, 2 minutes each; 95% alcohol, 2 changes, 2 minutes each; then rinse in distilled water.
2. Mordant in 3% acetic acid solution for 3 minutes.
3. Place in 1% alcian blue solution for 30 minutes.

4. Wash in running water for 5 minutes.
5. Rinse in distilled water.
6. Counterstain in Kernechtrot solution for 5 minutes.
7. Wash in running water for 1 minute.
8. Dehydrate slides in 95% alcohol and absolute alcohol 3 changes each. Clear in xylene, 3 changes.
9. Mount coverglass with appropriate medium (refractive index 1.48-1.56).

RESULTS
Weakly acidic mucopolysaccharides, sulfomucins, hyaluronic acid and sialomucins - dark blue.

Alcian Blue Method, pH 1.0

(See Fig. 156)

FIXATION 10% Buffered neutral formalin.
PROCESS Paraffin, frozen, or celloidin.
MICROTOMY Cut sections at 6 micrometers
CONTROL Use control number 10A.
SOLUTIONS

1% Alcian Blue

Alcian blue, 8GX	1.0	gm
0.1N Hydrochloric acid	100.0	ml

STAINING PROCEDURE
1. Decerate slides in xylene, 2 changes, 2 minutes each. Place in absolute alcohol, 2 changes, 2 minutes each; 95% alcohol, 2 changes, 2 minutes each; then rinse in distilled water.
2. Mordant in 3% acetic acid for 3 minutes.
3. Place in alcian blue solution for 30 minutes.
4. Wash in running water for 10 minutes.
5. Counterstain in Kernechtrot for 5 minutes.
6. Wash in running tap water for 1 minute.
7. Dehydrate slides in 95% alcohol and absolute alcohol 3 changes each. Clear in xylene, 3 changes.
8. Mount coverglass with appropriate medium (refractive index 1.48-1.56).

RESULTS
Sulfated mucosubstances .... dark blue

Alcian Blue Method, pH 0.4

FIXATION 10% Buffered neutral formalin.
PROCESS Paraffin, frozen, or celloidin.
MICROTOMY Cut sections at 6 micrometers.
CONTROL Use control number 10A.

SOLUTIONS

Phosphate-Hydrochloric Acid

Concentrated hydrochloric acid	42.0	ml
Sodium phosphate, monobasic	13.8	gm
Distilled water to make	1000.0	ml

1% Alcian Blue

Alcian blue 8GX	2.5	gm
Phosphate-hydrochloric acid	250.0	ml

Kernechtrot (Nuclear Fast Red)
(See Page 84)

STAINING PROCEDURE

1. Decerate slides in xylene, 2 changes, 2 minutes each. Place in absolute alcohol, 2 changes, 2 minutes each; 95% alcohol, 2 changes, 2 minutes each; then rinse in distilled water.
2. Place in phosphate-hydrochloric acid solution for 3 minutes.
3. Place in alcian blue solution for 30 minutes.
4. Wash in running tap water for 5 minutes.
5. Rinse in distilled water.
6. Counterstain in Kernechtrot for 5 minutes.
7. Wash in running water for 1 minute.
8. Dehydrate slides in 95% alcohol and absolute alcohol 3 changes each. Clear in xylene, 3 changes.
9. Mount coverglass with appropriate medium (refractive index 1.48-1.56).

RESULTS

Strongly acidic sulfated mucopolysaccharides .... blue

Aldehyde-Fuchsin Method, pH 1.7

(See Fig. 157 and 160)

FIXATION 10% Buffered neutral formalin.
PROCESS Paraffin, frozen, or celloidin.
MICROTOMY Cut sections at 6 micrometers.
CONTROL Use control number 10A.
SOLUTIONS

Aldehyde-Fuchsin, pH 1.7

Basic fuchsin	1.0	gm
70% Alchohol	200.0	ml
Concentrated hydrochloric acid	2.0	ml
Paraldehyde	2.0	ml

Let stand at room temperature for 2 to 5 days until stain is deep purple in color. Store in refrigerator. This solution has a pH of 1.7. To adjust the pH of the aldehyde fuchsin to 1.0 use concentrated hydrochloric acid.

STAINING PROCEDURE

1. Decerate slides in xylene, 2 changes, 2 minutes each. Place in absolute alcohol, 2 changes, 2 minutes each; 95% alcohol, 2 changes, 2 minutes each; then rinse in distilled water.
2. Place in 70% alcohol for 1 minute.
3. Place in aldehyde-fuchsin solution for 30 minutes.
4. Rinse in 70% alcohol.
5. Dehydrate slides in 95% alcohol and absolute alcohol 3 changes each. Clear in xylene, 3 changes.
6. Mount coverglass with appropriate medium (refractive index 1.48-1.56).

RESULTS

Sulfated mucopolysaccharides	purple
Mucins of epithelial origin	purple
Elastica	purple
Mast cells	purple
Paget's cell cytoplasm	purple

Aldehyde Fuchsin (pH 1.7)-Alcian Blue (pH 2.5) Method

(See Fig. 158)

FIXATION 10% Buffered neutral formalin.
PROCESS Paraffin, frozen, or celloidin.
MICROTOMY Cut sections at 6 micrometers.
CONTROL Use control number 39B.
SOLUTIONS

Alcian Blue pH 2.5

Alcian blue, 8GX	1.0	gm
3% Glacial acetic acid	100.0	ml

Aldehyde Fuchsin pH 1.7
(See Page 379)

STAINING PROCEDURE

1. Decerate slides in xylene, 2 changes, 2 minutes each. Place in absolute alcohol, 2 changes, 2 minutes each; 95% alcohol, 2 changes, 2 minutes each; then rinse in distilled water.
2. Rinse in 70% alcohol.
3. Place in aldehyde-fuchsin solution for 30 minutes.
4. Rinse in 70% alcohol, rinse in distilled water, several changes, for 5 minutes.
5. Place in alcian blue solution for 30 minutes.
6. Rinse in distilled water.
7. Dehydrate slides in 95% alcohol and absolute alcohol 3 changes each. Clear in xylene, 3 changes.
8. Mount coverglass with appropriate medium (refractive index 1.48-1.56).

RESULTS

Elastica .... purple
Weakly acidic sulfated mucopolysaccharides or mixed
sulfated and carboxylic mucosubstances .... blue to purple
Nonsulfated acid mucosubstances .... blue

Aldehyde Fuchsin, pH 1.0

FIXATION 10% Buffered neutral formalin.
PROCESS Paraffin, frozen, or celloidin.
MICROTOMY Cut sections at 6 micrometers.
CONTROL Use control number 10A.
SOLUTIONS

Aldehyde Fuchsin

Basic fuchsin	1.0	gm
70% Alcohol	200.0	ml
Concentrated hydrochloric acid....................................	2.0	ml
Paraldehyde	2.0	ml

Let stand at room temperature for 2 to 5 days until stain is deep purple in color. Store in refrigerator. This solution has a pH of 1.7. To adjust the pH of the aldehyde fuchsin to 1.0 use concentrated hydrochloric acid.

0.25% Metanil Yellow

Metanil yellow	0.25	gm
Distilled water	100.0	ml
Glacial acetic acid	0.25	ml

80% Alcohol Solution, pH 1.0

To adjust the pH of the 80% alcoholic rinse, use concentrated hydrochloric acid.

STAINING PROCEDURE

1. Decerate slides in xylene, 2 changes, 2 minutes each. Place in absolute alcohol, 2 changes, 2 minutes each; 95% alcohol, 2 changes, 2 minutes each; then rinse in distilled water.
2. Rinse in 2 changes of 80% alcohol (pH 1.0) for 30 seconds each.
3. Place in aldehyde-fuchsin solution (pH 1.0) for 30 minutes.
4. Rinse off excess stain in 80% alcohol solution (pH 1.0)
5. Place in metanil yellow solution 1 minute or until background is a light yellow.
6. Rinse in distilled water.
7. Dehydrate slides in 95% alcohol and absolute alcohol 3 changes each. Clear in xylene, 3 changes.
8. Mount coverglass with appropriate medium (refractive index 1.48-1.56).

RESULTS

Highly acidic sulfated mucopolysaccharides ....purple

Mowry's Method for Acid Mucopolysaccharides

(See Fig. 159)

FIXATION 10% Buffered neutral formalin.
PROCESS Paraffin, frozen, or celloidin.
MICROTOMY Cut sections at 6 micrometers.
CONTROL Use control number 23.
SOLUTIONS

29% Ferric Chloride

Ferric chloride	29.0	gm
Distilled water	100.0	ml

Muller's Colloidal Iron Oxide (Stock)

Bring 250 ml of distilled water to a boil. While the water is still boiling, add 4.4 ml of 29% ferric chloride solution and stir. When the solution has turned red, remove from heat and allow to cool. Label: Stock Muller's Colloidal Iron. This reagent is stable for one year. In the staining procedures, the stock solution of colloidal iron is diluted just before use as follows:

Colloidal Iron (Working)

Muller's colloidal iron (stock)	20.0	ml
Distilled water	15.0	ml
Glacial acetic acid	5.0	ml

Mix just before use.

12% Acetic Acid

Glacial acetic acid	12.0	ml
Distilled water	88.0	ml

5% Potassium Ferrocyanide

Potassium ferrocyanide	5.0	gm
Distilled water	100.0	ml

5% Hydrochloric Acid

Concentrated hydrochloric acid	5.0	ml
Distilled water	95.0	ml

Potassium Ferrocyanide-Hydrochloric Acid

5% Potassium ferrocyanide	50.0	ml
5% Hydrochloric acid	50.0	ml

Mix just before use.

1% Acid Fuchsin (Stock)

Acid fuchsin	1.0	gm
Distilled water	100.0	ml

1.7% Picric Acid (Stock)

Picric acid	1.7	gm
Distilled water	100.0	ml

Van Gieson (Working)

1% Acid fuchsin (stock)	2.5	ml
1.7% Picric acid (stock)	97.5	ml

STAINING PROCEDURE

1. Decerate slides in xylene, 2 changes, 2 minutes each. Place in absolute alcohol, 2 changes, 2 minutes each; 95% alcohol, 2 changes, 2 minutes each; then rinse in distilled water.
2. Rinse in 12% glacial acetic acid solution for 3 minutes.
3. Place in working colloidal iron solution for 1 hour.
4. Rinse in 12% glacial acetic acid 4 changes, 3 minutes each.
5. Place in potassium ferrocyanide-hydrochloric acid solution for 20 minutes.
6. Wash in tap water for 5 minutes.
7. Rinse in distilled water.
8. Place in Van Gieson's solution for 5 minutes.
9. Dehydrate slides in 95% alcohol and absolute alcohol 3 changes each. Clear in xylene, 3 changes.
10. Mount coverglass with appropriate medium (refractive index 1.48-1.56).

RESULTS

Acidic mucopolysaccharides and acidic epithelial mucins	Prussian blue
Collagen	red
Background	yellow

PAS-Alcian Blue Method for pH 2.5 or 1.0

(See Fig. 161)

FIXATION 10% Buffered neutral formalin.
PROCESS Paraffin, frozen, or celloidin.
MICROTOMY Cut sections at 6 micrometers.
CONTROL Use control number 39B.
SOLUTIONS

1% Periodic Acid
(See Page 306)

Schiff's Reagent
(See Page 128)

Alcian Blue pH 2.5
(See Page 380)

Alcian Blue pH 1.0

Alcian blue, 8GX	1.0	gm
0.1N Hydrochloric acid	100.0	ml

0.5% Sodium Metabisulfite

Sodium metabisulfite	0.5	gm
Distilled water	100.0	ml

STAINING PROCEDURE

1. Decerate slides in xylene, 2 changes, 2 minutes each. Place in absolute alcohol, 2 changes, 2 minutes each; 95% alcohol, 2 changes, 2 minutes each; then rinse in distilled water.
2. Place in alcian blue solution (pH 2.5 or 1.0) for 30 minutes.
3. Blot section dry with filter paper if alcian blue solution pH 1.0 is used. Wash in water after alcian blue pH 2.5 for 5 minutes.
4. Oxidize in 1% periodic acid solution for 10 minutes.
5. Wash in running tap water for 5 minutes.
6. Place in Schiff's solution for 10 minutes.
7. Rinse in sodium metabisulfite solution 3 changes, 2 minutes each.
8. Wash in running tap water for 10 minutes.
9. Dehydrate slides in 95% alcohol and absolute alcohol 3 changes each. Clear in xylene, 3 changes.
10. Mount coverglass with appropriate medium (refractive index 1.48-1.56).

RESULTS

PAS alcian blue pH 2.5: All polysaccharides and mucosubstances containing hexoses or deoxyhexoses with vicinal glycol groups stain magenta to red. Those mucosubstances staining red include neutral mucosubstances. Hyaluronic acid, sialomucins and all but the most strongly acidic sulfated mucosubstances stain blue.

PAS alcian blue pH 1.0: PAS positive same as above. Acidophilic substances at this pH include only the sulfated mucosaccharides.

Amylase Digestion Method

The digestion of glycogen is best accomplished with the use of amylase in place of the long and often used diastase. This is suggested since numerous problems have surfaced in recent years, with the use of diastase to digest glycogen. This problem is directly related to the fact that for some unknown reason, diastase does not digest all the glycogen from a given section. Amylase, on the other hand, works very well and therefore should become the enzyme of choice for digesting glycogen from tissue sections. It should be pointed out that glycogen is removed completely from sections, when saliva is used. It may not be difficult to understand why amylase and saliva work so well if one considers the following fact. Amylase is an enzyme that catalyzes the hydrolysis of starch into smaller molecules. The a-amylases are found in animals and include salivary amylase and pancreatic amylase. These are the a-amylase of the pancreas and a-amylase occurring in the saliva.

On the other hand, diastase is a white amorphous soluble enzyme produced during the germination of seeds, and contained in malt. Diastase of malt converts starch into maltose and then into dextrose, which in turn should dissolve in water.

Personal experience strongly suggests that commercially available amylase and amylase contained in one's saliva, work equally well for dissolving (removing) glycogen from tissue sections. The practice of using saliva to digest glycogen should be used only in emergencies; i.e., when amylase is not available in the laboratory to perform the test. The use of saliva is not sanitary; sections often contain bacteria from the technician's mouth and in general the practice of expectorating on a slide is not appealing to those in the same environment.

SOLUTIONS

Phosphate Buffer Solution pH 6.0

Sodium chloride	8.0	gm
Sodium phosphate, dibasic	0.282	gm
Sodium phosphate, monobasic	1.97	gm
Distilled water	1000.0	ml

Amylase Digestion Solution

Amylase of malt	0.1	gm
Phosphate buffer solution	100.0	ml

Mix just before use. Note: Distilled water can also be used to dissolve the amylase powder.

DIGESTION PROCEDURE

1. Decerate slides in xylene, 2 changes, 2 minutes each. Place in absolute alcohol, 2 changes, 2 minutes each; 95% alcohol, 2 changes, 2 minutes each; then rinse in distilled water.
2. Place slides in amylase solution at 37°C for 30 minutes or room temperature.
3. Wash in running water for 5 minutes.
4. Stain with McManus' method for glycoproteins (PAS) or selected method.

RESULTS

Staining attributable to glycogen is selectively eliminated.

Hyaluronidase Digestion Method

The following provides information for compounding solutions of hyaluronidase of different TR Units.

Man has long taken advantage of enzymatic reactions to improve his well-being without understanding the exact nature of the process; as, for example, the process in baking and in the manufacturing of cheese and alcoholic beverages. It is only in comparatively recent years that specific enzymes have been isolated and their mode of action studied.

The determination of enzyme activity in serum or other body tissues and fluids, has become an indispensable tool of modern diagnosis. Practical experience has shown that expert knowledge of enzyme chemistry is not essential, in order to benefit from its use.

Enzymes are biological catalysts. Like other catalysts, they accelerate chemical reactions or make them possible by lowering the reaction energy requirement. They are protein in nature and can be isolated or sometimes crystallized like other proteins. Like all proteins, enzymes may be denatured and thereby inactivated by physical or chemical means. They are susceptible to heat, pH and microorganisms.

Many factors influence the rate of enzyme reactions. The most important of these are pH ionic strength, temperature, buffer and technical factors, such as inhibitors from contaminated glassware or vigorous shaking which may cause denaturation.

Most enzyme assays utilize 37°C as a reaction temperature. This assures a more natural or physiological condition.

There are various simple precautions which can frequently improve accuracy and precision in clinical enzymology:

1. Glassware contamination is a frequent cause of error.
2. Sample handling presents problems: Since enzymes may be denatured at air-liquid interfaces (due to distortion of the protein molecules), mixing should be done by gentle inversion rather than vigorous shaking.
3. Accurate temperature control during incubation is essential.

With this general background information on enzymes, the following point needs emphasized as it relates to histotechnology. Enzyme activity is expressed in units of activity, not necessarily in grams of enzymes. Therefore, when making a solution utilizing an enzyme, the concentration should not be expressed in terms of weight (grams) but units of activity.

Hyaluronidase can be purchased in different units of activity from 300 to 2,000 units of activity per milligram (300 units/mg or 2,000 units/mg). The activity in terms of units for removing hyaluronic acid from sections should be 150-200 units per milligram. This being the case, one should purchase hyaluronidase in units of activity which are easy to calculate. For example, if the available hyaluronidase is 400 units of activity per milligram, one should weigh out 0.05 gm and dissolve in 100 milliliters (ml) of normal saline or distilled water. The concentration in terms of activity is 200 units per milliliter.
(400 units/mg = 400,000 units/gm); 0.05 gm = 20,000 units.

Example:
(400 units/mg) (1000 mg/gm) 0.05 gm = 20, 000 units
20,000 units/100 ml = 200 units/ml

This general formula can be used to calculate the amount of milligrams or grams necessary for mixing working hyaluronidase solutions with a unit of activity between 150-200 units per milligram.

Hyaluronidase from bovine testes lyophilized powder should be purchased for use in this procedure. The use of hyaluronidase containing various agents such as protein or lactose should not be used.

SOLUTIONS

0.1M Potassium Phosphate Monobasic

Potassium phosphate, monobasic	13.61	gm
Distilled water	1000.0	ml

0.1M Sodium Phosphate, Dibasic

Sodium phosphate, dibasic	14.20	gm
Sodium phosphate, dibasic, 0.1M	1000.0	ml

Buffer

Potassium phosphate, monobasic, 0.1M	94.0	ml
Sodium phosphate, dibasic, 0.1M	6.0	ml

Hyaluronidase

Hyaluronidase	0.5	gm
Buffer solution	100.0	ml

Mix just before use.

DIGESTION PROCEDURE: Use two slides from same block; one for digestion with hyaluronidase and one with buffer solution.

1. Decerate slides in xylene, 2 changes, 2 minutes each. Place in absolute alcohol, 2 changes, 2 minutes each; 95% alcohol, 2 changes, 2 minutes each; then rinse in distilled water.
2. Digest slides in hyaluronidase solution at 37°C for 1 hour. Incubate the duplicate slides in buffer solution at 37°C for 1 hour.
3. Wash both sets of slides in running tap water for 5 minutes.
4. Stain as desired.

Sialidase Digestion Method

SOLUTIONS

Buffer #1

Sodium acetate	26.2	gm
Calcium chloride	8.9	gm
Distilled water	200.0	ml

Buffer #2

Sodium acetate	2.6	gm
Calcium chloride	0.89	gm
Distilled water	200.0	ml

Sialidase Digestion

Sialidase (neuraminidase)*	1.0	ml
Buffer solution #1	0.1	ml

*General Biochemical Co., Chagrin Falls, OH 44022

DIGESTION PROCEDURE: Use two slides from the same block. One for digestion and one for treatment with buffer solution #2 (control).

1. Decerate slides in xylene, 2 changes, 2 minutes each. Place in absolute alcohol, 2 changes, 2 minutes each; 95% alcohol, 2 changes, 2 minutes each; then rinse in distilled water.
2. Circle sections to be digested with diamond hydrophobic pencil or marking pencil to contain enzyme solution.
3. Air-dry slides for 1 hour.
4. Place slides in humidity chamber, or on glass rods just above a layer of water in a petri dish. Place sialidase digestion solution on one slide in sufficient quantity to completely cover section. Cover the second slide, to be used as "control", with buffer solution #2. The humidity chamber or petri dishes are then incubated for 3 hours at 37-40°C.
5. Remove from oven and rinse in distilled water.
6. Stain as desired.

RESULTS
Staining attributable to the sialomucins will be selectively eliminated.

Ono-Katsuyama-Hotchi Method for Elimination of Mucin-Type Glycoprotein[125]

FIXATION 10% Buffered neutral formalin or 4% sodium phosphate buffered paraformaldehyde.
PROCESS Paraffin, frozen, or celloidin.
MICROTOMY Cut section at 3 micrometers; use poly-L-lysine or silanated slides.
CONTROL Use control number 19 of 39B.
SOLUTIONS

Amylase Digestion
(See Page 385)

Periodic Acid-Schiff

Periodic acid	1.0	gm
Distilled water	100.0	ml

Schiff's Reagent
(See Page 128)

0.5M Sodium Hydroxide

Sodium hydroxide	20.0	ml
70% Ethyl Alcohol	1000.0	ml

STAINING PROCEDURE
1. Decerate slides in xylene, 2 changes, 2 minutes each. Place in absolute alcohol, 2 changes, 2 minutes each; 95% alcohol, 2 changes, 2 minutes each; then rinse in distilled water.
2. Place in amylase digestion at 37°C for 8 minutes.
3. Wash in running tap water for 5 minutes.
4. Place in 70% ethyl alcohol for 3 minutes.
5. Place in 0.5M sodium hydroxide in 70% ethyl alcohol at 4°C for 3-7 days.
6. Rinse in 70% cold ethyl alcohol for 3 times.
7. Rinse in distilled water.
8. Stain with periodic acid-Schiff (PAS) technique.
9. Rinse in 0.5% sodium metabisulate, 3 changes, 2 minutes each.
10. Rinse in tap water and distilled water.
11. Place in 95% ethyl alcohol for 1 minute.
12. Place in 100% ethyl alcohol for 1 minute.
13. Place in ethyl ether-100% alcohol solution for 1 minute.
14. Place in absolute ethyl alcohol for 10 minutes.
15. Dehydrate slides in 95% alcohol and absolute alcohol 3 changes each. Clear in xylene, 3 changes.
16. Mount coverglass with appropriate medium (refractive index 1.48-1.56).

RESULTS
O-glycosidically linked glycoproteins (mucin-type glycoprotein) is selectively eliminated.

Culling-Reid-Burton-Dunn Method for Differentiating Lower Gastrointestinal Tract Mucins[126]

Epithelial mucins of the normal terminal ileum, caecum, colon, and rectum of man are unique in that they alone exhibit staining following the periodate-borohydride/KOH/PAS technique. Application of this technique enables one to differentiate those mucin-producing metastases arising from adenocarcinoma of the lower gastrointestinal tract from those arising elsewhere, and may occasionally be useful in determining the site of the primary tumor when it is in doubt. Furthermore, it was found to be especially useful in distinguishing between primary adenocarcinoma of the lung and metastases in the lung from adenocarcinoma of the lower gastrointestinal tract.

FIXATION 10% Buffered neutral formalin.
PROCESS Paraffin, frozen, or celloidin.
MICROTOMY Cut section at 6 micrometers.
CONTROL Use control number 39B.
SOLUTIONS

1% Periodic Acid

Periodic acid	1.0	gm
Distilled water	100.0	ml

Schiff's Reagent
(See Page 128)

2.45 Boric Acid (Solution A)

Boric acid	2.45	gm
Distilled water	100.0	ml

This solution may be stored in the refrigerator.

Sodium Borohydrate (Solution B)

Sodium borohydrate	1.89	gm
Distilled water	167.0	ml

This solution must be made immediately before use. (See step 4 of staining procedure.)

0.5% Potassium Hydroxide

Potassium hydroxide	0.5	gm
70% ethyl alcohol	100.0	ml

STAINING PROCEDURE
1. Decerate slides in xylene, 2 changes, 2 minutes each. Place in absolute alcohol, 2 changes, 2 minutes each; 95% alcohol, 2 changes, 2 minutes each; then rinse in distilled water.
2. Oxidize with 1% periodic acid at room temperature for 60 minutes.
3. Wash in running tap water for 1 minute.
4. Reduce in sodium borohydrate as follows:

Slides are placed in a glass or stainless steel slide holder (of the bottomless variety, not a coplin jar). This is placed in a container which holds 300 ml, but narrow enough for the sections to be completely covered by the addition of 100 ml of solution A. This container must be kept cool by surrounding it with crushed ice. Solution B is then added gradually over a period of 30 minutes and the sections are left for an additional hour; this step should be carried out under a fume hood.

5. Slides are washed in running water for 10-15 minutes.
6. Sections are treated in 0.5% potassium hydroxide in 70% alcohol at room temperature for 30 minutes.
7. Wash in 70% alcohol, then in water.
8. Stain with periodic acid Schiff (PAS) technique.
9. Dehydrate slides in 95% alcohol and absolute alcohol 3 changes each. Clear in xylene, 3 changes.
10. Mount coverglass with appropriate medium (refractive index 1.48-1.56).

RESULTS
Mucin showing any red coloration is interpreted as being KOH/PAS positive, provided the slide treated by the PB/PAS technique (see below) is unstained. Optionally, sections may be lightly counterstained with hematoxylin, or tartrazine in cellosolve, which will visualize the tissue and will aid the orientation of the section; we preferred the former.

Rapid Periodate-Borohydride/KOH/PAS Technique

This method, which utilizes a modification of Lillie and Pizzalato's borohydride aldehyde blocking method, can be used as a routine method. However, if a particular section treated by the periodate borohydride/PAS technique (see below), which is used as a negative control, shows any red color, then the long method (above) should be used.

1. Sections are brought to water and then treated with 1% periodic acid for 30 minutes.
2. Wash in running tap water for 10 minutes.
3. Treat in 0.1% sodium borohydride in 1% disodium hydrogen phosphate for 30 minutes,
4. Wash slides in water.

See steps 5, 6, 7 and 8 and results, in method above.

Periodate Borohydride/PAS Technique (PB/PAS)

This technique is used to insure that normal PAS reactivity has been abolished. The method consists of steps 1, 2, 3 and 4 above, followed by the routine PAS technique performed in parallel with the above method (PB/KOH/PAS). Any staining present in this section will indicate that the periodic acid-borohydride treatment is incomplete due to either (1) insufficient time of oxidation with periodic acid, or (2) incomplete reduction by sodium borohydride. The substitution of simple Schiff's reagent for 30 minutes in place of a full PAS method will help to determine the cause, since a positive result by the latter indicates that the reduction by sodium borohydride is incomplete.

Churukian-Schenk Method for Keratin and Mucin[127]

FIXATION 10% Buffered neutral formalin.
PROCESS Paraffin, frozen, or celloidin.
MICROTOMY Cut section at 5 micrometers.
CONTROL Use control number 23.
SOLUTIONS

Alcian Blue, pH 2.5
(See Page 380)

Alkaline Alcohol
(See Page 366)

Weigert's Iron Hematoxylin, Solution A

Hematoxylin	2.0	gm
95% Alcohol	100.0	ml

Solution B

Ferric chloride, 62% aqueous	4.0	ml
Distilled water	95.0	ml
Hydrochloric acid	1.0	ml

Working Solution
Mix equal parts of solution A and B.

1% Acid Alcohol
(See Page 77)

Erythrosin B

Erythrosin B	1.0	gm
Distilled water	100.0	ml

Alcoholic Metanil Yellow

Metanil yellow	1.5	gm
100% Alcohol	100.0	ml
Filter and add:		
Acetic acid	1.0	ml

STAINING PROCEDURE

1. Decerate slides in xylene, 2 changes, 2 minutes each. Place in absolute alcohol, 2 changes, 2 minutes each; 95% alcohol, 2 changes, 2 minutes each; then rinse in distilled water.
2. Stain in alcian blue solution for 15 minutes.
3. Wash well with tap water and rinse in distilled water.
4. Place in alkaline-alcohol solution for 10 minutes.
5. Wash well with tap water and rinse in distilled water.
6. Stain in Weigert's iron hematoxylin for 1 minute.
7. Wash well with tap water and rinse in distilled water.
8. Decolorize in acid alcohol solution 10 seconds.
9. Wash well with tap water and rinse in distilled water.

10. Stain in erythrosin B solution for 5 minutes.
11. Rinse quickly in distilled water.
12. Dip in 2 changes of 95% alcohol followed by 2 changes of absolute alcohol.
13. Counterstain in alcoholic metanil yellow for 30 seconds.
14. Absolute alcohol, 4 changes.
15. Clear in xylene, 3 changes each.
16. Mount coverglass with appropriate medium (refractive index 1.48-1.56).

RESULTS

Keratin .... red
Acid mucopolysaccharides .... blue to bluish-green
Nuclear chromatin .... black
Collagen .... orange
Muscle .... various shades of red and orange
Erythrocytes .... red

Gowali's Method for Sulfated Mucosubstances[128]

FIXATION 10% Buffered neutral formalin.
PROCESS Paraffin, frozen, or celloidin.
MICROTOMY Cut section at 5 micrometers.
CONTROL Use control number 10A.
SOLUTIONS

0.5% Pyronin Y (Stock)

Pyronin Y	0.5	gm
Distilled water	100.0	ml

10% Ferric Ammonium Sulphate (Stock)

Ferric ammonium sulfate	10.0	gm
Distilled water	100.0	ml

Ferric chloride may be substituted for ferric ammonium sulfate.

Sulfate-Pyronin Y (Working)

0.5% Pyronin Y	300.0	ml
10% Ferric ammonium sulphate	20.0	ml

STAINING PROCEDURE

1. Decerate slides in xylene, 2 changes, 2 minutes each. Place in absolute alcohol, 2 changes, 2 minutes each; 95% alcohol, 2 changes, 2 minutes each; then rinse in distilled water.
2. Sections are immersed in working ferric ammonium sulfate-pyronin Y solution for 6 hours for sulfated acid mucin and for 5 to 10 minutes for mast cell granules.
3. Rinse well with distilled water.
4. Dehydrate slides in 95% alcohol and absolute alcohol 3 changes each. Clear in xylene, 3 changes.
5. Mount coverglass with appropriate medium (refractive index 1.48-1.56).

RESULTS

Sulfated acid mucin .... red
Mast cell granules .... bright red
Nuclear chromatin .... faint pink

Churukian-Schenk Congo Red Method for Amyloid[129]

FIXATION 10% Buffered neutral formalin.
PROCESS Paraffin or glycol methacrylate.
MICROTOMY Cut section at 6 micrometers (Paraffin) 2.5 for glycol methacrylate.
CONTROL Use control number 2.
SOLUTIONS

Modified Weigert's Iron Hematoxylin
Solution A

Hematoxylin (CI 75290)	2.0	gm
95% Alcohol	100.0	ml

Solution B

Ferric chloride, 62% aqueous	4.0	ml
Distilled water	95.0	ml
Concentrated hydrochloric acid	1.0	ml

To make working solution mix equal parts of solutions A and B.

0.5% Acid Alcohol

70% Alcohol	100.0	ml
Concentrated hydrochloric acid	0.5	ml

0.1M Glycine (Stock)

Glycine	7.51	gm
Distilled water	1000.0	ml

0.1M Sodium Chloride (Stock)

Sodium chloride	5.85	gm
Distilled water	1000.0	ml

0.1M Sodium Hydroxide (Stock)

Sodium hydroxide	4.0	gm
Distilled water	1000.0	ml

Buffer Solution, pH 10.0

0.1M Glycine (stock)	30.0	ml
0.1M Sodium chloride (stock)	30.0	ml
0.1M Sodium hydroxide (stock)	40.0	ml

Adjust the pH to 10.0 with 1M sodium hydroxide or 1M hydrochloric acid.

Buffered Congo Red Solution

Congo red	0.5	gm
Buffer solution, pH 10.0	50.0	ml
Absolute alcohol	50.0	ml

Dissolve the congo red in the buffer solution; then add the absolute alcohol. This solution is stable for at least six months at room temperature.

STAINING PROCEDURE

1. Decerate slides in xylene, 2 changes, 2 minutes each. Place in absolute alcohol, 2 changes, 2 minutes each; 95% alcohol, 2 changes, 2 minutes each; then rinse in distilled water. For glycol methacrylate sections, start at step 2.
2. Stain in modified Weigert's iron hematoxylin for 1 minute.
3. Wash well with running tap water and rinse in distilled water.
4. Decolorize in acid alcohol solution for 30 seconds.
5. Wash well with running tap water and rinse in distilled water.
6. Place in 50% alcohol for 30 seconds.
7. Stain in buffered congo red solution for 20 seconds.
8. Place in 70% alcohol for 2 seconds.
9. Dehydrate slides in 95% alcohol and absolute alcohol 3 changes each. Clear in xylene, 3 changes.
10. Mount coverglass with appropriate medium (refractive index 1.48-1.56).

RESULTS

Amyloid .... pink to red
Nuclear chromatin .... black

When examined with polarization optics:

Amyloid .... green birefringence

When viewed with a fluorescence microscope:

Amyloid .... fluoresce orange-yellow to red

REMARKS

The optimal way to look for congo red stained amyloid is by fluorescence.

Attwood's Method for Amniotic Fluid Embolism

(See Fig. 164)

FIXATION	10% Buffered neutral formalin.
PROCESS	Paraffin, frozen, or celloidin.
MICROTOMY	Cut sections at 6 micrometers.
SOLUTIONS	

1% Alcian Green

Alcian green	1.0	gm
Distilled water	100.0	ml
Glacial water	2.0	ml

Phloxine-Calcium Chloride

Phloxine	0.5	gm
Calcium chloride	0.5	gm
Distilled water	100.0	ml

Tartrazine-Cellosolve

Tartrazine	2.5	gm
Cellosolve (ethylene glycol monoethyl ether)	100.0	ml

Scott's Tap Water
(See Page 78)

0.5% Oxalic Acid
(See Page 242)

STAINING PROCEDURE

Use control slide.

1. Decerate slides in xylene, 2 changes, 2 minutes each. Place in absolute alcohol, 2 changes, 2 minutes each; 95% alcohol, 2 changes, 2 minutes each; then rinse in distilled water.
2. Place in alcian green solution for 2 minutes.
3. Rinse in distilled water.
4. Place in potassium permanganate solution for 2 minutes.
5. Rinse in distilled water.
6. Place in oxalic acid solution for 2 minutes.
7. Rinse in distilled water.
8. Place in Mayer's hematoxylin solution for 3 minutes.
9. Rinse and blue in Scott's water solution for 5 minutes.
10. Place in phloxine solution for 30 minutes.
11. Rinse in distilled water, then in cellosolve.
12. Differentiate each slide with tartrazine-cellosolve solution, checking until the phloxine is removed from erythrocytes.
11. Dehydrate slides in 95% alcohol and absolute alcohol 3 changes each. Clear in xylene, 3 changes.
12. Mount coverglass with appropriate medium (refractive index 1.48-1.56).

RESULTS

Nuclei	blue-black
Mucin, ground substance of cartilage, granules of certain mast cells	green
Squamous epithelium, fibrin	red
Erythrocytes, plasma, collagen	yellow
Inclusion bodies	red

REMARKS

This method has been used very successfully for the demonstration of inclusion bodies. However, differentiation at step 12 must be rigidly controlled.

Note: The permanganate-oxalic bleach assists in the subsequent differentiation of the phloxine and tends to remove any background staining by alcian green. Any excess staining with tartrazine can be removed by rinsing in water until the desired intensity is obtained.

Hayama-Katsuyama-Nakayama-Akamatsu-Honda Method for Epithelial Cells of the Gastric Mucosa[130]

FIXATION Cold phosphate buffered 4% paraformaldehyde for 24 hours.
PROCESS Paraffin.
MICROTOMY Cut section at 3 micrometers.
CONTROL Use control number 48 and 52.
SOLUTIONS

1N Hydrochloric Acid (Stock)

Hydrochloric acid	83.5	ml
Distilled water	916.5	ml

2% Schiff's

Concentrated hydrochloric acid	8.0	ml
Sodium sulfite	5.0	gm
Basic fuchsin	4.0	gm
Distilled water	192.0	ml

Stir for 3 to 4 hours until solution is pale brown. Add 0.5 gm activated charcoal. Filter through double filter paper.

2.5% Sodium Metabisulfate

Sodium metabisulfite	2.5	gm
1N Hydrochloric acid	100.0	ml

1% Biebrich Scarlet

Biebrich scarlet	2.5	gm
Distilled water	250.0	ml

2% Ethyl Violet

Ethyl violet	10.0	gm
Distilled water	500.0	ml

Bowie Stain (Stock)

Biebrich scarlet-ethyl violet powder	0.2	gm
95% Ethyl alcohol	20.0	ml

Bowie Stain (Working)

Bowie stain (stock)	5	drops
20% Ethyl alcohol	50.0	ml

1% Periodic Acid

Periodic acid	1.0	gm
Distilled water	100.0	ml

Sodium Borohydride-Sodium Metaphosphate

Sodium borohydride	0.2	gm
Sodium metaphosphate	1.0	gm
Distilled water	100.0	ml

PBS Buffer

0.1M phosphate buffer, pH 7.4; 0.2M with respect to NaCl

Concanavalin A

Concanavalin A*	0.5	mg
PBS buffer	1.0	ml

Be sure to make sufficient solution needed for staining.

*Type IV, Sigma, St. Louis, MO, 63178.

Horseradish Peroxidase

Horseradish peroxidase*	0.5	mg
PBS buffer	100.0	ml

*Type VI, Sigma, St. Louis, MO, 63178.

Diaminobenzidine (DAB)

2 mg 3,3-diaminobenzidine tetrahydrochloride/10 ml in PBS containing 1 ul/10 ml of hydrogen peroxide

STAINING PROCEDURE

I.Modified Bowie's Stain (Feulgen-Bowie)

1. Decerate slides in xylene, 2 changes, 2 minutes each. Place in absolute alcohol, 2 changes, 2 minutes each; 95% alcohol, 2 changes, 2 minutes each; then rinse in distilled water.
2. Place slides in preheated 1N hydrochloric acid at 60°C for 10 minutes.
3. Wash in running tap water for 5 minutes and place in distilled water.
4. Place in 2% Schiff's reagent for 10 minutes.
5. Wash with 2.5% sodium metabisulfite 3 times, 3 minutes each.
6. Wash in running tap water for 5 minutes.
7. Immerse in 20% alcohol for 5 minutes.
8. Place in Bowie's stain for 15 hours at room temperature.
9. Wipe off excess fluid with filter paper and rinse in 100% acetone.
10. Differentiate in 98% acetone until only zymogen granules stand out (this usually takes 2-3 minutes).
11. Dehydrate with 100% acetone and clear in xylene.
12. Mount coverglass with appropriate medium (refractive index 1.48-1.56).

II.Feulgen Hydrolysis - PC5 - Bowie Stain

1. Decerate and hydrate sections.
2. Place slides in preheated 1N hydrochloric acid 60°C for 10 minutes.
3. Wash in running tap water for 5 minutes and place in distilled water.
4. Oxidize in 1% periodic acid for 30 minutes.
5. Rinse in running tap water for 5 minutes and place in distilled water.
6. Reduce in sodium borohydride-sodium metaphosphate for 2 minutes.
7. Wash in running tap water for 5 minutes and place in distilled water.
8. Rinse briefly in 0.1M phosphate buffer, pH 7.4; 0.2M with respect to NaCl (PBS).
9. Incubate in concanavalin A for 30 minutes at room temperature.
10. Rinse in PBS 3 times, 5 minutes each.
11. Incubate in horseradish peroxidase for 20 minutes.
12. Rinse in PBS 3 times, 5 minutes each.

13. Incubate for 5 minutes in DAB (3,3-diaminobenzidine tetrahydrochloride 2 mg/10 ml in 0.1M phosphate buffer, pH 7.4 containing 1 ul/10 ml of hydrogen peroxide).
14. Rinse in running tap water for 5 minutes and then place in Bowie's solution.
15. Follow steps 8-11 of Procedure I.

III. Feulgen Hydrolysis-PAS-Concanavalin A-Bowie Stain
1. Decerate, hydrate, hydrolyze and oxidize as in procedure II.
2. Place in 2% Schiff reagent for 10 minutes.
3. Rinse with 2.5% sodium metabisulfite 3 times, 3 minutes each.
4. Follow steps 7-15 of Procedure II.

RESULTS

I. Feulgen-Bowie Stain

Zymogen granules of chief cells	deep blue
Surface mucous cells, mucous neck cells, parietal cells	unstained
Granules of Paneth and mast cells	deep blue
Nuclei	purplish-red

II. Feulgen Hydrolysis-PCS-Bowie Stain

Zymogen granules of chief cells	deep blue
Mucous granules of mucous neck cells	dark brown
Surface mucous cells and parietal cells	unstained
Granules of Paneth and mast cells	deep blue
Nuclei	pale blue or unstained

III. Feulgen Hydrolysis-PAS-Concanavalin A-Bowie Stain

Zymogen granules of chief cells	deep blue to purplish-blue
Mucous granules of mucous neck cells	reddish-brown
Surface mucous cells	purplish-red
Parietal cells	unstained
Granules of Paneth and mast cells	deep blue
Nuclei	purplish-red

Chapter 13

Connective Tissues, Muscle and Muscle Entities

Puchtler-Sweat Method for Basement Membrane[55ai]

FIXATION	Modified Carnoy's solution (see below).
PROCESS	Paraffin, frozen, or celloidin.
MICROTOMY	Cut sections at 6 micrometers.
CONTROL	Use control number 6.
SOLUTIONS	

Modified Carnoy's Solution

100% Alcohol	60.0	ml
Chloroform	30.0	ml
Glacial acetic acid	10.0	ml

0.5% Periodic Acid
(See Page 307)

Sodium Bisulfite

Sodium bisulfite	20.0	gm
Distilled water	40.0	ml
100% Alcohol	10.0	ml

Resorcin-Fuchsin (Stock)

Basic fuchsin	2.0	gm
Resorcinol	4.0	gm
Distilled water	200.0	ml

Mix in a porcelain dish, bring to a boil and let boil for 1 minute, then add 25 ml of 29% ferric chloride solution. Cool, filter and leave precipitate on filter paper until dry. Return to porcelain dish, which should be dry but still contains whatever part of the precipitate remains adherent to it. Add 200 ml of 95% alcohol and heat very carefully dissolving all precipitate from filter paper before discarding it. Add 4 ml of hydrochloric acid. This solution keeps for several months.

Resorcin-Fuchsin (Working)

Resorcin-fuchsin (stock)	10.0	ml
70% Alcohol	100.0	ml
Concentrated hydrochloric acid	2.0	ml

Kernechtrot (Nuclear Fast Red)
(See Page 84)

STAINING PROCEDURE

1. Decerate slides in xylene, 2 changes, 2 minutes each. Place in absolute alcohol, 2 changes, 2 minutes each; 95% alcohol, 2 changes, 2 minutes each; then rinse in distilled water.
2. Oxidize in periodic acid for 5 minutes.
3. Rinse in distilled water, 2 changes.
4. Place in sodium bisulfite solution for 5 minutes.
5. Rinse in distilled water, 5 changes.
6. Stain in resorcin-fuchsin solution for 4 hours.
7. Rinse in distilled water, 3 changes.
8. Stain nuclear chromatin in nuclear fast red (Kernechtrot) for 5 minutes.
9. Rinse in distilled water, 3 changes.
10. Dehydrate slides in 95% alcohol and absolute alcohol 3 changes each. Clear in xylene, 3 changes.
11. Mount coverglass with appropriate medium (refractive index 1.48-1.56).

RESULTS

Basement membranes .... black in cross sections, gray in tangential sections
Nuclear chromatin .... pink to red

Jones' Method for Basement Membrane[55aj]

(See Fig. 165)

FIXATION 10% Buffered neutral formalin, Bouin's or Zenker's solution.
PROCESS Paraffin, frozen, or celloidin.
MICROTOMY Cut sections at 2 micrometers.
CONTROL Use control number 6.
SOLUTIONS

0.5% Periodic Acid
(See Page 307)

3% Methenamine (Stock)

Methenamine	3.0	gm
Distilled water	100.0	ml

5% Silver Nitrate (Stock)

Silver nitrate	5.0	gm
Distilled water	100.0	ml

Borate Buffer (Stock)
0.2M Boric Acid (Solution A)

Boric acid	12.36	gm
Distilled water	1000.0	ml

0.25M Sodium Borate (Solution B)

Sodium borate	19.07	gm
Distilled water	1000.0	ml

Borate Buffer, pH 8.2 (Working)

Solution A	6.5	ml
Solution B	3.5	ml

Methenamine-Silver, pH 8.2 (Working)

3% Methenamine (stock)	42.5	ml
5% Silver nitrate (stock)	2.5	ml
Borate buffer, pH 8.2 (working)	12.0	ml

Prepare fresh just before use and filter. This solution is stable for approximately 60-75 minutes. After this time, it begins to break down, producing a black precipitate which is picked up on the slides.

1% Gold Chloride (Stock)

Gold chloride	1.0	gm
Distilled water	100.0	ml

Gold Chloride (Working)

Gold chloride (stock)	10.0	ml
Distilled water	40.0	ml

Solution is stable for approximately 100 slides.

3% Sodium Thiosulfate (Hypo)

Sodium thiosulfate	3.0	gm
Distilled water	100.0	ml

STAINING PROCEDURE

Chemically clean glassware must be used.

Note: It is absolutely essential that all glassware be acid cleaned with concentrated nitric acid and rinsed in several changes of chloride-free distilled water. Distilled water may be checked for free chloride by the addition of several drops of 5% silver nitrate solution. If a white cloud appears upon the addition of the silver nitrate, the sample contains chlorides and must not be used. (Water for injection, USP is chloride free.)

1. Decerate slides in xylene, 2 changes, 2 minutes each. Place in absolute alcohol, 2 changes, 2 minutes each; 95% alcohol, 2 changes, 2 minutes each; then rinse in distilled water.
2. Oxidize in periodic acid solution for 10 minutes.
3. Rinse in chloride-free distilled water.
4. Filter freshly prepared methenamine-silver solution (working) into coplin jar or suitable staining dish.
5. Place slides in methenamine-silver solution and then place coplin jar in prewarmed 70°C water bath. Start timing at this point, approximately 60-75 minutes. Check under microscope when slides show a medium brown color.

Note: Solution and slides should be allowed to come to 70°C together. While slides are in the silver solution they may be examined after they begin to show macroscopically, a medium brown color reaction. Before checking under the microscope, rinse slides in hot 70°C chloride-free distilled water; check slides, return to hot-water rinse and then return to hot staining solution. Slides should be checked every 10 minutes once they have reached the dark or medium brown stage. Slides should be checked as rapidly as possible because uneven staining will occur if the sections cool. When the desired staining time has been reached, the slides should be checked as above every 1 to 2 minutes. Strict adherence to the timing is essential in order to obtain a uniform consistency in staining. A properly stained section should have a dark brownish-yellow background; the reticulum fibers will be an intense black, as will the basement membranes. An overstained section will be too black, making differentiation very difficult, as the black will be so intense as to obsure many or all of the tissue elements. The section may be destained with an extremely dilute solution of potassium ferricyanide, one or two dips.

6. Rinse section well in distilled water.
7. Tone in working gold chloride solution for 1 minute.
8. Rinse well in distilled water.
9. Place in sodium thiosulfate solution for 1 minute.
10. Wash in running tap water for 10 minutes.
11. Rinse well in distilled water.
12. Counterstain with a routine progressive hematoxylin and eosin stain or desired stain.
13. Dehydrate slides in 95% alcohol and absolute alcohol 3 changes each. Clear in xylene, 3 changes.
14. Mount coverglass with appropriate medium (refractive index 1.48-1.56).

RESULTS

Basement membrane.. black
Reticulum fibers.. black
Nuclei .... blue
Cytoplasm... pink to orange

Avallone's Modifications of Jones' Method for Basement Membranes[131]

FIXATION 10% Buffered neutral formalin, Bouin's or Zamboni's solution.
PROCESS Paraffin, frozen, or celloidin.
MICROTOMY Cut sections at 2 micrometers.
CONTROL Use control number 6.
SOLUTIONS

0.5% Periodic Acid
(See Page 307)

3% Methenamine (Stock)
(See Page 197)

5% Silver Nitrate (Stock)
(See Page 197)

0.05M Sodium Borate (Stock)

Sodium borate	19.07	gm
Distilled water, q.s. to	1000.0	ml

1% Gold Chloride (Stock)

Gold chloride	1.0	gm
Distilled water	100.0	ml

Gold Chloride (Working)

1% Gold chloride (stock)	10.0	ml
Distilled water	40.0	ml

Solution is stable for approximately 50 slides.

3% Sodium Thiosulfate

Sodium thiosulfate	3.0	gm
Distilled water	100.0	ml

Methenamine Silver pH 9.15 - 9.2 (Working)

3% Methenamine (stock)	42.5	ml
5% Silver nitrate (stock)	2.5	ml
0.05M Sodium borate (stock)	7.0	ml

Prepare this solution just before use; adjust pH with 1N sodium hydroxide, (approximately 0.5-1 ml) and check with glass electrode pH meter. This solution is stable for approximately 60-70 minutes; after this point, breakdown of silver will ensue, producing a black precipitate on the slides.

STAINING PROCEDURE

Note: It is absolutely essential that all glassware be acid-cleaned with concentrated nitric acid and rinsed in several changes of chloride-free distilled water. Distilled water may be checked for chloride ions by the addition of several drops of 5% silver nitrate solution. If a white cloud appears upon the addition of the silver nitrate, the sample contains chlorides and must not be used. (Water for injection, USP is chloride free.)

1. Place several coplin jars filled with distilled water into a water bath filled with tepid water. Allow the temperature of water bath to reach 68° to 70°C and hold at that temperature. Place a thermometer in one coplin jar to monitor water temperature. Coplin jars will crack if placed directly into hot water.
2. Decerate slides in xylene, 2 changes, 2 minutes each. Place in absolute alcohol, 2 changes, 2 minutes each; 95% alcohol, 2 changes, 2 minutes each; then rinse in distilled water.
3. Oxidize in periodic acid for 10 minutes.
4. Rinse well in 3 changes of chloride-free distilled water, 20 seconds each.
5. Place the slides in the coplin jars in the heated distilled water. Bring them to water bath temperature; approximately 3 minutes. (Slides may be left for longer periods with no danger to sections.)
6. This is the most critical step. Temperature, time and pH determine the results. Therefore, the directions must be followed conscientiously. Prepare a working batch of methenamine-silver solution. Use a glass electrode, pH meter and adjust the pH from 9.15 to 9.2 with 1N sodium hydroxide. Prepare a sufficient quantity for the volume of slides; approximately 35 ml of solution is needed per coplin jar containing 9 slides. Filter solutions before use.
7. Pour 35 ml of the methenamine-silver working solution into a 250 ml Erlenmeyer flask and swirl it around in the water bath to bring solution to the required working temperature; approximately 1-2 minutes. Pour off the distilled water from a coplin jar filled with slides. Replace with hot methenamine-silver solution. Start timer. If there are several coplin jars filled with slides, allow a 5 minute interval before adding methenamine-silver to each succeeding coplin jar. Repeat for each unit of

slides. Allow ample time for checking slides. Do not attempt to stain by mass production techniques since a quality control cannot be maintained. Each slide must be checked under the microscope.

8. After approximately 8 minutes in methenamine-silver, the sections will appear yellow. Times may vary slightly. Check microscopically. CAUTION. Rinse the slide in HOT DISTILLED water before examining and also before returning the slide to the HOT methenamine-silver solution. This prevents contamination and brings the section back to proper staining temperature. The first component to stain will be the capsular basement membrane. It will appear yellow. The glomerular basement membrane will also become yellow, 1 to 2 minutes later. After 12 minutes exposure to methenamine-silver, check the slide microscopically at intervals of 1 minute. When the capsular basement membrane is black and clearly outlined and the glomerular basement membrane is a finely delineated black line, remove the section from methenamine-silver and place in distilled water. Times may vary slightly due to variations in pH, temperature, section thickness, and fixation. It may be necessary to pull some slides earlier or later depending on different staining characteristics of the tissues.
9. Tone in the working gold chloride solution for 10-15 seconds or until background is clear. Check each slide under the microscope.
   CAUTION: DO NOT OVERTONE. If sections are overtoned, place in 3% sodium metabisulfite for 1 to 3 minutes, checking periodically.
10. Rinse well in distilled water.
11. Place in sodium thiosulfate solution for 2-3 minutes.
12. Wash in tepid running tap water for 10 minutes.
13. Rinse well in distilled water.
14. Counterstain with Harris' hematoxylin and eosin (approximately 2 minutes each) or desired stain.
15. Dehydrate slides in 95% alcohol and absolute alcohol 3 changes each. Clear in xylene, 3 changes.
16. Mount coverglass with appropriate medium (refractive index 1.48-1.56).

RESULTS

Basement membrane.. black
Reticulum fibers.. black
Nuclear chromatin .... blue
Cytoplasm, RBC's, collagen, connective tissue................ pink to orange

Lillie's Method for Basement Membrane, Muscle and Arteriosclerotic Lesions[132]

FIXATION 10% Buffered neutral formalin.
PROCESS Paraffin, frozen, or celloidin.
MICROTOMY Cut sections at 6 micrometers.
CONTROL Use control number 6.
SOLUTIONS

Schiff's Solution
(See Page 128)

0.5% Periodic Acid

Periodic acid	0.5	gm
Distilled water	100.0	ml

0.5% Sodium Metabisulfite

Sodium metabisulfite	0.5	gm
Distilled water	100.0	ml

Weigert's Iron-Hematoxylin
(See Page 79)

Picric Acid-Methyl Blue

Picric acid, saturated aqueous	100.0	ml
Methyl blue	40.0	mg

STAINING PROCEDURE

1. Decerate slides in xylene, 2 changes, 2 minutes each. Place in absolute alcohol, 2 changes, 2 minutes each; 95% alcohol, 2 changes, 2 minutes each; then rinse in distilled water.
2. Oxidize in periodic acid for 10 minutes.
3. Wash in running tap water for 5 minutes.
4. Stain in Schiff's solution for 10 minutes.
5. Rinse in sodium metabisulfite solution, 2 changes, 2 minutes each.
6. Wash in running tap water for 10 minutes.
7. Stain in working Weigert's iron hematoxylin for 2 minutes.
8. Wash in running tap water for 10 minutes.
9. Stain in picric acid-methyl blue for 6 minutes.
10. Dehydrate slides in 95% alcohol and absolute alcohol 3 changes each. Clear in xylene, 3 changes.
11. Mount coverglass with appropriate medium (refractive index 1.48-1.56).

RESULTS

Nuclear chromatin	black, gray, or brown
Cytoplasm, muscle cells	gray, green to greenish-yellow
Reticulum and collagen	blue
Basment membranes	red

REMARKS

This procedure is valuable for distinguishing between basement membrane and reticulum fibers and the demonstration of arteriosclerotic lesions.

Luna-Ishak Method for Bile Canaliculi[55ak]

(See Fig. 178)

FIXATION	10% Buffered neutral formalin.
PROCESS	Paraffin, frozen, or celloidin.
MICROTOMY	Cut sections at 6 micrometers.
CONTROL	Use control number 7.

SOLUTIONS

Celestine Blue B

Celestine blue B	0.5	gm
Ferric ammonium sulfate (iron alum)	5.0	gm
Concentrated hydrochloric acid	2.0	ml
Glycerin	15.0	ml
Distilled water	85.0	ml

Dissolve the iron alum in the distilled water, then dissolve the celestine blue in this solution and bring to a boil for 3-5 minutes. Cool and add the glycerin and hydrochloric acid.

1% Phosphotungstic Acid

Phosphotungstic acid	1.0	gm
Distilled water	100.0	ml

1% Acid Fuchsin (Stock)

Acid fuchsin	1.0	gm
Distilled water	100.0	ml

1% Oxalic Acid (Stock)

Oxalic acid	1.0	gm
Distilled water	100.0	ml

Acid Fuchsin-Oxalic Acid (Working)

Acid fuchsin, (stock)	10.0	ml
Oxalic acid, (stock)	1.0	ml
Distilled water	80.0	ml

STAINING PROCEDURE

1. Decerate slides in xylene, 2 changes, 2 minutes each. Place in absolute alcohol, 2 changes, 2 minutes each; 95% alcohol, 2 changes, 2 minutes each; then rinse in distilled water.
2. Stain in celestine blue B for 30 minutes.
3. Rinse in distilled water.
4. Immerse in 1% phosphotungstic acid solution for 15 minutes.
5. Wash in running tap water for 3 minutes.
6. Stain in working acid fuchsin solution for 8 minutes.
7. Dehydrate slides in 95% alcohol and absolute alcohol 3 changes each. Clear in xylene, 3 changes.
8. Mount coverglass with appropriate medium (refractive index 1.48-1.56).

RESULTS

Bile canaliculi	pink to red
Nuclear chromatin	blue
Bile	yellowish-green
Iron	brown to red
Background	bluish-pink

REMARKS

Water and alcohol will remove red dye on prolonged exposure; thus, the reason for rapid dehydration.

Van Gieson's Method for Collagen Fibers[55al]

FIXATION 10% Buffered neutral formalin.
PROCESS Paraffin, frozen, or celloidin.
MICROTOMY Cut sections at 6 micrometers.
CONTROL Use control number 11.
SOLUTIONS

Weigert's Iron-Hematoxylin
(See Page 79)

Van Gieson's

1% Aqueous acid fuchsin	2.5	ml
Saturated aqueous picric acid	97.5	ml

STAINING PROCEDURE

1. Decerate slides in xylene, 2 changes, 2 minutes each. Place in absolute alcohol, 2 changes, 2 minutes each; 95% alcohol, 2 changes, 2 minutes each; then rinse in distilled water.
2. Stain nuclear chromatin in Weigert's hematoxylin solution for 10 minutes.
3. Wash in distilled water for 1 minute.
4. Stain in Van Gieson's for 3 minutes.
5. Dehydrate slides in 95% alcohol and absolute alcohol 3 changes each. Clear in xylene, 3 changes.
6. Mount coverglass with appropriate medium (refractive index 1.48-1.56).

RESULTS

Collagen	red
Muscle	yellow
Cornified epithelium, if present	yellow
Nuclear chromatin	blue to black

Miller's Method for Elastica[133]

(See Fig. 167)

This method fulfills the following requirements.

1. Precise demonstration of both coarse and fine elastica fibers with high selectivity, eliminating the need for differentiation.
2. Intense black staining to provide maximum contrast with the usual counterstains, especially Van Gieson.
3. A stable staining solution.
4. Good staining results possible within a short time.

FIXATION 10% Buffered formol saline; formol sublimate; Heidenhain's Susa; Helly, Zenker, Clarke and Bouin's solution. Cryostat sections fixed in Wolman's fixation also give good results.

SOLUTIONS

Elastica Staining Solution

Victoria blue 4R	1.0	gm
New fuchsin	1.0	gm
Crystal violet	1.0	gm

Dissolve above dyes in 200 ml of hot distilled water, then add in the following order:

Resorcin	4.0	gm
Dextrin	1.0	gm
30% Ferric chloride (freshly prepared)	50.0	ml

Boil solution for 5 minutes, then filter while hot. Transfer precipitate plus filter paper to original beaker and redissolve in 200 ml of 95% alcohol. Boil (on a hot plate or in a water bath) for 15 to 20 minutes. Filter and make up to 200 ml with 95% alcohol. Finally, add 2 ml of concentrated hydrochloric acid.

0.5% Potassium Permanganate

Potassium permanganate	0.5	gm
Distilled water	100.0	ml

5% Oxalic Acid

Oxalic acid	5.0	gm
Distilled water	100.0	ml

Van Gieson

Acid fuchsin	0.1	gm
Picric acid, saturated solution	100.0	ml

STAINING PROCEDURE

1. Decerate slides in xylene, 2 changes, 2 minutes each. Place in absolute alcohol, 2 changes, 2 minutes each; 95% alcohol, 2 changes, 2 minutes each; then rinse in distilled water.
2. Oxidize in 0.5% potassium permanganate for 5 minutes.
3. Rinse in distilled water for 1 minute.
4. Decolorize in 5% oxalic acid for 2 minutes.
5. Rinse in distilled water.
6. Rinse in 95% alcohol.
7. Place in elastica staining solution for 3 hours.
8. Wash slides in 95% alcohol to remove excess stain for 1 minute.
9. Wash slides in distilled water.
10. Stain cytoplasm in Van Gieson for 2 minutes.
11. Dehydrate slides in 95% alcohol and absolute alcohol 3 changes each. Clear in xylene, 3 changes.
12. Mount coverglass with appropriate medium (refractive index 1.48-1.56).

RESULTS

Elastica	black
Mast cell granules	black
Muscle	yellow
Collagen	red

REMARKS

The method described stains elastica selectively and precisely and no differentiation is required. The staining reagent does not deteriorate and may be used repeatedly. Reliable results are obtained on tissues fixed in a wide variety of fixatives. Counterstaining with Van Gieson or Masson trichrome yields excellent results.

Luna's Method for Demonstrating Bone Canaliculi[134]

(See Fig. 170)

The following method can be used successfully for demonstrating a variety of entities, including bone canaliculi and cement lines. Bone specimens which have been exposed to formalin fixation for extended periods of time (2 months or longer) may not demonstrate the canaliculi well. As with other Giemsa-type stains, this procedure demonstrates other entities, including bacteria, mast cells and cartilage.

SOLUTIONS

Giemsa (Stock)

Giemsa powder	1.0	gm
Glycerin, A.C.S.	66.0	ml
Methyl alcohol	66.0	ml

Mix glycerin and Giemsa powder. Place in a 60°C oven for 2 hours. Let cool and add the methyl alcohol.

Buffer, pH 6.0

Potassium phosphate, monobasic	5.0	gm
Sodium phosphate, dibasic	1.0	gm
Distilled water	1000.0	ml

Giemsa (Working)

Giemsa solution (stock)	1.0	ml
Buffer solution, pH 6.0	50.0	ml

Alcoholic Ammonia

60% Methyl alcohol	95.0	ml
Concentrated ammonium hydroxide.	5.0	ml

STAINING PROCEDURE

1. Decerate slides in xylene, 2 changes, 2 minutes each. Place in absolute alcohol, 2 changes, 2 minutes each; 95% alcohol, 2 changes, 2 minutes each; then rinse in distilled water.
2. Place slides in alcoholic ammonia solution for 5 minutes.
3. Stain slides in working Giemsa solution for 45 minutes.
4. Rinse slides briefly in distilled water.
5. Dehydrate slides in 95% alcohol and absolute alcohol 3 changes each. Clear in xylene, 3 changes.
6. Mount coverglass with appropriate medium (refractive index 1.48-1.56).

RESULTS

Bone background	bluish-purple
Canaliculi	reddish-purple
Cement lines	blue
Cartilage	reddish-purple

Fraser-Lendrum Method for Fibrin, Keratin and Cytoplasmic Granules[55am]

FIXATION	Zenker's solution or 10% buffered neutral formalin (see step 2).
PROCESS	Paraffin, frozen, or celloidin.
MICROTOMY	Cut sections at 6 micrometers.
CONTROL	Use control number 17 or 27.
SOLUTIONS	

Celestine Blue B

Ferric ammonium sulfate	2.5	gm
Distilled water	50.0	ml

Dissolve overnight at room temperature.
Add:

Celestine blue B	0.25	gm

Boil for 3 minutes. Filter when cool and add 7 ml glycerin.

Mayer's Hematoxylin
(See Page 81)

Orange G-Picric Acid (Stock)

Saturated picric acid in 80% alcohol	200.0	ml
Orange G	0.4	gm

1% Acid Fuchsin

Acid fuchsin	1.0	gm
Distilled water	99.0	ml
Glacial acetic acid	1.0	ml

Differentiating Solution

Orange G - picric acid (stock)	30.0	ml
80% Alcohol	70.0	ml

Phosphotungstic-Picric Acid (Stock)

Phosphotungstic acid	25.0	gm
Picric acid	2.5	gm
95% Alcohol	100.0	ml

Phosphotungstic-Picric Acid (Working)

Phosphotungstic-picric acid (stock)	40.0	ml
95% Alcohol	40.0	ml
Distilled water	20.0	ml

2% Light Green

Light green, SF yellowish	2.0	gm
Distilled water	100.0	ml
Glacial acetic acid	1.0	ml

STAINING PROCEDURE

1. Decerate slides in xylene, 2 changes, 2 minutes each. Place in absolute alcohol, 2 changes, 2 minutes each; 95% alcohol, 2 changes, 2 minutes each; then rinse in distilled water.
2. Mordant in Zenker's solution overnight if tissue is formalin fixed.
3. Wash in running tap water for 3 minutes.
4. Remove mercuric chloride crystals with iodine and clear with sodium thiosulfate.
5. Wash in running tap water for 10 minutes.
6. Stain in celestine blue B for 5 minutes.
7. Wash in running tap water for 4 minutes.
8. Stain in Mayer's hematoxylin for 5 minutes.
9. Wash in running tap water for 5 minutes.
10. Stain in orange G-picric acid for 5 minutes.
11. Wash in running tap water for 1 minute.
12. Stain in acid fuchsin solution for 5 minutes.
13. Wash in running tap water for 1 minute.
14. Rinse in differentiating solution for 10 to 15 seconds.
15. Wash in running tap water for 1 minute.
16. Place in phosphotungstic-picric acid working solution for 5 minutes.
17. Wash in running tap water.
18. Counterstain in light green solution for 1 minute.
19. Wash in running tap water for 1 minute.
20. Dehydrate slides in 95% alcohol and absolute alcohol 3 changes each. Clear in xylene, 3 changes.
21. Mount coverglass with appropriate medium (refractive index 1.48-1.56).

RESULTS

Fibrin, keratin	red
Cytoplasmic granules	red
Erythrocytes	orange
Collagen	green

Ayoub-Shklar Method for Keratin and Prekeratin[55an]

FIXATION 10% Buffered neutral formalin.
PROCESS Paraffin, frozen, or celloidin.
MICROTOMY Cut sections at 6 micrometers.
CONTROL Use control number 27.
SOLUTIONS

5% Acid Fuchsin

Acid fuchsin	5.0	gm
Distilled water	100.0	ml

Aniline Blue-Orange G		
Aniline blue, water soluble	0.5	gm
Orange G	2.0	gm
Phosphotungstic acid	1.0	gm
Distilled water	100.0	ml

STAINING PROCEDURE

1. Decerate slides in xylene, 2 changes, 2 minutes each. Place in absolute alcohol, 2 changes, 2 minutes each; 95% alcohol, 2 changes, 2 minutes each; then rinse in distilled water.
2. Stain in acid fuchsin for 3 minutes.
3. Transfer slides directly to aniline blue-orange G solution for 45 minutes.
4. Transfer slides directly to 95% alcohol, 3 changes, 1 minute each.
5. Dehydrate slides in 95% alcohol and absolute alcohol 3 changes each. Clear in xylene, 3 changes.
6. Mount coverglass with appropriate medium (refractive index 1.48-1.56).

RESULTS

Keratin	brilliant red
Stratified squamous epithelium	gray
Stratum spinosum (prekeratin)	orange
Connective tissue	blue
Erythrocytes	red

Heidenhain's Method for Osteoid Collagen and Muscle[55ao]

FIXATION Zenker's, Helly's, Bouin's, or Carnoy's solution.
PROCESS Paraffin, frozen, or celloidin.
MICROTOMY Cut sections at 6 micrometers.
CONTROL Use control number 9 or 11.
SOLUTIONS

1% Azocarmine B		
Azocarmine B	1.0	gm
Distilled water	100.0	ml

Bring to a boil, filter at 56°C, cool and add 1 ml glacial acetic acid. Must be kept in refrigerator and filtered before use.

1% Aniline-Alcohol		
Aniline	1.0	ml
95% alcohol	100.0	ml

5% Phosphotungstic Acid		
Phosphotungstic acid	5.0	gm
Distilled water	100.0	ml

Aniline Blue-Orange G

Aniline blue, water-soluble	0.5	gm
Orange G	2.0	gm
Distilled water	300.0	ml
Concentrated glacial acetic acid	8.0	ml

1% Acetic Acid-Alcohol

Glacial acetic acid	1.0	ml
95% Alcohol	100.0	ml

STAINING PROCEDURE

1. Decerate slides in xylene, 2 changes, 2 minutes each. Place in absolute alcohol, 2 changes, 2 minutes each; 95% alcohol, 2 changes, 2 minutes each; then rinse in distilled water.
2. If formalin fixed, mordant in Zenker's solution for 4 hours. Then remove mercuric chloride crystals with iodine and clear with sodium thiosulfate.
3. Wash in running tap water for 5 minutes.
4. Rinse in distilled water for 30 seconds.
5. Stain in preheated azocarmine B solution at 56°C for 15 minutes.
6. Rinse in distilled water for 30 seconds.
7. Differentiate in aniline-alcohol until cytoplasm and connective tissue are pale pink and nuclei stand out sharply.
8. Rinse in 1% acetic acid-alcohol for 45 seconds.
9. Mordant in phosphotungstic acid for 15 minutes.
10. Rinse in distilled water for 30 seconds.
11. Stain in aniline blue for 15 minutes.
12. Rinse in distilled water for 30 seconds.
13. Dehydrate slides in 95% alcohol and absolute alcohol 3 changes each. Clear in xylene, 3 changes.
14. Mount coverglass with appropriate medium (refractive index 1.48-1.56).

RESULTS

Nuclear chromatin, osteocytes, neuroglial cells.................................. red
Collagen and reticulum.. blue
Muscle... red to yellow
Osteoid material in decalcified sections.. red

Gomori's Method for Reticulum[55ap]

FIXATION	10% Buffered neutral formalin.
PROCESS	Paraffin, frozen, or celloidin.
MICROTOMY	Cut sections at 6 micrometers.
CONTROL	Use control number 38.

SOLUTIONS

Ammoniacal Silver

To 10 ml of 10% silver nitrate solution add 2.5 ml of a 10% aqueous solution of potassium hydroxide. Add 28% ammonium hydroxide, drop by drop, while shaking the container continously, until precipitate is completely dissolved. Add again four drops of silver nitrate solution for every 10 ml of silver nitrate used. Add sufficient distilled water to double the volume of this mixture.

0.5% Potassium Permanganate

Potassium permanganate	0.5	gm
Distilled water	100.0	ml

2% Potassium Metabisulfite

Potassium metabisulfite	2.0	gm
Distilled water	100.0	ml

2% Ferric Ammonium Sulfate

Ferric ammonium sulfate	2.0	gm
Distilled water	100.0	ml

20% Formalin

37% Formaldehyde	20.0	ml
Distilled water	80.0	ml

0.2% Gold Chloride

Gold chloride	0.2	gm
Distilled water	100.0	ml

2% Sodium Thiosulfate

Sodium thiosulfate	2.0	gm
Distilled water	100.0	ml

STAINING PROCEDURE

1. Decerate slides in xylene, 2 changes, 2 minutes each. Place in absolute alcohol, 2 changes, 2 minutes each; 95% alcohol, 2 changes, 2 minutes each; then rinse in distilled water.
2. Oxidize in 0.5% potassium permanganate for 1 minute.
3. Wash in running tap water for 2 minutes.
4. Differentiate with 2% potassium metabisulfite for 1 minute.
5. Wash in running tap water for 2 minutes.
6. Sensitize in ferric ammonium sulfate for 1 minute.
7. Wash in running tap water for 2 minutes followed by 2 changes of distilled water for 30 seconds each.
8. Impregnate in the ammoniacal silver for 1 minute.
9. Rinse in distilled water for 20 seconds.
10. Reduce in 20% formalin for 3 minutes.
11. Wash in running tap water for 3 minutes.
12. Tone in gold chloride for 10 minutes.
13. Rinse in distilled water for 30 seconds.
14. Reduce in 2% potassium metabisulfite for 1 minute.

15. Fix in 2% sodium thiosulfate for 1 minute.
16. Dehydrate slides in 95% alcohol and absolute alcohol 3 changes each. Clear in xylene, 3 changes.
17. Mount coverglass with appropriate medium (refractive index 1.48-1.56).

RESULTS

Reticulum fibers .... black
Background .... gray

REMARKS

This technique is especially useful for demonstrating reticulum in the central nervous system. Use a recently opened bottle of ammonium hydroxide for making up the ammoniacal silver.

Manuel's Method for Reticulum

(See Fig. 168 and 169)

FIXATION	10% Buffered neutral formalin.
PROCESS	Paraffin, frozen, or celloidin.
MICROTOMY	Cut sections at 6 micrometers.
CONTROL	Use control number 38.
SOLUTIONS	

1% Uranium Nitrate

Uranium nitrate	1.0	gm
Distilled water	100.0	ml

Ammoniacal Silver

In 1000 ml of distilled water dissolve 100 gm of silver nitrate. Pour off 70 ml and save. To the 930 ml of 10% silver nitrate add 60 ml of 28% ammonium hydroxide, shaking the flask vigorously (solution becomes dark, then clears). Of the 70 ml saved, add slowly 50 ml to the ammonium hydroxide-silver nitrate mixture. Solution should become slightly cloudy; if not, add part or all of the remaining 20 ml of the silver nitrate. Let stand overnight and filter before using. Store in the refrigerator and use as needed.

1% Formalin

Concentrated formalin	1.0	ml
Distilled water	100.0	ml

0.2% Gold Chloride

(See Page 205)

5% Sodium Thiosulfate

Sodium thiosulfate	5.0	gm
Distilled water	100.0	ml

Kernechtrot (Nuclear Fast Red)

(See Page 84)

STAINING PROCEDURE

1. Decerate slides in xylene, 2 changes, 2 minutes each. Place in absolute alcohol, 2 changes, 2 minutes each; 95% alcohol, 2 changes, 2 minutes each; then rinse in distilled water.
2. Sensitize in uranium nitrate for 2 minutes.
3. Dip 3 times in running tap water.
4. Impregnate in ammoniacal silver for 1 minute.
5. Wash in running water, 2 or 3 quick dips, until no more white precipitate appears in the water.
6. Develop in formalin for 1 minute.
7. Wash in running tap water for 2 minutes.
8. Tone in gold chloride for 1 minute.
9. Wash in running tap water for 2 minutes.
10. Reduce in 5% sodium thiosulfate for 1 minute.
11. Wash in running tap water for 3 minutes.
12. Stain nuclear chromatin with nuclear fast red for 5 minutes.
13. Rinse in distilled water, 3 changes, 30 seconds each.
14. Dehydrate slides in 95% alcohol and absolute alcohol 3 changes each. Clear in xylene, 3 changes.
15. Mount coverglass with appropriate medium (refractive index 1.48-1.56).

RESULTS

Reticulum fibers.. black
Nuclear chromatin.. red

REMARKS

Use a recently opened bottle of ammonium hydroxide for making up the ammoniacal silver.

Wilder's Method for Reticulum

FIXATION 10% Buffered neutral formalin.
PROCESS Paraffin, frozen, or celloidin.
MICROTOMY Cut sections at 6 micrometers.
CONTROL Use control number 38.
SOLUTIONS

10% Phosphomolybdic Acid

Phosphomolybdic acid	10.0	gm
Distilled water	100.0	ml

1% Uranium Nitrate
(See Page 415)

Ammoniacal Silver

To 5 ml of 10% aqueous solution of silver nitrate add 28% ammonium hydroxide, drop by drop, until the precipitate which forms is almost dissolved. Add 5 ml of 3.1% sodium hydroxide and barely dissolve the resulting precipitate with a few drops of ammonium hydroxide. Make solution up to 50 ml with distilled water. Use at once. Glassware must be acid-clean.

Reducing Solution

Distilled water	50.0	ml
Neutral formalin	0.5	ml
1% Uranium nitrate	1.5	ml

Make fresh just before use.

Concentrated Neutral Formalin

Concentrated formalin, 37%	40.0	ml
Distilled water	100.0	ml
Calcium carbonate	10.0	gm

0.2% Gold Chloride

Gold chloride solution, 1%	10.0	ml

(Break glass vial [15 grains] in graduated cylinder with 100.0 ml distilled water for 1% solution.)

Distilled water	40.0	ml

5% Sodium Thiosulfate
(See Page 134)

Kernechtrot (Nuclear Fast Red)
(See Page 84)

STAINING PROCEDURE

1. Decerate slides in xylene, 2 changes, 2 minutes each. Place in absolute alcohol, 2 changes, 2 minutes each; 95% alcohol, 2 changes, 2 minutes each; then rinse in distilled water.
2. Oxidize in phosphomolybdic acid for 1 minute.
3. Rinse well in running tap water or cells will remain yellow.
4. Sensitize in 1% uranium nitrate for 1 minute.
5. Rinse in distilled water for 20 seconds.
6. Place in ammoniacal silver for 1 minute (change solution frequently).
7. Dip very quickly in 95% alcohol and go immediately to step 8.
8. Place in concentrated neutral formalin for 1 minute.
9. Rinse in distilled water for 30 seconds.
10. Tone in 0.2% gold chloride for 1 minute or until sections lose their yellow color and turn lavender. Too much toning will make sections red. Check individually under microscope.
11. Rinse in distilled water for 30 seconds.
12. Place in sodium thiosulfate solution for 1 minute.
13. Wash in running tap water for 3 minutes.
14. Counterstain in nuclear fast red for 5 minutes. Rinse well in distilled water.
15. Dehydrate slides in 95% alcohol and absolute alcohol 3 changes each. Clear in xylene, 3 changes.
16. Mount coverglass with appropriate medium (refractive index 1.48-1.56).

RESULTS

Reticulum fibers	black
Collagen	rose
Nuclear chromatin	red

REMARKS
Use a recently opened bottle of ammonium hydroxide for making up the ammoniacal silver. Well-stained reticulum exhibits black, well-defined fibers without beading effect and relatively clear background. If beading effect and dark background are noticed, replace ammoniacal silver and reducing solution. Change the neutral formalin (step 8) each time 5 slides have been exposed to this solution.

Gomori's Method for Muscle and Collagen[55aq]

(See Fig. 171)

FIXATION 10% Buffered neutral formalin.
PROCESS Paraffin, frozen, or celloidin.
MICROTOMY Cut sections at 6 micrometers.
CONTROL Use control number 11.
SOLUTIONS

0.5% Glacial Acetic Acid

Glacial acetic acid	0.5	ml
Distilled water	100.0	ml

Weigert's Hematoxylin
(See Page 79)

Bouin's Solution
(See Page 13)

Chromotrope-Light Green

Chromotrope 2R	0.6	gm
Light green, SF yellowish	0.3	gm
Glacial acetic acid	1.0	ml
Phosphotungstic acid	0.8	gm
Distilled water	100.0	ml

Aniline blue may be substituted for light green if it is more desirable to have collagen stained blue.

STAINING PROCEDURE

1. Decerate slides in xylene, 2 changes, 2 minutes each. Place in absolute alcohol, 2 changes, 2 minutes each; 95% alcohol, 2 changes, 2 minutes each; then rinse in distilled water.
2. Mordant in Bouin's solution in an oven at 56°C for 1 hour.
3. Wash well in running tap water or until yellow color disappears.
4. Stain nuclear chromatin with Weigert's or Gomori's chromium-hematoxylin solution for 10 minutes.
5. Wash in running water for 10 minutes.
6. Stain in chromotrope-light green for 15 minutes.
7. Place in 0.5% glacial acetic acid for 2 minutes.
8. Dehydrate slides in 95% alcohol and absolute alcohol 3 changes each. Clear in xylene, 3 changes.
9. Mount coverglass with appropriate medium (refractive index 1.48-1.56).

RESULTS

Muscle fibers .... red
Fibrin .... red
Collagen .... green (blue, if aniline blue is used)
Nuclear chromatin .... blue to black

REMARKS

If sections are too dark, differentiate in 1% glacial acetic water to which 0.7 gm of phosphotungstic acid has been added. Rinse in distilled water. Elastica can be demonstrated with aldehyde fuchsin if one introduces aldehyde fuchsin at step 4 in place of the hematoxylin. Exposure should be for 30 minutes followed by a 2 minute wash in 95% ethyl alcohol; follow with step 5.

Gomori's Method for Elastica and Sulfated Mucosaccharides

(See Fig. 172 and 179)

FIXATION	10% Buffered neutral formalin.
PROCESS	Paraffin, frozen, and celloidin.
MICROTOMY	Cut paraffin sections at 6 micrometers.
CONTROL	Use control number 16.
SOLUTIONS	

Aldehyde Fuchsin

Basic fuchsin	1.0	gm
70% Alcohol	200.0	ml
Concentrated hydrochloric acid	2.0	ml
Paraldehyde	2.0	ml

Let stand at room temperature for 2 to 3 days or until stain is deep purple in color.

Van Gieson's
(See Page 362)

0.25% Metanil Yellow

Metanil yellow	0.25	gm
Distilled water	100.0	ml
Glacial acetic acid	0.25	ml

STAINING PROCEDURE

1. Decerate slides in xylene, 2 changes, 2 minutes each. Place in absolute alcohol, 2 changes, 2 minutes each; 95% alcohol, 2 changes, 2 minutes each; then rinse in distilled water.
2. Stain in aldehyde fuchsin for 30 minutes.
3. Rinse off excess stain in 95% alcohol.
4. Rinse in distilled water.
5. Counterstain as desired with Van Gieson's solution, or metanil yellow, or Gomori's trichrome stain.
6. Dehydrate slides in 95% alcohol and absolute alcohol 3 changes each. Clear in xylene, 3 changes.
7. Mount coverglass with appropriate medium (refractive index 1.48-1.56).

RESULTS

Elastic fibers and mucosaccharides ....deep purple

Appearance of other tissues is determined by the counterstain used.

REMARKS

Aldehyde fuchsin has been used quite successfully in the study of mucosaccharides. The details for these methods may be found in chapter 12.

Masson's Method for Collagen and Muscle[55ar]

(See Fig. 173, 174 and 175)

FIXATION 10% Buffered neutral formalin.
PROCESS Paraffin, frozen, or celloidin.
MICROTOMY Cut sections at 6 micrometers.
CONTROL Use control number 11.
SOLUTIONS

Bouin's Solution

(See Page 13)

Weigert's Iron-Hematoxylin

(See Page 79)

Biebrich Scarlet-Acid Fuchsin

1% Biebrich scarlet, aqueous	90.0	ml
1% Acid fuchsin, aqueous	10.0	ml
Glacial acetic acid	1.0	ml

Phosphotungstic-Phosphomolybdic Acid

Phosphotungstic acid	5.0	gm
Phosphomolybdic acid	5.0	gm
Distilled water	200.0	ml

Aniline Blue

Aniline blue	2.5	gm
Glacial acetic acid	2.0	ml
Distilled water	100.0	ml

2% Light Green

Light green, SF yellowish	2.0	gm
Distilled water	100.0	ml

1% Glacial Acetic Acid

Glacial acetic acid	1.0	ml
Distilled water	100.0	ml

STAINING PROCEDURE

1. Decerate slides in xylene, 2 changes, 2 minutes each. Place in absolute alcohol, 2 changes, 2 minutes each; 95% alcohol, 2 changes, 2 minutes each; then rinse in distilled water.

2. Mordant in Bouin's solution for 1 hour at 56°C or overnight at room temperature, if sections are formalin fixed.
3. Cool and wash in running tap water until yellow color disappears.
4. Rinse in distilled water.
5. Stain nuclear chromatin in Weigert's iron-hematoxylin for 10 minutes. Wash in running tap water for 10 minutes.
6. Rinse in distilled water.
7. Place in Biebrich scarlet-acid fuchsin solution for 2 minutes. Save solution.
8. Rinse in distilled water for 30 seconds.
9. Place in phosphomolybdic acid-phosphotungstic acid solution for 15 minutes if using aniline blue solution. Place in aqueous phosphotungstic acid 5% for 15 minutes if using light green counterstain. Discard solution.
10. Place in aniline blue solution for 5 minutes or light green solution for 1 minute (for central nervous system, 15 to 20 minutes). Save solution.
11. Rinse in distilled water for 30 seconds.
12. Place in 1% glacial acetic acid solution for 3 to 5 minutes. Discard solution.
13. Dehydrate slides in 95% alcohol and absolute alcohol 3 changes each. Clear in xylene, 3 changes.
14. Mount coverglass with appropriate medium (refractive index 1.48-1.56).

RESULTS

Nuclear chromatin .... black
Keratin, muscle fibers .... red
Collagen .... blue

REMARKS

To insure proper mordanting Bouin's solution should be preheated to 58-60°C.

Picrosirius Red Method for Collagen

(See Fig. 189)

FIXATION 10% Buffered neutral formalin.
PROCESS Paraffin.
MICROTOMY Cut sections at 5 micrometers.
CONTROL Use control number 11.
SOLUTIONS

Picrosirius Red

Sirius red F3BA	0.1	gm
Saturated aqueous picric acid	100.0	ml

0.01N Hydrochloric Acid

Concentrated hydrochloric acid	0.83	ml
Distilled water, q.s. to	1000.0	ml

STAINING PROCEDURE

1. Decreate slides in xylene, 2 changes, 2 minutes each. Place in absolute alcohol, 2 changes, 2 minutes each; 95% alcohol, 2 changes, 2 minutes each; then rinse in distilled water.

2. Stain in picrosirius red solution.
3. Wash in 0.01N hydrochloric acid for 2 minutes.
4. Dehydrate in 95% alcohol and absolute alcohol 3 changes each. Clear in xylene, 3 changes.
5. Mount coverglass with appropriate medium (refractive index 1.48-1.56).

RESULTS
With light microscopy, this staining procedure produces a red color in collagenous fibers. All other tissue elements are stained bright yellow. In plain polarized light, collagen fibrils are manifested in a striking manner, as vivid orange-red bands against a black background.

Barbeito-Lopez Trichrome Stain[135]

FIXATION 10% Buffered neutral formalin.
PROCESS Paraffin.
MICROTOMY Cut section at 6 micrometers.
CONTROL Use control number 11 or 20.
SOLUTIONS

Acetic Fuchsin Solution

Ziehl's fuchsin solution	10.0	ml
Glacial acetic acid	0.2	ml
Distilled water	100.0	ml

Acetic Formalin

37% Formalin	4.0	ml
Glacial acetic acid	0.2	ml
Distilled water	100.0	ml

Molybdic Aniline Blue-Methyl Orange
To 100 ml of distilled water in a 250 ml Erlenmeyer flask add 1 gram of molybdic acid, shake vigorously and heat gently for 5 minutes. Add 0.5 gm of water soluble aniline blue and dissolve. Add methyl orange in excess until a deposit forms in the bottom of the flask. Filter.

STAINING PROCEDURE
1. Decerate slides in xylene, 2 changes, 2 minutes each. Place in absolute alcohol, 2 changes, 2 minutes each; 95% alcohol, 2 changes, 2 minutes each; then rinse in distilled water.
2. Stain in acetic fuchsin solution for 5 minutes.
3. Wash vigorously in running tap water to remove excess solution.
4. Fix color in acetic-formalin solution, up to 3 minutes. Place sections in coplin jar containing the solution, agitate and renew solution after four or five uses. (The color changes from red to pale violet.)
5. Wash in running tap water for 1 minute.
6. Stain in molybdic aniline blue-methyl orange for 1 minute.

7. Wash in running tap water for 10 seconds.
8. Dehydrate slides rapidly in 95% alcohol and absolute alcohol. Clear in xylene, 3 changes.
9. Mount coverglass with appropriate medium (refractive index 1.48-1.56).

RESULTS

Nuclear chromatin	violet red
Cytoplasm	green to pale blue
Reticulum fibers	deep blue
Collagen	green
Red blood cells	orange
Bacteria	violet red

REMARKS

Disregard the color clouds that appear in the 95% alcohol (step 8).

Ehrenreich-Espinosa Method for Basement Membrane and Protein Deposits[136]

FIXATION	10% Buffered neutral formalin or Bouin's solution.
PROCESS	Paraffin or celloidin.
MICROTOMY	Cut sections at 2 to 3 micrometers.
CONTROL	Use control number 6.
SOLUTIONS	

Silver-Methenamine

3% Methenamine (see page 197)	100.0	ml
5% Silver nitrate (see page 197)	5.0	ml
5% Sodium borate (see page 200)	12.0	ml

A stock solution of 3% methenamine and 5% silver nitrate in the proportions indicated may be kept in a dark bottle in the refrigerator for a month. Sodium borate is added before use.

0.5% Periodic Acid

Periodic acid	0.5	gm
Distilled water	100.0	ml

0.2% Gold Chloride

Gold chloride	0.2	gm
Distilled water	100.0	ml

1% Phosphotungstic Acid

Phosphotungstic acid	1.0	gm
Distilled water	100.0	ml

2% Chromotrope 2R

Chromotrope 2R	2.0	gm
0.1N Hydrochloric acid	100.0	ml

This solution may be kept for 6 months at room temperature.

2% Sodium Thiosulfate

Sodium thiosulfate	2.0	gm
Distilled water	100.0	ml

STAINING PROCEDURE

1. Decerate slides in xylene, 2 changes, 2 minutes each. Place in absolute alcohol, 2 changes, 2 minutes each; 95% alcohol, 2 changes, 2 minutes each; then rinse in distilled water.
2. Oxidize in 0.5% periodic acid for 30 minutes. Wash thoroughly in tap water; before proceeding, rinse twice in distilled water.
3. Place in silver-methenamine at 55°C until the sections turn dark mahogany (about 1 1/2 hour). Check slides with the microscope to prevent overstaining. Rinse well in distilled water (several changes).
4. Dip 5 times in 0.2% gold chloride or until sections turn gray. Rinse well in distilled water (several changes).
5. Dip 5 times in 2% sodium thiosulfate. Wash well in running tap water for 5 minutes.
6. Place in Bouin's solution at 55°C for 30 minutes. Rinse well in running tap water for 5 minutes.
7. Place in 1% phosphotungstic acid for 1 minute.
8. Stain in 2% chromotrope 2R for 15 minutes. Rinse in 2 changes of distilled water.
9. Dehydrate slides in 95% alcohol and absolute alcohol 3 changes each. Clear in xylene, 3 changes.
10. Mount coverglass with appropriate medium (refractive index 1.48-1.56).

RESULTS

Basement membrane	black
Cytoplasm	light red or pink
Erythrocytes	red
Protein deposits	bright red or dark red (the shade is different from that of erythocytes)
Nuclear chromatin	faint gray
Nucleoli	faint red
Hyaline	red
Fibrin	red

Luna's Method for Rhabdomyoblast and Nemaline Rods

(See Fig. 176 and 177)

The term rhabdomyoblast is used here to describe the characteristics of the cell cytoplasm, which tends to be acidophilic and differentiates to form cross striations or longitudinal myofibrils. The term nemaline rod is used to describe a rod-shaped structure which is found in the cytoplasm of skelatal muscle in nemaline myopathy.

FIXATION	10% Buffered neutral formalin.
PROCESS	Paraffin, frozen, or celloidin.
MICROTOMY	Cut sections at 4 micrometers.
CONTROL	Use control number 14.

SOLUTIONS

5% Ferric Ammonium Sulfate

Ferric ammonium sulfate	5.0	gm
Concentrated hydrochloric acid	2.0	ml
Distilled water	100.0	ml

1% Phosphotungstic Acid

Phosphotungstic acid	1.0	gm
Distilled water	100.0	ml

1% Acid Fuchsin (Stock)

Acid fuchsin	1.0	gm
Distilled water	100.0	ml

1% Oxalic Acid (Stock)

Oxalic acid	1.0	gm
Distilled water	100.0	ml

Acid Fuchsin (Working)

1% Acid fuchsin (stock)	1.0	ml
1% Oxalic acid (stock)	1.0	ml
Distilled water	98.0	ml

20% Alcohol (Stock)

Absolute alcohol	20.0	ml
Distilled water	100.0	ml

2% Picric Acid (Stock)

Picric acid	2.0	gm
Distilled water	100.0	ml

1% Fast Green (Stock)

Fast green FCF	1.0	gm
Distilled water	100.0	ml

Picric Acid-Fast Green

20% Alcohol (stock)	50.0	ml
2% Picric acid (stock)	20.0	ml
1% Fast green (stock)	1.0	ml

STAINING PROCEDURE

1. Decerate slides in xylene, 2 changes, 2 minutes each. Place in absolute alcohol, 2 changes, 2 minutes each; 95% alcohol, 2 changes, 2 minutes each; then rinse in distilled water.
2. Place slides in 5% ferric ammonium sulfate solution for 30 minutes.
3. Rinse slides in 3 changes of distilled water (pour on and off).
4. Place slides in 1% phosphotungstic acid solution for 15 minutes.
5. Wash slides for 3 minutes in running tap water, then rinse in distilled water.
6. Stain slides in working acid fuchsin for 1 hour in a 55-60°C oven.
7. Remove slides from oven and cool for 15 minutes.

8. Counterstain and differentiate in picric acid-fast green for 6 minutes. Agitate once every minute.
9. Dehydrate slides in 95% alcohol and absolute alcohol 3 changes each. Clear in xylene, 3 changes.
10. Mount coverglass with appropriate medium (refractive index 1.48-1.56).

REMARKS

Step 8 is critical since it is possible to overdifferentiate the cross bond of rhabdomyomblast and nemaline rods by overexposure to the differentiation solution. Slides may be restained in the acid fuchsin working solution if overdifferentiation occurs. Mallory's phosphotungstic acid hematoxylin (PTAH) should be performed in conjunction with this procedure when rhabdomyoma and nemaline rod demonstration is desired.

RESULTS

Rhabdomyomblast and nemaline rods
(Z band material).. bright red
Erythrocytes.. red (formalin fixed)
yellow (Bouin's or Zenker's fixed)
Background .... light green

Clark-Spooner-Dodds-Churukian Method for Fine Connective Tissue[137]

There are considered to be at least three types of fine connective tissues: basement membranes, reticular fibers forming the supporting system of some organs, and the "ring fibers" in the spleen, (see results).

FIXATION	10% Buffered neutral formalin.
PROCESS	Paraffin.
MICROTOMY	Cut section at 6 micrometers.
CONTROL	Use control number 6 or 38.
SOLUTIONS	

0.03M Silver Nitrate (Stock)

Silver nitrate	0.510	gm
Distilled water	100.0	ml

0.07M Ammonium Acetate (Stock)

Ammonium acetate ($NH_4CH_3C0_2$)	0.540	gm
Distilled water	100.0	ml

0.07M Sodium Hydroxide (Stock)

Sodium hydroxide	0.280	gm
Distilled water	100.0	ml

Silver Solution (Working)		
0.03M Silver nitrate	5.0	ml
0.07M Ammonium acetate	5.0	ml
0.07M Sodium hydroxide	5.0	ml
Distilled water	35.0	ml

Make fresh just before use. Use only once.

20% Formalin		
37% Concentrated formalin	20.0	ml
Distilled water	80.0	ml

1% Periodic Acid		
Periodic acid	1.0	gm
Distilled water	100.0	ml

1% Gold Chloride		
Gold chloride	1.0	gm
Distilled water	100.0	ml

STAINING PROCEDURE

1. Decerate slides in xylene, 2 changes, 2 minutes each. Place in absolute alcohol, 2 changes, 2 minutes each; 95% alcohol, 2 changes, 2 minutes each; then rinse in distilled water.
2. Oxidize in 1% periodic acid for 10 minutes.
3. Rinse well in running tap water and then distilled water. Let slides set in distilled water until working silver solution is hot.
4. Preheat working silver solution in 60°C oven or 55°C water bath for 45 minutes. Add the slides to this solution and return to oven or water bath for 1 hour. At this point sections should have a golden brown color.
5. Rinse very briefly in distilled water.
6. Reduce in 20% formalin for 5 minutes. Save solution to reduce gold (step 10).
7. Rinse well in running tap water and then at least 2 changes of distilled water.
8. Place in 1% gold chloride for 5 minutes.
9. Rinse briefly in distilled water.
10. Reduce in 20% formalin for 5 minutes.
11. Rinse well in running tap water.
12. Dehydrate slides in 95% alcohol and absolute alcohol 3 changes each. Clear in xylene, 3 changes.
13. Mount coverglass with appropriate medium (refractive index 1.48-1.56).

RESULTS

This stain demonstrates basement membranes well in the kidney, but it may not stain them in the skin. Similarly, reticular fibers are well-stained in the kidney and in the liver, but inconsistently stained in lymph nodes and spleen. Ring fibers do not stain. Fine connective tissue fibers stain black, but as the size increases, they stain reddish-black. The larger connective tissue fibers stain red. Fungi stain black to grey-black.

Garvey-Fathi-Bigelow-Carpenter-Jimenez Method for Elastica, Fibrin and Collagen[138]

FIXATION 10% Buffered neutral formalin.
PROCESS Paraffin.
MICROTOMY Cut section at 3 micrometers.
CONTROL Use control number 12, 17 or 11.
SOLUTIONS

Mordanting Solution

Mercuric acid	3-5.0	gm
Saturated aqueous picric acid	100.0	ml

Modified Verhoeff's
Solution A (Stock)

Hematoxylin	3.0	gm
100% Alcohol	100.0	ml

Solution B (Stock)

Ferric chloride (hexahydrate)	2.3	gm
Distilled water	100.0	ml

This solution is unstable and should be replaced when a precipitate has formed.

Solution C (Stock)

Potassium iodide	4.0	gm
Iodine	2.0	gm
Distilled water	100.0	ml

Working solution consists of three parts of solution A to two parts of solution B to one part of solution C.

1% Lissamine Fast Yellow

Lissamine Fast Yellow	1.0	gm
Distilled water	100.0	ml
Add:		
Glacial acetic acid	0.5	ml

Biebrich Scarlet-Acid Fuchsin-Ponceau 2R

Biebrich scarlet	0.75	gm
Acid fuchsin	0.75	gm
Ponceau GR or 2R	0.75	gm
Distilled water	100.0	ml
Add:		
Glacial acetic acid	0.5	ml

This solution remains useful for at least six months.

0.5% Aniline Blue

Aniline blue WS	0.5	gm
Distilled water	100.0	ml
Add:		
Glacial acetic acid	1.0	ml

2% Light Green

Light green SF yellowish	2.0	gm
Distilled water	100.0	ml
Add:		
Glacial acetic acid	2.0	ml

2% Phosphotungstic Acid

Phosphotungstic acid	2.0	gm
Distilled water	100.0	ml

STAINING PROCEDURE

1. Decerate slides in xylene, 2 changes, 2 minutes each. Place in absolute alcohol, 2 changes, 2 minutes each; 95% alcohol, 2 changes, 2 minutes each; then rinse in distilled water.
2. Stain nuclei in working modified Verhoeff's stain for 9 minutes.
3. Wash in warm tap water for 5 minutes.
4. Stain in 1% lissamine fast yellow for 2 minutes.
5. Rinse in 0.5% acetic acid for 15 seconds.
6. Stain in Biebrich scarlet-acid fuchsin-ponceau 2R for 5 minutes.
7. Rinse in distilled water for 15 seconds.
8. Place in 2% phosphotungstic acid for 10 minutes.
9. Rinse in distilled water for 15 seconds.
10. Stain in 0.5% aniline blue or 2% light green for 2 minutes.
11. Rinse in 0.5% acetic acid, 3 changes for 15 seconds each.
12. Dehydrate slides in 95% alcohol and absolute alcohol 3 changes each. Clear in xylene, 3 changes.
13. Mount coverglass with appropriate medium (refractive index 1.48-1.56).

RESULTS

Nuclear chromatin	blue-black
Elastic fibers	black
Fibrin (newly formed)	orange-yellow
Fibrin (mature)	red
Fibrin (old)	blue or green
Muscle	red
Collagen	blue or green and some red
Erythrocytes	yellow

Garvey's Method for Elastica, Muscle and Collagen

FIXATION 10% Buffered neutral formalin.
PROCESS Paraffin.
MICROTOMY Cut section at 3 mirometers.
CONTROL Use control number 16 or 11.
SOLUTIONS

Alcoholic-Picric Acid

Picric acid	2.5	gm
Alcohol	100.0	ml

Modified Verhoeff's
Solution A

Hematoxylin	3.0	gm
Absolute alcohol	100.0	ml

Solution B

Ferric chloride (hexahydrate)	2.0	gm
Distilled water	100.0	ml

This solution is unstable and should be replaced when a precipitate forms.

Solution C

Potassium iodine	4.0	gm
Iodine	2.0	gm
Distilled water	100.0	ml

Working solution consists of three parts A to two parts B and one part C.

5% Phosphotungstic Acid

Phosphotungstic acid	5.0	gm
Distilled water	100.0	ml

Biebrich Scarlet-Acid Fuchsin

0.5% Aqueous biebrich scarlet	90.0	ml
0.5% Aqueous acid fuchsin	10.0	ml
Glacial acetic acid	1.0	ml

2% Light Green

Light green SF yellowish	2.0	gm
Hot distilled water	100.0	ml
Cool, filter and add:		
Glacial acetic acid	2.0	ml

STAINING PROCEDURE

1. Decerate slides in xylene, 2 changes, 2 minutes each. Place in absolute alcohol, 2 changes, 2 minutes each; 95% alcohol, 2 changes, 2 minutes each; then rinse in distilled water.
2. Mordant sections in alcoholic picric acid for 5 minutes.
3. Wash sections in running tap water.
4. Stain in modified Verhoeff's stain for 6 minutes.
5. Wash in running tap water for 10 minutes.
6. Stain in Biebrich scarlet-acid fuchsin for 3 minutes.
7. Rinse in distilled water.
8. Place in 5% phosphotungstic acid, 3 changes, 5 minutes each.
9. Stain in 2% light green for 3 minutes.
10. Differentiate in 1% acetic acid for 3 minutes.
11. Dehydrate slides in 95% alcohol and absolute alcohol 3 changes each. Clear in xylene, 3 changes.
12. Mount coverglass with appropriate medium (refractive index 1.48-1.56).

RESULTS

Nuclear chromatin	reddish-brown to blue-black
Elastica	blue-black
Muscle	red
Collagen	green or blue

Buk's Method for Elastica and Fibrin

FIXATION 10% Buffered neutral formalin.
PROCESS Paraffin.
MICROTOMY Cut section at 3 to 4 micrometers.
CONTROL Use control number 16 and 17.
SOLUTIONS

Bouin's Solution

Saturated aqueous picric acid	75.0	ml
37% Formalin	25.0	ml
Glacial acetic acid	5.0	ml

5% Hematoxylin (Stock)

Hematoxylin	5.0	gm
100% Alcohol	100.0	ml

10% Ferric Chloride (Stock)

Ferric chloride	10.0	gm
Distilled water	100.0	ml

5% Sodium Thiosulfate

Sodium thiosulfate	5.0	gm
Distilled water	100.0	ml

Lugol's Iodine (Stock)

(See Page 81)

Verhoeff's Stain

5% Hematoxylin (stock)	20.0	ml
10% Ferric chloride (stock)	8.0	ml
Lugol's iodine (stock)	8.0	ml

Martius Yellow Stain

Martius yellow	0.5	gm
95% Alcohol	50.0	ml
Phosphotungstic acid	2.0	gm

Mix with magnetic stirrer.

1% Crystal Ponceau 6R

Crystal ponceau 6R	1.0	gm
1% Glacial acetic acid	100.0	ml

0.5% Soluble Blue

Soluble blue	0.5	gm
1% Aqueous glacial acetic acid	100.0	ml

0.5% Fast Green FCF

Fast green FCF	0.5	gm
1% Glacial acetic acid	100.0	ml

STAINING PROCEDURE

1. Decerate slides in xylene, 2 changes, 2 minutes each. Place in absolute alcohol, 2 changes, 2 minutes each; 95% alcohol, 2 changes, 2 minutes each; then rinse in distilled water.
2. Mordant sections in Bouin's solution preheated to 56°C for 40 minutes or overnight at room temperature.
3. Wash well in running tap water until yellow disappears (8-10 minutes).
4. Place in freshly prepared Verhoeff's stain for 15 minutes.
5. Wash in running tap water for 1 minute.
6. Differentiate in fresh 2% aqueous ferric chloride until both coarse and fine elastic fibers are clearly visible against a slate blue-gray background.
7. Place in 5% aqueous sodium thiosulfate for 3 minutes.
8. Wash in running tap water for 3 minutes.
9. Stain in Martius yellow for 2 minutes.
10. Wash briefly in distilled water.
11. Stain in crystal ponceau for 10 minutes.
12. Wash briefly in distilled water.
13. Place in fresh 1% phosphotungstic acid for 10 minutes with agitation to remove red dye from all structures except fibrin and muscle fiber.
14. Rinse in distilled water.
15. Stain in soluble blue for 5 minutes. Examine after 2 minutes since overstaining with soluble blue is hard to remove.
16. Blot sections carefully and rinse in 1% glacial acetic acid for 10-20 seconds.
17. Dehydrate slides in 95% alcohol and absolute alcohol 3 changes each. Clear in xylene, 3 changes.
18. Mount coverglass with appropriate medium (refractive index 1.48-1.56).

RESULTS

Muscle	red
Erythrocytes	yellow
Young fibrin	yellow to weak red
Mature fibrin	bright red
Older fibrin	blue to green
Nuclear chromatin	blue to black

Musto's Method for Muscle, Elastica and Mucosubstances[139]

FIXATION	10% Buffered neutral formalin or Bouin's solution.
PROCESS	Paraffin.
MICROTOMY	Cut section at 5 to 6 micrometers.
CONTROL	Use control number 11 and 39B.
SOLUTIONS	

Hematoxylin
Solution A (Stock)

Hematoxylin	2.0	gm
95% Ethyl alcohol	100.0	ml

Ferric Chloride
Solution B (Stock)

Ferric chloride hexahydrate	12.4	gm
Distilled water	500.0	ml
Concentrated hydrochloric acid	5.0	ml

Potassium Iodide-Iodine
Solution C (Stock)

Potassium iodide	20.0	gm
Iodine crystals	10.0	gm
Distilled water	20.0	ml
Shake until iodine dissolves and add:		
Distilled water	480.0	ml

Elastic Stain (Working)

Solution A	30.0	ml
Solution B	20.0	ml
Solution C	10.0	ml

2% Alcian Blue

Alcian blue	2.0	gm
Distilled water	100.0	ml
Glacial acetic acid	1.0	ml

Alkaline-Alcohol

95% Ethyl alcohol	95.0	ml
Concentrated ammonium hydroxide	5.0	ml

1% Biebrich Scarlet (Stock)

Biebrich scarlet	1.0	gm
Distilled water	100.0	ml

1% Crocein Scarlet Moo (Stock)

Crocein scarlet moo	1.0	gm
Distilled water	100.0	ml

Biebrich-Crocein Scarlet (Working)

1% Crocein scarlet moo	40.0	ml
Phosphotungstic acid	0.8	gm
Dissolve and then add:		
1% Biebrich scarlet (stock)	30.0	ml

5% Phosphotungstic Acid

Phosphotungstic acid	5.0	gm
Distilled water	100.0	ml

Alcoholic Safran

Safran du Gantinais	2.0	gm
Absolute alcohol	100.0	ml

Use a tightly stoppered bottle to prevent contamination with atmospheric water. Keep in suspension at 55-60°C for 24 hours, shaking occasionally. Cool to room temperature. Filter rapidly through coarse filter paper. Store in a tightly stoppered container.

1% Acetic Acid

Glacial acetic acid	3.0	ml
Distilled water	300.0	ml

STAINING PROCEDURE

1. Decerate slides in xylene, 2 changes, 2 minutes each. Place in absolute alcohol, 2 changes, 2 minutes each; 95% alcohol, 2 changes, 2 minutes each; then rinse in distilled water.
2. Mordant in Bouin's solution (if formalin fixed) at 55-60°C for 20 minutes or overnight at room temperature.
3. Wash in running tap water until the yellow color of the picric acid dissappears; rinse in distilled water.
4. Stain in 2% alcian blue for 10 minutes. Wash in running tap water for 1 minute; rinse in distilled water.
5. Place in alkaline alcohol for 10 minutes. Wash in running tap water for 5 minutes; rinse in distilled water.
6. Stain in working elastic stain for 45 minutes. Wash briefly in running tap water.
7. Stain in Biebrich-crocein scarlet for 10 minutes. Wash in 2 changes of distilled water.
8. Decolorize in 5% phosphotungstic acid for 5 minutes.
9. Transfer slides directly to 1% acetic acid for 5 minutes. Rinse well in distilled water.
10. Dehydrate through 1 change of 95% ethyl alcohol and 2 changes of absolute ethyl alcohol, 1 minute each.
11. Stain in alcoholic safran for 10 minutes. Rinse in 2 changes of absolute ethyl alcohol, 1 minute each.
12. Clear in xylene, 3 changes each.
13. Mount coverglass with appropriate medium (refractive index 1.48-1.56).

RESULTS

Nuclear chromatin	purple
Elastic fibers	black
Acid mucosubstances and mucins	blue to bluish-green
Muscle	red
Collagen	yellow

Lascane-Gomez-Berria Silver Impregnation of Cultured Cardiomyocytes[140]

FIXATION 5% Methyl alcohol in acetic acid.
PROCESS Paraffin, cell cultures.
MICROTOMY Cut sections at 9 micrometers.
CONTROL Use control number 11.
SOLUTIONS

0.05% Potassium Permanganate

Potassium permanganate	0.05	gm
Distilled water	100.0	ml

5% Oxalic Acid

Oxalic acid	5.0	gm
Distilled water	100.0	ml

2% Potassium Bichromate

Potassium bichromate	2.0	gm
Distilled water	100.0	ml

Prepare fresh and filter.

Silver Impregnating Solution (Stock)

Distilled water	100.0	ml
Silver nitrate	1.0	gm
Potassium carbonate	0.4	gm

Dissolve resulting precipitate by the addition of ammonium hydroxide solution, drop by drop.

Silver Impregnating Solution (Working)

Silver impregnating solution (stock)	40.0	ml
Pure pyridine	80	drops

Filter before use.

Reducing Solution

Gelatin	0.25	gm
Warm distilled water	100.0	ml
Concentrated formalin	2.0	ml

0.2% Gold Chloride

Gold chloride	0.2	gm
Distilled water	100.0	ml

5% Sodium Thiosulfate

Sodium Thiosulfate	5.0	gm
Distilled water	100.0	ml

STAINING PROCEDURE

Cultured cells start at step 2.

Paraffin embedded sections:

1. Decerate slides in xylene, 2 changes, 2 minutes each. Place in absolute alcohol, 2 changes, 2 minutes each; 95% alcohol, 2 changes, 2 minutes each; then rinse in distilled water.
2. Place preparations in 0.05% potassium permanganate for 30 seconds at room temperature.
3. Wash in distilled water, 2 changes.
4. Decolorize in 5% oxalic acid for 1 minute.
5. Wash in distilled water.
6. Sensitize in 2% potassium bichromate for 1 minute.
7. Wash in distilled water, three 2 minute washings.
8. Place in working impregnating solution for 15 minutes at 60°C.
9. Place in reducing solution for 10 minutes. If too weak, repeat steps 8-10 as often as required.
10. Wash in distilled water.
11. Tone in 0.2% gold chloride for 5 minutes.

12. Wash in distilled water.
13. Place in 5% sodium thiosulfate for 2 minutes.
14. Wash in running tap water.
15. Dehydrate slides in 95% alcohol and absolute alcohol 3 changes each. Clear in xylene, 3 changes.
16. Mount coverglass with appropriate medium (refractive index 1.48-1.56).

RESULTS

Fibrillar bundles....................black
Nuclei....................black
Collagen fibers....................black

REMARKS

At the end of the staining process, cultured cells attached to plastic surfaces should be allowed to dry at room temperature, then mounted in cedarwood oil.

Meloan-Puchtler Phosphotungstic Acid Hematoxylin (PTAH)[141]

FIXATION 10% Buffered neutral formalin, methacarn, Zenker-formol.
PROCESS Paraffin.
MICROTOMY Cut sections at 5 micrometers.
CONTROL Use control number 14.
SOLUTIONS

Langeron's Iodine

Iodine	1.0	gm
Potassium iodide	2.0	gm
Distilled water	200.0	ml

5% Sodium Thiosulfate

Sodium thiosulfate	5.0	gm
Distilled water	100.0	ml

Modified PTAH Solution
Solution A

Hematoxylin	1.0	gm
Distilled water	500.0	ml

Dissolve with the aid of gentle heat but do not exceed 60°C. Let cool for 10 minutes and add 0.177 gm potassium permanganate. Allow to stand for at least 30 minutes.

Solution B

Phosphotungstic acid	20.0	gm
Distilled water	500.0	ml

Working PTAH

Combine solutions A and B and let stand for 1 hour. This solution is stable for several months. If stored in a dark bottle away from strong light, stock solutions can remain good for 18 months to several years.

STAINING PROCEDURE

1. Decerate slides in xylene, 2 changes, 2 minutes each. Place in absolute alcohol, 2 changes, 2 minutes each; 95% alcohol, 2 changes, 2 minutes each; then rinse in distilled water.
2. Wash in running tap water for 10 minutes.
3. Treat sections with Langeron's iodine for 20 minutes.
4. Place in 5% sodium thiosulfate until sections are decolorized; approximately 1 minute.
5. Wash in running tap water for 5 minutes.
6. Stain in PTAH solution in a 55-60°C oven for 2 hours.
7. Dehydrate slides in 95% alcohol and absolute alcohol 3 changes each. Clear in xylene, 3 changes.
8. Mount coverglass with appropriate medium (refractive index 1.48-1.56).

RESULTS

"A" bands of striated muscle..blue
Smooth muscle cells, fibrin, nuclei ....blue
Collagen, reticulum, basement membranes ....red
Amyloid..bright red
Elastin, membrane..dark blue

REMARKS - CHEMISTRY OF PTAH

During preparation of hemalum solutions, hematoxylin is oxidized, then the hematein and alum solutions are combined. Acid is added last, after hemalum chelates have been formed. In contrast, in the traditional PTAH technique, unoxidized hematoxylin is mixed with the strongly acid phosphotungstic acid (PTA) solution. The oxidant potassium permanganate is then added, removing two hydrogens during oxidation of hematoxylin. These reactions proceed smoothly in aqueous or alcoholic solutions. However, strongly acid solutions contain a large number of protons that react rapidly with electron-donating $_{-}O^{-}$ groups of hematein. Consequently, hematein in acid solutions becomes pale to colorless. Thus, in traditional PTAH solutions as permanganate oxidizes hematoxylin to hematein, the numerous protons of the PTA solution apparently convert many hematein ions into hematoxylin-like products before they can form chelates with tungstic atoms. Therefore, it takes as long as six months for enough chelates to form for optimal staining.

Owing to steric conditions in PTA, red and blue PTAH compounds are formed. The 1:1 hematein:metal complexes are red; the 2:1 chelates are blue.

Martius-Scarlet-Blue Method for Fibrin, Muscle and Collagen

This method has been used for determining the age of gunshot wounds in tissue specimens. It is unfortunate that a reference stating that fact could not be found. Although the method is being presented here for the demonstration of the entities cited in the title, one should not discount its use for determining the age of gunshot wounds.

FIXATION	10% Buffered neutral formalin.
PROCESS	Paraffin, frozen, or celloidin.
MICROTOMY	Cut sections at 6 micrometers.
CONTROL	Use control number 17 or 11.
SOLUTIONS	

Weigert's Hematoxylin
(See Page 79)

1% Acid Alcohol
(See Page 77)

Martius Yellow

Martius yellow	0.5	gm
Phosphotungstic acid	2.0	gm
Water	5.0	ml
Alcohol	95.0	ml

0.5% Aniline Blue

Aniline blue	0.5	gm
Glacial acetic acid	1.0	ml
Distilled water	100.0	ml

Brilliant Crystal Scarlet 6R

Brilliant crystal scarlet 6R	1.0	gm
Glacial acetic acid	2.5	ml
Distilled water	97.5	ml

1% Phosphotungstic Acid

Phosphotungstic acid	1.0	gm
Distilled water	100.0	ml

STAINING PROCEDURE

1. Decerate slides in xylene, 2 changes, 2 minutes each. Place in absolute alcohol, 2 changes, 2 minutes each; 95% alcohol, 2 changes, 2 minutes each; then rinse in distilled water.
2. Stain nuclear chromatin in Weigert's hematoxylin for 15 minutes.
3. Wash in running tap water for 5 minutes.
4. Place in 1% hydrochloric acid alcohol for 10 seconds.
5. Wash in running tap water for 5 minutes.
6. Rinse in 95% alcohol.
7. Stain in Martius yellow for 2 minutes.
8. Rinse in running tap water.
9. Stain in brilliant crystal scarlet 6R for 10 minutes.
10. Rinse in running tap water.
11. Place in 1% phosphotungstic acid for 5 minutes.
12. Rinse in running tap water.
13. Stain in 0.5% aniline blue for 10 minutes.
14. Rinse in running tap water.
15. Dehydrate slides in 95% alcohol and absolute alcohol 3 changes each. Clear in xylene, 3 changes.
16. Mount coverglass with appropriate medium (refractive index 1.48-1.56).

RESULTS

Fibrin	bright red
Muscle	red
Erythrocytes	yellow
Nuclear chromatin	gray
Collagen, basement membranes, reticulin, elastic fibers	blue

Mallory's Method for Collagen and Pituitary Cells[55as]

FIXATION Zenker's solution.
PROCESS Paraffin or celloidin.
MICROTOMY Cut paraffin sections at 6 micrometers.
CONTROL Use control number 11.
SOLUTIONS

0.5% Acid Fuchsin Solution

Acid fuchsin	0.5	gm
Distilled water	100.0	ml

Aniline Blue-Orange G Solution

Aniline blue, water soluble	0.5	gm
Orange G	2.0	gm
Phosphotungstic acid	1.0	gm
Distilled water	100.0	ml

STAINING PROCEDURE

1. Decerate slides in xylene, 2 changes, 2 minutes each. Place in absolute alcohol, 2 changes, 2 minutes each; 95% alcohol, 2 changes, 2 minutes each; then rinse in distilled water.
2. Remove mercuric chloride crystals with iodine and clear with sodium thiosulfate.
3. Wash in running tap water for 10 minutes.
4. Rinse in distilled water.
5. Stain in acid fuchsin for 3 minutes. Staining with acid fuchsin solution may be omitted if it is desired that collagen fibers stand out more sharply.
6. Transfer directly to aniline blue-orange G solution for 30 minutes.
7. Dehydrate slides in 95% alcohol and absolute alcohol 3 changes each. Clear in xylene, 3 changes.
8. Mount coverglass with appropriate medium (refractive index 1.48-1.56).

RESULTS

Nuclear chromatin	red
Collagen fibrils	blue
Cartilage	varying shades of blue
Mucin	varying shades of blue
Amyloid	varying shades of blue
Erythrocytes	yellow
Basophils (beta cell)	blue
Acidophils (alpha cell)	red
Chromophobes	gray
Myelin	yellow

REMARKS
This stain is not especially useful for collagen. However, it is very useful for pituitary cell differentiation.

Pinkus' Method for Elastica and Collagen[142]

FIXATION 10% Buffered neutral formalin.
PROCESS Paraffin, frozen, or celloidin.
MICROTOMY Cut sections at 6 micrometers.
CONTROL Use control number 23.
SOLUTIONS

0.2% Orcein

Orcein, synthetic	0.2	gm
70% Alcohol	100.0	ml
Concentrated hydrochloric acid	0.6	ml

Solution is ready for staining immediately and improves on standing. It is stable for many months.

Dilute Giemsa Solution

Giemsa, any good stock solution	1	drop
Distilled water, adjusted to pH 7.0 with phosphate buffer	20.0	ml

Alcohol-Eosin Solution

1% Alcoholic eosin	2.0	ml
95% Alcohol	100.0	ml

STAINING PROCEDURE
1. Decerate slides in xylene, 2 changes, 2 minutes each. Place in absolute alcohol, 2 changes, 2 minutes each; 95% alcohol, 2 changes, 2 minutes each; then rinse in distilled water.
2. Place in 70% alcohol for 1 minute.
3. Stain in 2% orcein for 30 minutes.
4. Wash in running tap water for 15 minutes.
5. Stain overnight in dilute Giemsa. Do not hurry this step.
6. Drain off excess fluid. If collagen in sections does not have the desired rose-pink color, decolorize slightly and counterstain with eosin. Personal preference and experience controls this step.
7. Dehydrate slides in 95% alcohol and absolute alcohol 3 changes each. Clear in xylene, 3 changes.
8. Mount coverglass with appropriate medium (refractive index 1.48-1.56).

RESULTS

Nuclear chromatin	deep blue
Cytoplasm of epidermis, smooth muscle	light blue
Collagen	rose-pink
Elastic fibers	dark brown
Mast cell granules and mucoid substances	purple
Eosinophilic granules and erythrocytes	bright red

REMARKS
Sections cut at 10 micrometers are preferable, since they give a better three-dimensional picture. Step 3 stains elastic fibers specifically. If background becomes overstained, decolorize by dipping in absolute alcohol or 0.1% hydrochloric acid alcohol.

Hart's Method for Elastica

FIXATION	10% Buffered neutral formalin.
PROCESS	Paraffin, frozen, or celloidin.
MICROTOMY	Cut paraffin sections at 6 micrometers.
CONTROL	Use control number 16.

SOLUTIONS

0.25% Potassium Permanganate

Potassium permanganate	0.25	gm
Distilled water	100.0	ml

5% Oxalic Acid

Oxalic acid	5.0	gm
Distilled water	100.0	ml

Resorcin-Fuchsin (Stock)

Basic fuchsin	2.0	gm
Resorcinol	4.0	gm
Distilled water	200.0	ml

Mix in a porcelain dish, bring to boil and let boil for 1 minute, then add 25 ml of 29% ferric chloride solution. Cool, filter, and leave precipitate on filter paper until dry. Return to porcelain dish, which should be dry but still contains whatever part of the precipitate remains adherent to it. Add 200 ml of 95% alcohol and heat very carefully, dissolving all precipitate from filter paper before discarding it. Add 4 ml of hydrochloric acid. This solution keeps for several months.

Resorcin-Fuchsin (Working)

Resorcin-fuchsin (stock)	10.0	ml
70% Alcohol	100.0	ml
Concentrated hydrochloric acid	2.0	ml

Van Gieson's
(See Page 362)

STAINING PROCEDURE
1. Decerate slides in xylene, 2 changes, 2 minutes each. Place in absolute alcohol, 2 changes, 2 minutes each; 95% alcohol, 2 changes, 2 minutes each; then rinse in distilled water.
2. Oxidize in potassium permanganate for 5 minutes.
3. Rinse in distilled water for 30 seconds.
4. Place in oxalic acid until sections are clear.
5. Wash in running tap water for 5 minutes, then place in distilled water.
6. Stain in working resorcin-fuchsin overnight.

7. Wash in running tap water for 10 minutes.
8. Rinse in distilled water.
9. Stain in Van Gieson's solution for 1 minute.
10. Dehydrate slides in 95% alcohol and absolute alcohol 3 changes each. Clear in xylene, 3 changes.
11. Mount coverglass with appropriate medium (refractive index 1.48-1.56).

RESULTS

Elastic fibers	blue-black to black
Nuclear chromatin	blue to black
Collagen	pink to red
Other tissue elements	yellow

REMARKS
This method is especially useful in the demonstration of fine elastic fibers and elacin (degenerated elastic fibers).

Luna's Modification of Shorr's Method for Bone, Cartilage, Collagen, Muscle, Keratin and RNA Inclusion Bodies

(See Fig. 180)

Shorr's method was originally used for the demonstration of inclusion bodies, but it is useful in the interpretation of other pathologic entities, as well. Besides inclusion bodies, Shorr's method demonstrates collagen, muscle, keratin and erythrocytes. The modification provided below (see remarks) was made to differentiate bone from cartilage and other entities.

FIXATION	10% Buffered formalin or Bouin's solutions. If formalin is used mordant slide as indicated in step 2.
PROCESS	Paraffin or celloidin.
MICROTOMY	Cut sections at 6 micrometers.
CONTROL	Use control number 43, 45 or bone specimen.

SOLUTIONS

Aldehyde Fuchsin

Basic fuchsin	1.0	gm
70% Alcohol	200.0	ml
Concentrated hydrochloric acid	2.0	ml
Paraldehyde	2.0	ml

Let stand at room temperature for 2 to 3 days or until stain is deep purple in color. Store in refrigerator.

1% Acid Alcohol

Concentrated hydrochloric acid	1.0	ml
70% Alcohol	99.0	ml

Shorr's

50% Alcohol	100.0	ml
Biebrich scarlet	0.5	gm
Orange G	0.25	gm
Fast green, FCF	0.075	gm
Phosphotungstic acid	0.5	gm
Phosphomolybdic acid	0.5	gm
Glacial acetic acid	1.0	ml

Harris' Hematoxylin

Hematoxylin crystals	5.0	gm
95% Alcohol	50.0	ml
Ammonium or potassium alum	100.0	gm
Distilled water	1000.0	ml
Mercuric oxide	2.5	gm

Dissolve the hematoxylin in the alcohol and the alum in water with the aid of heat. Mix the two solutions. Bring the mixture to a boil as rapidly as possible; then remove from the heat and add the mercuric oxide. Reheat the solution until it becomes a dark purple (about 1 minute) and promptly plunge the container into a basin of cold water. The solution is ready to use when cool. Add 2 to 4 ml of glacial acetic acid for every 100 ml of solution.

Ammonia Water

Tap water	1000.0	ml
Ammonium hydroxide	3.0	ml

STAINING PROCEDURE

1. Decerate slides in xylene, 2 changes, 2 minutes each. Place in absolute alcohol, 2 changes, 2 minutes each; 95% alcohol, 2 changes, 2 minutes each; then rinse in distilled water.
2. Stain in Harris' hematoxylin for 3 minutes.
3. Rinse in running tap water for 30 seconds.
4. Differentiate in 1% acid alcohol until there is no hematoxylin in the cytoplasm of the cells. Check with microscope.
5. Dip in tap water, 5 times.
6. Dip sections 5 times in ammonia water to blue. Check with microscope.
7. Wash slides for 10 minutes in running tap water. (See remarks for differentiating cartilage from bone.)
8. Place slides in Shorr's staining solution for 1 minute.
9. Wash slides in 95% alcohol. Check with microscope. Connective tissue will be a clear light green when differentiation is complete.
10. Rinse several times in absolute alcohol and clear in 3 changes of xylene.
11. Mount coverglass with appropriate medium (refractive index 1.48-1.56).

RESULTS (when staining for bone and cartilage)

Bone	reddish-purple
Cartilage	purple
Collagen	green
Muscle	red
Nuclear chromatin	purplish-black

RESULTS (when staining for inclusion bodies)

Inclusion bodies	orange-red
Collagen	green
Muscle	red
Keratin	orange-red
Erythrocytes	red
Nuclear chromatin	grayish-purple

REMARKS

The staining procedure for the differentiation of bone from cartilage requires the introduction of aldehyde fuchsin between steps 7 and 8. Following step 7 rinse the slides in 95% alcohol and place in aldehyde fuchsin for 30 minutes. Rinse slides in 2 changes of 95% ethyl alcohol, followed by a water rinse, before placing slides in Shorr's solution.

Kellogg-McAuliffe Method for Myoepithelial Cells

FIXATION	10% Buffered neutral formalin.
PROCESS	Paraffin.
MICROTOMY	Cut sections at 3 to 5 micrometers.
CONTROL	Use control cross section of artery.
SOLUTIONS	

Bouin's Solution
(See Page 13)

5% Tannic Acid

Tannic acid	5.0	gm
Distilled water	100.0	ml

Allow to ripen for 48 hours before use.

1% Phosphomolybdic Acid

Phosphomolybdic acid	1.0	gm
Distilled water	100.0	ml

Amido Black

Methyl alcohol	90.0	ml
Glacial acetic acid	10.0	ml

Saturate with amido black.

Phloxine B

Methyl alcohol	90.0	ml
Glacial acetic acid	10.0	ml

Saturate with phloxine B.

Methyl Alcohol-Acetic Acid

Methyl alcohol	90.0	ml
Glacial acetic acid	10.0	ml

STAINING PROCEDURE

1. Decerate slides in xylene, 2 changes, 2 minutes each. Place in absolute alcohol, 2 changes, 2 minutes each; 95% alcohol, 2 changes, 2 minutes each; then rinse in distilled water.
2. Rinse in distilled water.
3. Postfix for 1 hour in a covered coplin jar in Bouin's fixative preheated to 56°C. Allow slides to cool in Bouin's for 10 to 20 minutes. Slides may be left at this stage for 48 hours.
4. Wash in running tap water until sections are colorless; 5 to 10 minutes.
5. Rinse in 2 changes of distilled water, 1 minute each.
6. Mordant in 5% aqueous tannic acid for 5 to 20 minutes.
7. Rinse slides in distilled water and blot.
8. Mordant in 1% aqueous phosphomolybdic acid for 10 to 15 minutes. This solution turns green and cannot be reused. Only 3 to 4 slides should be processed in a coplin jar.
9. Rinse slides for 30 seconds in distilled water.
10. Rinse slides in several dips of methyl alcohol.
11. Stain in saturated solution of amido black or phloxine B for 30 minutes to 1 hour.
12. Rinse in several changes of methyl alcohol-glacial acetic acid.
13. Finish dehydration in 2 changes of absolute ethyl alcohol.
14. Clear in 2 changes of xylene and mount with a nonfluorescent medium.

RESULTS

With amido black:
Myoepithelial cells..deep blue-green
Muscles and erythrocytes...blue
Collagen and basement membranes....................................golden yellow

With Phloxine B:
Myoepithelial cells and muscle...intensely pink
Background..golden yellow

REMARKS

If the background is too heavy, the time in tannic acid should be increased.

Roush-Breur-Wilson Method for Dental Structures[143]

FIXATION 10% Buffered neutral formalin.
PROCESS Paraffin.
MICROTOMY Cut sections at 5 micrometers.
CONTROL Use control containing tooth and jaw bone.
SOLUTIONS

1% Sirius Red (Stock)

Sirius red	1.0	gm
Distilled water	100.0	ml

Picrosirius Red (Working)

Sirius red (stock)...	10.0	ml
Saturated aqueous picric acid	90.0	ml

Mayer's Hematoxylin
(See Page 81)

STAINING PROCEDURE

1. Decerate slides in xylene, 2 changes, 2 minutes each. Place in absolute alcohol, 2 changes, 2 minutes each; 95% alcohol, 2 changes, 2 minutes each; then rinse in distilled water.
2. Stain for 1 hour in picrosirius red.
3. Wash briefly in running tap water.
4. Stain for 15 minutes in Mayer's hematoxylin.
5. Wash in running tap water for 15 minutes.
6. Dehydrate slides in 95% alcohol and absolute alcohol 3 changes each. Clear in xylene, 3 changes.
7. Mount coverglass with appropriate medium (refractive index 1.48-1.56).

RESULTS

Cementum	red
Connective tissues	red
Bone	red
Stratified squamous epithelium	light blue
Keratohyalin granules of the stratum corneum	red
Epineurium and perineurium	red
Dentin	light red
Korff's fibers	dark red
Dentinal tubules	dark red
Odontiblasts, fibroblasts, mesenchymal cells	light blue
Nuclei	dark blue
Sharpey's fibers	red

Gordon-Sweet's Method for Reticulin Fibers[144]

FIXATION 10% Buffered neutral formalin or Bouin's solution.
PROCESS Paraffin or celloidin.
MICROTOMY Cut sections at 6 micrometers.
CONTROL Use control number 38.
SOLUTIONS

Acidified Potassium Permanganate

0.5% Aqueous potassium permanganate	47.5	ml
3% Sulfuric acid	2.5	ml

1% Oxalic Acid

Oxalic acid	2.0	gm
Distilled water	200.0	ml

2.5% Ferric Ammonium Sulfate

Ferric ammonium sulfate	5.0	gm
Distilled water	200.0	ml

3% Sodium Hydroxide (Stock)

Sodium hydroxide	1.5	gm
Distilled water	50.0	ml

10.2% Aqueous Silver Nitrate

Silver nitrate	5.1	gm
Distilled water	50.0	ml

10% Formalin

Concentrated formalin, 37%	20.0	ml
Distilled water	200.0	ml

0.2% Gold Chloride

1% Gold chloride	4.0	ml
Distilled water	200.0	ml

5% Sodium Thiosulfate

Sodium thiosulfate	5.0	gm
Distilled water	100.0	ml

Ammonical Silver Solution

To 5 ml of 10.2% aqueous silver nitrate solution add strong ammonium hydroxide solution, drop by drop, until the precipitate which forms is just dissolved. Add 5.0 ml stock 3% sodium hydroxide to the ammoniated silver solution, redissolve the resultant preceipitate with a drop or two of strong ammonium solution and dilute to 50.0 ml with distilled water.

STAINING PROCEDURE

1. Decerate slides in xylene, 2 changes, 2 minutes each. Place in absolute alcohol, 2 changes, 2 minutes each; 95% alcohol, 2 changes, 2 minutes each; then rinse in distilled water.
2. Oxidize in acidified potassium permanganate for 5 minutes.
3. Wash in running tap water for 2 minutes.
4. Bleach in 1% oxalic acid for 5 minutes.
5. Wash in running tap water and then 2 to 3 changes of distilled water.
6. Mordant in 2.5% ferric ammonium sulfate for 10 minutes.
7. Wash in 3 changes of distilled water.
8. Impregnate with 10.2% aqueous silver nitrate until sections are transparent (30 seconds).
9. Wash well in 4 changes of distilled water.
10. Reduce in 10% formalin. Move each section back and forth in this solution. Reduction is complete almost immediately.
11. Wash in running tap water and then in distilled water.
12. Tone in 0.2% gold chloride for 10 to 15 minutes or until section turns a purplish color.
13. Rinse in distilled water for 30 seconds.

14. Fix in 5% sodium thiosulfate for 5 minutes.
15. Wash in running tap water for 30 seconds.
16. Dehydrate slides in 95% alcohol and absolute alcohol 3 changes each. Clear in xylene, 3 changes.
17. Mount coverglass with appropriate medium (refractive index 1.48-1.56).

RESULTS

Reticulin fibers	dark purple
Cartilage and bone	purplish-gray
Elastic	black
Nerve fibers	black
Background	almost colorless

Luna's PTAH Method for Muscle Entities

(See Fig. 186 and 187)

FIXATION 10% Buffered neutral formalin.
PROCESS Paraffin or celloidin.
MICROTOMY Cut sections at 6 micrometers.
CONTROL Use control number 14.
SOLUTIONS

Phosphotungstic Acid-Hematoxylin

Hematoxylin	1.0	gm
Phosphotungstic acid	20.0	gm
Distilled water	1000.0	ml

Dissolve the solid ingredients in separate portions of distilled water, the hematoxylin with the aid of gentle heat. When cool, combine solutions. Preservatives are not necessary. Ripening requires several weeks but the addition of 0.2 gm of potassium permanganate to each 1000 ml of solution will ripen the stain immediately.

Lugol's Iodine

Iodine	1.0	gm
Potassium iodide	2.0	gm
Distilled water	100.0	ml

2.5% Potassium Dichromate (Stock)

Potassium dichromate	2.5	gm
Distilled water	100.0	ml

Potassium Dichromate (Working)

2.5% Potassium dichromate (stock)	95.0	ml
Glacial acetic acid	5.0	ml

STAINING PROCEDURE

1. Decerate slides in xylene, 2 changes, 2 minutes each. Place in absolute alcohol, 2 changes, 2 minutes each; 95% alcohol, 2 changes, 2 minutes each; then rinse in distilled water.
2. Mordant sections in working potassium dichromate overnight.

3. Wash slides in running tap water for 10 minutes.
4. Place slides in Lugol's iodine for 15 minutes.
5. Place slides in 95% alcohol until sections are devoid of iodine (brownish) coloration; minimum time, 1 hour.
6. Rinse slides in 3 rapid changes of distilled water.
7. Stain slides in phosphotungstic acid-hematoxylin solution for 1 hour or more. Check sections microscopically every hour until muscle striations are well demonstrated.
8. Rinse slides briefly in 95% alcohol to remove excess stain.
9. Dehydrate slides quickly in absolute alcohol 3 changes each. Clear in xylene, 3 changes.
10. Mount coverglass with appropriate medium (refractive index 1.48-1.56).

RESULTS

Muscle fibers	bluish-purple
Muscle striations	dark blue
Collagen	reddish-orange
Nuclear chromatin	blue

REMARKS

This is the method of choice for staining muscle, particularly muscle biopsies. The staining results are light, producing a delicate demonstration of muscle elements. Sodium thiosulfate (hypo) is not used to clear the Lugol's iodine (step 5), since it will impair staining.

Russell's Modification of Movat's Pentachrome Stain

(See Fig. 183, 184 and 185)

FIXATION	10% Buffered neutral formalin or formol-sublimate.
PROCESS	Paraffin.
MICROTOMY	Cut sections at 6 micrometers.
CONTROL	Use control number 27.
SOLUTIONS	

1% Alcian Blue

Alcian blue, 8 GS	1.0	gm
Distilled water	100.0	ml
Glacial acetic acid	1.0	ml

Alkaline Alcohol

Ammonium hydroxide	10.0	gm
95% Alcohol	90.0	ml

Iodine-Iodide (Stock)

Iodine	2.0	gm
Potassium iodide	4.0	gm
Distilled water	100.0	ml

10% Hematoxylin (Stock)

Hematoxylin	10.0	gm
100% Alcohol	100.0	ml

10% Ferric Chloride (Stock)

Ferric chloride	10.0	gm
Distilled water	100.0	ml

Hematoxylin (Working)

10% Hematoxylin (stock)	25.0	ml
Absolute alcohol	25.0	ml
10% Ferric chloride (stock)	25.0	ml
Iodine-iodide (stock)	25.0	ml

2% Ferric Chloride (Working)

10% Ferric chloride	20.0	ml
Distilled water	80.0	ml

0.5% Acetic Acid Water

Glacial acetic acid	0.5	ml
Distilled water	100.0	ml

5% Phosphotungstic Acid

Phosphotungstic acid	5.0	gm
Distilled water	100.0	ml

5% Sodium Thiosulfate

Sodium thiosulfate	5.0	gm
Distilled water	100.0	ml

Crocein Scarlet-Acid Fuchsin
Solution A (Stock)

Crocein scarlet	1.0	gm
Distilled water	99.5	ml
Glacial acetic acid	0.5	ml

Solution B (Stock)

Acid fuchsin	0.1	gm
Distilled water	99.5	ml
Glacial acetic acid	0.5	ml

Working Solution

Solution A	8	parts
Solution B	2	parts

Alcoholic Safran

Safran du Gantinais	6.0	gm
Absolute alcohol	100.0	ml

Keep tightly closed to prevent hydration.

STAINING PROCEDURE

1. Decerate slides in xylene, 2 changes, 2 minutes each. Place in absolute alcohol, 2 changes, 2 minutes each; 95% alcohol, 2 changes, 2 minutes each; then rinse in distilled water.
2. Stain in 1% alcian blue for 20 minutes.
3. Wash in running tap water for 5 minutes.
4. Place slides in alkaline alcohol for 30 minutes.
5. Wash in running tap water for 10 minutes.
6. Rinse in distilled water.
7. Stain in working hematoxylin for 15 minutes.
8. Rinse in several changes of distilled water.
9. Differentiate in 2% ferric chloride for 1 minute. Rinse in distilled water.
10. Place slides in 5% sodium thiosulfate for 1 minute.
11. Wash in running tap water for 5 minutes; rinse in distilled water.
12. Stain in working crocein scarlet-acid fuchsin for 1 minute.
13. Rinse in several changes of distilled water.
14. Rinse in 0.5% acetic acid water for 30 seconds.
15. Place slide in 5% phosphotungstic acid, 2 changes for 5 minutes each.
16. Rinse in 0.5% acetic acid water for 30 seconds.
17. Rinse in 3 changes of absolute alcohol.
18. Stain in alcoholic safran for 15 minutes.
19. Dehydrate slides in 95% alcohol and absolute alcohol 3 changes each. Clear in xylene, 3 changes.
20. Mount coverglass with appropriate medium (refractive index 1.48-1.56).

RESULTS

Nuclei and elastic fibers .... black
Collagen .... yellow
Ground substance, mucin .... blue
Fibrinoid, fibrin .... intense red
Muscle .... red

REMARKS

Differentiation of the elastica is usually accomplished in 2 to 3 minutes and is complete when the elastic fibers remain black and the background becomes relatively clear. Most sections contain blood vessels and the elastica within them is easily identified, which is helpful in differentiation. A control section should be included each time the stain is performed. (We usually use a section of skin.) The complete removal of alkaline alcohol with running water (step 5 of the staining procedure) is important. Failure to remove all alkaline alcohol will inhibit the stains that follow.

Goldner's Trichrome Method

FIXATION 10% Buffered neutral formalin or Bouin's solution.
PROCESS Paraffin, frozen, or celloidin.
MICROTOMY Cut paraffin sections at 6 micrometers.
CONTROL Use control number 11 or selected specimen.

SOLUTIONS

Bouin's Solution
(See Page 13)

Ponceau 2R-Acid Fuchsin

Ponceau 2R	0.07	gm
Acid fuchsin	0.03	gm
Glacial acetic acid	0.2	ml
Distilled water	100.0	ml

Phosphomolybdic Acid-Orange G

Phosphomolybdic acid	4.0	gm
Orange G	2.0	gm
Distilled water	100.0	ml

1% Acetic Acid

Acetic acid	1.0	ml
Distilled water	99.0	ml

Light Green

Light green	0.2	gm
Glacial acetic acid	0.2	ml
Distilled water	100.0	ml

Weigert's Hematoxylin
(See Page 79)

STAINING PROCEDURE

1. Decerate slides in xylene, 2 changes, 2 minutes each. Place in absolute alcohol, 2 changes, 2 minutes each; 95% alcohol, 2 changes, 2 minutes each; then rinse in distilled water.
2. Mordant in Bouin's solution for 1 hour at 56°C, or overnight at room temperature, if formalin fixed.
3. Cool and wash in running tap water until yellow color disappears.
4. Rinse in distilled water.
5. Place in Weigert's hematoxylin for 10 minutes.
6. Wash in running tap water for 10 minutes.
7. Rinse in distilled water.
8. Stain in Ponceau 2R-acid fuchsin for 5 minutes.
9. Wash in 1% acetic acid.
10. Place in phosphomolybdic acid-orange G solution until collagen is decolorized.
11. Rinse in 1% acetic acid for 30 seconds.
12. Stain in light green for 5 minutes.
13. Rinse in 1% acetic acid for 5 minutes.
14. Blot sections but do not allow to dry.
15. Dip quickly in absolute alcohol.
16. Dehydrate slides in 95% alcohol and absolute alcohol 3 changes each. Clear in xylene, 3 changes.
17. Mount coverglass with appropriate medium (refractive index 1.48-1.56).

RESULTS

Nuclear chromatin	brownish-black
Cytoplasm	bright red
Erythrocytes	orange
Muscle	red
Collagen	green

Method for Differentiating Precollagen from Collagen

FIXATION	10% Buffered neutral formalin.
PROCESS	Paraffin.
MICROTOMY	Cut sections at 6 micrometers.
CONTROL	Use control number 11.
SOLUTIONS	

Celestine Blue

Celestine blue	0.5	gm
Iron alum	5.0	gm
Distilled water	100.0	ml
Glycerin	20.0	ml

Dissolve the celestine blue and iron alum in distilled water. Heat and let boil for 3 minutes. When cool, add glycerin.

Hematoxylin

Aluminum sulfate, 5% aqueous	100.0	ml
Hematoxylin, 1% alcoholic	50.0	ml
Distilled water	100.0	ml
Ferric chloride, 4% aqueous	10.0	ml
Concentrated hydrochloric acid	1.0	ml

Heat the aluminum sulfate solution. At the boiling point add very slowly the 1% alcoholic hematoxylin. Let boil for 3 minutes. When cool, add the distilled water, aqueous solution of ferric chloride and hydrochloric acid.

0.25% Metanil Yellow

Metanil yellow	0.25	gm
Distilled water	100.0	ml
Glacial acetic acid	8	drops

Acidified Water

Distilled water	100.0	ml
Glacial acetic acid	8	drops

Lithium Carbonate

Lithium carbonate, saturated aqueous	3	drops
Distilled water	100.0	ml

Picro-polychrome (Stock)

Solution A

Methyl blue	0.1	gm
Distilled water	100.0	ml

Solution B

Acid fuchsin	0.2	gm
Picric acid, saturated aqueous	100.0	ml

Picro-polychrome (Working)

Combine equal parts of Solution A and Solution B. Add 20 ml of glycerin and 1 ml of saturated aqueous lithium carbonate solution.

1% Acetic Acid

Distilled water	99.0	ml
Glacial acetic acid	1.0	ml

STAINING PROCEDURE

1. Decerate slides in xylene, 2 changes, 2 minutes each. Place in absolute alcohol, 2 changes, 2 minutes each; 95% alcohol, 2 changes, 2 minutes each; then rinse in distilled water.
2. Stain in celestine blue for 5 minutes.
3. Rinse in distilled water.
4. Stain in hematoxylin for 2 minutes.
5. Wash slides for 15 to 30 minutes in running tap water.
6. Rinse in distilled water for 30 seconds.
7. Stain slides in 0.25% metanil yellow for 2 minutes.
8. Differentiate in acidified water for 30 seconds.
9. Rinse in distilled water.
10. Place slides in lithium carbonate for 2 minutes.
11. Stain in working picro-polychrome for 2 minutes.
12. Dip the slides in 1% acetic acid for 2 minutes.
13. Dehydrate slides in 95% alcohol and absolute alcohol 3 changes each. Clear in xylene, 3 changes.
14. Mount coverglass with appropriate medium (refractive index 1.48-1.56).

RESULTS

Nuclei	black
Cytoplasm	green-yellow
Keratin, muscle and red blood cells	bright yellow
Young collagen	blue
Mature collagen	red
Hyaline	blue-green in non-fibrillary spots

Luna's Method for Heart Muscle Intercalated Discs

FIXATION	10% Buffered neutral formalin or Bouin's solution (preferred).
PROCESS	Paraffin, frozen, or celloidin.
MICROTOMY	Cut sections at 6 micrometers.
CONTROL	Use control number 14 (cardiac muscle).
SOLUTIONS	

Weigert's Hematoxylin
(See Page 79)

Bouin's Solution
(See Page 13)

Chromotrope 2R-Light Green

Chromotrope 2R	0.1	gm
Light green	0.3	gm
Acetic acid	1.0	ml
Phosphotungstic acid	0.8	gm
Distilled water	100.0	ml

0.5% Acetic Acid

Glacial acetic acid	0.5	ml
Distilled water	100.0	ml

STAINING PROCEDURE

1. Decerate slides in xylene, 2 changes, 2 minutes each. Place in absolute alcohol, 2 changes, 2 minutes each; 95% alcohol, 2 changes, 2 minutes each; then rinse in distilled water.
2. Mordant in Bouin's solution in a 58-60°C oven for 2 hours. (This step is necessary only if tissues were fixed in formalin or other fixative not containing picric acid.)
3. Wash in running tap water for 15 minutes.
4. Stain with Weigert's or Gomori's hematoxylin for 10 minutes.
5. Wash in running tap water for 10 minutes.
6. Stain in chromotrope 2R-light green for 10 minutes.
7. Differentiate in 0.5% aqueous acetic acid for 2 minutes.
8. Dehydrate slides in 95% alcohol and absolute alcohol 3 changes each. Clear in xylene, 3 changes.
9. Mount coverglass with appropriate medium (refractive index 1.48-1.56).

RESULTS

Muscle fiber	purplish-red
Intercalated discs	(center) bright green, (peripheral) red
Nuclear chromatin	blue-black
Erythrocytes	orange

Safranin O Method for Cartilage and Mast Cells

FIXATION 10% Neutral formalin.
PROCESS Paraffin.
MICROTOMY Cut sections at 6 micrometers.
CONTROL Use control specimen of cartilage.
SOLUTIONS

Weigert's Iron Hematoxylin
(See Page 79)

Fast Green

Fast green	1.0	gm
Distilled water	5000.0	ml

Note: Although the amount of distilled water used in the fast green seems excessive, the formulation is correct.

1% Acetic Acid

Glacial acetic acid	1.0	ml
Distilled water	99.0	ml

0.1% Safranin O

Safranin O	0.1	gm
Distilled water	100.0	ml

STAINING PROCEDURE

1. Decerate slides in xylene, 2 changes, 2 minutes each. Place in absolute alcohol, 2 changes, 2 minutes each; 95% alcohol, 2 changes, 2 minutes each; then rinse in distilled water.
2. Stain in working Weigert's hematoxylin for 7 minutes.
3. Wash in running tap water for 10 minutes.
4. Counterstain in aqueous fast green for 3 minutes.
5. Rinse and agitate slides in 1% acetic acid for 15 seconds.
6. Stain in 0.1% safranin O for 5 minutes.
7. Dehydrate slides in 95% alcohol and absolute alcohol 3 changes each. Clear in xylene, 3 changes.
8. Mount coverglass with appropriate medium (refractive index 1.48-1.56).

RESULTS

Nuclear chromatin	black
Cytoplasm	gray-green
Mucus, cartilage, and mast cell granules	orange-red

A Method for Myofibrils and Intermyofibrillary Material (Nemaline Rods)

FIXATION 10% Buffered neutral formalin.
PROCESS Frozen.
MICROTOMY Cut sections at 8 micrometers.
CONTROL Use material containing nemaline rods.

SOLUTIONS

Mayer's Hematoxylin
(See Page 81)

Gomori's Solution

Chromotrope 2R	0.8	gm
Fast green FCF	0.3	gm
Phosphotungstic acid	0.6	gm
Glacial acetic acid	1.0	ml

Distilled water to make 100 ml of solution; pH should be 3.4.

0.2% Glacial Acetic Acid

Glacial acetic acid	0.2	ml
Distilled water	100.0	ml

STAINING PROCEDURE

1. Stain in Mayer's hematoxylin for 10 minutes.
2. Rinse in distilled water for 30 seconds.
3. Stain in Gormori's solution for 10 minutes.
4. Differentiate in 0.2% acetic acid for 30 seconds.
5. Rinse in distilled water.
6. Dehydrate slides in 95% alcohol and absolute alcohol 3 changes each. Clear in xylene, 3 changes.
7. Mount coverglass with appropriate medium (refractive index 1.48-1.56).

RESULTS

Myofibrils	green
A bands	dark green
I bands	light green
Intermyofibrillary material (nemaline rods)	bright red

Hinds-Murray-Kitynsky Method for Glomerular Basement Membranes[145]

FIXATION 10% Buffered neutral formalin.
PROCESS Paraffin.
MICROTOMY Cut section at 2 micrometers
CONTROL Use control number 6.
SOLUTIONS

3% Methenamine (Stock)

Methenamine	3.0	gm
Distilled water	100.0	ml

5% Silver Nitrate (Stock)

Silver nitrate	5.0	gm
Distilled water	100.0	ml

5% Sodium Tetraborate (Borax)

Sodium tetraborate	5.0	gm
Distilled water	100.0	ml

Methenamine-Silver Nitrate (Stock)

5% Silver nitrate (stock)	5.0	ml
3% Methenamine (stock)	100.0	ml

Shake well to dissolve any precipitate.

Methenamine-Silver Nitrate (Working)

Methenamine - silver (stock)	20.0	ml
Distilled water	30.0	ml
Borax	2.0	ml

0.5% Periodic Acid

Periodic acid	0.5	gm
Distilled water	100.0	ml

5% Sodium Thiosulfate

Sodium thiosulfate	5.0	gm
Distilled water	100.0	ml

0.1% Gold Chloride

Gold chloride	0.1	gm
Distilled water	100.0	ml

STAINING PROCEDURE

1. Decerate slides in xylene, 2 changes, 2 minutes each. Place in absolute alcohol, 2 changes, 2 minutes each; 95% alcohol, 2 changes, 2 minutes each; then rinse in distilled water.
2. Place tissue in 0.5% periodic acid for 7 minutes at room temperature.
3. Rinse in 4 changes distilled water, leave in the final change for 2 minutes.
4. Place the slides in an unused, vertical, plastic slide mailer. Add methenamine-silver working solution to the mailer to cover 2/3 of the slides, or to just below the frosted tip. Tissue sections must be completely immersed. Press the lid down securely and place mailer into a 200 ml pyrex beaker with distilled water to the same level as the methenamine-silver solution in the mailer.
5. Microwave for 2 minutes.
6. Remove the beaker from the microwave and place in a conventional oven at 70°C for 1 minute.
7. Remove slides from oven after 1 minute and rinse in distilled water. Check slides microscopically to see if they have reached the desired intensity. If the slides are not sufficiently stained, rinse them well in distilled water. Wipe the backs with lint-free tissue paper and place them into a fresh working methenamine-silver solution in another unused slide mailer. Replace the mailer in the beaker and put into the conventional oven for 30 seconds. Rinse slides in distilled water and re-examine them microscopically. Alternately, the sections can be kept in the working methenamine-silver solution at room temperature for 1 to 2 minutes, after impregnation in the microwave. However, they must be examined microscopically at 30 second intervals.

8. Rinse in distilled water and tone in 0.1% gold chloride for 30 seconds or until sections turn gray.
9. Rinse in distilled water and examine microscopically. If necessary, bleach by dipping in 5% sodium thiosulfate for no more than 3 seconds.
10. Wash in running water for 2 minutes.
11. Rinse in distilled tap water. Counterstain with hematoxylin and eosin, if desired.
12. Dehydrate slides in 95% alcohol and absolute alcohol 3 changes each. Clear in xylene, 3 changes.
13. Mount coverglass with appropriate media (refractive index 1.48-1.56).

RESULTS

Glomerular basement membranes..black
Tubular basement membranes ....black
Background .... depends on counterstain used

REMARKS

All glassware and other utensils should be acid cleaned.

Chapter 14

Fats and Lipids

Schultz's Method for Cholesterol

(See Fig. 191)

FIXATION 10% Buffered neutral formalin.
PROCESS Frozen sections.
CRYOTOMY Cut sections at 10 micrometers.
CONTROL Use control number 10.
SOLUTIONS

2.5% Ferric Ammonium Sulfate

Ferric ammonium sulfate	2.5	gm
Distilled water	100.0	ml

Acetic-Sulfuric Acid

Glacial acetic acid	50.0	ml
Sulfuric acid, concentrated	50.0	ml

Constantly stirring, add the sulfuric acid very slowly to the acetic acid in a test tube or flask cooled with ice. Prepare fresh just before using.

STAINING PROCEDURE

1. Collect cut frozen sections in distilled water.
2. Place sections in 2.5% ferric ammonium sulfate solution at room temperature for 3 days.
3. Rinse briefly in distilled water, 3 changes.
4. Float sections onto the slides and blot dry. Slides may be held at this point until the pathologist is ready for wet reading.
5. Place 1 drop of freshly prepared acetic-sulfuric acid solution on each section.
6. Apply coverglass immediately, but carefully, on the drop of acetic-sulfuric acid.
7. Examine microscopically within a few minutes. Some residual faint acid smell may be noticed.

RESULTS

Cholesterol	green, blue-green or blue
Background	colorless

Oil Red O Propylene Glycol for Fat

(See Fig. 190)

FIXATION 10% Buffered neutral formalin.
PROCESS Frozen sections.
CRYOTOMY Cut sections at 10 micrometers.
CONTROL Use control specimen containing fat.

SOLUTIONS

0.5% Oil Red O
(See Page 352)

85% Propylene Glycol
(See Page 352)

Mayer's Hematoxylin
(See Page 81)

Glycerin Jelly

Gelatin	10.0	gm
Distilled water	60.0	ml
Heat until gelatin is dissolved.		
Add:		
Glycerin	70.0	ml
Phenol	1.0	ml

STAINING PROCEDURE
1. Mount cut sections on slides containing proper adhesive and allow to dry for 1 hour or more.
2. Place in 100% propylene glycol for 2 minutes.
3. Stain in 0.5% oil red O solution for 30 minutes.
4. Differentiate in 85% propylene glycol solution for 1 minute.
5. Rinse in distilled water, 2 changes.
6. Stain in Mayer's hematoxylin solution for 5 minutes.
7. Rinse under running tap water, 5 minutes.
8. Mount coverglass with glycerin jelly.

RESULTS

Fat	red
Nuclear chromatin	blue

Fischler's Method for Fatty Acids

FIXATION 10% Buffered neutral formalin.
PROCESS Frozen sections.
CRYOTOMY Cut sections at 8 micrometers.
CONTROL Use control specimen containing fat.
SOLUTIONS

Alcohol-Ethyl Ether Extraction Solution

100% Alcohol	50.0	ml
Ethyl ether	50.0	ml

Saturated Cupric Acetate

Cupric acetate, normal ($Cu[CH_3COO]_2H_20$)	10.0	gm
Distilled water	100.0	ml

Lithium-Hematoxylin

Hematoxylin	1.0	gm
100% alcohol	10.0	ml
Dissolve hematoxylin, then add:		
Distilled water	90.0	ml
Saturated aqueous lithium carbonate	1.0	ml

Borax-Potassium Ferricyanide

Borax	2.0	gm
Potassium ferricyanide	2.5	gm
Distilled water	100.0	ml

To use, dilute 1 part of this solution with 10 parts of distilled water.

STAINING PROCEDURE

Fatty Acid Extraction Procedure

1. Treat one control section as follows. Place in alcohol-ethyl ether extraction solution for 3 changes, 2 minutes each. Continue with all sections as follows:
2. Place in cupric acetate solution for 4 hours at room temperature.
3. Rinse in distilled water for 3 minutes.
4. Stain in lithium-hematoxylin for 20 minutes.
5. Decolorize in diluted borax-potassium-ferricyanide solution until erthrocytes are decolorized.
6. Rinse in distilled water for 1 minute.
7. Mount coverglass with glycerin jelly.

RESULTS

Fatty acids .... deep blue-black

REMARKS

Iron, hemaglobin and calcium may also stain but are not removed by the extraction procedure. Free fatty acid will be removed by the extraction procedure. Following the extraction test, only the calcium salts will remain.

Osmium Tetroxide Method for Fat

Osmium tetroxide blackens fats containing double bonds. Actually any double bond, whether lipid or not, will be blackened by osmium tetroxide.

FIXATION	10% Buffered neutral formalin.
PROCESS	Frozen sections.
CRYOTOMY	Cut sections at 8 micrometers.
CONTROL	Use control specimen containing fat.
SOLUTIONS	

1% Osmic Acid (Osmium Tetroxide)

Osmium tetroxide	1.0	gm
Distilled water	100.0	ml

Score a 1 gm ampule of osmium tetroxide with a file and drop into a bottle containing the distilled water. Vigorous shaking will break the ampule safely. Note: Use osmium tetroxide under a hood or in a well-ventilated room. Do not inhale fumes.

Glycerin Jelly
(See Page 461)

Kernechtrot (Nuclear Fast Red)
(See Page 84)

STAINING PROCEDURE

1. Mount cut frozen sections on slide containing section adhesive, if necessary.
2. Stain sections in osmium tetroxide for 2 hours.
3. Wash in several changes of tap water over a period of 1 hour.
4. Counterstain, if desired, with Kernechtrot or hematoxylin and eosin.
5. Place in absolute alcohol for 1 hour to obtain secondary staining of fat.
6. Rinse well in distilled water for 2 minutes.
7. Mount coverglass with glycerin jelly.

RESULTS

Fat .... black
Nuclear chromatin .... depends on counterstain
Background .... depends on counterstain

Luna's Method for Quantitative Demonstration of Fat (Paraffin Sections)

FIXATION	10% Buffered neutral formalin.
PROCESS	Paraffin sections.
MICROTOMY	Cut sections at 6 micrometers, but first, note step 1 of staining procedure.
CONTROL	Use control specimen containing fat.
SOLUTIONS	

5% Potassium Dichromate (Stock)

Potassium dichromate	5.0	gm
Distilled water	100.0	ml

2% Osmium Tetroxide (Stock)

Osmium tetroxide	2.0	gm
Distilled water	100.0	ml

Score, if necessary, two 1 gm ampules of osmium tetroxide with a file and drop into bottle containing distilled water. Vigorous shaking will break ampule safely. Note: Use osmium tetroxide under a hood or in a well-ventilated room. Do not inhale fumes.

Potassium Dichromate-Osmium Tetroxide (Working)

5% Potassium dichromate (stock)	50.0	ml
2% Osmium tetroxide (stock)	50.0	ml

STAINING PROCEDURE

1. Place formalin fixed specimens, which are no thicker than 4 mm, in working potassium dichromate-osmium tetroxide solution for 8 hours or overnight.
2. Wash specimens in running tap water for 2 hours.

3. Process specimens in desired tissue processing system, embed in paraffin, and cut sections at 6 micrometers. (Dry sections in conventional manner.)
4. Decerate sections and hydrate to distilled water.
5. Counterstain with Kernechtrot (nuclear fast red) or Mayer's hematoxylin and eosin.
6. Dehydrate slides in 95% alcohol and absolute alcohol, 3 changes each. Clear in xylene, 3 changes.
7. Mount coverglass with appropriate medium (refractive index 1.48-1.56).

RESULTS

Fat .... black
Nuclei .... depends on counterstain

REMARKS

The following modification to the previous method has proven extremely valuable in the author's hands, since it eliminates the many artifacts produced by conventional cryotomy and staining. This method utilizes the fat-preserving qualities of osmium tetroxide and the staining qualities of oil red O. Using the following procedure, it is possible to demonstrate small intracytoplasmic fat droplets not visible with conventional fat stains. This results in both a more quantitative and qualitative demonstration of fat.

0.5% Oil Red O
(See Page 352)

STAINING PROCEDURE

1. Prepare tissue as above (steps 1 - 4).
2. Remove osmium tetroxide by placing sections in 3% hydrogen peroxide solution; normally 4 hours exposure is adequate.
3. Wash sections in running tap water for 15 minutes.
4. Stain in oil red O for 30 minutes.
5. Wash in running tap water for 2 minutes.
6. Counterstain with Mayer's hematoxylin for 10 minutes.
7. Wash in running tap water for 10 minutes.
8. Mount coverglass with glycerin jelly.

RESULTS

Fat .... red
Nuclei .... blue

REMARKS

It may be necessary to expose slides to a 5% aqueous solution of periodic acid for 5 minutes following step 4 to ensure section backgrounds are clear enough to accept the hematoxylin counterstain.

Sudan Black B Method for Fat

(See Fig. 195)

FIXATION	10% Buffered neutral formalin.
PROCESS	Frozen sections.
CRYOTOMY	Cut sections at 10 micrometers.
CONTROL	Use control specimen containing fat.

SOLUTIONS

Sudan Black B

Dissolve 0.7 gm sudan black B in 100 ml of propylene glycol by gently heating to 100-110°C and stirring thoroughly for a few minutes. Do not exceed 110°C. While hot, filter through Whatman No. 2 paper to remove excess dye. After cooling to room temperature, filter through a Seitz filter with the aid of vacuum.

85% Propylene Glycol
(See Page 352)

Kernechtrot (Nuclear Fast Red)
(See Page 84)

STAINING PROCEDURE

1. Cut frozen sections and pick up on chrome alum-gelatin coated slides (or use any appropriate section adhesive).
2. Place in absolute propylene glycol for 10 minutes.
3. Place in sudan black B solution for 30 minutes. Note: If sections are not stained sufficiently, allow them to remain in staining solution for an additional 30 minutes.
4. Differentiate in 85% propylene glycol solution for 3 minutes. Agitate sections several times.
5. Rinse in distilled water.
6. Counterstain in nuclear fast red for 5 minutes.
7. Rinse well in distilled water.
8. Mount coverglass with glycerin jelly.

RESULTS

Fat .... blue-black
Nuclei .... red

Landing's Method for Globosides, Sphingomyelin and Glucocerebrosides

FIXATION 10% Buffered neutral formalin.

PROCESS Paraffin, frozen, or celloidin.

MICROTOMY Cut paraffin and celloidin sections at 6 micrometers. Cut frozen sections at 10 micrometers.

CONTROL Use control of human serum or section of spleen, liver or erythrocytes.

SOLUTIONS

Acetone-Ethyl Ether

Acetone	50.0	ml
Ethyl ether	50.0	ml

1% Phosphomolybdic Acid

Phosphomolybdic acid	1.0	gm
100% Alcohol	50.0	ml
Chloroform	50.0	ml

Use chloroform with extreme care since it is a suspected carcinogen.

3N Hydrochloric Acid (Stock)

Concentrated hydrochloric acid	10.0	ml
Distilled water	30.0	ml

1% Stannous Chloride

Stannous chloride	1.0	gm
3N Hydrochloric acid (stock)	100.0	ml

Make this solution fresh just before use.

Aqueous Eosin

Eosin Y, water soluble	1.0	gm
Distilled water	100.0	ml

STAINING PROCEDURE

1. (a) Frozen: Place sections on chrome-gelatin subbed slide. Allow sections to dry for 2 hours.
   (b) Paraffin: Decerate slides in 2 changes of xylene, 2 minutes each; absolute alcohol, 2 changes, 2 minutes each; and 95% alcohol, 2 changes, 2 minutes each.
2. Dip slides in acetone-ethyl ether solution.
3. Place slides in 1% phosphomolybdic acid solution for 15 minutes.
4. Rinse in equal parts ethyl alcohol and chloroform.
5. Rinse in chloroform, 2 changes.
6. Immerse in 1% stannous chloride solution for 5 seconds.
7. Wash in distilled water for 2 minutes.
8. Counterstain with aqueous eosin solution for 2 minutes.
9. Dehydrate slides in 95% alcohol and absolute alcohol, 3 changes each, clear in xylene, 3 changes.
10. Mount coverglass with appropriate medium (refractive index 1.48-1.56).

RESULTS

Glucocerebroside (Kerasin - Gaucher disease lipid)	deep blue
Sphingomyelin (Nieman-Pick lipid)	light blue
Globoside (ganglioside)	deep blue
Normal myelin	deep blue

Baker's Method for Phospholipids

(See Fig. 192)

(See Pyridine Extraction Procedure Below)

FIXATION 10% Buffered neutral formalin.
PROCESS Frozen sections. (See below for specimen processing.)
CRYOTOMY Cut sections at 8 micrometers.
CONTROL Use control number 5.
SOLUTIONS

Potassium Dichromate-Calcium Chloride

Potassium dichromate	5.0	gm
Calcium chloride	1.0	gm
Distilled water	100.0	ml

Borax-Potassium Ferricyanide

Borax	0.25	gm
Potassium ferricyanide	0.25	gm
Distilled water	100.0	ml

Store in refrigerator.

1% Sodium Iodate (Stock)

Sodium iodate	1.0	gm
Distilled water	100.0	ml

Acid-Hematein

Hematoxylin crystals	0.05	gm
1% Sodium iodate (stock)	1.0	ml
Distilled water	48.0	ml

Bring to a boil, then cool and add 1 ml of glacial acetic acid. Make fresh just before use.

Glycerin Jelly
(See Page 461)

STAINING PROCEDURE

1. Cut frozen sections and mount on chrome alum-gelatin coated slides (or use other selected section adhesive).
2. Place in potassium dichromate-calcium chloride solution at room temperature for 8 hours.
3. Transfer to a second change of potassium dichromate-calcium chloride solution at 60°C for 4 hours.
4. Wash in running tap water for 1 hour, then distilled water for 5 minutes.
5. Place in acid-hematein solution at 37°C for 2 hours.
6. Rinse in distilled water, then transfer to borax-ferricyanide solution at 37°C for 8 hours.
7. Rinse in distilled water for 10 minutes.
8. Mount coverglass with glycerin jelly.

RESULTS

Phospholipids (lecithin, kephalin, sphingomyelin)	blue black
Galactolipids	blue black to pale blue
Background	grayish brown

Method for Differentiating Phospholipids and Galactolipids from Protein

Pyridine extraction procedure is used to distinguish between phospholipids and certain proteins which give a positive reaction with the use of this acid hematein method.

SOLUTIONS

Modified Bouin's Solution

Saturated aqueous picric acid	50.0	ml
37% Formalin	10.0	ml
Glacial acetic acid	5.0	ml
Distilled water	35.0	ml

Pyridine

Pyridine can be purchased at most laboratory supply houses.

STAINING PROCEDURE

1. Place frozen sections in modified Bouin's solution for 16 hours.
2. Place in 70% alcohol for 1 hour.
3. Place in 50% alcohol for 30 minutes.
4. Wash in running tap water for 10 minutes.
5. Place in pyridine at room temperature, 2 changes, 1 hour each.
6. Place in pyridine at 60°C overnight.
7. Wash in running tap water for 1 hour.
8. Proceed at step 2 of Baker's Method for Phospolipids above.

RESULTS

With pyridine extraction the phospholipids and galactolipids remain unstained. Red cells stain black, both with and without pyridine extraction. Nuclear chromatin stains after extraction but not before.

REMARKS

The time suggested at each step in the staining procedure can be extended considerably to accommodate overnight non-working hours. This procedure was originally developed for staining thin gross tissue blocks. However, frozen cut sections can be employed satisfactorily. The block method should be used whenever possible. Only by using the method in this way can one be absolutely certain of the results.

Pearse's Method for Phospholipids, Lipofucsin and Eosinophilic Granules

FIXATION Baker's calcium-cadmium formol.
PROCESS Paraffin, frozen, or celloidin.
MICROTOMY Cut sections at 6 micrometers.
CONTROL Use control number 5 or 27A.
SOLUTIONS

Baker's Calcium-Cadmium

Calcium chloride (anhydrous)	1.0	gm
Cadmium chloride (anhydrous)	1.0	gm
Concentrated 37% formalin	10.0	ml
Distilled water	90.0	ml

0.1% Methasol Fast Blue 2G

Methasol fast blue 2G	0.1	gm
100% Alcohol	100.0	ml

Luxol fast blue MBS may be substituted for methasol fast blue 2G.

0.05% Lithium Carbonate

Lithium carbonate	0.05	gm
Distilled water	100.0	ml

1% Neutral Red

Neutral red	1.0	gm
Distilled water	100.0	ml

STAINING PROCEDURE

1. Decerate slides in xylene, 2 changes, 2 minutes each. Place in absolute alcohol, 2 changes, 2 minutes each; 95% alcohol, 2 changes, 2 minutes each; then rinse in distilled water.
2. Stain in 1% methasol fast blue solution at 58°C overnight.
3. Rinse in 70% alcohol for 30 seconds.
4. Rinse in distilled water for 30 seconds.
5. Decolorize in lithium carbonate solution for 30 minutes. (Decolorization may take as long as 2 hours.)
6. Place in distilled water for 1 minute.
7. Counterstain in neutral red solution for 30 minutes.
8. Rinse in distilled water for 30 seconds.
9. Dehydrate slides in 95% alcohol and absolute alcohol, 3 changes each. Clear in xylene, 3 changes.
10. Mount coverglass with appropriate medium (refractive index 1.48-1.56).

RESULTS

Phospholipids	blue
Eosinophilic granules, lipofucsin, keratin, keratohyalin, human elastic tissues	blue to purple
Nuclear chromatin	red

Mukherji's Method for Unsaturated Lipids

FIXATION 10% Buffered neutral formalin.
PROCESS Frozen.
CRYOTOMY Cut sections at 8 micrometers.
CONTROL Use control specimen containing fat.
SOLUTIONS

3.5% Bromine

Bromine	3.5	gm
Distilled water	100.0	ml

5% Sodium Thiosulfate
(See Page 134)

10% Silver Nitrate

Silver nitrate	10.0	gm
50% alcohol	100.0	ml

10% Formalin (Reducer)

Concentrated 37% formalin	10.0	ml
Distilled water..	90.0	ml

STAINING PROCEDURE

1. Cut frozen sections and pick up on gelatinized slide. (Any good section adhesive may be used.)
2. Wash in running tap water for 5 minutes, followed by several changes of distilled water.
3. Carefully blot slides dry.
4. Expose sections to 3.5% bromine solution fumes in a closed jar at 37°C for 1 hour. (A coplin jar containing 15 ml of bromine may serve the purpose.) Do not allow section to come in contact with solution.
5. Place in 5% sodium thiosulfate, 2 changes, 5 minutes each.
6. Rinse in distilled water, 3 changes for 10 seconds each.
7. Place in 10% silver nitrate solution at 37°C in the dark until sections are faint brown or pale yellow.
8. Rinse in distilled water.
9. Place in 10% formalin reducing solution until section is deep brown.
10. Rinse in distilled water.
11. Place in 5% sodium thiosulfate solution for 2 minutes.
12. Rinse in distilled water for 30 seconds.
13. Dehydrate slides in 95% alcohol and absolute alcohol, 3 changes each. Clear in xylene, 3 changes.
14. Mount coverglass with appropriate medium (refractive index 1.48-1.56).

RESULTS

Unsaturated lipids .... brown to black

REMARKS

If a critical study of unsaturated lipid is desired, sections should be treated with amylase after step 2 and before bromination. This will eliminate glycogen staining interference with the unsaturated lipid.

Seligman-Ashbel Method for Active Carbonyl Groups (Lipids)

FIXATION 10% Buffered neutral formalin (not more than 48 hours).

PROCESS Frozen.

CRYOTOMY Cut sections at 15 micrometers. Mount sections with Poly-L-Lysine or other good section adhesive.

CONTROL Use control specimen containing fat.

SOLUTIONS

Glycerin Jelly

(See Page 461)

0.1% Hydroxy-Naphthoic Acid Hydrazide

2-hydroxy-3-naphthoic acid hydrazide	0.1	gm
Glacial acetic acid	5.0	ml

Dissolve powder carefully in hot glacial acetic acid, then add:

50% Alcohol	95.0	ml

5% Acetic Acid-Alcohol

Glacial acetic acid	5.0	ml
50% Alcohol	95.0	ml

0.67M Phosphate Buffer, pH 7.4

1/15M Sodium phosphate, dibasic (9.465 gm made up to 1000 ml with distilled water)	80.0	ml
1/15M Potassium acid phosphate (9.07 gm made up to 1000 ml with distilled water)	20.0	ml

Phosphate Buffer-Alcohol (Stock)

0.67M Phosphate buffer solution (stock)	50.0	ml
100% Alcohol..	50.0	ml

0.05% Fast Black B

Fast black B	0.05	gm
Phosphate buffer-alcohol (stock).....................................	100.0	ml

0.1% Safranin

Safranin	0.1	gm
Glacial acetic acid, 1% aqueous	100.0	ml

STAINING PROCEDURE

1. Pick up cut frozen sections on Poly-L-Lysine coated slides. (Any other good section adhesive may also be used.)
2. Dry slides in 37°C oven for 4 hours to overnight.
3. Wash slide in running tap water for 2 hours to remove formalin.
4. Place in freshly made 0.1% hydroxy-naphthoic acid hydrazide for 2 hours at 25°C.
5. Wash in 5% acetic-alcohol solution, 4 changes, 2 minutes each.
6. Place in phosphate buffer-alcohol for 2 minutes.
7. Immerse directly into fast black B for 2 minutes.
8 Rinse in 5% acetic acid alcohol.
9. Counterstain with 0.1% safranin for 2 minutes.
10. Dip 5 times in distilled water.
11. Mount coverglass with glycerin jelly or other aqueous mounting medium.

RESULTS

Active carbonyl lipids................................... dark blue or greenish-blue

Background.. pink

Menschik's Method for Phospholipids

FIXATION Baker's formol-calcium.
PROCESS Frozen.
CRYOTOMY Cut sections at 10 micrometers.
CONTROL Use control number 37.
SOLUTIONS

Nile Blue Sulphate

Nile blue sulphate, saturated aqueous	100.0	ml
Sulphuric acid, 0.05% aqueous	10.0	ml

Boil for 30 minutes using a reflux condenser.

5% Glacial Acetic Acid

Glacial acetic acid	5.0	ml
Distilled water	95.0	ml

0.5% Hydrochloric Acid

Concentrated hydrochloric acid	0.5	ml
Distilled water	99.5	ml

Glycerin Jelly
(See Page 461)

STAINING PROCEDURE

1. Pick up cut frozen sections on Poly-L-Lysine coated slides. (Any other good section adhesive may also be used.)
2. Place in Nile blue sulphate at 60°C for 90 minutes.
3. Rinse in distilled water.
4. Preheat acetone to 50°C, then remove from source of heat and immediately immerse sections into it for 30 minutes.
5. Differentiate in 5% glacial acetic acid solution for 30 minutes.
6. Rinse in distilled water.
7. Differentiate in 0.5% hydrochloric acid solution for 3 minutes.
8. Rinse in distilled water.
9. Mount coverglass with glycerin jelly.

RESULTS

Phospholipids .... blue

Luna-Verfuerth Method for STAT H&E

FIXATION Formol methyl alcohol.
CRYOTOMY Cut frozen sections at 8 micrometers.
SOLUTIONS

Formol-Methyl Alcohol

10% Buffered neutral formalin	80.0	ml
Absolute methyl alcohol	20.0	ml

Bullard's Hematoxylin

Hematoxylin	8.0	gm
80% Alcohol	144.0	ml
Glacial acetic acid	16.0	ml
Dissolve by gentle heat and add mixture of:		
Ammonium or potassium alum	20.0	gm
Distilled water	250.0	ml
Heat to boiling, remove from flame and slowly add:		
Mercuric oxide	8.0	gm
Cool quickly, filter and add mixture of:		
95% Alcohol	275.0	ml
Glycerin	330.0	ml
Glacial acetic acid	18.0	ml
Ammonium or potassium alum	40.0	gm

Eosin Y-Phloxine B

1.0% Eosin Y (Stock)

Eosin Y, water soluble	1.0	gm
Distilled water	100.0	ml

1.0% Phloxine B (Stock)

Phloxine B	1.0	gm
Distilled water	100.0	ml

Eosin Y-Phloxine B Solution (Working)

1.0% Eosin Y (stock)	100.0	ml
1.0% Phloxine B (stock)	10.0	ml
95% Alcohol	780.0	ml
Glacial acetic acid	4.0	ml

1.0% Acid Alcohol

70% Alcohol	100.0	ml
Hydrochloric acid, concentrated	1.0	ml

1.0% Lithium Carbonate

Lithium carbonate	1.0	gm
Distilled water	100.0	ml

STAINING PROCEDURE

1. Cut frozen sections and pick up on acid cleaned slides. Keep sections in cryostat until ready to stain. This will inhibit autolysis.
2. Fix in formol-methyl alcohol for 30 seconds.
3. Dip 5 times in tap water (change frequently).
4. Place in Bullard's hematoxylin for 1 minute.
5. Tap water (change frequently).
6. Decolorize in 1% acid alcohol, 1 dip.
7. Tap water (change frequently).
8. Dip once in 1% lithium carbonate.
9. Tap water (change frequently).
10. Dip 3 times in 80% alcohol.

11. Place in eosin Y-phloxine B working solution for 30 seconds.
12. Dip 3 times in 95% alcohol.
13. Dip 3 times in 100% alcohol.
14. Dip 3 times in 100% alcohol.
15. Dip 3 times in xylene.
16. Dip 3 times in xylene.
17. Mount coverglass with appropriate medium (refractive index 1.48-1.56).

RESULTS

Nuclear chromatin..........blue
Cytoplasm ....pink to red

Terry's STAT Method for Unfixed Tissue[146]

FIXATION Fresh tissue.

CRYOTOMY Cut frozen sections at 10 micrometers. Mount sections on clean slides containing proper adhesive.

SOLUTIONS

1% Methylene Blue (Stock)

Methylene blue..........	1.0	gm
Distilled water..........	100.0	ml

1% Potassium Carbonate (Stock)

Potassium carbonate	1.0	gm
Distilled water..........	100.0	ml

Terry's Methylene Blue (Working)

1% Methylene blue (stock)	6.0	ml
1% Potassium carbonate (stock)..........	6.0	ml
Distilled water..........	18.0	ml

Combine above ingredients and boil over a flame for exactly 2 1/2 minutes, (see remarks). Cool solution under running tap water. After the solution has cooled, q.s. to 30 ml to replace any water lost during boiling. Filter solution before use.

STAINING PROCEDURE

1. Drain excess water from slide but do not allow sections to dry.
2. Cover the sections with a small amount of Terry's working methylene blue for 10 seconds.
3. Drain the stain off the slides and rinse in distilled water.
4. Examine sections while still wet.

RESULTS

Nuclei and epithelial cells..........blue
Connective tissue fibrils..........red to colorless
Smooth muscle ....purple

REMARKS

This stain gives its best results when used on frozen sections of living, unfixed tissue. As soon as the cells die, the staining is unsatisfactory. Sections may be coverslipped by employing a drop of water on the under surface of the coverglass. The time the Terry's working solution is boiled is very important in developing polychrome. If the alkaline stain is boiled longer than 2 1/2 minutes or if it is slowly cooled, more red is present, more crystalline precipitate forms, and the stain is less intensely active.

Terry's STAT Method for Fixed Tissue[146]

FIXATION 10% Buffered neutral formalin.

CRYOTOMY Cut frozen sections at 10 micrometers. Mount sections on glass slides containing proper adhesive.

SOLUTIONS

1% Methylene Blue (Stock)
(See Page 100)

1% Potassium Carbonate (Stock)
(See Page 100)

10% Glacial Acetic Acid (Stock)

Glacial acetic acid	10.0	ml
Distilled water	90.0	ml

Terry's Acid Polychrome Methylene Blue (Working)

1% Methylene blue (stock)	20.0	ml
1% Potassium carbonate (stock)	20.0	ml
Distilled water	60.0	ml

Combine and heat to boiling. Boil gently for 2 1/2 minutes. Cool rapidly under running tap water. Add 10 ml of 10% glacial acetic acid (stock) to cooled solution, (see Remarks). Shake the solution for 1 minute and q.s. up to 100 ml with distilled water to replace any water lost during boiling. Filter solution before use.

STAINING PROCEDURE

1. Drain excess water, if any, from slide but do not allow to dry.
2. Cover the sections with a small amount of Terry's working acid polychrome methylene blue solution for 3 to 5 seconds.
3. Drain the stain off the slide and rinse in distilled water.
4. Remount sections on a clean glass slide and examine while still wet.

RESULTS

Nuclear chromatin and epithelial cells	blue
Connective tissue	bright red to colorless
Smooth muscle	purple

REMARKS

If less acetic acid is added to the stain, connective tissue staining will be a brighter red. Conversely, more acetic acid results in pale to colorless staining of connective tissue. The time the solution is boiled is very important in developing polychrome. Boiling for 2 to 2 1/2 minutes seems to be best, although good acid stains may be made from solutions boiled only 30 seconds or as long as 5 minutes.

Luna's Method for Staining Fat in Brain Sections

This method is especially valuable for the demonstration of the lipoid degeneration products in the CNS. Since it stains the myelin sheaths quite intensely, brain sections should be differentiated for at least half an hour with pyridine.

FIXATION 10% Buffered neutral formalin.

CRYOTOMY Cut frozen sections at 8 micrometers. Mount sections on slide containing appropriate section adhesive.

SOLUTIONS

Oil Red O-Pyridine

Oil red O	5.0	gm
70% Pyridine	100.0	ml

Add stain to 70% pyridine (70 ml pyridine and 30 ml of distilled water). After standing 1 hour at room temperature with occasional shaking, the solution is saturated and ready for use. Store in well-stoppered bottle in cool, dark cabinet, and filter before use.

STAINING PROCEDURE

1. Place section in 50% pyridine for 5 minutes.
2. Stain in oil red O-pyridine for 30 minutes.
3. Differentiate in 50% pyridine for 2 minutes and dip in distilled water for 5 dips.
4. Counterstain in Harris's hematoxylin for 2 minutes.
5. Blue in running tap water for 5 minutes.
6. Wash in distilled water for 5 minutes.
7. Mount coverglass with gum acacia or glycerol-gelatin.

RESULTS

Fat	orange-red
Nuclear chromatin	blue

REMARKS

Pyridine, which is slightly alkaline, does not alter the cellular structures and in a 70% solution, will not dissolve fat or lipoids if used at room temperature. A further advantage of pyridine as a solvent is that it does not cause precipitation of the stain.

Luna's Method for Lecithins

FIXATION	10% Buffered neutral formalin.
PROCESS	Frozen.
CRYOTOMY	Cut section at 8 micrometers.
CONTROL	Use control specimen of brain, liver, semen or egg yolk.
SOLUTIONS	

Acetone-Ether

Acetone	50.0	ml
Ether	50.0	ml

Ethyl Alcohol-Chloroform (Stock)

100% Ethyl alcohol	50.0	ml
Chloroform	50.0	ml

1% Phosphomolybdic Acid

Phosphomolybdic acid	1.0	gm
Ethyl alcohol-chloroform (stock)	100.0	ml

3N Hydrochloric Acid (Stock)

Concentrated hydrochloric acid	40.0	ml
Distilled water	60.0	ml

1% Stannous Chloride

Stannous chloride	1.0	gm
3N Hydrochloric acid (stock)	100.0	ml

STAINING PROCEDURE

1. Mount frozen section on gelatinized slide (or use any appropriate section adhesive).
2. Allow sections to dry thoroughly.
3. Dip in acetone-ether.
4. Place in 1% phosphomolybdic acid for 15 minutes.
5. Rinse in ethyl alcohol chloroform, followed by 2 changes of chloroform, 10 seconds each.
6. Dip for 10 seconds in 1% aqueous stannous chloride.
7. Wash in running tap water for 2 minutes.
8. Counterstain for 30 seconds in eosin. Positive areas stain blue; negative areas stain red.
9. Dehydrate slides in 95% alcohol and absolute alcohol, 3 changes each. Clear in xylene, 3 changes.
10. Mount coverglass with appropriate medium (refractive index 1.48-1.56).

RESULTS

Lecithin	blue
Unreactive sites	pink

Berenbaum's Method for Bound Lipids[147]

Many tissue lipids are bound in varying degrees to other substances, mainly protein, and thus are resistant to extraction by organic solvents. Such bound lipids are still present in paraffin sections, but they cannot be satisfactorily demonstrated by conventional use of the Sudans. An essential condition for such demonstration is their prior disassociation from their complexes by hydration, heat or the action of proteolytic enzymes.

After routine fixation, bound lipids in paraffin sections or in smears, may be partially dissociated from their complexes and then demonstrated by Sudan black with Berenbaum's method.

This method shows lipids to be present constantly in association with nuclear and cytoplasmic nucleoprotein, in basement membranes, epithelial brush borders, neurones, and many other structures. There is evidence that these lipids do not form a homogeneous class, but vary widely in their physico-chemical natures.

FIXATION 10% Buffered neutral formalin.
PROCESS Paraffin or frozen.
MICROTOMY Cut sections at 6 micrometers.
CONTROL Use section of myelin sheath for moderately bound lipids and epithelial brush border (lung) or reticulum fibers for strongly bound lipids.

SOLUTIONS

2% Sudan Black B-Acetone

Sudan black B	2.0	gm
Acetone	100.0	ml

STAINING PROCEDURE

1. Decerate slides in xylene, 2 changes, 2 minutes each. Place in absolute alcohol, 2 changes, 2 minutes each; 95% alcohol, 2 changes, 2 minutes each; then rinse in distilled water.
2. Wash slides in running tap water for 1 hour.
3. Rinse in acetone for 15 seconds.
4. Stain in sudan black B-acetone at 50°C for 4 hours.
5. Place in xylene until all coloring agent is extracted; usually 15 minutes is sufficient.
6. Mount coverglass with appropriate medium (refractive index 1.48-1.56).

RESULTS

The colors obtained are usually black or gray, but a polychromatic effect is often observed, some structures becoming blue or brown. The following entities can be demonstrated with this method:

- — Chromatin
- — Nuclear membrane
- — Leucocytic inclusions (seen in disseminated lupus erythematosus)
- — Cytoplasmic granules (seen in macrophages, in the granular cell myoblastoma granules of eosinophilic leucocytes)
- — Thyroid colloid
- — Mucin

— Cartilage
— Protein cast in renal tubules
— Bacteria and lipofuscin
— Amyloid
— Fibrin and fibrinoid of rheumatic nodules
— Corpora amylacea
— Morula bodies of cerebral trypanosomiasis

Chapter 15

Nerve Cells, Fibers and Cell Inclusions

Ramon-Cajal's Method for Astrocytes

(See Fig. 196)

FIXATION	Formalin-ammonium bromide for 1 day at 37°C or 10% buffered neutral formalin (see remarks).
PROCESS	Frozen.
CRYOTOMY	Cut sections at 15 micrometers; collect sections in distilled water.
CONTROL	Use control number 5.
SOLUTIONS	

Formalin-Ammonium Bromide

37% Concentrated formalin solution	15.0	ml
Ammonium bromide	2.0	gm
Distilled water	85.0	ml

1% Gold Chloride (Stock)

Gold chloride (brown)	1.0	gm
Distilled water	100.0	ml

Mercuric Chloride (Stock)

Mercuric chloride	0.5	gm
Distilled water	10.0	ml

Dissolve mercuric chloride using gentle heat.

Gold-Mercuric Chloride (Working)

1% Gold chloride (stock)	10.0	ml
Mercuric chloride (stock)	10.0	ml

Store solution in brown bottle. It will be necessary to dissolve the mercuric chloride with the aid of gentle heat. The 1% aqueous gold chloride is added while the mercuric chloride solution is warm. A precipitate will form if the mercuric chloride solution is not warm enough when mixed. In this case, repeat the preparation of gold working solution.

5% Sodium Thiosulfate
(See Page 134)

STAINING PROCEDURE

Use acid-cleaned glassware. Carry frozen sections on glass rods.

1. Rinse sections in distilled water, 2 changes.
2. Place sections flat in freshly prepared (working) gold-mercuric chloride solution for 4 to 6 hours in the dark at room temperature.
3. Rinse in distilled water, 5 minutes.
4. Place in sodium thiosulfate for 5 minutes.

5. Rinse well in 3 changes of distilled water.
6. Float sections onto slides containing a good section adhesive and air-dry.
7. Dehydrate slides in 95% alcohol and absolute alcohol 3 changes each. Clear in xylene, 3 changes.
8. Mount coverglass with appropriate medium (refractive index 1.48-1.56).

RESULTS

Astrocytes and their processes	black
Background	light brownish-purple
Nerve cells	pale red
Axon fibers	unstained

REMARKS

Formalin-fixed frozen tissues may be used provided sections are washed well and placed in formalin ammonium bromide solution for 24 hours prior to staining. Use mercuric bichloride crystals and brown gold chloride only.

Guillery's Method for Degenerating Axons

FIXATION	10% Formol-saline by perfusion (see remarks for processing schedule). After fixation by perfusion, specimens are macrosectioned (trimmed) to 4 mm thickness or less. They are then placed in the following solution until ready to process: 37% concentrated formalin solution, 100.0 ml. Filter before use. Add lithium carbonate in excess.
PROCESS	Paraffin, frozen, or celloidin.
MICROTOMY	Cut sections at 15 micrometers.
CONTROL	Use control section of brain of known degenerative disease.
SOLUTIONS	

Alcohol-Ammonium Hydroxide

28% Ammonium hydroxide	1.0	ml
50% Alcohol	100.0	ml

1.5% Silver Nitrate (Stock)

Silver nitrate	1.5	gm
Distilled water	100.0	ml

Silver Nitrate-Pyridine

Pyridine*	5.0	ml
1.5% Silver nitrate solution (stock)	95.0	ml

*Pyridine can be replaced by higher methylated derivatives of pyridine, or the concentration of these substances can be increased from 5% up to 20%, if necessary, to improve differentiation; 10% 2, 4, 6-trimethypyridine (collidine) has been found particularly useful.

4.5% Silver Nitrate (Stock)

Silver nitrate	4.5	gm
Distilled water	100.0	ml

2.5% Sodium Hydroxide (Stock)

Sodium hydroxide	2.5	gm
Distilled water	100.0	ml

Ammoniacal Silver

4.5% Silver nitrate (stock)	20.0	ml
100% Ethyl alcohol	10.0	ml
28% Ammonium hydroxide	1.8	ml
2.5% Sodium hydroxide (stock)	1.5	ml

This solution should be placed in a staining dish on a 40-45°C hot plate just before use and covered.

1% Citric Acid (Stock)

Citric Acid	1.0	gm
Distilled water	100.0	ml

10% Formalin (Stock)

Concentrated formalin, 37%	10.0	ml
Distilled water	100.0	ml

Reducing Solution

1% Citric acid (stock)	17.5	ml
10% Formalin (stock)	17.5	ml
10% Alcohol	450.0	ml

1% Sodium Thiosulfate

Sodium thiosulfate	1.0	gm
Distilled water	100.0	ml

STAINING PROCEDURE

1. Decerate slides in xylene, 2 changes, 2 minutes each. Place in absolute alcohol, 2 changes, 2 minutes each; 95% alcohol, 2 changes, 2 minutes each; then rinse in distilled water.
2. Place in alcohol-ammonium hydroxide for 6 hours.
3. Rinse in distilled water, 3 changes.
4. Place in silver nitrate-pyridine for 24 hours in the dark. Do not wash.
5. Preheat working ammoniacal silver solution to 45°C. Place three slides in this solution for 3 minutes. Solution should be changed frequently. Do not wash.
6. Place in reducing solution for 1 minute.
7. Rinse in distilled water.
8. Place in 1% sodium thiosulfate solution for 2 minutes.
9. Wash in distilled water for 1 minute.
10. Dehydrate slides in 95% alcohol and absolute alcohol 3 changes each. Clear in xylene, 3 changes.
11. Mount coverglass with appropriate medium (refractive index 1.48-1.56).

RESULTS

Degenerating axons	dark black
Normal fibers	light black

REMARKS
Processing Schedule: Wash specimens 8 hours in 3 changes of distilled water. Dehydrate the specimens in 80%, 95%, and 100% alcohol as usual. Clear in cedarwood oil (2 days) and transfer to paraffin (melting point 45°C) at 37°C for 3 hours. Embed in 54°C paraffin.

Luna's Method for Neurofibrils, Axons and Dendrites

FIXATION 10% Buffered neutral formalin.
PROCESS Paraffin, frozen, or celloidin.
MICROTOMY Cut paraffin sections at 8 micrometers. Cut frozen sections at 15 micrometers.
CONTROL Use control number 5.
SOLUTIONS

20% Silver Nitrate

Silver nitrate	20.0	gm
Distilled water	100.0	ml

Warm solution to 37°C just before use.

1% Ammonium Hydroxide

Ammonium hydroxide (NH_4OH, approximately 58%)	1.0	ml
Distilled water	100.0	ml

5% Sodium Thiosulfate

Sodium thiosulfate	5.0	gm
Distilled water	100.0	ml

Developer

Formalin, 10% unbuffered	20.0	ml
Distilled water	100.0	ml
Nitric acid, concentrated	1	drop
Citric acid	0.5	gm

STAINING PROCEDURE

1. Decerate slides in xylene, 2 changes, 2 minutes each. Place in absolute alcohol, 2 changes, 2 minutes each; 95% alcohol, 2 changes, 2 minutes each; then rinse in distilled water.
2. Place slides in pre-warmed 20% silver nitrate solution for 15 minutes in a 37°C oven.
3. Pour silver nitrate solution into clean container and save.
4. Wash slides in distilled water for 10 minutes.
5. Place slides in the following solution for 10 minutes in a 37°C oven. To the silver nitrate solution used in step 3 add concentrated ammonium hydroxide, drop by drop, agitating the container continuously until the precipitate which forms is completely dissolved. Since approximately 7.0 to 7.5 ml are required, 6.0 ml may be added at once; thereafter, add the ammonium hydroxide drop by drop until the end point is reached. If more than 10.0 ml of ammonium hydroxide is required to dissolve the precipitate, discard the silver nitrate solution and, using a fresh

20% silver nitrate solution, titrate with a new bottle of ammonium hydroxide. An excess of ammonia may cause a precipitate or poor impregnation of fibers.

6. Pour above solution into a clean container and save.
7. Place slides in 1% ammonium hydroxide solution for 2 minutes.
8. To the ammoniacal silver solution saved at step 6, add 3-10 drops of developer (see remarks). Place slides in this solution and agitate 3-5 minutes. It is important to check slides microscopically at this point for proper development of fibers. To do this, remove slides individually and dip in 1% aqueous ammonium hydroxide solution. This solution stops silver development. If fibers are not dark enough, return slides to the silver nitrate solution.
9. Wash slides in 1% ammonium hydroxide solution for 3 minutes.
10. Wash slides in distilled water for 1 minute.
11. Place slides in 5% sodium thiosulfate solution for 5 minutes.
12. Wash slides in 3 changes of distilled water, 5 minutes each.
13. Dehydrate slides in 95% alcohol and absolute alcohol 3 changes each. Clear in xylene, 3 changes.
14. Mount coverglass with appropriate medium (refractive index 1.48-1.56).

RESULTS

Neurofibrils and senile plaques	black
Fibrillary tangles	black
Background	yellow to brown

REMARKS

Three to 10 drops of developer may be used in step 7. The more developer used, the quicker the reaction, but it also results in darker background staining. In addition, there is a greater chance of precipitate formation when larger quantities of developer are used. Overstained slides may be corrected if they are dipped in 0.5% potassium ferricyanide after step 13.

A Modified Bielschowsky for Nerve Entities (Senile Plaques and Fibrillary Tangles)

(See Fig. 197, 198, 199, 200 and 201)

FIXATION	10% Buffered neutral formalin.
PROCESS	Paraffin, frozen, or celloidin.
MICROTOMY	Cut paraffin sections at 6 micrometers. Cut frozen sections at 15-25 micrometers.
CONTROL	Use control number 5 or known case of Alzheimer's disease.

SOLUTIONS

18% Silver Nitrate

Silver nitrate	18.0	gm
Distilled water	100.0	ml

1% Ammonium Hydroxide

Ammonium hydroxide	1.0	ml
Distilled water	100.0	ml

5% Sodium Thiosulfate (Hypo)

Sodium thiosulfate	5.0	gm
Distilled water	100.0	ml

Developing Solution

10% Unbuffered formalin	20.0	ml
Distilled water	100.0	ml
Nitric acid, concentrated	1	drop
Citric acid	0.5	gm

Make developing solution a week before using.

STAINING PROCEDURE

1. Decerate slides in xylene, 2 changes, 2 minutes each. Place in absolute alcohol, 2 changes, 2 minutes each; 95% alcohol, 2 changes, 2 minutes each.
2. Wash well in distilled water for 15 minutes.
3. Place slides in 18% silver nitrate for 15 minutes in the dark at 37°C.
4. Save the silver solution and transfer slides to distilled water for 10 minutes.
5. Add ammonium hydroxide to the saved silver solution using constant agitation, until the precipitate that forms dissolves. It should not take more than 14-15 ml/100 ml solution. Do not add more than 18-20 ml ammonium hydroxide, since an excess may cause a precipitate to form or cause the fibers to impregnate lightly.
6. Return the slides to this ammoniacal silver solution for 10 minutes in the dark at 37°C.
7. Save the silver solution and transfer the slides to 1% ammonium hydroxide for 1-3 minutes.
8. Add 3-10 drops, as needed, of developing solution to the saved ammoniacal silver solution. Return the slides to this solution and agitate for 3-5 minutes. Check microscopically for black fibers and plaques. If an excess of developing solution is needed to develop fibers or plaques or a heavy photographic film forms on slides, make up a new developing solution.
9. Wash in 1% ammonium hydroxide for 1-3 minutes.
10. Wash in distilled water for 1 minute.
11. Place slides in 5% sodium thiosulfate for 5 minutes.
12. Wash well in distilled water.
13. Dehydrate slides in 95% alcohol and absolute alcohol 3 changes each. Clear in xylene, 3 changes.
14. Mount coverglass with appropriate medium (refractive index 1.48-1.56).

RESULTS

Fibers, senile plaques	black
Fibrillary tangles	black
Background	yellow to gold

REMARKS

Use acid cleaned glassware. Do not crowd slides in staining container. Use the same silver solution throughout the technique. Three to 10 drops of developing solution may be used in step 8; however, the more used the quicker the reaction takes place and the more the background stains. Also, there is a greater chance of precipitation.

Luna's Modification of Bodian's Method for Axons and Nerve Endings[148]

(See Fig. 202, 203 and 204)

FIXATION	10% Buffered neutral formalin.
PROCESS	Paraffin, frozen, or celloidin.
MICROTOMY	Cut sections at 6 micrometers.
CONTROL	Use control number 5.
SOLUTIONS	

1% Silver Proteinate (Protargol)

Protargol	1.0	gm
Distilled water	100.0	ml

Sprinkle the protargol on the surface of the water and allow it to remain undisturbed until it dissolves.

Hydroquinone-Formalin

Hydroquinone	1.0	gm
Concentrated formalin, 37%	5.0	ml
Distilled water	100.0	ml

1% Gold Chloride

(See Page 203)

2% Oxalic Acid

Oxalic acid	2.0	gm
Distilled water	100.0	ml

5% Sodium Thiosulfate

(See Page 134)

Aniline Blue-Phosphomolybdic Acid

Aniline blue	0.1	gm
Oxalic acid	2.0	gm
Phosphomolybdic acid	2.0	gm
Distilled water	300.0	ml

1% Lissamine Fast Red

Lissamine fast red	1.0	gm
Glacial acetic acid	1.0	ml
Distilled water	100.0	ml

1% Phosphomolybdic Acid

Phosphomolybdic acid	1.0	gm
Distilled water	100.0	ml

1.5% Tartrazine

Tartrazine	1.5	gm
Glacial acetic acid	1.5	ml
Distilled water	100.0	ml

STAINING PROCEDURE

1. Decerate slides in xylene, 2 changes, 2 minutes each. Place in absolute alcohol, 2 changes, 2 minutes each; 95% alcohol, 2 changes, 2 minutes each; then rinse in distilled water.
2. Place slides in 1% silver proteinate solution and add 6 gm of clean copper shot per 100 ml of solution. Let stand at 37°C overnight.
3. Rinse in distilled water, 3 changes.
4. Place in hydroqinone-formalin reducing solution for 10 minutes.
5. Rinse in distilled water, 3 changes.
6. Tone in 1% gold chloride solution for 10 minutes.
7. Rinse in distilled water, 3 changes.
8. Develop in 2% oxalic acid solution under the microscope until background is gray and nerve fibers appear clearly; approximately 5 minutes.
9. Rinse in distilled water, 3 changes.
10. Place in 5% sodium thiosulfate for 5 minutes.
11. Rinse in distilled water.
12. Counterstain, if desired, with aniline blue-phosphomolybic acid solution (2 or 3 quick dips, to give a light blue background) or 1% lissamine fast red solution (see procedure below).
13. Dehydrate slides in 95% alcohol and absolute alcohol 3 changes each. Clear in xylene, 3 changes.
14. Mount coverglass with appropriate medium (refractive index 1.48-1.56).

RESULTS

If aniline blue-phosphomolybdic acid solution is used for the counterstain:

Axon and nerve endings.. black
Background .... blue
Nuclear chromatin.. black

Luna's Method for Counterstaining with Lissamine Fast Red

1. At step 12 above place slides in 1% lissamine fast red for 5 minutes.
2. Place in 1% aqueous phosphomolybdic acid solution for 5 minutes.
3. Rinse briefly in distilled water.
4. Place in 1.5% tartrazine solution for 5 minutes.
5. Rinse briefly in absolute alcohol and clear in xylene, 2 changes each.
6. Dehydrate slides in 95% alcohol and absolute alcohol 3 changes each. Clear in xylene, 3 changes.
7. Mount coverglass with appropriate medium (refractive index 1.48-1.56).

RESULTS

Axons .... black
Melanin .... black
Myelin .... red
Nuclei .... black
Muscle .... red
Erythrocytes .... red
Background .... yellow

REMARKS
Copper wire may be used in place of copper shot, both of which may be reused if cleaned with three alternating changes of concentrated nitric acid and distilled water. The cleaning of copper with nitric acid should be performed with a great deal of care and under the hood.

Holmes' Method for Nerve Cells and Axons

(See Fig. 206)

FIXATION 10% Buffered neutral formalin.
PROCESS Paraffin, frozen, or celloidin.
MICROTOMY Cut sections at 6 micrometers.
CONTROL Use control number 5.
SOLUTIONS

20% Silver Nitrate

Silver nitrate	20.0	gm
Distilled water	100.0	ml

1.2% Boric Acid (Stock)

Boric acid	12.0	gm
Distilled water	1000.0	ml

1.9% Borax (Stock)

Sodium borate, tetra (borax)	19.0	gm
Distilled water	1000.0	ml

1% Silver Nitrate (Stock)

Silver nitrate	1.0	gm
Distilled water	100.0	ml

10% Pyridine (Stock)

Pyridine	10.0	ml
Distilled water	90.0	ml

Silver Nitrate Impregnating Solution

1.2% Boric acid (stock)	55.0	ml
1.9% Borax (stock)	45.0	ml
Distilled water	394.0	ml
1% Silver nitrate	1.0	ml
10% Pyridine	5.0	ml

Hydroquinone-Sodium Sulfite

Hydroquinone	1.0	gm
Sodium sulfite, anhydrous	10.0	gm
Distilled water	100.0	ml

This solution may be used repeatedly to reduce tissue sections but does not have a long shelf life (generally good for 5-10 days).

5% Sodium Thiosulfate		
Sodium thiosulfate	5.0	gm
Distilled water	100.0	ml

0.2% Gold Chloride		
Gold chloride, 1% aqueous	2.0	ml
Distilled water	100.0	ml

2% Oxalic Acid		
Glacial acetic acid	4	drops
Oxalic acid	2.0	gm
Distilled water	100.0	ml

STAINING PROCEDURE

1. Decerate slides in xylene, 2 changes, 2 minutes each. Place in absolute alcohol, 2 changes, 2 minutes each; 95% alcohol, 2 changes, 2 minutes each; then rinse in distilled water.
2. Place in 20% silver nitrate for 2 hours.
3. Rinse in 3 changes of distilled water during a 10 minute period.
4. Place in silver nitrate impregnating solution at 37°C in a covered vessel overnight.
5. Remove slides, drain excess solution, and place in hydroqinone-sodium sulfite reducing solution for at least 2 minutes.
6. Wash in running water for 3 minutes, then rinse in distilled water.
7. Tone in 0.2% gold chloride for 3 minutes.
8. Rinse briefly in distilled water.
9. Place in 2% oxalic acid for 3 minutes. Examine microscopically until axons are blue-black.
10. Rinse in distilled water for 30 seconds.
11. Place in 5% sodium thiosulfate solution for 5 minutes, then wash in running tap water for 5 minutes.
12. Dehydrate slides in 95% alcohol and absolute alcohol 3 changes each. Clear in xylene, 3 changes.
13. Mount coverglass with appropriate medium (refractive index 1.48-1.56).

RESULTS

Axons	blue to black
Nerve endings	black
Background	gray to rose

Hirano-Zimmerman Method for Axons, Nerve Endings, Senile Plaques and Fibrillary Tangles

FIXATION	10% Buffered neutral formalin.
PROCESS	Paraffin, frozen, or celloidin.
MICROTOMY	Cut sections at 12-15 micrometers.
CONTROL	Use control number 5.

SOLUTIONS

10% Silver Nitrate

Silver nitrate	10.0	gm
Distilled water	100.0	ml

Ammonia Water

28% Ammonium hydroxide	2	drops
Distilled water	100.0	ml

Formalin-Tap Water

Concentrated formalin, 37%	50.0	ml
Tap water	50.0	ml

1% Gold Chloride (Stock)

Gold chloride	1.0	gm
Distilled water	100.0	ml

Gold Chloride (Working)

1% Gold chloride (stock)	10.0	ml
Distilled water	200.0	ml

5% Sodium Thiosulfate
(See Page 134)

STAINING PROCEDURE

1. Decerate slides in xylene, 2 changes, 2 minutes each. Place in absolute alcohol, 2 changes, 2 minutes each; 95% alcohol, 2 changes, 2 minutes each; then rinse in distilled water.
2. Place in 10% silver nitrate solution at room temperature for 2 hours.
3. Place sections directly in ammonia water solution for 3 minutes.
4. Place sections directly in formalin-tap water solution for 3 minutes. Tissue becomes brown to black.
5. Rinse in distilled water, 2 changes for 15 seconds.
6. Place sections again in 10% silver nitrate solution for 3 to 5 minutes.
7. Repeat steps 3 through 5. Tissue will become darker in color.
8. Tone in 1% gold chloride for 20 minutes.
9. Place sections directly in 5% sodium thiosulfate solution for 1 minute.
10. Rinse in distilled water for 30 seconds.
11. Dehydrate slides in 95% alcohol and absolute alcohol 3 changes each. Clear in xylene, 3 changes.
12. Mount coverglass with appropriate medium (refractive index 1.48-1.56).

RESULTS

Nucleolus and nuclear membrane	black
Axons and dendrites	black
Cytoplasm of astrocytes and the cytoplasmic membranes of macrophages	gray
Senile plaques	black
Fibrillary tangles	black

Luna's Modification of PTAH Method for CNS and Other Tissue Elements

(See Fig. 208)

FIXATION 10% Buffered neutral formalin.

PROCESS Paraffin or celloidin.

MICROTOMY Cut sections at 6 micrometers.

CONTROL Use control number 5 or selected control depending on the use of the stain.

SOLUTIONS

Phosphotungstic Acid-Hematoxylin (PTAH)

Hematoxylin	1.0	gm
Phosphotungstic acid	20.0	gm
Distilled water	1000.0	ml

Dissolve the solid ingredients in separate portions of the distilled water, the hematoxylin with the aid of gentle heat. When cool, combine solutions. Preservatives are not necessary. Ripening requires several weeks, but the addition of 0.2 gm of potassium permanganate will ripen the stain immediately. Or, add 5.0 ml of 0.25% potassium permanganate to each 100.0 ml of stock solution.

0.25% Potassium Permanganate

Potassium permanganate	0.25	gm
Distilled water	100.0	ml

Make fresh each time; 0.25% potassium permanganate is not stable. Alternatively, this can be made by diluting 1% potassium permanganate.

Lugol's Iodine

Iodine	1.0	gm
Potassium iodide	2.0	gm
Distilled water	100.0	ml

5% Oxalic Acid

Oxalic acid	5.0	gm
Distilled water	100.0	ml

2.5% Potassium Dichromate (Stock)

Potassium dichromate	2.5	gm
Distilled water	100.0	ml

2.5% Potassium Dichromate (Working)

2.5% Potassium dichromate (stock)	95.0	ml
Glacial acetic acid	5.0	ml

STAINING PROCEDURE

1. Decerate slides in xylene, 2 changes, 2 minutes each. Place in absolute alcohol, 2 changes, 2 minutes each; 95% alcohol, 2 changes, 2 minutes each; then rinse in distilled water.

2. Mordant sections in 2.5% potassium dichromate working solution overnight
3. Wash slides in running tap water for 15 minutes.
4. Place slides in Lugol's iodine for 15 minutes.
5. Place slides in 95% alcohol until sections are devoid of iodine (brownish) coloration. Minimum of 1 hour.
6. Rinse slides in 3 rapid changes of distilled water.
7. Place slides in 0.25% potassium permanganate solution for 5 minutes.
8. Rinse slides in 3 rapid changes of distilled water.
9. Place slides in 5% oxalic acid solution for 5 minutes.
10. Rinse slides in 3 rapid changes of distilled water.
11. Wash slides in running tap water for 5 minutes.
12. Rinse slides in 3 rapid changes of distilled water.
13. Place slides in phosphotungstic acid hematoxylin solution overnight.
14. Rinse sections in 95% alcohol to remove excess stain.
15. Dehydrate quickly in 3 changes of absolute alcohol.
16. Clear in xylene, 3 changes each.
17. Mount coverglass with appropriate medium (refractive index 1.48-1.56).

RESULTS

Nuclei, centrosomes, spindles, neuro, myo, fibro, epithelioglial fibers, kidney and liver mitochondria, myelin sheath .... blue
Fibrin .... darker blue
Collagen, ground substance of bone, cartilage ....pinkish-red
Cytoplasm .... terra-cotta pink

REMARKS

Sodium thiosulfate is not used to clear the Lugol's iodine (step 5), since it will impair staining. It is generally believed that iodine is used solely for removal of mercuric chloride deposits. In this technique, however, the iodine prepares the tissue for the stain in some way that is not clear but is essential for good results.

The combination of potassium permanganate and oxalic acid is often referred to as the Mallory bleach. After the original mordanting of the sections with potassium dichromate, the permanganate probably acts by oxidizing compounds which the tissues have formed with salts of chromium, thus liberating chromic acid. Essentially, it acts as a differentiating agent before staining with phosphotungstic acid hematoxylin. The purpose of the oxalic acid is to remove brown precipitate manganous salts. An essential feature of this stage is that the time of exposure of the section to oxalic acid should be the same as its exposure to the permanganate. The characteristics of the final stain by the phosphotungstic acid hematoxylin can be regulated, to a large extent, by varying the length of time the sections are exposed to the Mallory bleach. Lengthening the time tends to result in feeble staining of those elements which normally should appear blue. Shortening the time, on the other hand, leads to an increase in the density of the blue components and ultimately to an absence of differential staining if the Mallory bleach is eliminated altogether. Therefore, a decision to reduce or increase the time in potassium permanganate and oxalic acid must be based on the effect ultimately required and the results obtained with the particular batch of stain in use.

Holzer's Method for Neuroglial Cells

(See Fig. 209)

FIXATION	10% Buffered neutral formalin.
PROCESS	Paraffin, frozen, or celloidin.
MICROTOMY	Cut sections at 6 micrometers.
CONTROL	Use control number 5.
SOLUTIONS	

0.5% Phosphomolybdic Acid (Stock)

Phosphomolybdic acid	0.5	gm
Distilled water	100.0	ml

Phosphomolybic Alcohol

0.5% Phosphomolybdic acid (stock)	50.0	ml
95% Alcohol	100.0	ml

Alcohol-Chloroform

100% Alcohol	40.0	ml
Chloroform	160.0	ml

Chloroform has been declared a potential carcinogen; therefore, use with care and under a hood.

5% Crystal Violet

Crystal violet	5.0	gm
100% Alcohol	20.0	ml
Chloroform	80.0	ml

10% Potassium Bromide

Potassium bromide	10.0	gm
Distilled water	100.0	ml

Aniline-Chloroform

Aniline	120.0	ml
Chloroform	180.0	ml
Ammonium hydroxide, 28%	20	drops

STAINING PROCEDURE

1. Decerate slides in xylene, 2 changes, 2 minutes each. Place in absolute alcohol, 2 changes, 2 minutes each; 95% alcohol, 2 changes, 2 minutes each; then rinse in distilled water.
2. Place in phosphomolybdic alcohol solution for 3 minutes. Drain slides well.
3. Place in alcohol-chloroform solution until section becomes translucent.
4. Place slides on staining rack and flood sections with 5% crystal violet solution for 30 seconds. Blot dry.
5. Flood sections with 10% potassium bromide solution for 1 minute. Blot dry.
6. Differentiate slides in aniline-chloroform for 30 seconds. If overdifferentiated, restain with 5% crystal violet.
7. Clear in xylene, 3 changes.
8. Mount coverglass with appropriate medium (refractive index 1.48-1.56).

RESULTS

Neuroglial cell cytoplasm	violet
Neuroglial processes	deep violet
Background	pale violet

REMARKS

If a crystal violet precipitate forms on slide, it may be removed with full strength aniline oil.

Casper-Wolman Fluorescent Method for Myelin Figures

FIXATION Fresh, unfixed tissue must be used.
PROCESS Frozen.
MICROTOMY Cut sections at 4-6 micrometers. (See procedure, step 1.)
SOLUTIONS

Bohmer's Hematoxylin
(See Page 83)

1% Thioflavine TFS

Thioflavine, TFS	1.0	gm
Distilled water	100.0	ml

1% Glacial Acetic Acid
(See Page 308)

Apathy's Mounting Medium
(See Page 367)

STAINING PROCEDURE

1. Mount frozen sections on glass slides and allow sections to dry for 2 hours.
2. Stain in Bohmer's hematoxylin solution for 2 minutes.
3. Wash in running tap water for 5 minutes.
4. Place in thioflavine TFS solution for 3 minutes.
5. Rinse in distilled water.
6. Differentiate in 1% glacial acetic acid solution for 6 minutes.
7. Wash in running tap water for 2 minutes, then rinse in distilled water.
8. Mount coverglass with Apathy's mounting medium. Examine sections with fluorescent microscope.

RESULTS

Myelin figures	bluish-white fluorescence

Klüver-Barrera Method for Myelin Sheath

FIXATION 10% Buffered neutral formalin.
PROCESS Paraffin, frozen, or celloidin.
MICROTOMY Cut sections at 15 micrometers.
CONTROL Use control number 5.

SOLUTIONS

0.1% Luxol Fast Blue

Luxol fast blue, MBS	0.1	gm
95% Alcohol	100.0	ml

Dissolve dye in the alcohol. Add 0.5 ml of 10% glacial acetic acid to each 100 ml. Solution is stable for months.

0.1% Cresyl Echt Violet

Cresyl echt violet	0.1	gm
Distilled water	100.0	ml

Just before using, add 15 drops 10% glacial acetic acid. Filter.

0.05% Lithium Carbonate

Lithium carbonate	0.05	gm
Distilled water	100.0	ml

70% Alcohol

100% Alcohol	70.0	ml
Distilled water	30.0	ml

STAINING PROCEDURE

1. Decerate slides in xylene, 2 changes, 2 minutes each. Place in absolute alcohol, 2 changes, 2 minutes each; 95% alcohol, 2 changes, 2 minutes each; then rinse in distilled water.
2. Stain in 0.1% luxol fast blue solution at 56-60°C overnight.
3. Rinse in 95% alcohol to remove excess stain.
4. Rinse in distilled water.
5. Begin differentiation by 3 quick dips in 0.05% lithium carbonate.
6. Continue differentiation in 70% alcohol solution until gray and white matter can be distinguished from each other.
7. Dip in distilled water 5 times.
8. Complete differentiation by rinsing briefly in 0.05% lithium carbonate solution and then placing in several changes of 70% alcohol solution until the greenish-blue of the white matter contrasts sharply with the colorless gray matter.
9. Rinse thoroughly in distilled water.
10. Place in 0.1% cresyl echt violet solution for 6 minutes. (Filter and preheat cresyl echt violet solution to 57°C just before use.)
11. Differentiate in several changes of 95% alcohol.
12. Dehydrate slides in 95% alcohol and absolute alcohol 3 changes each. Clear in xylene, 3 changes.
13. Mount coverglass with appropriate medium (refractive index 1.48-1.56).

RESULTS

Myelin sheath	blue
Nuclei and Nissl substances	violet

REMARKS

Luxol fast blue has been successfully combined with other stains; for these combinations see: Margolis, G., and Pickett, J.P.: Lab. Invest. 5: 459-473, 1956.

Woelcke's Method for Myelin Sheath

FIXATION — 10% Buffered neutral formalin.
PROCESS — Paraffin, frozen, or celloidin.
MICROTOMY — Cut sections at 15 micrometers.
CONTROL — Use control number 5.
SOLUTIONS

2.5% Ferric Ammonium Sulfate

Ferric ammonium sulfate	2.5	gm
Distilled water	100.0	ml

10% Alcoholic Hematoxylin (Stock)

Hematoxylin	10.0	gm
100% Alcohol	100.0	ml

Solution should be ripened for 6 months.

Saturated Lithium Carbonate (Stock)

Lithium carbonate	0.75	gm
Distilled water	100.0	ml

Hematoxylin (Working)

Distilled water	45.0	ml
10% Hematoxylin (stock)	10.0	ml
Saturated lithium carbonate (stock)	7.0	ml
Distilled water	45.0	ml

Mix ingredients in order listed just before use.

STAINING PROCEDURE

1. Decerate slides in xylene, 2 changes, 2 minutes each. Place in absolute alcohol, 2 changes, 2 minutes each; 95% alcohol, 2 changes, 2 minutes each; then rinse in distilled water.
2. Mordant in 2.5 ferric ammonium sulfate solution overnight.
3. Rinse in distilled water, 2 changes, 30 seconds each.
4. Place in working hematoxylin solution for 2 hours.
5. Rinse in distilled water, 2 changes, 30 seconds each.
6. Differentiate in 80% alcohol until background is clear and myelin stands out sharply.
7. Dehydrate slides in 95% alcohol and absolute alcohol 3 changes each. Clear in xylene, 3 changes.
8. Mount coverglass with appropriate medium (refractive index 1.48-1.56).

RESLUTS

Myelin sheath	blue
Background	clear
Glial cells	black
Nucleoli of neurons	black

Vogt's Method for Nissl Substance

FIXATION 10% Buffered neutral formalin.
PROCESS Paraffin, frozen, or celloidin.
MICROTOMY Cut sections at 6 micrometers.
CONTROL Use control number 5.
SOLUTIONS

2% Cresyl Echt Violet (Stock)

Cresyl echt violet	2.0	gm
Distilled water	100.0	ml

Sodium Acetate Buffer (Stock)

Sodium acetate	2.0	gm
Distilled water	1000.0	ml
Glacial acetic acid	3.0	ml

1% Cresyl Echt Violet (Working)

2% Cresyl echt violet (stock)	1.0	ml
Sodium acetate buffer (stock)	100.0	ml

STAINING PROCEDURE

1. Decerate slides in xylene, 2 changes, 2 minutes each. Place in absolute alcohol, 2 changes, 2 minutes each; 95% alcohol, 2 changes, 2 minutes each; then rinse in distilled water.
2. Let stand in absolute alcohol for 30 minutes.
3. Place in working 1% cresyl echt violet for 40 minutes.
4. Dehydrate slides in 95% alcohol and absolute alcohol 3 changes each. Clear in xylene, 3 changes.
5. Mount coverglass with appropriate medium (refractive index 1.48-1.56).

RESULTS

Nissl substance	intense purple
Nuclear chromatin	purple
Background	clear

Luna's Method for Nissl Substance, Aldehyde Reactive Groups and Pituitary Cells

FIXATION 10% Buffered neutral formalin.
PROCESS Paraffin, frozen, or celloidin.
MICROTOMY Cut sections at 6 micrometers.
CONTROL Use control number 35 and pituitary cells.
SOLUTIONS

0.5% Sulfuric Acid (Stock)

Sulfuric acid, concentrated	0.5	ml
Distilled water	99.5	ml

0.5% Potassium Permanganate (Stock)

Potassium permanganate	0.5	gm
Distilled water	100.0	ml

Sulfuric Acid-Potassium Permanganate (Working)

0.5% Sulfuric acid (stock)	50.0	ml
0.5% Potassium permanganate (stock)	50.0	ml

2% Potassium Metabisulfite

Potassium metabisulfite	2.0	gm
Distilled water	100.0	ml

Aldehyde-Thionin

Thionin	0.5	gm
50% Alcohol	91.5	ml
Paraldehyde	7.5	ml
Concentrated hydrochloric acid	1.0	ml

This solution should ripen in a tightly stoppered container for 5 days at room temperature before use.

0.5% Periodic Acid

Periodic acid	0.5	gm
Distilled water	100.0	ml

Schiff's Solution
(See Page 128)

1% Orange G

Orange G	1.0	gm
Distilled water	100.0	ml

1% Glacial Acetic Acid

Glacial acetic acid	1.0	ml
Distilled water	100.0	ml

5% Phosphotungstic Acid

Phosphotungstic acid	5.0	gm
Distilled water	100.0	ml

STAINING PROCEDURE

1. Decerate slides in xylene, 2 changes, 2 minutes each. Place in absolute alcohol, 2 changes, 2 minutes each; 95% alcohol, 2 changes, 2 minutes each; then rinse in distilled water.
2. Place in working sulfuric acid-potassium permanganate for 2 minutes.
3. Bleach with 2% potassium metabisulfite for 1 minute.
4. Rinse in distilled water for 1 minute.
5. Place in aldehyde-thionin solution in an air-tight container for 1 hour.
6. Rinse in distilled water for 30 seconds.
7. Place in periodic acid for 5 minutes.
8. Rinse in distilled water for 30 seconds.
9. Place in Schiff's reagent for 15 minutes.
10. Wash in running tap water for 15 minutes.

11. Place in 1% orange G for 3 minutes.
12. Place in 5% phosphotungstic acid for 1 minute.
13. Rinse briefly in 1% glacial acetic acid for 20 seconds.
14. Dehydrate slides in 95% alcohol and absolute alcohol 3 changes each. Clear in xylene, 3 changes.
15. Mount coverglass with appropriate medium (refractive index 1.48-1.56).

RESULTS

Nissl substances .... blue-black
PAS-positive materials.. red
Background .... various shades depending on amount of differentiation

Gwyn-Heardman Method for Motor End Plates

FIXATION 10% Buffered neutral formalin.
PROCESS Frozen sections.
CRYOTOMY Cut frozen sections at 50 micrometers. Collect sections in distilled water.
CONTROL Use control number 14.
SOLUTIONS

0.1M Copper Sulfate (Stock)

Cupric sulfate ($CuSO_4.5H_20$)	24.9	gm
Distilled water	100.0	ml

0.5M Aminoacetic Acid (Stock)

Aminoacetic acid	3.7	gm
Distilled water	100.0	ml

Acetate Buffer, pH 5 (Stock)

0.1N Glacial acetic acid, (6 ml to 1000 ml distilled water)	3.0	ml
0.1N Sodium acetate, (13.6 gm to 1000 ml distilled water)	6.0	ml

2% Acetylthiocholine Iodide (Stock)

Acetylthiocholine iodide	2.0	gm
Distilled water	100.0	ml

Copper Sulfate-Acetylthiocholine Iodine

0.1M Copper sulfate (stock)	10.0	ml
2% Acetylthiocholine iodide (stock)	30.0	ml

Centrifuge this mixture until solution is clear; about 5-10 minutes.

Copper Sulfate-Aminoacetic Acid (Working)

0.1M Copper sulfate (stock)	0.2	ml
0.5M Aminoacetic acid (stock)	0.2	ml
Acetate buffer pH 5 (stock)	5.0	ml
Copper sulfate-acetylthiocholine iodine (stock)	0.8	ml

5% Ammonium Sulphide

Ammonium sulphide, yellow	5.0	ml
Distilled water	100.0	ml

10% Silver Nitrate

Silver nitrate	10.0	gm
Distilled water	100.0	ml

Ammoniacal Silver

To 10 ml of 10% silver nitrate add 28% ammonium hydroxide, drop by drop, shaking the flask vigorously until the brown precipitate disappears. Add 3 drops ammonium hydroxide in excess.

1% Ammonia Water

28% Ammonium hydroxide	1.0	ml
Distilled water	99.0	ml

3% Sodium Thiosulfate

Sodium thiosulfate	3.0	gm
Distilled water	100.0	ml

1% Glacial Acetic Acid

Glacial acetic acid	1.0	ml
Distilled water	100.0	ml

Formalin-Pyridine

Pyridine	2.0	ml
10% Neutralized formalin (stock)	98.0	ml

10% Neutralized Formalin (Stock)

Magnesium carbonate	10.0	gm
10% Unbuffered formalin	100.0	ml

STAINING PROCEDURE

1. Place sections in working copper sulfate-aminoacetic acid solution at 37°C for 15 minutes.
2. Wash in distilled water for 30 seconds.
3. Place in 5% ammonium sulphide solution 2 minutes.
4. Wash in distilled water, 3 changes, for 20 minutes each.
5. Place in formalin-pyridine solution for 15 days.
6. Wash in tap water, 3 changes, 20 minutes each.
7. Place in 10% silver nitrate at 37°C in the dark for 1 1/2 hours.
8. Wash in neutralized formalin until the white precipitate no longer forms; about 6 changes.
9. Place in ammoniacal silver until impregnation of the terminal portion of the nerve fibers has occurred. (May take several hours.)
10. Immerse in 1% ammonia water for 2 minutes.
11. Immerse in 1% glacial acetic acid for 2 minutes.
12. Place in distilled water for 2 minutes.

13. Place in 3% sodium thiosulfate for 5 minutes.
14. Place in distilled water for 30 minutes.
15. Dehydrate slides in 95% alcohol and absolute alcohol 3 changes each. Clear in xylene, 3 changes.
16. Mount coverglass with appropriate medium (refractive index 1.48-1.56).

RESULTS

Motor end plates and associated nerve fibers.................................. black
Background.. pale grey

D'Amelios' Method for Neuronal Bodies and Neuroglial Cells[149]

FIXATION 10% Buffered neutral formalin, paraformaldehyde, glutaraldehyde. Select slices of brain no more than 3-4 mm thick and fix in fixative of choice.

PROCESS Specimens are processed in toto (see procedure, step 1).

CRYOTOMY Cut section at 50-100 micrometers.

CONTROL Use control number 5.

SOLUTIONS

del Rio Hortega

Potassium dichromate	3.0	gm
Chloral hydrate	2.5	gm
10% Unbuffered formalin	50.0	ml

1.5% Silver Nitrate

Silver nitrate	1.5	gm
Distilled water	100.0	ml

Lavilla's Mixture

5% Sodium carbonate	1	volume
10% Formalin	1	volume

5% Sodium Thiosulfate

Sodium thiosulfate	5.0	gm
Distilled water	100.0	ml

STAINING PROCEDURE

1. Place specimens in del Rio Hortega solution for 4 days with gentle agitation. Change mixture daily.
2. Rinse in distilled water.
3. Blot specimens with filter paper.
4. Place in 1.5% silver nitrate solution in brown glass bottles for 2-3 days, again with gentle agitation but without fluid change.
5. Rinse in distilled water.
6. Section specimens on a cryostat or freezing microtome.

7. Collect sections, 50-100 micrometers thick, in distilled water for 5 minutes.
8. Place in Lavilla's mixture for 20 minutes.
9. Place in 5% sodium thiosulfate for 2 minutes.
10. Rinse in distilled water and dehydrate in 95% and 100% alcohol, 2 changes each. Clear in creosote.
11. Mount sections on glass slides and blot gently.
12. Mount coverglass using appropriate medium (refractive 1.48-1.56).

RESULTS

Neurons, astrocytes and dendrites .... black
Background .... gray

deBold-deBold-Kraicer Method for Stellate Cells of the Pituitary Gland[150]

FIXATION 10% Buffered neutral formalin.
PROCESS Paraffin.
MICROTOMY Cut sections at 15 micrometers (see procedure, step 1).
CONTROL Use control number 46.
SOLUTIONS

1% Ammonia Water

28% Ammonium hydroxide	1.0	ml
Distilled water	100.0	ml

10% Silver Nitrate (Stock)

Silver nitrate	10.0	ml
Distilled water	100.0	ml

Silver Nitrate (Working)

10% Silver nitrate (stock)	1.0	ml
Distilled water	25.0	ml

Adjust solution to pH 8.2-8.3 by adding 28% ammonium hydroxide, drop by drop. Prepare fresh.

Reducer

Hydroquinone	1.0	gm
Sodium sulfate	10.0	gm
Distilled water	100.0	ml

STAINING PROCEDURE

1. Float 15 micrometer thick sections in toluene. Hydrate through graded alcohol to water.
2. Immerse sections for 10 minutes in 1% ammonia water in a petri dish.
3. Wash for 3 minutes.
4. Stain for 4 minutes in silver nitrate working solution in the dark, at room temperature. Filter slightly gray solution into a brown glass bottle. Solution is stable up to 48 hours.

5. Agitate sections in reducer for 30 seconds.
6. Wash in water, mount sections on glass slides and blot dry.
7. Dehydrate slides in 95% alcohol and absolute alcohol 3 changes each. Clear in xylene, 3 changes.
8. Mount coverglass with appropriate medium (refractive index 1.48-1.56).

RESULTS

Cytoplasm body of stellate cells .... dark brown
Cytoplasm of pars intermedia cells.......................................light yellow

Zimny-Onge-Schutte Fresh Frozen Method for Nerve Endings[151]

This is a gold chloride method for the demonstration of nerve endings in skeletal muscle. The ligament is dissected, kept oriented as to its bony attachments and cut into pieces 3-5 mm thick. The tissue used must be fresh, obtained at autopsy. Following autopsy, tissue samples can be placed in saline for transport. Tissue samples in saline may be frozen for storage and future processing.

FIXATION None.
PROCESS Frozen (see step 5).
CRYOTOMY Cut sections at 100 micrometers.
CONTROL Use control number 5.
SOLUTIONS

1% Aqueous Gold Chloride

Gold chloride	1.0	gm
Distilled water	100.0	ml

25% Formic Acid

Formic acid, concentrated	25.0	ml
Distilled water	100.0	ml

STAINING PROCEDURE

1. Place tissue in 1 part 88% formic acid to 3 parts filtered, freshly squeezed lemon juice and place in the dark for 15 minutes. Hereafter, handle the tissue with paraffin-coated forceps.
2. The lemon juice is decanted off the tissue and the tissue blotted. A 1% solution of aqueous gold chloride is added and the tissue returned to the dark for 20 minutes. The tissue turns a uniform golden yellow.
3. The gold chloride solution is decanted off (it can be filtered and reused) and 25% formic acid added. The tissue is kept in the dark for 15-16 hours.
4. Wash the tissue in water and store in the refrigerator.
5. Cut frozen sections on a sliding microtome at 100 microns and hold in water.
6. Mount the sections on chrome gelatin or Poly-L-Lysine coated slides and allow to dry for a minimum of 5 hours.
7. Immerse the slides in absolute alcohol for 2 minutes and 2 changes of xylene, 3 minutes each.
8. Mount coverglass with appropriate medium (refractive index 1.48-1.56).

RESULTS

Nerve fibers and nerve endings .... black

Degirolami-Zvaigzne Method for Pineocyte Processes[152]

FIXATION	10% Formalin neutralized with calcium carbonate.
PROCESS	Paraffin.
MICROTOMY	Cut sections at 6 micrometers.
CONTROL	Use section of tumor arising from pineal parenchymal cells (pineocytes).
SOLUTIONS	

10% Formalin Neutralized with Calcium Carbonate

Concentrated 37% formalin	10.0	ml
Distilled water	100.0	ml
Calcium carbonate	1.0	gm

There should be an excess of calcium carbonate. Let the solution stand overnight. If the calcium carbonate dissolves, add another gram.

5% Sodium Carbonate (Stock)

Sodium carbonate	5.0	gm
Distilled water	100.0	ml

If the sodium carbonate resists going into solution because of hydrate formation, use heat to dissolve it and then cool to room temperature.

10% Silver Nitrate (Stock)

Silver nitrate, ACS	15.0	gm
Distilled water	150.0	ml

Solution A

10% Silver nitrate (stock)	20.0	ml
Distilled water	80.0	ml
Pyridine	1.2	ml

Solution B

10% Silver nitrate (stock)	10.0	ml
5% Sodium carbonate (stock)	30.0	ml

Mix well. A precipitate will form. Add concentrated ammonium hydroxide drop by drop while constantly shaking the solution until the precipitate is barely dissolved. (Avoid excess ammonium hydroxide. The solution should have no smell of ammonia.) Add a sufficient quantity of distilled water to bring the volume to 148.0 ml and then filter. Add:

Pyridine	2.0	ml

Solution C

10% Silver nitrate (stock)	100.0	ml

Add concentrated ammonium hydroxide drop by drop while shaking the solution until the precipitate which first forms is almost dissolved. Filter.

1% Neutral Formalin Reducing Solution

10% Formalin neutralized with $CaCO_3$	30.0	ml
Distilled water	270.0	ml
Pyridine	0.6	ml

0.2% Gold Chloride

Remove all extraneous material from a 1 gm vial of gold chloride. Place in a heavy, glass-stoppered, 500 ml container. Shake vigorously until the vial breaks. Add:

Distilled water	500.0	ml

5% Sodium Thiosulfate

Sodium Thiosulfate	5.0	gm
Distilled water	100.0	ml

STAINING PROCEDURE

1. Decerate slides in xylene, 2 changes, 2 minutes each. Place in absolute alcohol, 2 changes, 2 minutes each; 95% alcohol, 2 changes, 2 minutes each; then rinse in distilled water.
2. Transfer to solution A for 2 hours at 56-58°C.
3. Impregnate in solution B for 16 hours in the dark at room temperature.
4. Reinforce the impregnation in solution C for 2 minutes.
5. Reduce for 5 minutes in the 1% neutral formalin reducing solution, changing the solution 3 times. Make the first change after 5 seconds.
6. Wash in running tap water for 5 minutes. Rinse once with distilled water.
7. Tone in 0.2% gold chloride for 10 minutes at room temperature.
8. Rinse 2 times in distilled water for 5 seconds each.
9. Place in 5% sodium thiosulfate for 1 minute.
10. Wash in running tap water for 5 minutes. Rinse once in distilled water.
11. Dehydrate slides in 95% alcohol and absolute alcohol 3 changes each. Clear in xylene, 3 changes.
12. Mount coverglass with appropriate medium (refractive index 1.48-1.56).

RESULTS

Club shaped process ends of pineocytes	black

Garvey-Fathi-Bigelow-Jimenez-Carpenter Silver Technique for the Nervous System[153]

FIXATION 10% Buffered neutral formalin for 10 to 14 days.
PROCESS Paraffin.
MICROTOMY Cut sections at 8 micrometers.
CONTROL Use control number 5.
SOLUTIONS

1% Silver Nitrate

Silver nitrate	1.0	gm
Deionized water	100.0	ml

3% Gum Mastic

Gum mastic	3.0	gm
100% ethyl alcohol	100.0	ml

Filter before use.

Hydroquinone (Stock)

Hydroquinone	0.5	gm
Deionized water	25.0	ml

Make just before use.

0.1% Glacial Acetic Acid (Stock)

Glacial acetic acid	0.1	ml
Deionized water	100.0	ml

Developing Solution

Hydroquinone (stock)	25.0	ml
3% Gum mastic (stock)	15.0	ml

Filter and add 1.2 ml of 1.0% silver nitrate just before slides are immersed.

0.25% Gold Chloride

Gold chloride	0.25	gm
0.1% Glacial acetic acid (stock)	100.0	ml

1% Sodium Thiosulfate

Sodium thiosulfate	1.0	gm
Deionized water	100.0	ml

STAINING PROCEDURE

1. Decerate slides in xylene, 2 changes, 2 minutes each. Place in absolute alcohol, 2 changes, 2 minutes each; 95% alcohol, 2 changes, 2 minutes each; then rinse in distilled water.
2. Place sections in 1% silver nitrate at 4°C for 30 minutes. (Refrigerating at 4-8°C avoids variation in background staining.)
3. Rinse in 3 changes of distilled water.
4. Rinse in 95% alcohol, 2 changes.
5. Rinse in 100% alcohol, 2 changes.
6. Infiltrate with 3% gum mastic for 5 minutes. Avoid drying of the sections to prevent improper impregnation.
7. Drain slides and place in the developing solution at 60°C in floatation water bath for 20 minutes.
8. Rinse in 100% alcohol, 3 changes over a period of 2 minutes.
9. Rinse in distilled water.
10. Place in 0.25% gold chloride solution for 5 minutes.
12. Rinse in distilled water.
13. Place in 1% sodium thiosulfate for 2 minutes.
14. Dehydrate slides in 95% alcohol and absolute alcohol 3 changes each. Clear in xylene, 3 changes.
15. Mount coverglass with appropriate medium (refractive index 1.48-1.56).

RESULTS

Axons	black
Dendrites	black
Neurofibrils	black
Neurofibrillary tangles	black

Braak-Braak-Ohm-Bohl Method for Alzheimer's Neurofibrillary Changes[154]

FIXATION 10% Buffered neutral formalin.
PROCESS Paraffin.
MICROTOMY Cut section at 10 micrometers.
CONTROL Use control from a known Alzheimer's disease.
SOLUTIONS

1% Silver Nitrate

Silver nitrate	1.0	gm
Distilled water	100.0	ml

Lanthanum Nitrate

Lanthanum nitrate hexahydrate	0.4	gm
Sodium acetate	2.0	gm
Distilled water	100.0	ml

Alkaline Silver Iodide

Sodium hydroxide	4.0	gm
Distilled water	50.0	ml
Potassium iodide	10.0	gm

Stir and wait until dissolved. Then add slowly:

1% Aqueous solution of silver nitrate	3.5	ml

Stir until clear. Add distilled water to make final volume of 100 ml.

Developer Solution A (Stock)

Distilled water	1000.0	ml
Ammonium nitrate	2.0	gm
Silver nitrate	2.0	gm
Tungstosilicic acid	10.0	gm
Concentrated 37% formalin	4.3	ml

Combine agents in order given above.

Developer Solution B (Stock)

Distilled water	1000.0	ml
Anhydrous sodium carbonate	50.0	gm

Solutions A and B are stable for 8 months if stored in brown bottles.

0.5% Glacial Acetic Acid

Glacial acetic acid	0.5	ml
Distilled water	100.0	ml

1% Gold Chloride

Gold chloride	1.0	gm
Distilled water	100.0	ml

1% Sodium Thiosulfate

Sodium thiosulfate	1.0	gm
Distilled water	100.0	ml

Concentrated Performic Acid (Stock)

30% Hydrogen peroxide	3.0	ml
98-100% Formic acid	27.0	ml

Let ripen for 1 hour. Use solution for 1 day only.

Dilute Performic Acid (Working)

70% Ethyl alcohol	70.0	ml
Concentrated performic acid (stock)	30.0	ml

Aldehyde Fuchsin (Stock)

Pararosaniline chloride	0.5	gm
70% Ethyl alcohol	100.0	ml
Add:		
25% Hydrochloric acid	1.0	ml
and wait until clear.		
Add:		
100% Crotonaldehyde	1.0	ml

Shake and let ripen for 1 week at room temperature. Solution is stable for 1 month.

Aldehyde Fuchsin (Working)

Aldehyde fuchsin (stock)	6.0	ml
96% Ethyl alcohol	400.0	ml
Distilled water	100.0	ml
98-100% Formic acid	50.0	ml

Darrow Red

Darrow red	0.1	gm
Distilled water	500.0	ml
Glacial acetic acid	5.0	ml
Boil for 10 minutes.		
Cool and add:		
Sodium acetate	3.6	gm

Filter solution before use.

STAINING PROCEDURE

1. Decerate slides in xylene, 2 changes, 2 minutes each. Place in absolute alcohol, 2 changes, 2 minutes each; 95% alcohol, 2 changes, 2 minutes each; then rinse in distilled water.
2. Place sections in a freshly prepared solution of lanthanum nitrate for 1 hour to suppress non-specific background staining.
3. Wash in distilled water for 5 minutes.
4. Place sections in the alkaline silver iodide solution for 1 minute.
5. Neutralize in 0.5% glacial acetic acid for 3 minutes.

6. Prepare developer using equal volumes of solutions A and B. Stir vigorously and wait until clear. Develop sections for 5 to 30 minutes. Staining is adequate when sections turn pale brown. View neurofibrillary changes microscopically.
7. Place in working developer solution for 20 minutes or until sections turn pale brown.
8. Wash in 0.5% glacial acetic acid for 3 minutes.
9. Tone sections in 1% gold chloride for 5 minutes. Gently agitate sections.
10. Wash in distilled water and place in 1% sodium thiosulfate for 5 minutes.
11. Wash in distilled water for 5 minutes.
12. Place sections in working dilute performic acid for 4 hours.
13. Wash sections in 70% ethyl alcohol for 5 minutes.
14. Stain in working aldehyde fuchsin for 16 hours (overnight).
15. Wash sections several times in 70% ethyl alcohol for 5 minutes.
16. Rinse in distilled water.
17. Counterstain sections with Darrow red for 60 minutes at 60°C.
18. Dehydrate slides in 95% alcohol and absolute. Transfer to a mixture of 1 part terpineol and 2 parts xylene, 3 changes. Clear in xylene, 3 changes.
19. Mount coverglass with appropriate medium (refractive index 1.48-1.56).

RESULTS

Neurofibrillary changes of Alzheimer's ....bluish-black
Basophilic material (nissl substances) ....red
Lipofuscin deposits ....purple

REMARKS

In most silver techniques the structures appear black against a clear background. Therefore, it is difficult to delineate the boundaries of subcortical nuclei or of cortical areas and layers. This technique, with a combined stain for lipofuscin pigment and Nissl material, allows for recognition of the boundaries of layers, areas, and nuclei in the brain of the human adult.

Feeback-Brumback Method for Degenerating Nerve Terminals[155]

FIXATION — 10% Buffered neutral formalin. Then place in a 30% sucrose solution made up in 10% buffered neutral formalin for 1-2 days or until tissue sinks to the bottom of the solution.

PROCESS — Frozen.

MICROTOMY — Cut sections at 10 micrometers. Mount sections on Poly-L-Lysine coated slides (see procedure, step 1).

CONTROL — Use control specimens from Alzheimer's disease.

SOLUTIONS

0.025% Potassium Permanganate

Potassium permanganate	0.025	gm
Distilled water	100.0	ml

Bleaching Solution

1% Oxalic acid, aqueous	50.0	ml
1% Hydroquinone, aqueous	50.0	ml

2.5% Uranyl Nitrate

Uranyl nitrate	2.5	gm
Distilled water	100.0	ml

0.2% Silver Nitrate

Silver nitrate	0.2	gm
Distilled water	100.0	ml

Ammoniacal Silver Nitrate Solution

1.5% Silver nitrate	20.0	ml
95% Ethyl alcohol	12.0	ml
2.5% Sodium hydroxide	1.7	ml
28.0% Ammonium hydroxide	2 to 5	ml

Add the ammonium hydroxide last, beginning with 2.0 ml and adding just enough to dissolve the silver nitrate. This solution should be prepared fresh.

Reducing Solution

Distilled water	91.0	ml
95% Ethyl alcohol	12.0	ml
10% Formalin (unbuffered)	3.0	ml
1% Citric acid	3.0	ml

0.5% Sodium Thiosulfate

Sodium thiosulfate	0.5	gm
Distilled water	100.0	ml

0.25M Ammonium Acetate

Ammonium acetate	1.93	gm
Distilled water	100.0	ml

PROCEDURES

POLY-L-LYSINE SLIDE - COATING PROCEDURE

1. Place slides in glass slide racks.
2. Rinse slides in distilled water for 10 minutes.
3. Dip slides 2 times in 0.25M ammonium acetate.
4. Dip slides 2 times in distilled water.
5. Immerse slides in Poly-L-Lysine working solution for 10 minutes at room temperature.
6. Air dry slides in relatively dust-free area at room temperature or in a 50°C oven. Once dry, slides may be stored in boxes until ready for use. To mount tissue sections, the slide should be slightly moistened to activate the Poly-L-Lysine.

STAINING PROCEDURE

1. Place chucks with frozen tissue in a cryostat set to a temperature of 8-10°C and allow frozen tissue to equilibrate to this temperature in the cryostat for at least 1 hour prior to sectioning.
2. Cut sections at a thickness of 10 mm and pick them up on Poly-L-Lysine coated slides. Allow to air dry.

3. Fix slides in 95% ethyl alcohol for 1-2 minutes at room temperature.
4. Rinse briefly in distilled water.
5. Soak sections for 10 minutes in 0.025% potassium permanganate.
6. Place sections in bleaching solution until clear.
7. Rinse slides thoroughly in distilled water.
8. Incubate slides for 10 minutes at room temperature in 2.5% uranyl nitrate.
9. Rinse slides thoroughly in distilled water.
10. Incubate slides for 90 minutes at room temperature in a solution of 0.2% silver nitrate.
11. Without rinsing, transfer slides to freshly prepared ammoniacal silver nitrate solution for 5 minutes at room temperature.
12. Rinse slides briefly in distilled water.
13. Transfer slides to reducing solution for 2 minutes.
14. Rinse sections well in distilled water.
15. Place slides in 0.5% sodium thiosulfate for 1-2 minutes.
16. Rinse slides in distilled water.
17. Dehydrate slides in 95% alcohol and absolute alcohol 3 changes each. Clear in xylene, 3 changes.
18. Mount coverglass with appropriate medium (refractive index 1.48-1.56).

RESULTS

Axons ....black
Senile plaques ....black

Nauta-Gygax Method for Degenerating Nerve Terminals[156]

FIXATION — 10% Buffered neutral formalin.
PROCESS — Frozen.
CRYOTOMY — Cut sections at 15 micrometers. Mount sections on glass slides which have been treated with Poly-L-Lysine adhesive.
CONTROL — Use control specimen from Alzheimer's disease.
SOLUTIONS

15% Alcohol

100% Alcohol	15.0	ml
Distilled water	85.0	ml

0.05% Potassium Permanganate

Potassium permanganate	0.05	gm
Distilled water	100.0	ml

0.5% Phosphomolybdic Acid

Phosphomolybdic acid	0.5	gm
Distilled water	100.0	ml

Hydroquinone-Oxalic Acid

1% Aqueous hydroquinone	25.0	ml
1% Aqueous oxalic acid	25.0	ml

1.5% Silver Nitrate

Silver nitrate	1.5	gm
Distilled water	100.0	ml

Ammoniacal Silver Nitrate

Dissolve 0.9 gm silver nitrate in 20 ml of distilled water. Add 10 ml absolute alcohol, 1.8 ml ammonium hydroxide, and 1.5 ml of 2.5% sodium hydroxide. Mix well. Prepare fresh, just before use.

Reducing Solution

Distilled water	400.0	ml
100% Alcohol	45.0	ml
10% Formalin (non-neutralized)	13.5	ml
1% Aqueous citric acid	13.5	ml

1% Sodium Thiosulfate

Sodium thiosulfate	1.0	gm
Distilled water	100.0	ml

STAINING PROCEDURE

1. Place slides in 15% alcohol solution for 15 minutes.
2. Rinse in distilled water.
3. Place in 0.5 phosphomolybdic acid for 1 hour.
4. Without washing, place sections in 0.05% potassium permanganate for 4-10 minutes (see remarks).
5. Rinse in distilled water.
6. Decolorize in hydroquinone-oxalic acid for 2 minutes.
7. Rinse well in distilled water, 3 or 4 changes.
8. Place in 1.5% silver nitrate solution for 30 minutes.
9. Rinse in distilled water.
10. Place in ammoniacal silver nitrate for 1 minute.
11. Place in reducing solution for 1 minute.
12. Rinse in distilled water.
13. Place in 1% sodium thiosulfate for 1 minute.
14. Rinse in distilled water, 3 changes.
15. Dehydrate slides in 95% alcohol and absolute alcohol 3 changes each. Clear in xylene, 3 changes.
16. Mount coverglass with appropriate medium (refractive index 1.48-1.56).

RESULTS

Degenerating axons stain black and have fragmented and swollen fibers.

REMARKS

Step 4 is extremely critical. Excessive exposure will inhibit subsequent silver impregnation of all fibers, inadequate exposure will cause impregnation of too many normal fibers. The optimum duration varies from specimen to specimen and can best be established in a trial run, treating 3 sections for 5, 7, and 9 minutes respectively. During treatment, sections must be agitated from time to time in order to insure an even reaction.

Penfield's Method for Oligodendroglia and Microglia[157]

FIXATION	Formalin-ammonium bromide or 10% buffered neutral formalin.
PROCESS	Frozen.
CRYOTOMY	Cut sections at 20 micrometers. Mount sections on Poly-L-Lysine adhesive-treated slides.
CONTROL	Use control number 5.

SOLUTIONS

10% Silver Nitrate (Stock)

Silver nitrate	10.0	ml
Distilled water	100.0	ml

5% Hydrobromic Acid

Hydrobromic acid	5.0	ml
Distilled water	100.0	ml

5% Sodium Carbonate

Sodium carbonate	5.0	gm
Distilled water	100.0	ml

Silver-Carbonate

10% Silver nitrate (stock)	5.0	ml
5% Sodium carbonate (stock)	20.0	ml

Add 28% ammonium hydroxide, drop by drop, in sufficient quantity to dissolve the precipitate which forms. Shake vigorously during this step.

Distilled water to make	75.0	ml

1% Formalin Solution

37% Formalin	1.0	ml
Distilled water	100.0	ml

0.2% Gold Chloride

Gold chloride	0.2	gm
Distilled water	100.0	ml

5% Sodium Thiosulfate

Sodium thiosulfate	5.0	gm
Distilled water	100.0	ml

Hortega's Carbol-Xylene-Creosote Mixture

Creosote	10.0	ml
Phenol crystals, melted	10.0	ml
Xylene	80.0	ml

STAINING PROCEDURE

1. Decerate slides in xylene, 2 changes, 2 minutes each. Place in absolute alcohol, 2 changes, 2 minutes each; 95% alcohol, 2 changes, 2 minutes each; then rinse in distilled water.
2. Place slides in the following solution for 30 minutes: 100 ml distilled water, 15 drops 28% ammonium hydroxide.

3. Place in 5% hydrobromic acid at 37°C for 1 hour.
4. Wash in distilled water, 3 changes, 30 seconds each.
5. Place in 5% sodium carbonate for 1 hour.
6. Place in silver-carbonate 3-5 minutes (see remarks).
7. Develop in 1% formalin until sections turn a uniform gray color.
8. Rinse in distilled water for 20 seconds.
9. Place in 0.2% gold chloride until sections turn bluish-gray.
10. Place in 5% sodium thiosulfate for 30 seconds.
11. Wash thoroughly in several changes of distilled water.
12. Dehydrate slides in 95% alcohol and absolute alcohol 3 changes each. Clear in xylene, 3 changes.
13. Mount coverglass with appropriate medium (refractive index 1.48-1.56).

RESULTS

Oligodendroglial processes	black
Microglia	black
Background	yellowish-brown

REMARKS

Faint staining of astrocytes may occur, especially if sections are left too long in the silver solution. Leave in stain 3-5 minutes, or until sections turn a uniform gray color when transferred to the reducer. Control staining by taking a section out of the silver-carbonate at intervals of 1-2 minutes and examining with the microscope.

Lapham's Method for Myelin and Glial Fibers[158]

FIXATION 10% Buffered neutral formalin.
PROCESS Paraffin.
MICROTOMY Cut sections at 15 micrometers.
CONTROL Use control number 5.
SOLUTIONS

Chromium-Gallocyanin

Chromium potassium sulfate (chrome alum)	10.0	gm
Distilled water	200.0	ml
Gallocyanin	0.3	gm

Heat slowly to boiling; boil for 12-25 minutes with frequent stirring. Cool slowly to room temperature. Filter before using.

0.5% Phloxine B Aqueous

Phloxine B	0.5	gm
Distilled water	100.0	ml

5% Phosphotungstic Acid

Phosphotungstic acid	5.0	gm
Distilled water	100.0	ml

1% Glacial Acetic Acid

Glacial acetic acid	1.0	ml
Distilled water	100.0	ml

0.05% Fast Green FCF Solution		
Fast green, FCF	0.05	gm
Distilled water	100.0	ml

STAINING PROCEDURE

1. Decerate slides in xylene, 2 changes, 2 minutes each. Place in absolute alcohol, 2 changes, 2 minutes each; 95% alcohol, 2 changes, 2 minutes each; then rinse in distilled water.
2. Stain in chromium-gallocyanin overnight.
3. Wash in running tap water for 5 minutes.
4. Stain in 0.5% phloxine B for 5 minutes.
5. Wash in running tap water for 3 minutes.
6. Place in 5% phosphotungstic acid for 1 minute.
7. Wash in running tap water for 3 minutes.
8. Place in 1% glacial acetic acid for 2 minutes.
9. Differentiate in 80% alcohol for 2 minutes.
10. Stain in 1% glacial acetic acid for 1 minute.
11. Stain in 0.05% fast green FCF for 1 minute.
12. Place in 1% glacial acetic acid for 1 minute.
13. Dehydrate slides in 95% alcohol and absolute alcohol 3 changes each. Clear in xylene, 3 changes.
14. Mount coverglass with appropriate medium (refractive index 1.48-1.56).

RESULTS

Myelin	magenta
Glial fibers, cytoplasm of astrocytes	bright green
Nuclei	blue-black to black
Collagen	blue-green
Erythrocytes	orange to red

REMARKS

The phloxine is differentiated at step 9. All tissue structures are decolorized at this time except myelin and erythrocytes. Over-differentiation does not occur within limits.

Luna's Method for Cells of the Pituitary and CNS

FIXATION	10% Buffered neutral formalin.
PROCESS	Paraffin.
MICROTOMY	Cut sections at 6 micrometers.
CONTROL	Use control number 5 or 46.
SOLUTIONS	

0.5% Sulfuric Acid (Stock)		
Sulfuric acid, concentrated	0.5	ml
Distilled water	99.5	ml

0.5% Potassium Permanganate (Stock)		
Potassium permanganate	0.5	gm
Distilled water	100.0	ml

Sulfuric Acid-Potassium Permanganate (Working)

0.5% Sulfuric acid (stock)	25.0	ml
0.5% Potassium permanganate (stock)	25.0	ml

2% Potassium Metabisulfite

Potassium metabisulfite	2.0	gm
Distilled water	100.0	ml

Aldehyde-Thionin

Thionin	0.5	gm
70% Alcohol	92.0	ml
Paraldehyde	7.5	ml
Hydrochloric acid, concentrated	1.0	ml

This solution should ripen before use in a tightly stoppered container for 3-5 days at room temperature.

0.5% Periodic Acid Solution

Periodic acid	0.5	gm
Distilled water	100.0	ml

Schiff's Solution
(See Page 128)

1% Orange G

Orange G	1.0	gm
Distilled water	100.0	ml

1% Glacial Acetic Acid

Glacial acetic acid	1.0	ml
Distilled water	100.0	ml

5% Phosphotungstic Acid

Phosphotungstic acid	5.0	gm
Distilled water	100.0	ml

STAINING PROCEDURE

1. Decerate slides in xylene, 2 changes, 2 minutes each. Place in absolute alcohol, 2 changes, 2 minutes each; 95% alcohol, 2 changes, 2 minutes each; then rinse in distilled water.
2. Place in working sulfuric acid-potassium permanganate for 2 minutes.
3. Bleach with 2% potassium metabisulfite for 1 minute.
4. Rinse well in distilled water.
5. Stain in aldehyde-thionin (in an air tight stoppered container) for approximately 50 minutes.
6. Rinse in distilled water.
7. Oxidize in 0.5% periodic acid for 5 minutes.
8. Rinse in distilled water.
9. Stain in Schiff's solution for 15 minutes.
10. Wash in running tap water for 15 minutes.

11. Stain in 1% orange G for 3 minutes.
12. Place in 5% phosphotungstic acid for 1 minute.
13. Rinse briefly in 1% glacial acetic acid.
14. Dehydrate slides in 95% alcohol and absolute alcohol 3 changes each. Clear in xylene, 3 changes.
15. Mount coverglass with appropriate medium (refractive index 1.48-1.56).

RESULTS

Alpha cells	red
Beta cells	orange
Delta cells	light blue
Gamma cells	gray
Nissl substance	blue-black
PAS positive material	red
Background	various shades depending on amount of differentiation

Sevier-Munger Method for Neural Tissues[159]

FIXATION 10% Buffered neutral formalin. Avoid chromate fixatives.
PROCESS Paraffin.
MICROTOMY Cut section at 6 micrometers.
CONTROL Use control number 5.
SOLUTIONS

20% Silver Nitrate

Silver nitrate	20.0	gm
Distilled water	100.0	ml

10% Silver Nitrate

Silver nitrate	10.0	gm
Distilled water	100.0	ml

2% Formalin

37% Formalin	2.0	ml
Tap water	98.0	ml

5% Sodium Thiosulfate

Sodium thiosulfate	5.0	gm
Distilled water	100.0	ml

Sodium Carbonate

Sodium carbonate (hydrated)	8.0	gm
Distilled water	30.0	ml

Ammoniacal Silver Solution

Place 50 ml of 10% silver nitrate solution in a 200 ml beaker, add ammonium hydroxide (28-30%) drop by drop until the dark brown precipitate which forms has almost disappeared. Shake vigorously between drops. Avoid complete decolorization. The end point is a slightly cloudy solution. Add to this solution 0.5 ml of sodium carbonate solution. Shake well. Add 25 drops of ammonium hydroxide. Shake well. Solution should now be crystal clear. Filter and place in a well stoppered container.

STAINING PROCEDURE

1. Decerate slides in xylene, 2 changes, 2 minutes each. Place in absolute alcohol, 2 changes, 2 minutes each; 95% alcohol, 2 changes, 2 minutes each; then rinse in distilled water.
2. Preheat 20% silver nitrate to 60°C for 15 minutes. Add slides to warm silver nitrate and let them remain in the 60°C oven for 15 minutes.
3. Rinse 1 slide at a time in distilled water, then place in a clean coplin jar.
4. To working ammoniacal silver solution add 10 drops of the 2% formalin solution, shaking gently while adding formalin. Pour quickly over slides. Let develop for 5 to 30 minutes until golden brown. Keep in motion to avoid precipitate.
5. Examine microscopically. Do not wash. Developing should be neither too fast nor too slow.
6. Rinse well in fresh tap water, 3 changes.
7. Place in 5% sodium thiosulfate for 2 minutes.
8. Wash in running tap water for 3 minutes.
9. Dehydrate slides in 95% alcohol and absolute alcohol 3 changes each. Clear in xylene, 3 changes.
10. Mount coverglass with appropriate medium (refractive index 1.48-1.56).

RESULTS

Peripheral neurities	black
Axons	black
Myelin sheath	light brown
Collagen and muscle	brown
Argentaffin granules	black

REMARKS

Sections of the central nervous system have also been stained with this technique. The cerebellar basket cells and axons of the Purkinje cell become beautifully impregnated. Within the brain stem and spinal cord, large fibers are stained but small fibers are not visualized. The optic and olfactory nerves do not stain.

Hirsch-Pfeiffer Method for Metachromatic Leukodystrophy

(See Fig. 193 and 194)

FIXATION 10% Buffered neutral formalin.

MICROTOMY Sections may be cut at 12 to 20 microns depending on the material's ability to section. More distinct intracellular sulfatide detail can be seen in 12 micron sections. Conversely, more overall sulfatides reaction can be seen on the 20 micron sections. Use control material if available. Stain works well in floating sections, as well as sections placed on albuminized slides. See staining procedures.

SOLUTION

0.04% Cresyl Violet V (Stock)

Distilled water	85.0	ml
Glacial acetic acid	15.0	ml
Cresyl violet V	0.04	gm

15% Triethanolamine (Stock)

Distilled water	85.0	ml
Triethanolamine	15.0	ml

Cresyl Violet V (Working)

Add the 0.04% stock cresyl violet V to the 15% triethanolamine stock solution.

STAINING PROCEDURE. The following method is used for staining free floating sections.

1. Frozen cut sections are placed in distilled water for 10 minutes.
2. Preheat cresyl violet-triethanolamine solution in 60°C oven for 1 hour.
3. Transfer sections from distilled water to preheated cresyl violet-triethanolamine solution and allow to stain for 10 minutes in 60°C oven.
4. Remove staining solution, containing the sections, from oven and allow to cool to room temperature (20-25 minutes).
5. Transfer sections to distilled water, 2 changes, 5 minutes each.
6. Float sections on albuminized slides and allow to air dry for a minimum of 4 hours.
7. Dip slides in distilled water several times.
8. Mount coverslip with glycerin jelly and ring coverslip with clear fingernail polish.

RESULTS

Areas containing abnormal sulfatides	brown
Other tissue structures	blue to purple

STAINING PROCEDURE. The following procedure is used for sections which have been mounted on glass slides.

1. Preheat cresyl violet-triethanolamine solution in 60°C oven for 1 hour.
2. Transfer slides from distilled water to preheated cresyl violet-triethanolamine solution for 10 minutes in 60°C oven.
3. Remove staining dish from oven and allow to cool to room temperature (20-25 minutes).
4. Wash slides in distilled water for 15 minutes.
5. Mount coverslip with glycerin jelly and ring coverslip with clear fingernail polish.

RESULTS

Areas containing abnormal sulfatides	brown
Other tissue structures	blue to purple

REMARKS. Slides or loose sections may be left in distilled water overnight if necessary at step 5 (floating sections procedure), or step 4 (mounted section procedure).

Sirkin's Method for Defatting Frozen Brain Sections[160]

Cresyl violet is perhaps the most frequently used stain in clinical and experimental neurology. The need to defat frozen sections in alcohols or xylenes before staining has been noted frequently. To achieve consistently good results, the amount of defatting is critical. If this step is omitted, the stain will be poor and uneven. On the other hand, too much defatting prior to staining can sometimes make it difficult or impossible to adequately stain the cell bodies of neurons without the white matter also retaining appreciable stain after differentiation.

FIXATION	10% Buffered neutral formalin.
PROCESS	Frozen.
CRYOTOMY	Cut sections at 40 micrometers.
CONTROL	Use control number 5.
SOLUTIONS	

0.25% Cresyl Violet Acetate

Cresyl violet acetate	0.25	gm
Distilled water	100.0	ml

STAINING PROCEDURE

1. Dry slides well.
2. Defat in ascending concentrations of ethyl alcohol: 1 dish of 70%, 2 dishes of 95%, and 2 dishes of 100%, 10 dips each. Immediately start descending through the same dishes in reverse order, again 10 dips each.
3. Place the slides in water, for 1 minute. Then place in 0.25% cresyl violet acetate staining solution for 1.5 minutes.
4. Rinse in running tap water.
5. Dehydrate slides in 95% alcohol and absolute alcohol 3 changes each. Clear in xylene, 3 changes.
6. Mount coverglass with appropriate medium (refractive index 1.48-1.56).

REMARKS

If slides are kept in alcohol for longer periods, or if xylenes are used, as is common practice, they will be overdefatted. 24 hours, 20 minutes or even 1 minute in xylenes will result in overdefatting and excessive staining of white matter.

Chapter 16

Methods for Microwave Technology

Most of the methods utilized in this chapter were provided by Mr. Charles Churukian, Department of Pathology and Laboratory Medicine, The University of Rochester Medical Center, Rochester, New York, 14626. I am pleased that Mr. Churukian has allowed publication of his modifications utilizing microwave technology. Many of these procedures were tried in our laboratory and all performed well. It is because of this degree of quality and reproducibility that these methods are included in this book.

Theory of Microwave Ovens

The microwave oven operates with a heating source that is much different from that of the conventional resistive electrical heating element. Microwave ovens are equipped with a special radio tube called a magnetron. The magnetron generates electromagnetic waves called microwaves (radar is a form of microwave). The energy of the microwave is transformed into heat when absorbed by water. Polarized water molecules and other dipolar molecules are flipped back and forth in the electromagnetic field produced by the magnetron. This results in tremendous molecular friction which produces heat.

When water is exposed to microwaves, oscillation and collision of water molecules occur. Oscillation of the water molecules cause friction which results in heat. The collision of molecules adds energy to random motion and also results in heat.

The microwave energy generated by the magnetron is fed into the metal-lined oven cavity by a wave guide. A metal fan called a mode stirrer is usually positioned between the cavity and the wave guide to help distribute the energy and reduce hot and cold spots, the microwave oven. Good oven design minimizes hot and cold spots in the microwave oven, allowing for more even heating. Microwaves with carousels will help minimize hot and cold spots also.

Microwave ovens are rated by watts. An increase in watts produces a faster response time for the production of energy and a decrease in watts produces a slower response time for the production of energy. Most domestic microwave ovens are 500 to 600 watts, although lower wattage microwaves ovens are available. As one can see the wattage of a microwave can influence the time necessary for a specific fixation or staining procedure to occur.

Practical Effects of Microwave in Fixation and Staining

The effect of microwave on biological material is definitely one of heat generation. One certainly would wonder if there are in addition, non-thermal effects.

Microwave fixation can occur through both non-additive and additive fixation, depending on the medium in which the tissue is placed. In a non-fixative solution such as saline, non-additive fixation occurs by the removal of bound water through microwave radiation. In the presence of an additive fixative the rapid molecular movement of the fixative accelerates crosslinking by allowing the fixative to find its binding sites faster. For a complete understanding of additive and non-additive fixation see chapter 1.

Microwaves decrease staining times by increasing molecular movement of the dye molecule. As discussed earlier, this is accomplished by oscillation and collision of water molecules and other dipolar molecules resulting in rapid molecular movement and the production of heat. When rapid molecular movement occurs, dye molecules find their binding sites much faster. Note: heat by itself regardless of the generating source (microwave, electrical resistance or flame) will result in rapid molecular movement.

Health and Safety of Microwave

In 1968, The Bureau of Radiologic Health, (an FDA agency) set the limit of 1 mw/cm as the maximum standard of allowable leakage of microwave from an oven as it leaves the manufacturer. This can rise to no more than 5 mw/cm during the life of the unit. The intensity of microwave radiation falls off drastically with distance from the oven. Two major concerns may be what hazards are associated with low levels of radiation not yet commonly recognized and how often do leak rates exceed the standards?

Microwave Hazards in the Histology Laboratory

The following is a list of hazards that one will/may encounter while fixing tissue, making solutions or staining in a microwave oven.

1. Explosion of liquid may occur if lids on containers are tightened (caps and/or lids should be loosely fitted on containers).
2. Super heated water (steam) can cause burns.
3. Containers that are not fully microwavable may cause burns when handled.
4. Heating of some solutions may cause fumes that may explode (xylene, alcohol) or produce toxic fumes (formalin, mercury ions, osmium tetroxide, lead and uranium nitrate).
5. Metals (non microwavable) should not be used in the microwave oven. The use of metals in microwave ovens radically changes the electromagnetic field. This change causes arcing (sparking) and damage to the oven. The arcing may be harmful to the operator if in close proximity to the oven.

To determine if containers are microwavable (plastic cassettes, etc.), place them in the microwave and heat them for 30 seconds. Any detectable warmth of the container indicates that it is not suitable for microwave use. A beaker of saline or water should always be placed in the oven during such a test to provide for absorption of microwaves and to prevent damage to the microwave oven. (Jerry L. Fredenburgh)

Note: The staining temperature for the following methods are designated by power levels on the microwave oven from 1-10. It is therefore suggested that one purchase a 500 to 600 watt microwave oven with power levels 1-10 for use with these methods.

Warthin-Starry Method for Bacteria (Microwave)

FIXATION	10% Buffered neutral formalin. Do not use dichromate fixatives.
PROCESS	Paraffin.
MICROTOMY	Cut sections at 5 micrometers.
CONTROL	Use control number 22A, 29 or 40.
SOLUTIONS	

Glycine-Acetic Acid (Stock)

Glycine	2.4	gm
Acetic acid, glacial	0.3	ml
Distilled water	100.0	ml

The pH of this solution is approximately 3.9. Store in a refrigerator at 3-6°C.

Glycine-Acetic Acid (Working)

Glycine-acetic acid (stock)	5.0	ml
Distilled water	500.0	ml

The pH of this solution is approximately 4.2.

0.5% Silver Nitrate Solution*

Silver nitrate	0.5	gm
Glycine-acetic acid (working)	100.0	ml

2% Silver Nitrate Solution (Stock)*

Silver nitrate	2.0	gm
Glycine-acetic acid (working)	100.0	ml

4% Gelatin Solution (Stock)*

Gelatin	4.0	gm
Glycine-acetic acid (working)	100.0	ml

Dissolve by placing on a magnetic stirrer and applying gentle heat.

0.1% Hydroquinone Solution (Stock)*

Hydroquinone	0.1	gm
Glycine-acetic acid (working)	100.0	ml

*Place the 2% silver nitrate, 4% gelatin, and the 0.1% hydroquinone in separate flasks in a 56-60°C. oven. These solutions and the 0.5% silver nitrate are prepared immediately before use.

Silver Nitrate-Gelatin-Hydroquinone Developer

2% Silver nitrate (stock)	10.0	ml
4% Gelatin (stock)	25.0	ml
0.1% Hydroquinone (stock)	15.0	ml

Combine in order given, making certain the solutions are mixed well after each addition. Prepare immediately.

STAINING PROCEDURE

1. Decerate slides in xylene, 2 changes, 2 minutes each. Place in absolute alcohol, 2 changes, 2 minutes each; 95% alcohol, 2 changes, 2 minutes each; then rinse in distilled water.
2. Place in working glycine-acetic acid solution.

3. Place in 40 ml of 0.5% silver nitrate in a plastic coplin jar with the cap loosely applied. Place in a microwave oven and microwave at power level 1 (the lowest power setting) for 5 minutes. Allow the slides to remain in the solution for 2 more minutes.
4. Place in freshly prepared silver nitrate-gelatin-hydroquinone developer in a plastic coplin jar with the cap loosely applied. Place in a microwave oven and microwave at power level 1 for one minute. Allow the slides to remain in the solution until the sections appear a grayish-brown. This usually takes 30-60 seconds but may take longer depending upon the type of gelatin used.
5. Wash quickly and thoroughly in hot running water.
6. Rinse in 2 changes of distilled water.
7. Dehydrate slides in 95% alcohol and absolute alcohol 3 changes each. Clear in xylene, 3 changes.
8. Mount coverglass with appropriate medium (refractive index 1.48-1.56).

RESULTS

Spirochetes ....black
Legionnaires' bacteria ....black
Nocardia ....black
Klebsiella ....brown to black
Campylobacter ....gray to black
Nuclear chromatin ....brown
Erythrocytes ....brown

REMARKS

We prefer this microwave oven method over the regular Warthin-Starry and Dieterle methods for demonstrating bacteria, because it is easier to do, less time consuming, and gives equally good, if not superior, results. Remember that all bacteria are stained by these silver methods, and they are not specific for Legionella.

The type of gelatin used in this method is important in obtaining satisfactory results. We prefer using Fisher type A gelatin because there is less background staining when this gelatin is used.

We have observed that the staining results are more consistent and reliable when the impregnation and developing is done at the lowest power level. This is because there is a more uniform distribution of heat in the solutions as compared to the use of higher power levels, which can result in as much as a 10°C difference in temperature between the top and bottom portions. The use of xylene substitutes in this method is not recommended.

Steiner's Method for Spirochetes and Bacteria (Microwave)

FIXATION	10% Buffered neutral formalin. Do not use fixatives which contain chromium or mercury.
PROCESS	Paraffin.
MICROTOMY	Cut sections at 5 micrometers.
CONTROL	Use control number 22, 29 or 40.

SOLUTIONS

1% Uranyl Nitrate

Uranyl nitrate	1.0	gm
Distilled water	100.0	ml

0.5% Silver Nitrate (Stock)

Silver nitrate	0.5	gm
Distilled water	100.0	ml

2.5% Gum Mastic (Stock)

Gum mastic	2.5	gm
Absolute alcohol	100.0	ml

Allow the gum mastic to dissolve for 24 hours. Filter through Whatman #4 filter paper and store in a refrigerator at 3-6°C.

2% Hydroquinone (Stock)

Hydroquinone	0.5	gm
Distilled water	25.0	ml

Prepare fresh each time.

Reducing Solution

2.5% Gum mastic (stock)	10.0	ml
2% Hydroquinone (stock)	25.0	ml
Absolute alcohol	5.0	ml
0.5% Silver nitrate (stock)	2.5	ml

Prepare just before use. The solution will have a milky appearance but does not need to be filtered.

STAINING PROCEDURE

1. Decerate slides in xylene, 2 changes, 2 minutes each. Place in absolute alcohol, 2 changes, 2 minutes each; 95% alcohol, 2 changes, 2 minutes each; then rinse in distilled water.
2. Place in 40 ml of 1% uranyl nitrate in a plastic coplin jar with the cap loosely applied and microwave at power level 7 for 1 minute. Allow the slides to remain in the hot solution for 1 more minute.
3. Rinse with 5 changes of distilled water.
4. Place in 40 ml of 0.5% silver nitrate in plastic coplin jar with the cap loosely applied and microwave at power level 2 for 3 minutes. Allow the slides to remain in this hot solution for 5 more minutes.
5. Rinse with 5 changes of distilled water.
6. Dehydrate in 2 changes of 95% ethyl alcohol and 2 changes of absolute alcohol
7. Place in 2.5% gum mastic for 1 minute.
8. Allow to air-dry for 1 minute.
9. Rinse with 3 changes of distilled water. The slides may remain in distilled water while the reducing solution is being prepared.
10. Place slides in freshly prepared reducing solution in a plastic coplin jar with the cap loosely applied and microwave at power level 1 for 3 minutes. Allow the slides to remain in this warm solution for 3-5 minutes or until the sections are properly developed. Adequately developed sections should appear light brown. Care needs to be exercised to avoid under-or-overdeveloping the sections.

RESULTS

Spirochetes	dark brown to black
Legionella pneumophila	dark brown to black
Campylobacter	dark brown to black
Other nonfilamentous bacteria	dark brown to black

REMARKS

The most commonly used staining methods for demonstrating spirochetes and some gram-negative bacteria which stain poorly with tissue gram stains, are the Dieterle, Warthin-Starry and Steiner methods. All of these methods require the use of heat to impregnate tissue sections with silver nitrate, followed by reducing solution which converts the silver nitrate to deposits of black metallic silver on the spirochetes and bacteria. The staining time required to perform these methods has been greatly reduced by the use of the microwave oven. We have tried all of these methods and found that all will give satisfactory staining results. However, for demonstrating spirochetes, the Steiner method produces the most consistent and reliable staining results.

Truant's Fluorescent Method for Acid-Fast Bacteria (Microwave)

FIXATION 10% Buffered neutral formalin.
PROCESS Paraffin.
MICROTOMY Cut sections at 5 micrometers.
CONTROL Use control number 41.
SOLUTIONS

Auramine-Rhodamine Solution

Auramine 0	1.5	gm
Rhodamine B	0.03	gm
Glycerine	75.0	ml
Phenol, liquid, approximately 90%	11.0	ml
Distilled water	50.0	ml

Combine the liquids in a 250 ml flask and place on a magnetic stirrer. Add the dyes to the solution, apply gentle heat and allow to mix for about 30 minutes. Before using, filter through Whatman #4 filter paper while warm.

0.5% Acid Alcohol Solution

70% Alcohol	1000.0	ml
Hydrochloric acid, concentrated	5.0	ml

0.5% Potassium Permanganate

Potassium permanganate	0.5	gm
Distilled water	100.0	ml

STAINING PROCEDURE

1. Decerate slides in xylene, 2 changes, 2 minutes each. Place in absolute alcohol, 2 changes, 2 minutes each; 95% alcohol, 2 changes, 2 minutes each; then rinse in distilled water.

2. Place in 45 ml of auramine-rhodamine solution in a glass coplin jar. Place in a microwave oven and microwave at power level 1 (the lowest power setting) for 5 minutes. Allow to set in the hot solution for 2 more minutes.
3. Rinse with 3 changes of distilled water.
4. Differentiate in 2 changes of 0.5% acid alcohol, 1 minute in each change.
5. Rinse in 4 changes of distilled water.
6. Place in 0.5% potassium permanganate solution for 1 1/2 minutes.
7. Rinse with 3 changes of distilled water.
8. Stand slide on end and thoroughly air-dry.
9. Dip in 2 changes of xylene.
10. Mount coverglass with appropriate medium (refractive index 1.48-1.56).

RESULTS

Acid fast bacteria, including leprosy and mycobacterium avium intracellular ....fluorescent orange-yellow on a dark background

REMARKS

Always use a positive and a negative control slide. The negative control slide is important because tap water and, on occasion, staining solutions may contain acid-fast bacilli. Look at the slide away from the tissue section for such contaminants. False positive staining in our laboratory was found on culture to be due to Runyon Group II scotochromogenic acid-fast bacteria present in tap water.

Remember that this stain is not specific for mycobacterium tuberculosum; it will stain several types of acid-fast organisms. Also, it does not appear to be affected by the state of viability of the organism. The use of heat is necessary to obtain satisfactory staining results. Most texts recommend staining for 10 minutes in auramine-rhodamine solution which has been preheated to 60°C. For many years we did the staining in a 58°C water bath for 30 minutes with good results. This microwave oven method saves time and produces consistently good staining results.

Usually a plastic coplin jar is used when staining in the microwave oven. This is because there is a tendency for glass coplin jars to break when solutions in them are rapidly heated with microwaves. However, when the lowest power setting is used the solution heats up slowly and there is no danger of the coplin jar cracking. The main reason for using a glass coplin jar in this method is that it is easier to clean than a plastic coplin jar. Cleaning is accomplished with the acid alcohol solution. The amount of rhodamine B in the staining solution is much less (0.03 gm versus 0.75 gm) than indicated in Truant's method. The reason for this is that rhodamine B, even though it is a fluorescent dye, can sometimes act as a fluorescence quencher when used even in relatively low concentrations. Studies completed in our laboratory on several samples of rhodamine B showed that all of them acted as fluorescence quenchers when used in the concentration prescribed by Truant. There was only weak fluorescence of acid-fast bacilli. By greatly reducing the amount of rhodamine B, the fluorescence of acid-fast bacilli was significantly intensified.

A Silver Method for Fungi and Pneumocystis Carinii in Unfixed Specimens (Microwave)

FIXATION Unfixed specimens on slides are briefly heat fixed.
PROCESS Cytologic preparations of lung aspirates, bronchial washings, sputum specimens, and other cytologic preparations.
CONTROL Use control number 18 or 37A.
SOLUTIONS

4% Periodic Acid

Periodic acid	4.0	gm
Distilled water	100.0	ml

Ammoniacal Silver Solution

To 10 ml of 10% silver nitrate, add 5.0 ml of 4% lithium hydroxide, monohydrate. Then add concentrated ammonium hydroxide, drop by drop with constant shaking, until the precipitate dissolves. Make up the solution to 1000 ml with distilled water and store in a refrigerator at 3°C. This solution is stable for at least 3 months.

0.2% Gold Chloride

Gold chloride	0.2	gm
Distilled water	100.0	ml

2% Sodium Thiosulfate

Sodium thiosulfate	2.0	gm
Distilled water	100.0	ml

Fast Green (Stock)

Fast green FCF	0.2	gm
Distilled water	100.0	ml
Acetic acid	0.2	ml

Fast Green (Working)

Fast green (stock)	10.0	ml
Distilled water	40.0	ml

STAINING PROCEDURE

1. Oxidize slides in 4% periodic acid for 5 minutes.
2. Rinse in 5 changes of distilled water.
3. Microwave 45 ml of refrigerated cold ammoniacal silver in a plastic coplin jar with the cap loosely applied, at power level 7 for 1 minute. Immediately place the slides in this hot solution, return to the microwave oven and microwave at power level 1 for 1 minute. Allow the slides to remain in the solution for 1 more minute or until the material appears yellow-brown.
4. Rinse in 4 changes of distilled water.
5. Tone in 0.2% gold chloride for 30 seconds.
6 Rinse in 2 changes of distilled water.
7. Place in 2% sodium thiosulfate for 30 seconds.
8. Rinse in 4 changes of distilled water.

9. Counterstain with working fast green solution for 30 seconds.
10. Rinse in 3 changes of distilled water.
11. Dehydrate slides in 95% alcohol and absolute alcohol 3 changes each. Clear in xylene, 3 changes.
12. Mount coverglass with appropriate medium (refractive index 1.48-1.56).

RESULTS

Fungi and pneumocystis carinii ....sharply delineated in black
Background ....green

REMARKS

In this staining method 4% periodic acid is used as the oxidizing agent in place of 5% chromic acid. The reason for this, is that chromic acid oxidation sometimes results in mucin staining so intensely that the organisms are difficult to find. Using 4% periodic acid eliminates mucin staining, making Pneumocystis carinii and fungi easy to locate and identify.

We use this method only on cytologic preparations. It is not performed on paraffin sections because there is a tendency for the sections to become detached from the slides.

Luxol Fast Blue-Cresyl Violet Method for Myelin and Nerve Cells (Microwave)

FIXATION 10% Buffered neutral formalin.
PROCESS Paraffin.
MICROTOMY Cut sections at 10 micrometers.
CONTROL Use control number 34.
SOLUTIONS

0.1% Luxol Fast Blue

Luxol fast blue, MBS	0.1	gm
95% Alcohol	100.0	ml

Dissolve dye in alcohol. Add 0.5 ml of 10% glacial acetic acid to each 100 ml.

Cresyl Violet Acetate

Cresyl violet acetate*	0.04	gm
Distilled water..	40.0	ml
1% Oxalic acid	0.2	ml

*Cresyl echt violet may be substituted for cresyl violet acetate.

0.05% Lithium Carbonate

Lithium carbonate...	0.05	gm
Distilled water..	100.0	ml

70% Alcohol

100% Alcohol..	70.0	ml
Distilled water..	30.0	ml

STAINING PROCEDURE

1. Decerate slides in xylene, 2 changes, 2 minutes each. Place in absolute alcohol, 2 changes, 2 minutes each; 95% alcohol, 2 changes, 2 minutes each.
2. Place in 40 ml of luxol fast blue solution in a glass coplin jar and microwave at power level 1 (the lowest power setting) for 4 minutes. Allow the slides to remain in the hot solution for 30 minutes.
3. Rinse off excess stain in 95% alcohol.
4. Rinse in 3 changes of distilled water.
5. Begin differentiation by placing in 0.05% lithium carbonate solution for 5 seconds.
6. Continue differentiation in 70% alcohol for 10 seconds.
7. Rinse in 4 changes of distilled water. Check with a microscope and repeat steps 5 and 6, if necessary.
8. Place in cresyl violet acetate solution for 5 minutes.
9. Rinse in 3 changes of distilled water.
10. Dehydrate slides in 95% alcohol and absolute alcohol 3 changes each. Clear in xylene, 3 changes.
11. Mount coverglass with appropriate medium (refractive index 1.48-1.56).

RESULTS

Myelin ....blue
Cell products ....pink to violet

REMARKS

The use of the microwave oven greatly reduces the time required to perform the method. The power output at power level 1 is 50 watts on the oven we use. Other models of microwave ovens can be used in this method, but it may be necessary to change the time required to heat the luxol fast blue solution. The temperature of the solution should be 75-80°C.

Rhodanine Method for Copper (Microwave)

FIXATION 10% Buffered neutral formalin.
PROCESS Paraffin.
MICROTOMY Cut paraffin sections at 5 micrometers.
CONTROL Use control number 12.
SOLUTIONS

Rhodanine Solution (Stock)

Rhodanine 5-p-dimethylaminobenzylidene.....................	0.2	gm
Absolute alcohol..	50.0	ml

Filter through Whatman #1 filter paper and store in a refrigerator. The solution is stable for about 1 week.

Rhodanine Solution (Working)

Rhodanine solution (stock)	5.0	ml
Distilled water..	45.0	ml

0.5% Sodium Borate

Sodium borate, $Na_2B_4O_7 \cdot 10H_2O$	5.0	gm
Distilled water	1000.0	ml

STAINING PROCEDURE

1. Decerate slides in xylene, 2 changes, 2 minutes each. Place in absolute alcohol, 2 changes, 2 minutes each; 95% alcohol, 2 changes, 2 minutes each; then rinse in distilled water.
2. Place slides in 50 ml of working rhodanine solution in a plastic coplin jar with the cap loosely applied and microwave at power level 1 for 6 minutes. Care should be taken not to boil the solution. Allow the slides to remain in this hot solution for 4 more minutes.
3. Wash with 6 changes of distilled water.
4. Stain in Mayer's hematoxylin for 7 minutes.
5. Rinse with 2 changes of distilled water.
6. Blue nuclei with 0.5% sodium borate for 5 seconds.
7. Rinse with 4 changes of distilled water.
8. Dehydrate slides in 95% alcohol and absolute alcohol 3 changes each. Clear in xylene, 3 changes.
9. Mount coverglass immediately. If there is a delay in mounting the sections, there will be a loss of some or all of the copper.

RESULTS

Copper	bright red intracytoplasmic granules
Nuclei	blue

REMARKS

We have found this stain to be both sensitive and specific and allows for semiquantitative evaluation of the amount of copper present. This microwave method gives more intense positive staining results than the overnight method which is done in a 37°C oven. A good positive control for this method is any fetal liver fixed in 10% buffered neutral formalin, for not longer than 24 hours prior to paraffin processing. Fetal liver always contain 3 to 4+ copper within hepatocytes.

Care must be taken not to cover counterstain with hematoxylin, which will mask the positive staining of copper.

Giemsa Method for Plastic Bone Marrow Sections (Microwave)

PROCESS Undecalcified glycol methacrylate embedded bone marrow.
MICROTOMY Cut sections at 3-4 micrometers.
CONTROL Use section of bone marrow.
SOLUTIONS

0.5% Hydrochloric Acid-Alcohol

70% Alcohol	1000.0	ml
Hydrochloric acid, concentrated	5.0	ml

10% Triton X-100 (Stock)

Triton X-100	10.0	ml
Distilled water	90.0	ml

Phosphate Buffer, pH 6.4

Potassium phosphate, monobasic, KH_2PO_4	7.25	gm
Sodium phosphate, dibasic, Na_2HPO_4	2.75	gm
Distilled water	1000.0	ml
10% Triton X-100 (stock)	10.0	ml

Store in refrigerator at 3-5°C.

Giemsa Solution (Stock)

Azure A, C.I. 52005	0.15	gm
Methylene blue, C.I. 52015	0.30	gm
Eosin Y, C.I. 45380	0.36	gm
Phloxine B, C.I. 45410	0.04	gm
Glycerin	60.0	ml
Methyl alcohol	40.0	ml

Pour the glycerin into a beaker and add the dyes. Place in a 55-58°C oven and mix occasionally with a glass stirring rod. Allow to remain in the oven overnight. Remove from the oven, add the methyl alcohol and thoroughly mix on a magnetic stirrer using gentle heat. Allow to cool and store in a tightly stoppered brown bottle. This Giemsa solution is stable for at least 1 year.

Giemsa Solution (Working)

Giemsa solution (stock)	4.0	ml
Phosphate buffer, pH 6.4	40.0	ml

1% Acetic Acid

Acetic Acid	1.0	ml
Distilled water	99.0	ml

STAINING PROCEDURE

1. Place in 0.5% hydrochloric acid-alcohol for 10 minutes.
2. Wash in running tap water for 5 minutes.
3. Rinse in 2 changes of distilled water.
4. Place slides in the working Giemsa solution in a glass coplin jar and microwave at power level 1 (the lowest power setting) for 3 1/2 minutes. Allow the slides to remain in the warm solution for 45 minutes. (The stain may be done at room temperature, by staining in the working Giemsa solution for 2 hours.)
5. Rinse in distilled water, 3 dips.
6. Place in 1% acetic acid for 5 seconds.
7. Rinse in distilled water, 3 dips.
8. Dehydrate slides in 95% alcohol and absolute alcohol 3 changes each. Clear in xylene, 3 changes.
9. Mount coverglass with appropriate medium (refractive index 1.48-1.56).

RESULTS

All the colors of the hemopoietic bone marrow are similar to the colors of bone marrow air-dried aspirates stained with Giemsa.

Eosinophil granules	red refractile granules
Mature erythrocytes	red

Immature erythrocytes ....purple (azuraphilic) granules
Mast cells ....deep purple
Megakaryocytes ....pale purple
Immature megakaryocytes ....eosinophilic cytoplasm
Mature neutrophils ....tan to pink
Nuclear chromatin ....blue
Plasma cells ....purple with the characteristic perinuclear clear region

REMARKS

The use of the microwave oven reduces the time required to perform the method. Also, the staining results appear to be better than when the stain is done at room temperature. The power output at power level 1 is 50 watts in the oven we use. Other models of microwave ovens can be used in this method, but it may be necessary to change the time required to heat the Giemsa solution. The final temperature of the solution should be approximately 60°C.

We have found considerable variation in the quality of commercial Giemsa stains in both solution and powder form. The Giemsa formula used in this method compares favorably with the best commercial Giemsa stains and can be used in any Giemsa staining method.

The addition of the detergent Triton X-100 to the staining solution results in more distinct staining of the various cell types found in hemopoietic bone marrow. Because morphological details are more distinct, microscopic detection and identification of the cells is facilitated. Another advantage gained by the addition of Triton X-100 is that it stabilizes the working Giemsa solution. If the Triton X-100 is omitted, a fine precipitate forms in the solution. This precipitate also forms when Triton X-100 is omitted from solutions prepared with commercial Giemsa stains.

This Giemsa method differs from Wolbach's method in that the working Giemsa solution contains more stock Giemsa and the staining time is much shorter.

Wolbach's Giemsa Method (Microwave)

FIXATION 10% Buffered neutral formalin.
PROCESS Paraffin.
MICROTOMY Cut sections at 5 micrometers.
CONTROL Use bone marrow or entity one desires to stain (see controls, chapter 17).
SOLUTIONS

0.5% Hydrochloric Acid-Alcohol

70% Alcohol	995.0	ml
Hydrochloric acid, concentrated	5.0	ml

Giemsa Solution (Stock)

Azure A, C.I. 52005	0.15	gm
Azure B, C. I. 52010	0.15	gm
Methylene blue, C.I. 52015	0.15	gm
Eosin Y, C.I. 45380	0.36	gm
Phloxine B, C.I. 45410	0.04	gm
Glycerol	60.0	ml
Methyl alcohol	40.0	ml

Pour the glycerol into a beaker and add the dyes. Place in a 55-60°C oven and mix occasionally with a glass stirring rod. Allow to remain in the oven overnight. Remove from the oven, add the methyl alcohol and thoroughly mix on a magnetic stirrer using gentle heat. Allow to cool and store in a tightly stoppered amber bottle. The solution is stable for at least 1 year.

Giemsa Solution (Working)

Giemsa solution (stock)	1.0	ml
Methyl alcohol	1.0	ml
Distilled water	40.0	ml

Rosin-Alcohol Solution (Stock)

Rosin	10.0	gm
Absolute alcohol	100.0	ml

Rosin-Alcohol Solution (Working)

Rosin solution (stock)	5.0	ml
95% Alcohol	45.0	ml

STAINING PROCEDURE

1. Decerate slides in xylene, 2 changes, 2 minutes each. Place in absolute alcohol, 2 changes, 2 minutes each; 95% alcohol, 2 changes, 2 minutes each; then rinse in distilled water.
2. Place in 0.5% hydrochloric acid-alcohol for 10 minutes.
3. Wash well in running tap water and rinse in 2 changes of distilled water.
4. Place in the working Giemsa solution in a glass coplin jar. Place in a microwave oven and microwave at power level 1 (the lowest power setting) for 3 1/2 minutes. Allow the slides to remain in this warm solution for 1 hour. 5. Differentiate in working rosin-alcohol for 5 to 10 seconds.
6. Rinse in 2 changes of 95% alcohol and check with a microscope. If too dark, repeat step 5 until nuclei are well differentiated against a pink to rose background.
7. Rinse in 2 changes of 95% alcohol and 3 changes of absolute alcohol.
8. Dehydrate slides in 95% alcohol and absolute alcohol 3 changes each. Clear in xylene, 3 changes.
9. Mount coverglass with appropriate medium (refractive index 1.48-1.56).

RESULTS

Nuclei	blue
Collagen and other tissue elements	pink to rose
Bacteria	blue
Mast cells	purple
Eosinophils	pink
Rickettsia	purple

REMARKS
We have found some variation in the quality of commercial Giemsa stains in both solution and powder form. The Giemsa formula used in this method compares favorably with the best commercial Giemsa stains and can be used in any Giemsa staining method.

The use of the microwave oven has greatly reduced the time required to perform this method. Also, the staining results appear to be better than the original overnight staining method. In the oven we use, the power output at level 1 is 50 watts. Microwave ovens with different wattages can be used in this method, but it may be necessary to change the time required to heat the Giemsa solution. The final temperature of the solution should be approximately 60°C.

Churukian-Schenk's Method for Pneumocystis Carinii and Fungi (Microwave)

FIXATION	Unfixed specimens on slides are briefly heat fixed.
PROCESS	Cytologic preparations of lung aspirates, bronchial washings, sputum specimens.
CONTROL	Use control number 18 or 37A.

SOLUTIONS

5% Periodic Acid

Periodic acid	5.0	gm
Distilled water	100.0	ml

Ammoniacal Silver

To 10 ml of 10% silver nitrate, add 5.0 ml of 4% lithium hydroxide, monohydrate. Then add concentrated ammonium hydroxide, drop by drop with constant shaking, until the precipitate dissolves. Make up the solution to 1000.0 ml with distilled water and store in the refrigerator at 3-6°C. This solution is stable for at least 3 months.

0.2% Gold Chloride

Gold chloride	0.2	gm
Distilled water	100.0	ml

Sodium Thiosulfate

Sodium thiosulfate	2.0	gm
Distilled water	100.0	ml

0.2% Fast Green (Stock)

Fast green FCF	0.2	gm
Distilled water	100.0	ml
Glacial acetic acid	0.2	ml

Fast Green (Working)

0.2% Fast green (stock)	10.0	ml
Distilled water	40.0	ml

STAINING PROCEDURE

1. Oxidize slides in 5% periodic acid for 5 minutes.
2. Rinse in 5 changes of distilled water.
3. Microwave 45 ml of refrigerated ammoniacal silver in a plastic coplin jar, with cap loosely applied, at power level 7 for 1 minute. (If 5 or more slides are to be done, use power level 8.) Mix the solution with a glass rod or one of the slides. Immediately place the slides in this hot solution and return to the microwave oven at power level 1 for 1 minute. Allow the slides to remain in the solution for 1 more minute or until the material appears medium brown.
4. Rinse in 4 changes of distilled water.
5. Tone in 0.2% gold chloride for 30 seconds.
6. Rinse in 2 changes of distilled water.
7. Place in sodium thiosulfate for 30 seconds.
8. Rinse in 3 changes of distilled water.
9. Counterstain with working fast green solution for 30 seconds.
10. Rinse in 3 changes of distilled water.
11. Dehydrate slides in 95% alcohol and absolute alcohol 3 changes each. Clear in xylene, 3 changes.
12. Mount coverglass with appropriate medium (refractive index 1.48-1.56).

RESULTS

Pneumocystis carinii ....black
Fungi ....black
Glycogen, mucin, lipofuscin, melanin ....black
Background ....green

REMARKS

Since unfixed tissue is a biohazard, the following precautions should be taken:

1. Wear rubber gloves.
2. Use metal forceps throughout the procedure.
3. Containers used to transport the specimens can be placed in zip-lock bags and discarded in a biohazardous waste container.
4. After use, the rubber gloves are also placed in zip-lock bags.
5. Wash hands thoroughly with soap and water.

Hind's Method for Pneumocystis Carinii and Fungi (Microwave)

This modification of Grocott's methenamine-silver is relatively stable at the high temperature of the microwave oven. Silver impregnation is initiated in the microwave oven to accelerate heating. The solution is then transferred to a conventional oven where the process of metallic silver deposition in the tissue can be easily controlled. The method drastically reduces the time required for the staining of pneumocystis carinii and fungi and assures consistency of results. It is rapid, easy to perform, and does not present any of the problems encountered with the use of other microwave oven techniques. The staining results are as good as those obtained with Grocott's original method or other modifications.

FIXATION	10% Buffered neutral formalin or DuBoscq Brasil (alcoholic Bouin's) solution.
PROCESS	Paraffin.
MICROTOMY	Cut sections at 4 micrometers.
CONTROL	Use control number 18 or 37A.

SOLUTIONS

5% Chromic Acid

Chromic acid	5.0	gm
Distilled water	100.0	ml

3% Methenamine (Stock)

Methenamine (hexamethylenetetramine)	3.0	gm
Distilled water	100.0	ml

5% Silver Nitrate (Stock)

Silver nitrate	5.0	gm
Distilled water	100.0	ml

5% Borax (Stock)

Borax (sodium borate)	5.0	gm
Distilled water	100.0	ml

Methenamine-Silver Nitrate (Stock)

5% Silver nitrate (stock)	5.0	ml
3% Methenamine (stock)	100.0	ml

Shake well to dissolve any precipitate formed.

Methenamine-Silver Nitrate (Working)

Methenamine-silver nitrate (stock)	20.0	ml
Distilled water	30.0	ml
5% Borax (stock)	2.0	ml

2% Sodium Thiosulfate

Sodium thiosulfate	2.0	gm
Distilled water	100.0	ml

0.1% Gold Chloride

Gold chloride	0.1	gm
Distilled water	100.0	ml

0.02% Light Green

Light green, SF yellowish	0.02	gm
Distilled water	100.0	ml

1% Sodium Metabisulfite

Sodium metabisulfite	1.0	gm
Distilled water	100.0	ml

STAINING PROCEDURE

1. Decerate slides in xylene, 2 changes, 2 minutes each. Place in absolute alcohol, 2 changes, 2 minutes each; 95% alcohol, 2 changes, 2 minutes each; then rinse in distilled water.

2. Place in 5% chromic acid for 5 minutes at room temperature.
3. Rinse in 4 changes of distilled water and leave in the final change for 2 minutes.
4. Place in 1% sodium metabisulfite for 30 seconds and wash in running tap water for 1 minute.
5. Rinse in 4 changes of distilled water.
6. Place slides in a new (unused) vertical plastic slide mailer or similar container. Add working methenamine-silver solution to cover 2/3 of the slides. Be sure tissue sections are immersed. Press the lid down securely on the slide-mailer and place in a 200 ml pyrex beaker containing distilled water at the same level as the methenamine-silver solution in the mailer.
7. Place the beaker in the microwave oven and heat for 2 minutes at power level 2.
8. Remove from microwave and place in a conventional oven for 1 minute at 60°C.
9. Remove slides from the oven and rinse in distilled water. Examine slides microscopically to be sure the organisms are stained to the desired intensity. If they are not sufficiently stained, rinse the sections well in distilled water and place them in fresh methenamine-silver working solution in another new plastic mailer. Put the mailer into the original beaker and place in the conventional oven for 30 seconds.
10. Rinse in distilled water.
11. Tone in 0.1% gold chloride for 30 seconds or until sections turn gray.
12. Rinse in distilled water and examine microscopically. If necessary, bleach by dipping in 5% sodium thiosulfate for 3 seconds or less.
13. Wash in running tap water for 1 to 2 minutes.
14. Rinse in distilled water and counterstain in light green for 30 seconds to 1 minute.
15. Dehydrate slides in 95% alcohol and absolute alcohol 3 changes each. Clear in xylene, 3 changes.
16. Mount coverglass with appropriate medium (refractive index 1.48-1.56).

RESULTS

Pneumocystis carinii ....sharp black
Fungi ....black
Background ....light green

Pascual's Method for Argyrophil Granules

FIXATION 10% Buffered neutral formalin.
PROCESS Paraffin.
MICROTOMY Cut sections at 5 micrometers.
CONTROL Use control number 1A, 3 or 6A.
SOLUTIONS

Citric Acid - Glycine (Stock)

Citric acid	0.06	gm
Glycine	1.2	gm
Distilled water	100.0	ml

The pH of this solution is approximately 3.8. Store in a refrigerator at 3-6°C.

Citric Acid - Glycine (Working)

Citric acid - glycine (stock)	5.0	ml
Distilled water	500.0	ml

The pH of this solution is approximately 4.2.

0.5% Silver Nitrate

Silver nitrate	0.5	gm
Citric acid - glycine (working)	100.0	ml

Bodian's Developer

Sodium sulfite (anhydrous)	5.0	gm
Hydroquinone	1.0	gm
Distilled water	100.0	ml

Prepare fresh.

Kernechtrot (Nuclear Fast Red)
(See Page 84)

STAINING PROCEDURE

1. Decerate slides in xylene, 2 changes, 2 minutes each. Place in absolute alcohol, 2 changes, 2 minutes each; 95% alcohol, 2 changes, 2 minutes each; then rinse in distilled water.
2. Place in citric acid - glycine working solution for 30 seconds.
3. Place in 40 ml of 0.5% silver nitrate solution in a plastic coplin jar and microwave at power level 8 (400W) for 1 minute. Dip the slides up and down several times and allow them to remain in the hot solution for 2 minutes.
4. Rinse in 4 changes of distilled water.
5. Place 40 ml of Bodian's developer in a plastic coplin jar and microwave at power level 8 for 40 seconds. Immediately place the slides in this hot solution and allow them to remain in the solution for 1 minute.
6. Rinse in 4 changes of distilled water.
7. Place the slides in the same 0.5% silver nitrate solution for 1 minute.
8. Rinse in 4 changes of distilled water.
9. Place the slides in the same Bodian's developer for 1 minute.
10. Rinse in 4 changes of distilled water.
11. Place the slides in the same 0.5% silver nitrate solution and microwave at power level 8 for 15 seconds. Dip the slides up and down several times and allow them to remain in the hot solution for 1 minute.
12. Rinse in 4 changes of distilled.
13. Microwave the same Bodian's developer at power level 8 for 15 seconds. Immediately place the slides in the solution and allow them to remain in the solution for 1 minute.
14. Rinse in 4 changes of distilled water.
15. Counterstain with 0.1% nuclear fast red for 1 minute.
16. Rinse in 3 changes of distilled water.
17. Dehydrate slides in 95% alcohol and absolute alcohol 3 changes each. Clear in xylene, 3 changes.
18. Mount coverglass with appropriate medium (refractive index 1.48-1.56).

RESULTS

Neuroendocrine cell granules (argentaffin and argyrophil) in various organs stain dark brown to black. Nuclei are orange to red and the background a light yellow-orange.

REMARKS

Argyrophil cells are found in the gastrointestinal mucosa, pancreas, trachea, bronchi, prostate and ovary. Tumors derived from these cells are called apudomas or neuroendocrine tumors. Argyrophil cells, as their name indicates, have an affinity for silver ions but lack the ability to convert the silver ions to the metallic state. This is why a reducing agent is needed in the staining procedure.

In this staining method, a triple treatment with silver nitrate and the reducing solution is necessary in order to obtain optimal staining results.

When solutions are heated with microwave irradiation, there can be up to a 15°C difference in temperature between the top and bottom portion of the solutions. Therefore, in order to equalize the temperature of the solutions, the slides are dipped up and down in steps 2 and 10 of the staining procedure. The temperature of the 0.5% silver nitrate will be about 92°C and that of the developer about 72°C after they have been microwaved.

The shelf life of silver nitrate and hydroquinone can be greatly increased by storing the reagents in a refrigerator at 3-6°C.

Noble's Method for Pneumocystis Carinii in Cytoprep Smears

Grocott's methenamine silver method is modified for use in demonstrating Pneumocystis carinii in cytoprep smears. The microwave oven is used to accelerate silver impregnation, thus providing a faster diagnostic turn-around time. This particular modification allows a better recovery rate of material on the slide than any method tried previously.

SOLUTIONS

10% Chromic Acid

Chromic acid (chromium trioxide)	50.0	gm
Distilled water, q.s. to	500.0	ml

1% Sodium Metabisulfite

Sodium metabisulfite	5.0	gm
Distilled water	500.0	ml

5% Silver Nitrate (Stock)

Silver nitrate	25.0	gm
Distilled water, q.s. to	500.0	ml

3% Methenamine (Stock)

Methenamine (hexamethylenetetramine)	15.0	gm
Distilled water	500.0	ml

5% Borax (Stock)

Borax (sodium borate)	25.0	gm
Distilled water	500.0	ml

1% Gold Chloride		
Gold chloride	1.0	gm
Distilled water	100.0	ml

5% Sodium Thiosulfate		
Sodium thiosulfate	25.0	gm
Distilled water	500.0	ml

0.5% Light Green		
Light green	2.5	gm
Distilled water	500.0	ml

Methenamine-Silver Nitrate (Stock)		
5% Silver nitrate (stock)	25.0	ml
3% Methenamine (stock)	500.0	ml

When mixing, shake to dissolve white precipitate. This solution will remain stable for several months if kept refrigerated in a brown bottle.

Methenamine-Silver Nitrate (Working)		
Methenamine-silver nitrate (stock)	25.0	ml
Distilled water	25.0	ml
5% Borax (stock)	2.0	ml

STAINING PROCEDURE

1. Decerate slides in xylene, 2 changes, 2 minutes each. Place in absolute alcohol, 2 changes, 2 minutes each; 95% alcohol, 2 changes, 2 minutes each; then rinse in distilled water.
2. Place 500 ml of distilled water in a beaker and heat in the microwave oven until the water is 60°C. Place the beaker in a 60°C oven for rinse purposes.
3. Place slides in 10% chromic acid for 5 minutes at room temperature.
4. Rinse slides in 3 changes of distilled water or until chromic acid is removed.
5. Place slides in 1% sodium metabisulfite for 1 minute.
6. Rinse slides in 3 changes of distilled water.
7. Prepare working methenamine-silver nitrate and pour into an acid-cleaned plastic coplin jar. Pour 300 ml of distilled water into a 600 ml beaker. Place the uncovered coplin jar in the beaker.
8. Place beaker with the coplin jar containing the working methenamine silver into the microwave oven and heat for 2 1/2 to 3 minutes until the water in the beaker reaches 70°C. (The actual time will have to be calibrated in each laboratory, because not all microwave ovens have the same wattage.) Check the temperature of the water in the beaker with a centigrade thermometer. Do not check the temperature of the contents in the coplin jar with the thermometer!

   Stir the working methenamine-silver nitrate solution with an acid cleaned glass rod and put slides into the solution. Place the beaker and coplin jar back in the microwave oven and heat for 1 minute. Check sections to see if they are a medium golden brown color. If not, leave slides in the methenamine-silver nitrate until the desired color is obtained.
9. Take slides out of the methenamine-silver nitrate and place in a coplin jar containing hot distilled water. Check slides microscopically. The Pneumocystis carinii cyst forms should be dark brown.
10. Rinse slides in 3 changes of distilled water.

11. Place slides in 1% gold chloride for 10 seconds.
12. Rinse slides in 3 changes of distilled water.
13. Place slides in 5% sodium thiosulfate for 3 minutes.
14. Rinse slides in 3 changes of distilled water.
15. Counterstain in 0.5% light green for 30 seconds.
16. Quickly agitate slides in 95% ethyl alcohol for 20 seconds.
17. Dehydrate with absolute ethyl alcohol, clear in xylene.
18. Mount coverglass with appropriate medium (refractive index 1.48-1.56).

RESULTS
Pneumocystis carinii cyst forms are sharply delineated in black. The background is light green.

Gomori's Method for Fungi (Microwave)

FIXATION	10% Buffered neutral formalin.
PROCESS	Paraffin, frozen, smears and touch preparations may be used.
MICROTOMY	Cut sections at 5 micrometers.
CONTROL	Use control number 18.
SOLUTIONS

5% Chromic Acid

Chromium trioxide	5.0	gm
Distilled water	100.0	ml

5% Sodium Borate (Stock)

Sodium borate (Borax)	5.0	gm
Distilled water	100.0	ml

Methenamine-Silver Nitrate (Stock)

5% Silver nitrate	5.0	ml
3% Methenamine	100.0	ml

This solution is good for about 1 month when stored in a refrigerator.

Methenamine-Silver Nitrate (Working)

Methenamine-silver nitrate (stock)	20.0	ml
Distilled water	20.0	ml
5% Sodium borate (stock)	1.6	ml

Prepare fresh.

1% Sodium Bisulfite

Sodium bisulfite	1.0	gm
Distilled water	100.0	ml

0.2% Gold Chloride

1% Gold chloride	20.0	ml
Distilled water	80.0	ml

2% Sodium Thiosulfate		
Sodium thiosulfate	2.0	gm
Distilled water	100.0	ml

0.2% Fast Green (Stock)		
0.2% Fast green FCF	0.2	gm
Distilled water	100.0	ml
Glacial acetic acid	0.2	ml

Fast Green (Working)		
0.02% Fast green (stock)	10.0	ml
Distilled water	40.0	ml

STAINING PROCEDURE

1. Decerate slides in xylene, 2 changes, 2 minutes each. Place in absolute alcohol, 2 changes, 2 minutes each; 95% alcohol, 2 changes, 2 minutes each; then rinse in distilled water.
2. Place in 40 ml of 5% chromic acid solution in a glass coplin jar and microwave at power level 2 (100W) for 2 minutes. Dip the slides up and down several times and allow them to remain in the warm solution for 1 1/2 minutes.
3. Wash in running tap water for 30 seconds.
4. Place in 1% sodium bisulfite solution for 30 seconds to remove any residual chromic acid.
5. Wash in running tap water for 30 seconds.
6. Rinse in 3 changes of distilled water.
7. Place freshly prepared working methenamine-silver nitrate solution in a plastic coplin jar and microwave at power level 8 (400W) for 45 seconds. Immediately place the slides in this hot solution, return to the microwave oven and microwave at power level 2 (100W) for 30 seconds. Dip the slides up and down several times and allow them to remain in the hot solution for about 30 seconds or until the sections turn a medium brown color.
8. Rinse in 4 changes of distilled water.
9. Tone in 0.2% gold chloride solution for 30 seconds.
10. Rinse in 2 changes of distilled water.
11. Remove unreduced silver with 2% sodium thiosulfate solution for 30 seconds.
12. Wash in running tap water for 15 seconds and rinse in 2 changes of distilled water.
13. Counterstain with working fast green solution for 30 seconds.
14. Wash in running tap water for a few seconds and rinse in 2 changes of distilled water.
15. Dehydrate slides in 95% alcohol and absolute alcohol 3 changes each. Clear in xylene, 3 changes.
16. Mount coverglass with appropriate medium (refractive index 1.48-1.56).

RESULTS

Fungi and pneumocystis carinii	sharply delineated in black
Mucin and glycogen	gray to black
Melanin	black
Background	green

Note: When heating the 5% chromic acid and working methenamine-silver nitrate solutions with microwave irradiation, the top portion of the solution will be 10-15°C warmer than that near the bottom of the coplin jar. Therefore, in order to equalize the temperature of the solutions, the slides are dipped up and down. This assures uniformity of staining throughout the tissue sections. There are a number of other things which the rapid GMS will stain, including histiocytes, melanin, actinomyces, certain bacteria, starch, glycogen and mucin. Nuclei, collagen, and erythrocytes also will stain if the time in the working methenamine-silver nitrate solution is prolonged. It has been found that silver nitrate has a shelf life of several years when stored in a refrigerator.

Perl's Method for Ferric Iron (Microwave)

FIXATION	10% Buffered neutral formalin.
PROCESS	Paraffin.
MICROTOMY	Cut sections at 5 micrometers.
CONTROL	Use control number 25.
SOLUTIONS	

4% Hydrochloric Acid (Stock)

Hydrochloric acid, concentrated	4.0	ml
Distilled water	96.0	ml

2% Potassium Ferrocyanide (Stock)

Potassium ferrocyanide	2.0	gm
Distilled water	100.0	ml

10% Triton X-100 (Stock)

Triton X-100	10.0	ml
Distilled water	90.0	ml

Hydrochloric Acid-Potassium Ferrocyanide

4% Hydrochloric acid (stock)	20.0	ml
2% Potassium ferrocyanide (stock)	20.0	ml
10% Triton X-100 (stock)	0.4	ml

Prepare just before use.

Kernechtrot (Nuclear Fast Red)
(See Page 84)

STAINING PROCEDURES

1. Decerate slides in xylene, 2 changes, 2 minutes each. Place in absolute alcohol, 2 changes, 2 minutes each; 95% alcohol, 2 changes, 2 minutes each; then rinse in distilled water.
2. Stain slides in the hydrochloric acid-potassium ferrocyanide solution in a glass coplin jar and microwave at power level 2 (100W) for 1 3/4 minutes. Dip the slides up and down several times and allow them to remain in the warm solution for 2 minutes.
3. Rinse with 5 changes of distilled water.
4. Counterstain with nuclear fast red solution for 5 minutes.
5. Rinse with 3 changes of distilled water.

6. Dehydrate slides in 95% alcohol and absolute alcohol 3 changes each. Clear in xylene, 3 changes.
7. Mount coverglass with appropriate medium (refractive index 1.48-1.56).

RESULTS

Ferric iron	bright blue
Nuclei	red
Cytoplasm	pink

Note: Hemosiderin is a breakdown product of hemoglobin and is thought to be composed of ferric iron and protein. It may be present in tissues in certain pathologic conditions such as hemochromatosis. This yellow-brown pigment is insoluble in alkalis and water, but is soluble in acids even after fixation. Treating a tissue section with 10% sulfuric acid overnight will usually remove this pigment. The principle of Perl's Prussian blue reaction is that potassium ferrocyanide will form ferric ferrocyanide (Prussian blue) with reactive ferric salts in an acid solution. Dilute hydrochloric acid liberates loosely bound ferric iron from protein. According to Lillie, the optimal pH for the ferrocyanide reaction is around 1.5. Because 4% hydrochloric gives a final pH of about 1.5, we decided to use it rather than the 2% solution which most texts indicate. The pH of the working ferrocyanide solution made up with 2% hydrochloric acid is 1.75. In developing this staining method we found that good staining results are obtained when the solution is heated as described above. The final temperature of the solution will be about 60°C. However, the temperature of the solution is not uniform throughout the coplin jar. The top portion of the solution is warmer than that near the bottom of the coplin jar. Therefore, in order to equalize the temperature of the solution, the slides are dipped up and down. This assures uniformity of staining throughout the tissue sections. Triton X-100 is added to the hydrochloric acid-potassium ferrocyanide solution because it helps to keep the solution stable when it is heated with microwave irradiation. If the Triton X-100 is omitted, there is a tendency for the solution to break down, resulting in a fine blue precipitate forming in the coplin jar and sometimes on the tissue sections. The stain may be performed at room temperature by staining the sections for 30 minutes in the hydrochloric acid-potassium ferrocyanide solution.

Ziehl-Neelsen Method for Acid Fast Bacteria (Microwave)

FIXATION 10% Buffered neutral formalin.
PROCESS Paraffin or frozen.
MICROTOMY Cut sections at 5 micrometers.
CONTROL Use control number 41.
SOLUTIONS

Carbol Fuchsin

Pararosaniline or basic fuchsin	0.8	gm
Absolute alcohol	10.0	ml
90% Liquid phenol	6.0	ml
Distilled water	84.0	ml

Dissolve the pararosaniline or basic fuchsin in the absolute alcohol, then add the phenol and distilled water.

0.5% Acid Alcohol

70% Alcohol	1000.0	ml
Hydrochloric acid, concentrated	5.0	ml

Methylene Blue

Methylene blue	0.25	gm
Distilled water	100.0	ml
Glacial acetic acid	1.0	ml

STAINING PROCEDURE

1. Decerate slides in xylene, 2 changes, 2 minutes each. Place in absolute alcohol, 2 changes, 2 minutes each; 95% alcohol, 2 changes, 2 minutes each; then rinse in distilled water.
2. Place slides in carbol fuchsin in a glass coplin jar and microwave at power level 1 (approximately 50W) for 2 minutes. Dip the slides up and down several times and allow them to remain in the warm solution for 7 minutes.
3. Wash in running tap water to remove excess stain.
4. Decolorize with 0.5% acid alcohol until sections are pale pink.
5. Wash in running tap water for 1 minute and rinse in 2 changes of distilled water.
6. Counterstain with methylene blue solution for 15 seconds.
7. Rinse in 3 changes of distilled water.
8. Dehydrate slides in 95% alcohol and absolute alcohol 3 changes each. Clear in xylene, 3 changes.
9. Mount coverglass with appropriate medium (refractive index 1.48-1.56).

RESULTS

Acid-fast bacteria, including Mycobacterium avium intracellular	red
Erythrocytes	pink
Mast cells	blue
Other tissue elements	pale blue

REMARKS

Ziehl-Neelsen or similar carbol fuchsin solutions are usually prepared with basic fuchsin. The dye sold as "basic fuchsin" is usually a mixture of pararosaniline (C.I. 42500), rosanilin (C.I. 42510), magenta II and new fuchsin (C.I. 42520). Carbol fuchsin prepared with any of these dyes (or mixtures thereof) stains acid fast bacteria well with the above procedure. In developing this staining method we exposed the carbol fuchsin to varying amounts of microwave radiation. We observed that optimal staining results are obtained when the solution is heated as described above. The final temperature of the solution will be about 46°C. However, the temperature of the solution is not uniform throughout the coplin jar. The top portion of the solution is warmer than that near the bottom of the coplin jar. Therefore, in order to equalize the temperature of the solution, the slides are dipped up and down. This assures uniformity of staining throughout the tissue sections. The stain may be performed at room temperature by staining the sections for 45 minutes in the carbol fuchsin solution.

Razzaboni's Method for Demonstrating Bacillus Piliformis (Microwave)

The Warthin-Starry silver technique is modified for use in demonstrating Bacillus piliformis in paraffin sections. Microwave heating reduces staining time, and more importantly, dramatically improves the reliability of the technique when used in conjunction with Walpole's buffer.

Bacillus piliformis, the pathogenic bacterium that causes Tyzzer's disease, is a major concern for all rodent breeding laboratories as well as research facilities that use rodents. An outbreak of this disease often means that an entire colony must be broken down, equipment disinfected and the animals destroyed. The devastating effect of Tyzzer's disease, both financially and on long-term research, is immeasurable. Thus, the Warthin-Starry silver technique is a valuable tool in detection, prevention and eradication of this pathogen.

FIXATION	10% Buffered neutral formalin.
PROCESS	Paraffin.
MICROTOMY	Cut sections at 6 micrometers.
CONTROL	Use control number 40.
SOLUTIONS	

5M Glacial Acetic Acid (Stock)

Glacial acetic acid	11.8	ml
Deionized water	1000.0	ml

5M Sodium Acetate (Stock)

Sodium acetate	16.4	gm
Deionized water	1000.0	ml

Walpole's Buffer

5M Glacial acetic acid (stock)	17.6	ml
5M Sodium acetate (stock)	2.4	ml
Deionized water	480.0	ml

2% Silver Nitrate (Stock)

Silver nitrate	2.0	gm
Walpole's buffer	100.0	ml

1% Silver Nitrate

Silver nitrate	1.0	gm
Walpole's buffer	100.0	ml

5% Gelatin (Stock)

Gelatin	5.0	gm
Walpole's buffer	100.0	ml

3% Hydroquinone (Stock)

Hydroquinone	3.0	gm
Walpole's buffer	100.0	ml

Developer

2% Silver nitrate (stock)	1.5	ml
5% Gelatin (stock)	7.5	ml
3% Hydroquinone (stock)	0.5	ml

STAINING PROCEDURE

1. Decerate slides in xylene, 2 changes, 2 minutes each. Place in absolute alcohol, 2 changes, 2 minutes each; 95% alcohol, 2 changes, 2 minutes each; then rinse in distilled water.
2. Rinse in Walpole's buffer for 10 seconds.
3. Place slides in a plastic coplin jar, cover with 1% silver nitrate, and microwave for 1.5 minutes at a medium power setting. Remove slides from oven and let slides stand for 5 minutes in this solution.
4. Rinse in Walpole's buffer.
5. Flood slides with developer, which has been freshly prepared during the preceding 5 minutes standing time.
6. Wash in warm tap water.
7. Dehydrate slides in 95% alcohol and absolute alcohol 3 changes each. Clear in xylene, 3 changes.
8. Mount coverglass with appropriate medium (refractive index 1.48-1.56).

RESULTS

Bacillus piliformis .... black
Background .... light brown to golden brown

REMARKS

In our laboratory we make up a large volume of the gelatin solution in advance and then aliquot 7.5 ml into 15 ml screw top vials, which are then stored until needed in the refrigerator. We can then take one vial of gelatin and heat it in warm (56°C) tap water, add the solutions and then discard the vial after the procedure has been finished. Using a disposable vial greatly reduces the chance of contamination. Plastic coplin jars used in this procedure should be acid cleaned. As with most silver stains, the slides should never be touched with metal instruments.

Ammoniacal Silver Method for Demonstrating Fungi and Pneumocystis Carinii (Microwave)

FIXATION	10% Buffered neutral formalin.
PROCESS	Paraffin, frozen, smears and touch preparations may be done.
MICROTOMY	Cut sections at 5 micrometers.
CONTROL	Use control number 18 or 37A.
SOLUTIONS	

5% Chromic Acid

Chromium trioxide	5.0	gm
Distilled water	100.0	ml

1% Sodium Bisulfite

Sodium bisulfite	1.0	gm
Distilled water	100.0	ml

Ammoniacal Silver Solution

To 10 ml of 10% silver nitrate, add 5.0 ml of 4% lithium hydroxide monohydrate. Then add concentrated ammonium hydroxide drop by drop, with constant shaking until the precipitate dissolves. Make up the solution to 1000 ml with distilled water and store in a refrigerator at 3°C. This solution is stable for about 1 month.

0.2% Gold Chloride

Gold chloride	0.2	gm
Distilled water	100.0	ml

2% Sodium Thiosulfate

Sodium thiosulfate	2.0	gm
Distilled water	100.0	ml

Fast Green (Stock)

Fast green FCF	0.2	gm
Distilled water	100.0	ml
Acetic acid	0.2	ml

Fast Green (Working)

Fast green (stock)	10.0	ml
Distilled water	40.0	ml

STAINING PROCEDURE

1. Decerate slides in xylene, 2 changes, 2 minutes each. Place in absolute alcohol, 2 changes, 2 minutes each; 95% alcohol, 2 changes, 2 minutes each; then rinse in distilled water.
2. Place slides in 40 ml of 5% chromic acid in a glass coplin jar and microwave at power level 2 (100W) for 2 1/2 minutes. Dip the slides up and down several times and allow them to remain in the warm solution for 1 1/2 minutes.
3. Rinse with 3 changes of distilled water.
4. Treat with 1% sodium bisulfite for 30 seconds.
5. Wash in running tap water for 30 seconds.
6. Rinse in 3 changes of distilled water.
7. Microwave 40 ml of cold ammoniacal silver in a plastic coplin jar at power level 7 (350W) for 1 minute. Immediately place the slides in this hot solution, return to the microwave oven and microwave at power level 2 for 30 seconds. Dip the slides up and down several times and allow them to remain in the hot solution for 30-60 seconds, or until the sections turn a medium brown.
8. Rinse in 4 changes of distilled water.
9. Tone in 0.2% gold chloride for 30 seconds.
10. Rinse in 2 changes of distilled water.
11. Place in 2% sodium thiosulfate for 30 seconds.
12. Rinse in 4 changes of distilled water.
13. Counterstain with working fast green solution for 30 seconds.
14. Rinse in 3 changes of distilled water.

15. Dehydrate slides in 95% alcohol and absolute alcohol 3 changes each. Clear in xylene, 3 changes.
16. Mount coverglass with appropriate medium (refractive index 1.48-1.56).

RESULTS

Fungi and pneumocystis carinii	sharply delineated in black
Mucin and glycogen	black
Melanin	black
Background	green

REMARKS

We prefer this microwave ammoniacal silver method over the microwave methenamine-silver nitrate method for the following reasons:

1. Dilute ammoniacal silver is much more stable at high temperatures (above 80°C) than methenamine-silver nitrate.
2. When methenamine-silver nitrate is heated to high temperatures, the solution rapidly breaks down and sometimes produces a deposit of silver on the slides and the coplin jar. This never occurs when dilute ammoniacal silver is heated to temperatures even above 90°C.
3. Dilute ammoniacal silver is ready for immediate use and does not need to be diluted and buffered like methenamine-silver nitrate.

The volume and temperature of the ammoniacal silver used in this method is important. We remove the ammoniacal silver from the refrigerator and pour 40 ml of the cold solution into a plastic coplin jar immediately before use. The temperature of the ammoniacal silver is 3-6°C before and about 85°C after, the second exposure to microwave irradiation.

When heating a solution with microwave irradiation, the top portion of the solution is warmer by 10-15°C than that near the bottom of the coplin jar. Therefore, in order to equalize the temperature of the solution, the slides are dipped up and down in steps 2 and 7 of the staining procedure. This assures uniformity of staining throughout the tissue sections.

We have found that silver nitrate has a shelf life of several years when stored in a refrigerator.

Luxol Fast Blue-Periodic Acid-Schiff-Cresyl Violet Method (Microwave)

FIXATION	10% Buffered neutral formalin.
PROCESS	Paraffin.
MICROTOMY	Cut sections at 8-12 micrometers.
CONTROL	Use control number 34.

SOLUTIONS

0.1% Luxol Fast Blue

Luxol fast blue, MBS	0.1	gm
95% Alcohol	100.0	ml
10% Glacial acetic acid	0.5	ml

0.05% Lithium Carbonate

Lithium carbonate	0.05	gm
Distilled water	100.0	ml

70% Alcohol

100% Alcohol	70.0	ml
Distilled water	30.0	ml

0.5% Periodic Acid

Periodic acid	0.5	gm
Distilled water	100.0	ml

Schiff's Solution

Dissolve 1.0 gm basic fuchsin in 200 ml hot distilled water. Bring to near boiling point. Cool to 50°C and add 20 ml normal hydrochloric acid. Cool further to 25°C and add 1.0 gm anhydrous sodium bisulfate or metabisulfite. Keep in the dark for 48 hours. Add 0.2 gm of activated charcoal and agitate for 3 minutes. Filter through Whatman #4, then Whatman #1, filter papers. This keeps for about 6 months if stored in refrigerator.

Cresyl Violet Acetate

Cresyl violet acetate	0.04	gm
Distilled water	40.0	ml
1% Oxalic acid	0.2	ml

Cresyl echt violet may be substituted for cresyl violet acetate.

STAINING PROCEDURE

1. Decerate slides in xylene, 2 changes, 2 minutes each. Place in absolute alcohol, 2 changes, 2 minutes each; 95% alcohol, 2 changes, 2 minutes each; then rinse in distilled water.
2. Place in 40 ml of 0.1% luxol fast blue solution in a glass coplin jar and microwave at power level 1 (the lowest power setting) for 4 minutes. Allow the slides to remain in the hot solution for 30 minutes.
3. Rinse off excess stain in 95% alcohol.
4. Rinse in 3 changes of distilled water.
5. Begin differentiation by placing in 0.05% lithium carbonate for 5 seconds.
6. Continue differentiation in 70% alcohol for 5-10 seconds.
7. Rinse in 4 changes of distilled water. Check under microscope and repeat steps 5 and 6, if necessary.
8. Place in 0.5% periodic acid for 10 minutes.
9. Rinse in 4 changes of distilled water.
10. Place in Schiff's solution for 15 minutes.
11. Wash in running tap water for 10 minutes and rinse in 2 changes of distilled water.
12. Place in cresyl violet acetate solution for 5 minutes.
13. Rinse in 3 changes of distilled water

14. Dehydrate slides in 95% alcohol and absolute alcohol 3 changes each. Clear in xylene, 3 changes.
15. Mount coverglass with appropriate medium (refractive index 1.48-1.56).

RESULTS

Myelin	blue
PAS positive elements	rose to red
Capillaries	red
Nuclei	purple-blue

REMARKS

The use of the microwave oven greatly reduces the time required to perform this method. In the microwave oven which we use, the power output at power level 1 is 50 watts. Other microwave ovens can be used in this method, but it may be necessary to change the time required to heat the luxol fast blue solution. The temperature of the solution should be 75-80°C.

Alcian Blue-Periodic Acid Schiff Method for Kidney (Microwave)

FIXATION 10% Buffered neutral formalin.
PROCESS Paraffin or plastic sections.
MICROTOMY Cut sections at 2 to 3 micrometers.
CONTROL Use control number 28.
SOLUTIONS

3% Acetic Acid

Glacial acetic acid	3.0	ml
Distilled water	97.0	ml

1% Alcian Blue

Alcian blue 8GX	1.0	gm
3% Acetic acid	100.0	ml

Filter and add a few crystals of thymol.

0.5% Periodic Acid

Periodic acid	0.5	gm
Distilled water	100.0	ml

Lillie's Schiff Reagent

Basic fuchsin	1.0	gm
Distilled water	65.0	ml
1N Hydrochloric acid	35.0	ml
Sodium metabisulfate ($Na_2S_2O_5$)	1.9	gm

Place in a 250 ml flask and stir at intervals for 4 hours. Add 500 mg activated charcoal; shake for 2 minutes. Filter through Whatman #1 filter paper. The solution should be water clear. Store Schiff's reagent in refrigerator.

STAINING PROCEDURE

1. Decerate slides in xylene, 2 changes, 2 minutes each. Place in absolute alcohol,

2 changes, 2 minutes each; 95% alcohol, 2 changes, 2 minutes each; then rinse in distilled water.
2. Place in 3% acetic acid for 3 minutes.
3. Place in 40 ml of 1% alcian blue solution in a glass coplin jar and microwave at power level 2 (100W) for 2 1/2 minutes. Dip the slides up and down several times and allow them to remain in the hot solution for 2 minutes.
4. Wash briefly in running tap water and rinse in 2 changes of distilled water.
5. Oxidize in 0.5% periodic acid for 15 minutes.
6. Rinse with 4 changes of distilled water.
7. Place in Lillie's Schiff reagent for 30 minutes.
8. Rinse in 2 changes of distilled water and wash in running tap water for 10 minutes.
9. Rinse in 2 changes of distilled water.
10. Dehydrate slides in 95% alcohol and absolute alcohol 3 changes each. Clear in xylene, 3 changes.
11. Mount coverglass with appropriate medium (refractive index 1.48-1.56).

RESULTS
Normal and abnormal tubular and glomerular basement membranes, Bowman's capsule, elastic lamina of blood vessels, some renal casts, and juxtaglomerular granules are red. Epithelial cell mucoproteins and acid mucopolysaccharides in blood vessels are blue.

REMARKS
When heating the alcian blue solution with microwave irradiation, the top portion of the solution is warmer by 10-15°C than that near the bottom of the coplin jar. Therefore, in order to equalize the temperature of the solution, the slides are dipped up and down. This assures uniformity of staining throughout the tissue sections.

The stain may be performed at room temperature by staining the sections for 30 minutes in the alcian blue solution.

The formula for Lillie's Schiff reagent has been changed by increasing the amount of hydrochloric acid. The reason for this is that sometimes when pararosaniline is used, a white precipitate forms in the solution after 2 or 3 weeks. Increasing the amount of hydrochloric acid in the solution prevents the precipitate from forming. The solution is stable for 4 to 6 weeks when stored in a refrigerator.

Mowry's Colloidal Iron Method (Microwave)

FIXATION 10% Buffered neutral formalin. Avoid the use of dichromate fixatives.
PROCESS Paraffin.
MICROTOMY Cut sections at 5 micrometers.
CONTROL Use control number 23 or 39B.
SOLUTIONS

29% Ferric Chloride (Stock)

Ferric chloride, $FeCl_3\ 6H_20$	29.0	gm
Distilled water	100.0	ml

Colloidal Iron (Stock)

Distilled water	250.0	ml
29% Ferric chloride (stock)	4.4	ml

Bring the distilled water to a boil and add the 4.4 ml of ferric chloride. When the solution turns dark red, remove from heat and cool. This solution is stable for several months.

Colloidal Iron (Working)

Colloidal iron (stock)	20.0	ml
Distilled water	15.0	ml
Glacial acetic acid	5.0	ml

12% Glacial Acetic Acid

Glacial acetic acid	12.0	ml
Distilled water	88.0	ml

4% Hydrochloric Acid (Stock)

Hydrochloric acid, concentrated	4.0	ml
Distilled water	96.0	ml

2% Potassium Ferrocyanide (Stock)

Potassium ferrocyanide	2.0	gm
Distilled water	100.0	ml

10% Triton X-100 (Stock)

Triton X-100	10.0	ml
Distilled water	90.0	ml

Hydrochloric Acid-Potassium Ferrocyanide (Working)

4% Hydrochloric acid (stock)	20.0	ml
2% Potassium ferrocyanide (stock)	20.0	ml
10% Triton X-100 (stock)	0.4	ml

Prepare just before use.

Kernechtrot (Nuclear Fast Red)

(See Page 84)

STAINING PROCEDURE

1. Decerate slides in xylene, 2 changes, 2 minutes each. Place in absolute alcohol, 2 changes, 2 minutes each; 95% alcohol, 2 changes, 2 minutes each; then rinse in distilled water.
2. Place in 12% glacial acetic acid for 1 minute.
3. Place in working colloidal iron solution in a glass coplin jar and microwave at power level 2 (100W) for 2 minutes. Dip the slides up and down several times and allow them to remain in the warm solution for 8 minutes.
4. Rinse in 12% glacial acetic acid, 4 changes, 1 minute each.
5. Place in working hydrochloric acid-potassium ferrocyanide solution in a glass coplin jar and microwave at power level 2 (100W) for 1 3/4 minutes. Dip the slides up and down several times and allow them to remain in the warm solution for 2 minutes.
6. Wash in running tap water for 1 minute.
7. Rinse in 2 changes of distilled water.

8. Counterstain in nuclear fast red solution for 3 minutes.
9. Rinse in 3 changes of distilled water.
10. Dehydrate slides in 95% alcohol and absolute alcohol 3 changes each. Clear in xylene, 3 changes.
11. Mount coverglass with appropriate medium (refractive index 1.48-1.56).

RESULTS

Acidic mucopolysaccharides	deep blue
Acidic epithelial mucus	deep blue
Cryptococcus	deep blue
Nuclei	red

REMARKS

The colloidal iron method stains essentially the same acidic mucopolysaccharides that are stained with alcian blue, except that the color with colloidal iron is a more intense blue.

It appears that cryptococcus is the only fungus which can be demonstrated by methods for acid mucopolysaccharides, such as Mowry's alcian blue, Southgate's and Mayer's mucicarmine and Mowry's colloidal iron. Most texts recommend Southgate's or Mayer's mucicarmine for demonstrating cryptococcus, but based on my experience, this colloidal iron method gives the best results.

When heating a solution with microwave irradiation, the top portion of the solution is warmer by 10-15°C than that near the bottom of the coplin jar. Therefore, in order to equalize the temperature of the solution, the slides are dipped up and down in steps 3 and 5 of the staining procedure. This assures uniformity of staining throughout the tissue sections.

The stain may be performed at room temperature by staining the sections for 1 hour in the colloidal iron-acetic acid working solution and for 30 minutes in the hydrochloric acid-potassium ferrocyanide solution.

Alcian Blue-Periodic Acid Schiff (Microwave)

FIXATION	10% Buffered neutral formalin.
PROCESS	Paraffin.
MICROTOMY	Cut sections at 5 micrometers.
CONTROL	Use control number 39B.

SOLUTIONS

3% Acetic Acid

Glacial acetic acid	3.0	ml
Distilled water	97.0	ml

1% Alcian Blue

Alcian blue 8GX	1.0	gm
3% Acetic acid	100.0	ml

Filter and add a few crystals of thymol.

0.5% Periodic Acid

Periodic acid	0.5	gm
Distilled water	100.0	ml

Schiff's Reagent

Dissolve 1.0 gm basic fuchsin in 200 ml hot distilled water. Bring to near boiling point. Cool to 50°C and add 20 ml normal hydrochloric acid. Cool further to 25°C and add 1.0 gm anhydrous sodium bisulfite or metabisulfite. Keep in the dark for 24 - 48 hours. Add 0.2 gm of activated charcoal and agitate for 3 minutes. Filter through Whatman #4, then Whatman #1, filter papers. Keeps for about 6 months if stored in the refrigerator.

STAINING PROCEDURE

1. Decerate slides in xylene, 2 changes, 2 minutes each. Place in absolute alcohol, 2 changes, 2 minutes each; 95% alcohol, 2 changes, 2 minutes each; then rinse in distilled water.
2. Place in 3% acetic acid for 3 minutes.
3. Place in 40 ml of 1% alcian blue solution in a glass coplin jar and microwave at power level 2 (100W) for 2 1/2 minutes. Dip the slides up and down several times and allow them to remain in the hot solution for 2 minutes.
4. Wash in running tap water for 1 minute and rinse in distilled water.
5. Oxidize in 0.5% periodic acid for 10 minutes.
6. Rinse with 4 changes of distilled water.
7. Stain in Schiff's reagent for 15 minutes.
8. Dip in distilled water and wash with running tap water for 10 minutes.
9. Rinse in 2 changes of distilled water.
10. Dehydrate slides in 95% alcohol and absolute alcohol 3 changes each. Clear in xylene, 3 changes.
11. Mount coverglass with appropriate medium (refractive index 1.48-1.56).

RESULTS

Acidic sulfated mucosubstances	blue
Neutral polysaccharides	red

REMARKS

This stain is most useful for demonstrating transition zones and metaplastic changes in the gastrointestinal tract. Gastric mucins are exclusively PAS positive. Goblet cells in the small bowel are predominantly alcian blue positive. Metaplastic cells and glands in the esophagus and stomach stain blue and thus are easy to contrast with normal gastric mucins. The stain is also useful in a variety of mucin producing carcinomas.

When heating the alcian blue solution with microwave irradiation, the top portion of the solution is warmer by 10-15°C than that near the bottom of the coplin jar. Therefore, in order to equalize the temperature of the solution, the slides are dipped up and down. This assures uniformity of staining throughout the tissue sections.

The stain may be performed at room temperature by staining the sections for 30 minutes in the alcian blue solution.

Alcian Blue Method, pH 2.5 (Microwave)

FIXATION 10% Buffered neutral formalin.
PROCESS Paraffin.
MICROTOMY Cut sections at 5 micrometers.
CONTROL Use control number 39B.
SOLUTIONS

3% Glacial Acetic Acid

Glacial acetic acid	3.0	ml
Distilled water	97.0	ml

1% Alcian Blue

Alcian blue 8GX	1.0	gm
3% Acetic acid	100.0	ml

Filter and add a few crystals of thymol.

Kernechtrot (Nuclear Fast Red)
(See Page 84)

STAINING PROCEDURE

1. Decerate slides in xylene, 2 changes, 2 minutes each. Place in absolute alcohol, 2 changes, 2 minutes each; 95% alcohol, 2 changes, 2 minutes each; then rinse in distilled water.
2. Place in 3% glacial acetic acid for 3 minutes.
3. Place slides in 40 ml of 1% alcian blue solution in a glass coplin jar and microwave at power level 2 (100W) for 2 1/2 minutes. Dip the slides up and down several times and allow them to remain in the hot solution for 2 minutes.
4. Wash in running tap water for 1 minute and rinse in distilled water.
5. Counterstain in nuclear fast red for 3 minutes.
6. Rinse in 3 changes of distilled water.
7. Dehydrate slides in 95% alcohol and absolute alcohol 3 changes each. Clear in xylene, 3 changes.
8. Mount coverglass with appropriate medium (refractive index 1.48-1.56).

RESULTS

Acidic sulfated mucosubstances	bluish-green
Nuclear chromatin	red

REMARKS

No single staining solution of alcian blue colors all polyanions maximally, but the above procedure, performed at pH 2.5, stains all except the most strongly sulfated mucopolysaccharides. It is very useful for demonstrating goblet cell mucins in the respiratory tract and in small and large bowel, as well as various components of skin, heart valves, large arteries and cartilage.

Alcian blue probably finds its greatest usefulness in combination stains such as Alcian blue-PAS and the Kreyberg method.

When heating the alcian blue solution with microwave irradiation, the top portion of the solution is warmer by 10-15°C than that near the bottom of the coplin jar. Therefore, in order to equalize the temperature of the solution, the slides are dipped up and down. This assures uniformity of staining throughout the tissue sections.

The stain may be performed at room temperature by staining the sections for 30 minutes in the alcian blue solution.

Alcian Blue with Hyaluronidase Digestion (Microwave)

FIXATION 10% Buffered neutral formalin.
PROCESS Paraffin.
MICROTOMY Cut paraffin sections at 5 micrometers.
CONTROL Use control number 23.
SOLUTIONS

0.1M Potassium Phosphate, Monobasic (Stock)

Potassium phosphate, monobasic	6.80	gm
Distilled water to make a total of	500.0	ml

0.1M Sodium Phosphate, Dibasic (Stock)

Sodium phosphate, dibasic	7.10	gm
Distilled water to make a total of	500.0	ml

Buffer Solution

0.1M Potassium phosphate, monobasic (stock)	94.0	ml
0.1M Sodium phosphate, dibasic (stock)	6.0	ml

Hyaluronidase Digestion Solution

Hyaluronidase, testicular	0.04	gm
Buffer solution	40.0	ml

Mix just before use.

3% Acetic Acid

Glacial acetic acid	3.0	ml
Distilled water	97.0	ml

1% Alcian Blue

Alcian blue 8GX	1.0	gm
3% Acetic acid	100.0	ml

Filter and add a few crystals of thymol.

Kernechtrot (Nuclear Fast Red)
(See Page 84)

STAINING PROCEDURE

1. Decerate slides in xylene, 2 changes, 2 minutes each. Place in absolute alcohol, 2 changes, 2 minutes each; 95% alcohol, 2 changes, 2 minutes each; then rinse in distilled water.

2. Place 1 slide in the hyaluronidase digestion and place in a 37°C oven for 1 hour. Place the duplicate section, which is the control slide, in the buffer solution and place in a 37°C oven for 1 hour.
3. Wash both slides in running tap water for 5 minutes and rinse in distilled water.
4. Place both in 3% acetic acid for 3 minutes.
5. Place both in 40 ml of 1% alcian blue solution in a glass coplin jar and microwave at power level 2 (100W) for 2 1/2 minutes. Dip the slides up and down several times and allow them to remain in the hot solution for 2 minutes.
6. Wash in running tap water for 1 minute and rinse in 2 changes of distilled water.
7. Counterstain in nuclear fast red solution for 3 minutes.
8. Rinse in 3 changes of distilled water.
9. Dehydrate slides in 95% alcohol and absolute alcohol 3 changes each. Clear in xylene, 3 changes.
10. Mount coverglass with appropriate medium (refractive index 1.48-1.56).

RESULTS

Staining attributable to hyaluronic acid, chondroitin 4 sulfate, or chondroitin 6 sulfate is selectively eliminated.

REMARKS

The alcian blue method, with and without hyaluronidase digestion, has some value in differentiating mesotheliomas from carcinomas. Some mesotheliomas contain hyaluronic acid and therefore can be stained with alcian blue. If the alcian blue positively is destroyed following prior incubation with hyaluronidase, this helps in establishing the diagnosis of mesothelioma.

When heating the alcian blue solution with microwave irradiation, the top portion of the solution is warmer by 10-15°C than that near the bottom of the coplin jar. Therefore, in order to equalize the temperature of the solution, the slides are dipped up and down. This assures uniformity of staining throughout the tissue sections.

The stain may be performed at room temperature by staining the sections for 30 minutes in the alcian blue solution.

Fite's Method for Acid-Fast Microorganisms (Microwave)

FIXATION 10% Buffered neutral formalin.
PROCESS Paraffin.
MICROTOMY Cut sections at 5 micrometers.
CONTROL Use control number 41.
SOLUTIONS

Mineral Oil - Xylene

Mineral oil	20.0	ml
Xylene	80.0	ml

Carbol Fuchsin

Absolute alcohol	5.0	ml
Basic fuchsin or pararosaniline	0.5	gm
90% Liquid phenol	6.0	ml
Distilled water	94.0	ml

Dissolve the basic fuchsin in the absolute alcohol, then add the phenol and distilled water. Filter before use.

0.5% Acid Alcohol

70% Alcohol	1000.0	ml
Concentrated hydrochloric acid	5.0	ml

0.25% Methylene Blue

Methylene blue	0.25	gm
Distilled water	100.0	ml
Glacial acetic acid	1.0	ml

STAINING PROCEDURE

1. Decerate slides in mineral oil-xylene for 20 minutes.
2. Stand slides on end and allow to drain thoroughly. Wipe off excess oil and blot to opacity.
3. Place in carbol fuchsin in a glass coplin jar and microwave at power level 1 (50W) for 2 minutes. Dip the slides up and down several times and allow them to remain in the warm solution for 8 minutes.
4. Wash in running tap water for 2 minutes.
5. Decolorize with 0.5% acid alcohol until sections are faint pink. This may take 20-30 seconds.
6. Wash in running tap water for 1 minute and rinse in 2 changes of distilled water.
7. Counterstain in 0.25% methylene blue solution for 15 seconds.
8. Rinse in 3 changes of distilled water.
9. Stand slides on end and thoroughly air-dry.
10. Dip in xylene.
11. Mount coverglass with appropriate medium (refractive index 1.48-1.56).

RESULTS

Acid-fast bacilli	red
Lepra bacilli	red
Nocardia filaments	red
Background	pale blue

REMARKS

This method is especially recommended for mycobacterium leprae, which is more susceptible to over-decolorization with acid alcohol than are other acid-fast bacteria. Mineral-xylene coating around the bacilli helps avoid over-decolorization with acid alcohol. Most texts recommend using a weak solution of aqueous sulfuric acid as the decolorizer instead of the usual 1% hydrochloric acid-alcohol. We have found that 0.5% hydrochloric acid alcohol works more rapidly than the aqueous solution of sulfuric acid and has less tendency to over-decolorize the bacilli.

The temperature of the carbol fuchsin solution after exposure to microwave irradiation will be about 46°C. However, the temperature of the solution is not uniform throughout the coplin jar. The top portion of the solution is warmer by 10-15°C than that near the bottom of the coplin jar. Therefore, in order to equalize the temperature of the solution, the slides are dipped up and down. This assures uniformity of staining throughout the tissue sections.

The stain may be performed at room temperature by staining the sections for 30 minutes in the alcian blue solution.

Churukian's Modified Luna's for Melanin (Microwave)

FIXATION 10% Buffered neutral formalin.
PROCESS Paraffin.
MICROTOMY Cut sections at 5 micrometers.
CONTROL Use control number 32.
SOLUTIONS

Acidulated Water (Stock)

Distilled water	500.0	ml

Add enough 1% citric acid to bring water to pH 3.2.

0.5% Silver Nitrate

Silver nitrate	0.5	gm
Acidulated water (stock)	100.0	ml

2% Silver Nitrate (Stock)

Silver nitrate	2.0	gm
Acidulated water (stock)	100.0	ml

4% Gelatin (Stock)

Gelatin, BBL, Fisher Type A or Sigma Type III	4.0	gm
Acidulated water (stock)	100.0	ml

Place in a beaker, then on a magnetic stirrer and apply gentle heat until gelatin dissolves.

0.1% Hydroquinone (Stock)

Hydroquinone	0.1	gm
Acidulated water (stock)	100.0	ml

Silver Nitrate-Gelatin-Hydroquinone Developer

Silver nitrate, 2%	10.0	ml
Gelatin, 4%	25.0	ml
Hydroquinone	15.0	ml

Place the 2% silver nitrate, 4% gelatin, and the 0.1% hydroquinone in separate flasks in a 56-58°C oven. Allow the solutions to remain in the oven until thoroughly warmed before continuing. Then combine in order given, in a warm 125 ml flask, making certain the solutions are mixed well after each addition. Prepare immediately before use.

Kernechtrot (Nuclear Fast Red)
(See Page 84)

STAINING PROCEDURE

1. Decerate slides in xylene, 2 changes, 2 minutes each. Place in absolute alcohol, 2 changes, 2 minutes each; 95% alcohol, 2 changes, 2 minutes each; then rinse in distilled water.
2. Place in acidulated water, pH 3.2, for 1 minute.
3. Place slides in 40 ml of 0.5% silver nitrate in a plastic coplin jar and microwave at power level 8 (400W) for 45 seconds. Dip the slides up and down several times and allow them to remain in the hot solution for 5 minutes.
4. Place slides in freshly prepared silver nitrate-gelatin-hydroquinone developer in a plastic coplin jar and microwave at power level 1 (50W) for 1 minute. Dip the slides up and down several times and allow them to remain in the warm solution for 2-3 minutes.
5. Wash quickly and thoroughly in running hot water.
6. Rinse in 2 changes of distilled water.
7. Stain in nuclear fast red solution for 3 minutes.
8. Rinse in 2 changes of distilled water.
9. Dehydrate slides in 95% alcohol and absolute alcohol 3 changes each. Clear in xylene, 3 changes.
10. Mount coverglass with appropriate medium (refractive index 1.48-1.56).

RESULTS

Melanin .... black

Nuclei .... light pink or orange

Cytoplasm .... yellow

REMARKS

This staining method is a modification of the Warthin-Starry method for spirochetes and bacteria. The pH of the acidulated water in the Warthin-Starry is 4.0-4.4. In this method the pH of the acidulated water is 3.2. This lower pH results in less background staining, allowing for better visualization of melanin. We have found that this method is more sensitive and specific than the Fontana-Masson method for melanin. It will not stain argentaffin, chromaffin, lipofuscin and ferric iron, which all stain with the Fontana-Masson. Amelanotic melanomas, which stain poorly or not at all with the Fontana-Masson, are usually well stained with this method.

When solutions are heated with microwave irradiation, the top portion of the solution is warmer by 10-15°C than that near the bottom of the coplin jar. Therefore, in order to equalize the temperature of the solution, the slides are dipped up and down. This assures uniformity of staining throughout the tissue sections. The temperature of the 0.5% silver nitrate will be about 82°C and that of the developer about 60°C.

The shelf life of silver nitrate and hydroquinone can be greatly increased by storing the reagents in a refrigerator at 3-6°C.

Chapter 17

Staining Controls, Section Adhesives and Tissue Artifacts

Tissue Staining Controls

Tissue controls have the following important functions in the histotechnology laboratory:

1. They provide a means for determining the condition and quality of staining solutions.
2. They assist in determining if the fixation, processing, and staining methods are appropriate and/or functioning to optimum.
3. They provide excellent insight into how well the histotechnologist has performed the staining procedure.

Provided below are some of the most useful sources of tissue control material for identifying specific entities. The information is divided into three parts: the entity one wishes to demonstrate, source and location of the entity, and the preferred fixative. Also included are some general comments which apply to the demonstration of specific tissues or entities.

Important Considerations in the Selection and Use of Controls

In using normal tissue or tissue entity as controls one must remember that normal structures have a tendency to react faster and better to staining than abnormal structures or entities. For example, mucins from mucin producing tumors may be more difficult to demonstrate than mucin produced normally in goblet or similar cells. Likewise, in many instances, the granules of carcinoids do not react to silver as well as the normal argentaffin granules of the Kulchitsky cells of the small intestine.

Good quality control of staining reactions depends on more than the availability of control blocks. The real key to success in stain technology is the individual histotechnologist. I find that most reactions and/or stains provide a certain tinctorial quality to given tissue structures which is a good indicator to determine whether or not the reaction has worked properly. For example, there should always be some silver deposition or reaction on collagen in a well stained Grocott methenamine silver method for fungi. Blood vessels serve as a good built in control for the periodic acid Schiff reaction, since the vessel walls stain if the staining procedure is working well. These are examples of built-in controls which can be used to great advantage for controlling staining reactions. To utilize this system successfully one must:

1. Learn to recognize those structures in tissue which can be used as built-in controls.
2. Become tinctorially sensitive. That is, learn to recognize macroscopically and microscopically when the right shade of color reaction has been applied to a certain tissue structure or entity.

Control, Source and Preferred Fixative

(See General Comments for more information on specific controls.)

Control Number	Entity	Source	Fixative
1	Acid or Alkaline phosphatase	Kidney(rat) - epithelial cells of collecting tubules.	Formal-Calcium (4°C)
1A	Alpha (A) cell granules (glucagon)	Pancreas - usually found toward the peripheral margins of the islets of Langerhans.	10% BNF
2	Amyloid	Tissue containing amyloid. (See General Comments.)	10% BNF
3	Argentaffin granules	Small intestine - Kultschitzky's (argentaffin) cells in crypts of Lieberkühn of the terminal ileum. These cells are often referred to as "enteroendocrine".	10% BNF
4	Argyrophil granules	Pancreas - cells found toward the peripheral margins of the islet of Langerhans.	10% BNF
5	Axons and dendrites	Brain - neurons of the cerebrum, cerebellum and spinal cord.	10% BNF
6	Basement membrane	Kidney - glomerular tufts and collecting tubules of the kidney. Skin contains a basement membrane at the epidermal and dermal junction.	10% BNF
6A	Beta (B) cell granules (insulin)	Pancreas - concentrated toward the center of the islet of Langerhans.	10% BNF
7	Bile	Liver with obstructed bile duct (obstructive jaundice).	10% BNF
8	Calcium oxalate	Human kidney - calculus. Left colleteral gland of the cockroach Periplaneta Americana.	10% BNF

Control Number	Entity	Source	Fixative
9	Calcium	Bone from a young mouse, bone from calcified lesion or full term placenta.	10% BNF
9A	Calcium carbonate	Digestive gland of the snails, Helix aspersa and/or Helix pomatia.	10% BNF
9B	Calcium phosphate	Diverticulum of the crab, Carcinus maenas.	10% BNF
9C	Charcott-Leyden crystals	Bone - tissue where eosinophilic leukocytes are undergoing fragmentation (eosinophilic granuloma of bone); Bronchiole secretion in bronchiol asthma and in stools in some cases of intestinal parasitism.	10% BNF
10	Cholesterol	Gallbladder with cholesterolosis.	10% BNF
10A	Chondroitin sulfate A, B & C	Cartilage or tendon.	10% BNF
11	Collagen	Heart - normal heart muscle between muscle bundles. Lung is also a good organ to use as a control for collagen.	10% BNF
12	Copper	Liver - hepatic cell in Wilson's disease or fetal liver.	10% BNF
13	Chromaffin	Adrenal gland - chromaffin cells in the medulla of the adrenal gland.	Regaud's fixative or 2% potassium dichromate
14	Cross striations (muscle)	Cardiac or skeletal muscle, cross striations are the "A" band of the myofibril sarcomere.	10% BNF
14A	Delta (D) cell granules (somatostatin)	Pancreas - found throughout the islets of Langerhans.	10% BNF
15	Deoxyribose nucleic acid (DNA)	Tonsil - plasma cells in tonsil or lymph node.	Carnoy's

Control Number	Entity	Source	Fixative
16	Elastic fibers	Skin - thin fibers below the epidermis, as well as thicker fibers in the dermis; medium sized arteries also contain both thick and very thin elastic fibers.	10% BNF
17	Fibrin	Bone marrow aspirate, fibrillary tangles, or blood clot (see 39A).	10% BNF
18	Fungi	Tissue containing fungal organisms but particularly histoplasmosis capsulatum. (See General Comments)	10% BNF
19	Glycogen	Liver - cytoplasm in hepatic cells of the liver.	Methyl or Formal-alcohol
20	Gram-positive and Gram-negative organisms	Animal skin which has been injected with bacteria, infected appendix or sections of beef jerky. Alternatively, one can use the methods contained in General Comments.	10% BNF
21	Hepatitis B surface antigen	Liver with viral hepatitis.	10% BNF
22	Hemoglobin	Kidney - patient with hemoglobin uria.	10% BNF
22A	Helicobacter pylori	Gastric biopsy from patient with chronic gastritis or peptic ulcer.	10% BNF
23	Hyaluronic acid	Skin - normally found between collagen fibers and papillary layer of dermis. Umbilical cord also contains hyaluronic acid.	10% BNF
23A	Herpes	Inoculum of HSU-hand HSV-2. See in Vivo Productions of Controls.	10% BNF
24	Inflammatory infiltrate	A section of tissue containing an acute inflammatory infiltrate.	10% BNF
25	Iron	Liver - hepatic cell in a case of hemochromatosis, or produce your own control with the use of an iron nail. (See General Comments) Bone marrow specimens often contain iron deposits.	10% BNF

Control Number	Entity	Source	Fixative
26	Juxtaglo-merular cell granules	Kidney - afferent arteriole of the kidney glomeruli. The cells are found in the walls of the afferent arteriole.	Bouin's
27	Keratin	Skin - stratum, corneum. Palm of hand and sole of foot.	10% BNF
27A	Lipofuscin	Liver - lipofuscin can be found in the cytoplasm of hepatocyte of older persons.	10% BNF
28	H&E	Kidney - a wedge shaped, formalin fixed piece of kidney is ideal for controlling H&E staining results. This specimen would contain a diversity of cells in the glomeruli, Bowman's capsule, proximal and distal convoluted tubules, thin and thick limbs of Henle's loop and collecting tubules.	10% BNF
29	Legionella pneumophila or related bacteria	Lung from a patient with Legion-naires disease - bacterial clusters can be seen within macrophages. Individual bacteria can also be found free floating in the lung. For laboratory produced controls, see bacteria and fungi controls in General Comments.	10% BNF
30	Lepra bacilli	Skin - specimen containing lepra bacilli.	10% BNF
30A	Lyme's spirochete	Specimen from a patient with Lyme's disease or use spirochete control (see control 40).	10% BNF
30B	Malarial parasite	Blood smear or tissue specimen. Erythrocytes or patients with malaria.	10% BNF
31	Mast cells	Skin - near blood vessels throughout the dermis.	10% BNF
32	Melanin	Skin - in melanocytes in the stratum germinativum or a known case of malignant melanoma.	10% BNF
33	Mucin	Small intestine - goblet cell or a mucin producing tumor.	10% BNF

Control Number	Entity	Source	Fixative
34	Myelin	Cerebrum - myelin sheath envelopes the axon (nerve fibers) in the white matter (see phospholipid).	10% BNF
34A	Negri bodies	Cytoplasm of nerve cells of animal or human brain after death from rabies.	Zenker's
35	Nissl substance	Spinal cord - large cell such as motor neurons.	10% BNF
36	Nocardia asteroides	Lungs of patients with nocardiosis.	10% BNF
37	Phospho-lipid	Brain - myelin sheath a. Sphingomyelin-found in Gaucher's disease, Nieman-Pick disease, Tay-Sachs disease. b. Lecithin-found in egg yoke.	10% BNF
37A	Pneumocystis carinii	Lung from patient with Pneumocystic carinii pneumonia.	10% BNF
37B	Protozoa	Tissue containing the subphyla (amebae) protozoa are found in the digestive tract. Trypanosoma protozoa are found in blood macrophages. Balantidium organism found in digestive tract. Plasmodium-malaria parasites found on erythrocytes.	10% BNF
38	Reticulum fibers	Liver - abundant amount of fibers can be found in the walls of sinusoids. Lymph node - fibers found throughout the node.	10% BNF
38A	Rickettsia	Lymph node of infected person or animal.	Zenker's or 10% BNF
39	Ribonucleic acid (RNA)	Tonsil or lymph node - cytoplasm of plasma cells.	Carnoy's
39A	Senile plaques	Brain - patient with Alzheimer's disease.	10% BNF
39B	Sialic acid (sialomucin)	Sublingual and submaxillary glands.	10% BNF

Control Number	Entity	Source	Fixative
40	Spirochetes	Lymph node or lesions in organ of patient with venereal syphilis. For laboratory produced controls see spirochetes in General Comments.	10% BNF
41	Tubercle bacilli	Lung or lymph nodes - from patients with tuberculosis. Bacilli are usually found in giant cells or caseous (a form of necrosis in which the tissue is changed into an amorphous mass resembling cheese) material in patients with tuberculosis. For laboratory produced controls see bacteria and fungal controls in General Comments.	10% BNF
42	Urate	Gouty joint articulations.	Methyl alcohol crystals
43	Viral inclusions (RNA)	Tissue from patients with viral infections; e.g., cytomegalovirus.	See specific technique
44	Sex chromatin	Buccal smear - scrape the inside of the cheek of a female with tongue depressor. Smear scraped material on clear glass side.	Methyl alcohol or 10% BNF
45	Neutrophilic myeloid and blood cells	Bone marrow.	B-5
46	Basophils - R (TSH, LH, FSH) acidophils (prolactin) chromophobes basophils-S (ACTH, MSH)	Pituitary gland.	10% BNF
47	Spermatozoa	Testicle - human.	Bouin's or 10% BNF

Control Number	Entity	Source	Fixative
48	Paneth cell granules	Small intestine - cells are found at base of the glands in small intestine. Namely, the crypts of Leiberkühn.	10% BNF
49	Alcoholic hyaline (mallory bodies)	Liver - alcoholic hyaline is an amorphous eosinophilic material found in the cytoplasm of some hepatocytes of patients with alcoholic hepatitis.	10% BNF
50	Mitochondria	Cells - mitochondria are found in the cytoplasm of cells.	10% BNF
51	Blood cells	Blood smear.	Methyl alcohol
52	Zymogen granules	Stomach - sections of fundic, cardiac and pyloric regions.	10% BNF

In-Vivo Production of Controls and General Comments

AMYLOID

Amyloidosis of various types is fairly prevalent and therefore one should be able to obtain an amyloid control fairly easy. In the event one cannot find a control, the following procedure can be used to produce a control in the laboratory.

Sorenson-Wilson-Heefner's Method for Amyloid

1. Inject mice subcutaneously with 0.5 ml of the following solution:
   5% casein in 0.25% sodium hydroxide.
2. Repeat injections 5 times weekly for a period of 10-12 weeks.
3. Sacrifice mice using accepted and approved procedures.
4. Remove the spleen, liver and kidney and fix in 10% buffered formalin.
5. Gross to desired thickness and process in conventional manner and embed in paraffin wax.
6. Cut paraffin sections at desired thickness and stain with preferred method.

REMARKS
Amyloid deposits initially occur in the spleen and very shortly thereafter appear in the liver. Subsequently, similar accumulations can be observed in the kidney and other parts of the body.

REFERENCE

Cohen, A.S., et al: Studies on experimental amyloidosis. American Journal of Pathology 971-989, 1959.

CALCIUM OXALATE

Calcium oxalate can be found in abnormal concentrations occurring within the animal body. Collecting tubules of human kidney often contain calcium oxalate in cases of ethylene glycol poisoning, renal insufficiency and vitamin defiency.

REFERENCE

Chapin et al: Experimentally Produced calcium oxalate. Histochemistry 74: 149, 1982.

CHOLESTEROL AND CHOLESTEROL ESTERS

These entities can only be demonstrated on frozen cut sections. Various fixatives can be used but 10% buffered formalin works best. Although the Schultz reaction for cholesterol (page 460) does not provide permanent slides, in my opinion, it is the most specific method for cholesterol.

CHROMAFFIN CELL GRANULES

In order to preserve chromaffin cell granules, it is essential that tissue be fixed in Orth's solution.

Elston's Method for Bacteria or Fungi

BACTERIAL STRAINS

Fresh clinical isolates of mycobacterium scrofulaceum (for acid fast), Neisseria subflava (Gram negative), Escherichia coli ATCC 25922 (Gram negative) and Staphylococcus epidermidis ATCC 14990 (Gram-positive).

PREPARATION OF INOCULUM

Mycobacterium scrofulaceum is yielded on a Lowenstein-Jensen medium slant for 4 days. Neisseria subflava is yielded on heart infusion agar slant for 48 hours. Escherichia coli and staphylococcus epidermidis are yielded on heart infusion agar slants in 24 hours.

FUNGUS STRAINS

Aspergillus fumigate is yielded on a Sabouraud's agar slant. This control should be taken from a five day old aspergillus fumigatus culture.

PROCEDURE

1. Culture slants (24x125 mm test tube) are covered with 10% buffered formalin solution for 4 hours to kill the organisms.
2. Remove the colonies from the surface of the slant with a stiff wire loop. (This will form an organism-formalin mixture.)

3. Decant the organism-formalin mixture into a clean test tube.
4. Homogenize colonies as much as possible, centrifuge for 5 minutes at 2500 rpm, and decant the supernatant.
5. Pool equal amounts of sediment from Gram-positive and Gram-negative organisms.
6. Wash each pool of sediment by suspending the bacterial buttons in 5 ml of isotonic saline (1:5).
7. Further dilute (1:25) the mycobacterium inoculum by taking 1.0 ml of the 1:5 dilution and adding 5 ml of isotonic saline.

The entire button, which is 10-12 mm in diameter is used for the inoculum.

Preparation of Synthetic Control

PROCEDURE

1. Prepare thrombin solution from a vial containing 50 NIH units of lyophilized human thrombin by reconstituting the thrombin with 2 ml of isotonic saline (0.9% sodium chloride).
2. Reconstitute vials of human lyophilized standardized normal plasma with 1.0 ml of deionized water. Restopper vials and allow to stand until plasma dissolves.
3. Pipette 5 drops of the mycobacterium suspension or Gram-negative Gram-positive suspensions into the vials containing 2 ml of reconstituted normal human plasma.
4. Add 0.6 ml of the reconstituted human thrombin solution to the bacterial-plasma mixture.
5. Tilt mixture back and forth until a good clot forms (60 seconds).
6. Decant excess liquid; place clot in lens paper.
7. Place clot in labelled cassette and process in conventional manner.
8. Cut section at desired thickness and stain with preferred methods.

REMARKS
One should remember that the entire fungal button produced as the inoculum is used in the plasma-thrombin mixture. The staining results are similar to those seen on a tissue section containing these organisms. It should be pointed out that other bacterial and fungal organisms could be used for producing controls.

Bezahlers's Method for Bacterial Controls

PROCEDURE

1. Obtain pure cultures of Gram-positive (staphylococcus aureus) and Gram-negative (Escherichia coli) organisms from the microbiology laboratory. Use overgrown plates.
2. Flood plates with 10% formalin and let stand overnight.
3. The following day, gently scrape the surface of the plate to suspend the bacteria in the formalin and pour into 50 ml centrifuge tubes. (Be sure to label tubes positive and negative.)
4. Centrifuge for 5 minutes to concentrate the organisms; pour off supernatant, leaving behind a sediment of killed organism.
5. Add 10 ml of liquid agar (previously melted in a water bath), and stir to suspend the bacteria.

6. Pour the molten agar/bacteria suspension into a plastic disposable container and place in the refrigerator until solid.
7. Remove the chilled, solid agar from the container and slice into 2 mm thick slices. Place the slices of Gram-positive and Gram-negative organisms into tissue cassettes, process and embed in paraffin wax.
8. Cut sections at desired thickness and stain with preferred methods.

REMARKS
The paraffin blocks produced have a shelf life of several years. The controls are reliable and fairly easy to prepare. Fungus and acid fast bacilli controls are prepared in a similar fashion. For additional control methods for bacteria see the following references.
Tseng, C.H., and Renner, E.D.: Laboratory methods for producing tissue controls for bacterial stains. Laboratory Medicine, 14:298, 1983.
Jung, W.K.: In vitro positive controls for histochemical stains of bacteria and fungi. American Journal of Clinical Pathology, 84:342-345, 1985.

COPPER

The following reference may be helpful in obtaining a control for copper. Ludwig, J. et al: The liver in the inherited copper disease of Bedlington terriers. Laboratory Investigation, 43:1:82-87, 1980.

CROSS STRIATIONS-MUSCLE

Skeletal and cardiac muscle are commonly used as controls for the phosphotungstic acid hematoxylin (PTAH) staining procedure. I prefer to use a section of brain. In my experience, cardiac or skeletal muscle striations stain quite well with PTAH solutions which may not be reacting to optimum. These controls may produce the false impression that the staining solution is reacting well. A section of brain eliminates this problem since a very good working PTAH solution is required to stain the fibro, myo and neuroglial fiber dendrites blue and the cytoplasm of neurons salmon color. Fibrin (aspirate) is also a good control.

DEOXYRIBONUCLEIC (DNA) AND RIBONUCLEIC (RNA) ACIDS

The demonstration of DNA and RNA with the methyl green-pyronin Y (MGP) method is contingent upon tissue fixation in Carnoy's fluid. I have never seen a good reproducible methyl green-pyronin Y stain on formalin fixed material. When formalin fixative is used, it has been my experience that the quality of RNA and DNA staining is determined by the length of tissue exposure to the formalin fixative. For example, one obtains good MGP staining results if tissue is fixed for 6 hours in 10% buffered neutral formalin. Staining decreases considerably the longer the tissue remains in the formalin fixative until it reaches a low point after 24 hours, at which time staining results are poor. It should be pointed out that tissue fixation in unbuffered formalin provides better MGP staining potential than the normally used 10% buffered formalin.

ELASTIC FIBERS

A section of skin is the best control for demonstrating elastic fibers since skin contains coarse elastic fibers in the dermis and thin fibers immediately subjacent to the epidermis. A good demonstration of both the thin and coarse fibers indicates the staining solution is working to optimum.

FUNGI

Histoplasma capsulatum is one of the most difficult fungi to demonstrate with Grocott's methenamine silver method. It is for this reason that histoplasma capsulatum serves as the best control for paraffin embedded tissue sections. Generally speaking, if this fungus stains well, most likely all other types of fungi are going to stain to optimum.

IRON

An iron control can be produced simply in the laboratory on a piece of liver from an autopsy. Place a nail through a piece of unfixed liver and place in unbuffered 10% formalin for a week. Remove nail, process tissue in conventional manner and section. Iron deposits will radiate from the nail hole.

PHOSPHOLIPID

Phospholipids are one of a class of compounds widely distributed in living cells. Two major kinds of phospholipids are lecithin and sphingomyelin.

SPIROCHETES

The best control for all spirochetal organisms is a tissue specimen from a patient with venereal syphilis. The causitive agent of this disease is Treponema Pallidum. It has been my experience that this is the most difficult organism to demonstrate. That being the case, one must agree that this is the best control for the demonstration of all other types of spirochetes. Included in this group are Lyme disease spirochetes, Leptospira and Donovan bodies of the disease Donovania granulomatis.

Luna-Slayter-Sims Method for Gram Positive and Negative Bacteria

The following procedure for producing Gram-positive and Gram-negative controls has been used successfully by the author for many years. It provides tissue specimens which have been exposed to the animal's immune system and therefore, the staining reaction should resemble bacteria from a pathologic condition.

PROCEDURE

Staphylococcus aureus and pseudomonas aeruginosa are streaked on separate plates of 5% sheep's blood agar and incubated at 35°C for 24 hours. A small amount of each culture is retrieved with a cotton swab and suspended in 4 ml normal saline solution. One ml of each suspension is injected subcutaneously in the right flank of 3 mature Sprague Dawley rats. The rats are serially sacrificed at 24, 48, and 72 hours with an anesthetic overdose and from each rat a 2 centimeter round piece of skin at the injection sites is excised along with underlying muscle. The tissue is fixed in 10% neutral buffered formalin, processed for routine histologic examination, and embedded in paraffin so that each block contains Gram-positive and Gram-negative bacteria. Following microtomy, the slides can be stained using the method of choice.

RESULTS

Following 24 hours after injection of each bacterial suspension, all rats developed subcutaneous induration at the site which progressively enlarged to 3-4 cm in diameter by 72 hours.

The tissue harvested at 72 hours is generally the most suitable for the intended purpose. Microscopically, the tissues from both bacterial treatment groups exhibit edema and variable amounts of necrotic debris. Inflammatory infiltrates composed of neutrophils and lesser numbers of macrophages are generally present in the sample treated with staphylococci, the organisms being mostly intracellular. The pseudomonas organisms are generally not within phagocytes; in fact, an inflammatory cellular infiltrate was conspicuously absent in spite of an abundance of bacteria. The Gram-positive organisms stain intensely dark blue-black. The Gram-negative bacteria are distinctly red-purple and contrast well with the Gram-positive group.

Note: It is possible to mix and inject a suspension from both staphylococcus aureus and pseudomonas aeruginosa and thereby produce a specimen with both Gram-positive and Gram-negative bacteria.

Smith-Felton Method for Herpes Simplex Virus Types 1 and 2[162]

Controls specific for herpes simplex virus type 1 (HSV-1) and herpes simplex virus type 2 (HSV-2) are difficult, if not impossible, to obtain in the histopathology laboratory. In the procedure below, cell cultures of both types of HSV may be prepared and used to provide a control for the PAP stain.

MATERIALS AND METHODS

HSV-1 and HSV-2 are inoculated in separate 75 cm flasks containing 2-day old cultures of HeLa cells grown in medium EMEM-10% fetal bovine serum that has not reached monolayer confluency. One tissue culture flask containing no virus is processed as a negative control. Observe the cell cultures daily and harvest when a 3+ cytopathic effect is observed in the flasks containing the virus. The tissue culture media is gently washed once with phosphate-buffered saline (PBS), pH 7.6, to remove medium protein. The PBS is removed and 10 ml of 10% buffered neutral formalin (BNF) is added to the cell cultures. Cells are then scraped from the flasks using a rubber policeman and decanted into 15 ml conical centrifuge tubes. Centrifuge cell suspensions for three minutes at 1000 rpm to form pellets. Allow the pellets to fix in 10% NBF overnight and then process, embed in paraffin, and cut into 4 micrometer sections with a microtome. Dry the slides in an oven, deparaffinize, rehydrate, and then stain with the PAP stain specific for HSV-1 and HSV-2.

Immunohistochemistry Controls

The following list of controls for immunohistochemical stains are provided as an aid to those individuals performing the immunochemical procedures. Check procedure being performed for preferred fixative and recommended method of processing.

Antigen	Source
Adrenocorticotrophic hormone (ACTH)	Corticotroph cells of pituitary
Alpha-1-antitrysin	Hepatic cells of fetal liver
Alpha fetoprotein (AFP)	Hepatic cells of fetal liver
Calcitonin	Parafollicular "C" cells of thyroid
Carcinoembryonic antigen (CEA)	Epithelial cells of colonic cancer
Chorionic gonadotropin (HCG)	Trophoblast cells of placenta
Desmin	Skeletal muscle
Estradiol	Ovary
Factor VIII	Blood vessel of tonsil
Ferritin	Hepatic cells of fetal liver
Fibrinogen	Placenta
Follicle stimulating hormone (FSH)	Gonadotroph cells of pituitary
Gastrin	Normal cells of pyloric antrum
Glucagon	Alpha cells of pancreatic islet
Glial fibrillary acidic protein (GFAP)	Fibrillary astrocytes of brain
Growth hormone	Somatotroh cells of pituitary
Hepatitis B surface antigen (HBsAG)	Infected hepatocytes of liver
Hepatitis B core antigen (HBcAG)	Infected hepatocytes of liver
Herpes simplex virus 1 (HSV-1) and 2 (HSV-2)	Infected tissue, infected cell culture
Immunoglobulins G, A, M	Plasma cells of tonsil
Insulin	Beta cells of pancreatic islet

Antigen	Source
J chain	Plasma cells of tonsil
Kappa and Lambda	Plasma cells of tonsil
Keratin	Epidermis
Lactoferrin	Bone marrow
Luteinizing hormone (LH)	Gonadotroph cells of pituitary
Lysozyme	Histiocytes of tonsil
Mammary epithelial membrane (MEMA 70)	Breast
Melanocyte stimulating hormone (MSH)	Corticotroph cells of pituitary
Myoglobin	Skeletal muscle fibers
Neuron specific enolase (NSE)	Neural crest cells of pancreas
Neurophysin	Hypothalamus of brain
Papilloma virus	Infected tissue
Pregnancy specific beta-1-glycoprotein (SP1)	Trophoblast cells of placenta
Prolactin	Lactotroph cells of pituitary
Prostatic acid phosphatase	Prostate epithelium
Prostate epithelial antigen	Prostate epithelium
Secretory piece	Epithelial cells of colon
Somatostatin	Delta cells of pancreatic islet
Testosterone	Interstitial cells of testes
Thyroglobulin	Epithelial cells of thyroid
Thyroid stimulating hormone (TSH)	Thyrotroph cells of pituitary
Thyroxin (T4)	Epithelial cells of thyroid
Vasoactive intestinal peptide (VIP)	Nerve fibers and mucosal cells of colon

Tissue Section Adhesives

The Need for Section Adhesives

It is prudent to discuss the need for section adhesives before the various methods of adhering tissue sections to slides are presented. The need for section adhesives is simply a matter of mechanics; i.e., tissue sections which have been picked up on glass slides are going to float off unless there is some form of adhesive action between the section and the slide. The mere presence of water between these two objects allows for the section to separate from the slide unless the water is allowed to evaporate (Fig. 288d). The action of water evaporation allows the section to come directly in contact with the surface of the glass. This action forms a surface tension depressant. The greater the contact and more homogeneous the attachment, the more one can be assured of section adhesion. There are numerous reasons for poor, nonhomogeneous section contact to slides. For example, the presence of particulate matter on the slide (Fig. 288a). Obviously, such surface particulate matter will prevent complete and homogeneous section contact on the surface of the glass slide.

Another reason for poor adhesion has to do with the status of the undersurface of the section which comes in contact with the slide. Many section undersurfaces are uneven due to poor processing, inadequate knife sharpness or thick and thin areas due to excessively hard tissue (the result of overexposure to coagulant fixatives, processing agents or microvibrations). Uneven section undersurfaces can also result from excessive cooling of the paraffin block prior to cutting (Fig. 288b). In addition, tissues are exposed to a wide variety of environments during the staining process which may adversely affect their outer surfaces. For example, the performance of an H&E may expose tissue to as many as 8 solutions and 15 chemical agents. Sections may also fall off during staining if improperly dried. It is no wonder that sections fall off slides unless handled properly. It is my belief that were it not for the situations cited above, most tissues would remain on the slide during a staining process without the use of adhesives. However, some of the cited problems are unsolvable and therefore require the use of section adhesives in an attempt to negate all of the above section adhesion imperfections. The fact that a section adhesive is used, in no way guarantees that sections will remain on glass slides during the various environmental changes experienced in the course of a staining process. There are a number of reasons for sections falling off slides after an adhesive has been applied. Some of the most common are presented below.

Tissue Section Detachment

Inadequate Adhesion

Inadequate section adhesion often results from improper application of adhesives. For example, if one utilized granular gelatin by sprinkling it on the surface of the floatation bath, an excessive amount may be removed during the surface cleaning of the bath with a paper towel. This is particularly true if the gelatin is not mixed adequately with the water.

Section Overexpansion

Sections frequently fall off slides due to section overexpansion on the surface of the floatation bath. This occurs when the water temperature is excessive; 45°C or over. In this case, the section expands beyond its normal size and is picked up on a slide in this extended position. During the drying and/or staining steps, the section attempts to shrink to its original shape. Consequently, the bond between the adhesive and the section is broken at microscopic levels, allowing various liquids to seep between the section and the slide. These react with the adhesive and break the bonds further, allowing the section to detach.

Alkaline Solutions

Protein adhesive bonds between the section and the glass slide are broken when slides are exposed to alkaline solutions; e.g., ammoniacal silver, H&E blueing solutions, etc. Conversely, the bonds of protein adhesives do not break easily in acid solutions.

Enzyme Treatment

Sections which are treated with enzymes, whether for immunochemistry, histochemistry or mucosaccharide procedures, have a tendency to fall off slides. Therefore, it is imperative that one uses an appropriate section adhesive.

Heat Treatment

Sections which require heat for DNA probes, such as in situ hybridization, will come off slides if the proper adhesive has not been utilized. The method's instructions should be read carefully before an adhesive is used.

Adhesive Solutions and Methods

Glycerin-Egg Albumin

The glycerin used in compounding Mayer's egg albumin acts as a stabilizer, keeping the egg albumin (protein) in a liquid state. If one does not allow the glycerin to dissolve adequately in the water floatation bath, the protein will not gel or set when the slides are dried on the slide warming table or in the oven. The protein will remain dissolved in residual glycerin, thereby having no adhesive quality. This allows sections to become detached.

Mayer's Egg Albumin

Egg white	50.0	ml
Glycerin	50.0	ml

Mix well and filter through coarse filter paper, Buchner's funnel with vacuum or through several thicknesses of gauze. Add a crystal of thymol to preserve. Commercial egg albumin can be purchased ready for use. After microtomy, slides should be dried in a 58-60°C oven for 20 to 30 minutes for proper section adhesion. Here the egg white (albumin) coagulates forming a bond between the section and the slide. Inadequate heat results in a poor to nonexistent coagulant effect and subsequently, poor section adhesion. The glycerin is not a factor in the adhesive process, since most of it dissolves into the water floatation bath when the section is being picked up. The egg albumin solution is applied by smearing it on the slide with a small piece of fine sponge prior to section pickup.

Albumin-Sodium Chloride Method

(Powdered Albumin)

Albumin, dried	5.0	gm
Sodium chloride	0.5	gm
Distilled water	100.0	ml

Combine and filter through Buchner's funnel with vacuum to 50 ml of filtrate; add 50 ml of glycerin. Add a crystal of thymol to preserve.

The albumin-sodium chloride solution is applied by smearing it on the slide with a small piece of fine sponge prior to section pickup.

Gelatin Solution Method

5% Gelatin Type B

Gelatin, Type B	5.0	gm
Distilled water	100.0	ml

Dissolve the gelatin in the water with gentle heat. Add a small amount of thymol to prevent mold growth. Mix thoroughly 3 teaspoons of 5% solution of gelatin for every 1000 ml of water in bath used for floating sections. The gelatin adhesive added to the water bath has proven to be a good medium for affixing sections to slides. When sections are picked up on clean slides and processed through the stains, the gelatin, (if not in excess in the water bath), does not take up the stain.

Since the gelatin is a culture medium for bacteria and fungi, it is important that the floatation bath be cleaned and dried thoroughly after each day's use to insure against contamination of the sections, which might confuse the diagnosis.

Gelatin Type B Granular Method

The amount of gelatin which should be used in making this adhesive will depend upon the size of the floatation water bath in which it is made. If one is using a large bath (approximately 800 ml capacity), use enough gelatin granules to completely cover the surface of a quarter. In a smaller bath (approximately 400 ml capacity) use enough gelatin to cover the surface of a nickel. These amounts are merely estimates and may need to be adjusted to accommodate specific circumstances. It is extremely important that all the gelatin granules dissolve in the floatation bath in order to achieve proper section adhesion. This may be accomplished by performing the following steps in the order presented:

1. Sprinkle the gelatin granules on the bottom of the empty floatation bath.
2. Slowly pour approximately 100 ml hot (almost boiling) water into the floatation bath. Some of the gelatin granules will remain on the bottom surface while others will float to the top of the water. Let this stand for approximately ten minutes, then stir well with a camel's hair brush or similar tool.

3. Complete filling the floatation water bath with distilled water and allow the bath to reach desired temperature.
4. Wipe the surface of the floatation bath free of extraneous material by laying a piece of paper (newspaper works excellently) on top of the water and gently drawing it across the entire length of the container. It is important to clean the floatation bath (and any utensils used in it) very well each day to prevent the growth of bacteria.

Gelatin Type B-Formalin Method

Formalin hardens the protein contained in gelatin and egg albumin. This effect can be accomplished by exposing gelatinized or albuminized slides to formalin, thereby achieving a firmer bond between the section and the glass slide.

1. Gelatin or egg albumin is applied as described earlier.
2. Sections are placed on slides from the floatation bath and drained from 30 seconds to 1 minute.
3. Place slides in a coplin jar containing approximately 5 ml of formalin (full strength), making certain the formalin solution in no way comes in contact with the tissue sections. The tightly covered coplin jar is then placed in a 60°C oven for 1 hour.
4. Remove sections from the coplin jar and dry in the conventional way; i.e., either on slide warmer overnight or in a 60°C oven for 1 hour.
5. Treat and/or stain as desired.

Elmer's Glue Slide Dip Method

General purpose Elmer's glue has achieved some popularity as a section adhesive, particularly in those cases destined for enzyme treatment and subsequent immunochemical staining of paraffin embedded sections. The following procedure will give good results even for those specimens containing bone and keratin.

1. Make up a 15% aqueous solution of Elmer's glue in a Falcon conical tube or similar container. Store in refrigerator until ready for use.
2. Cut section ribbon and place on floatation bath.
3. Dip the slide into the tube container and wipe the excess glue from the back of the slide. Then dip the slide underneath the section in the floatation bath and gently tease it onto the slide.
4. Quickly draw the section up from the floatation bath and drain the slide in a vertical position for a few minutes. Allow the slide to dry in the conventional manner; i.e., on a slide warming table overnight or in a 60°C oven for 30 minutes. After drying, the slide will be white where excess glue remains, but this will not interfere with the staining of the tissue section. Finally, it is important to remember to keep the tube containing the glue solution in the refrigerator when not in use.

Elmer's Glue Solution Method

Elmer's Glue

Elmer's glue, general purpose	15.0	ml
Distilled water	100.0	ml

In this case add the suggested quantity of glue solution to the floatation bath and stir well. Use 4 tablespoons in an 800 ml floatation bath and 2 tablespoons in a 400 ml bath. Place sections on the surface of the water in the floatation bath and pick up in the conventional manner. Dry slides in a 60°C oven for 30 minutes or on a slide warmer overnight.

Elmer's Glue - Alcohol Solution Method

Alcohol	100.0	ml
Distilled water	2400.0	ml
Elmer's glue	6.0	ml

Fill water bath container with this solution. Pick up sections on glass slides and dry in a 58-60°C oven for 20 to 30 minutes.

Celloidin Section Coating Method

Paraffin sections which tend to become separated from the slide because of the action of enzymes and alkaline solutions used in the various staining procedures, or which are extremely dried, or which contain bone and keratinized material, are held more firmly to the slide if they are coated with a thin film of 0.5-1% celloidin solution before staining. This is an excellent method for retaining sections on slides because it is a coating rather than an adhesive.

1% Celloidin

Celloidin	1.0	gm
Absolute alcohol	50.0	ml
Ethyl ether	50.0	ml

Combine the ether and alcohol first and then add the celloidin to this solution. Stir until celloidin completely dissolves. Note: This solution is extremely flammable and should only be used far removed from potential fire hazards, such as Bunsen burners, sparks, etc.

1. Decerate slides in xylene and run through several changes of absolute alcohol.
2. Immerse in 1% celloidin solution for 2 minutes.
3. Air-dry sections for 30 minutes or until sections turn white.
4. Immerse sections in 80% alcohol for 2 minutes.
5. Rinse in distilled water and proceed with desired stain.

Note: Sections which are to be stained with the following procedures should not be coated with thin celloidin since the celloidin has a strong affinity for the stains used in these procedures: Aldehyde fuchsin, alcian blue, mucicarmine, Gridley fungus and McManus' periodic acid-Schiff stain.

Poly-L-Lysine Method

Poly-L-Lysine is one of the most popular and most useful tissue section adhesives available. Slides precoated with Poly-L-Lysine are for use with frozen sections, immunochemistry, in situ hybridization, and special stains. Poly-L-Lysine is a polycation which binds to DNA, red blood cell membranes and other negatively charged proteins. It is very useful for the immobilization of membranes in many applications.

Poly-L-Lysine

Poly-L-Lysine* 300,000 mol unit..................................	50.0	mg
0.01 Tris buffer (pH 8.0)	100.0	ml

Store in refrigerator.

*Sigma Cat #P1524

PROCEDURE

1. Use pre-cleaned slides. Alternatively, make sure slides have been cleaned properly before applying Poly-L-Lysine. Slides can be cleaned with potassium dichromate, sulfuric acid or a strong chlorine bleach (full strength Clorox). In both instances, submerge slides in cleaner for 30 minutes. This is followed by washing the slides in running tap water for 30 minutes. Slides are then rinsed well in deionized water and dried in 37°C oven overnight.
2. Place slides in staining rack and dip in Poly-L-Lysine solution for 15 minutes. Drain off excess solution.
3. Dry slides overnight in a 37°C oven. Coated slides will not dry at room temperature.
4. Store coated slides in slide boxes. Open boxes only when necessary to prevent accumulation of dust.
5. Slides can be used to mount tissue sections without further coating.

Edward-Price Butvar Method[160]

Butvar B-98 is used in electron microscopy as a support film. It has a sticky, hydrophilic surface ideal for use as a section adhesive. The resin contains polyvinyl alcohol, hence its hydrophilic qualities. The resin is a nonflammable granular powder, a definite advantage over celloidin.

The sections remain intact and adherent to the slides even after long exposures to heat, acid, alkalies and a saponification test (1% KOH in 70% ethyl alcohol for 30 minutes).

Slides prepared with this adhesive show its advantages over celloidin on several points. When used with hematoxylin and eosin, the periodic Schiff reaction, phosphotungstic acid hematoxylin, GMS, Masson's trichrome or luxol fast blue, slides remained water clear with no residual stain. Butvar B-98 provides a substantial cost savings over celloidin, and the working solution remains stable and useable for several months.

Butvar B-98 Resin*

Chloroform, preheated to 40°C	100.0	ml
Butvar B-98 resin	25.0	mg

Store at room temperature in a brown bottle.

*Butvar B-98 is a polyvinylbutynal resin, distributed by Sigma Chemical Co., St. Louis, MO.

PROCEDURE

1. Mount paraffin sections onto slides in the usual manner from a floatation water bath which does not contain a section adhesive.
2. Dry slides in the usual manner and decerate in xylene.

3. Remove xylene in absolute alcohol.
4. Immerse slides in the Butvar B-98 solution for 10 dips.
5. Allow sections to air-dry for 15 minutes.
6. Rehydrate and perform desired stain.

REMARKS
Most stains can be performed after the use of this adhesive. There is no residue on the slides in most instances. To my knowledge this section adhesive has not been tried on immunochemistry procedures. Chloroform has been declared a carcinogen. Therefore, all precautions should be taken to prevent skin contact or inhalation of fumes.

Gelatin-Chrome Alum Dip Method

This adhesive is very good for mounting sections from tissue which has been fixed and cut frozen on a cryostat or similar instrument and is destined for histochemical and immunofluorescence studies.

1. To 1 liter of distilled water, heated to approximately 60°C, add 3.0 gm gelatin.
2. Mix until completely dissolved, using magnetic stirrer. Gelatin must be completely dissolved before proceeding.
3. Add 0.5 gm chromium sulfate (chrome alum). Solution will appear bluish at this point.
4. Add several thymol crystals. Continue to agitate until alum and thymol are completely dissolved.
5. Filter into a 100 ml beaker. (Keep hot.)
6. Place several slides one at a time into the hot solution. Replenish solution as needed with hot solution. Keep beaker about 3/4 full at all times.
7. Withdraw slides one at a time. Blot edges on a towel. Place in vertical position in a drying rack, and place in dust free area to dry. A hot air oven is the most desirable.
8. When slides are completely dry, place in slide box.
9. Store at room temperature.

REMARKS
If slides appear dirty or oily, pre-cleaning in alcohol is essential to insure uniform coverage by adhesive solution.

Gelatin-Chrome Alum Subbing Method

With this method it is possible to dip an entire rack of microscopic slides into the gelatin-chrome alum. In addition, the slides may be used over an extended period of time.

Gelatin .... 1.0 gm
Distilled water .... 100.0 ml
Dissolve gelatin completely by gently heating above mixture.
Add:
Chromium sulfate (chrome alum) .... 0.1 gm
Mix solution well.

1. Rack clean dry slides in an appropriate slide rack.
2. Dip slides in gelatin-chrome alum solution and allow to remain for 3 minutes.
3. Drain slides through corner of rack and remove from solution.
4. Place slides in a 60°C oven for 15 minutes to dry.
5. Remove slides from rack and transfer to a clean slide box and store in clean area.

REMARKS
Subbed slides can be used for a multitude of purposes, especially frozen sections. This "adhesive" is very good at retaining tissue sections on slides, including bone, keratin, brain, fixed frozen sections and cytology preparations.

Frosted Slide Surface Method

Using slides with a frosted surface is an excellent method for maintaining sections or smears on slides through almost any treatment. This is equally true for cryostat and paraffin sections or cellular smears. The slides were originally developed for use in cytology but have been used for a variety of specimens including paraffin and cryostat sections, washed cell suspensions and cellular exudates. In terms of tissue binding to the slide surface, this method has no parallel.

PROCEDURE
Sections or cellular exudates are floated or placed and smears made directly onto the ground-glass surface of the slide. Drain paraffin and dry at 37°C or at 60°C for 1 to 4 hours. Drain (but do not dry) cryostat sections or smears, place in alcohol for 1 minute, following which they may be stained.

REMARKS
If cryostat sections are to be stained, it is often advantageous to treat them with xylol for 1 minute to remove lipids (which often prevent even staining), followed by hydration through absolute and 95% ethyl alcohol before staining. Depending upon the technique to be employed, cellular exudates and smears are usually treated in a manner similar to that described for cryostat sections.

Aminoalkylsilane (AAS) Adhesive[161]

Glass microscopic slides treated with 3-aminopropyltriethoxysilane (aminoalkylsilane) produce excellent tissue adhesion of paraffin and frozen material. This section adhesive prevents sections from falling off slides and also eliminates staining of the background (a common occurrence with many adhesives). This remarkable coating resists bleaching and enzymatic pre-digestion with protease K, collagenase, papain, pronase, sialidase, diastase, amylase and hyaluronidase. Even preparations heated to 92°C remain adherent. 3-aminopropyltriethoxysilane has been defined by different designations: Aminoalkylsilane (AAS), 3-aminopropyltriethoxysilane (AES), and APES solutions.

2% Aminoalkylsilane

3-aminopropyltriethoxysilane (aminoalkylsilane)*	2.0	ml
99.5% Acetone	100.0	ml

Because aminoalkylsilane is moisture sensitive, store in refrigerator (4°C) in a tightly closed container of anhydrous calcium sulfate.

Make up a 2% working solution just prior to use. This solution can be used 3 times. Slides should be cleaned well in acid or strong alkali (full strength chlorine bleach). Wash slides well in deionized or distilled water after exposure to cleaning solution.

*Sigma Cat #A-3648

PROCEDURE
Use clean slides and make certain slides are not touched with the fingers, since there is a strong possibility of the deposition of DNA, which may react with DNA probe procedures being performed.

1. Place slides in rack.
2. Place slides in 2% aminoalkylsilane for 2 minutes.
3. Dip slides 2 times in each of 2 changes of deionized water.
4. Dry slides overnight in a clean environment. Slides may be stored in a clean slide box until ready for use.

RESULTS
It is suspected that aminoalkylsilane reacts with silica glass to produce amino alkyl groups on the glass surface that bond either ionically or covalently with aldehyde or ketone groups in the tissue.

REMARKS
Slides which have been treated with aminoalkylsilane can be stored for as long as 6 months before use. We have taken a 50 slide staining rack and used the same solution for a total of 400 slides. The staining dishes handle approximately 800 ml of solution. AAS slides provide ideal, almost irreversible, adhesion of tissue for in situ hybridization of DNA probes.

Artifacts in Tissue Sections

Prefixation and Fixative Artifacts

Nuclear Distortion (Bubbling) Artifact

Tissue fixation in the paranitrophenol-cupric fixing solution produces the phenomenon of nuclear distortion in that the chromatin condenses at the end of the nuclei opposite to the exposed surface of the tissue. The mass of chromatin is so compact that in preparations stained with hematoxylin and eosin it appears deep blue, almost black, while the rest of the nucleus remains transparent due to the lack of chromatin in that region. This artifact may be similar to that produced by phosphate buffered formalin, Fig. 219 and 220.

Cupric nitrate alone is responsible for the peculiar appearance of the nuclei. Omission of the nitric acid increases the chromatin condensation effect. All salts of copper give the same result. It is interesting to note that aqueous solutions of all heavy metals in which the cation alone contains the metal, produce the same distorting effect on the nuclei. On the other hand, aqueous solutions in which the anion alone contains the metal, give normal fixation of nuclei. The same holds true for solutions of salts in which one metal forms part of the anion while another metal forms part of the decalcification.

Formalin-Acetic Acid Solution

(A solution to bubbling artifacts.)

Formalin, 10% aqueous, unbuffered	900.0	ml
Glacial acetic acid	10.0	ml

This solution is not used for initial fixation but rather to correct an artifact and is the last fixative step on the tissue processor. The time of exposure is approximately 1 hour. Paraffin embedded tissue specimens exhibiting artifactual bubbling can be run back through the various stages of deparaffinization to water (Fig. 219 and 220), treated with formalin acetic acid solution to eliminate this bubbling problem, and reprocessed.

Saprophytic Organism

A saprophyte is a microorganism which thrives on decaying matter. It is observed after autolysis commences. This organism is noted for its invasive characteristics (Fig. 2).

Gas Production Due to Microorganisms

Spaces observed in the tissue section are gas bubbles which were formed by saprophytic microorganisms (Fig. 289).

Nuclear Digestion Due to Improper Fixation

The liver section was stained with hematoxylin and eosin. The tissue was allowed to adhere to the bottom of the fixative container. The result is no exposure to the fixative and an autolytic reaction ensues. Enzymatic destruction of the nuclei is pronounced (Fig. 3).

Talcum Powder Artifact

A very common artifact is starch from surgical dusting powder. The classic appearance is an hexagonal crystal which is gray after staining with H&E. Talcum powder is PAS positive and stains green with the Giemsa stains. It possesses excellent polarization properties. When polarization is performed, a Maltese cross is observed. Powdered rubber gloves used for grossing tissue are the most frequent cause for introduction of the talcum powder. It may also be noted in GU specimens from infants (Fig. 210).

Ferrous Deposition from Fixative Container

Metal lids used on fixative containers are subject to corrosion. Ferrous depositions are noted in tissue when the diaphragm is loose or missing from the lid. Erroneous staining results are obtained in iron staining procedures. The deviations in appearance are most significant in specimen such as liver, spleen and bone marrow (Fig. 221, 222 and 223).

Effects of Unbuffered Formalin on Tissue

It must be recognized that staining reactions and tissue elements are altered and/or obliterated by the deleterious and insidious effects of unbuffered formalin fixative. Formalin is oxidized to formic acid, and the following ruinous effects are produced: (a) The so-called formalin pigment frequently will react during the staining procedure to mask or, in some cases, simulate microorganisms, pigments, or other elements suggesting disease. Formalin pigment has a notorious reputation for reducing silver in procedures for staining fungi (Grocott's methenamine-silver), melanin (Fontana-Masson), spirochetes (Warthin-Starry), and reticulum procedures. (b) Unbuffered formalin gradually destroys nuclear basophilia if tissue is exposed longer than three weeks. (c) Unbuffered formalin dissolves ante-mortem copper, iron, and calcium during the process of fixation. (d) It alters the staining of some cell granules. (e) It also will inhibit the aldehyde-fuchsin reaction for demonstrating Paget's cells. These are only a few of many examples but should be convincing evidence that unbuffered formalin should be used judiciously.

The use of 10% unbuffered formalin results in the formation of the so-called formalin pigment in tissue sections, Fig. 224, 225, 226, 229 and 230. The formation of this pigment is supposedly produced by the interaction of formic acid and hemaglobin. I do not believe this is entirely true since one can find formalin pigment in areas far removed from a bed of red blood cells (one would expect to find hemaglobin in or near red blood cells). Formic acid is produced in unbuffered formalin fixative solution by Cannizzaro's reaction (a reaction which aldehydes may undergo when brought in contact with tissue specimens). In this reaction one molecule of aldehyde (formalin) is reduced to the corresponding alcohol (methyl alcohol) and the other molecule is simultaneously oxidized to the corresponding acid (formic acid).It is true, however, that unbuffered formalin produces sharper staining of hematoxylin and eosin stained slides. This is also true of some special stains, such as the trichromes. There are some instances in which calcium carbonate (marble chips) has been used to neutralize formalin. In my opinion, this should never be done, since calcium carbonate produces artifactual crystals in tissue sections, Fig. 224, 225, 226, 229 and 230.

Neutral Formalin Artifacts

Neutralized formalin (not to be confused with buffered neutral formalin) creates a false sense of security. When tissue is placed in a neutralized fixative, the pH of the fixative is lowered rapidly from the formation of formic acid, and some of the irreversible alterations mentioned above occur. Unbuffered or neutralized formalins should not be used. Carnoy, Carnoy-Lebrun, formalin, formol-calcium, and osmic acid, which are stable in pure solution, all show rapid, substantial pH changes during tissue fixation. The speed at which pH changes occur may be correlated with the time necessary for adequate fixation. Carnoy's solution, which is known to fix rapidly, shows no further pH changes after 3 hours, and more than half of the total change takes place within 15 to 20 minutes. All fixatives show the most rapid pH changes during the first hours of fixation and a large group exhibit only slight changes after 12 hours.

Calcium Carbonate Crystals Artifacts

Calcium carbonate is frequently used as a neutralizer in some formalin fixative solutions. Calcium carbonate crystallization occurs within the specimen, resulting in cellular displacement, Fig. 11a and b. The appearance is misleading because it resembles an in vivo reaction. In vitro crystallization of calcium carbonate is dissimilar to in vivo. The reason for the difference is not clear, but it is possibly due to the resistance encountered by the tissue. Crystals polarize well. Calcium carbonate crystals are only slightly soluble in pure water. However, they dissolve readily in water containing dissolved carbon dioxide because of their conversion to the more soluble hydrogen carbonate. The presence of calcium carbonate crystals can be determined by exposing the slide to water containing a few small pieces of dry ice for 30 minutes. This is an artifact which warrants attention since it can simulate gouty tophi, and other normal or artifactual products in tissue.

Pseudocalcification (Ante-Mortem) Artifacts

Pseudocalcification artifacts are observed when calcium acetate is employed as a neutralizing agent in 10% formalin. The calcium deposits are difficult to distinguish from those produced during in vivo calcification. Hence, pseudocalcification can be mistaken for true calcification, Fig. 227 and 228. Staining properties with the various calcium stains are identical to true calcification. Pseudocalcium can be distinguished from true calcification by using mucosaccharide staining techniques (alcian blue or Mowry's colloidal iron). True calcification produces mucosaccharides outside the calcium peripheral margins while pseudocalcium does not.

Zenker's Fixative Artifacts

Diminished Hematoxylin Staining

Tissues overexposed to Zenker's fixative will not exhibit the usual hematoxylin staining characteristics (Fig. 8d). Overexposure produces an acidic staining appearance. Chromatin material cannot be differentiated and, in general, nuclei are not distinct. Restoration of staining properties can be accomplished by placing the slides in a 20% aqueous solution of sodium bicarbonate overnight, followed by routine H&E staining.

Mercuric Chloride Crystals

An ingredient of Zenker's fixative is mercuric chloride. In the classic form mercuric chloride crystals are spheroid, their periphery irregular. The crystals also have been observed in needle forms and many times are polymorphic, Fig. 7a-h. Mercuric chloride crystals present on tissue sections are insoluble in water and alcohol. Exposure to 20% aqueous ammonium hydroxide for 30 seconds results in color change. Chromatically, the crystals will be black, Fig. 7h. This is a fairly specific stain for these crystals and should prove helpful when confusion arises in the identification of mercuric chloride crystals.

Erythrocyte Crystallization

Zenker's fixative will often crystallize erythrocytes. Microscopically the RBCs will exhibit a transparent or a slightly translucent appearance. These erythrocytes will possess excellent polarization properties (Fig. 7e).

Mucosaccharide Staining Alterations

Zenker's fixative has an adverse effect on hyaluronic acid and cannot be demonstrated by staining (Fig. 235). It is not known if other mucosubstances are adversely affected. Buffered neutral formalin does not produce any gross alterations affecting the staining properties of hyaluronic acid.

It is generally believed that glacial acetic acid should not be added to Zenker's fixative solution until just before use. This is no longer true. Glacial acetic acid can be added to the mixture and stored for 3-5 months without deleterious effects. The omission of glacial acetic acid until just before use was necessary in 1894 when Zenker proposed this fixative mixture. At that time, glacial acetic acid was recovered from pyroligneous acid. Pyroligneous acid was obtained from the destructive distillation of wood, which was composed of nearly equal parts of methyl alcohol, acetone, and glacial acetic acid. Each of the major constituents commercially prepared in the past was considered contaminated with the other two. The contaminants contributed to a short shelf life. Today, modern chemistry ensures that the product meets purity requirements.

Urea Polymer Artifact

Urea polymers are formed in tissue which has been placed in a Zenker-type fixative containing urea and subsequently immersed in formalin for storage. Urea is added to a fixative to prevent hemolysis. Microscopically, the polymer is a round, clear-to-gray crystalline structure in hematoxylin and eosin preparations. It can be differentiated from the Zenker crystal since it does not react (turn black) when exposured to 20% ammonium water. The polymer polarizes easily, whereas Zenker crystals do not polarize after the ammonia water treatment. In addition, Zenker crystals appear to be slightly above the tissue surface, while the urea crystalline polymer is embedded in the tissue and appears as a flat structure (Fig. 7h).

Bouin's Fixative Artifacts

Picric Acid Effects on Tissue

It is absolutely imperative that there be no residual picric acid in Bouin's fixed tissue. Residual picric acid will alter the tissue and affect the staining properties. Eventually, almost all cellular structures will be distorted, unrecognizable, or obliterated. Remove picric acid by placing specimen in 60% ethyl alcohol until all yellow color disappears (Fig. 236, 237 and 238).

Mucopolysaccharide Alteration

Buffered neutral formalin does not produce any gross alterations affecting the staining properties of hyaluronic acid. Bouin's fixative has an adverse effect on hyaluronic acid and it cannot be demonstrated by staining (Fig. 236, 237 and 238).

Picric Acid Crystals

Bouin's fixative contains formalin, glacial acetic acid, and trinitrophenol (picric acid). Acids will break down hemoglobin and combine with the chromatin portions to produce acidic crystals similar to those produced by unbuffered formalin (Fig. 236, 237 and 238).

Processing Artifacts

Parched Earth Effect

Sections having inadequate paraffin impregnation resemble dry or parched earth (Fig. 239). This takes place because the section expands beyond normal proportions on the warm floatation bath. The section is picked up on a slide in this expanded condition. When it is exposed to the clearing xylene during H&E staining, the section contracts in an attempt to revert to its original size. The section then breaks away from the slide at the tissue section adhesive's weak points, which in turn, produces the parched earth effect.

Vacuoles in Rat Central Nervous System (CNS) Tissues

Holding rat CNS tissues in 70% alcohol for 20 hours produces artifactual vacuoles in the white matter (Fig. 245). Unfortunately, holding tissues in automatic tissue processors for this length of time for weekend runs, is routine in some laboratories. Tissues held in 10% neutral buffered formalin for the same length of time do not exhibit artifactual vacuoles. Thus, this troublesome artifact can be eliminated easily by using 10% neutral buffered formalin as the holding solution for automatic tissue processors.

Microtomy Artifacts

Three important indicators of a dull knife (lack of amorphous facet surface) are:

A. Paraffin block will not ribbon.

B. Sections stick to knife during the specimen block upstroke.

C. Microscopic compression of tissue section entities (cells) occurs.

See the chapter on Microtomy for information on amorphous vs. crystalline microtome bevel surfaces and their impact on microtomy

Venetian Blind Effect

The venetian blind effect, the thick and thin horizontal lines often seen on sections (Fig. 246 and 249), is due to poor embedding practices which ultimately affect the cutting of sections. Specimens of hard, fibrous tissue, like uterus, should never be embedded side by side, as it results in the knife having to travel through what amounts to a broader tissue specimen, thereby causing the knife to jump as it travels through the tissue, resulting in the thick and thin areas. This effect can also be produced if the knife is loose or if screws on the microtome have not been tightened adequately. Also, lack of oiling of all sliding microtome surfaces frequently causes venetian blind effect. Venetian blind effect is not to be confused with microvibration which is the result of failure to reintroduce moisture into tissue structures.

Moth Eaten Effect

This problem is characterized by the rough-edged holes seen in sections (Fig. 258). This artifact is the result of overexposure of tissue to dehydration, clearing, and paraffin impregnation, resulting in overhardening of the tissue specimen. Tissue is gouged out as the knife travels through the dried, paraffin embedded specimen. The problem can be eliminated by the application of a water soaked piece of cotton to the faced surface of the paraffin block. (See page 62 for information on reintroduction of moisture into tissue structures.)

H & E Staining Artifacts

Aluminum or Potassium Crystal Artifacts

Small crystalline structures often seen throughout sections may be crystals of aluminum potassium sulfate, a mordant in many hematoxylin solutions. The crystals indicate that the hematoxylin solution was not properly mixed; aluminum potassium sulfate retained its original crystalline form. Mixing hematoxylin solutions requires vigorous agitation. A magnetic stirrer is recommended for this purpose. Alum crystals can be visualized easily under polarized microscopy. When observed, the hematoxylin solution should be discarded. Several days after alum crystals are observed, a black pigment deposits throughout the section. Hematein (the coloring agent of hematoxylin) is precipitated due to the absence of the alum mordant. Alum and hematein are indicative of hematoxylin breakdown, resulting in poor section staining (Fig. 268).

Uneven Staining

Uneven staining in tissue sections is often due to a failure to agitate slides several times in the staining solution before allowing them to remain for the required exposure time.

Inadequate Dehydration

Inadequate dehydration of microscopic slides is characterized by a murky granular appearance of the tissue section. Microscopically, bubbles can be observed when the microscope condensor is lowered (Fig. 269).

Poor Coverslipping Practice

Slides should not be allowed to dry when applying a coverslip (Fig. 276 and 277). If that happens an artifact with two distinct features will be produced: (1) Stippling which resembles pigment; (2) Nuclei are seen as distinct, black, glossy structures (Fig. 277). Identification is accomplished by lowering the substage of the microscope.

Excessive Coverglass Mounting Medium

Excessive mounting medium or one which has thickened, will result in a foggy looking preparation when observed microscopically under high power (x450 or more) light microscopy (Fig. 274).

Artifactual Birefringent Artifacts

Intranuclear birefringent inclusions often occur in paraffin sections. This phenomenon occurs most often when paraffin impregnated sections are dehydrated in acetone instead of alcohol. This artifact is significant if one is performing polarized microscopic examinations, in which case some confusion with other birefringent material may occur.

Fig. 1
This section of liver stained with Best Carmine demonstrates glycogen displacement due to fixation. Note that glycogen has been forced against the wall of the cell by the action of the advancing fixative. (Chapter 1)

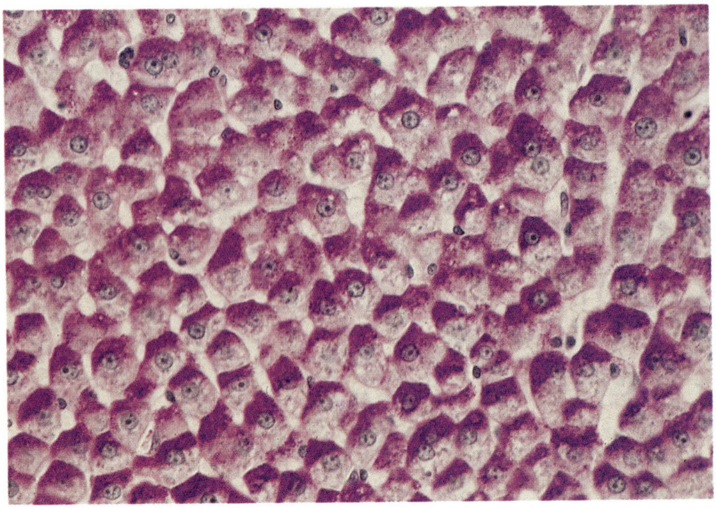

Fig. 2
Saprophytic bacteria can be seen throughout this section of a completely autolyzed specimen. (Chapter 1)

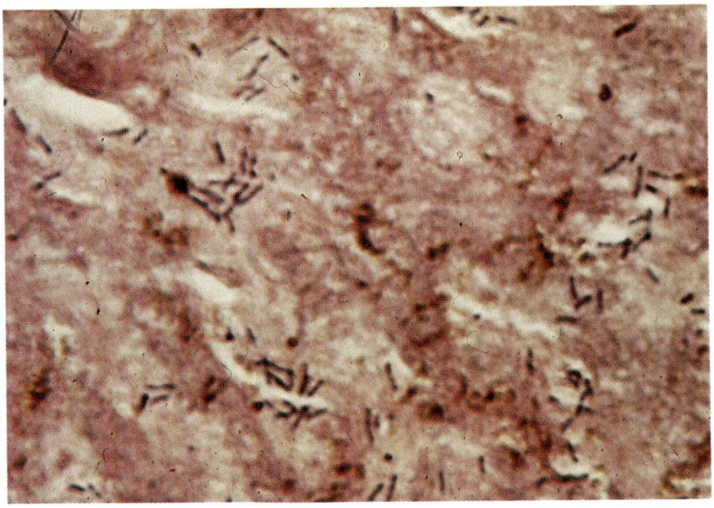

Fig. 3
This section of liver illustrates autolytic changes, primarily due to the specimen being placed in a container prior to the addition of the fixative. In this instance the portion seen toward the bottom left illustrates very extensive digestion of nuclear material. Tissue specimens should always be placed in the fixative and not in the container before the fixative. (Chapter 1)

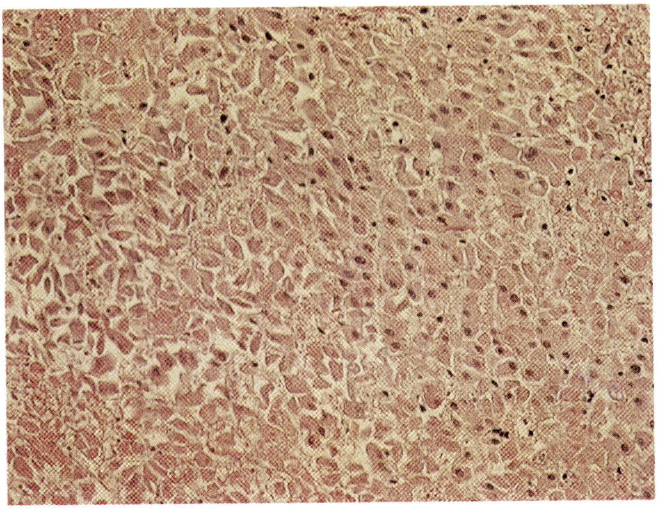

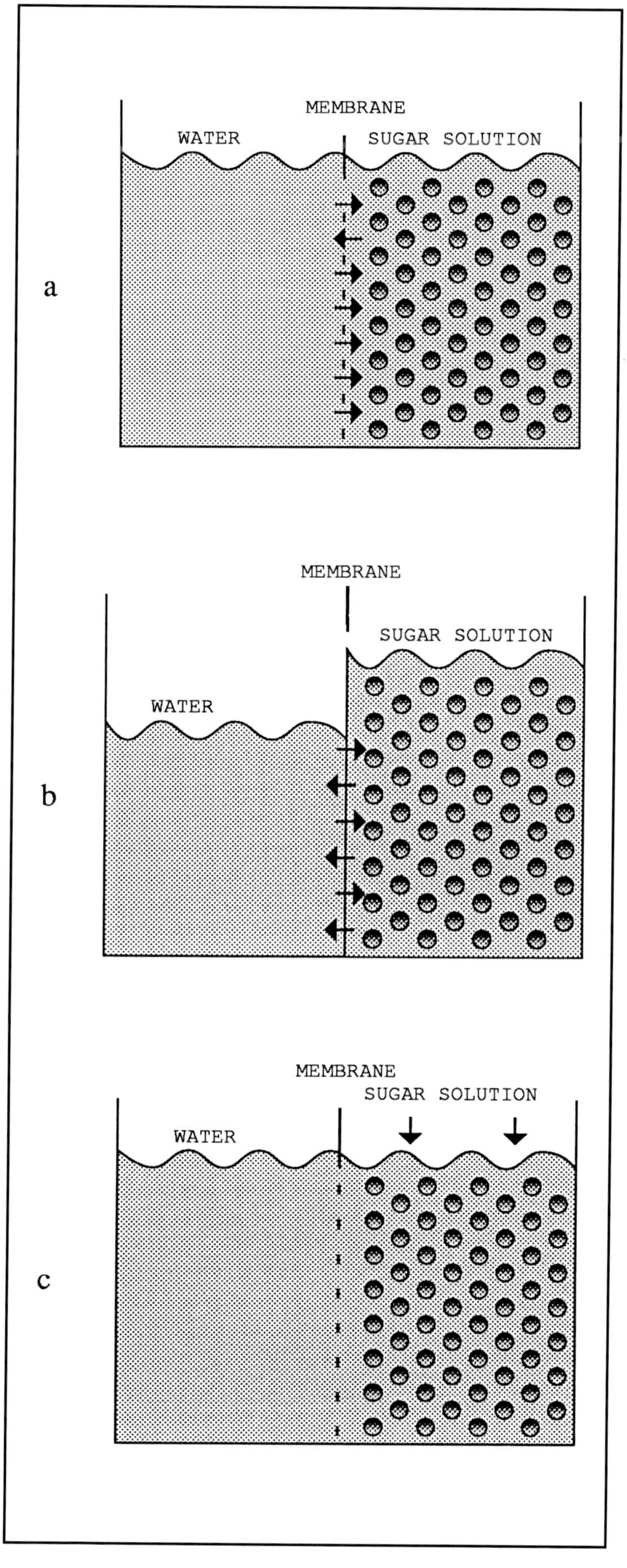

Fig. 4a
This schematic demonstrates the action of osmosis. The differentially permeable membrane will allow water molecules to pass into the sugar solution (circles), but it will not allow sugar molecules to pass into the water solution. In other words, water moves from the side of lower sugar concentration to the side of higher concentration. (Chapter 1)

Fig. 4b
Demonstrates that as water moves from the side of lower sugar concentration to the side of higher concentration, the side with the sugar solution ends up with more volume. (Chapter 1)

Fig. 4c
The osmotic pressure (arrows) equals the pressure that must be supplied to the sugar solution to prevent the flow of water (osmosis) into the sugar solution. (Chapter 1)

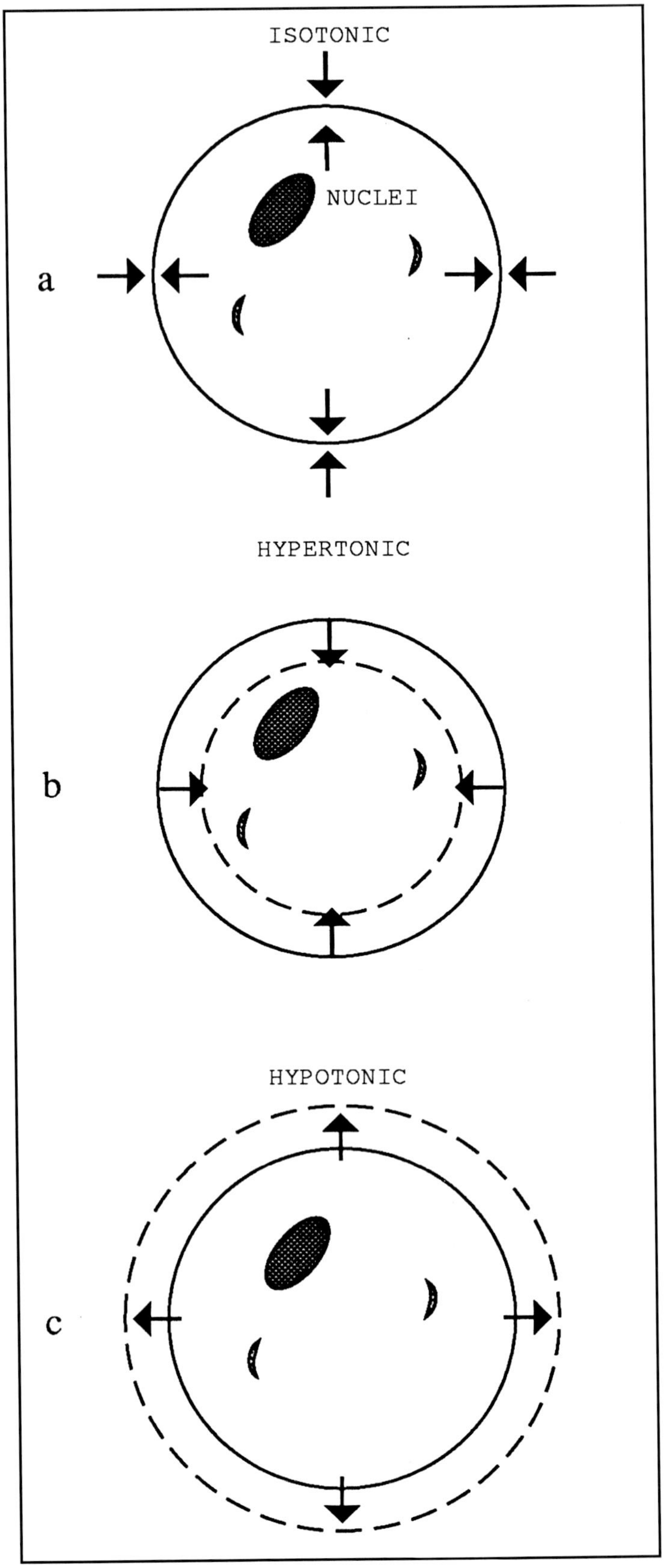

Fig. 5a
These schematics demonstrate that passage of water in and out of living cells is an important biological process. If the concentration of solute is equal on both sides of the cell membrane, the osmotic pressures will be equal and the solutions are then called isotonic. (a) Illustrates an isotonic solution. Since the osmotic pressure is equal on both sides, the cell will not shrink or swell.

Fig. 5b
This is an example of a hypertonic solution (concentration of solute is higher in the solution than in the cell). The osmotic pressure is unequal; therefore, water from the cell moves into the solution causing the cell to shrink (arrows).

Fig. 5c
This is an example of a hypotonic solution, (concentration of solute is higher in the cell). The unequal osmotic pressure causes the water to move into the cell, causing it to swell. (Chapter 1)

Fig. 6a, b, c and d
Demonstrate the results of testing for the presence of formalin (aldehyde groups) in solutions with the use of Coleman's leucofuchsin solution. (a) Notice vivid reaction of leucofuchsin with formalin. (b) Leucofuchsin. (c) Leucofuchsin combined with distilled water. (d) Leucofuchsin combined with tap water. Note the absence of positive reactions in Fig. 6b, c and d, indicating absence of formalin (aldehyde groups). (Chapter 1)

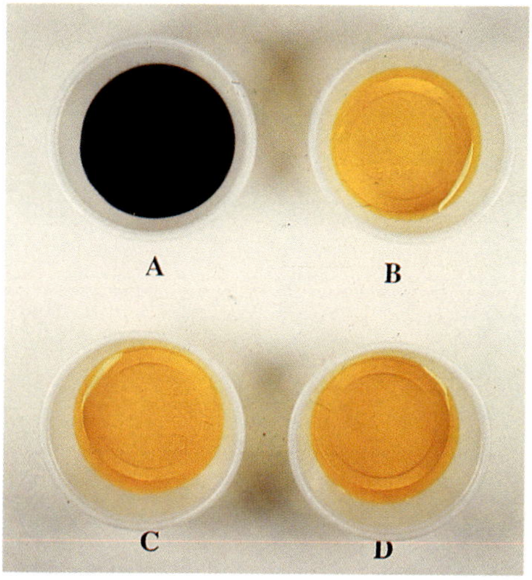

Fig. 7a
Demonstrate mercuric chloride crystals which have been deposited in tissue during fixation with a fixative containing mercuric chloride (e.g., Zenker's and B-5). (Chapter 1)

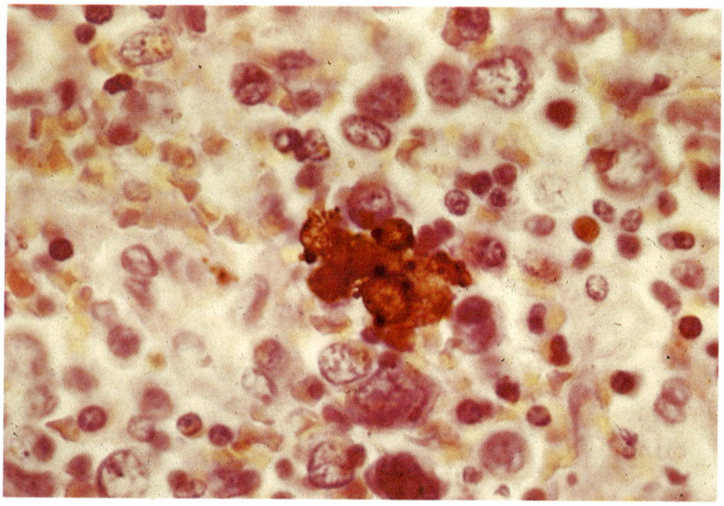

Fig. 7b
Needle-shaped clear mercuric chloride crystals can be seen on the left; section on right shows no mercuric chloride crystals since they have been removed. (Chapter 1)

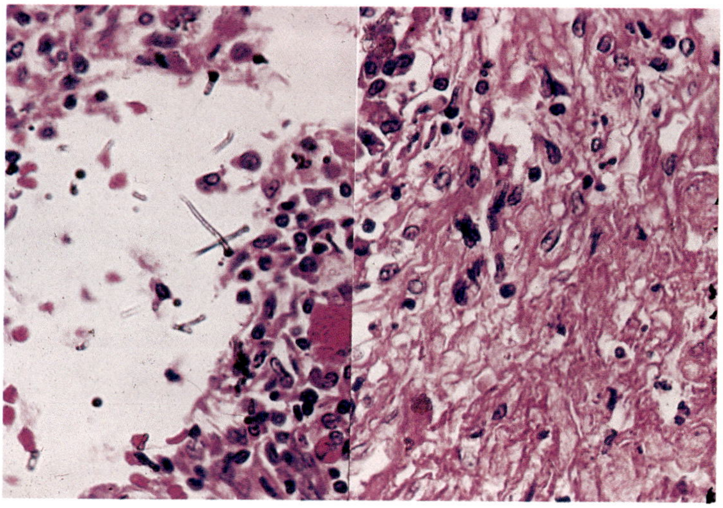

Fig. 7c
Shamrock shaped mercuric chloride crystal. (Chapter 1)

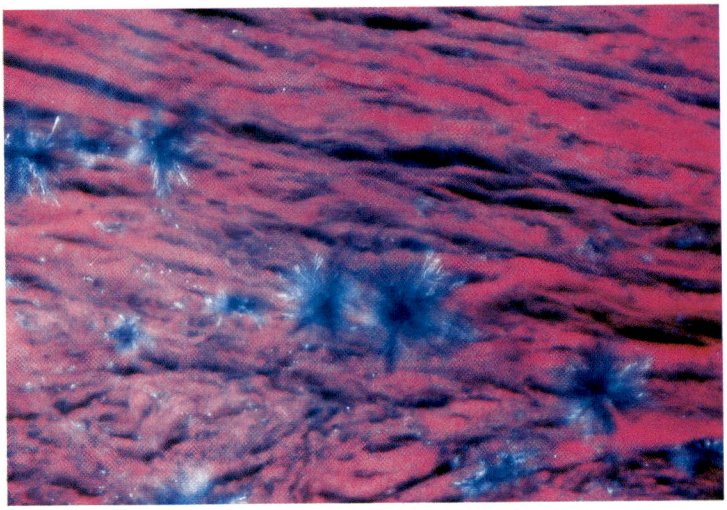

Fig. 7d
Fontana stained section demonstrating the strong silver reducing properties of mercuric chloride. In addition, this section demonstrates the more characteristic shape and pattern of mercuric chloride crystals. (Chapter 1)

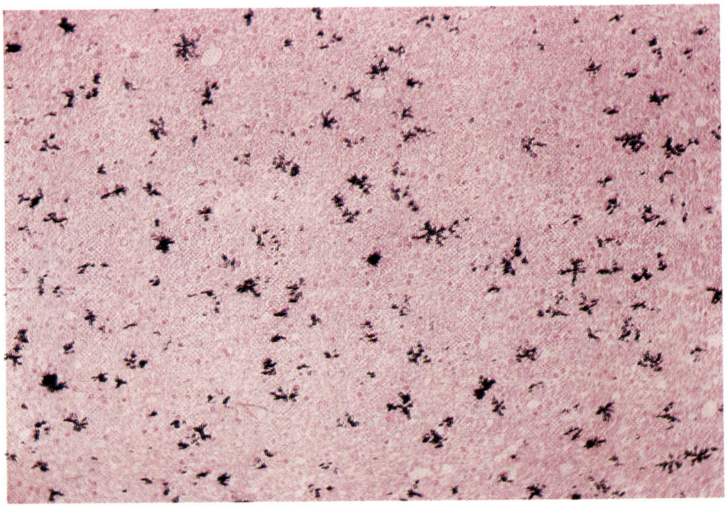

Fig. 7e
This section of liver under partial polarization demonstrates crystallized red blood cells. Crystallization of red blood cells is frequent with Zenker or Zenker-type fixatives. (Chapter 1)

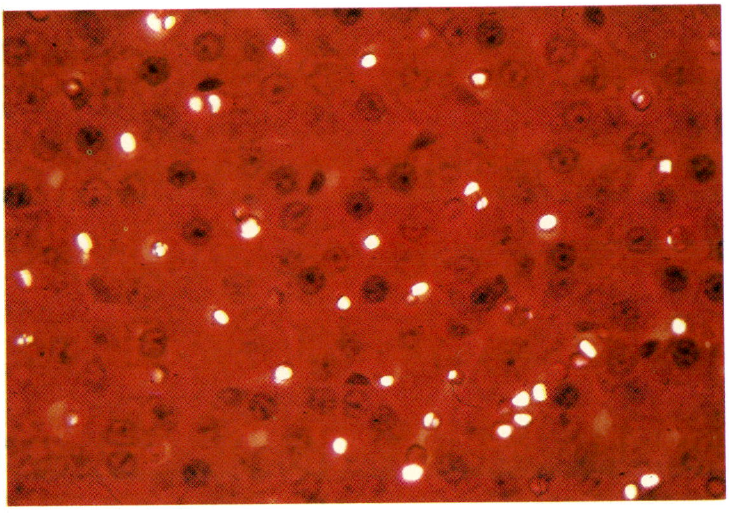

Fig. 7f
Section of liver demonstrating 3 black structures resembling nuclei that could be confused with histoplasmosis. These black structures are mercuric chloride deposition in a Zenker fixed specimen. (Chapter 1)

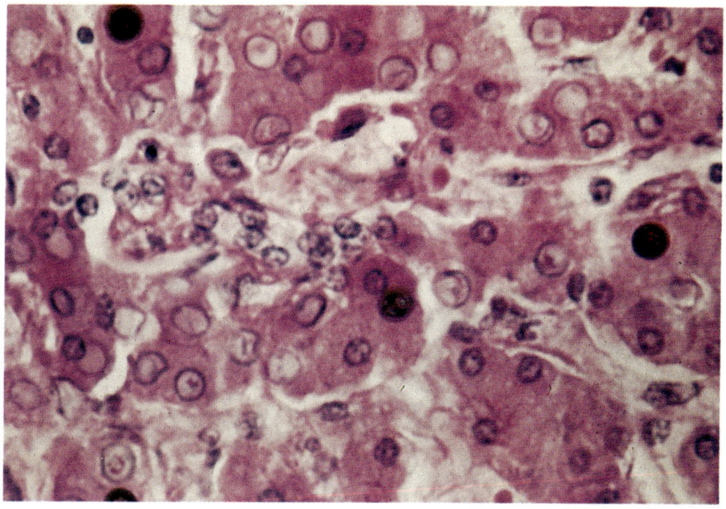

Fig. 7g
Section of bone fixed in Zenker's demonstrating another bizarre pattern of mercuric chloride deposition. This deposit simulates calcium salts. (Chapter 1)

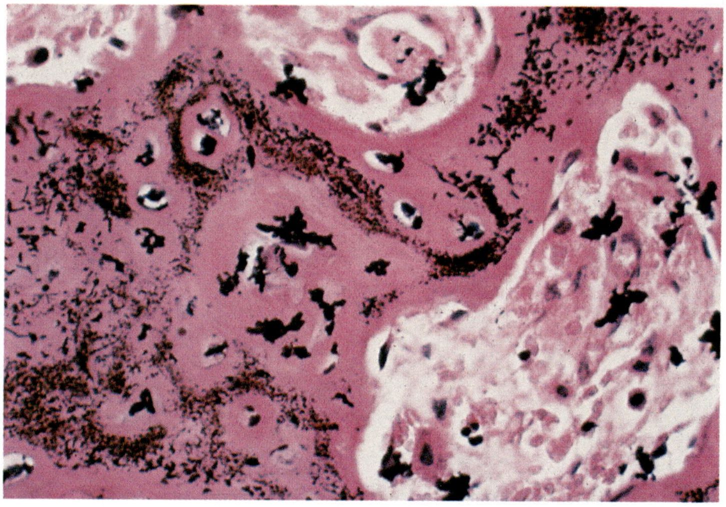

Fig. 7h
This composite of a section of heart demonstrates 2 artifactual depositions. Black structures are mercuric chloride from a Zenker fixed specimen. These sections have been stained with Luna's Method for Mercuric Chloride. This section also contains urea crystalline artifacts. Note the small, round, slightly colored structures seen on the left. The right side shows the urea crystals under partial polarization. (Chapter 1)

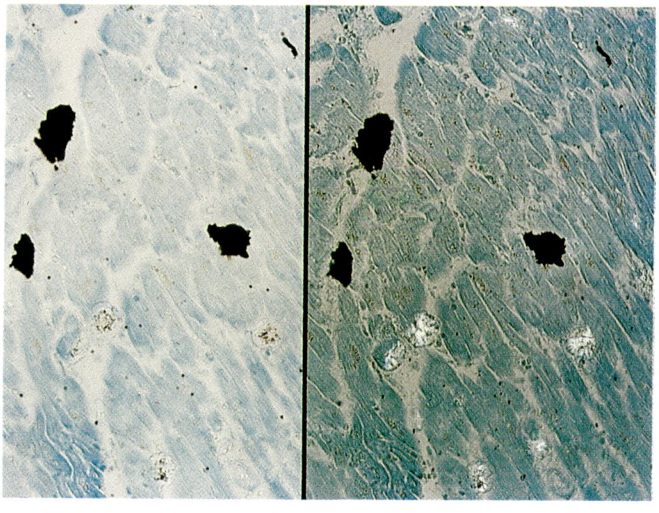

Fig. 8a, b, c and d
These photographs illustrate the three distinct changes which nuclei undergo during the process of fixation in Zenker's solution. (a) Demonstrates microscopically nuclear chromatin in a condensed and shrunken condition. (b) Demonstrates the effects of subsequent swelling of the nuclei after exposure of tissue to the fixative for several hours. It is at this stage that tissue should be removed from the fixative solution. (c) Demonstrates monochromatic staining as the results of tissue overexposure to Zenker's fixative solution. Notice lack of tinctorial differentiation between nuclei and cytoplasm. (d) This composite shows the deleterious effect of overexposure to Zenker's which can be seen on the right portion of the photograph. The left portion has been treated with sodium carbonate for 12 hours to restore the basophilic staining properties of the specimen. See page 79 for method for restoring staining properties. (Chapter 1)

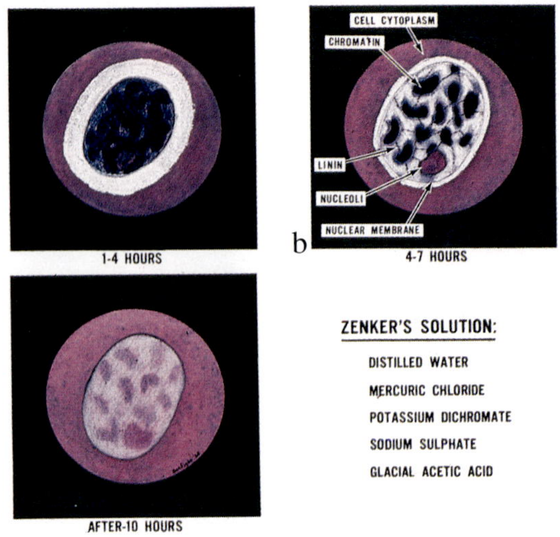

a b c

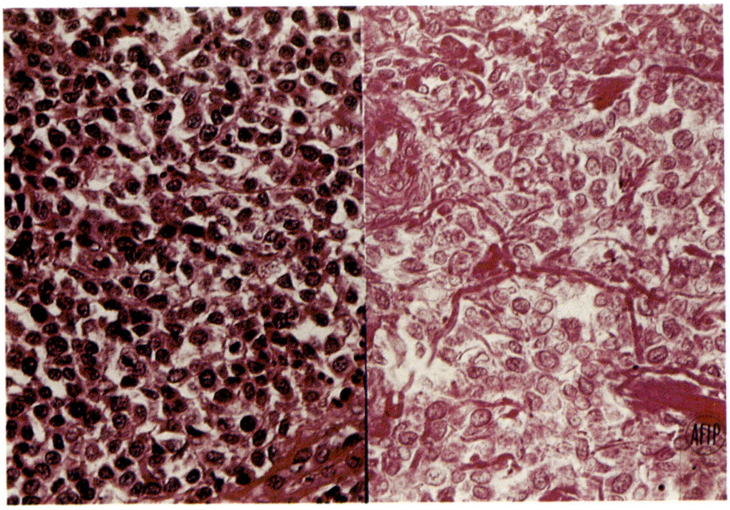

d

Fig. 9a and b
Bouin's fixed section of lymph node demonstrating the effects of picric acid on tissue after long term tissue storage. (a) This section was processed, sectioned and stained within 24 hours after removal from the host. (b) This section from the same case was cut and stained 6 months later. Notice the deleterious effects of picric acid on tissue morphology. This is even more dramatic if one compares the morphology of the Reed-Sternberg cell. (Chapter 1)

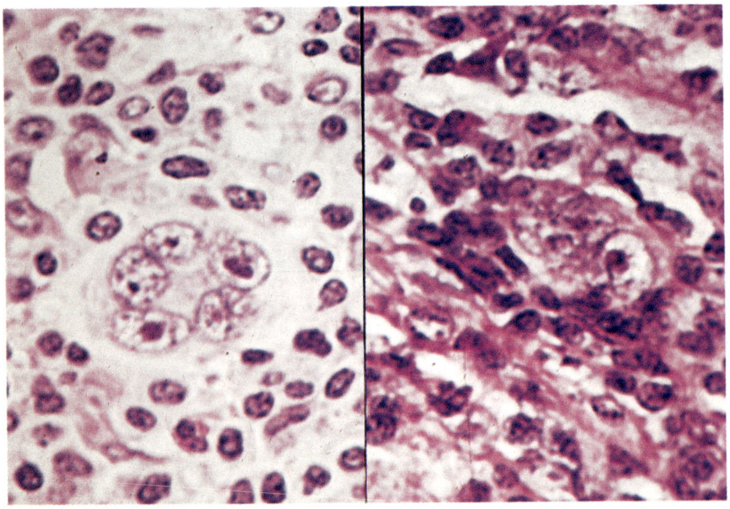

a b

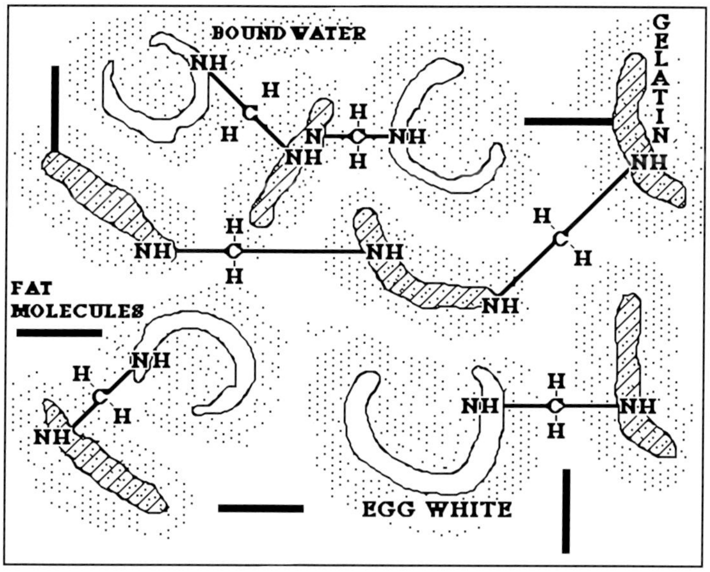

a

CROSS LINKING FIXATION

(SIMPLE ADDITION FIXATION)
TISSUE + FIXATIVE + TISSUE

TISSUE AMINO		FORMALIN FIXATIVE		TISSUE AMINO		
-N-H	+	$H_2C{=}O$	+	H-N-	⟶	TISSUE -N-CH_2-N- TISSUE + H_2O (METHYLENE BRIDGE, WATER)

TISSUE SULPHYDRYL		MERCURY FIXATIVE		TISSUE SULFHYDRYL		
-S-H	+	$HgCl_2$	+	H-S-	⟶	TISSUE -S-Hg-S- TISSUE + 2HC1 (MERCURY BRIDGE, HYDROCHLORIC ACID)

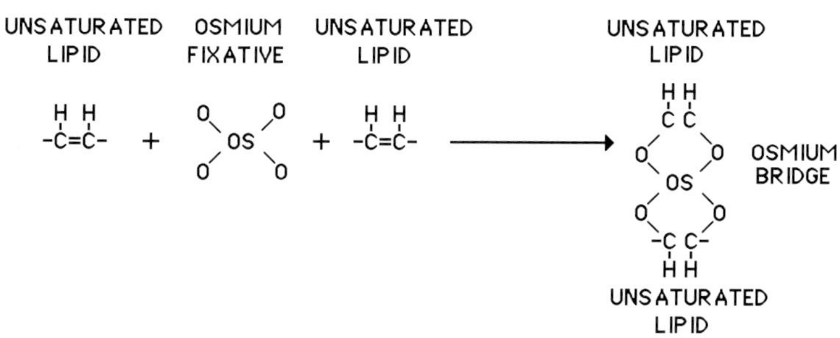

Fig. 10a
This schematic represents a model of protoplasm which is being used to demonstrate good (cross linking) fixation. The white horseshoe shaped structures (egg white) represents globular protein molecules. The linear gray structures (gelatin) represent fibrous protein molecules. The fine lines represent intracellular and intermixture bonds (cross linking). (a) This schematic represents chemical cross linking of formalin with tissue, resulting in good fixation. (Chapter 1)

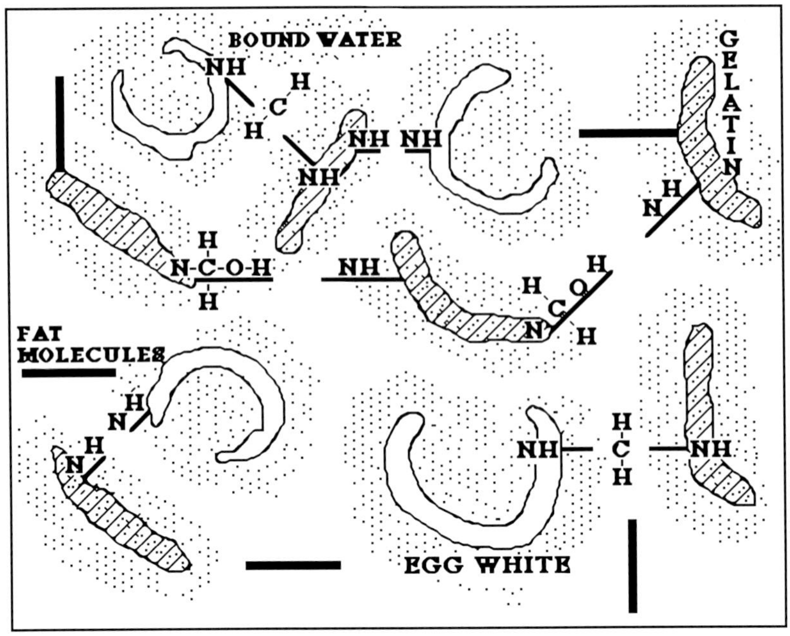

b

INADEQUATE CROSS LINKING FIXATION

(INCOMPLETE ADDITION FIXATION)

TISSUE + FIXATIVE

TISSUE AMINO		FORMALIN FIXATIVE		FORMALIN ADDITION PRODUCT
-N-H	+	H C=O H	⟶	TISSUE -N-C-O-H (H above and below C)

TISSUE SULPHYDRYL		MERCURY FIXATIVE		MERCURY ADDITION PRODUCT		HYDROCHLORIC ACID
-S-H	+	$HgCl_2$	⟶	TISSUE -S-HgCl	+	HCl

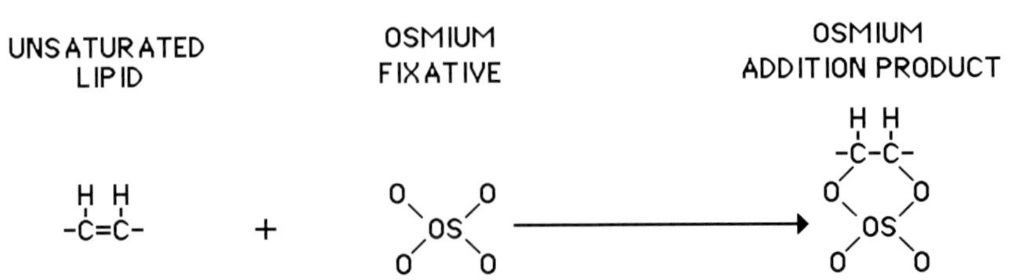

Fig. 10b

Poor or inadequate cross linking is evident in this schematic. This would result in poor fixation of tissue proteins. Note, in the native state these protein molecules would be closer to each other forming a compact structure. See page 19 for more information on this model of protoplasm. (Chapter 1)

Fig. 11a
These sections demonstrate artifactual calcium carbonate, (calcium carbonate is sometimes used to neutralize formalin fixative solutions). (a) H&E stained section of liver demonstrating large amounts of calcium carbonate. (Chapter 1)

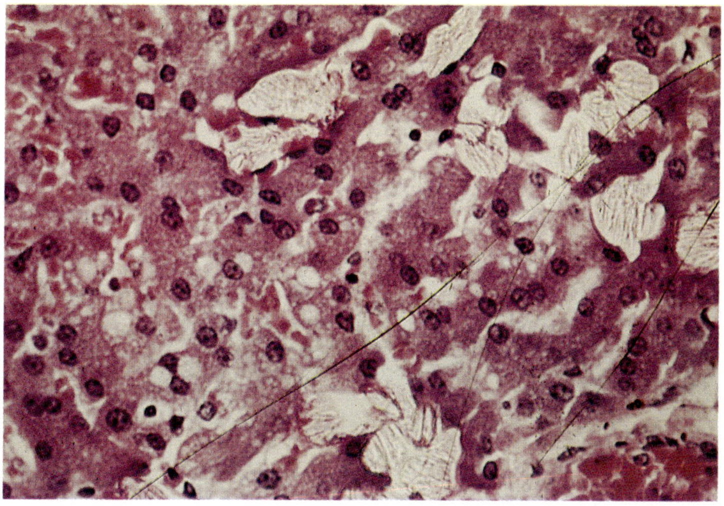

Fig. 11b
Section of brain showing that calcium carbonate polarizes quite well. (Chapter 1)

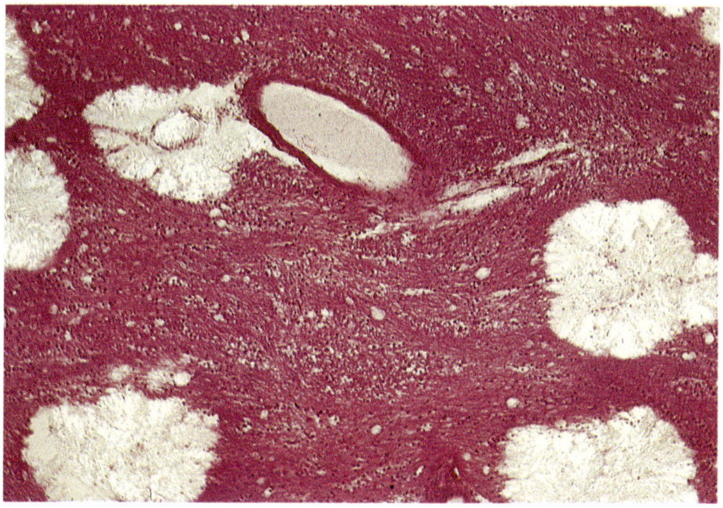

DEHYDRANT	BENZENE	XYLENE	CHLOROFORM
Methyl alcohol	6.5-7	3-3.5	17.5-18
Ethyl alcohol	12-14	7-8	13-14
Isopropyl alcohol	9.5-10	10	7-7.5
Acetone, technical	1.75-3	1.5-2	2-2.2
Dioxane, pure	1.5-2	1-1.5	1.5-2

Fig. 12
This water turbidity chart illustrates the percentage of added water required to render turbid equal mixtures of dehydrating and clearing agents. (Chapter 2)

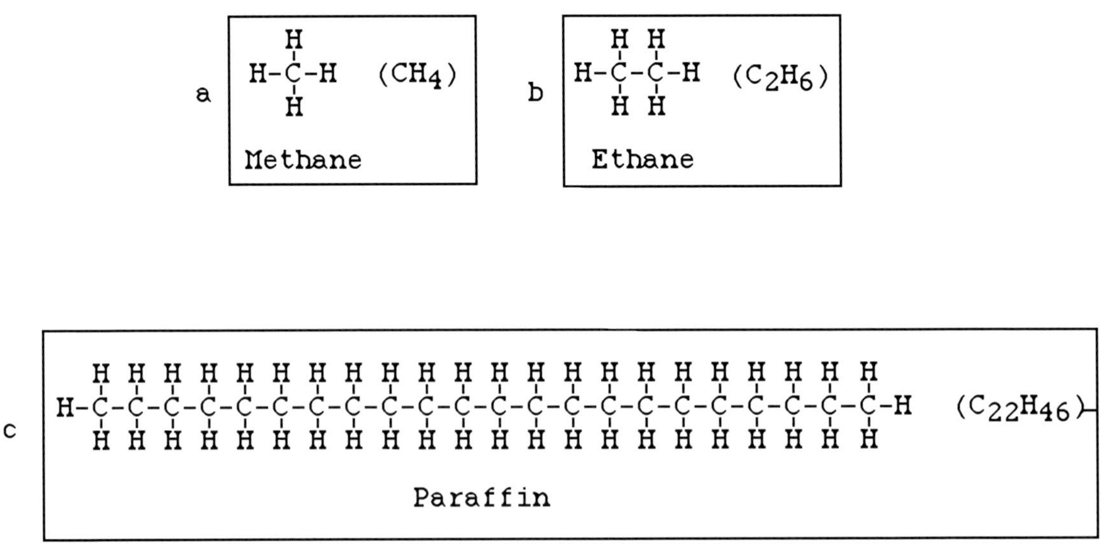

Fig. 13a, b and c
These formulas demonstrate the simplest hydrocarbon molecules, (a) methane and (b) an ethane molecule. (c) Paraffin wax long chain molecule compared with the simplest hydrocarbon molecules of methane and ethane. (Chapter 2)

Fig. 14a
This schematic demonstrates the effects of vacuum and pressure on specimens undergoing paraffin wax impregnation. (a) Air (arrows) being pulled from tissue interstices by the action of vacuum.

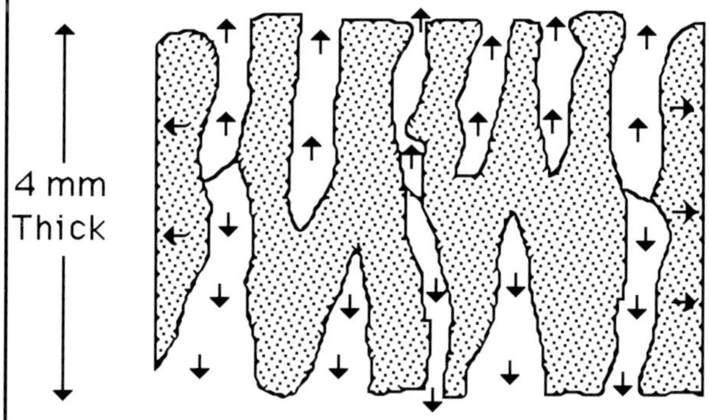

Fig. 14b
Long chain paraffin wax hydrocarbon molecules (small dashes) entering and filling voids (interstices) left by the effects of vacuum.

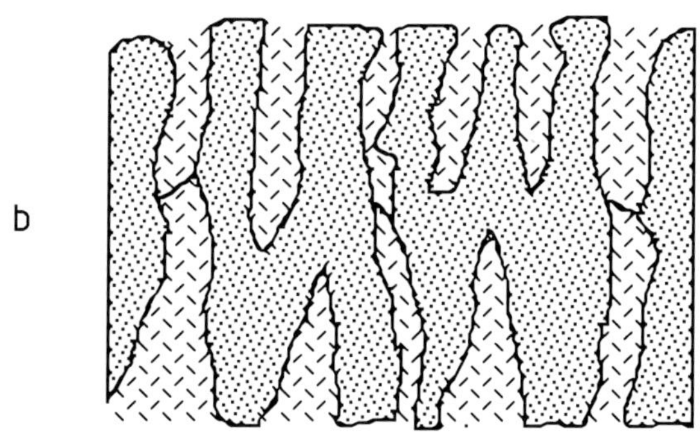

Fig. 14c
Compacted paraffin wax hydrocarbon molecule (dark areas) due to vacuum followed by pressure. This illustrates a specimen which has been impregnated well. (Chapter 2)

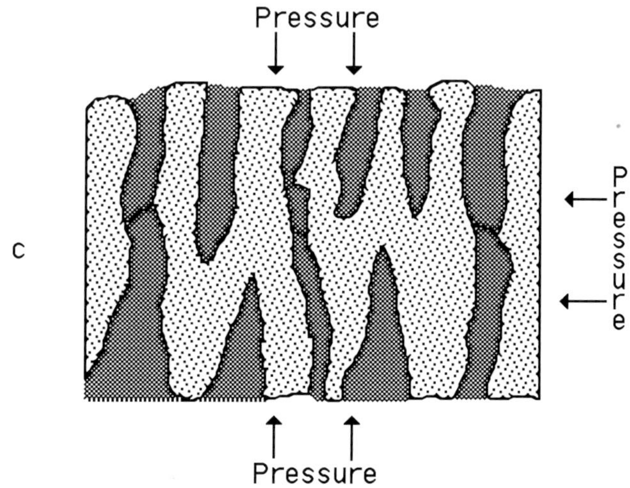

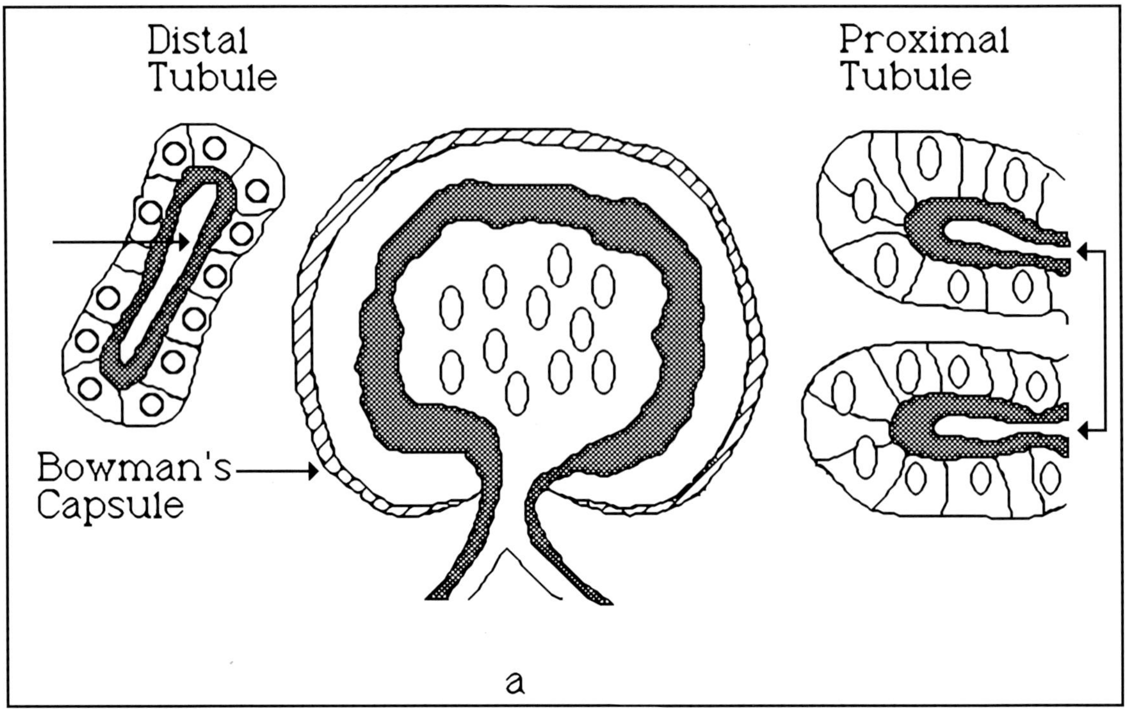

Fig. 15a and b
This schematic of a kidney glomerulus and adjacent area demonstrates (a) poor and (b) good paraffin impregnation. The gray area represents paraffin in tissue spaces. Notice that in (a) some areas have not been filled with paraffin. Schematic (b) demonstrates good paraffin impregnation; all spaces (gray areas) have been filled with paraffin. (Chapter 2)

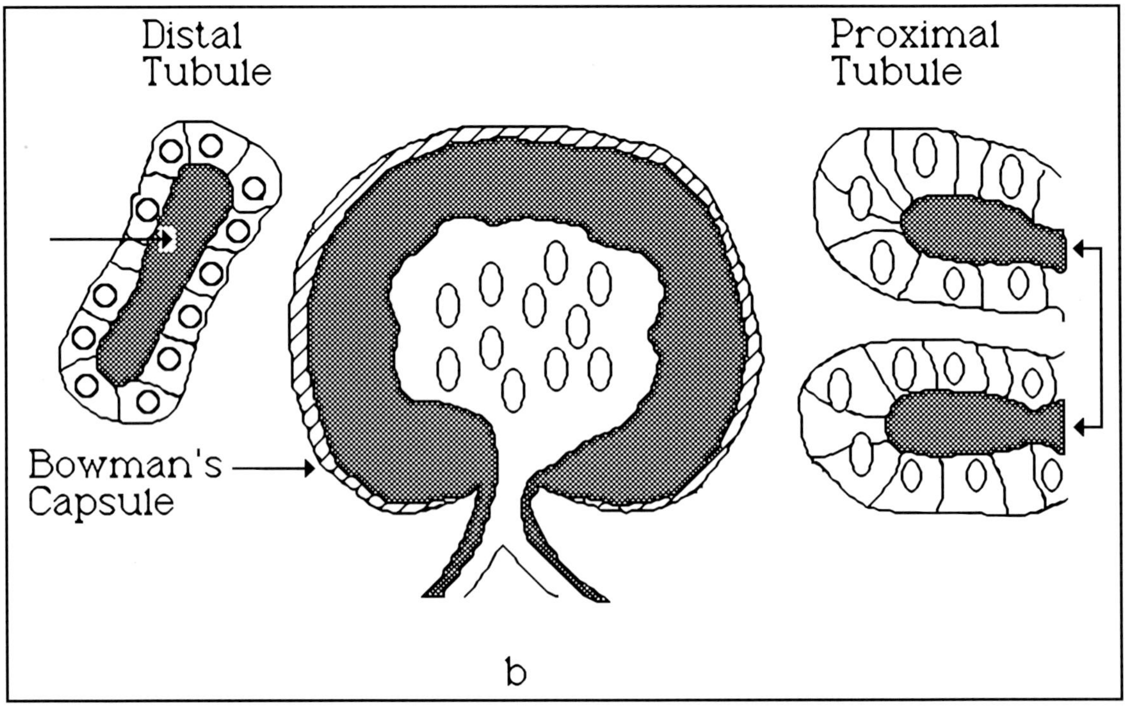

Fig. 16a
These figures illustrate how unstained, polarized sections of kidney can be used to evaluate the quality of paraffin impregnation. (a) Illustrates poor paraffin impregnation. Poor impregnation in this instance can be due to two factors: 1) insufficient time in the paraffin or 2) paraffin which has a shrinking factor beyond 12% as described on page 32. (Chapter 2)

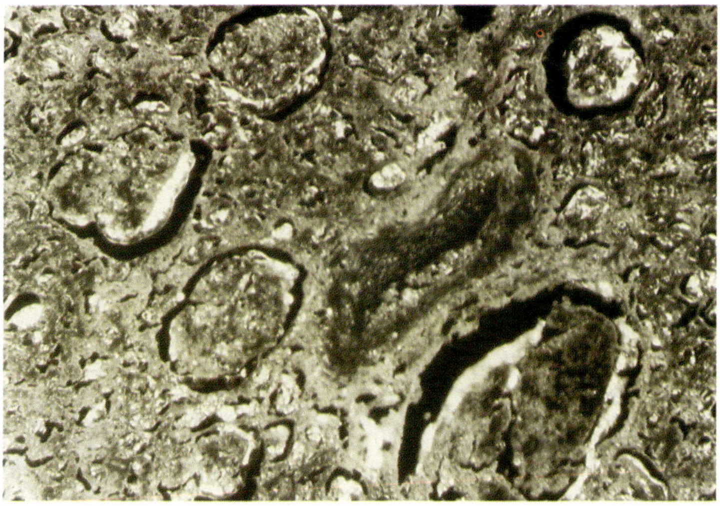

Fig. 16b
This section of a kidney which is under polarized microscopy shows a well impregnated specimen. Note that the spaces surrounding the glomeruli have been filled with paraffin which is illustrated by the complete white halo surrounding the glomeruli compared to Fig. 16a which is underprocessed. (Chapter 2)

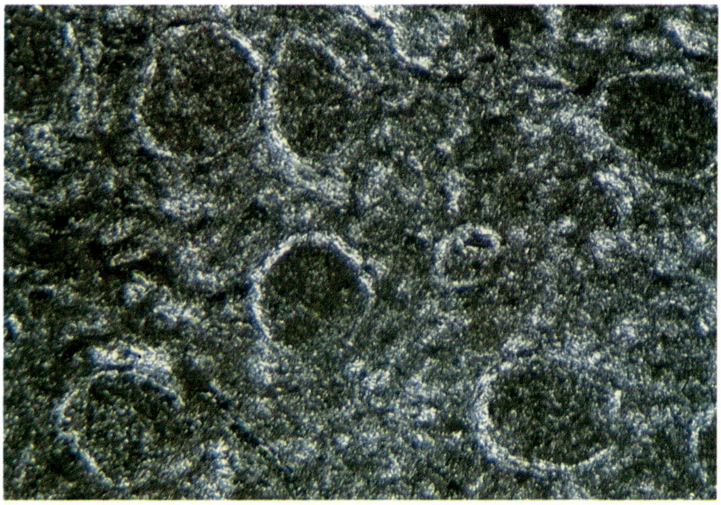

Fig. 17a - t
These schematics represent the right and wrong ways of blocking various tissue specimens. Tissue should not be placed too close to the edge of the paraffin block (a), but centered (b) so that a margin of paraffin completely surrounds it. Tissue placed too close to the edge of the paraffin block may not ribbon well. The long axis of a dome shaped tissue specimen should be embedded down (e). Specimens embedded at an angle (c) or vertical position (d) will produce wrinkling at certain pressure points (arrows) during microtomy. Narrow tubular pieces of tissue (h) should be straightened as much as possible without damaging the specimen. Twisted specimens, (f) and (g), will compress and produce wrinkles at pressure points (arrows) during microtomy. Cross sections of tubular specimens, such as appendix, blood vessels, intestine and trachea, may be blocked in two ways, (j) and (k). The wrong way is illustrated in (i) which shows the specimens embedded in a haphazard manner. This type of embedding causes some slide manipulation problems for the pathologist. Multiple small elongated specimens should be embedded horizontally and slightly separated (m) and (n). These specimens should not be embedded as illustrated in (l) since wrinkling will occur at pressure points (arrows) during microtomy. Embedding of long specimens with a bulge should be as illustrated in (p) and (r). This will prevent wrinkling during microtomy at pressure points (arrows) seen in (o) and (q). Also some compression of the specimen will be evident after microtomy. Hard specimens, such as bone, should be embedded in a slanting position (t). The correct degree of slant can be determined by lining up the specimen between the left bottom corner and the right upper corner of the embedding mold. This is especially important if the specimen contains both compact and spongy bone (t). The embedding of hard tissue in a slanting position will prevent, to some extent, the production of thick and thin sections of the hard tissue during microtomy. It will also reduce the separation of spongy from compact bone.

Remarks - For multiple tissue embedding: When more than one piece of tissue is to be included in one block, tissues should be placed carefully but as rapidly as possible in the embedding cassette so all the pieces will be at approximately the same level in the paraffin block. This is essential if the entire surface of each piece is to appear in the sections. The thicker specimens should be blocked first and the thinner ones last, in order that the thin pieces will not be completely cut away in trimming the paraffin block. (Chapter 2)

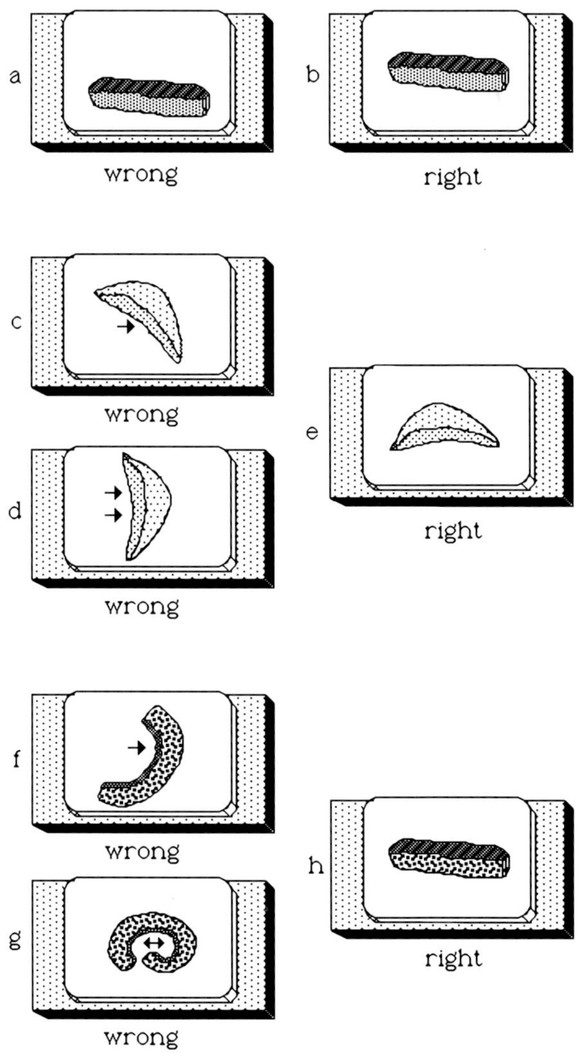

Fig. 17a - t continued

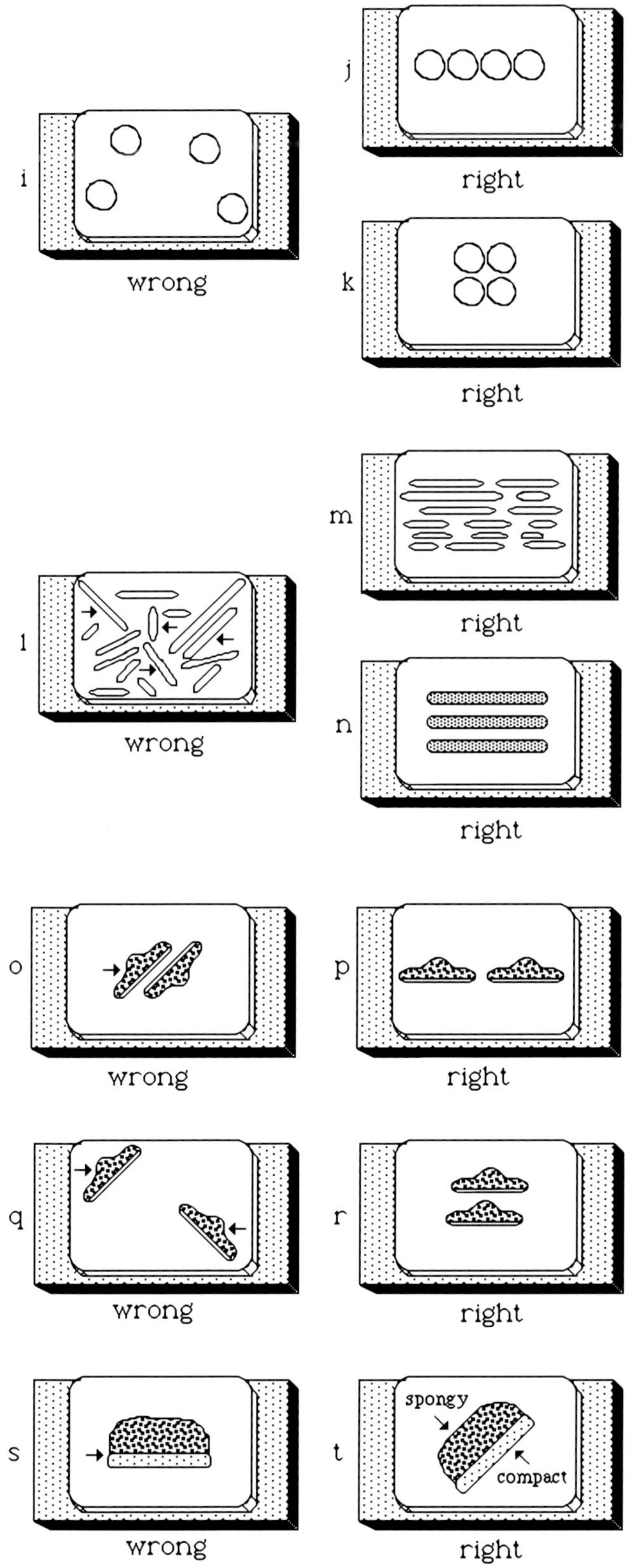

Fig. 18
This chart indicates how to determine the quality of tissue processing by the use of moisture. The suggested corrective actions are based on the utilization of proper processing solutions and grossing of specimens at suggested thickness; namely 4-5 micrometers thick. Moisture exposure must be performed with the use of a good quality piece of cotton. The cotton is soaked in warm water (floatation bath is excellent). The water soaked cotton is than applied against the surface of the specimens. The quality of processing is revealed by the number of specimen which turn white after exposure to moisture for the indicated time periods. (Chapter 2)

MOISTURE EXPOSURE	QUALITY OF IMPREGNATION	PROCESSING REMEDIAL STEPS
5 seconds	Extremely poor (all specimens turn white)	Requires drastic processing schedule changes in every step. (Change solutions. Increase time in all steps by 30 minutes each).
10 seconds	Very poor (90% of the specimens turn white).	Requires drastic processing changes in most steps. (Change solutions. Increase time in all steps by 30 minutes each).
20 seconds	Poor (50% of the specimens turn white).	Requires moderate processing changes in most steps. (Change solutions. Increase time in all steps by 15 minutes).
45 seconds	Good (20% of the specimens turn white).	Requires minor processing changes. (Usually an increase of 15 minutes in the last dehydrant, clearant and paraffin).
2 minutes	Excellent (5% of the specimens turn white).	Requires no changes.

These schematics demonstrate commonly encountered problems in cutting and their causes.

Fig. 19a
Ribbon is crooked or uneven - Paraffin upper and lower edges not parallel with knife edge. Block not square or rectangular. (Chapter 3)

Fig. 19b
Compressed wrinkled or jammed sections - Incomplete dehydration clearing and/or impregnations. Knife facet insufficiently polished. Block too warm. Low paraffin melting point. Excessive knife tilt. Loose microtome set screw. Paraffin formed on back of knife facet. (Chapter 3)

Fig. 19c
Section crumbles (saw dust) effect - Inadequate introduction of water into tissue prior to microtomy. Tissue excessively dry. Extremely over processed tissue. Over exposure to a coagulant fixative. (Chapter 3)

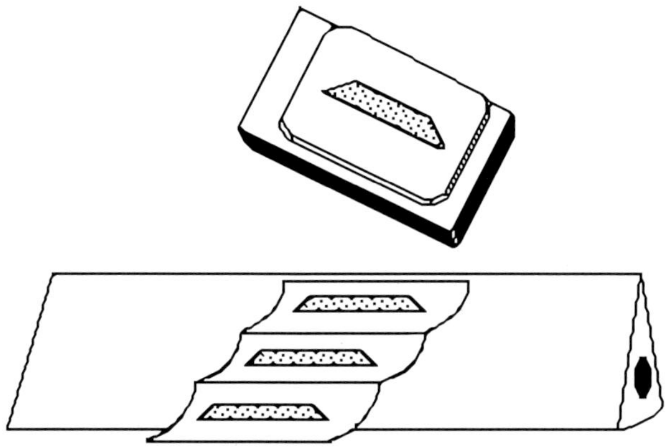

a

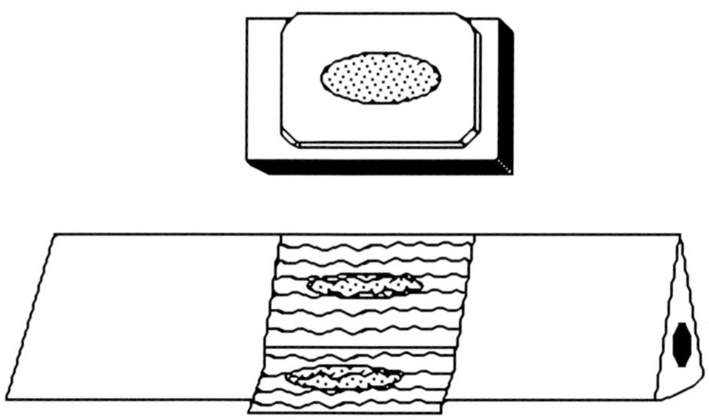

b

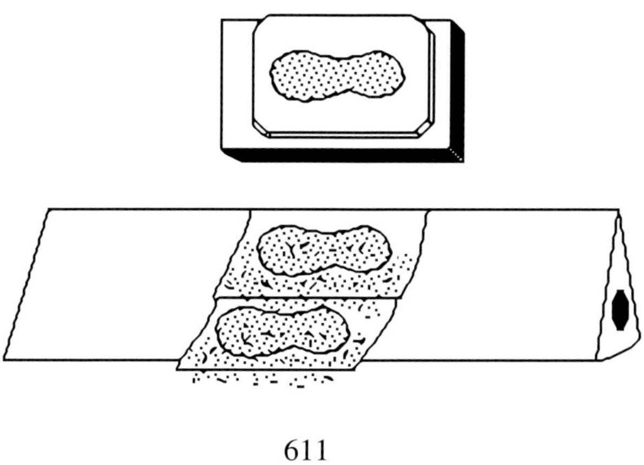

c

Fig. 19d
Sections split lengthwise or scratched - Hard particle in block, such as bone or crystal. Lint fiber or other foreign substance laying across the knife facet edge. Nicks in the knife. Paraffin or embedding material contain hard impurities (grit). (Chapter 3)

Fig. 19e
Section lifts on upstroke - Accumulation of paraffin on the knife facet. Upper and lower block edges not parallel with facet edges. Too vertical a knife tilt. Dirty knife. Insufficient cooling of block surface before section cutting. (Chapter 3)

Fig. 19f
Sections curl and do not adhere to one another - Block not cool enough. Too much bevel angle. Knife facet not polished enough. Dull knife. Block edges are not parallel with edge of knife. (Chapter 3)

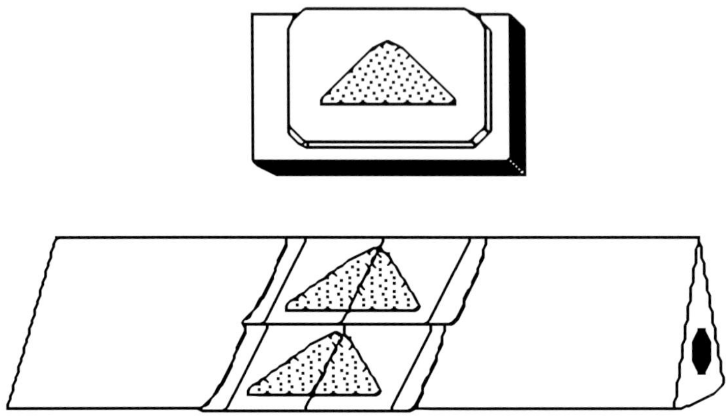
d

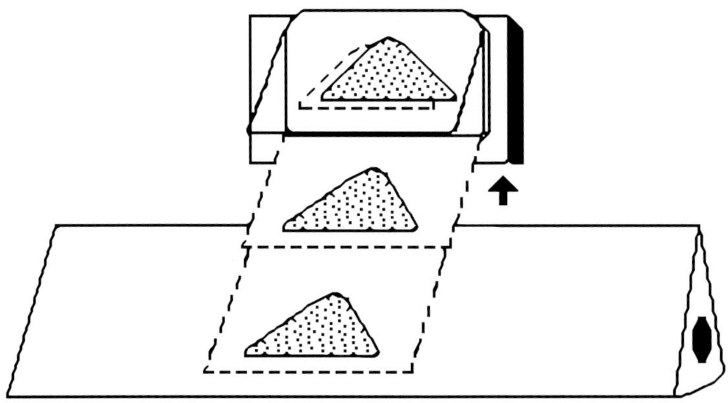
e

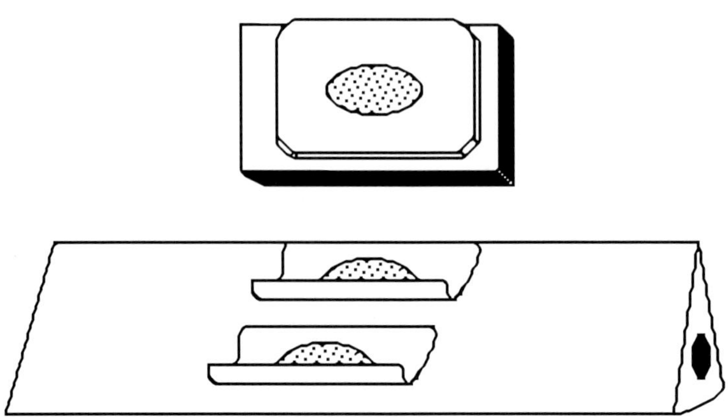
f

Fig. 19g
Tissue specimen soft and mushy - Incomplete processing most often associated with insufficient time in the clearing and/or impregnation steps. (Chapter 3)

Fig. 19h
Specimens cut thick and thin - Microtome needs repairing. Loose screw on blocks or knife holder. Loose knife. Specimen too hard and/or lacks sufficient moisture prior to microtomy. Insufficient knife tilt. Lack of microtome lubrication. Microtome wheel turned to fast or in a jerky motion. Block holding arm is advanced too far away from the microtome. (Chapter 3)

Fig. 19i
Section clings to facet - Knife facet contains a film of residual paraffin. Low humidity in the room causing static electricity. Dull knife. Insufficient facet polishing. Residual paraffin collected on back of microtome facet. (Chapter 3)

Fig. 19j
Tissue specimen turns chalky white with water - Incomplete dehydration clearing and/or impregnation. See chapter 2 for more information on this subject. (Chapter 3)

g

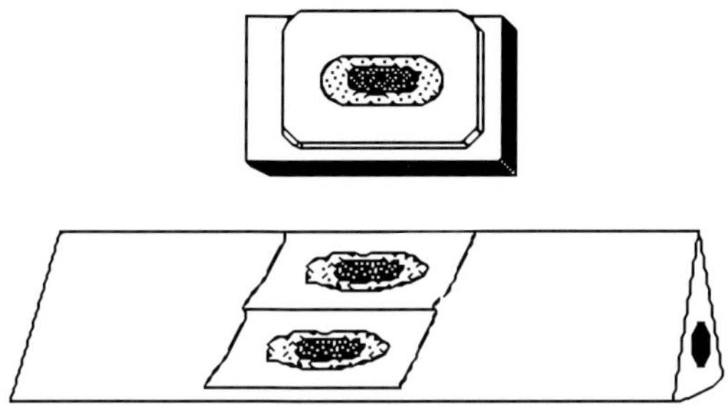

h

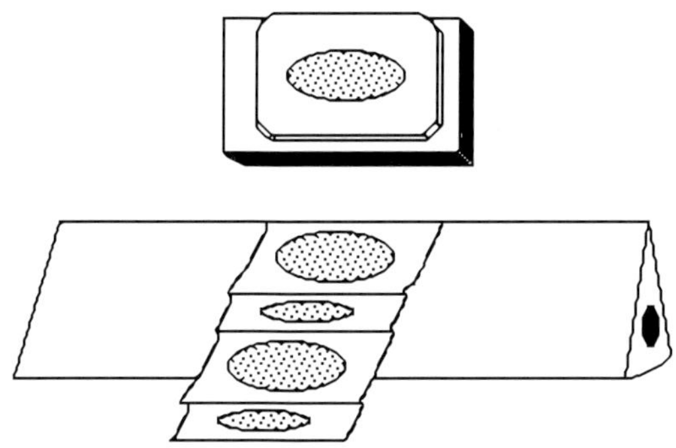

i

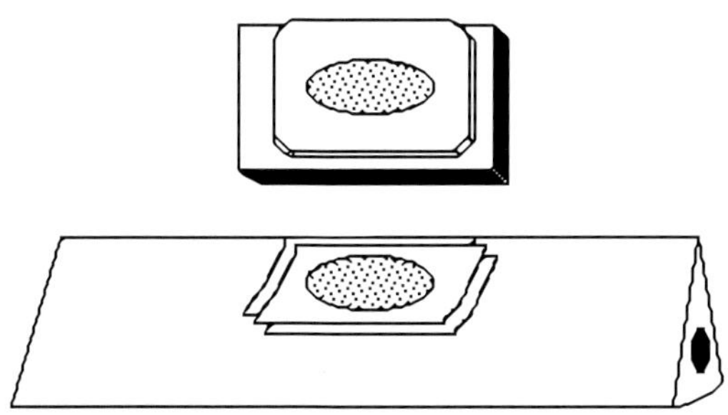

j

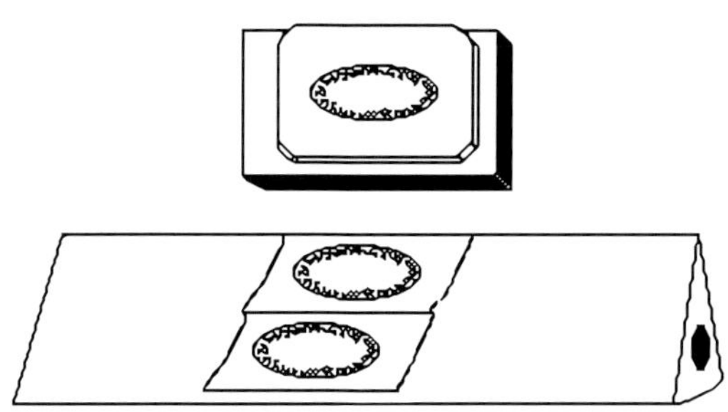

Fig. 20a-d

This schematic demonstrates tissue molecular bonds shown in gray (a) lying in the path of the advancing knife edge. As the knife advances (b), a wave of compression precedes the edge and exerts transverse tinsel stresses parallel to it. As the knife advances (c), the tinsel stresses increase and the molecular chains first undergo an elongation along the lines of tension. As the knife continues to advance (d), the molecular chains rupture and slip over each other (plastic flow) tearing out from the bottom of the material as the yield point is exceeded. (Chapter 3)

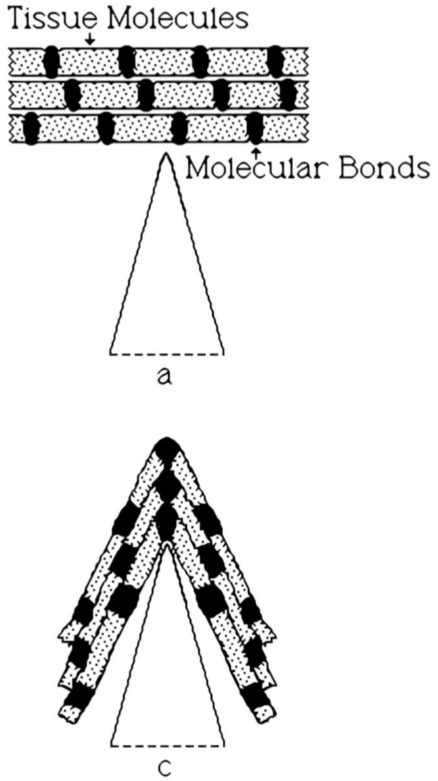

Fig. 21

This schematic demonstrates the appearance of crystalline structure similar to which make up a metal microtome blade. Notice that the crystalline structures results in a rough edge of the facet of the blade. It should be recognized that crystalline particles are arranged in a highly ordered fashion. A naturally produced angular solid of definite form in which the ultimate unit from which it is built up are systematically arranged. They are usually evenly spaced on a regular spaced lattice. This is quite evident in this schematic. (Chapter 3)

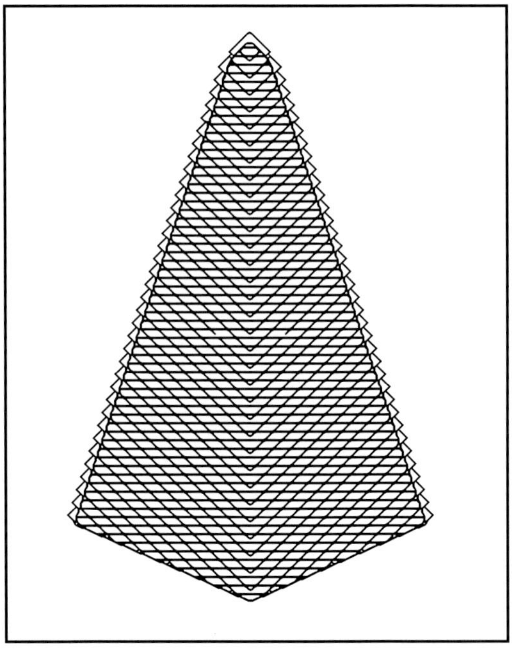

Fig. 22
This schematic illustrates the effects of polishing, seen in black, on the surface of the facet. As the polishing continues there is an amorphous layer which fills in the rough edge surfaces of the blade. (Chapter 3)

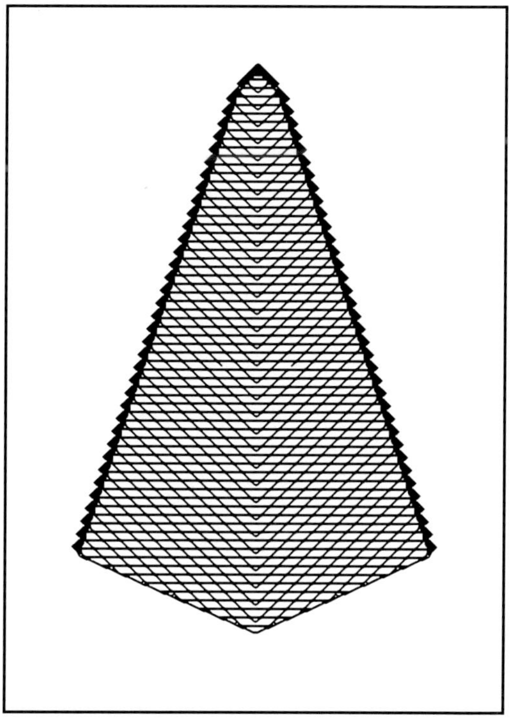

Fig. 23
This schematic illustrates the completion of the development of an amorphous surface on the facet of a blade. Note that the rough angular structure seen in Fig. 21 are no longer present. This is roughly what one would expect to see in what I referred to as a sharp blade, while in reality what it really implies is that the blade has an amorphous surface. Note that the metal is not effected by polishing in the inner structures of the blade. (Chapter 3)

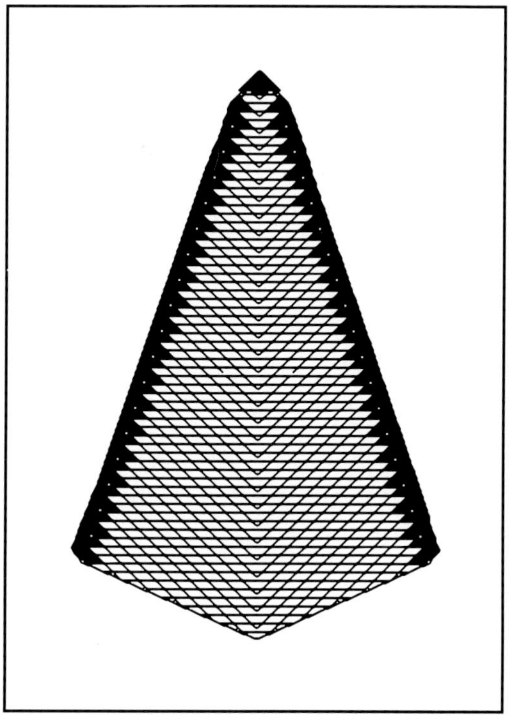

Fig. 24a
This schematic demonstrating a blade with an amorphous surface will reduce the drag and resistance as the blade travels through the specimen. (Chapter 3)

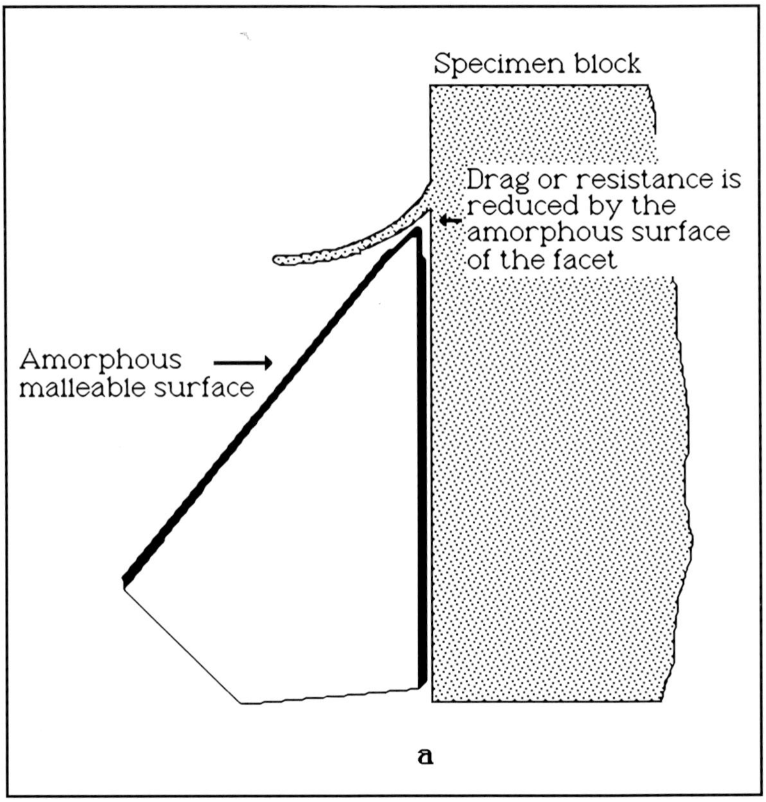

Fig. 24b
This schematic demonstrates a specimen block being cut with a blade which has a crystalline surface (b). Note that there is a fair amount of drag resistance produced at the surface of the blade as it attempts to section the specimen.

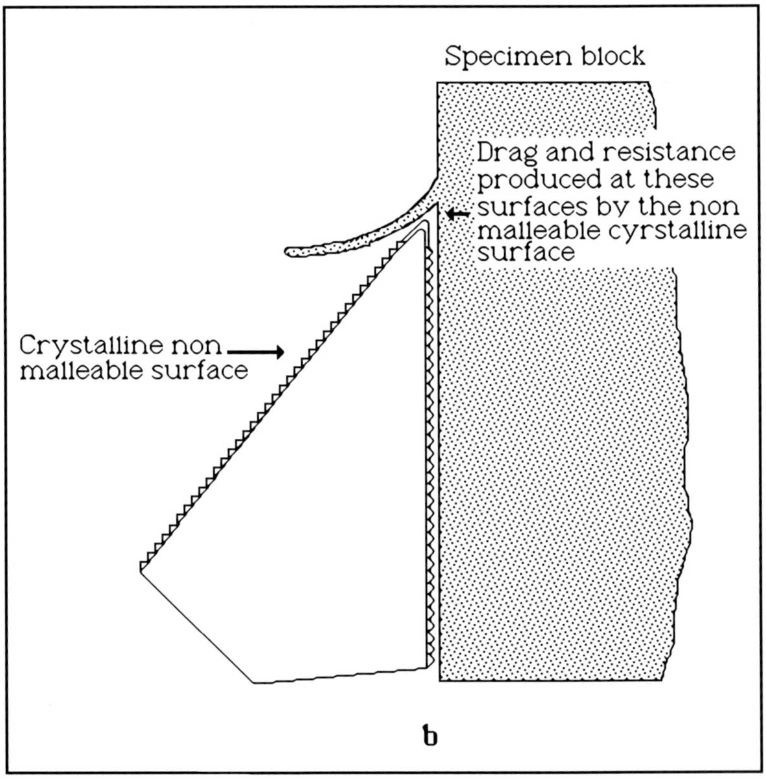

Fig. 25
This is a drawing of a linen strop holder which can be made in most hospital carpenter shops. Note that the sections with the large wingnut must be moveable in order for the linen to be made taught when the wingnut is tightened. (Chapter 3)

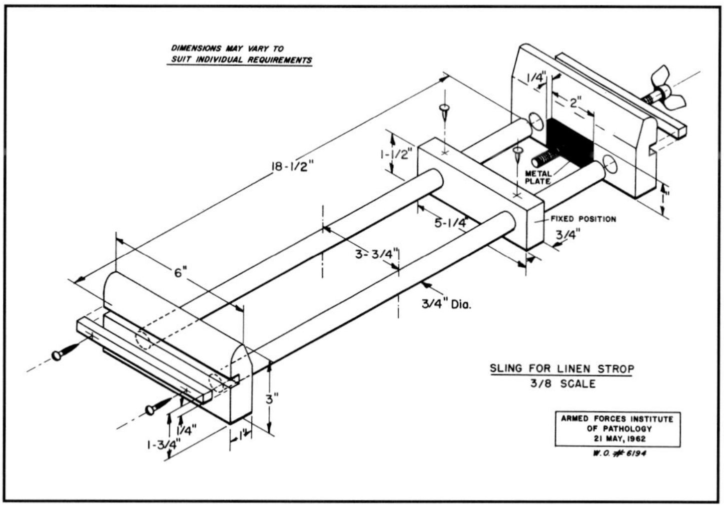

Fig. 26a and b
This diagram demonstrates the proper stropping pattern for both sides of the microtome knife. This produces a criss-crossing pattern of the sides of both facets of the blade which can be seen by a detailed study of a and b. (Chapter 3)

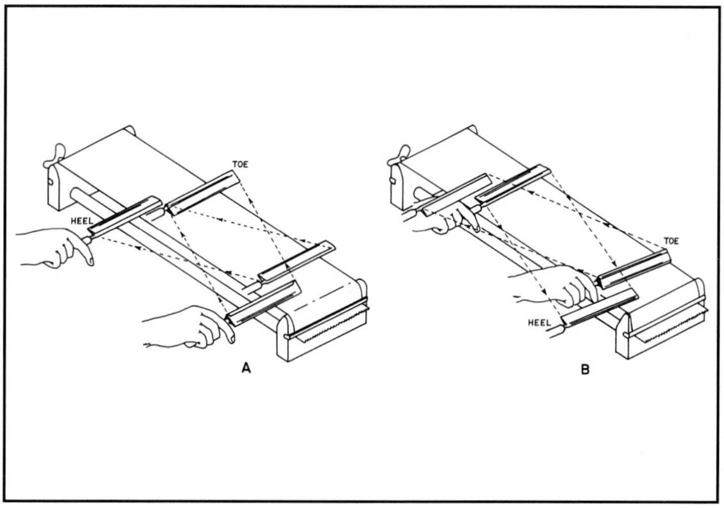

Fig. 27
This illustration demonstrates the proper angling of the knife against the paraffin block. Note that the facet angle should be approximately 27 to 32 degrees. The most important aspect of this is that the clearance angle between the knife facet and the paraffin block should be between 5 and 10 degrees. (Chapter 3)

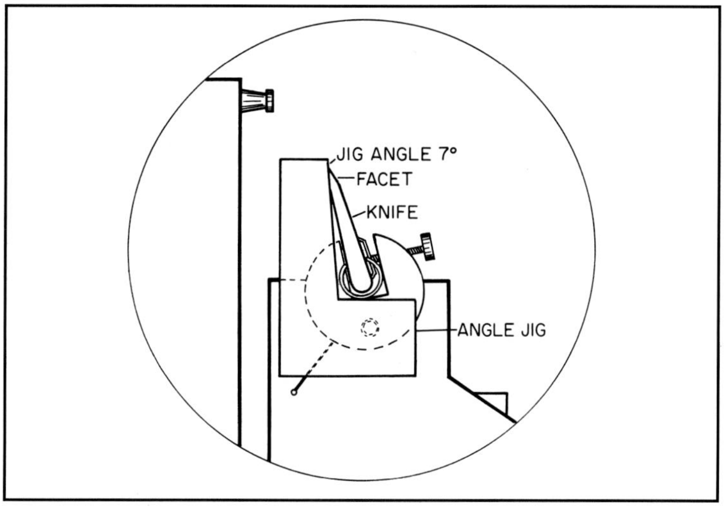

Fig. 28a

a) This illustration demonstrates the effects of cutting with properly prepared tissue and knife facet. Most importantly, one should note that there are certain compression points and wave compression areas on the specimen. One should be aware that these tissue forces are in two forms: Down toward the advancing knife blade and away from the surface of the specimen as can be seen in the upper portion of this schematic. Also, one should be aware that during the knife blade advance through fibrous specimens, an enormous amount of tissue forces are produced. This is in contrast to very little force produced where one has a highly cellular specimen. (Chapters 3 & 17)

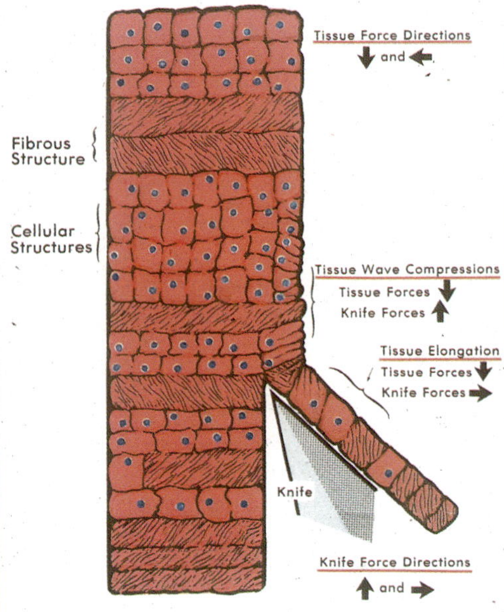

Effects of cutting with properly prepared tissue and knife facet.

Fig. 28b

In contrast to (a) notice effects of tissue artifacts produced when tissue is improperly prepared and the bevel of the microtome knife does not contain an amorphous surface. (Chapters 3 & 17)

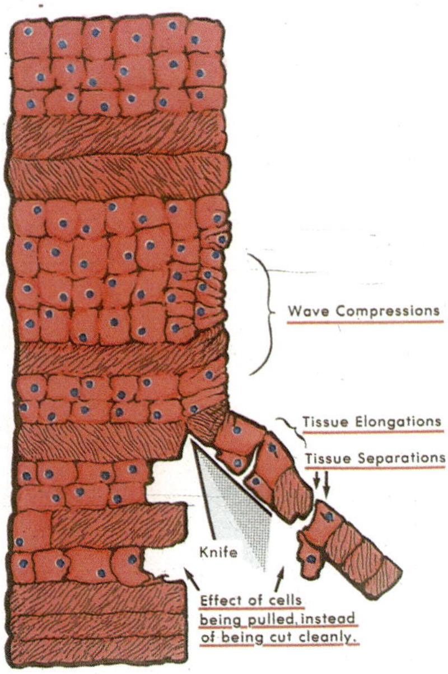

Effect of cutting with improperly prepared tissue or knife facet.

Fig. 29
This schematic demonstrates the cutting edge of a microtome blade which is estimated to measure 0.3 micrometers. (Chapter 3)

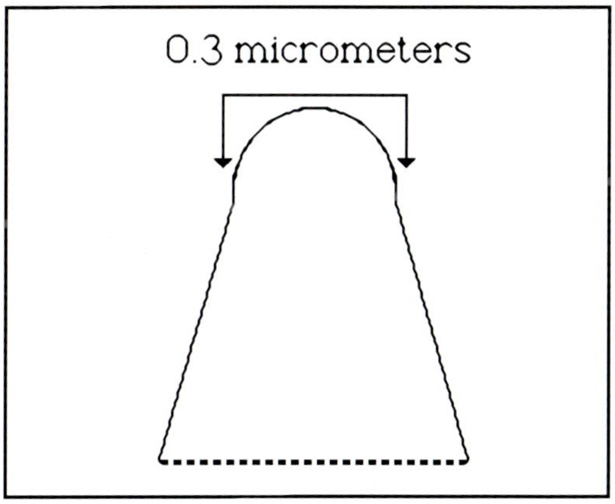

Fig. 30
This schematic demonstrates soaking depth which is often variable due to the type of tissue and/or length of soaking exposure. The illustration demonstrates the depth of water penetration in green. This particular area will not result in microvibrations during microtomy. Conversely, the portion of section in brown, where water did not penetrate sufficiently, will result in microvibrations. It should be understood that lack of proper water soaking will produce many of the artifacts often seen as a results of microtomy. One should ensure that proper reintroduction of moisture into the specimen has been accomplished. (Chapters 3 & 17)

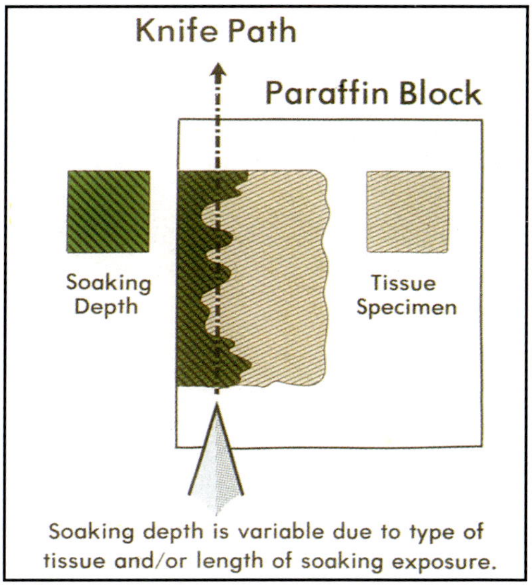

Fig. 31
This schematic demonstrates the effects of improper soaking. The green area illustrates proper reintroduction of moisture, while the gray area demonstrates an area which does not contain sufficient moisture prior to microtomy. This gray area will produce microvibrations as illustrated in the photograph, while the area that has been properly impregnated with moisture will be artifact free. (Chapters 3 & 17)

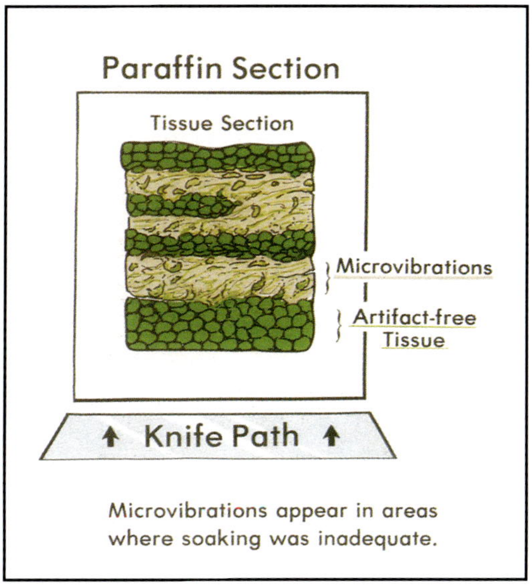

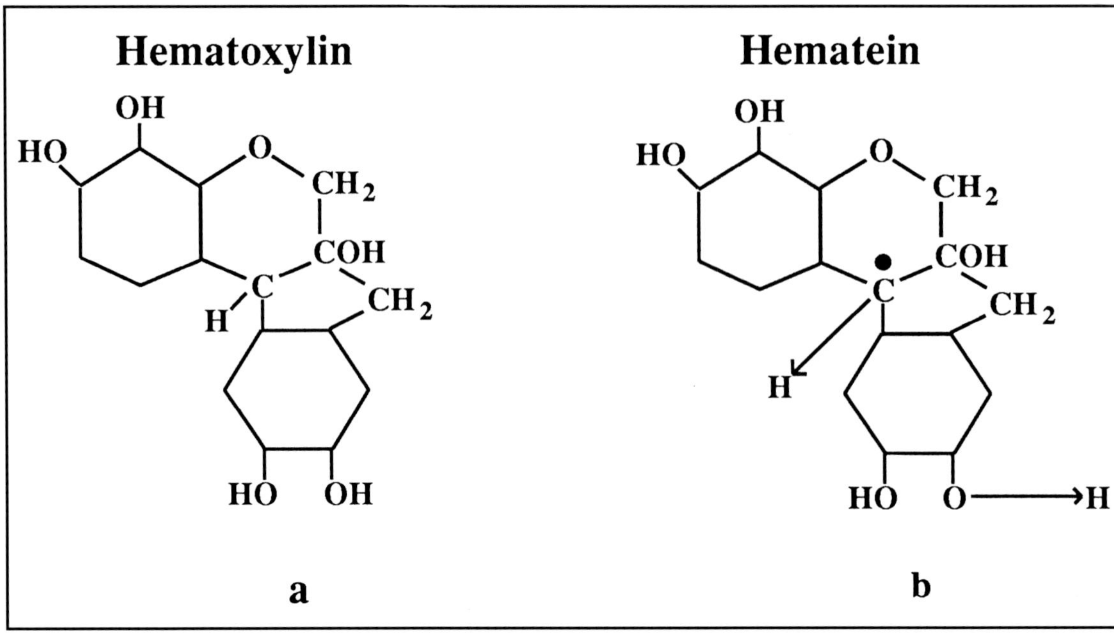

Fig. 32a and b
This schematic demonstrates the chemical formula for hematoxylin (a), followed by the hematein chemical formula (b). During oxidation of the hematein two hydrogens are removed (b) leaving a negatively charged phenolic $^{-}O_{-}$ group and a coordinately unsaturated carbon atom indicated by a dot. Hematein does not have a quinonoid structure. (Chapter 4)

Fig. 33
This illustration demonstrates that during formation of the hematoxylin two hematein ions form chelates with aluminum (Al^{+++}). Each $_{-}O^{-}$ group shares an electron pair with the metal forming a coordinate bond. The remaining coordination sites are occupied by two water molecules. (Chapter 4)

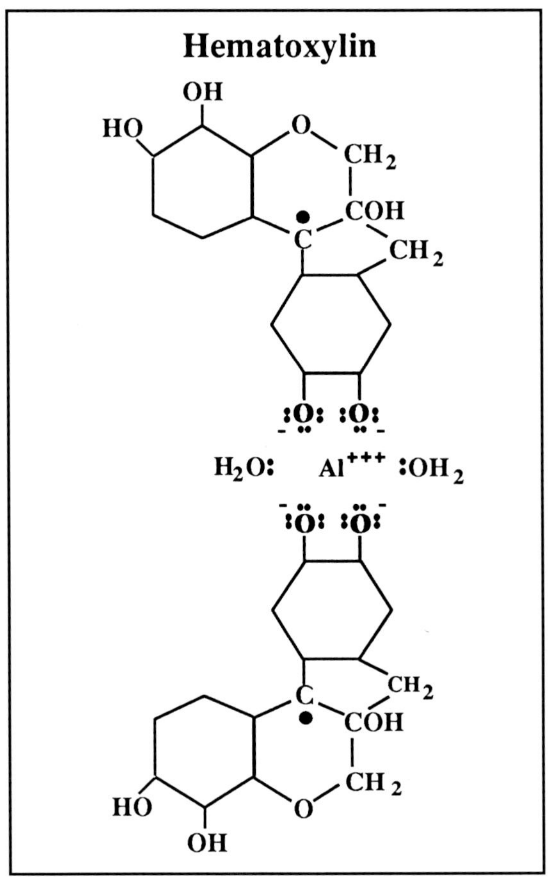

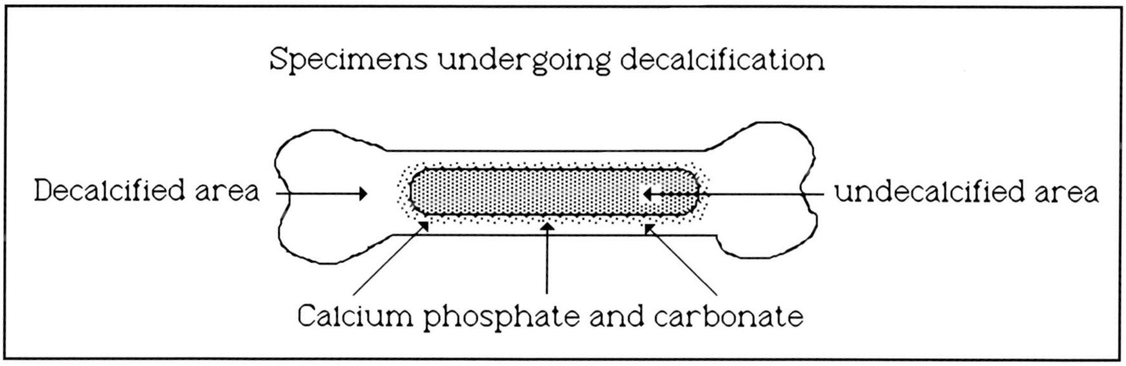

Fig. 34
This schematic illustrates the process of internal buffering by the release of calcium phosphate and carbonate. As long as this process is taking place, all surrounding tissues are protected from the deleterious effect of the decalcifying acids. It is important to recognize that once this decalcification process is complete and no calcium phosphate and/or carbonate are produced, then the decalcifying acids attack soft tissues very rapidly. See more information on page 103 related to this subject. (Chapter 5)

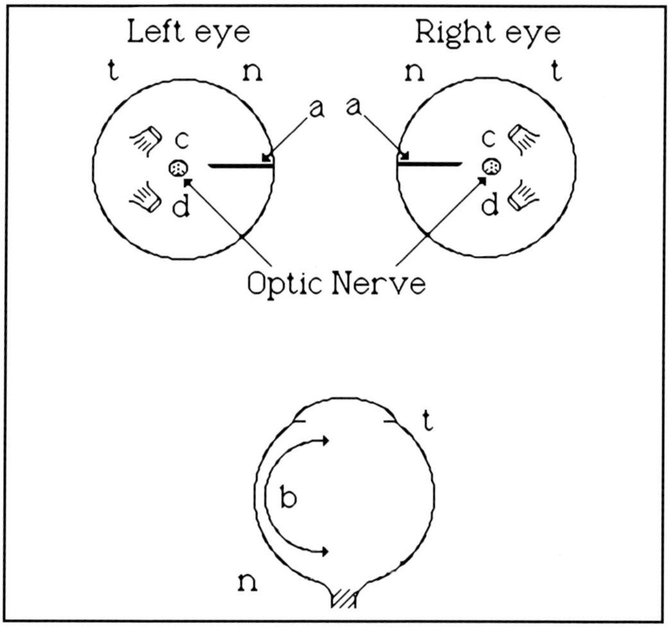

Fig. 35a-d
This schematic of the back of the eye illustrates the different structures and measurements for determining a left eye from a right eye. This information is important for proper identification of lesions if present. (Chapter 6)

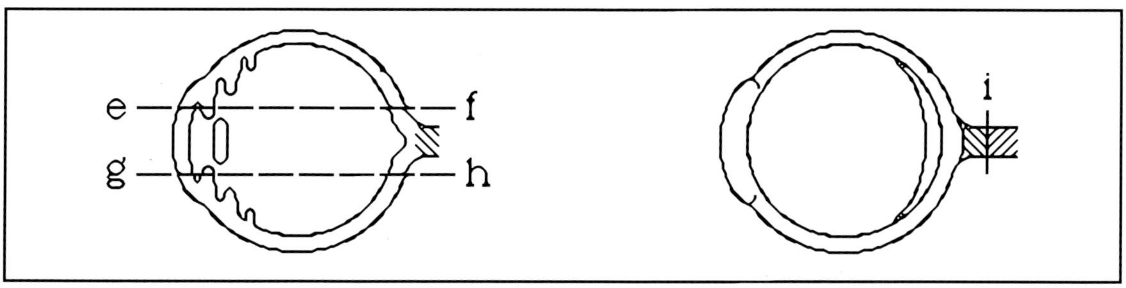

Fig. 35e-i
This schematic provides information on the proper way of cutting the eye to produce a specimen for processing, including the optic nerve. (Chapter 6)

Fig. 36
This demonstrates veins and arteries which branch and interconnect with capillary plexus. This is periodic acid Schiff method. (Chapter 6)

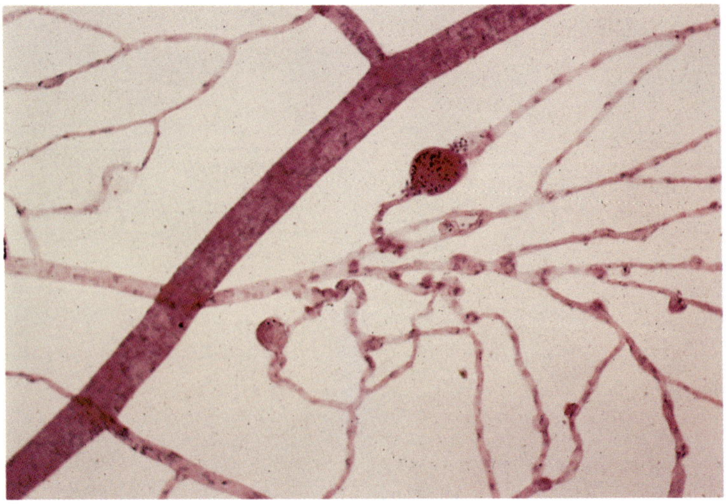

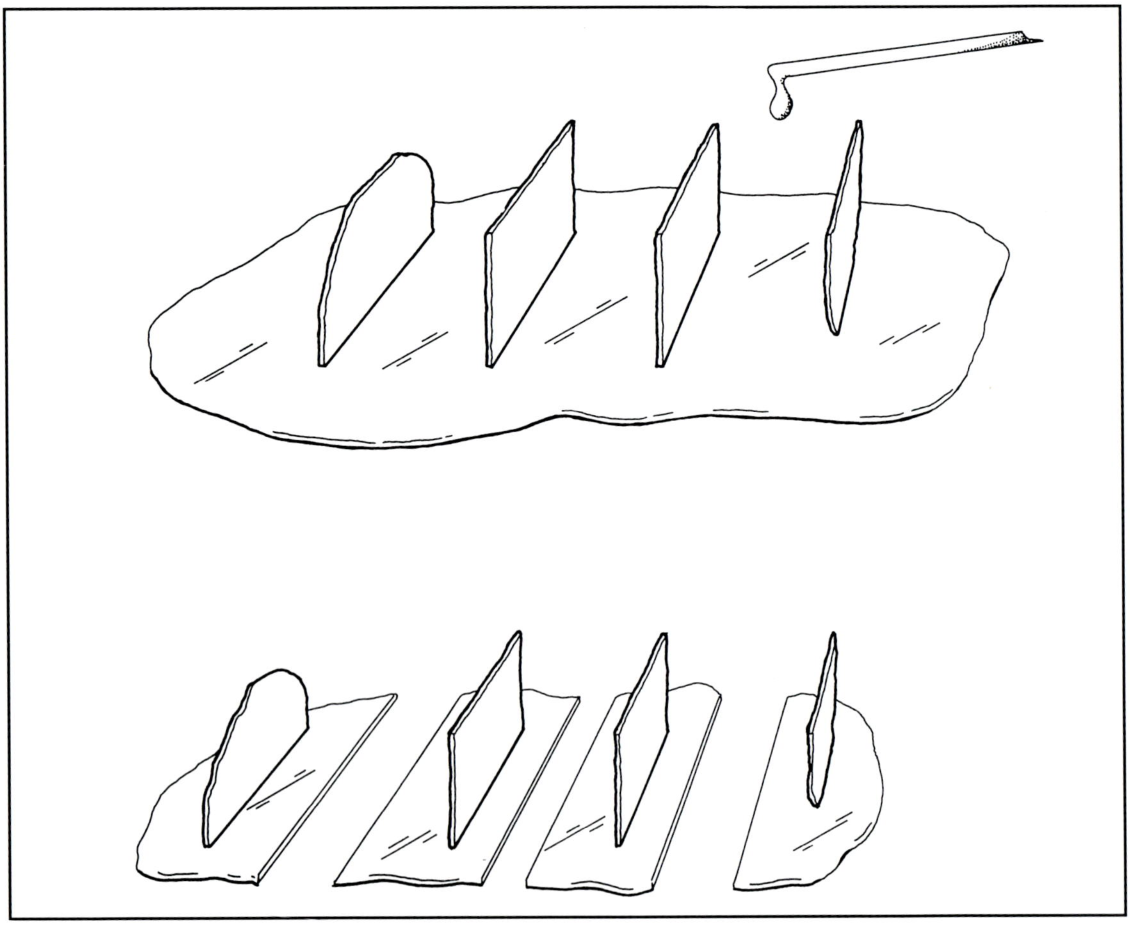

Fig. 37
The upper portion of this schematic shows the applicator stick application of agar to the specimens. The lower portion demonstrates the vertical orientation plane of the specimens. (Chapter 6)

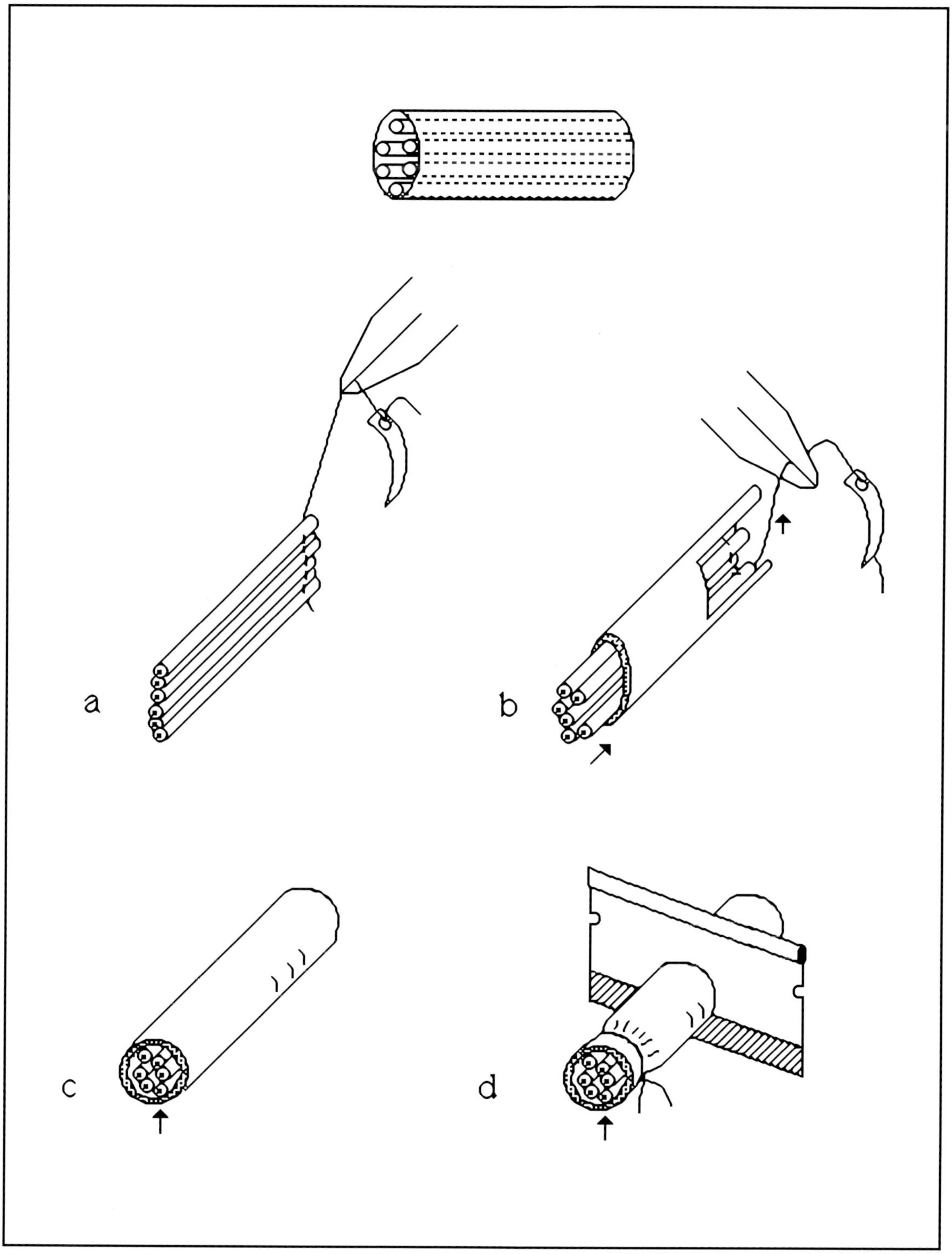

Fig. 38a-d
This schematic demonstrates the various steps for multiple specimen embedments. (Chapter 6)

Fig. 39a
Illustrates the effect of laminar flow on staining reactions. The stain molecules (clear circles) move considerably slower near the walls of the staining dish due to friction of the wall surface. Conversely, the molecules in or near the middle of the staining dish (dark circles) move faster. (Chapter 6)

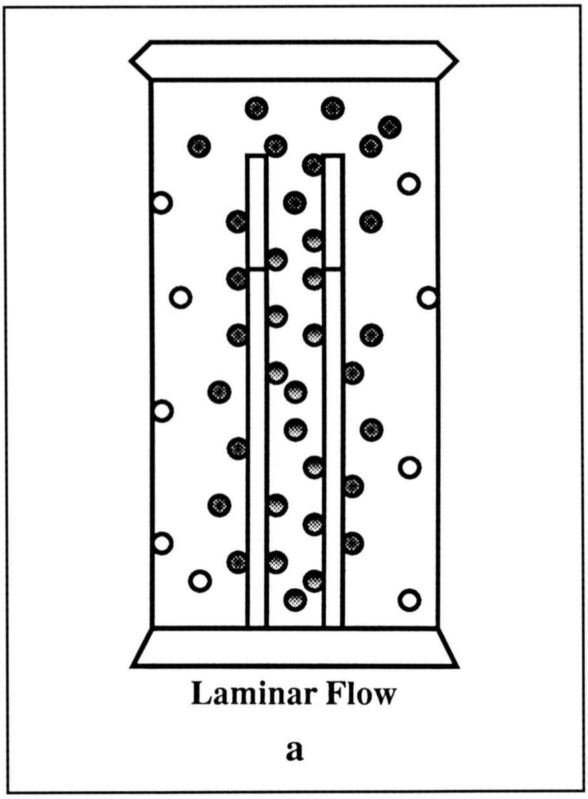

Fig. 39b
This schematic illustrates that microwave heating results in a 10-15°C difference in temperature between top and bottom levels of a staining container. (Chapter 6)

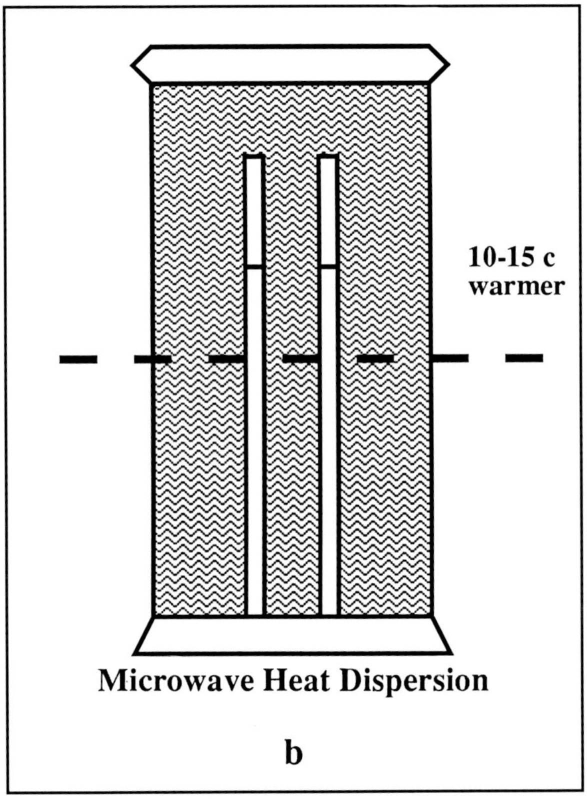

Fig. 39c
Convection heat radiates from all sides of a staining container when a conventional type oven is used. This results in variable staining between slides in the middle vs. those close to the outer walls of the staining dish. (Chapter 6)

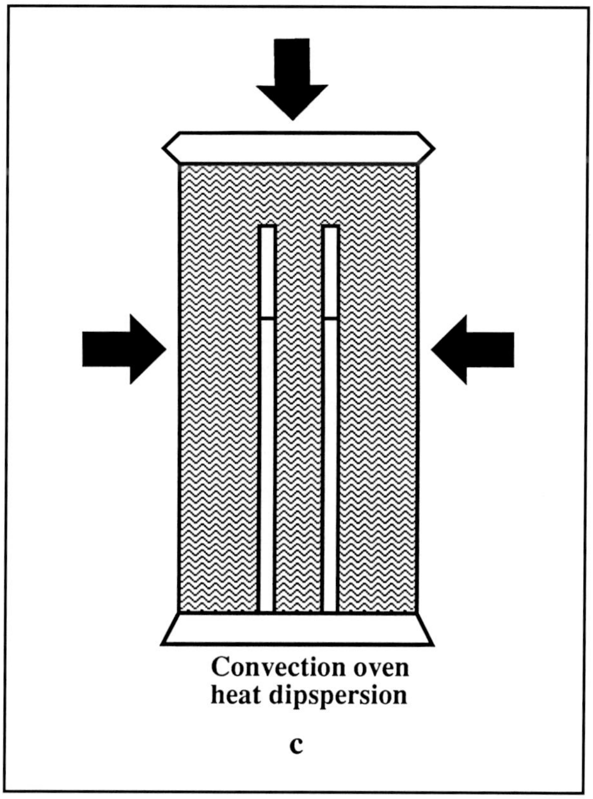

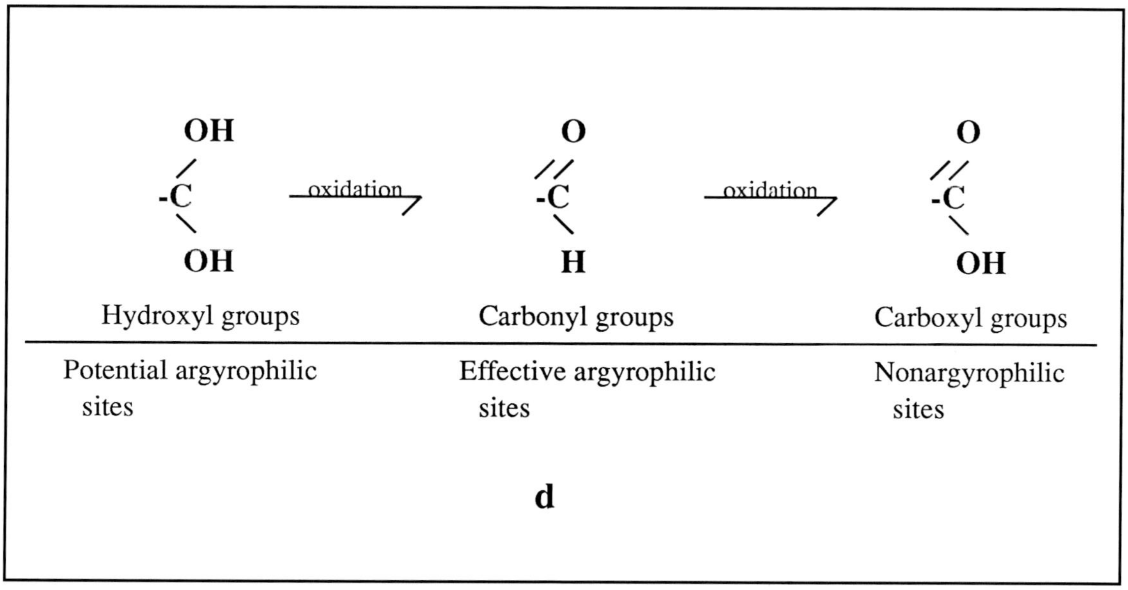

Fig. 39d
This formula demonstrates an example of the importance for proper utilization of oxidizers. (Chapter 6)

Fig. 40a
Is a section of skin with leprosy. Notice the well stained, brilliant demonstration of the leprae bacilli. Fite-Faraco Method (Chapter 7)

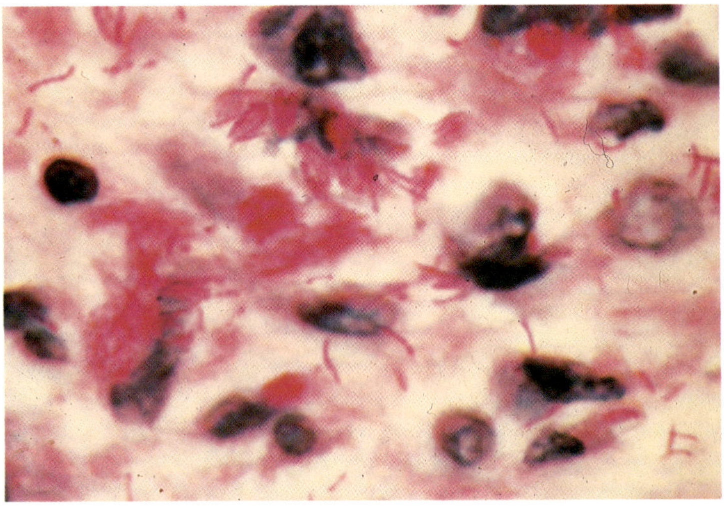

Fig. 40b
This composite demonstrates filaments of nocardia asteroides. Section is well stained. Fite-Faraco Method (Chapter 7)

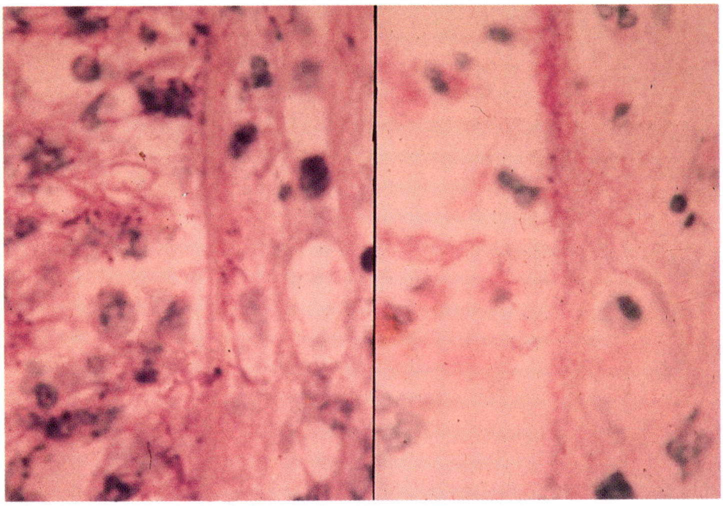

Fig. 41
Mycobacterium avian intercellular which is demonstrated by fluorescence microscopy. Truant's Method (Chapter 7)

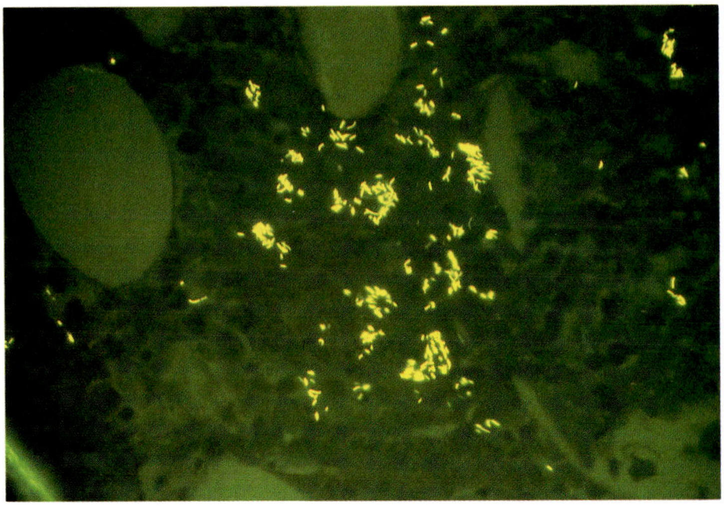

Fig. 42
This section shows a case of mycobacterium tuberculosis in an AIDS patient. Notice the enormous amount of tubercle bacilli throughout the section. Section is well stained. Kinyoun Method (Chapter 7)

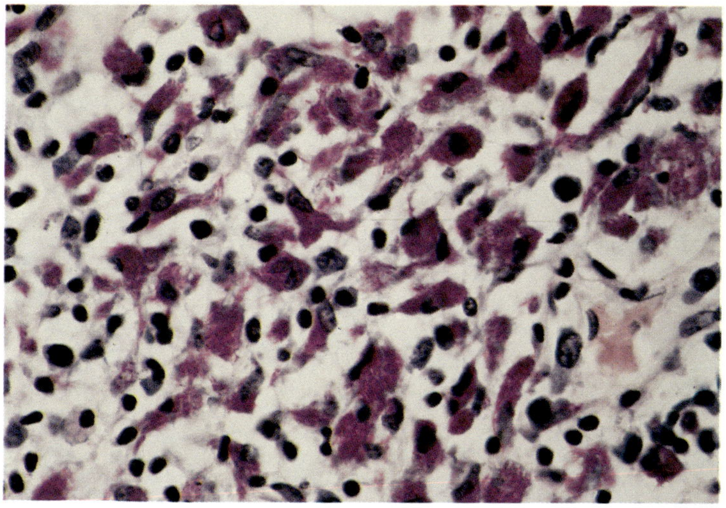

Fig. 43
This is Mycobacterium avian. Notice that in this case the background (methylene blue) is stained slightly darker than it should be for good demonstration of the bacilli. Kinyoun Method (Chapter 7)

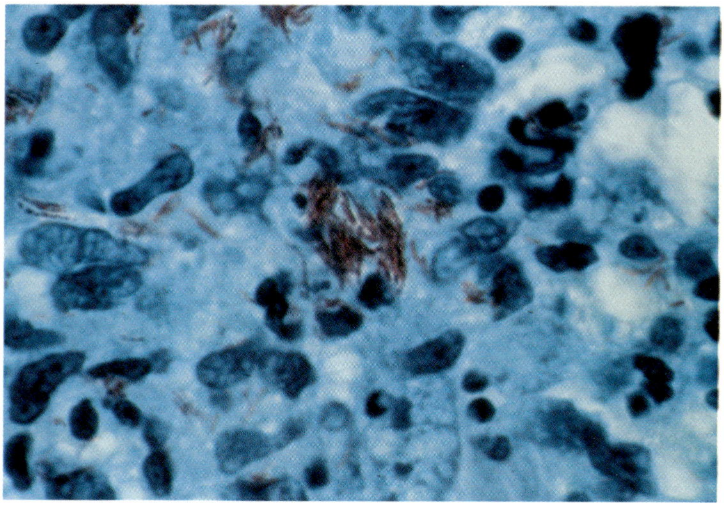

Fig. 44
This section contains gram-positive gram-negative bacteria. Notice the gram-positive bacteria in the middle of the photograph staining bluish-purple. The gram-negative bacteria can be seen surrounding the gram-positive organisms. Section is well stained. Brown-Hopps Method (Chapter 7)

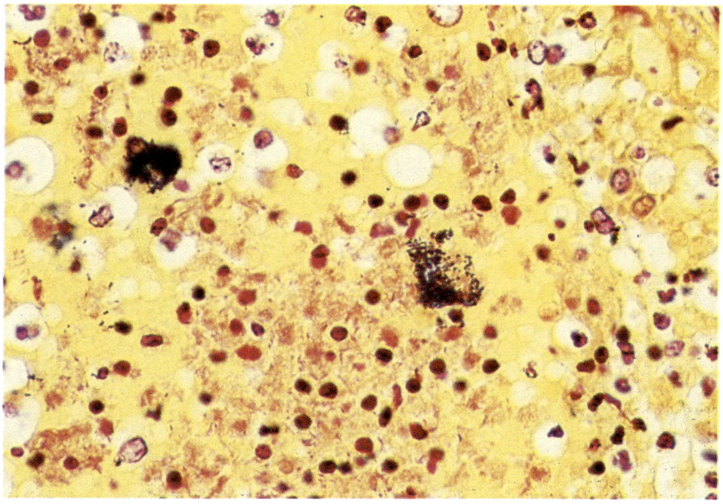

Fig. 45
This is a high power of Fig. 44 demonstrating again gram-positive cocci (upper portion of the photograph) and gram-negative rods. Brown-Hopps Method (Chapter 7)

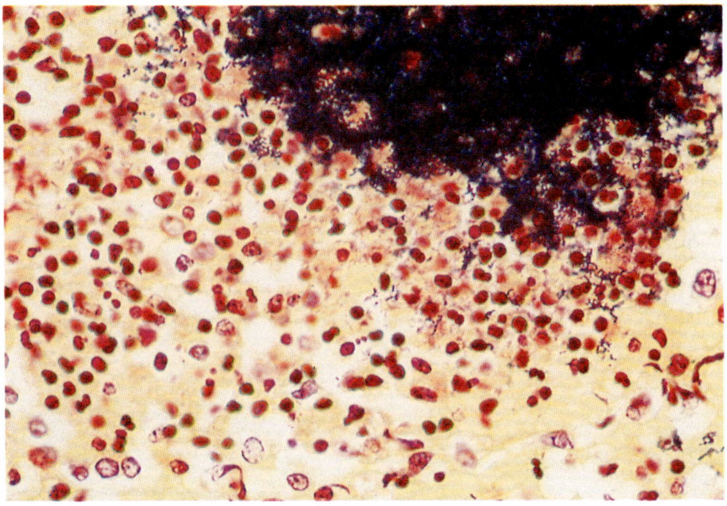

Fig. 46
Actinomycosis. Notice the large sulfur granules throughout the section. Section is well stained. Brown-Hopps Method (Chapter 7)

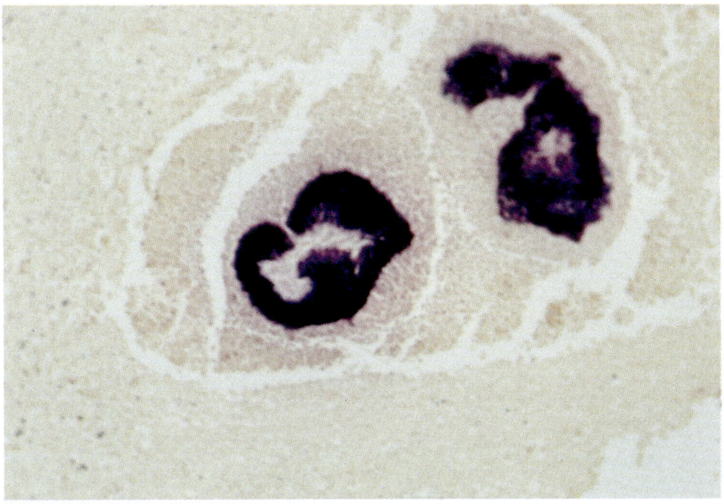

Fig. 47
This is a high power of Fig. 46 demonstrating the filaments of the actinomycosis staining gram-positive reaction. Section is well stained. Brown-Hopps Method (Chapter 7)

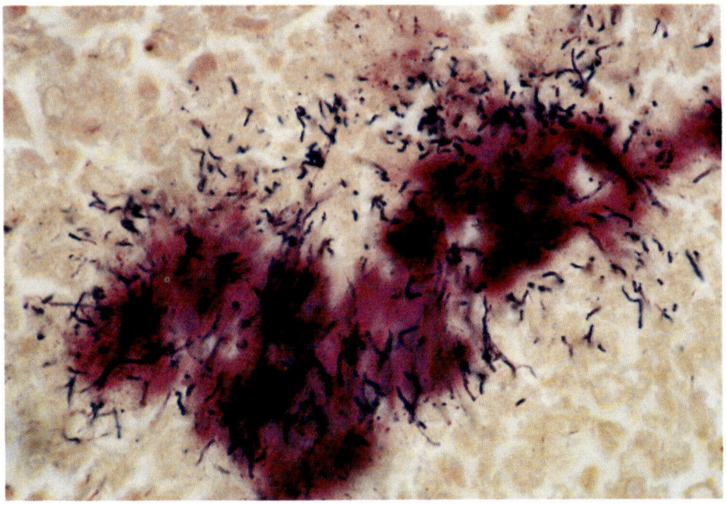

Fig. 48
This section which contains actinomycosis has been stained with Brown-Hopps Method and viewed under dark field microscopy. (Chapter 7)

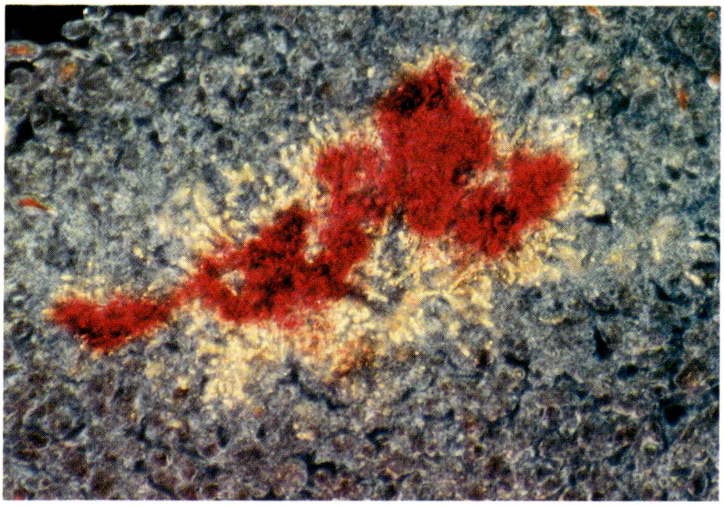

Fig. 49
This section demonstrates Legionella pneumophila or legionnaire disease bacilli. Notice the dark well demonstrated bacilli. Section is well stained. Greer-Chandler Method (Chapter 7)

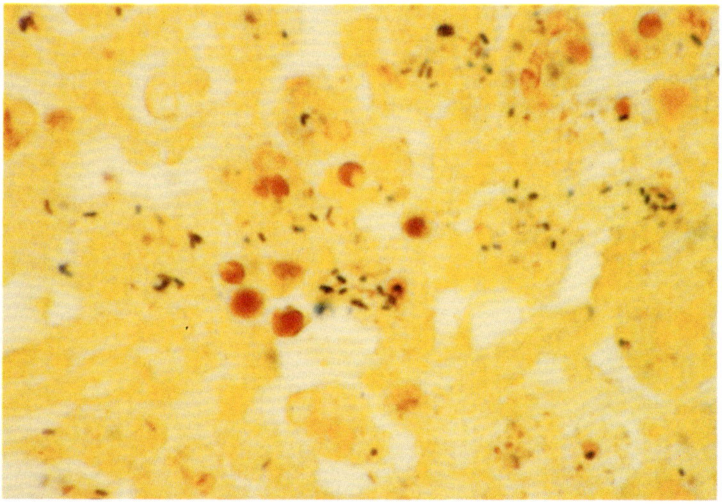

Fig. 50
This photograph demonstrates coccidiomycosis fungi. Section is well stained. Gridley's Method (Chapter 7)

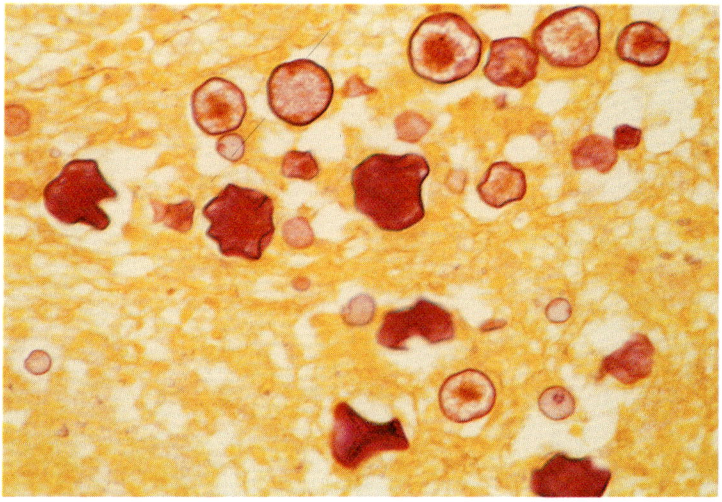

Fig. 51
This is a case of aspergillosis. Notice elongated conidiophores staining positive. Section is well stained. Gridley's Method (Chapter 7)

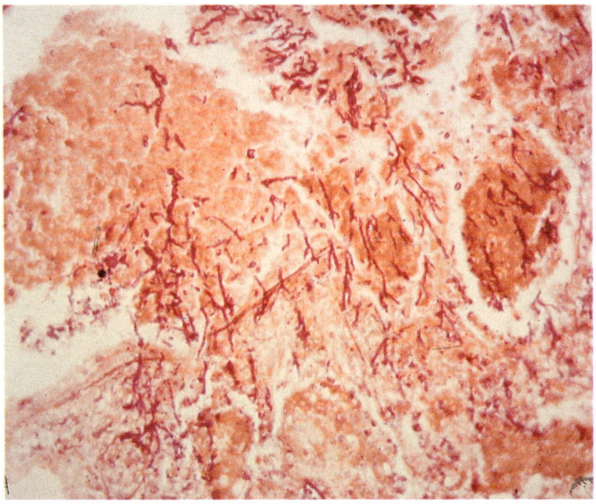

Fig. 52
Section contains actinomycosis filaments after staining with a silver reaction. Section is well stained. Grocott Methenamine Silver Method (Chapter 7)

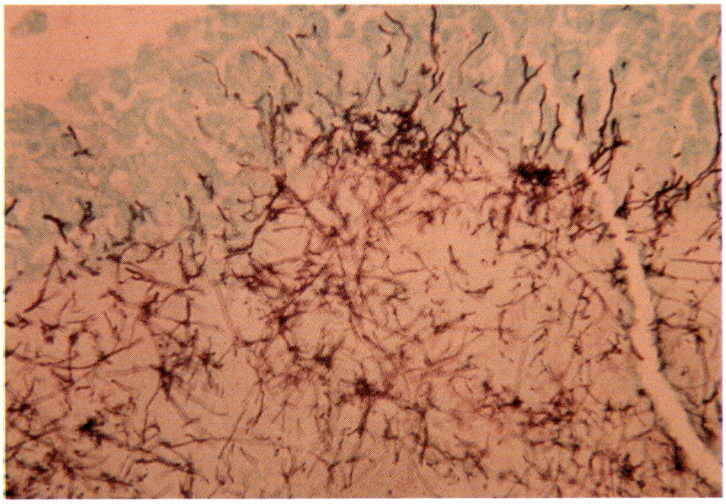

Fig. 53
Demonstrates South American blastomycosis. Section is well stained. Grocott's Methenamine Silver Method (Chapter 7)

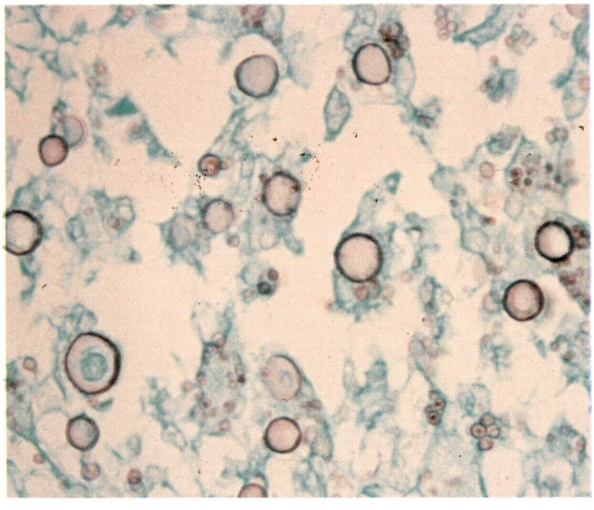

Fig. 54
Section of lymph node demonstrating histoplasma capsulatum. Section is well stained. Grocott's Methenamine Silver Method (Chapter 7)

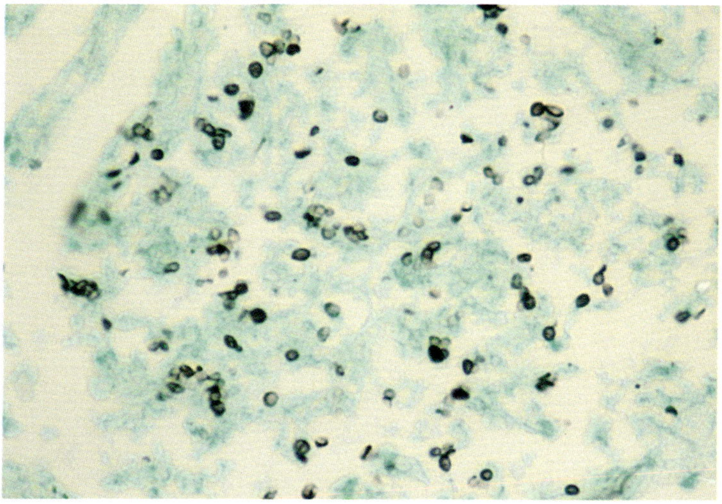

Fig. 55
Nocardia asteroides stained with a silver reaction. Section is well stained. Grocott's Methenamine Silver Method (Chapter 7)

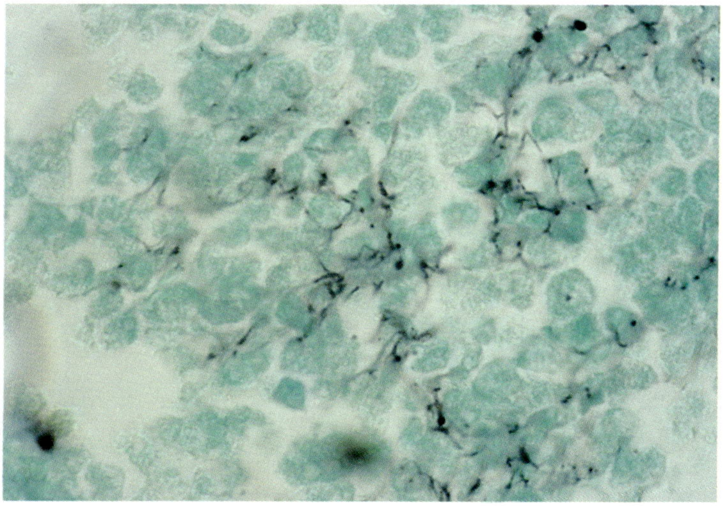

Fig. 56
Section of lung demonstrating Pneumocystis carinii. Notice the well differentiated kinetoplast. Section is well stained. Contributed by Leonard Noble. Grocott Methenamine Silver Method (Chapter 7)

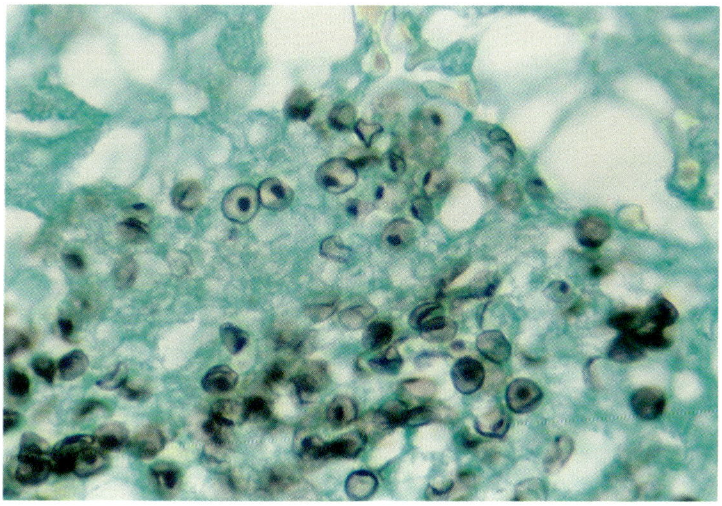

Fig. 57
This a bronchial alveolar lavage smear. Contributed by Leonard Noble. Grocott Methenamine Silver Method (Chapter 7)

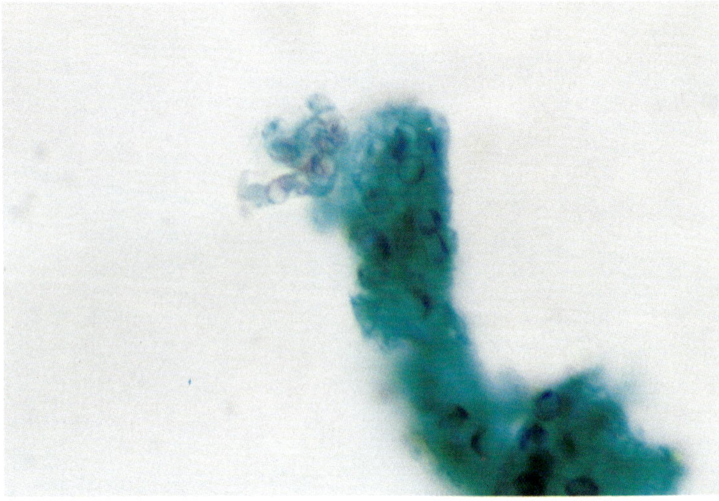

Fig. 58
A section of lung of an AIDS patient demonstrating an enormous amount of tubercle bacilli which is being demonstrated with a silver procedure. Grocott Methenamine Silver Method (Chapter 7)

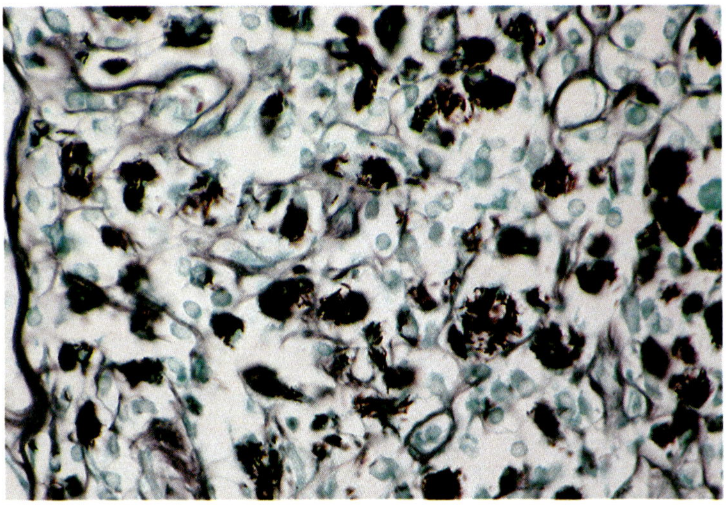

Fig. 59
Section of lung demonstrating Pneumocystis carinii. Section is well stained. Genaw's Method (Chapter 7)

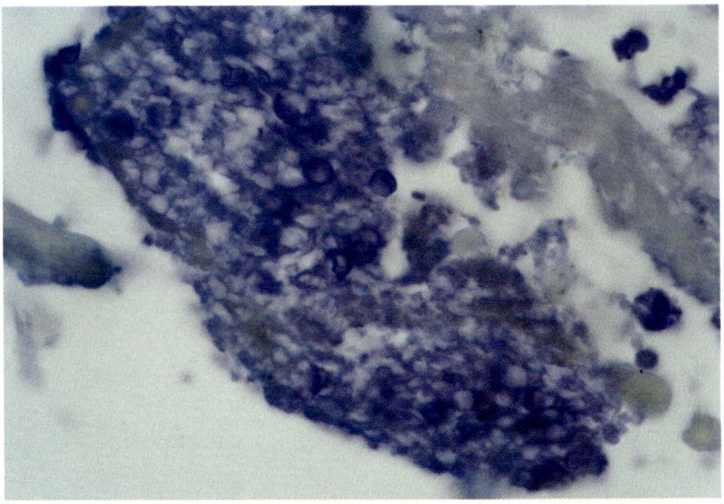

Fig. 60a
This is Treponema pallidum (a) which demonstrates isolated spirochetes. (Chapter 7)

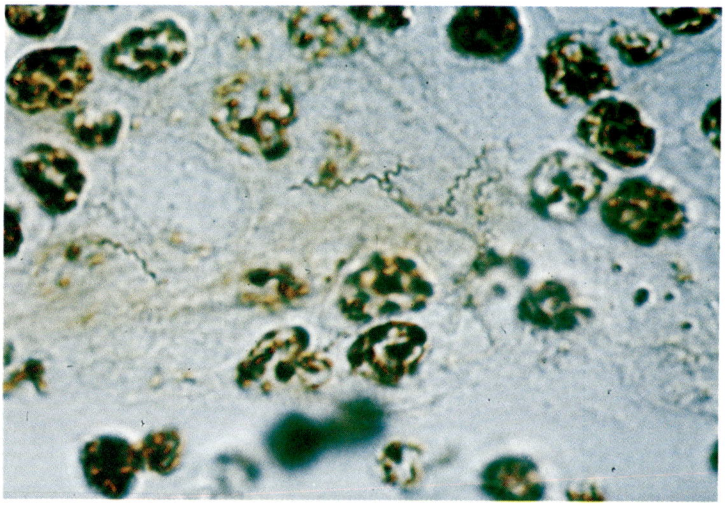

Fig. 60b
Demonstrates massive amounts of spirochetes. Warthin-Starry procedure. If this procedure is done correctly, the organisms that one is trying to demonstrate should be clearly differentiated upon a light yellow background. Warthin-Starry Method (Chapter 7)

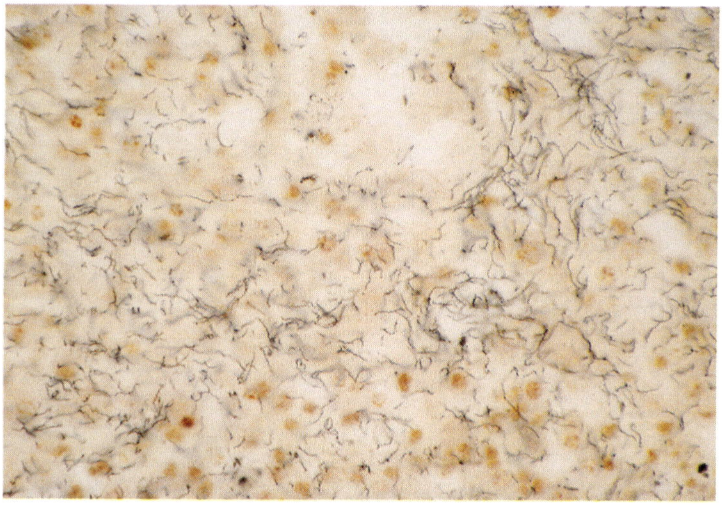

Fig. 61
This photograph demonstrates Donovan bodies which can be seen as safetypin-like structures in the middle of the photograph. Section is well stained. Warthin-Starry Method (Chapter 7)

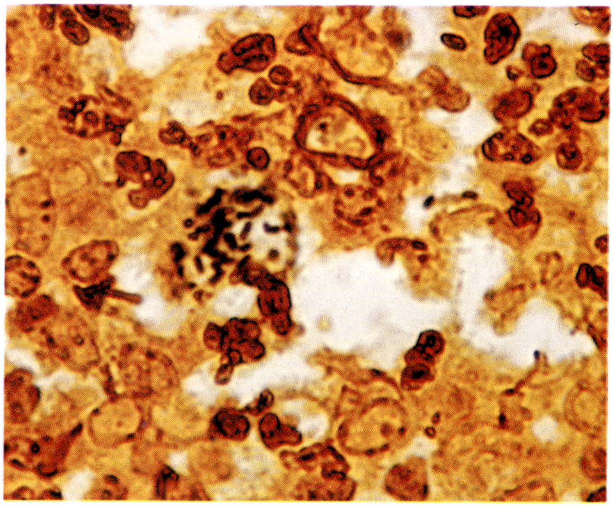

Fig. 62
Helicobacter pylori can be seen in this section of intestine. The photograph illustrates a slightly over-developed Warthin-Starry, varied by the dark staining cellular structures. However, notice that the organisms have been well demonstrated. Warthin-Starry Method (Chapter 7)

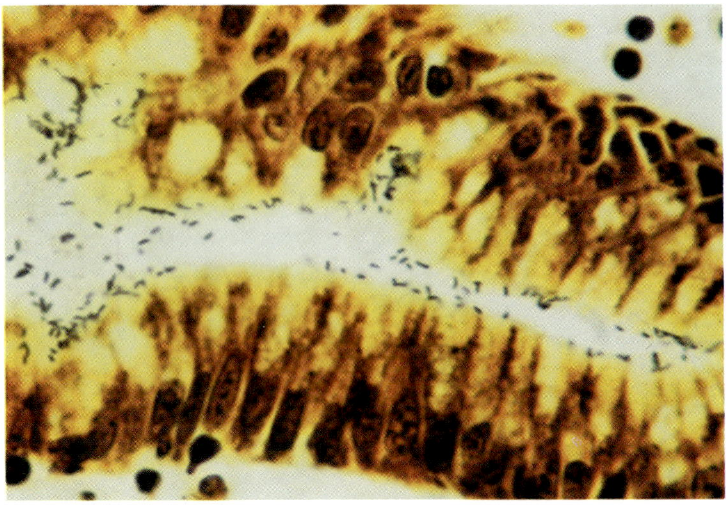

Fig. 63
Helicobacter pylori from an AIDS patient can be seen throughout this section. Section is well stained. Warthin-Starry Method (Chapter 7)

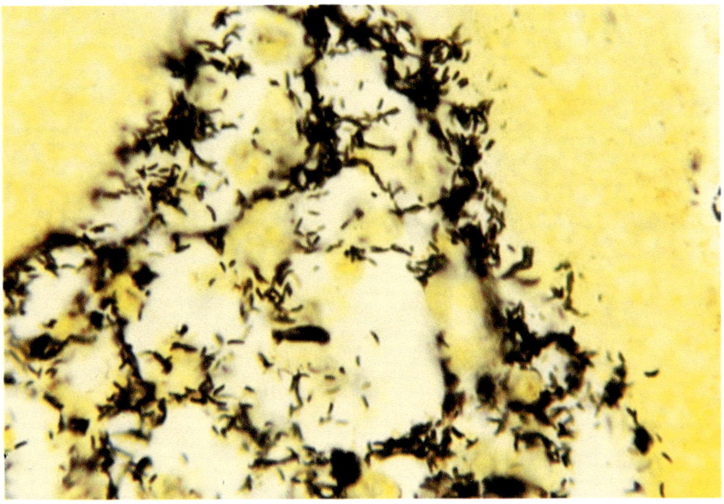

Fig. 64
A case of cat-scratch disease bacilli demonstrated extremely well throughout this section. Notice the clear light-brown background allowing for good demonstration of the organisms. Section is well stained. Warthin-Starry Method (Chapter 7)

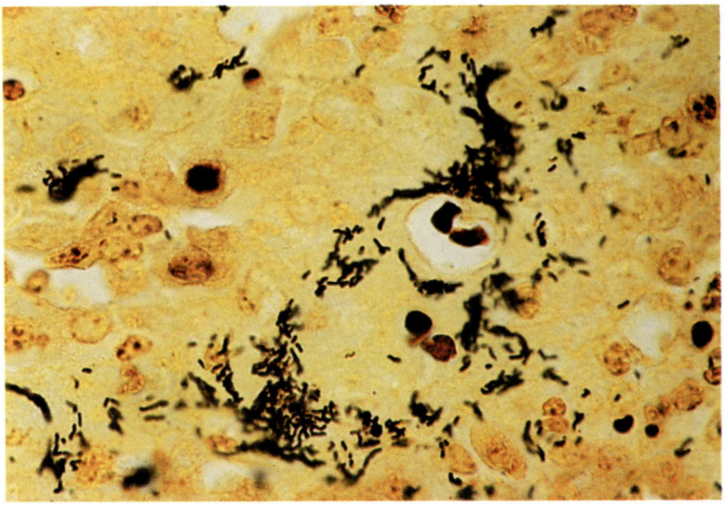

Fig. 65
This is a case of Lyme's disease demonstrating Borelia burgdorferi spirochete which is characteristic of Lymes disease. Warthin-Starry Method (Chapter 7)

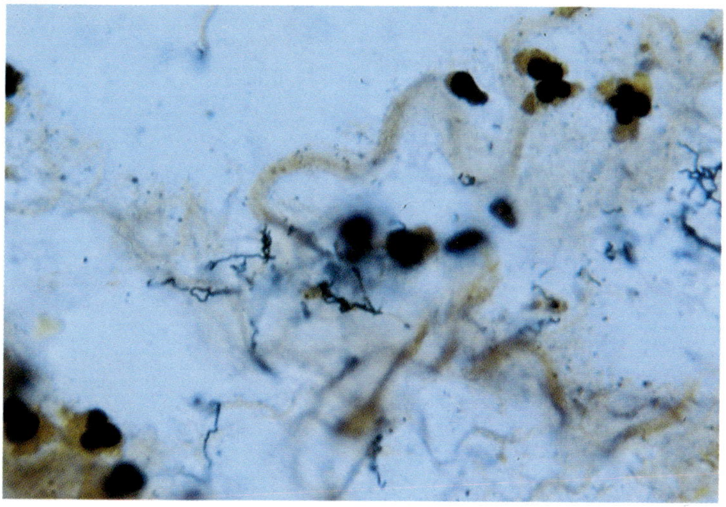

Fig. 66
This section demonstrates Bacillus piliformis seen as long rods. Section is well stained. Contributed by Frank Razzaboni. Warthin-Starry Method for Bacillus Piliformis (Chapter 7)

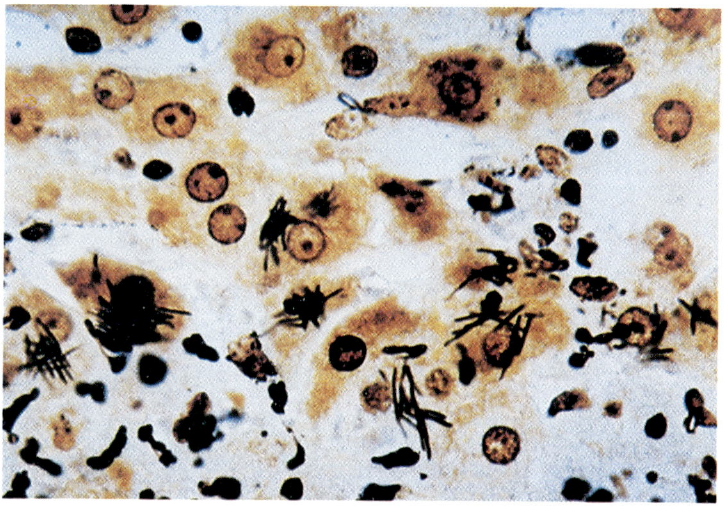

Fig. 67
Section contains filaments of Nocardia asteroides. Section is well stained. Luna-Parker Method (Chapter 7)

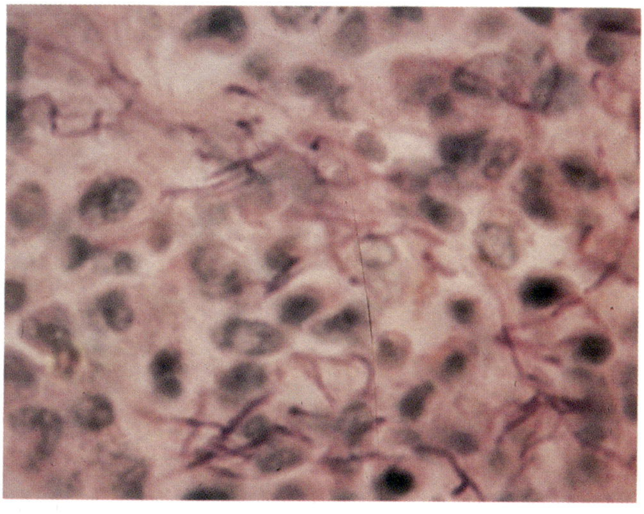

Fig. 68a
Section of lung demonstrating cytomegalovirus inclusion bodies. Lendrum's Method (Chapter 8)

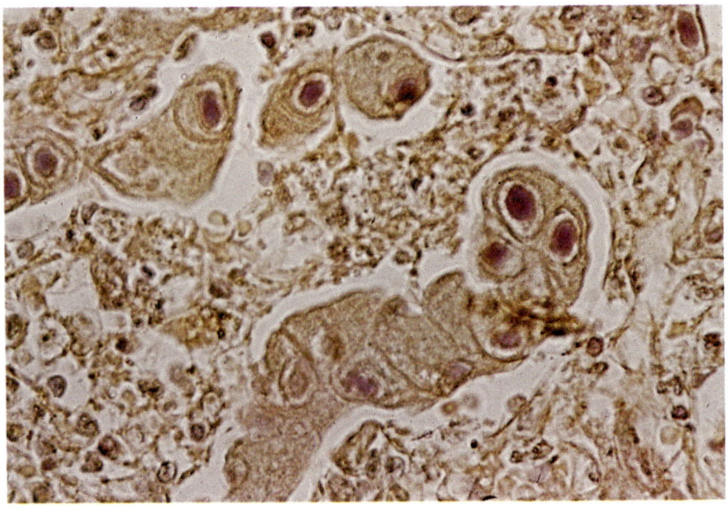

Cytomegalovirus may also be demonstrated with the following stains.

Fig. 68b
This section demonstrates cytomegalovirus in 3 large cells. Nuclear Fast Red Stain (Chapter 8)

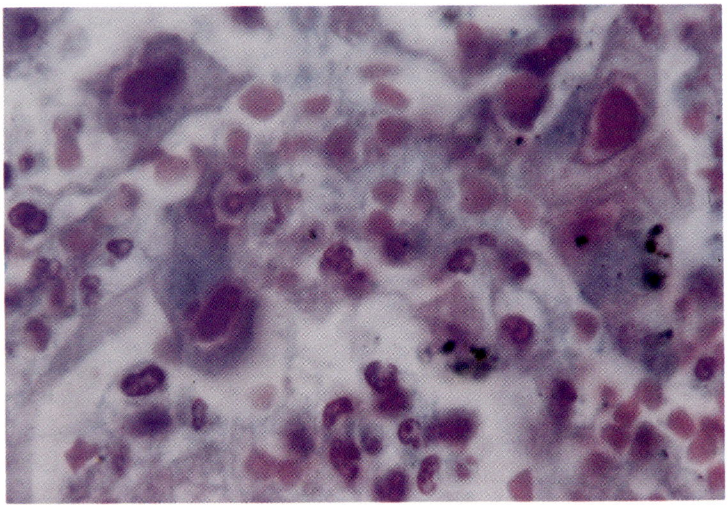

Fig. 68c
This is a section of lung demonstrating cytomegalovirus which can be seen stained bright red. Masson Trichrome Stain (Chapter 8)

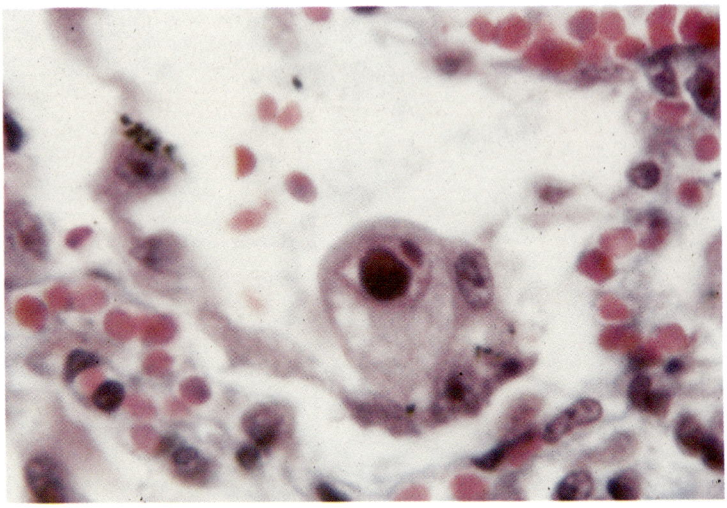

Fig. 69
This section of liver demonstrates Mallory bodies seen in red throughout the photograph. Lendrum's Method (Chapter 8)

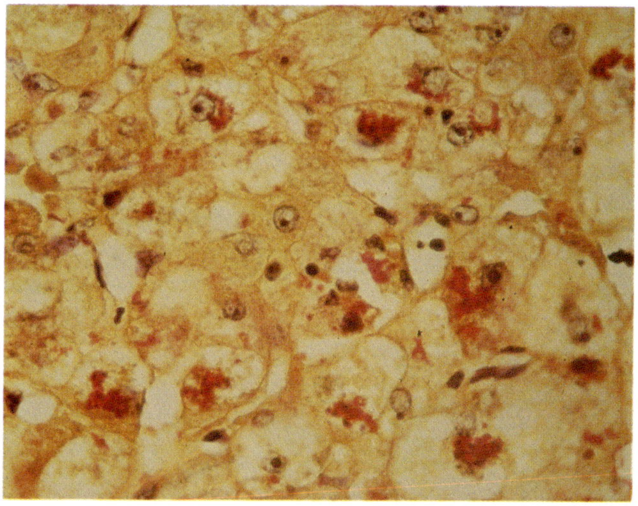

Fig. 70
This section of brain demonstrates rabies encephalitis. Note the Negri bodies in the cytoplasm of the large brain cells. Schleifstein's Method (Chapter 8)

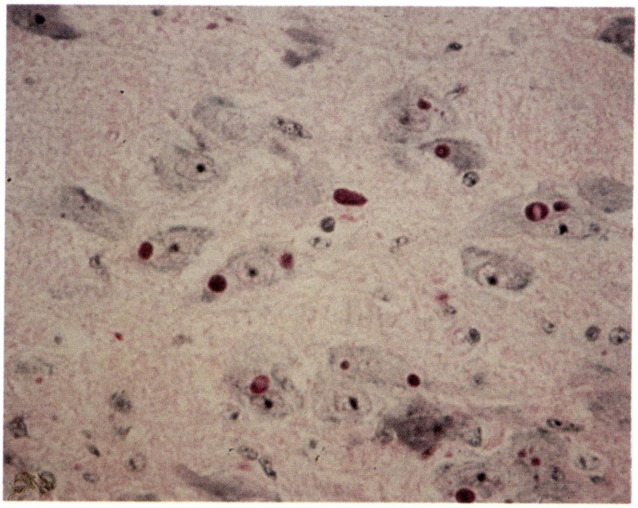

Fig. 71
Negri bodies can be seen more clearly in this higher power of the same section. Schleifstein's Method (Chapter 8)

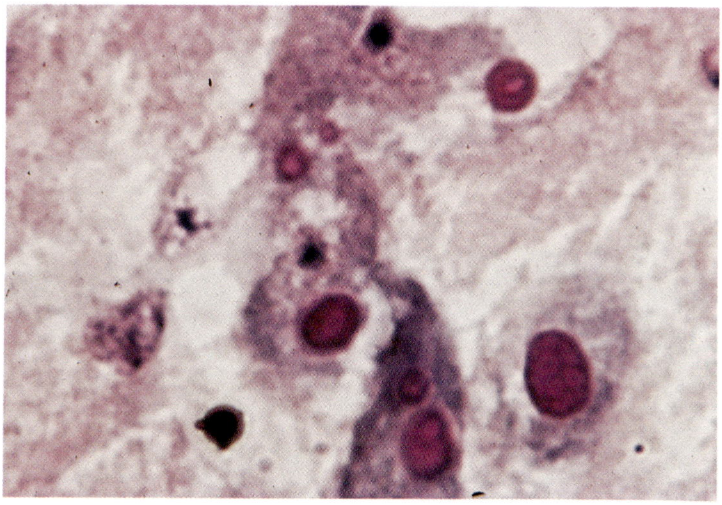

Fig. 72
This section of brain demonstrates rabies encephalitis. Notice the prominently stained negri bodies in these large cells. Luna's Method (Chapter 8)

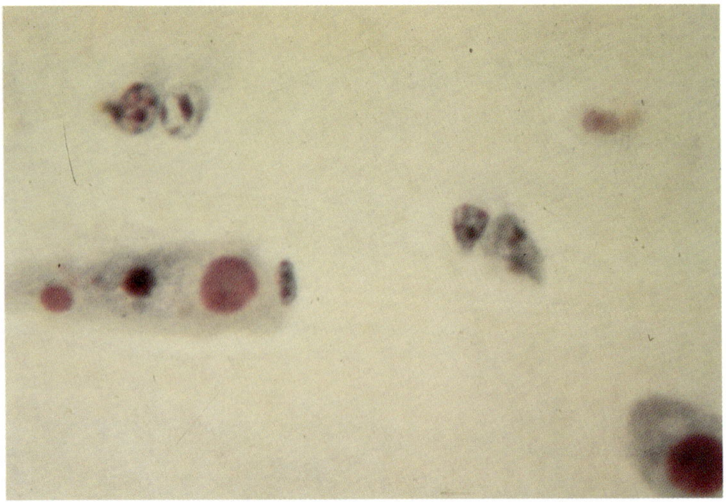

Fig. 73
This section demonstrates canine distemper inclusions. Inclusion bodies are seen as red structures. Luna's Method (Chapter 8)

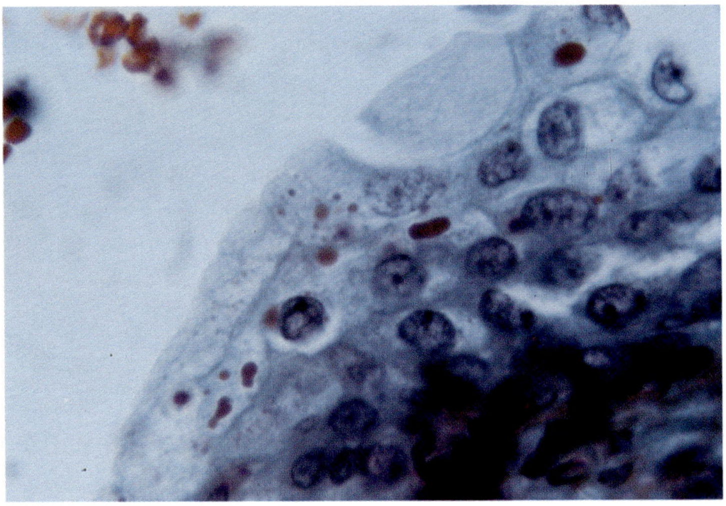

Fig. 74
This photograph demonstrates kala azar throughout the section as fine stippling. These organisms are stained well. Luna-Giemsa Method (Chapter 8)

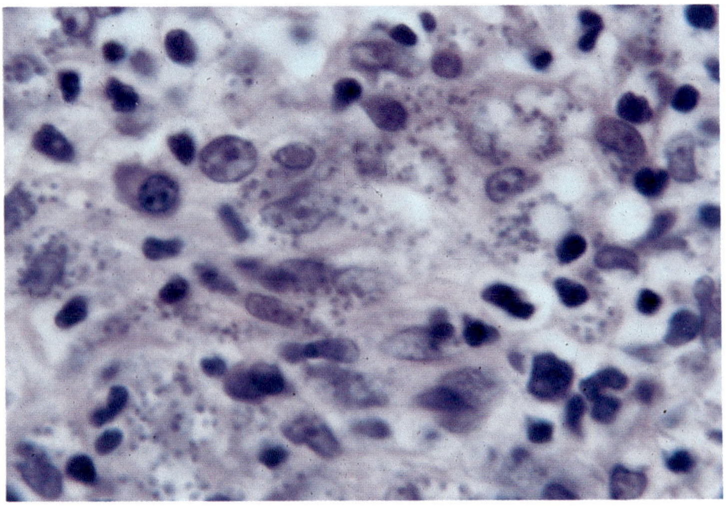

Fig. 75a
This section is demonstrating Entamoeba histolytica. Notice the ingested erythrocytes which resemble nuclei, particularly amoeba at mid bottom. Luna-Giemsa Method (Chapter 8)

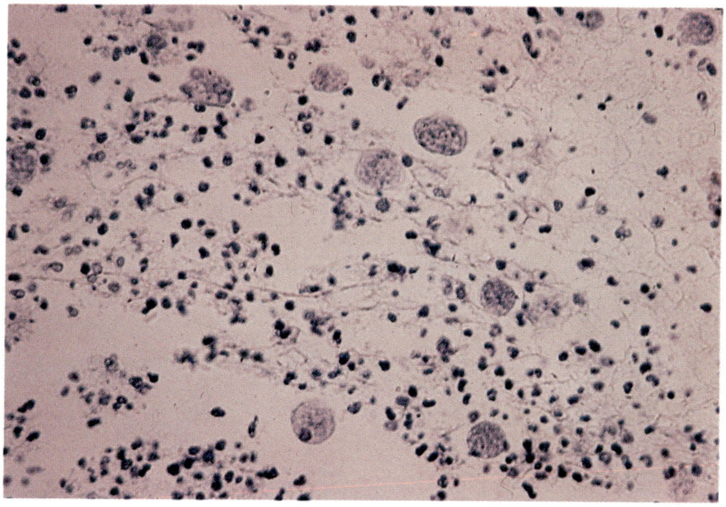

Fig. 75b
This is a section demonstrating Entamoeba histolytica with Grocott's Methenamine Silver. Notice the ingested erythrocytes in the section on the right. This particular section was stained in eosin before performing the GMS. This results (Chapter 8) in staining of the ingested erythrocytes. Grocott's Methenamine Silver Method

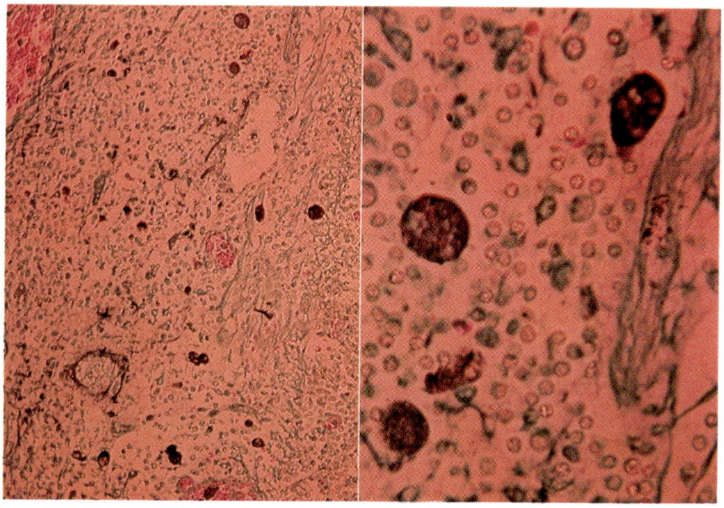

Fig. 75c
This is another example of Entamoeba histolytica using the PAS method. (Chapter 8)

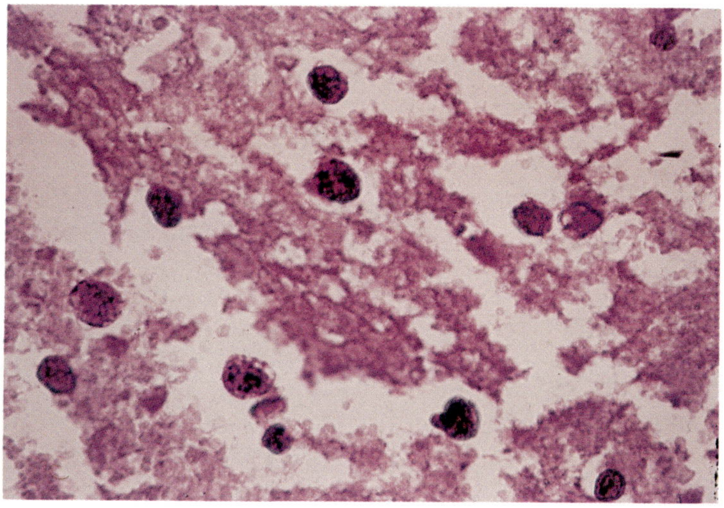

Fig. 75d
Fontana Masson stain may also be used for the demonstration of Entamoeba histolytica. (Chapter 8)

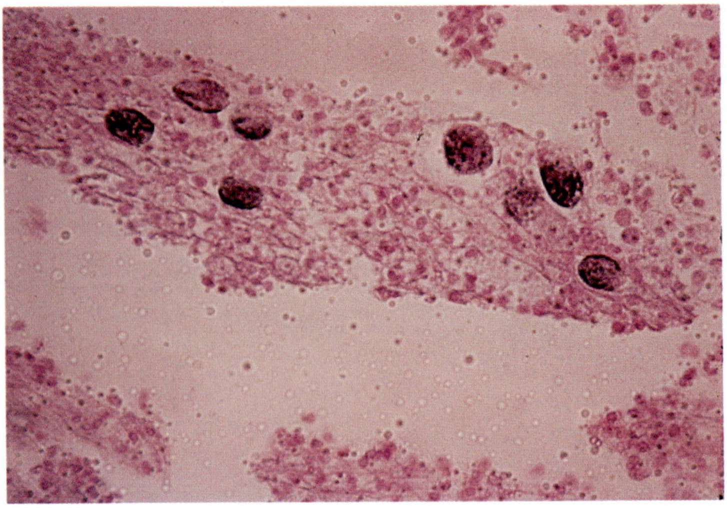

Fig. 76
This section demonstrates rickettsia (rocky mountain spotted fever). The organisms can be seen as very fine pink staining structures in the middle of the photograph. Luna Basic Fuchsin (Chapter 8)

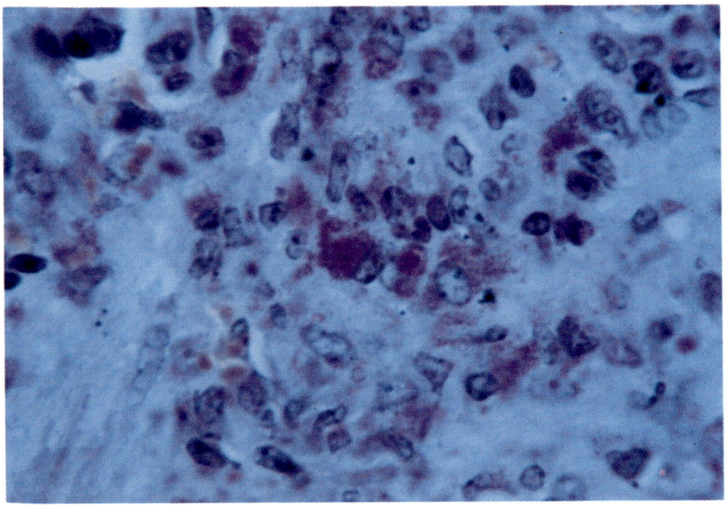

Fig. 77
This section is demonstrating rickettsiae in the middle of the section. Pinkerton's Method (Chapter 8)

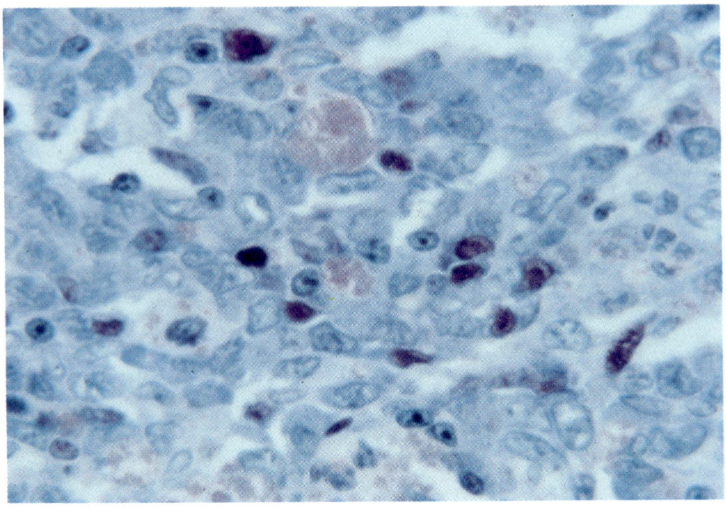

Fig. 78
This section is demonstrating rickettsiae in the middle of the section. Machivello's Method (Chapter 8)

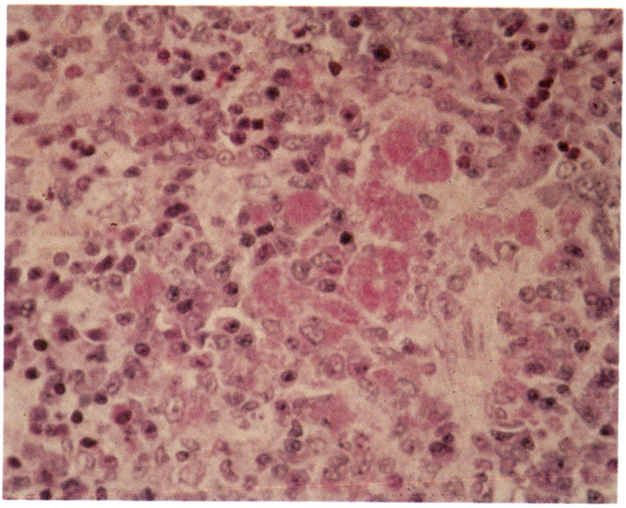

Fig. 79
Section of brain demonstrating malaria parasites in erythrocytes. The black structures are malaria pigment and subjacent to that lies the malaria parasite. Thomas' Method (Chapter 8)

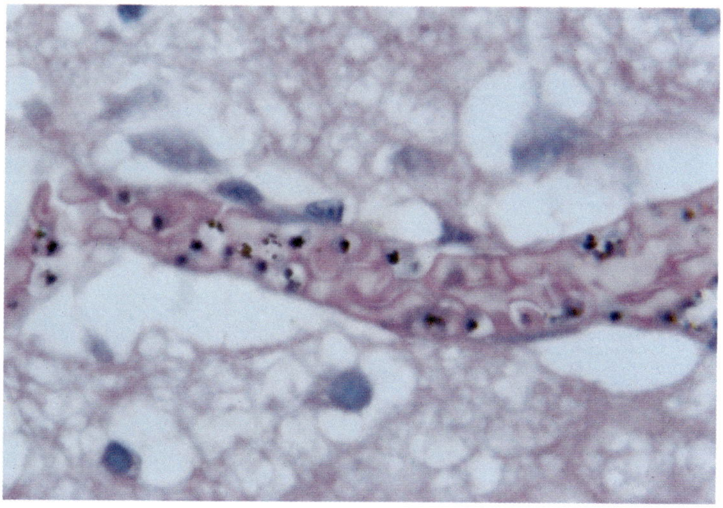

Fig. 80
This is a section taken under polarized microscopy to show the polarizing properties of malaria pigment which is somewhat identical to formalin pigment. Thomas' Method (Chapter 8)

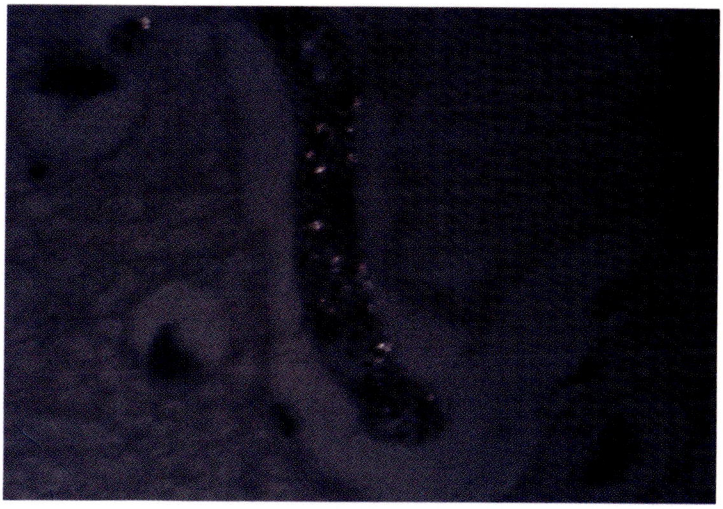

Fig. 81
This section of heart muscle is demonstrating trypanosoma cruzi in the center of the photograph. Luna-Bodian Stain (See Chapter 15)

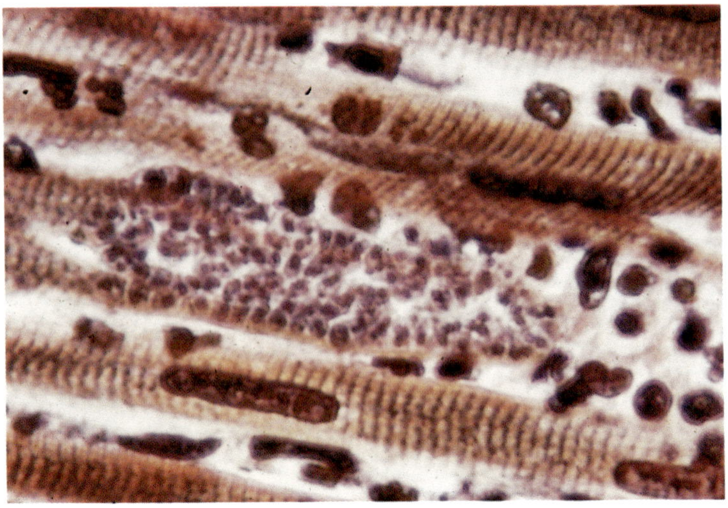

Fig. 82
This section of brain is demonstrating toxoplasmosis parasites. Hematoxylin and Eosin Method (Chapter 8)

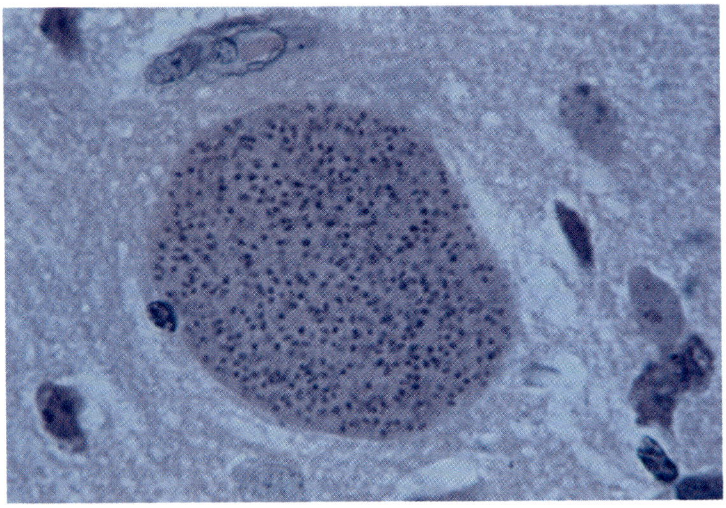

Fig. 83
This section of brain is demonstrating toxoplasmosis parasites. Notice the background was left lightly stained by overexposure to chromic acid to better identify the organisms. Grocott's Methenamine Silver Method (Chapter 8)

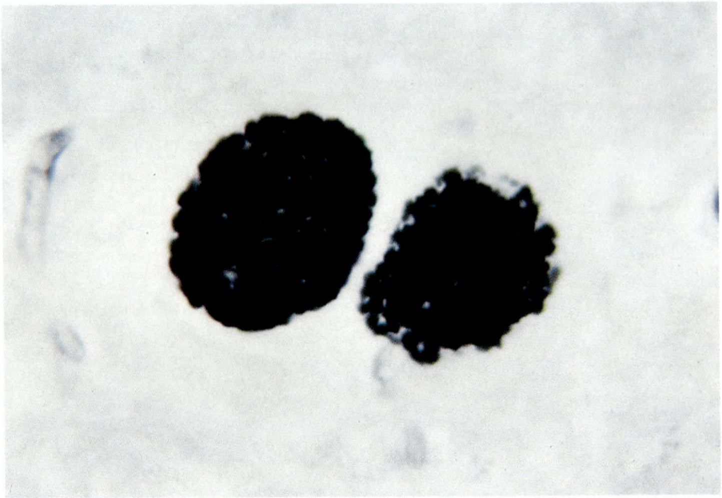

Fig. 84
Section demonstrating toxoplasmosis parasites. The interesting thing here is that the capsule is staining. Bodian Method (Chapter 8)

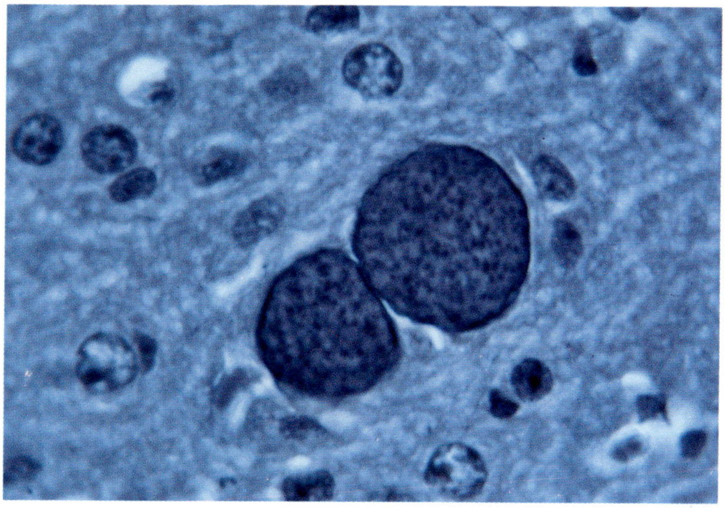

Fig. 85
Toxoplasmosis parasites demonstrated with PAS. (Chapter 8)

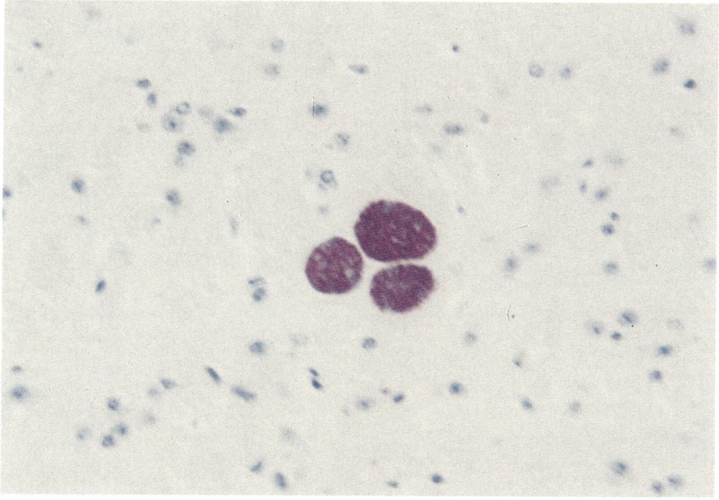

Fig. 86
This section demonstrates coarse round eosinophilic granules in the cytoplasm of each cell. Verfuerth-Luna Method (Chapter 9)

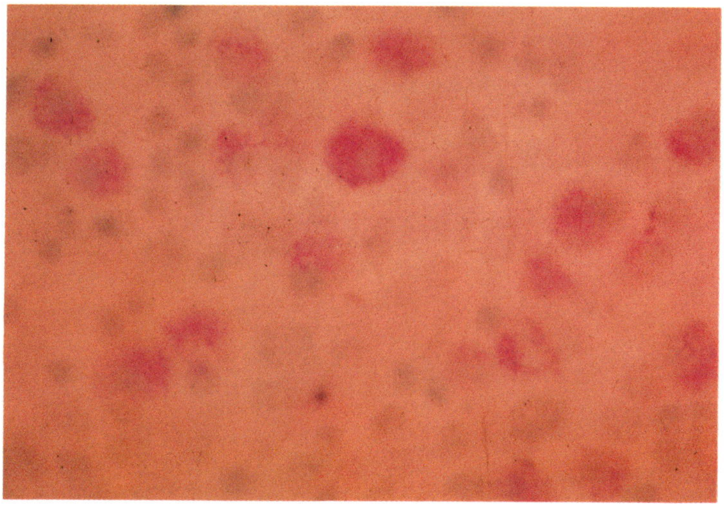

Fig. 87
Charcot-Leyden crystals can be seen in this section of bone marrow. The yellow background is too dark in this section. The base crystals are as bright as they should be. Luna-Brown-Brenn Method (Chapter 9)

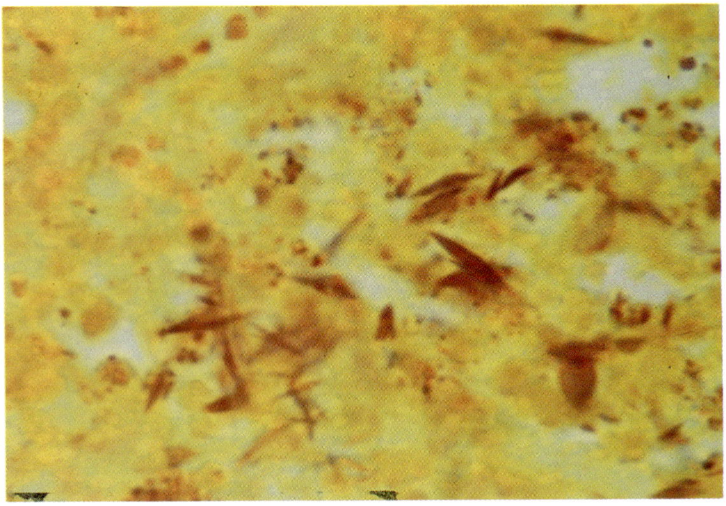

Fig. 88
Low power magnification of a section demonstrating Charcot-Leyden crystals. Hematoxylin is not blue enough in this section. Luna's Method (Chapter 9)

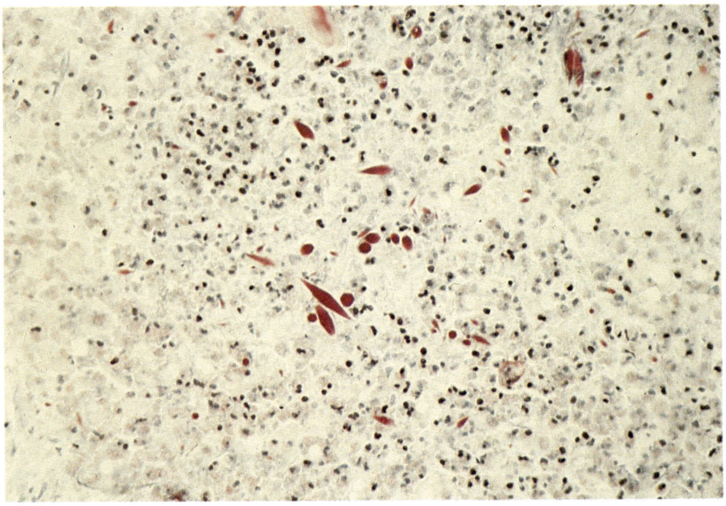

Fig. 89
Visceral larva migrams in a section of liver demonstrating Charcot-Leyden crystals. Section is well stained. Luna's Method (Chapter 9)

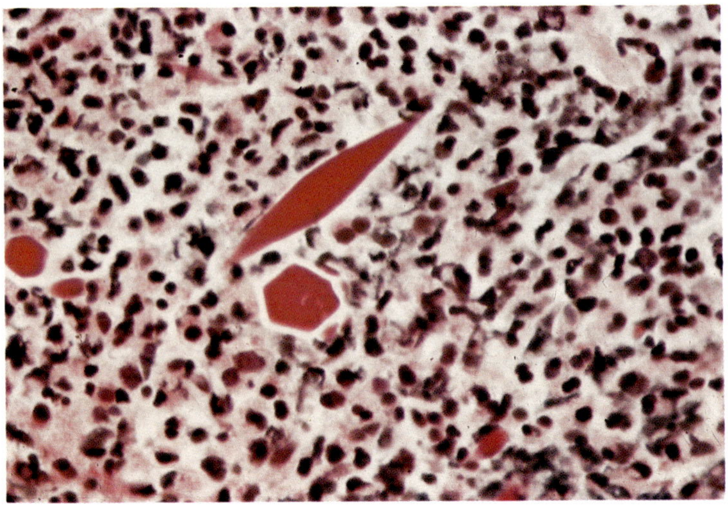

Fig. 90
Section of liver with visceral larva migrams demonstrating eosinophil granules. Section is well stained. Luna's Method (Chapter 9)

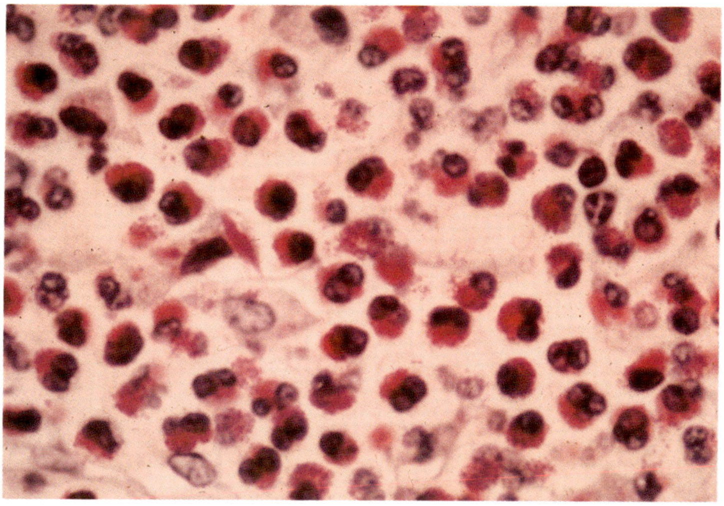

Fig. 91
This is a section of lymph node demonstrating RNA and DNA. The DNA can be seen as pink nuclear chromatin, the RNA stains green. Section is well stained. Lillie's Method (Chapter 9)

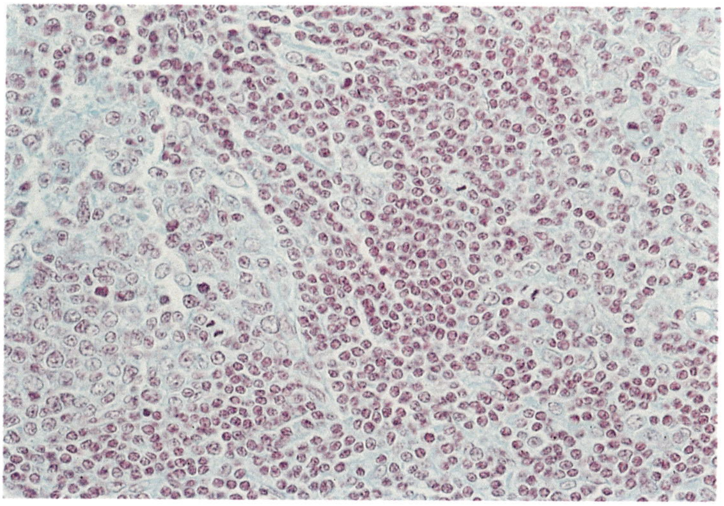

Fig. 92
Section of tonsil demonstrating ribonucleic acid (RNA) and deoxyribonucleic acid (DNA). The RNA can be seen as red cytoplasmic structures while the DNA is staining blue. Emanuele's Method (Chapter 9)

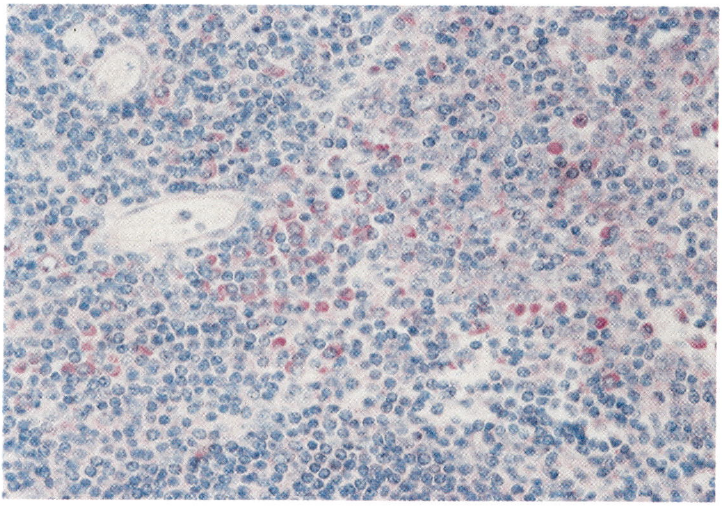

Fig. 93
Higher power of the same section seen in Fig. 92 demonstrating RNA in red and DNA in blue. Section is well stained. Emanuele's Method (Chapter 9)

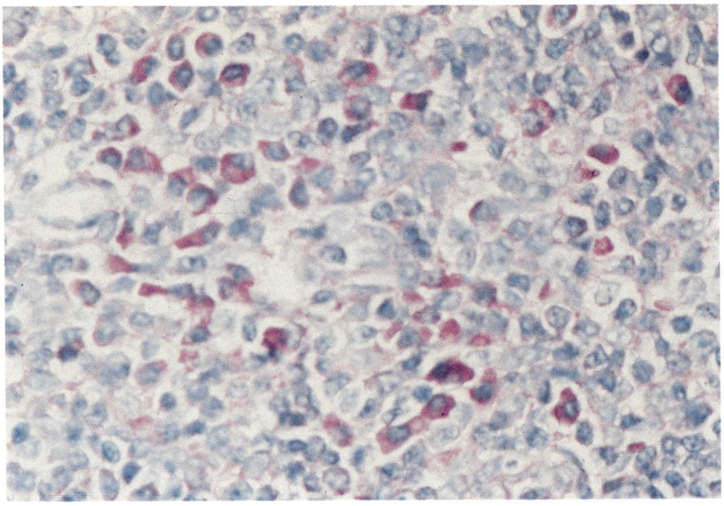

Fig. 94
Section of kidney demonstrating lead poisoning inclusion bodies. Lead poisoning inclusion bodies can be stained with Emanuele's RNA and DNA stain or any methyl green pyronin procedure. Lead poisoning inclusion bodies are demonstrated as red intranuclear round bodies. Emanuele's Method (Chapter 9)

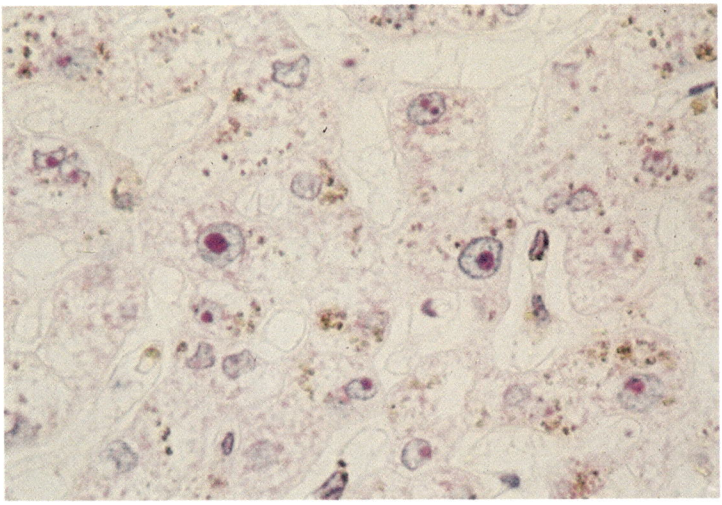

Fig. 95
A section demonstrating carcinoid granules stained reddish-brown. The section is well stained. Diazo's Method (Chapter 10)

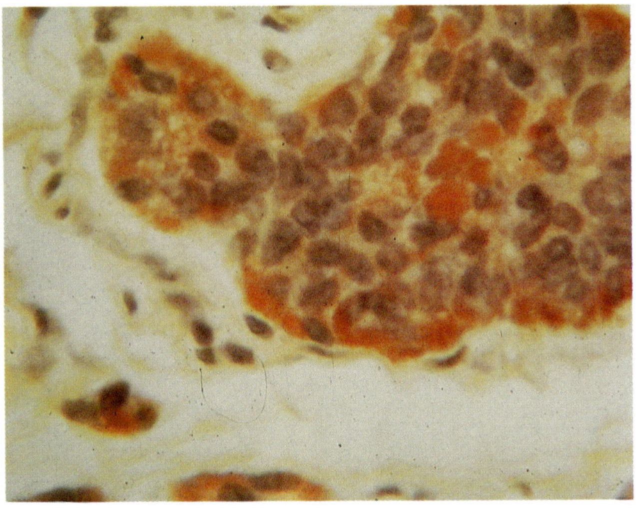

Fig. 96
Section of small intestine demonstrating the argentaffin cell granules in black. Section is well stained. Fontana-Masson Method (Chapter 10)

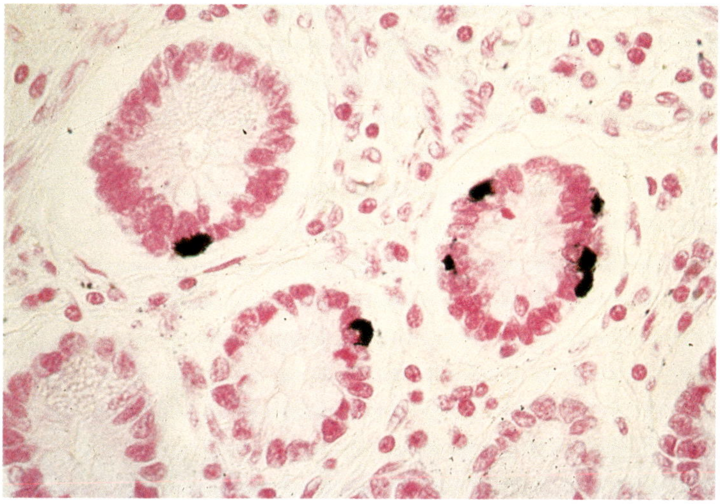

Fig. 97
Lymph node containing a carcinoid. Notice that not all the carcinoid granules are stained with this stain. This is fairly characteristic of staining with the Fontana-Masson Method. Fontana-Masson Method (Chapter 10)

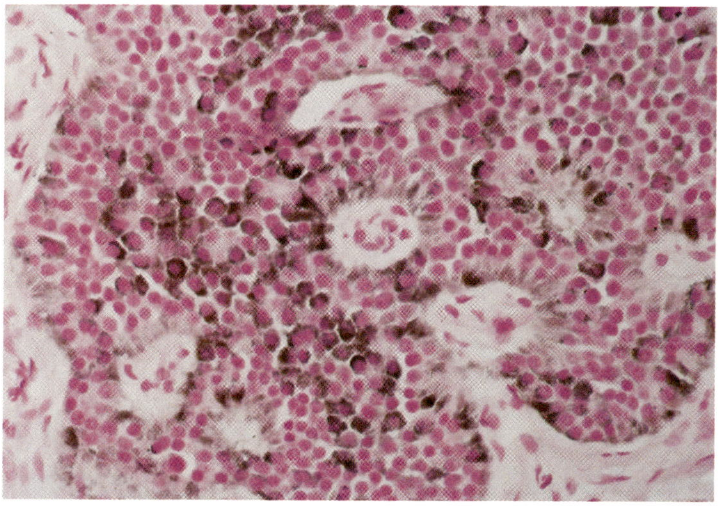

Fig. 98
Section of lymph node containing carcinoid granule. Notice the deep staining of the entire section which is characteristic of this particular staining procedure. Sevier-Munger Method (Chapter 10)

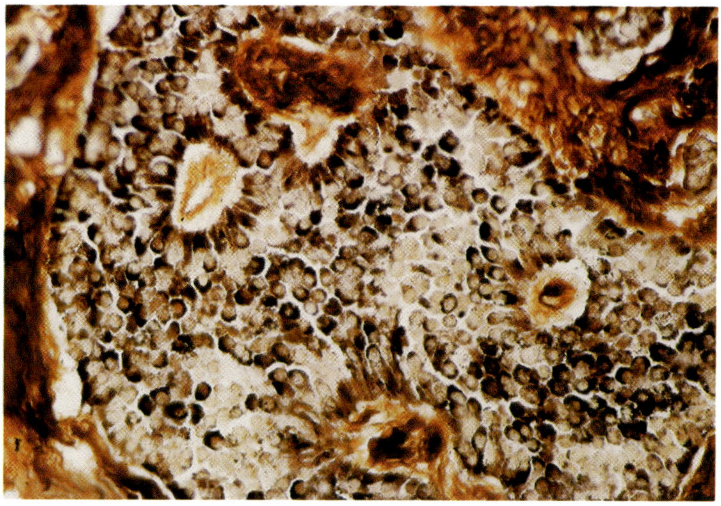

Fig. 99
Section of liver containing hepatitis B surface antigen. The staining reaction is well performed. Notice the clear background which should be characteristic of a good aldehyde fuchsin procedure for the demonstration of hepatitis B. Luna's Method (Chapter 10)

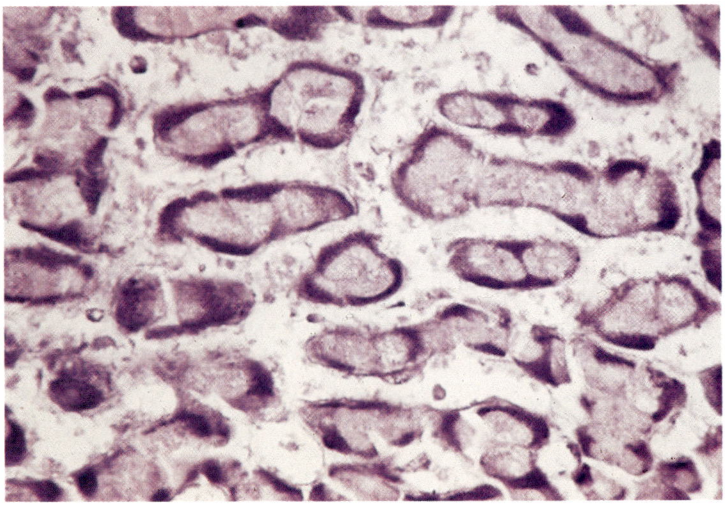

Fig. 100
Section of kidney embedded in Epon 2 micrometer thick section demonstrating juxtaglomerular granules. Notice the reddish-purple staining juxtaglomerular granules. Smith's Method (Chapter 10)

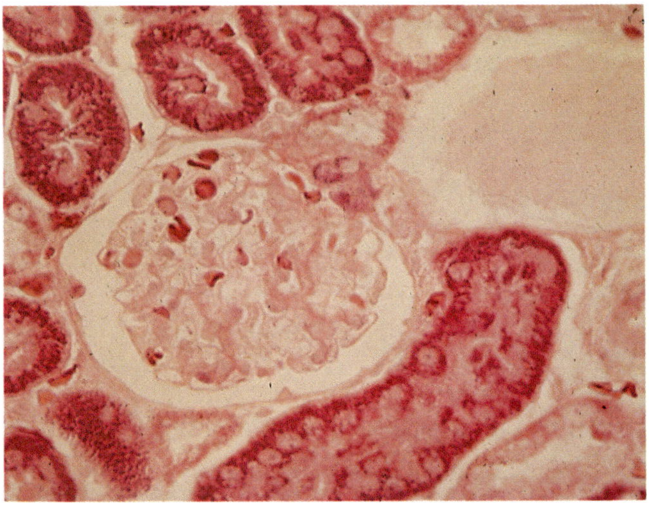

Fig. 101
Section demonstrating juxtaglomerular granules on the walls of an efferent and afferent vessel as they enter and leave the glomeruli. Section is well stained. This is an example of what a well fixed section can produce. In this instance the Bouin's fixation was properly handled and the juxtaglomerular granules stained well with the Masson procedure which normally does not stain them. Bouin's Fixed-Masson Method (Chapter 10)

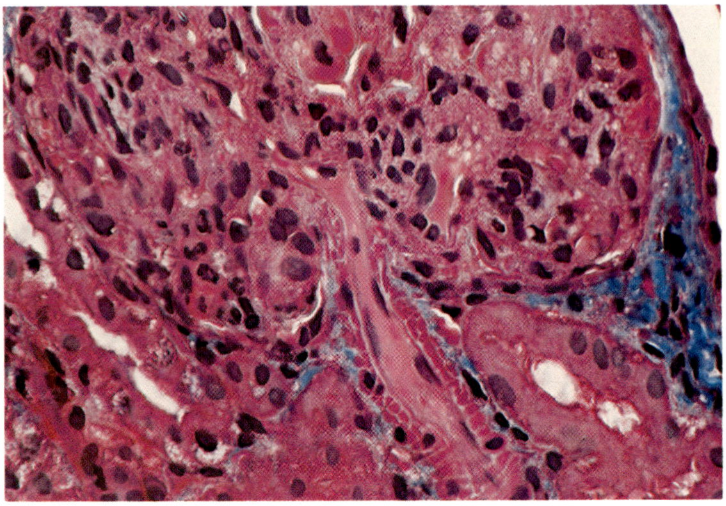

Fig. 102
Mast cell tumor demonstrating mast cell granules in the cytoplasm of various cells. Section is well stained. Luna Methyl Blue Method (Chapter 10)

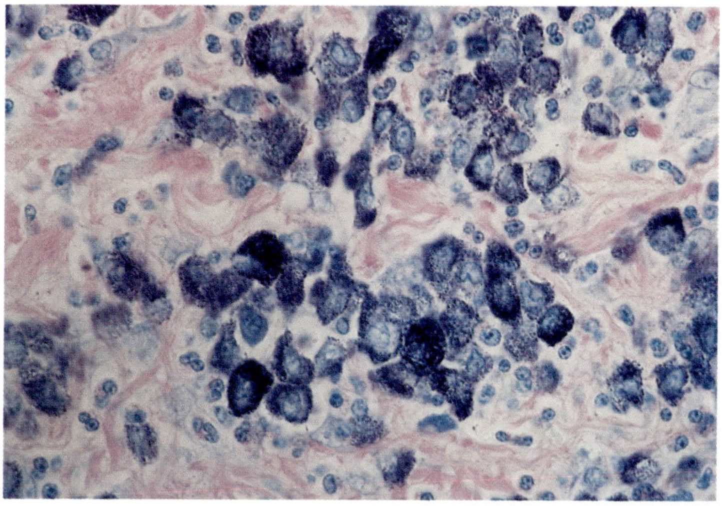

Fig. 103
Mast cell granules being demonstrated with the alcian blue (pH 2.5) method. Section is well stained. (Chapter 12)

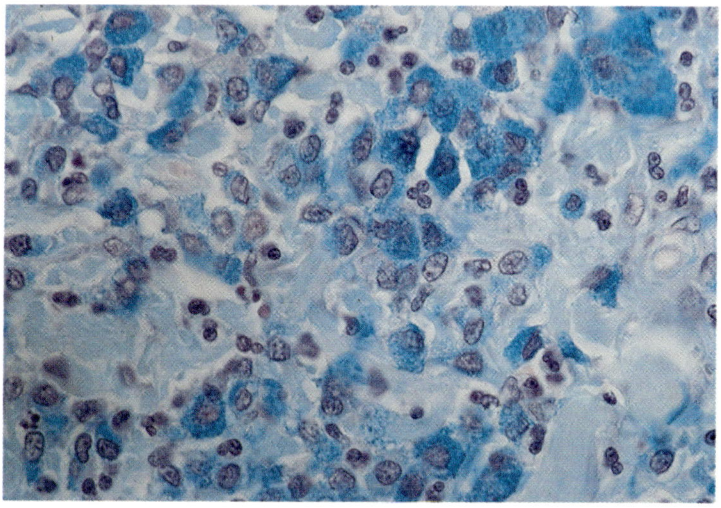

Fig. 104
Mast cell granules. Mast cell granules can be seen in aqua colored reaction. Section is well stained. Mowry's Colloidal Iron (Chapter 12)

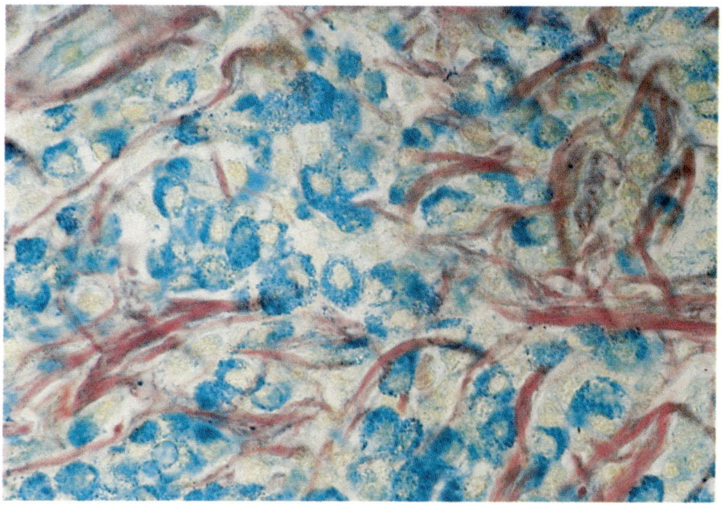

Fig. 105
Mast cell granules being demonstrated in red throughout the section. Section is well stained. Leder's Method (Chapter 9)

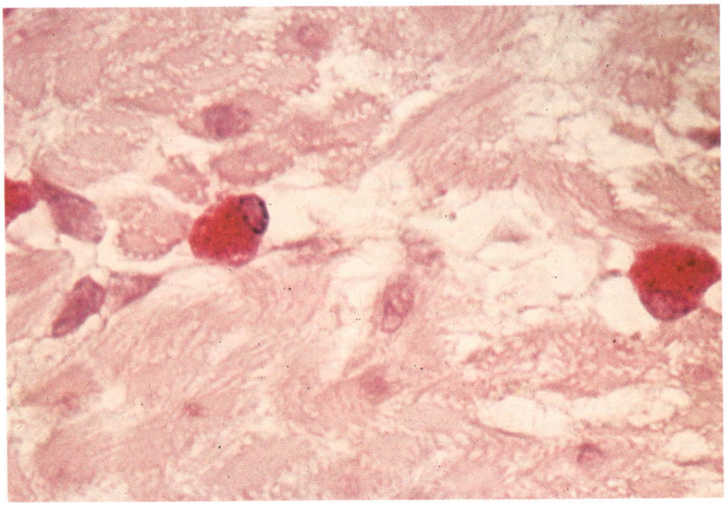

Fig. 106
This section demonstrates pancreatic beta cell granules. Section is well stained. Wenger's Aldehyde Fuchsin Method (Chapter 10)

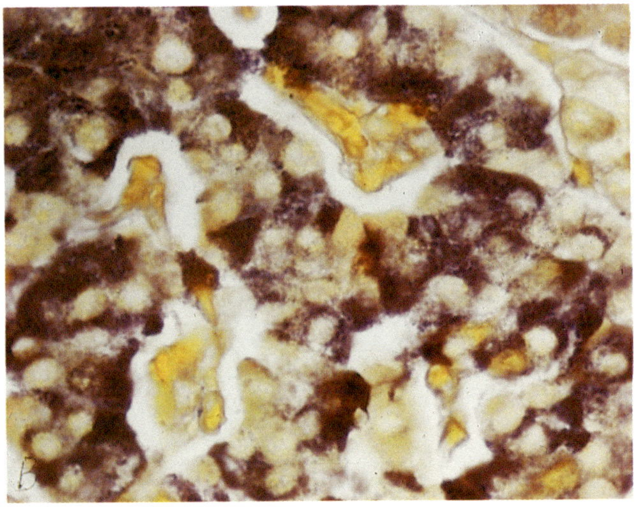

Fig. 107
Section of pancreas demonstrating pancreatic islet cells. Churukian-Schenk Method (Chapter 10)

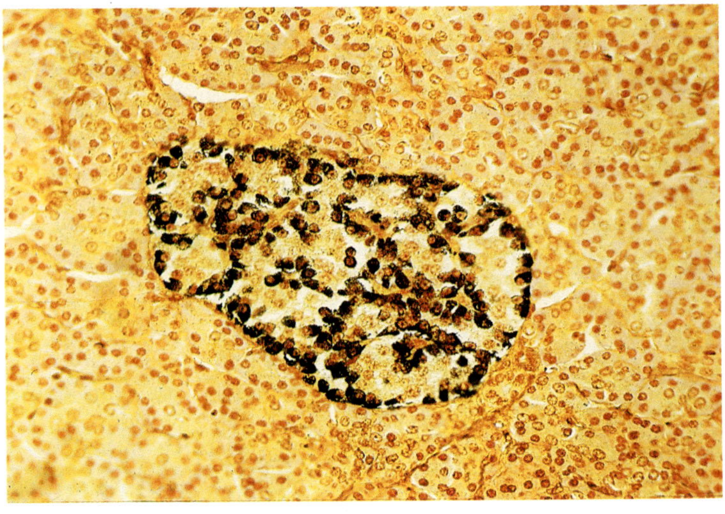

Fig. 108
Lymph node containing a carcinoid tumor demonstrating carcinoid granules. Section is well stained. Churukian-Schenk Method (Chapter 10)

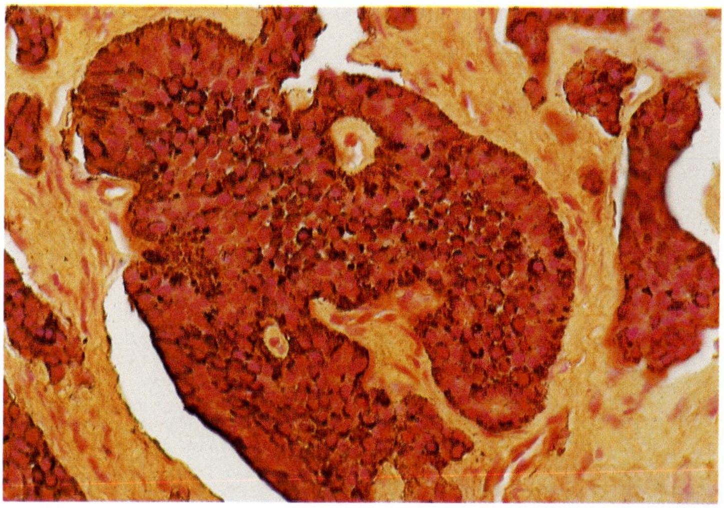

Fig. 109
This composite shows a section of a pheochromocytoma on the right side which has been fixed in a fixative containing a dichromate salt. In this particular instance it was potassium chromate. Notice the green staining reaction of the pheochrome cells (right side). The left photograph illustrates the same lesion stained with H&E. (Chapter 10)

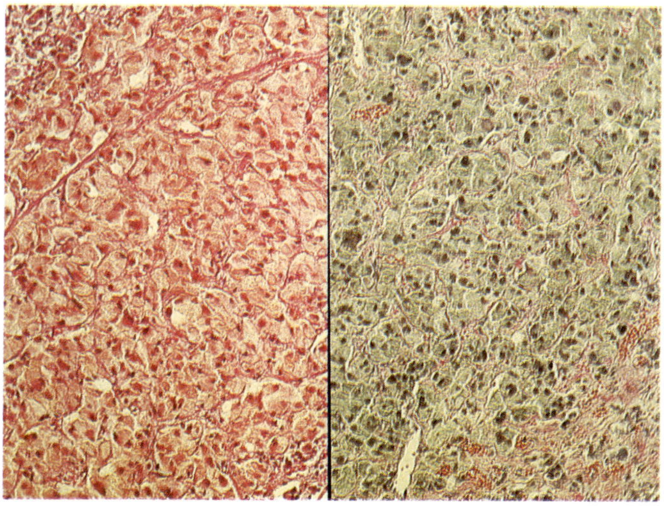

Fig. 110
Higher power of the pheochromocytoma. Again notice the green staining reaction of the pheochrome cell granules. Section is well stained. Gomori's Method for Chromophine Granules (Chapter 10)

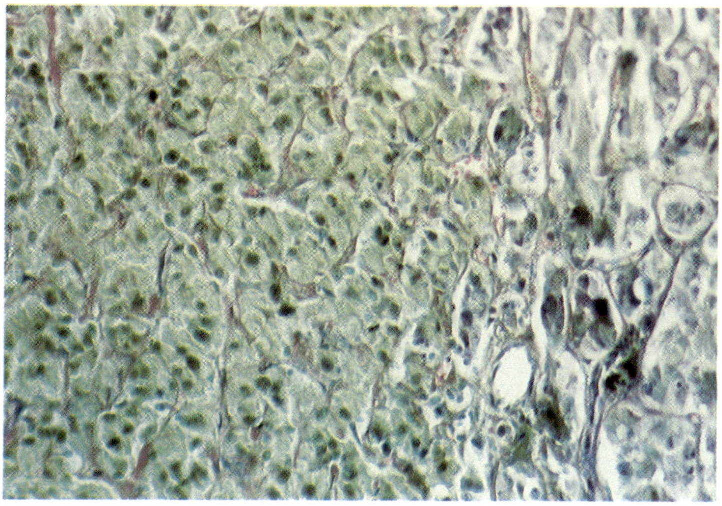

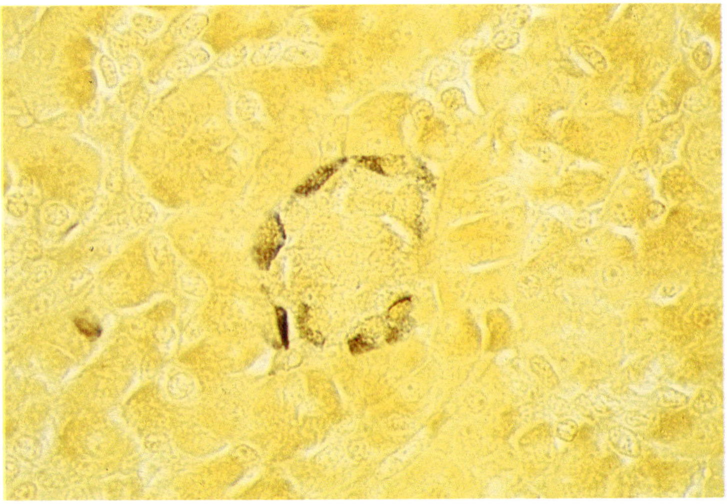

Fig. 111
This is a section of pancreas demonstrating $Alpha_2$ cell granules in the islet of Langerhan. Section is well stained. Grimelius Method (Chapter 10)

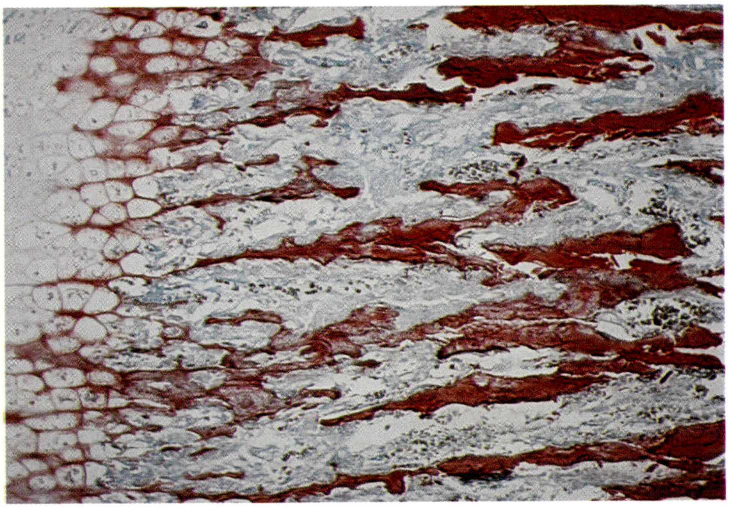

Fig. 112
Section of bone at the epiphyseal junction which demonstrates calcium stained with alizarin red S. Section is well stained. Dahl's Method (Chapter 11)

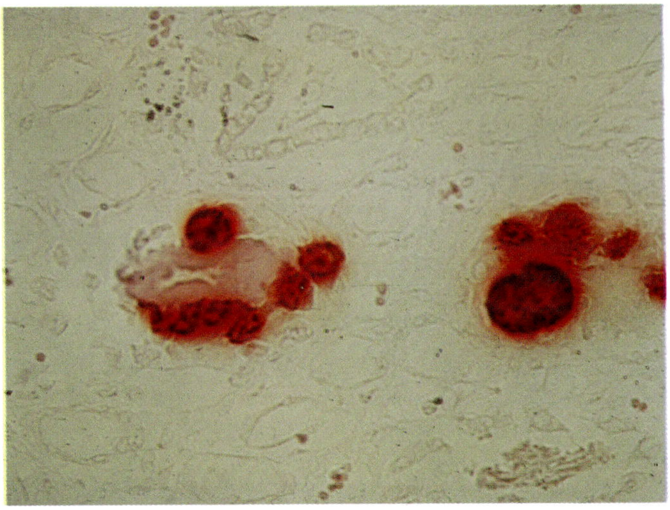

Fig. 113
Section of kidney demonstrating uric acid crystals stained with the alizarin red S. Section is well stained. Dahl's Method (Chapter 11)

Fig. 114
Section of bone at the epiphyseal junction which demonstrates calcium as brownish-black throughout the section. Section is well stained. Von Kossa's Method (Chapter 11)

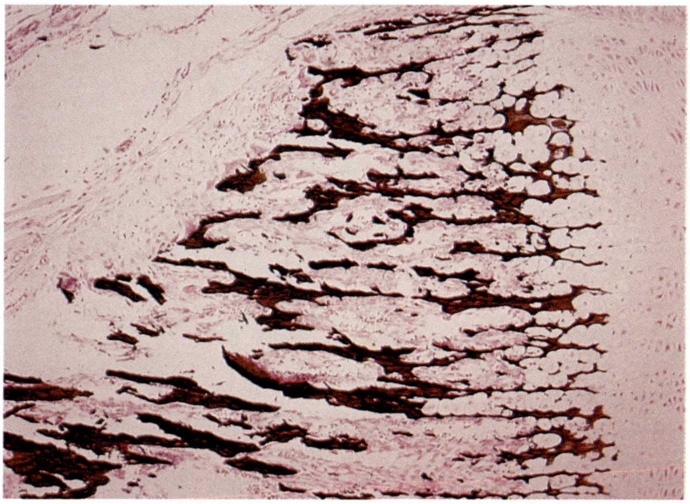

Fig. 115
Section of a malignant melanoma demonstrating ferric iron in blue. Section is well stained. Lillie's Stain for Ferric and Ferrous Iron (Chapter 11)

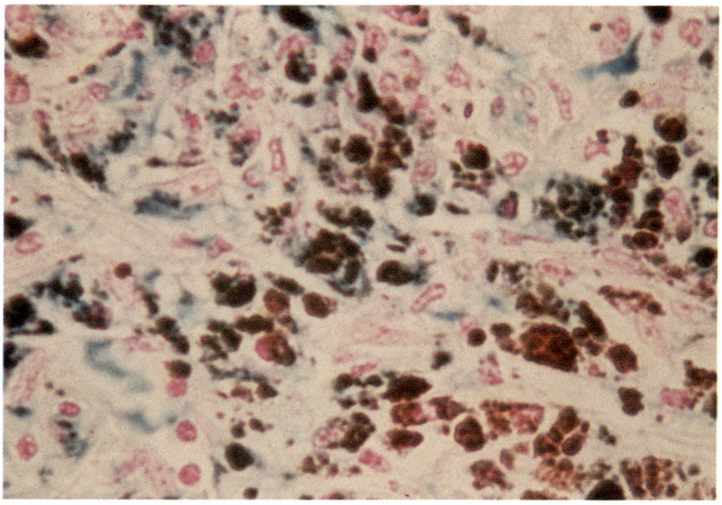

Fig. 116
Section of a malignant melanoma demonstrating ferrous iron in blue. The brown pigment is melanin. Lillie's Stain for Ferric and Ferrous Iron (Chapter 11)

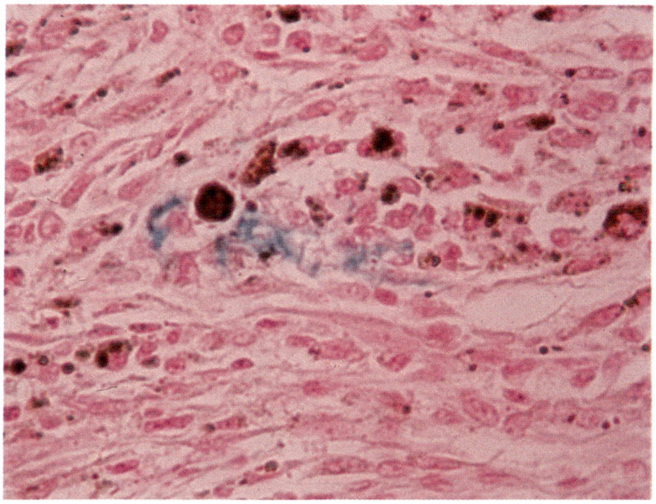

Fig. 117
Section of liver demonstrating iron. Section is well stained. Luna's Method (Chapter 11)

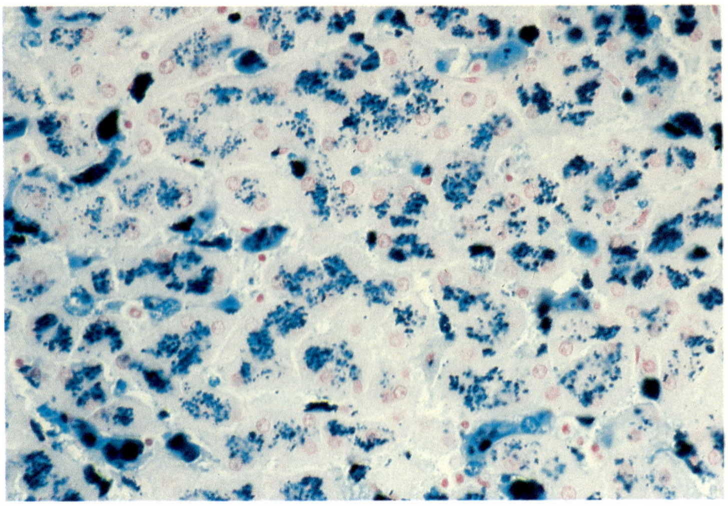

Fig. 118
A scanning electron micrograph of an asbestos fiber lying across the middle of the photograph. (Chapter 11)

Fig. 119
This photograph is similar to section shown in Fig. 117 demonstrating the iron content (right) which is a common occurrence when sections containing asbestos are stained with an iron reaction. Section at the left is an H&E. Luna's Method (Chapter 11)

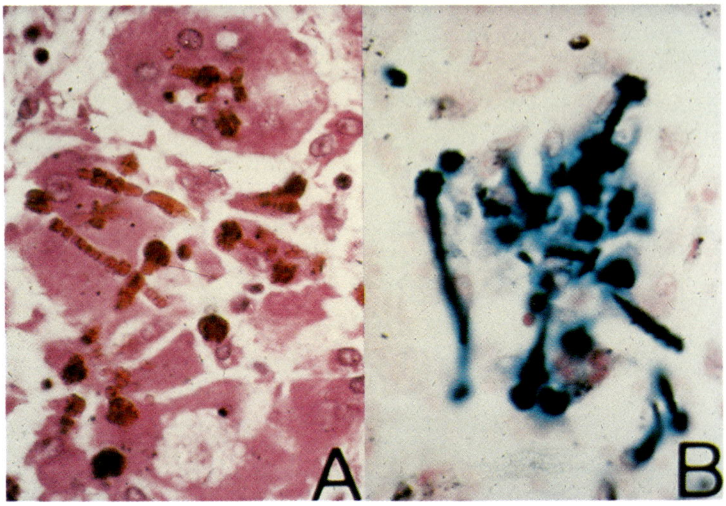

Fig. 120
Section of a malignant melanoma stained for both melanin (black pigment) and iron (blue pigment). Luna's Method for Ferric Iron and Melanin (Chapter 11)

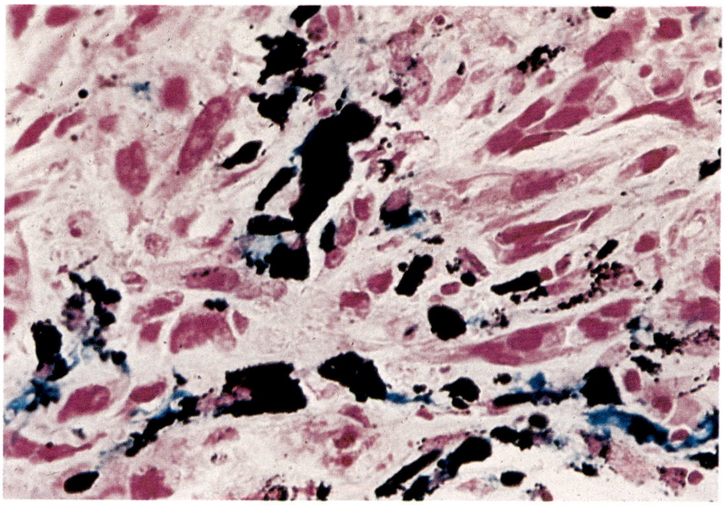

Fig. 121
Section of liver stained for both iron and bile canaliculi. Section is well stained. Luna's Method (Chapter 11)

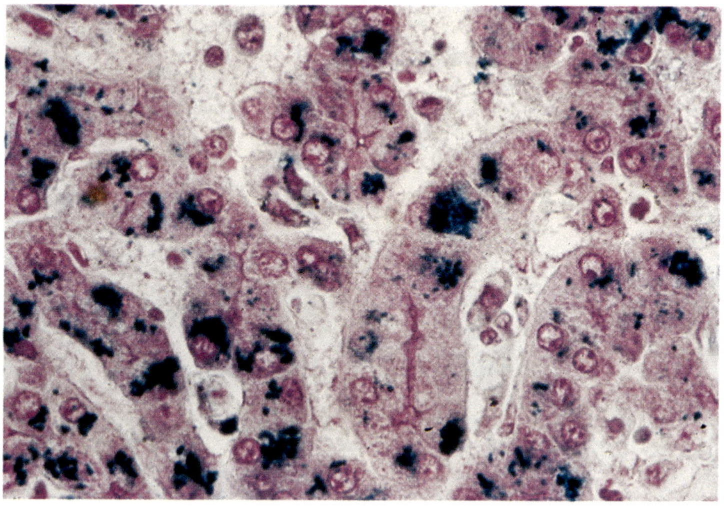

Fig. 122
Section of fetal liver demonstrating copper. Lindquist's Method (Chapter 11)

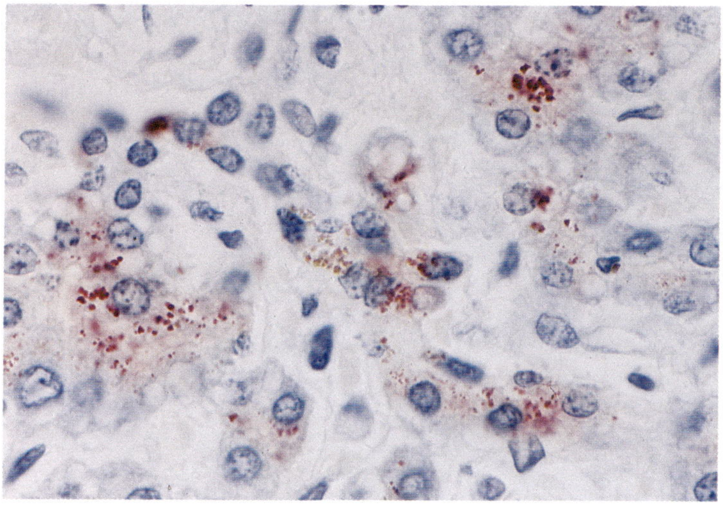

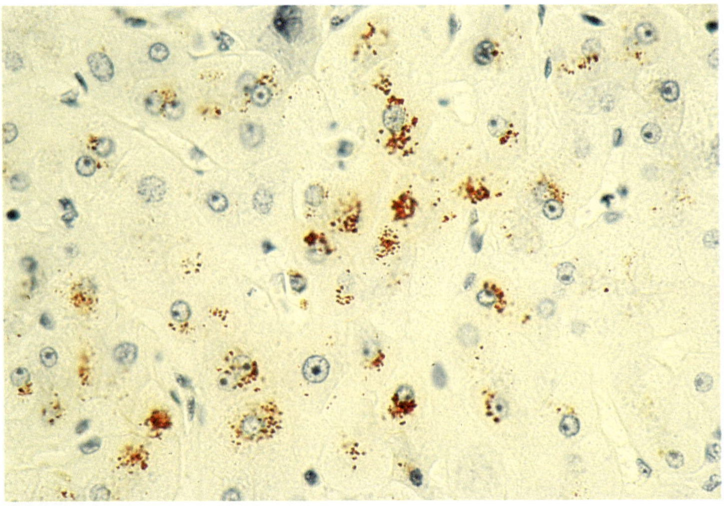

Fig. 123
Section of adult liver from a case of Wilson's disease. Section is well stained. Lindquist's Method (Chapter 11)

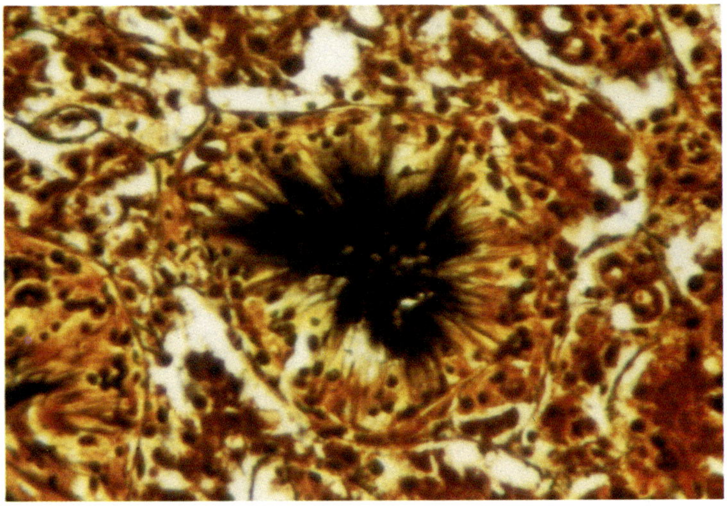

Fig. 124
Section demonstrating calcium oxalate reacting to silver nitrate. The reaction can be seen in black (middle). DeGalantha's Method (Chapter 11)

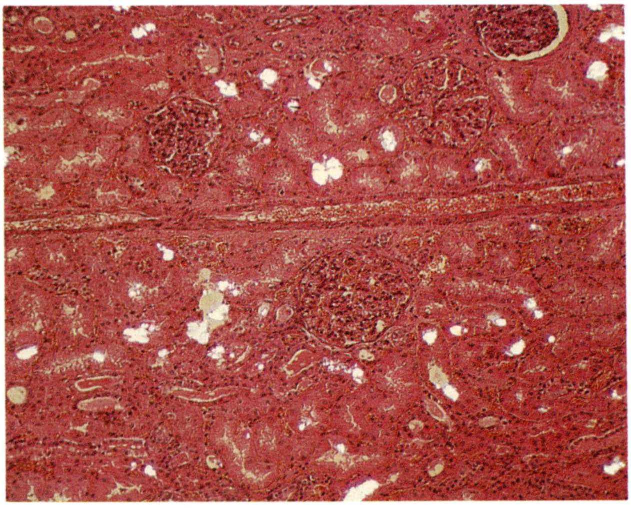

Fig. 125
Section demonstrating calcium oxalate crystals under partial polarized microscopy. (Chapter 11)

Fig. 126
Section containing uric acid crystals in a case of gout. Section is well stained. DeGalantha's Method (Chapter 11)

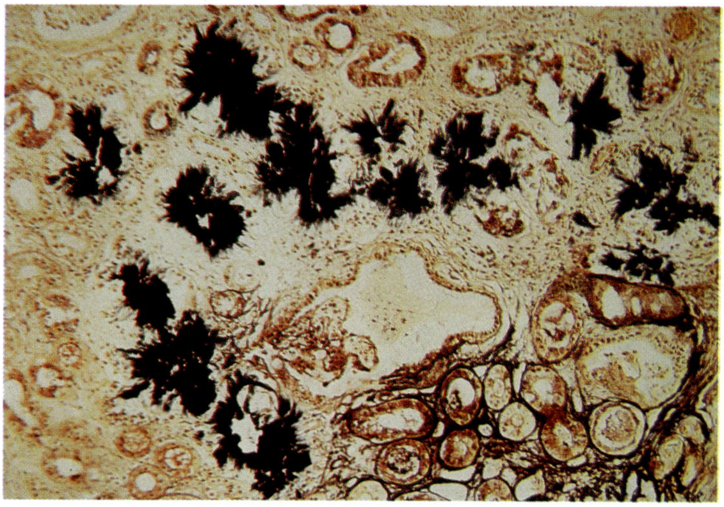

Fig. 127
Section of spleen demonstrating lipofuscin. Section is well stained. Carbol-Fuchsin Method for Lipofuscin (Chapter 11)

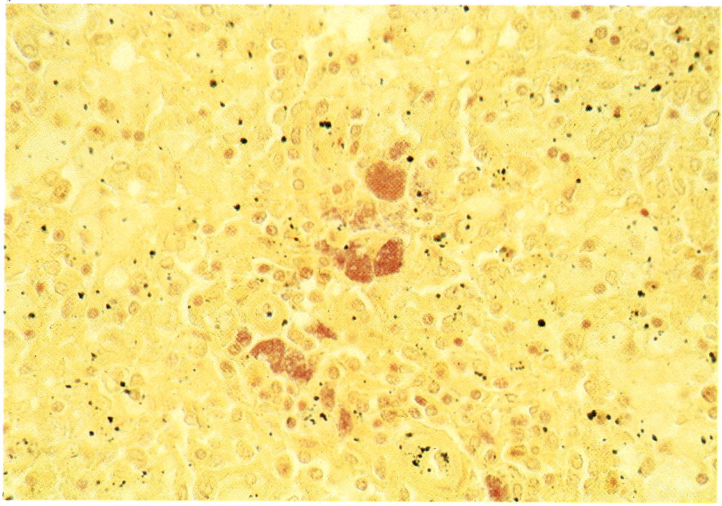

Fig. 128
Section of spleen demonstrating lipofuscin in red. Section is well stained. Oil Red O Stain for Lipofuscin (Chapter 11)

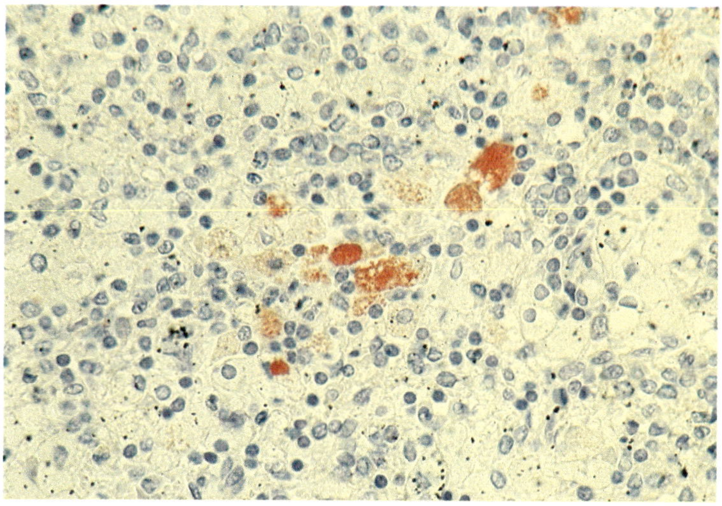

Fig. 129
Section of liver with a silver reaction demonstrating lipofuscin. It is not unusual for any silver reaction, particularly the Fontana-Masson, to demonstrate lipofuscin in tissue sections. Section is slightly over-developed. Churukian's Modification of Luna's Method for Melanin. (Chapter 11)

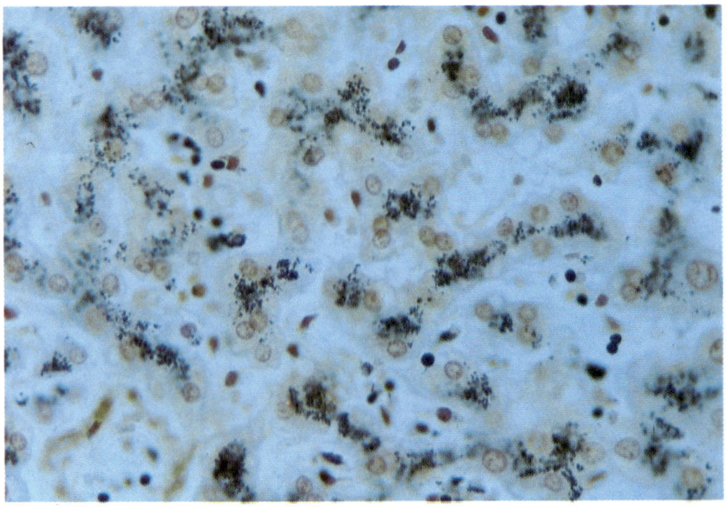

Fig. 130
Section of skin with a malignant melanoma. Notice the clearly defined melanin granules in this silver reaction. Warkel-Luna-Helwig Method (Chapter 11)

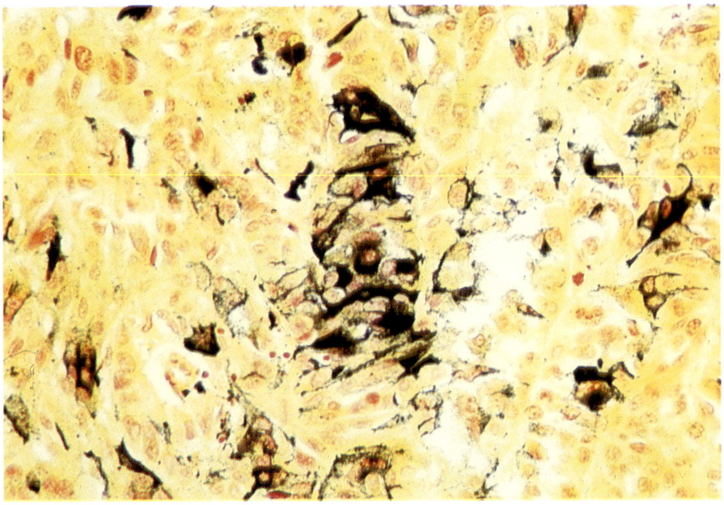

Fig. 131
Section of skin with a malignant melanoma. Melanin is staining green. Section is well stained. Luna's Method for Reducing Substances and Mucin (Chapter 11)

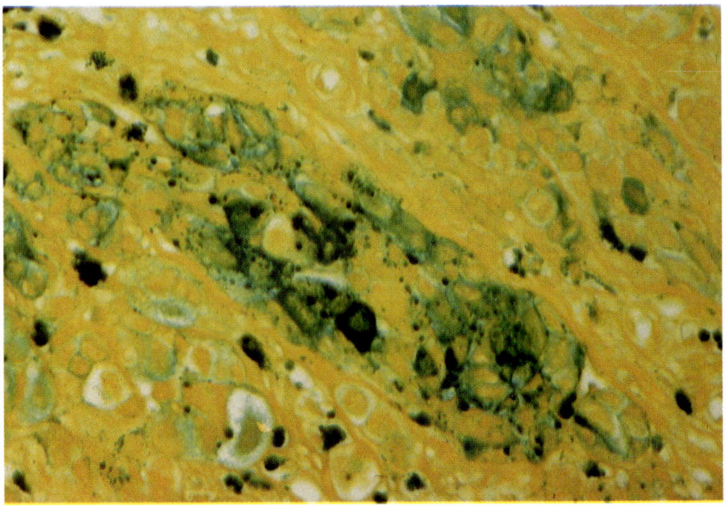

Fig. 132
Section of kidney demonstrating hemoglobin. Luna-Giemsa Method (Chapter 11)

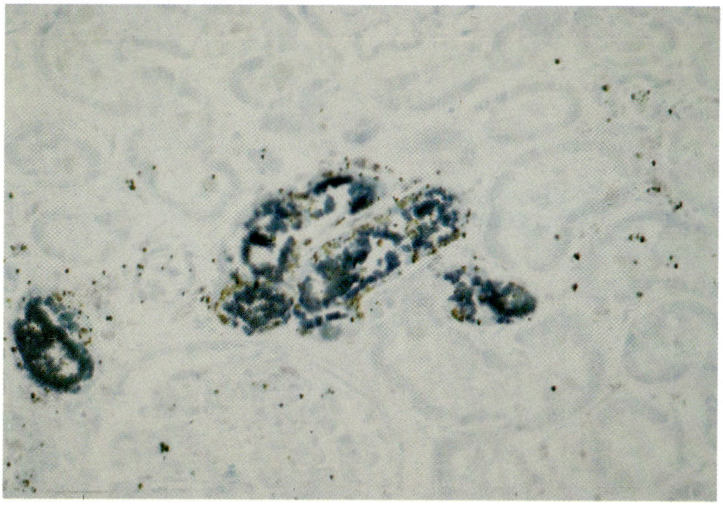

Fig. 133
Formalin pigment can be seen as black structures or black precipitate throughout this section. Hematoxylin and Eosin Method (Chapter 11)

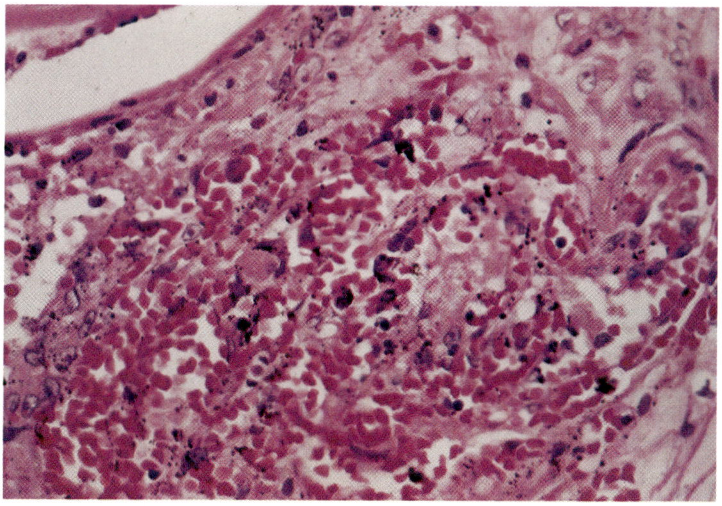

Fig. 134
Section of brain containing two blood vessels which contain malaria parasites upon which lies the dark staining malaria pigment. Section is well stained. Luna-Giemsa Method (Chapter 11)

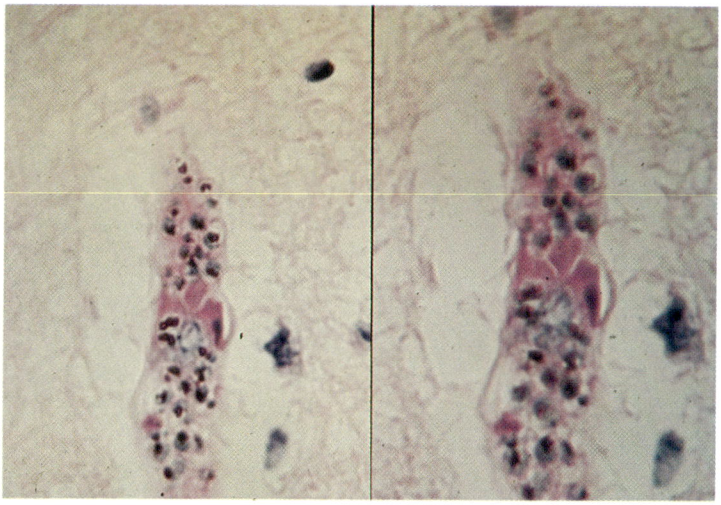

Fig. 135
H&E stained section of large intestine containing melanosis coli pigment in the center of section. (Chapter 11)

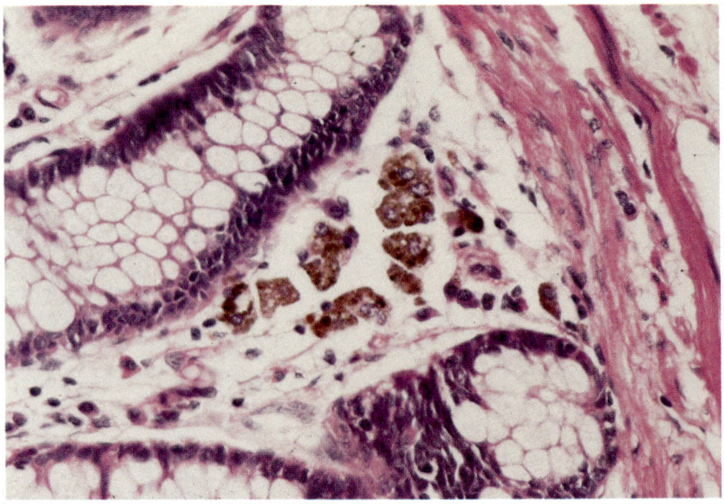

Fig. 136
Section similar to Fig. 135 containing melanosis coli pigment. Section is well stained. Periodic-Acid-Schiff Method (Chapter 11)

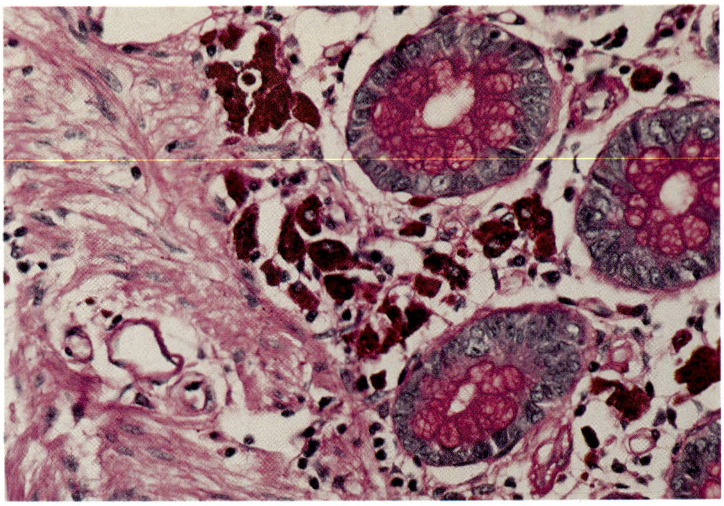

Fig. 137
Section containing melanosis coli pigment. Section is well stained. Luna-Giemsa Method (Chapter 11)

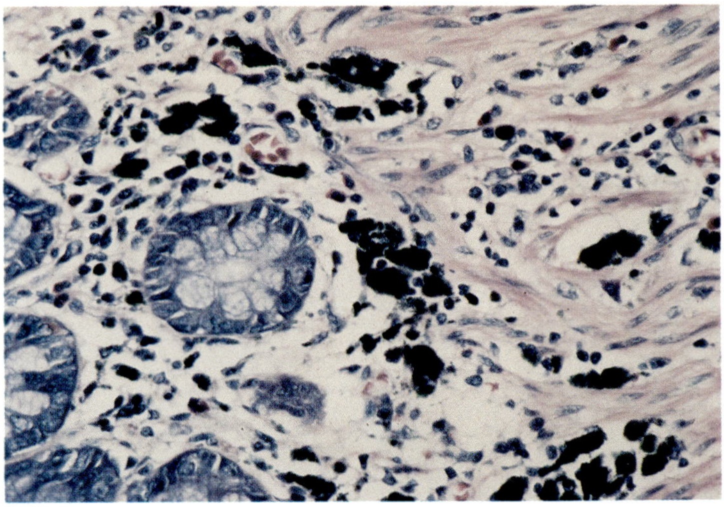

Fig. 138
This section of lung contains barium crystals. The crystals can be seen as white structures throughout the section as a result of partial polarized microscopy. (Chapter 11)

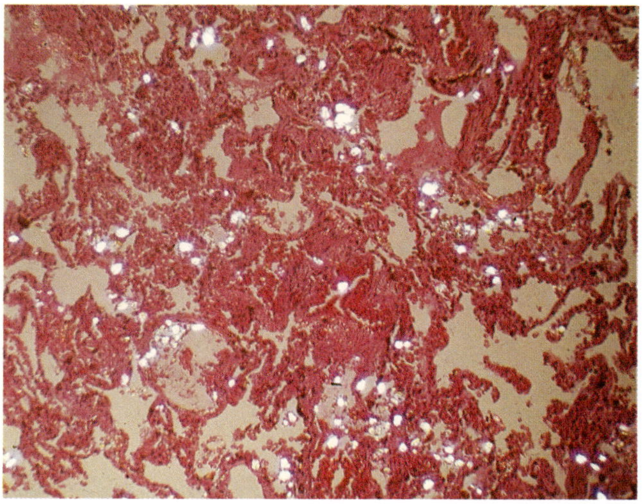

Fig. 139
H&E stained section of lung containing silicosis pigment which is identified as a dark precipitate type material throughout the section. (Chapter 11)

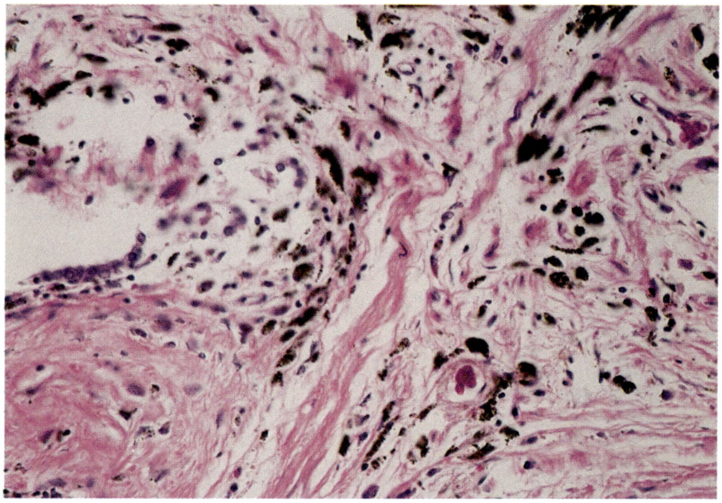

Fig. 140
Bronchial lymph node containing carbon particles (pigment) which can easily be identified throughout the section. Masson's Method (Chapter 11)

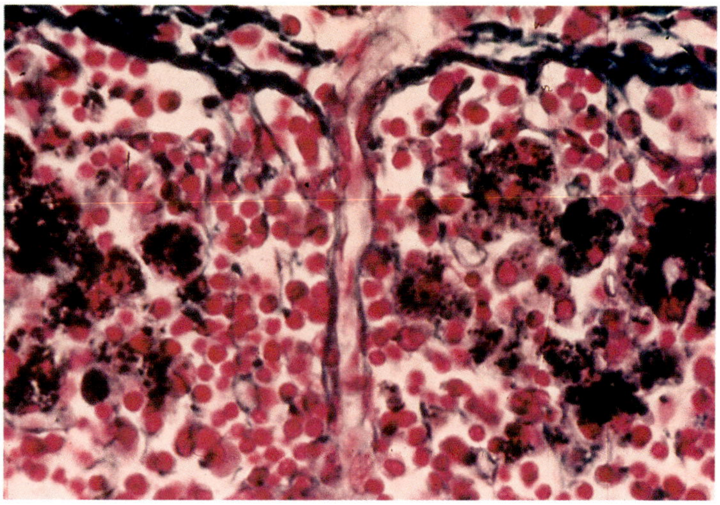

Fig. 141
This section of lung which is under partial polarization demonstrates talcum powder in the case of an IV drug user. (Chapter 11)

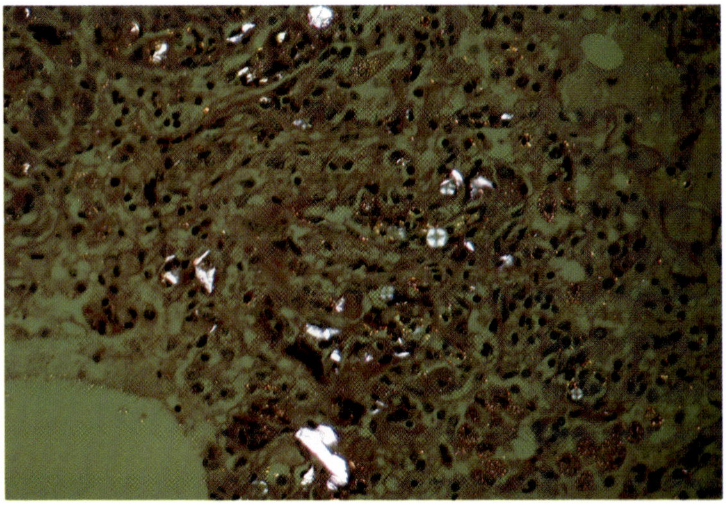

Fig. 142
The dark staining material in the dermis of this specimen demonstrates pigment which is the result of tattooing. Hematoxylin and Eosin Method (Chapter 11)

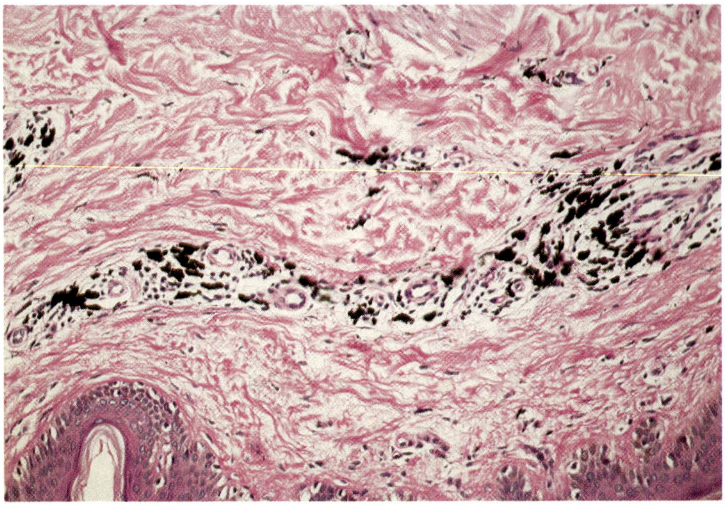

Fig. 143
Section of liver demonstrating bile pigment. Section is well stained. Hall's Method (Chapter 11)

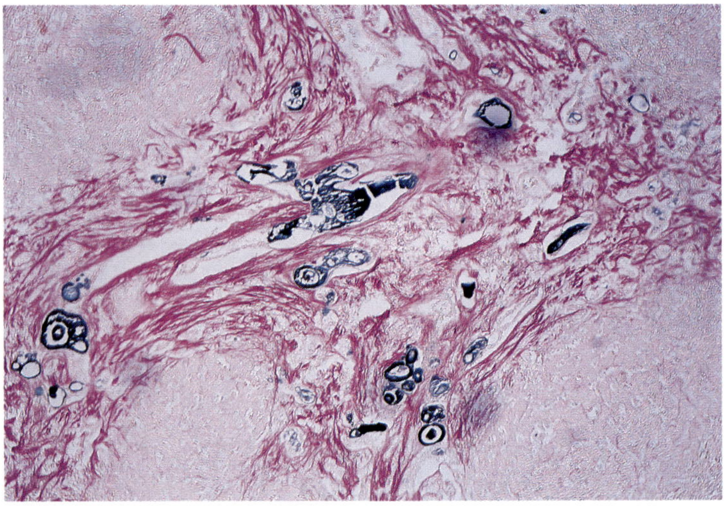

Fig. 144
Section of heart muscle demonstrating amyloid. The amyloid can be seen as a pinkish-red structure throughout the section. Section is well stained. Sweat-Puchtler Sirius Red S Method (Chapter 12)

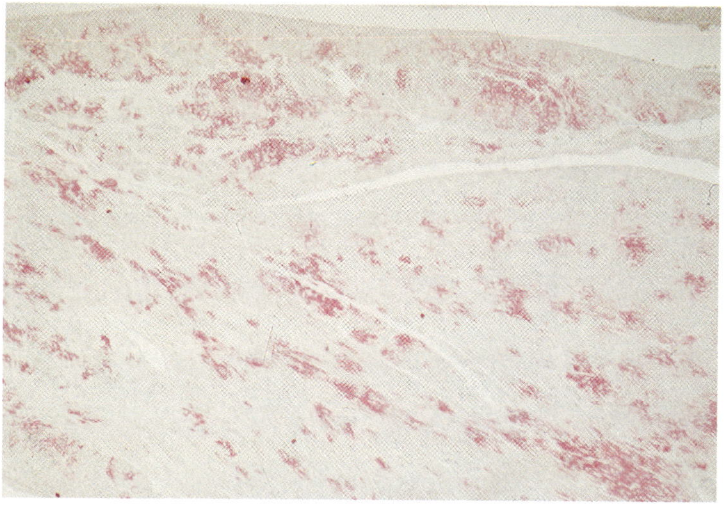

Fig. 145
This section is stained with Sweat-Puchtler sirius red S and is viewed under partial polarization to illustrate the polarizing properties of sirius red S which is somewhat similar to the properties seen with the congo red procedure. Section is well stained. Sweat-Puchtler Sirius Red S Method (Chapter 12)

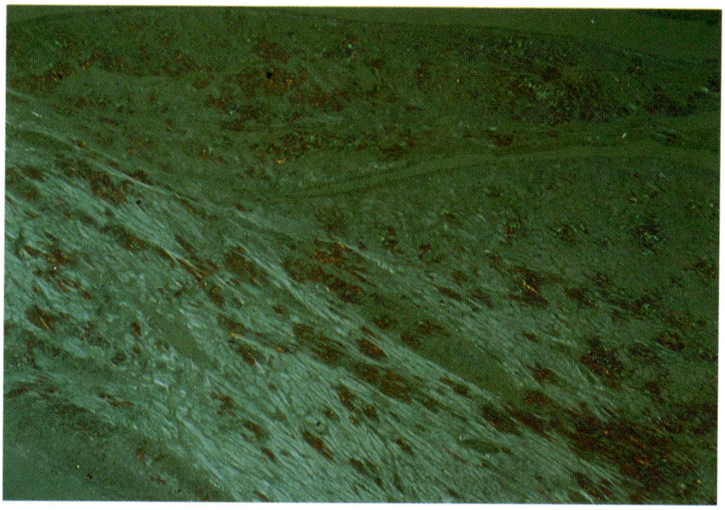

Fig. 146
Section of lung demonstrating large deposits of amyloid (middle). Section is well stained. Bennhold's Congo Red Method (Chapter 12)

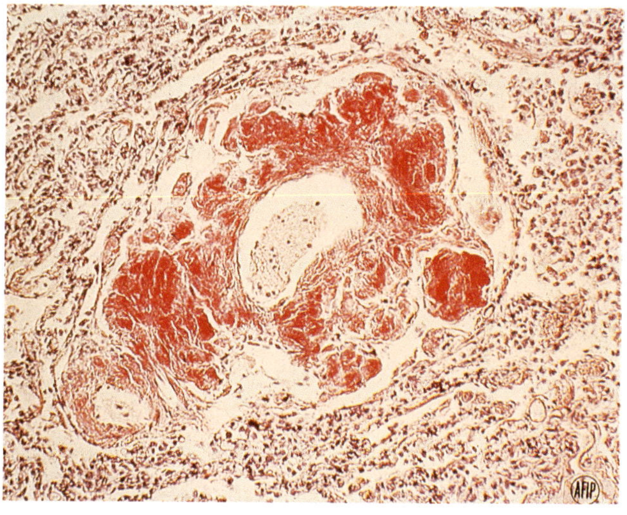

Fig. 147
Section of lung demonstrating large deposits of amyloid which are under partial polarization to show the polarizing properties of the combination of Congo Red and amyloid. Bennhold's Congo Red Method (Chapter 12)

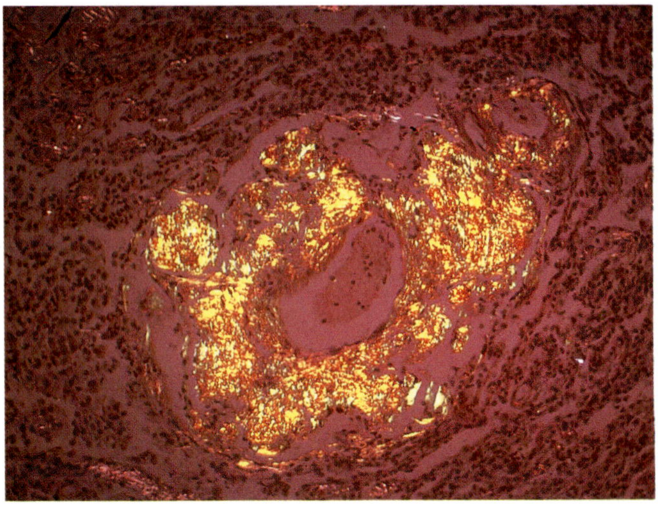

Fig. 148
Section of heart muscle showing deposits of amyloid throughout the section. The amyloid can be seen as red deposits. The background has a bluish tint. Section is well stained. Lieb's Method (Chapter 12)

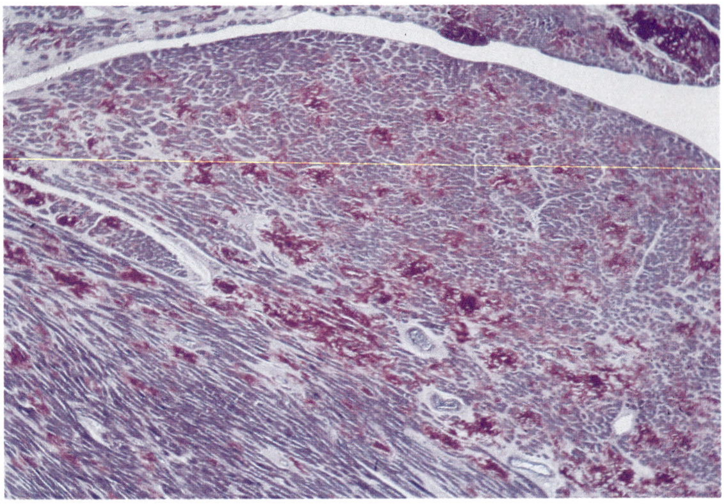

Fig. 149
Section of intestine demonstrating mucin staining in the goblet cells. Section is well stained. Mayer's Mucicarmine Method (Chapter 12)

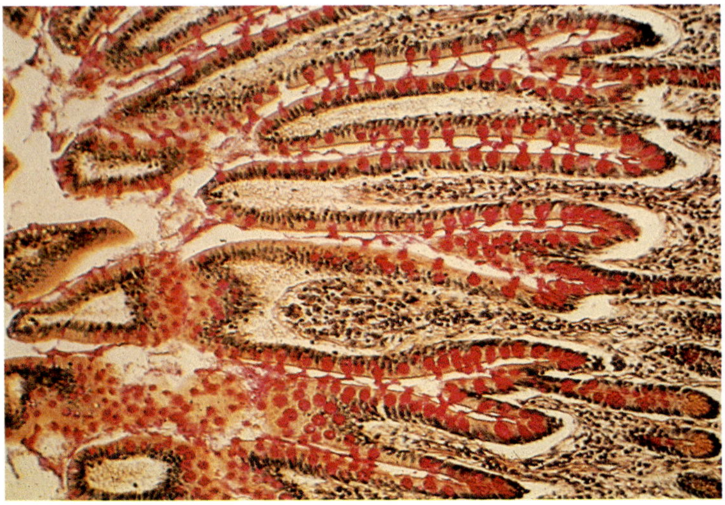

Fig. 150
This section demonstrates sublingual gland (top), submaxillary gland (left) and Brunner's gland (lower right). Notice the staining of cytoplasm of the cells in the sublingual gland. All sections were stained with mucicarmine. Stain is performed well. Mayer's Mucicarmine Method (Chapter 12)

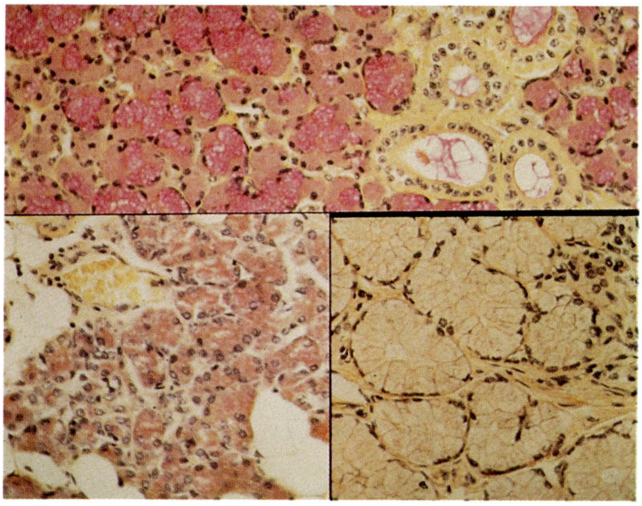

Fig. 151
Section of lung containing cryptococcus neoforms. Notice the reaction of the mucin type material surrounding the organism staining positive with the mucicarmine stain. Section is well stained. Mayer's Mucicarmine Method (Chapter 12)

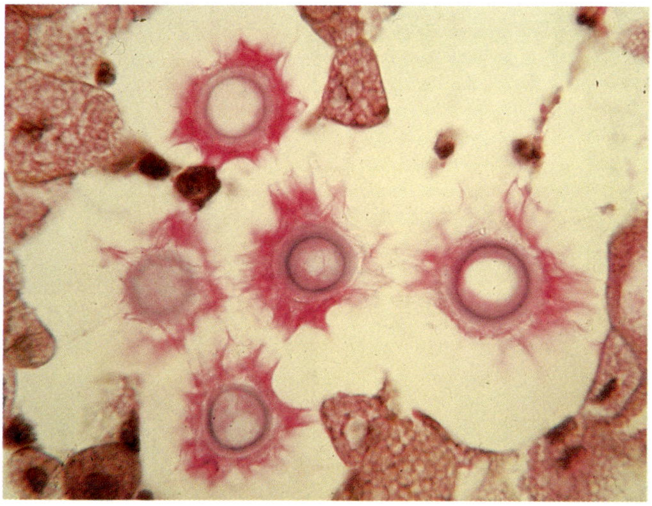

Fig. 152
Section of liver containing large amounts of glycogen and large vacuoles which probably represent fat globules. Section is well stained. McManus' Method for Glycoproteins (PAS) (Chapter 12)

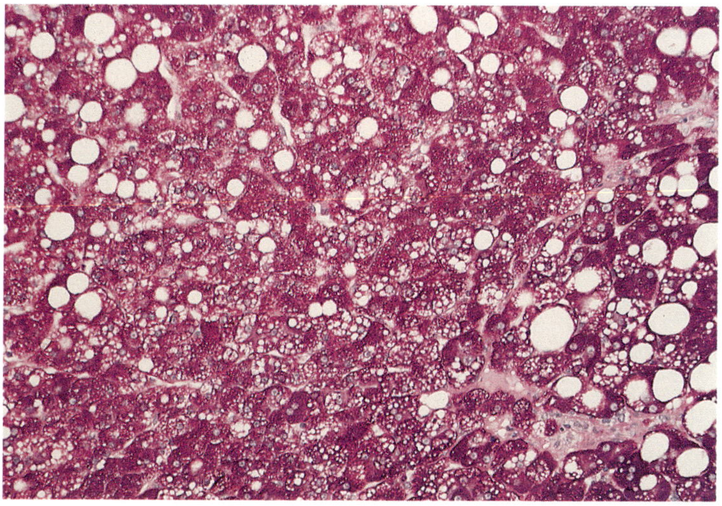

Fig. 152a
This is a case of coccidioidomycosis demonstrating the fungal organisms in red. Section is well stained. Periodic Acid Schiff Reaction with Light Green counterstain (Chapter 12)

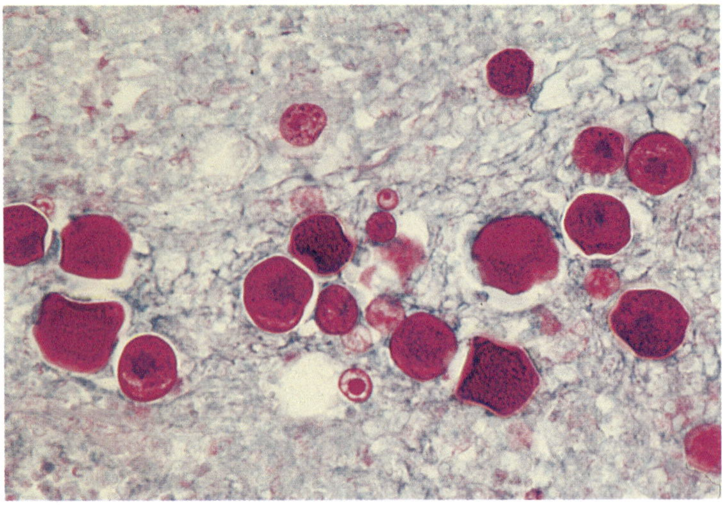

Fig. 152b
This section demonstrates cryptococcus neoforms stained with periodic acid Schiff and observed under dark field microscopy. Periodic Acid Schiff Method (Chapter 12)

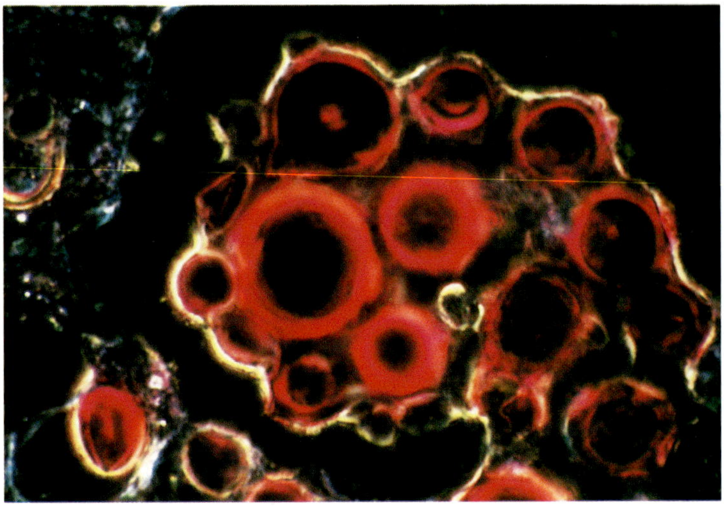

Fig. 153
Section of liver which has been treated with amylase digestion. Notice the complete absence of glycogen. Section is well stained. McManus' Method for Glycoproteins (PAS) (Chapter 12)

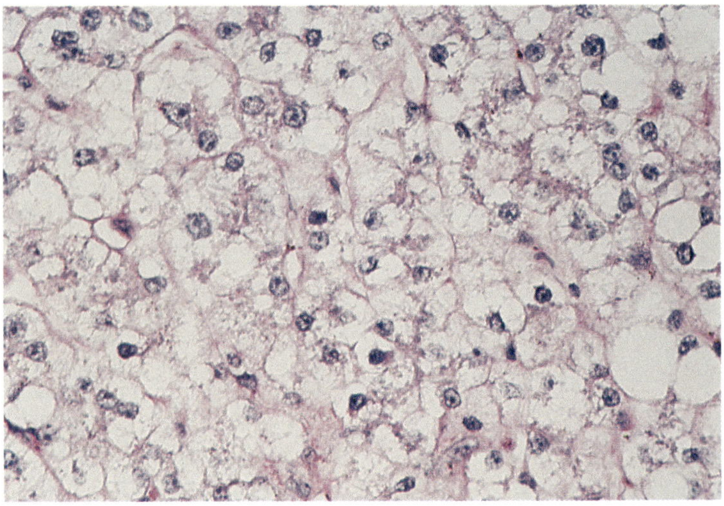

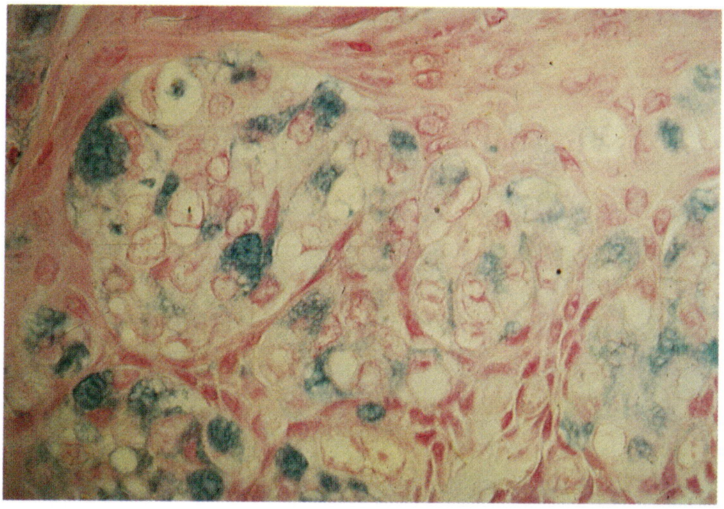

Fig. 154
Section of skin epidermis with Paget's disease. Notice the Paget's disease cells staining blue. Section is well stained. Alcian Blue Method pH 2.5 (Chapter 12)

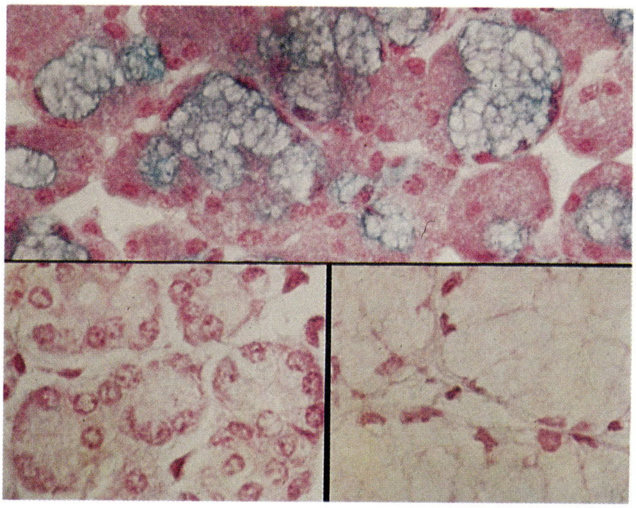

Fig. 155
This composite contains sublingual gland (top) submaxillary on (left) and Brunner's gland (lower right). Notice the reaction of the alcian blue at pH 2.5 of the mucoid material in the sublingual gland. Also notice the absence of alcian blue staining in the submaxillary and Brunner's gland. The counterstain here is Safranin O. Section is well stained. Alcian Blue Method pH 2.5 (Chapter 12)

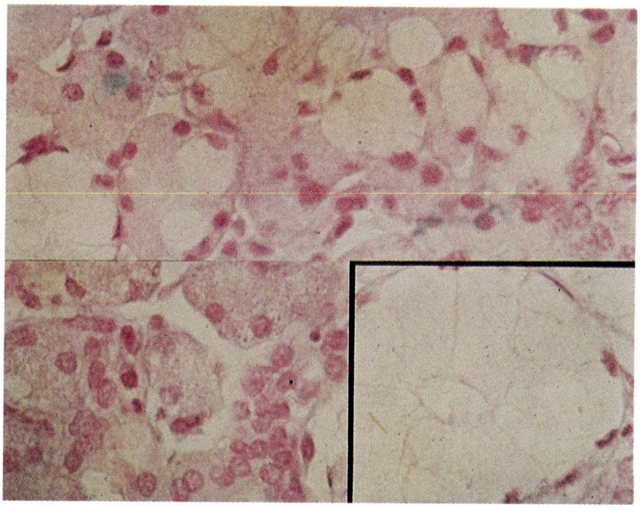

Fig. 156
This composite also shows sublingual gland (top), submaxillary (left) and Brunner's gland (right). Notice complete absence of staining. Section is well stained. Alcian Blue Method pH 1.0 (Chapter 12)

Fig. 157
Section of sublingual (top), submaxillary (left) and Brunner's gland (right). Notice the positive reaction of the sublingual gland while there is a negative reaction to both the submaxillary and Brunner's gland. Section is well stained. Aldehyde Fuchsin Method pH 1.7 (Chapter 12)

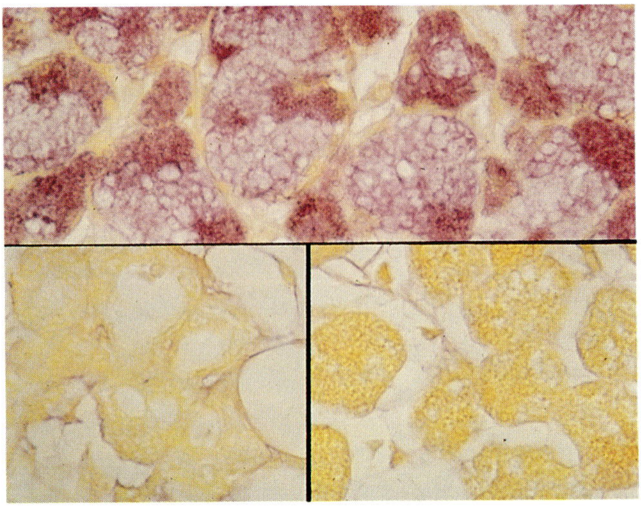

Fig. 158
Section of sublingual gland demonstrates a combination aldehyde fuchsin at pH 1.7 and alcian blue at pH 2.5. Notice the reaction to both the alcian blue and the aldehyde fuchsin. Section is well stained. Aldehyde Fuchsin (pH 1.7) and Alcian Blue (pH 2.5) Method (Chapter 12)

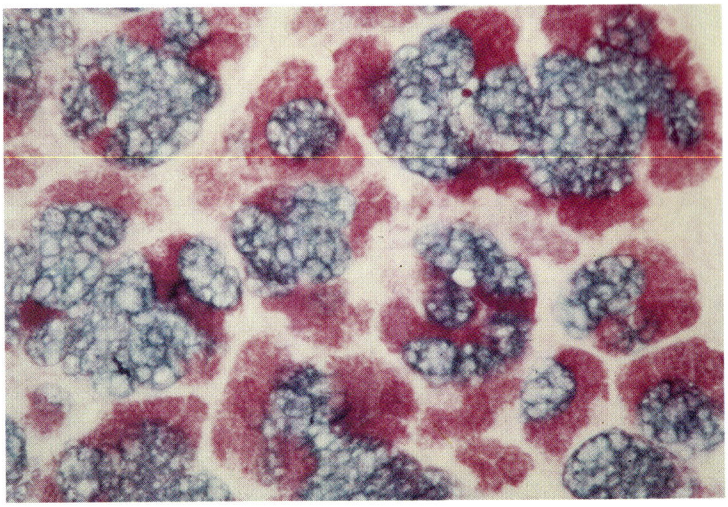

Fig. 159
Section of skin with Paget's disease. Notice the staining of the Paget's cells in the epidermis. Section is well stained. Mowry's Acid Mucopolysaccharide Method (Chapter 12)

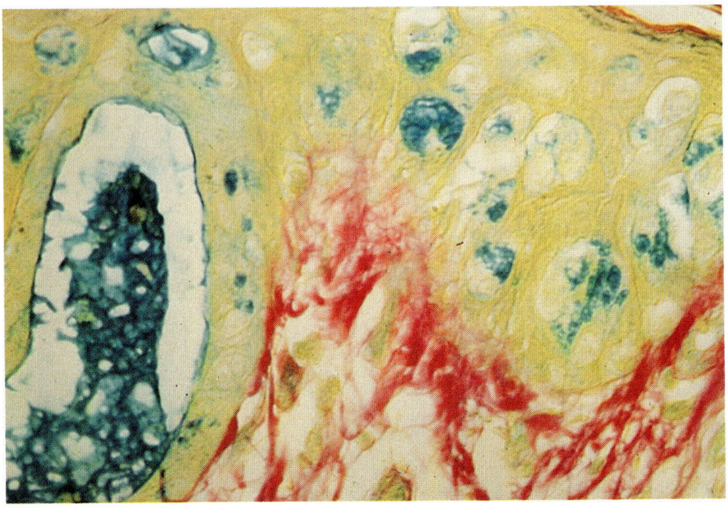

Fig. 160
Section of skin with Paget's disease demonstrating the epidermis which contains positive reaction to the aldehyde fuchsin. Aldehyde Fuchsin Method (pH 1.7) (Chapter 12)

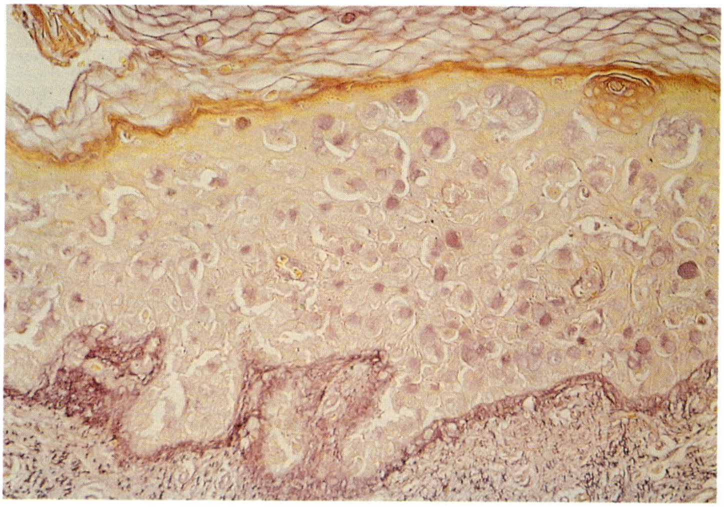

Fig. 161
Section of sublingual gland stained with alcian blue at pH 2.5 and the McManus' method for glycoproteins (PAS). Notice that the mucosubstances are staining positive with both the alcian blue and PAS method. Section is well stained. PAS-Alcian Blue (pH 2.5) Method (Chapter 12)

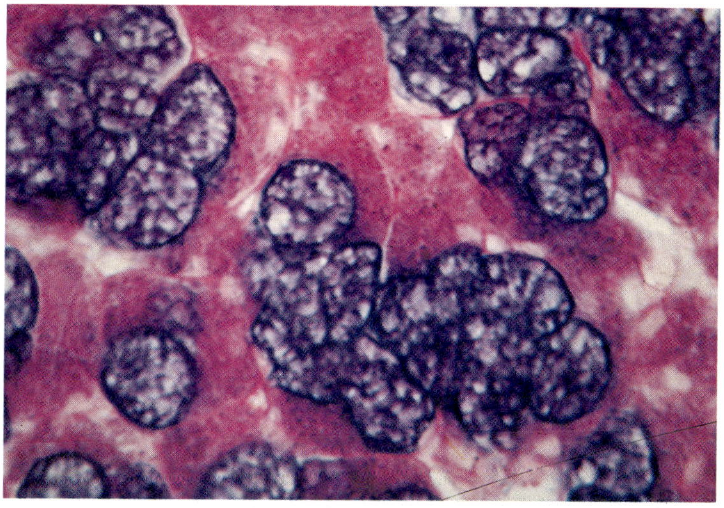

Fig. 162
Section of liver stained with Best's method for glycogen. Notice the positive reaction throughout the cytoplasm of the hepatocytes. Section is well stained. Best's Method for Glycogen (Chapter 12)

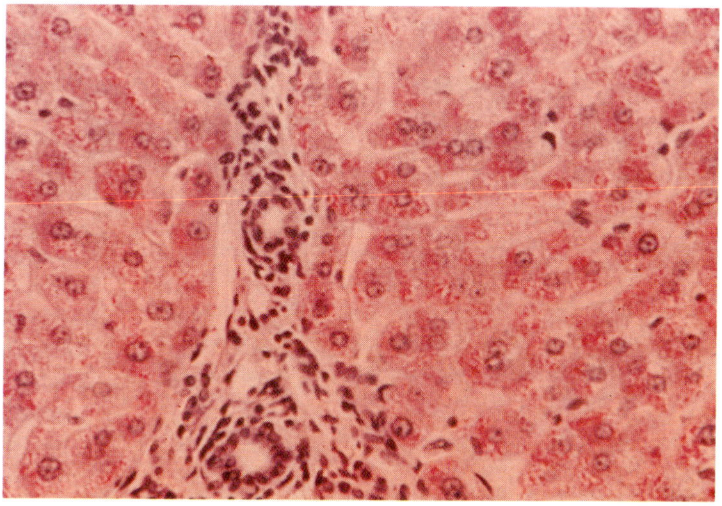

Fig. 163
Section of lung stained with Sweat-Puchtler-Valentine fluorescence method for amyloid. (Chapter 12)

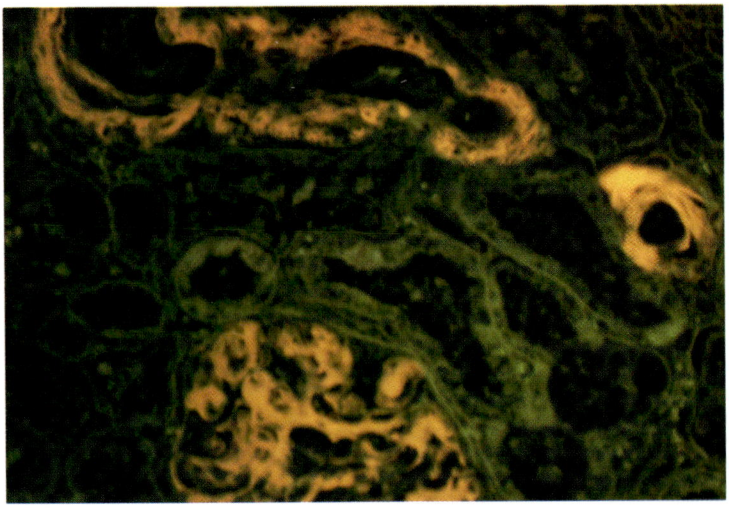

Fig. 164
This section demonstrates Kaposi sarcoma and more specifically the red staining granular material which can be found in the cytoplasma in some cells of this tumor. Notice the bright tinctorial color in several areas throughout. Attwood's Method for Amniotic Fluid Embolism (Chapter 12)

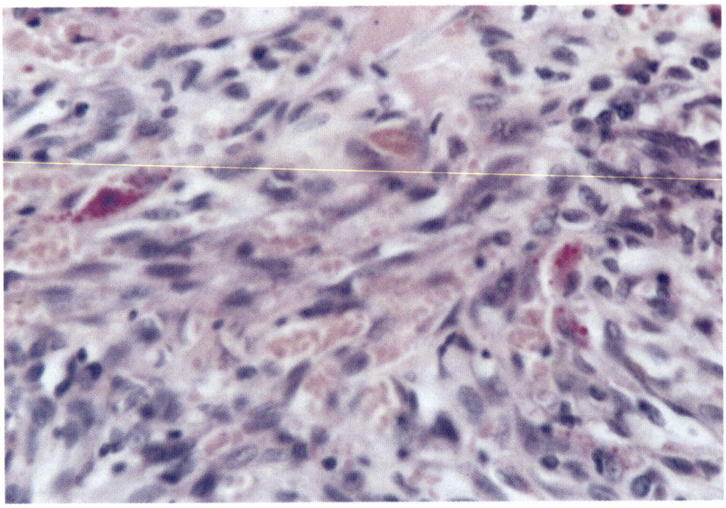

Fig. 165
Section of kidney glomeruli with basement membrane staining throughout. Section is well stained. Jones' Method for Basement Membrane (Chapter 13)

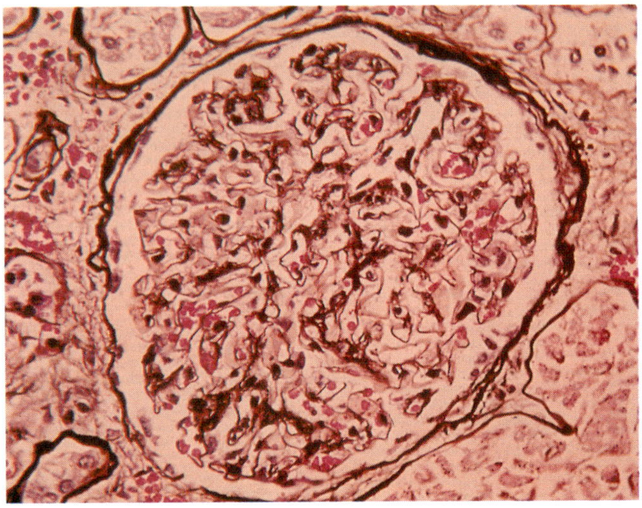

Fig. 166
Section of glomeruli with the basement membranes reacting positively. Section shows good staining of the PAS reaction but the hematoxylin is not dark enough. McManus' Method for Glycoprotein (PAS) (Chapter 13)

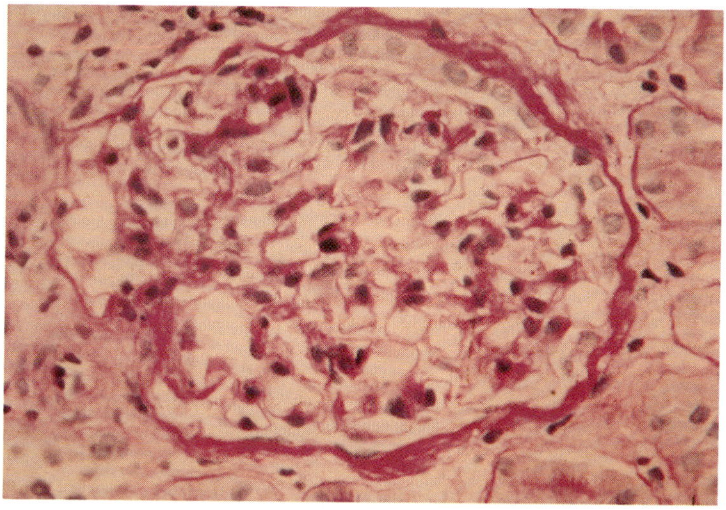

Fig. 167
This section demonstrates elastic fibers throughout. Notice the good demonstration of these fibers. Section is well stained. Miller's Method (Chapter 13)

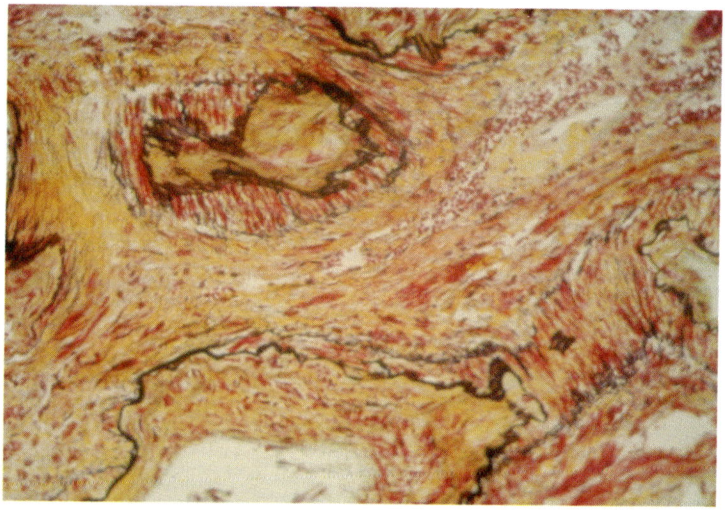

Fig. 168
Section of liver demonstrating reticulum fibers in the sinusoids. The stain has been counterstained with Kernechtrot (Nuclear Fast Red). Section is well stained. Manuel's Method (Chapter 13)

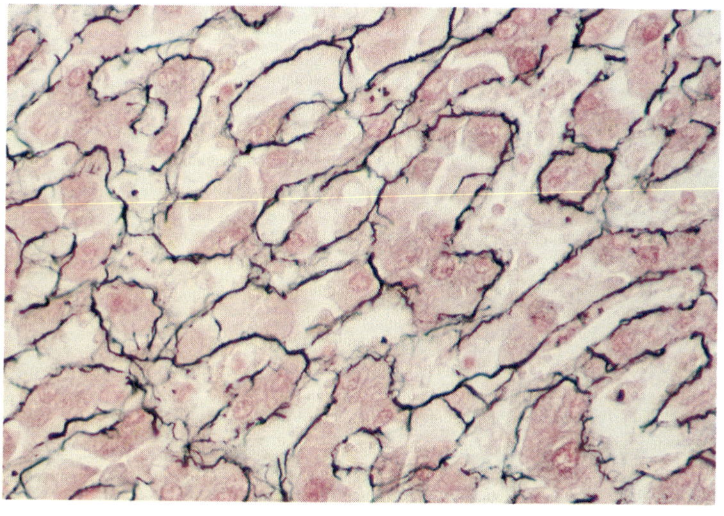

Fig. 169
Section of skin with a malignant melanoma demonstrating well delineated reticulum fibers. Section is well stained. Manuel's Method (Chapter 13)

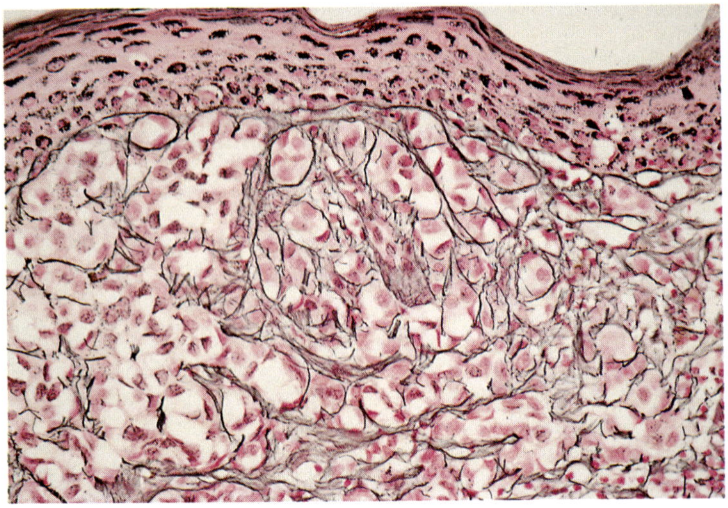

Fig. 170
Section of bone spicules demonstrating bone canaliculi projecting from the osteocytes. Section is well stained. Luna-Giemsa Method (Chapter 13)

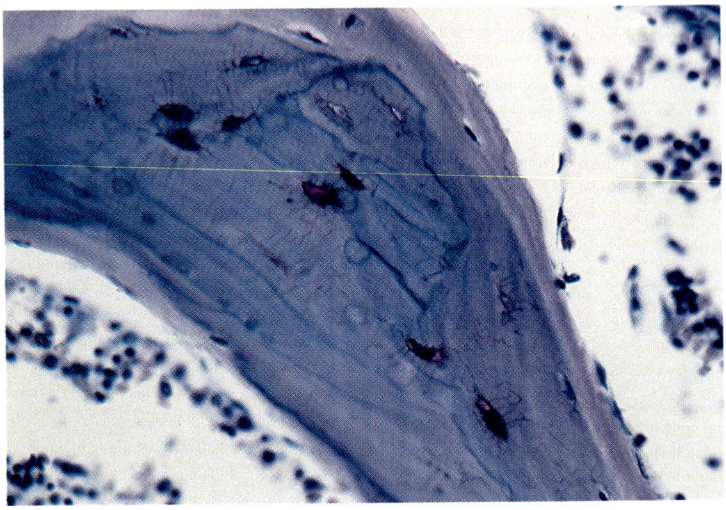

Fig. 171
Section demonstrating myoepithelial cells. The myoepithelial cell can be seen as elongated structures in the middle of the section (red). Section is well stained. Gomori's Method for Muscle and Collagen (Chapter 13)

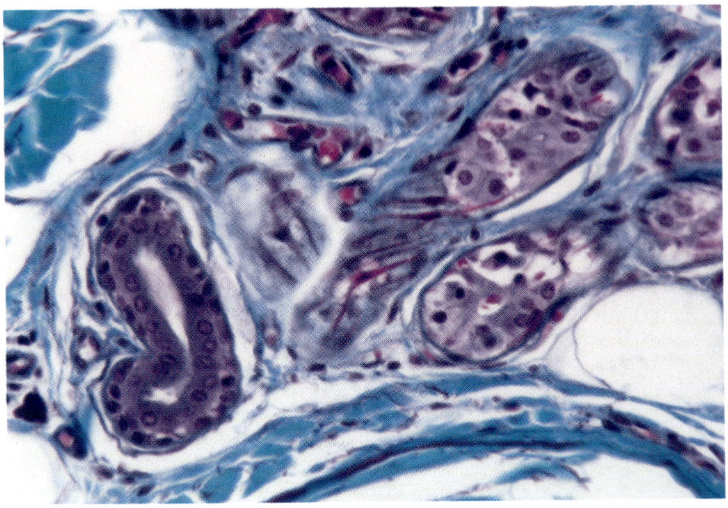

Fig. 172
Section of skin demonstrating solar elastosis. In this particular instance aldehyde fuchsin was used to demonstrate the elastic fibers. Section is well stained. Gomori's Method for Elastica and Sulfated Mucosaccharides (Chapter 13)

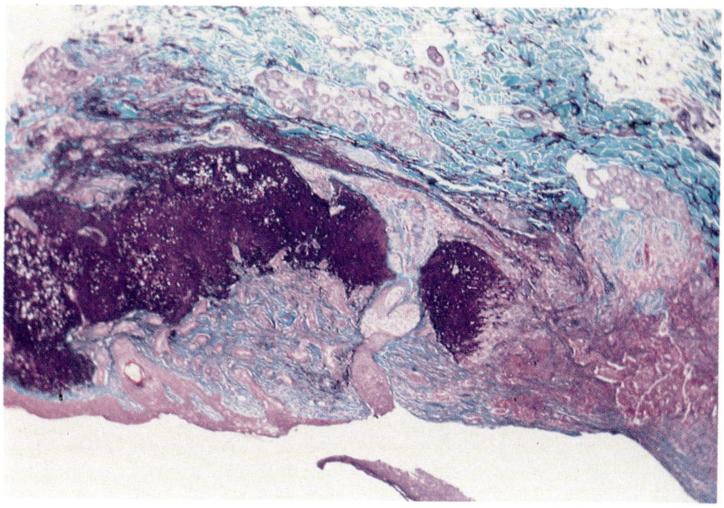

Fig. 173
Section of collecting tubule of the kidney demonstrating well delineated collagen fibers on the periphery of the tubule. Section is well stained. Masson's Trichrome Method (Chapter 13)

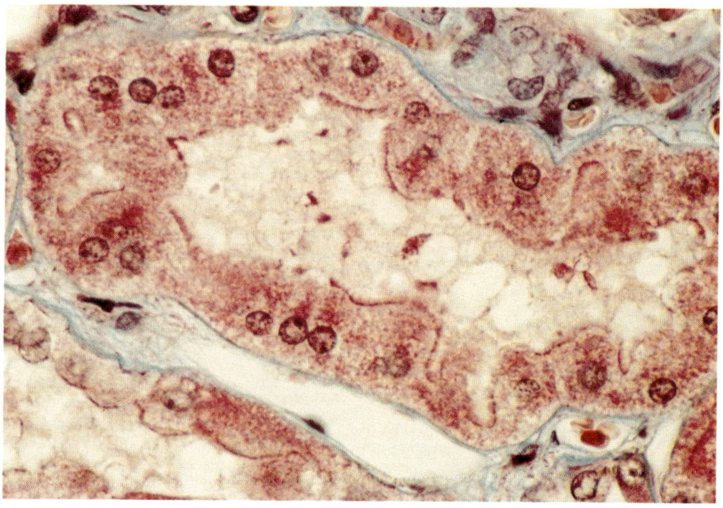

Fig. 174
Section of lung demonstrating collagen in blue and muscle fibers in red. Notice the good delineation between the collagen and the muscle fibers. Section is well stained. Masson's Trichrome Method (Chapter 13)

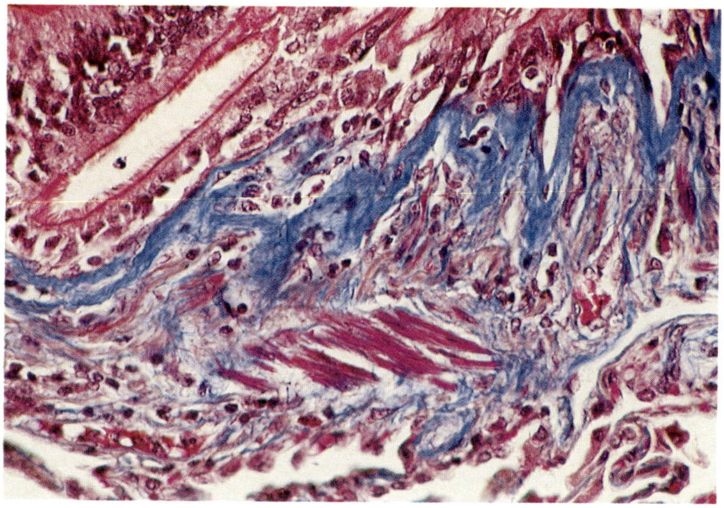

Fig. 175
Section of kidney demonstrating the staining of collagen around and through the glomeruli structure. Section is well stained. Masson's Trichrome Method (Chapter 13)

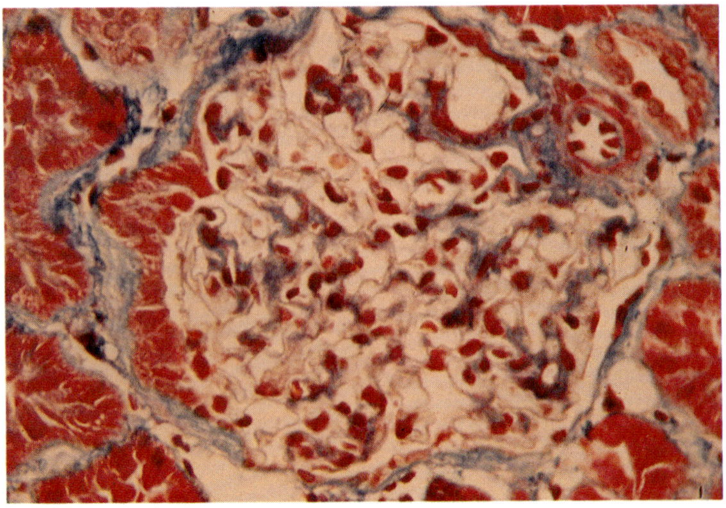

Fig. 176
Section of skeletal muscle demonstrating nemaline rods throughout the cytoplasm of the muscle fiber. Section is slightly understained. Luna's Method for Nemaline Rods (Chapter 13)

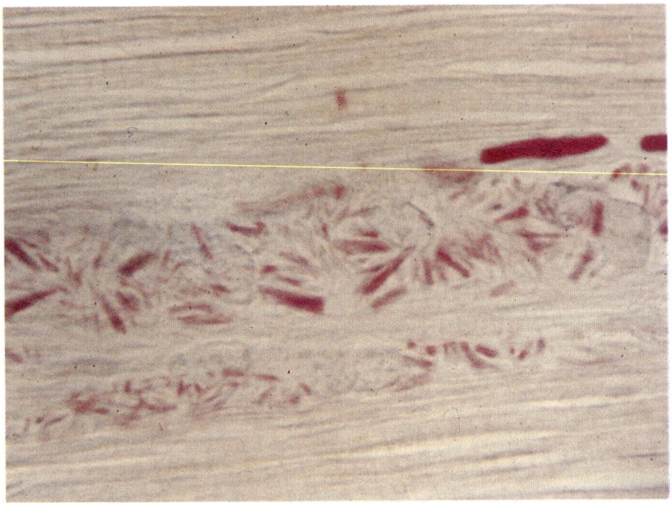

Fig. 177
Individual section of skeletal muscle demonstrating nemaline rods throughout the substance of the muscle fiber. Section is slightly understained. These rods are the results of nemaline myopathy which is a muscle disease. Luna's Method for Nemaline Rods (Chapter 13)

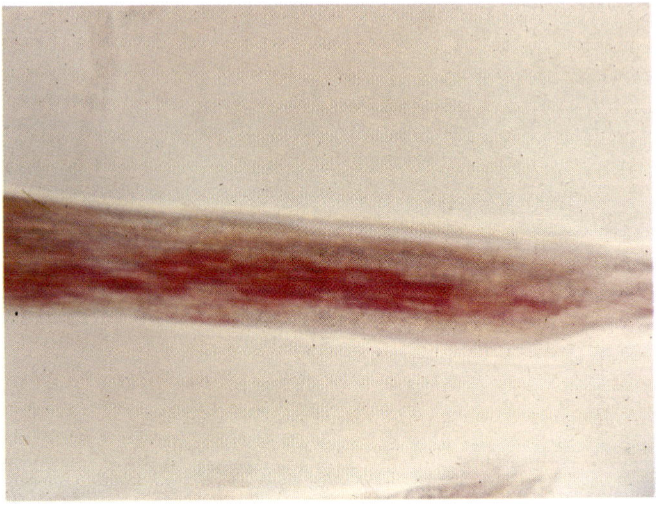

Fig. 178
This section of liver demonstrates bile/canaliculi. The hematoxylin nuclear chromatin stain is not blue enough in this section. Luna-Ishak Method (Chapter 13)

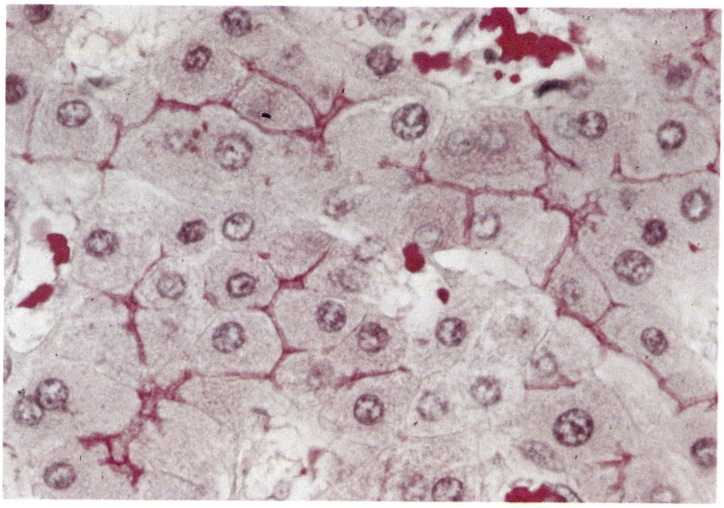

Fig. 179
Section of elastic cartilage showing good demonstration of the elastic fibers including some aldehyde fuchsin reaction in the cytoplasm of the section of cartilage. Section is well stained. Gomori's Method for Elastica and Sulfated Mucosaccharides (Chapter 13)

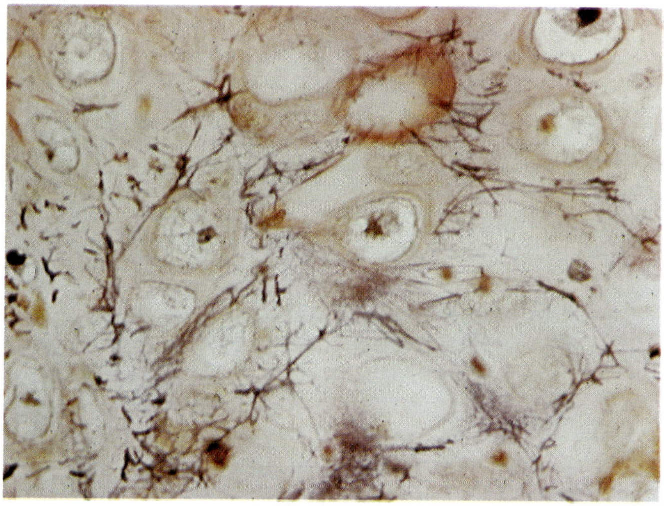

Fig. 180
Section of bone trabeculae demonstrating bone in reddish-purple and cartilage in purple. Also seen throughout this section is staining of soft tissues. Section is well stained. Luna's Modification of Shores' Method (Chapter 13)

Fig. 181
Section of bone demonstrating osteocytes from which bone canaliculi radiate. Luna-Bodian Method for Bone Canaliculi and Nerve Fibers (Chapter 15)

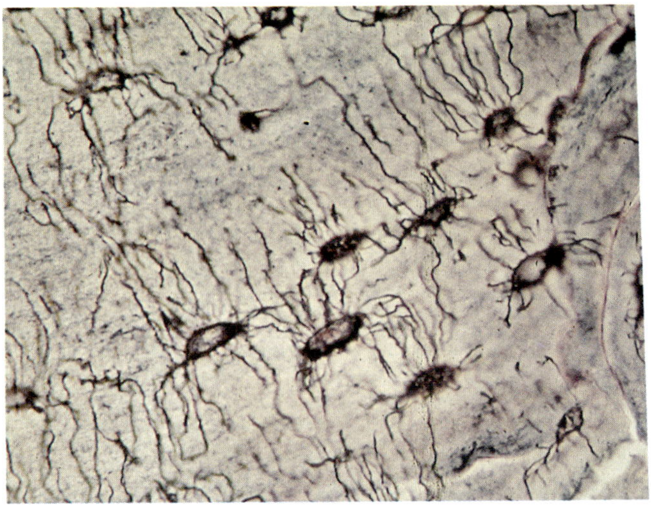

Fig. 182
Section of skin demonstrating stratum corneum, stratum lucidum and remaining three layers of the epidermis. Section is well stained. Luna-Bodian Method for Bone Canaliculi and Nerve Fibers (Chapter 15)

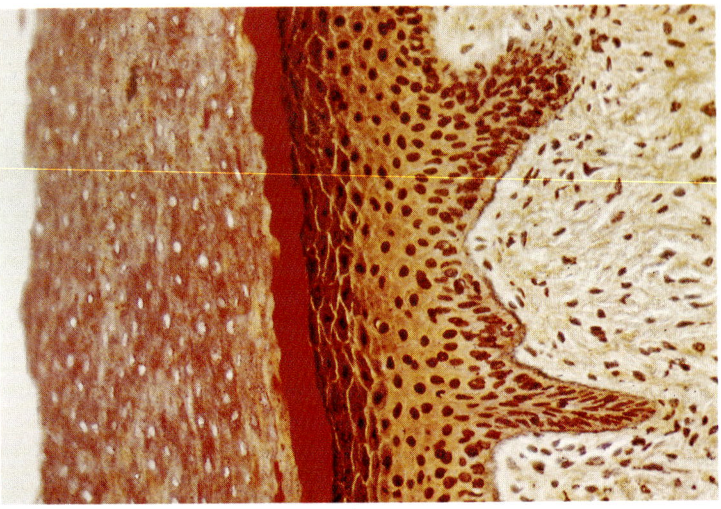

Fig. 183
Section of tracheal epithelium demonstrating some of the trachea gland's elastic fibers which can be seen as dark staining structures in the upper left hand corner. Muscle fibers can also be seen in the upper right hand corner. Russell's Modification of Movat's Method (Chapter 13)

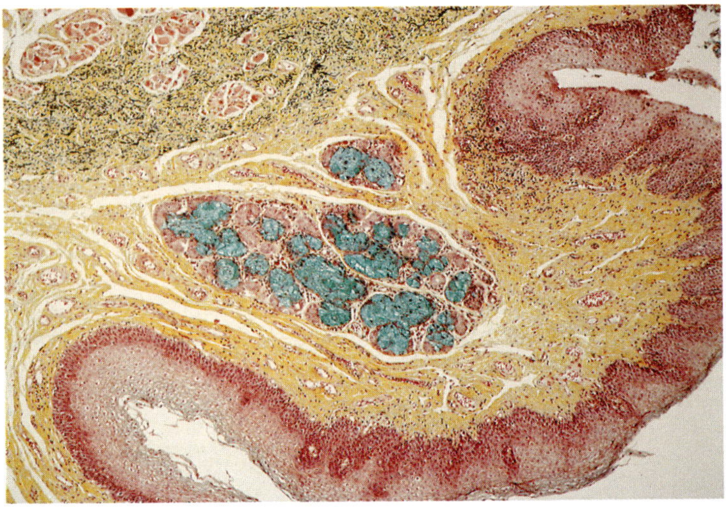

Fig. 184
Section of trachea demonstrating cartilage, muscle (lower right hand corner) and some of the tracheal glands (left side). Russell's Modification of Movat's Method (Chapter 13)

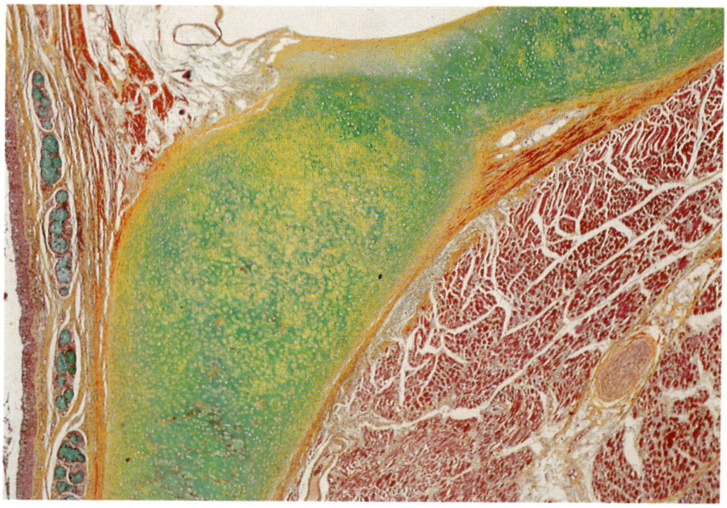

Fig. 185
Aorta containing Marfan's syndrome. Notice the well stained elastic fibers throughout the substance of the aortic specimen. Also, some mucosubstances can be seen as very light staining in the background. Section is well stained. Russell's Modification of Movat's Method (Chapter 13)

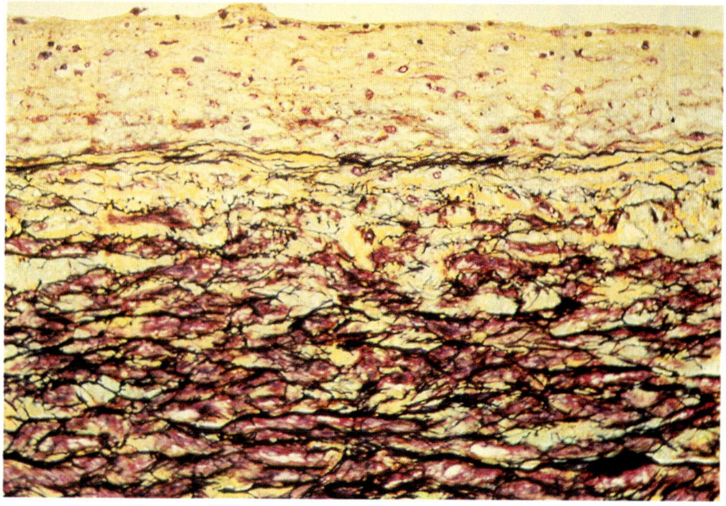

Fig. 186
Section of skeletal muscle in a case of rhabdomyosarcoma. Notice the fine cross striations in the muscle fibers. Section is well stained. Luna Modification of the PTAH Method (Chapter 13)

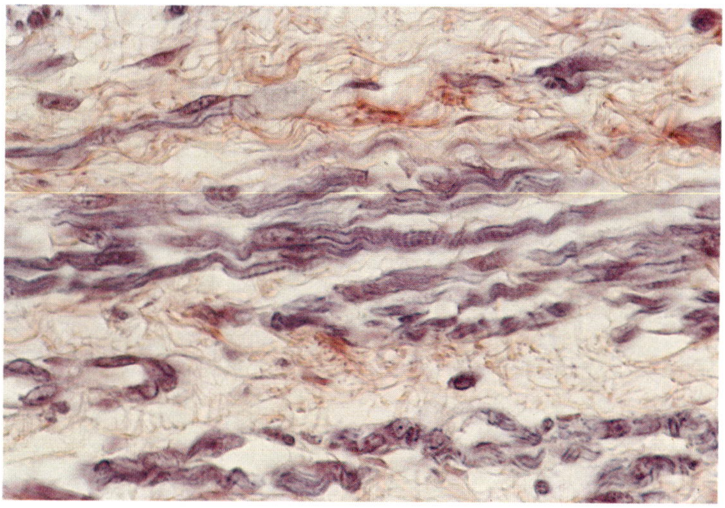

Fig. 187
This section demonstrates fibrin. Section is well stained. Luna Modification of the PTAH Method (Chapter 13)

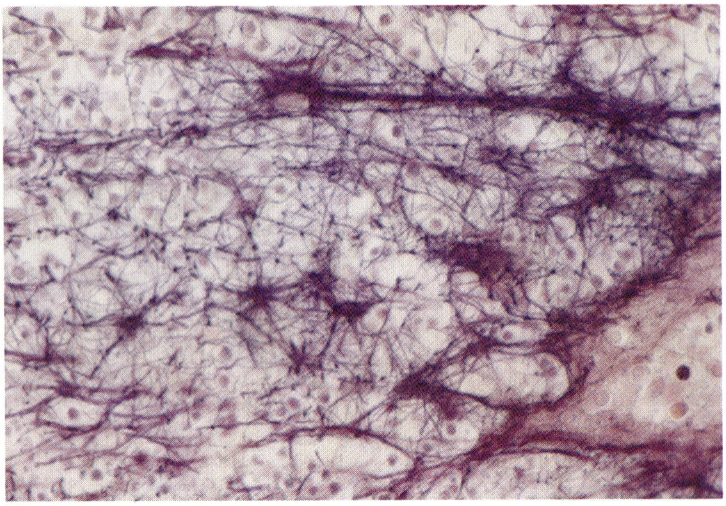

Fig. 188
Section of spleen demonstrating some connective tissue staining including muscle across the middle of the photograph. This photograph is being used to compare with Fig. 189. Masson's Trichrome Method (Chapter 13)

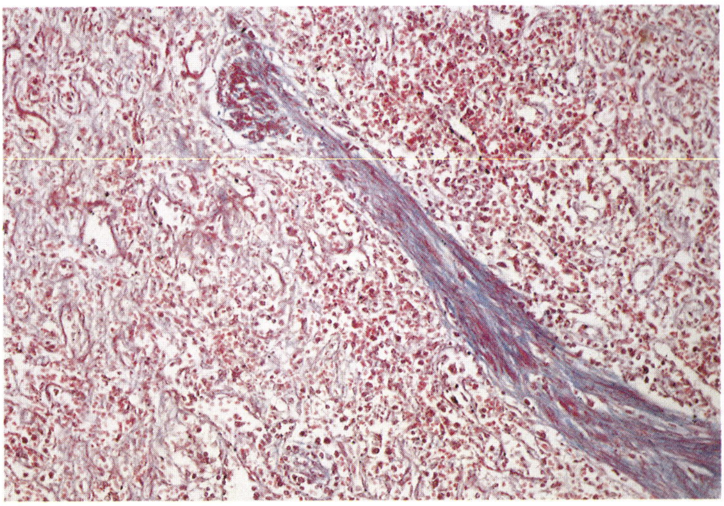

Fig. 189
Section of spleen demonstrating the same connective tissue island as seen in Fig. 188. Notice the well demonstrated collagen fibers. Picrosirius Red Method for Collagen (Chapter 13)

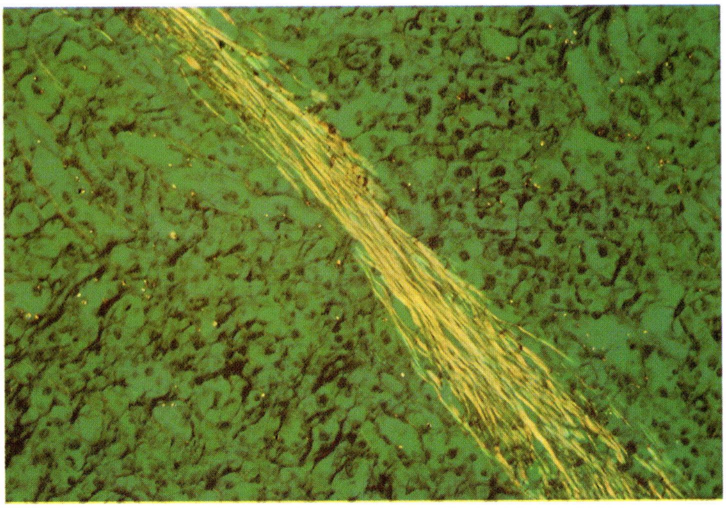

Fig. 190
This frozen section demonstrates fat throughout the section. Oil Red O Method (Chapter 14)

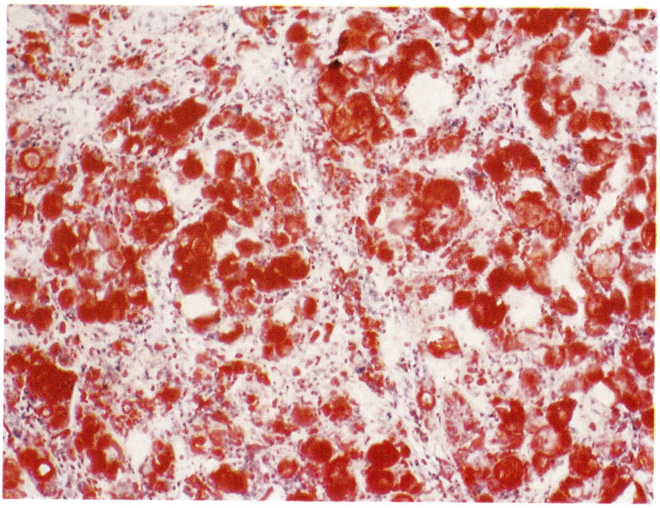

Fig. 191
This is a section of gall bladder demonstrating cholesterol in a case of cholesterosis. The reaction can be seen as green colored. The two dark circles represent air bubbles which are the results of the acid's reaction. Schultz's Method (Chapter 14)

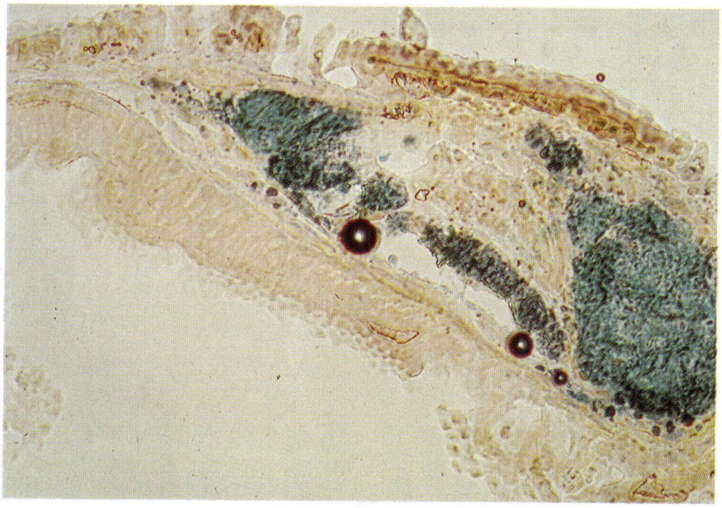

Fig. 192
This composite is a section of sebaceous cell adenoma demonstrating phospholipids. The section on the left shows the phospholipid being well demonstrated while the section on the right was treated with pyridine for phospholipid extraction. Baker's Method (Chapter 14)

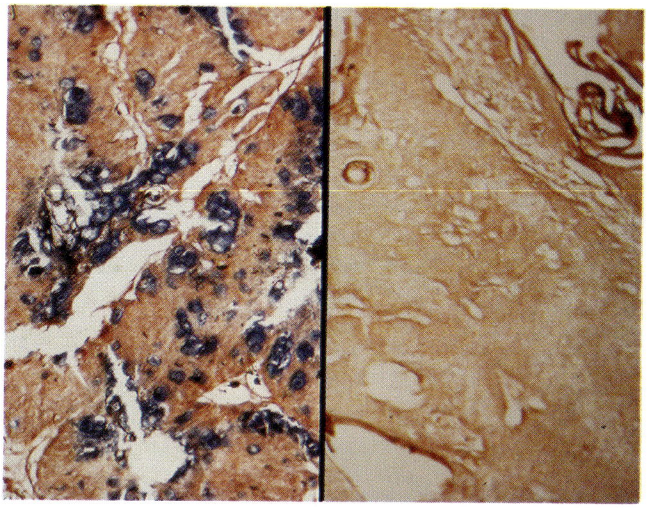

Fig. 193
This photograph demonstrates sphingolipid (sulfatide) in a case of metachromatic leukodystrophy. Notice the reaction of the lipid material. Hirsch-Pfeiffer's Method (Chapter 15)

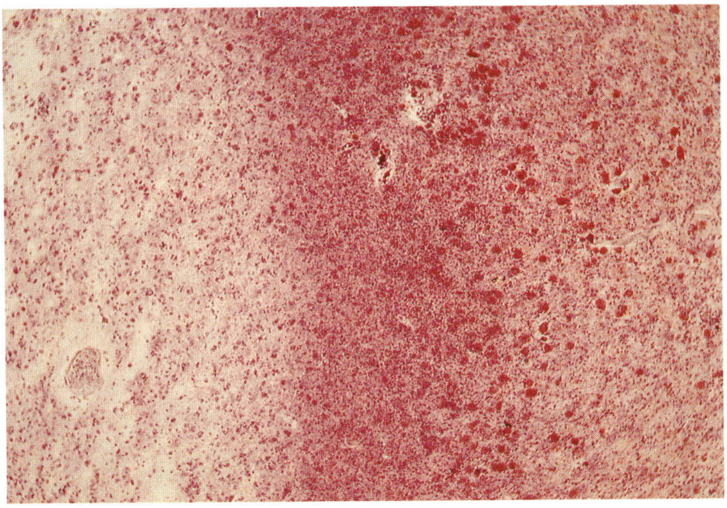

Fig. 194
This is a higher power of Fig. 193 to better illustrate the sphingolipid. Hirsch-Pfeiffer's Method (Chapter 15)

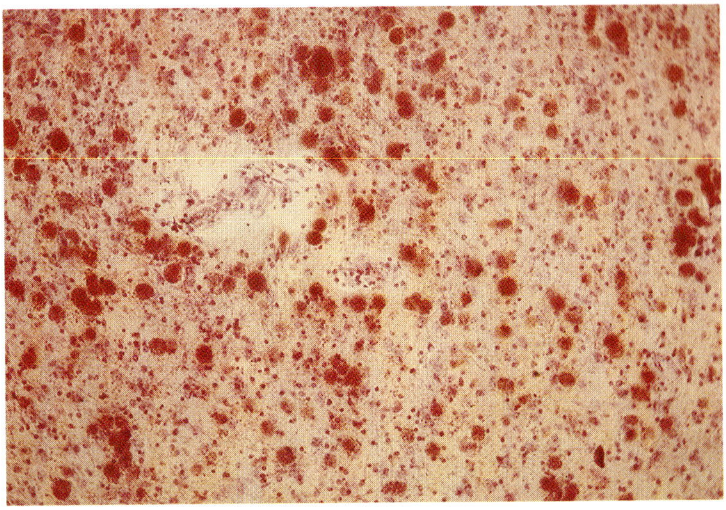

Fig. 195
Fatty accumulation in hepatocytes. Sudan Black B Method (Chapter 14)

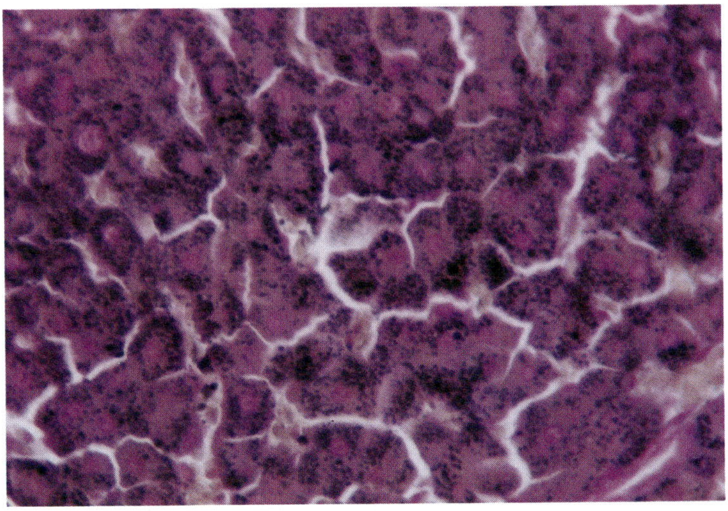

Fig. 196
This is a section of brain showing fibrous astrocytes throughout the photograph. Section is well stained. Ramon-Cajal's Method (Chapter 15)

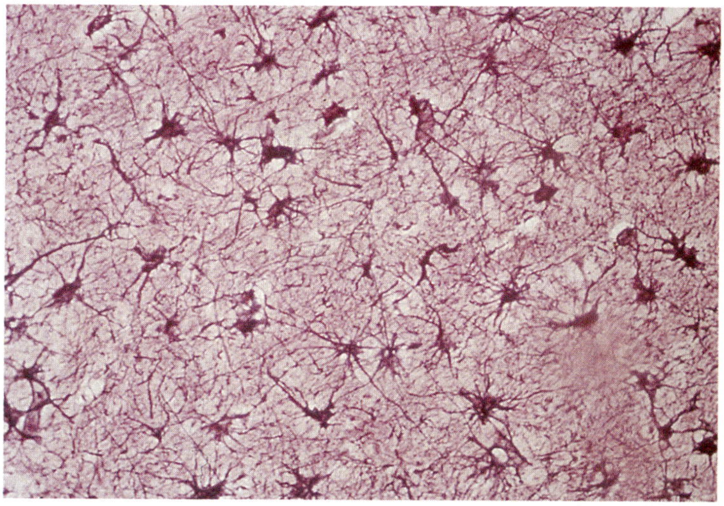

Fig. 197
A section of brain demonstrating dark splotches throughout representing senile plaques. Section is well stained. Bielschowsky's Method (Chapter 15)

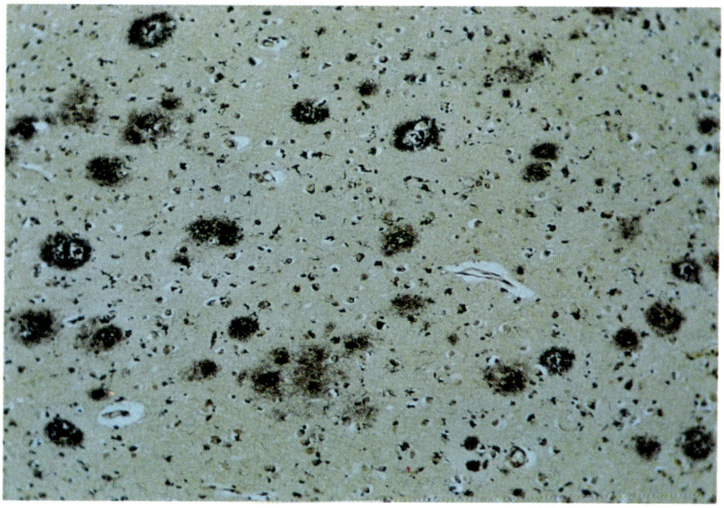

Fig. 198
This section of brain is a higher power of the senile plaques. Section is well stained. Bielschowsky's Method (Chapter 15)

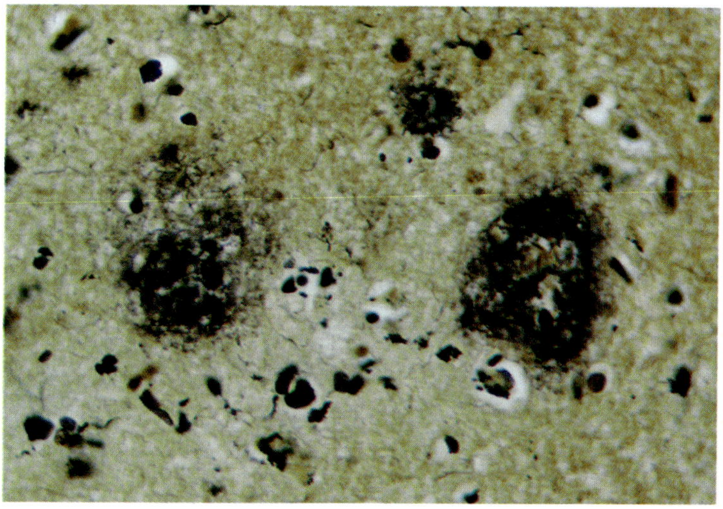

Fig. 199
Consistent with the pathology of Alzheimer's disease this photograph demonstrates fibrillary tangles. Bielschowsky's Method (Chapter 15)

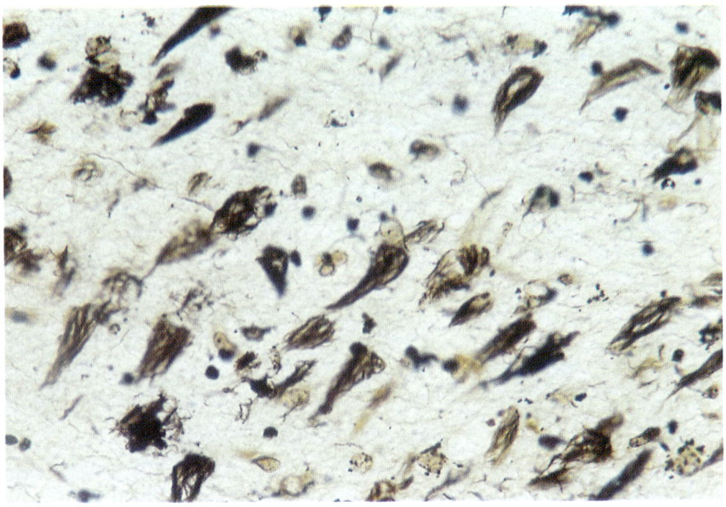

Fig. 200
This is a section of brain demonstrating nerve fibers (neurons). Section is well stained. Bielschowsky's Method (Chapter 15)

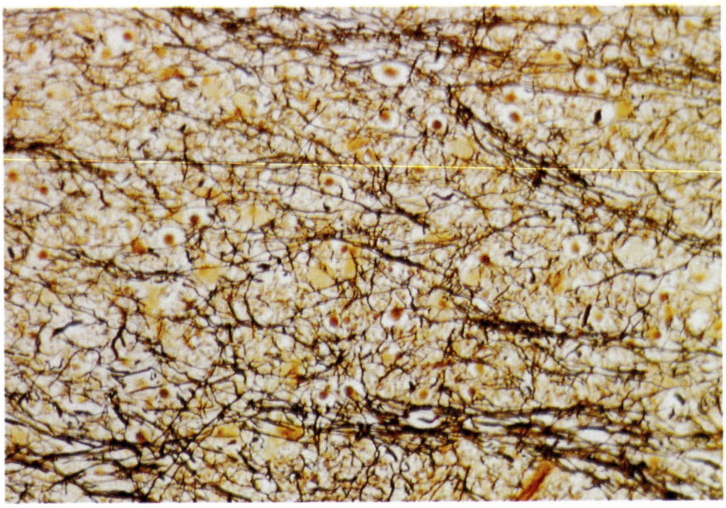

Fig. 201
This section demonstrates fibrillary fibers around the perkinje cells. Bielschowsky's Method (Chapter 15)

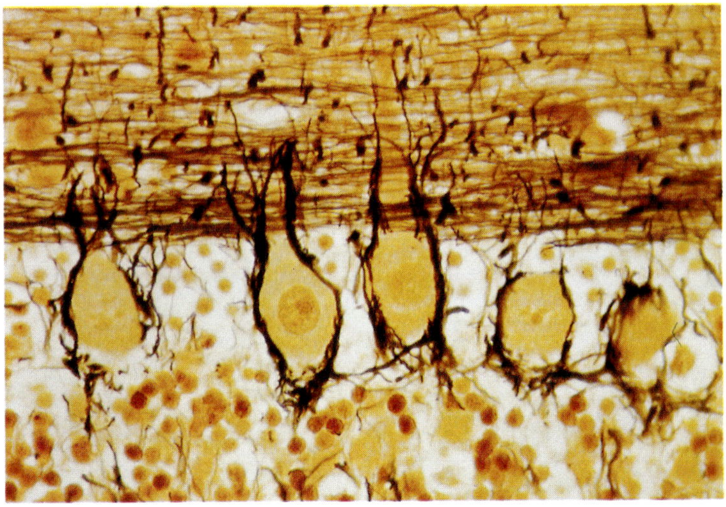

Fig. 202
Section of peripheral nerves seen throughout this photograph. Section is well stained. Luna's Modification of Bodian's Method (Chapter 15)

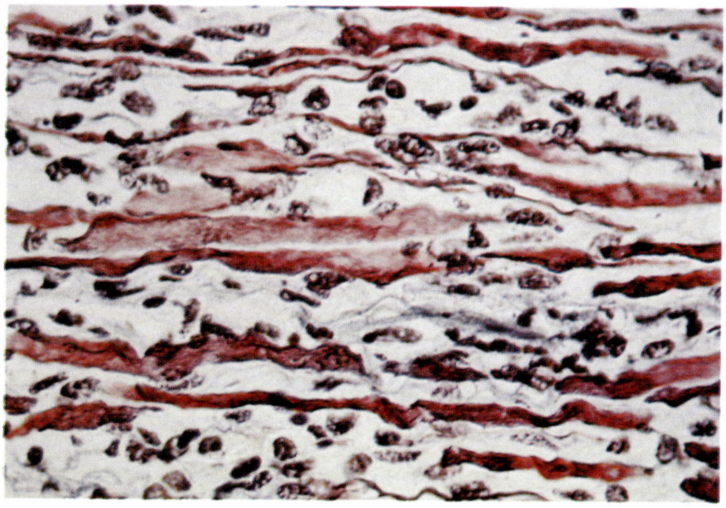

Fig. 203
Large neurons are seen in this section, as well as some of the smaller neurons. Section is well stained. Luna's Modification of Bodian's Method (Chapter 15)

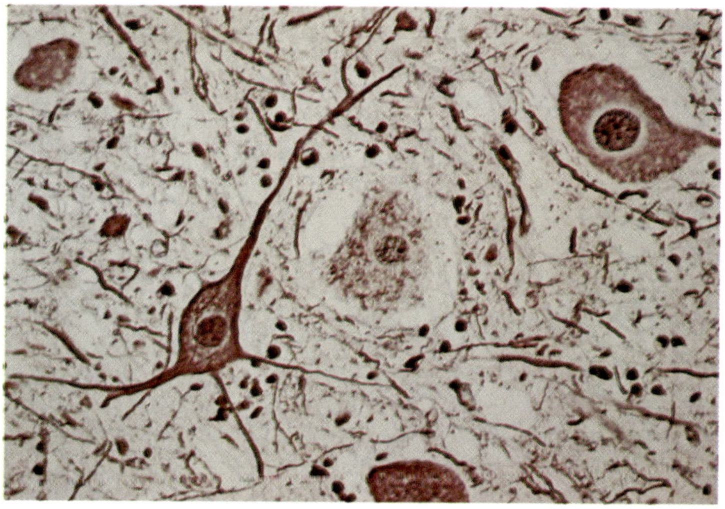

Fig. 204
This section demonstrates biscono bodies. Section is well stained. Luna's Modification of Bodian's Method (Chapter 15)

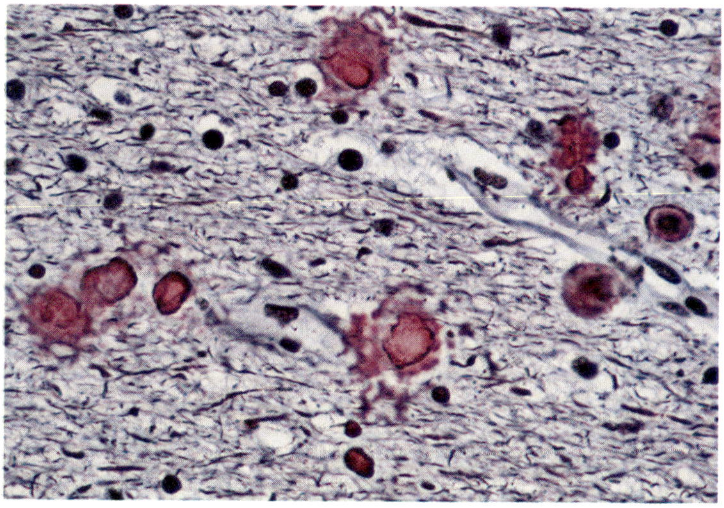

Fig. 205
Section of brain demonstrating nerve fibers and myelin sheath. Section is well stained. Bodian's Luxol Fast Blue Method (Chapter 15)

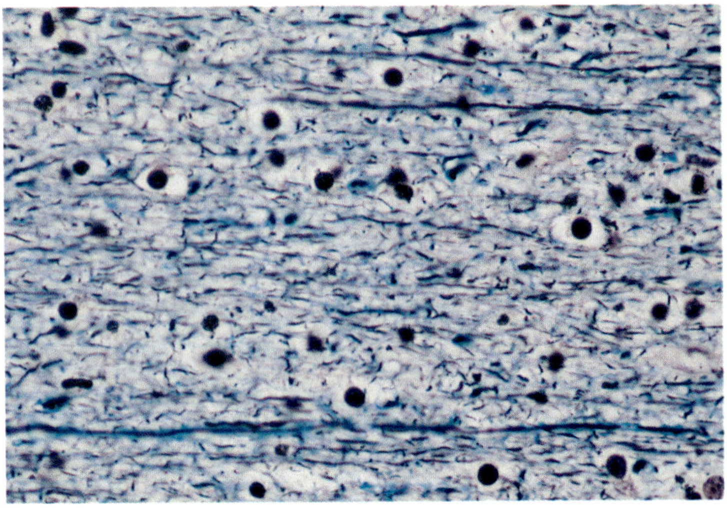

Fig. 206
This is a section of brain demonstrating neurons. Section is well stained. Holmes Method (Chapter 15)

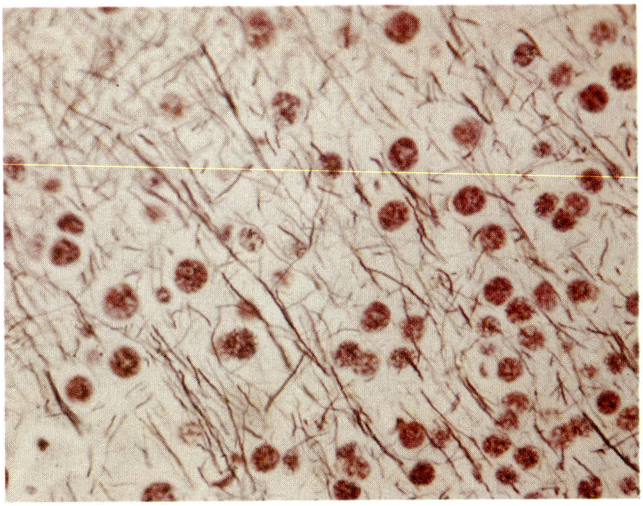

Fig. 207
This section of brain demonstrates Lafora bodies in the neurons. Section is well stained. McManus Periodic Acid Schiff Method (Chapter 12)

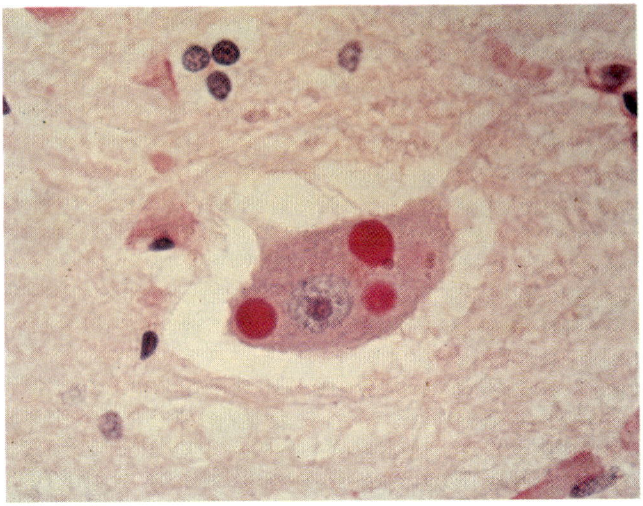

Fig. 208
This section of brain is demonstrating astrocytes. Section is well stained. Luna's Modification of PTAH Method (Chapter 15)

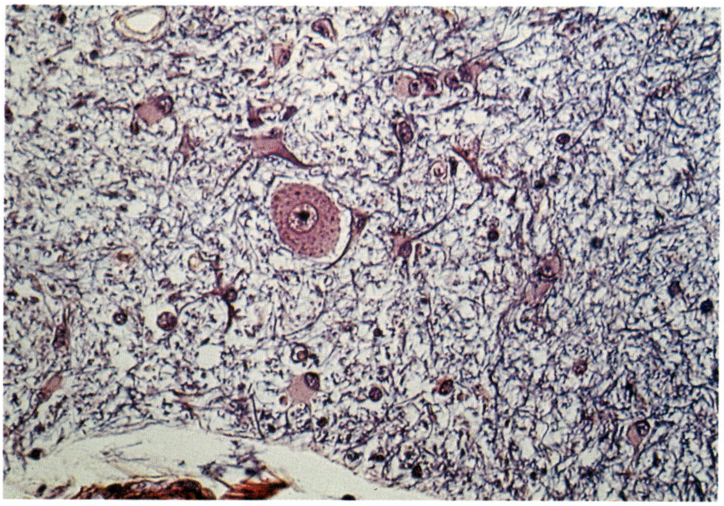

Fig. 209
This is a section of brain demonstrating astrocytes. Section is well stained. Holzer's Method (Chapter 15)

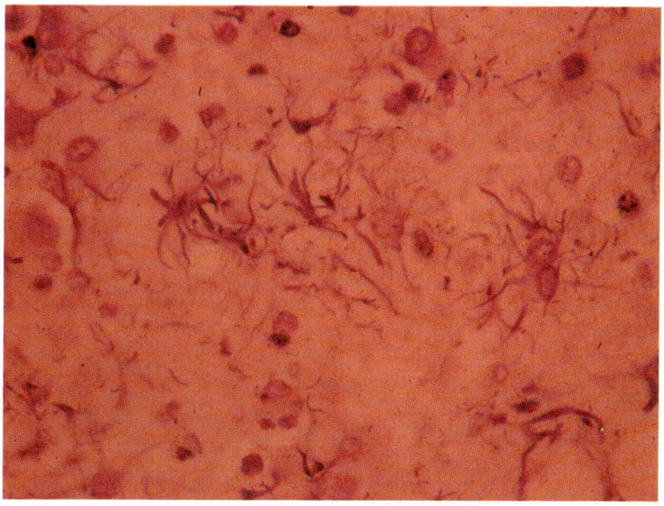

Pre-Fixation Artifacts

Fig. 210a and b
This composite demonstrates starch powder which is often used on surgical gloves. (a) H&E stained section which shows individual powder granules in the crevice. These structures stain gray with the H&E stain. (b) Section under polarized light which demonstrates that powder granules polarize well. (Chapter 17)

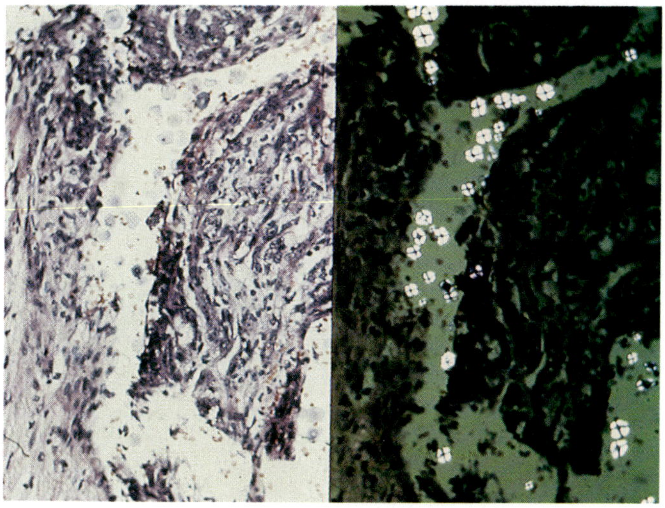

Fig. 211
This section of brain shows a great deal of cracks which are produced when an individual has been on a life support respirator for an extended period of time. (Chapter 17)

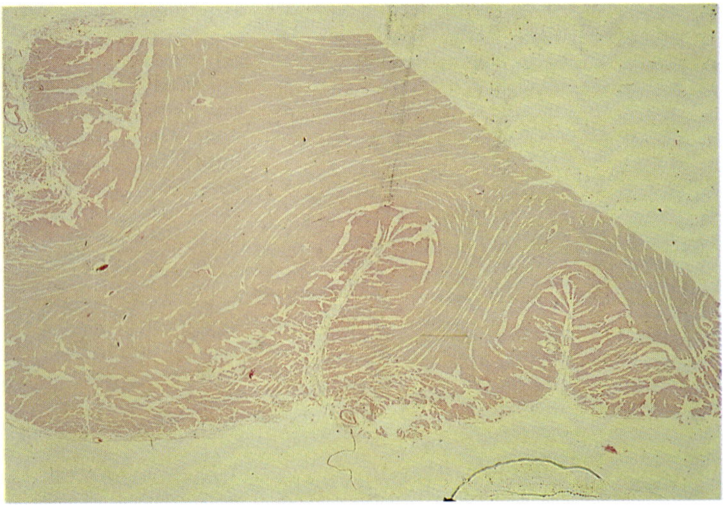

Fig. 212
Higher power of the section of brain seen in Fig. 211 illustrating the separation of the brain substance. (Chapter 17)

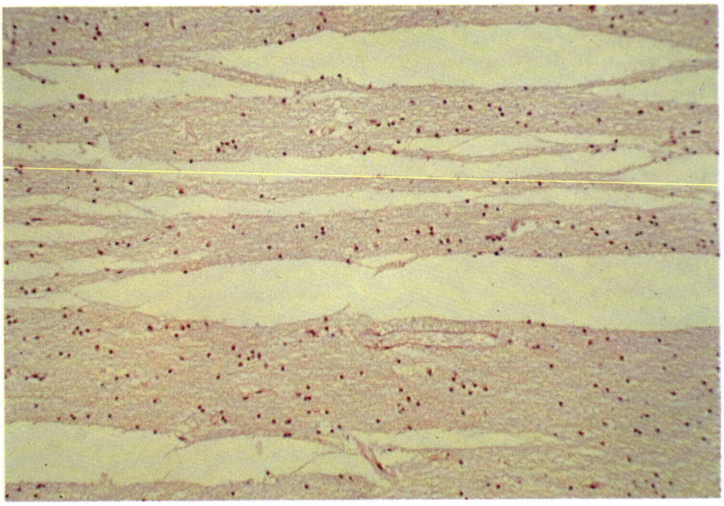

Fig. 213
This section of brain is to be compared with Fig. 211 and 212 since it simulates the condition mentioned above. However, it is a section of brain which has been blotted with filter paper. You can see where the filter paper picked out some of the brain substance. (Chapter 17)

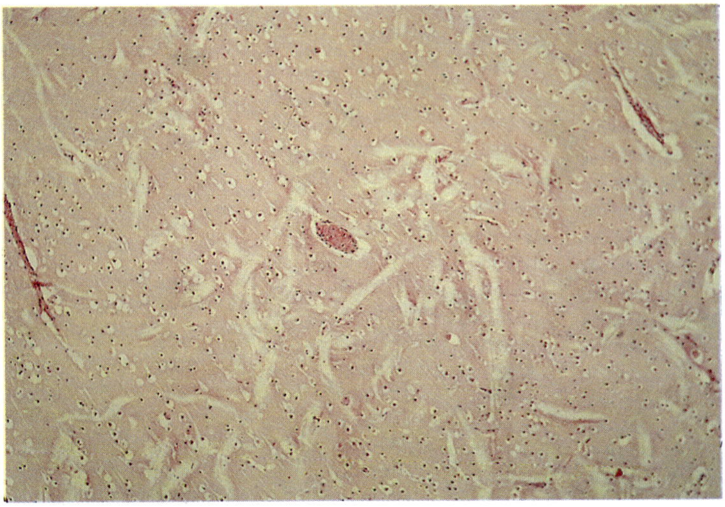

Fig. 214
This is a section of tissue which contains a gauze pad. In this particular instance the gauze pad was left in the individual after an operation. Several months later the tumor returned and upon removal it was determined that the tumor was the result of a reaction to the gauze pad. Note the giant cell reaction surrounding the gauze fiber. (Chapter 17)

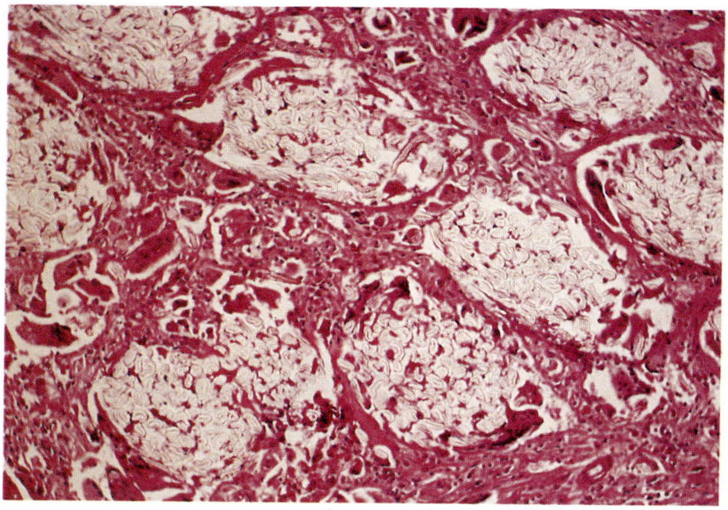

Fig. 215
Section of the same specimen as Fig. 214 except that in this case it is under partial polarized light showing the well delineated gauze pad fibers. (Chapter 17)

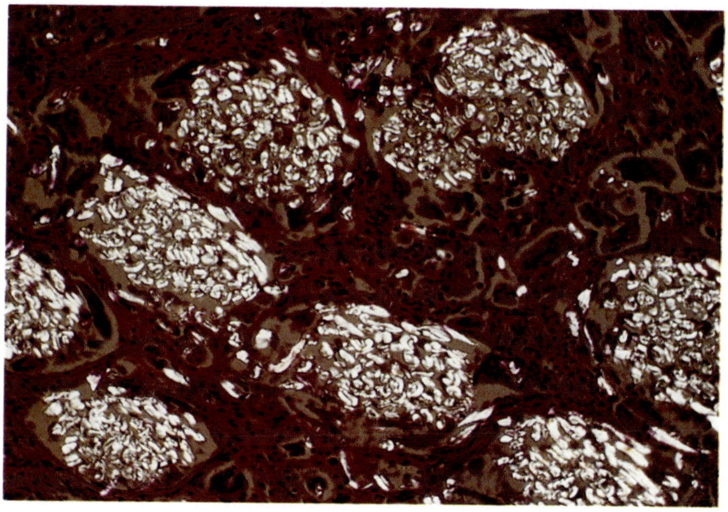

Fig. 216
This illustrates a modified Verhoeff's stain to demonstrate burnt areas in tissue sections. There is no indication of what burned the tissue, but the procedure can be used successfully to determine its presence. (Chapter 17)

Fig. 217
This section shows residual India ink which was used to identify a surface of the specimen for proper orientation. (Chapter 17)

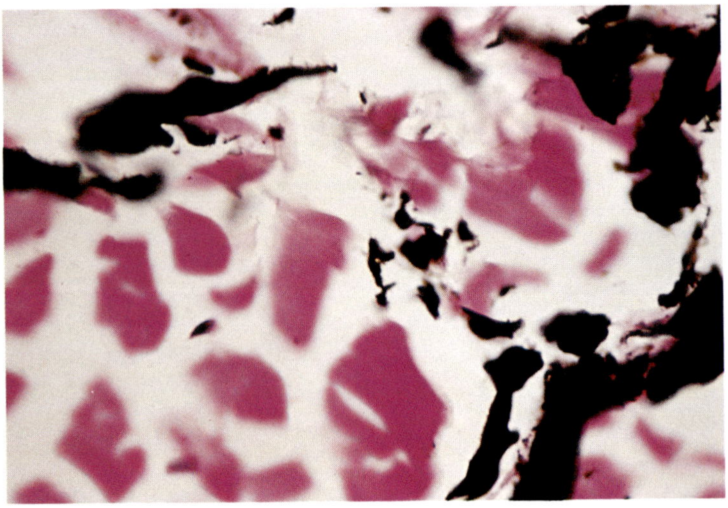

Fig. 218
Contraction artifact of muscle during freezing or fixation. This can be avoided by the use of an appropriate clamp which should be used when the specimen is frozen or fixed. (Chapter 17)

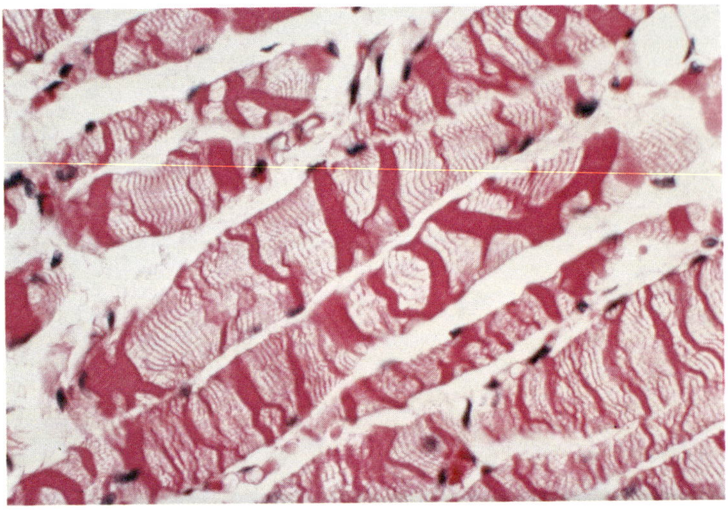

Formalin Fixative Artifacts

Fig. 219
This photograph illustrates vacuolated nuclei which is frequently produced by the use of buffered neutral formalin. In this case, it is assumed that the soft fixing effect of buffered formalin creates the vacuole by condensing the nuclear chromatin up against the wall of the nuclear capsule. This artifact can be corrected by use of the formalin alcohol-acetic acid method presented on page 11. (Chapters 1 & 17)

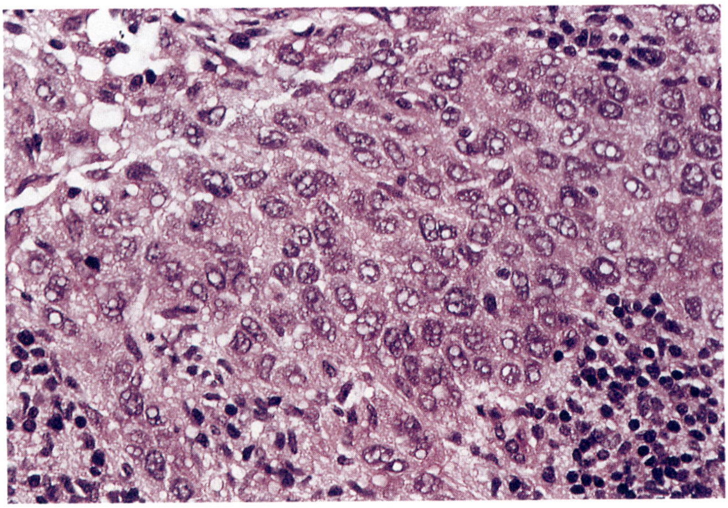

Fig. 220
A similar example of the vacuolated nuclear artifact illustrated in Fig. 219. Note that only the periphery is affected. (Chapters 1 & 17)

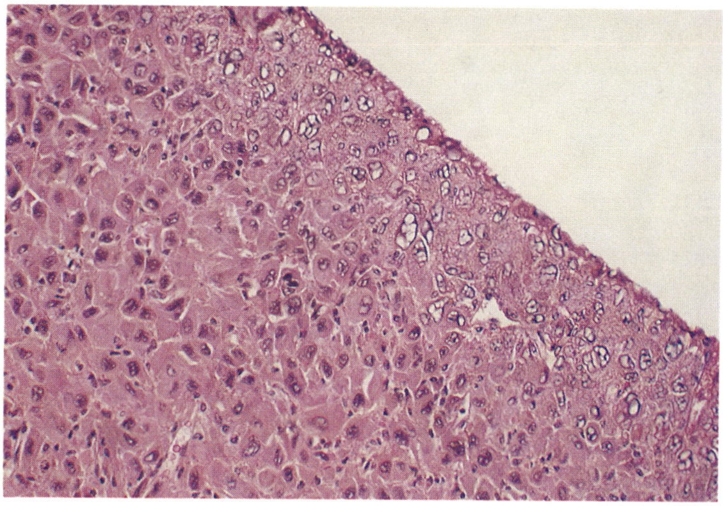

Fig. 221
This shows a container with a metal lid which did not have the plastic or cardboard diaphragm. Fixative was placed in the container and then an eye globe was placed in the container for fixation. Notice the rusty appearance of the lid of the container. The two eye specimens submitted to the laboratory were in different containers. (Chapter 17)

Fig. 222
The eye specimen seen on the left was fixed in the container with the rusty lid seen in Fig. 221. (Chapter 17)

Fig. 223
Note the extensive deposition of iron on the peripheral of the cornea after an iron stain was performed. (Chapter 17)

Fig. 224
This section shows a large vessel containing dark staining material which is formalin pigment. (Chapter 17)

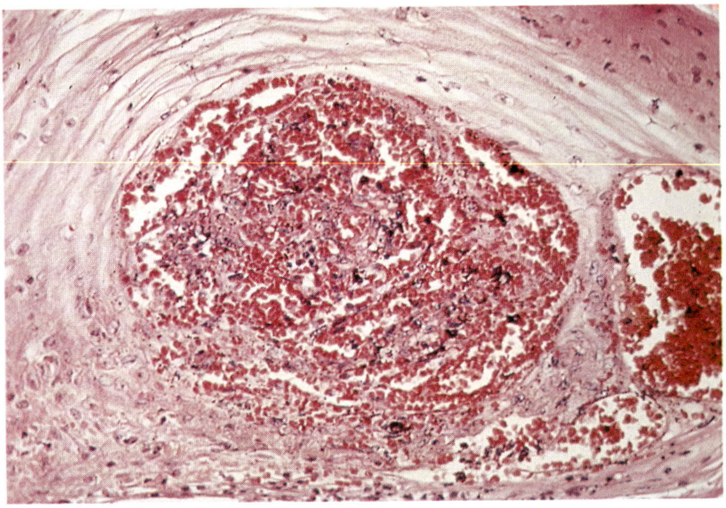

Fig. 225
This section of brain illustrates a potential problem dealing with formalin pigment. Note that the cytoplasm of these cells contain formalin pigment. This could erroneously be diagnosed as some other type of pigment. (Chapter 17)

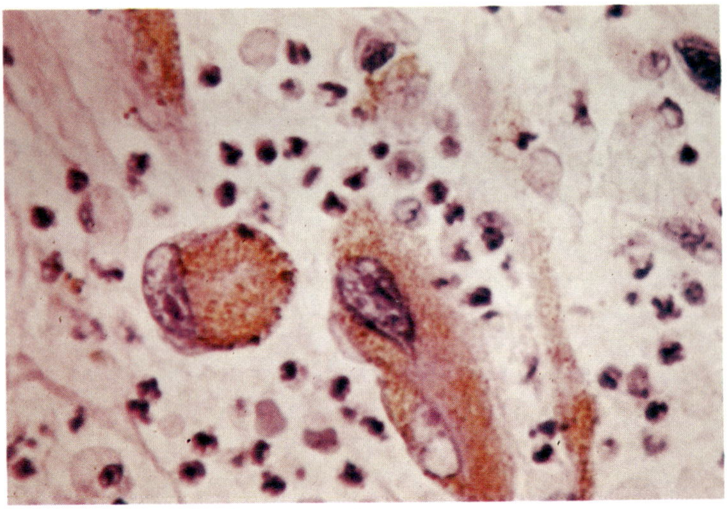

Fig. 226
This composite shows an additional problem related to formalin. Namely the fact that formalin pigment very often simulates other types of pigment. The photograph on the left shows a rhodanine copper stain. Note the red staining copper throughout. The section on the right is a partially polarized section of the same specimen showing polarized pigment. In this instance, the polarized pigment is formalin pigment. One can see in the background the red staining copper. (Chapter 17)

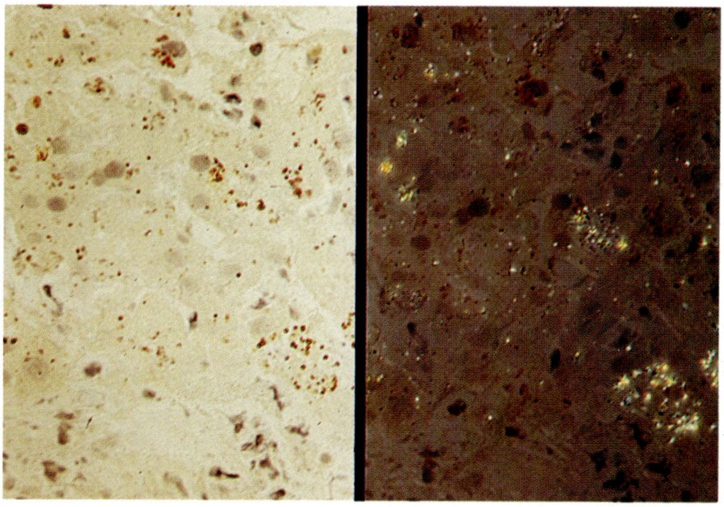

Fig. 227
This is a section of cardiac muscle demonstrating what appears to be calcium plaques. This specimen was fixed in 10% formalin buffered with calcium acetate. The plaques are pseudocalcification resulting from tissue fixation in formalin which contained a calcium salt.
(Chapter 17)

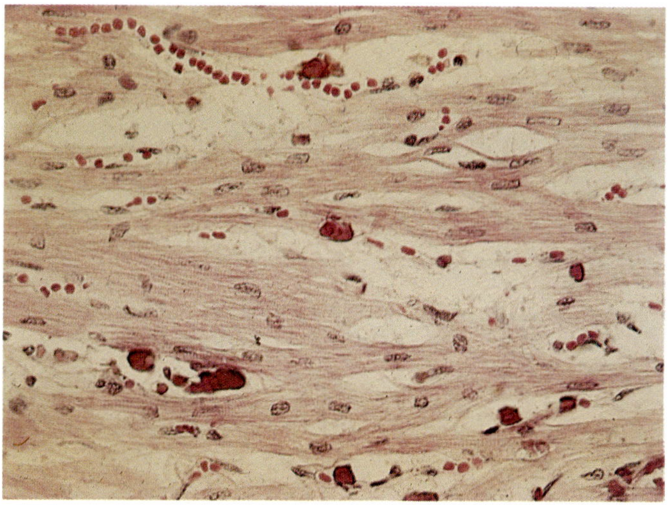

Fig. 228
Section of lung demonstrating a large vessel which contains pseudocalcified material. This specimen was fixed in 10% formalin buffered with calcium acetate. The Von Kossa stain was performed on this section. The exposure of tissue to any calcium salt will produce pseudocalcification. (Chapter 17)

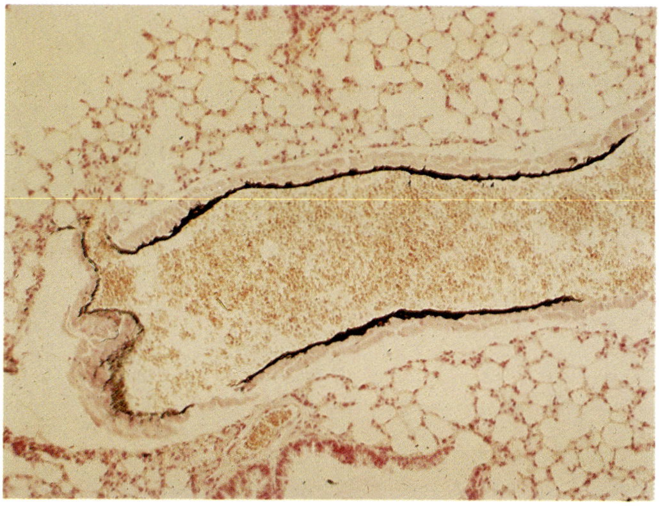

Fig. 229
This is a section of skin demonstrating Paget's cells stained with the aldehyde fuchsin method. This particular specimen was in the fixative solution for approximately 5 days. (Chapter 17)

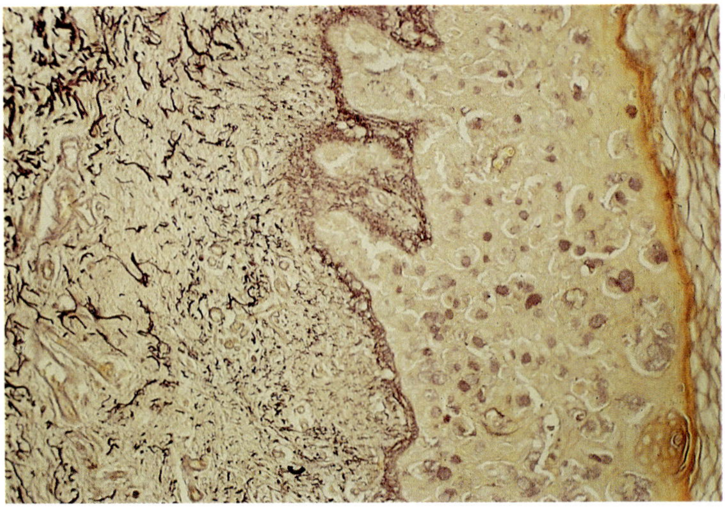

Fig. 230
This photograph is similar to the specimen shown in Fig. 229. This specimen was also stained with aldehyde fuchsin. Notice a lack of staining of the Paget cells. In this instance, extensive exposure (2 weeks) to unbuffered formalin reduced the staining properties of the aldehyde fuchsin. (Chapter 17)

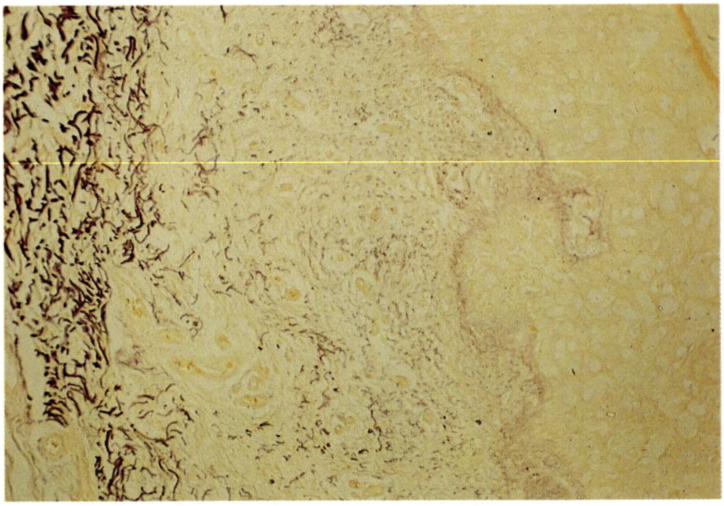

Fig. 231
This is a section of a kidney demonstrating extensive shrinkage of the glomeruli and surrounding tissue. In this instance the kidney specimen was extra thick. The section was taken from the middle of the specimen which was not fixed well. (Chapter 17)

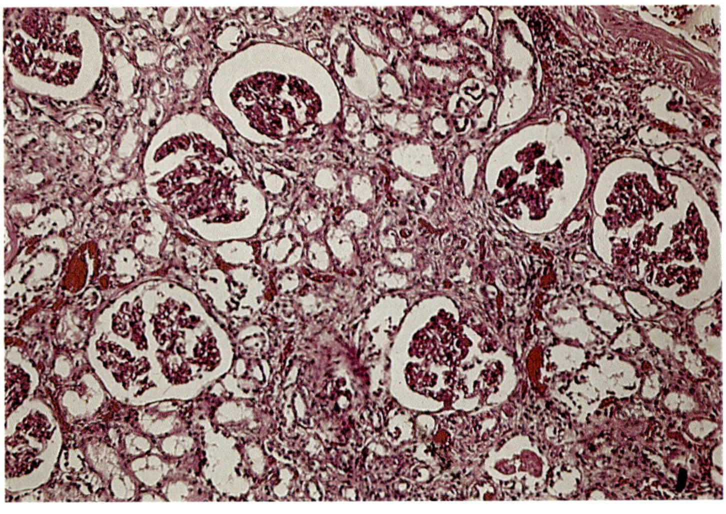

Fig. 232
This is a section taken from the periphery of the specimen seen in Fig. 231. Notice that there is less shrinkage of the glomeruli and surrounding tissues indicating better fixation. (Chapter 17)

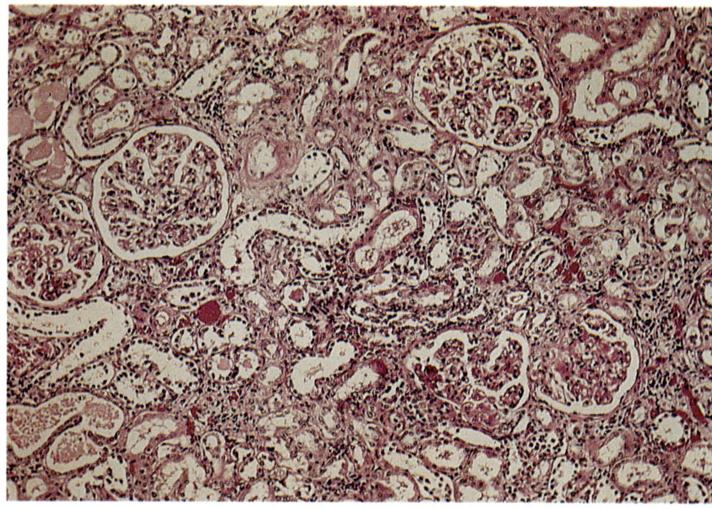

Fig. 233
This photograph demonstrates fungi artifact which can grow in tissue sections due to extensive washing subsequent to fixation. This fungi is deposited from rubber or plastic tubing which has not been replaced for an extended period of time. Note the extensive growth of fungi throughout the section. (Chapter 17)

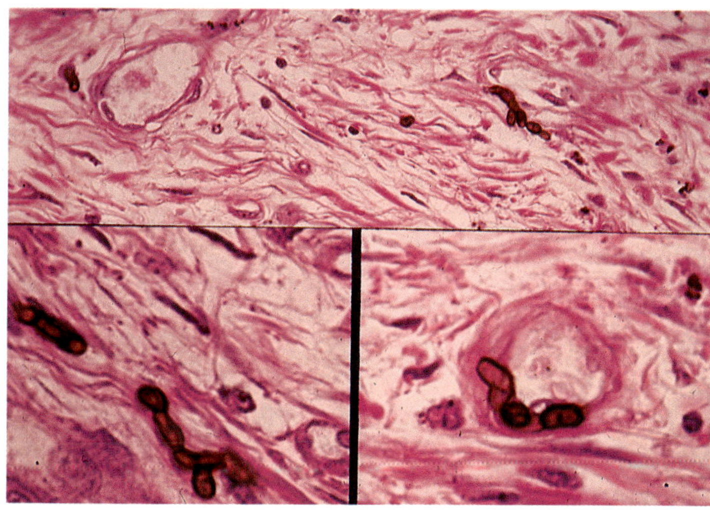

Fig. 234
Another illustration of artifactual bacteria growth in a tissue section (a result of extensive washing of tissue specimens in a sink which contained plastic or rubber tubing which had not been replaced frequently). (Chapter 17)

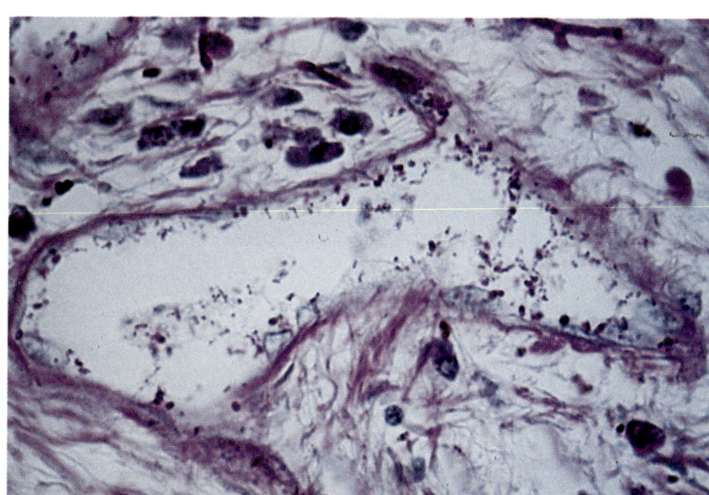

Fig. 235
This is a section of skin stained with the Mowry's colloidal iron stain. Note that the hyaluronic acid which is normally found between the collagen fibers is absent. This is due to extended exposure to Zenker's solution. In this instance the section remained in the Zenker's for 5 days. (Chapter 17)

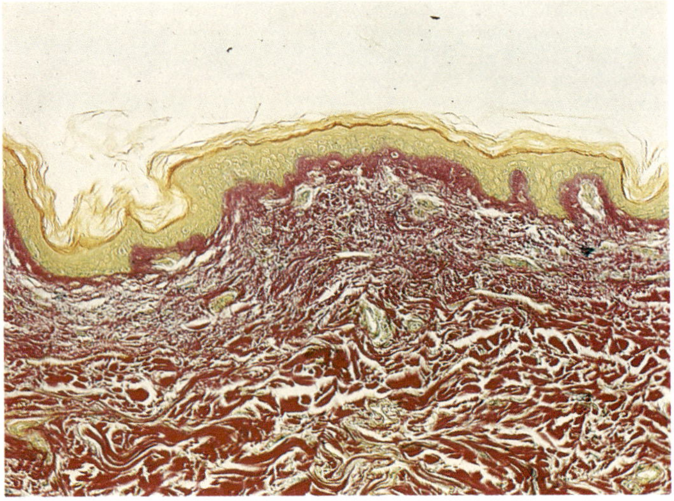

Fig. 236
This photograph is to illustrate the yellow effect produced by Bouin's fixative on tissue or slides exposed to the picric acid in this fixative. It should be understood that exposure to picric acid produces some deleterious effects on tissue. It should also be understood that unless most of the picric acid is removed from the section or slide, subsequent staining will be altered significantly. (Chapter 17)

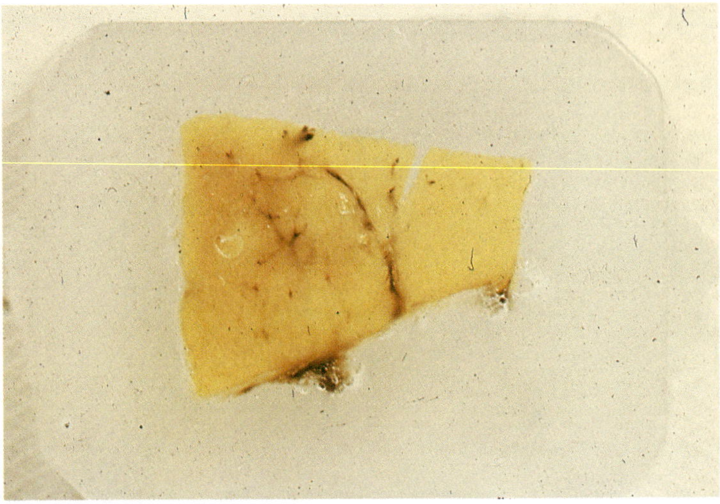

Fig. 237
This composite shows a section of skin. The photograph on the left was fixed in Bouin's. Note that most of the hyaluronic acid has been removed by fixation in this solution. In contrast, note the photograph on the right which contains most of the hyaluronic acid present in the section. In this instance this section was fixed in 10% buffered neutral formalin. (Chapter 17)

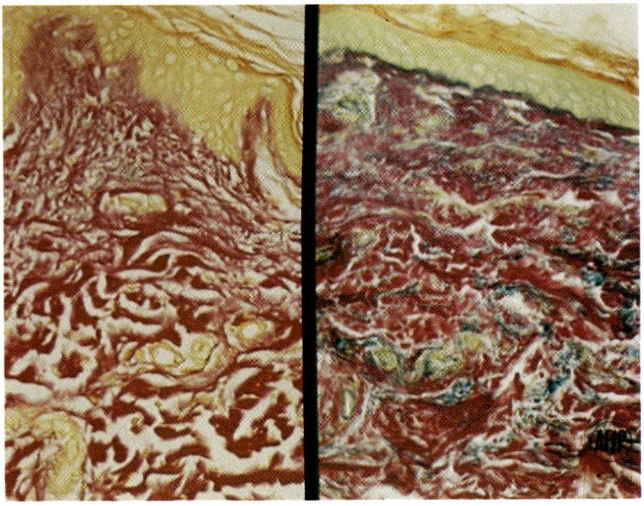

Fig. 238
This is a Bouin's fixed specimen. Note the H&E on the left which shows very little evidence of pigment while the section on the right shows a great deal of pigmentation due to the acidic effect of Bouin's fixation. (Chapter 17)

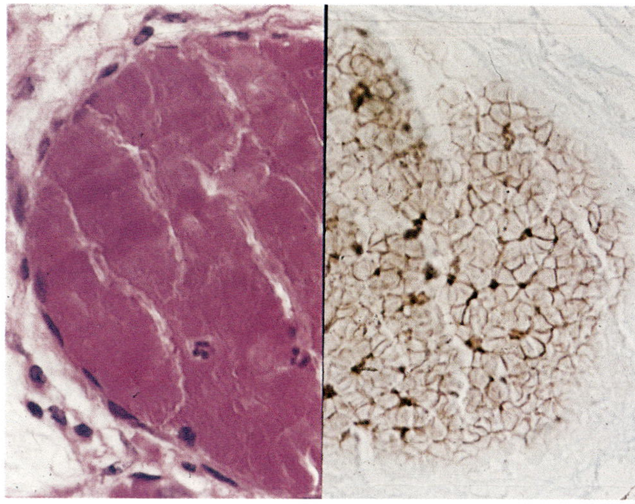

Processing Artifacts

Fig. 239
This section demonstrates what I refer to as the parched earth effect. Namely the cracking throughout the section. This is due to one of two factors; (1) underprocessed specimens, (underparaffin impregnation) and (2) a tissue floatation bath that is above the 45°C maximum temperature which creates an expansion factor beyond that which the tissue can withstand, creating tension which results in breaking of tissue in the section during subsequent drying or staining. This takes place because tissue attempts to return to its original form but is not allowed to do so by the section adhesive, resulting in tissue cracking. (Chapter 17)

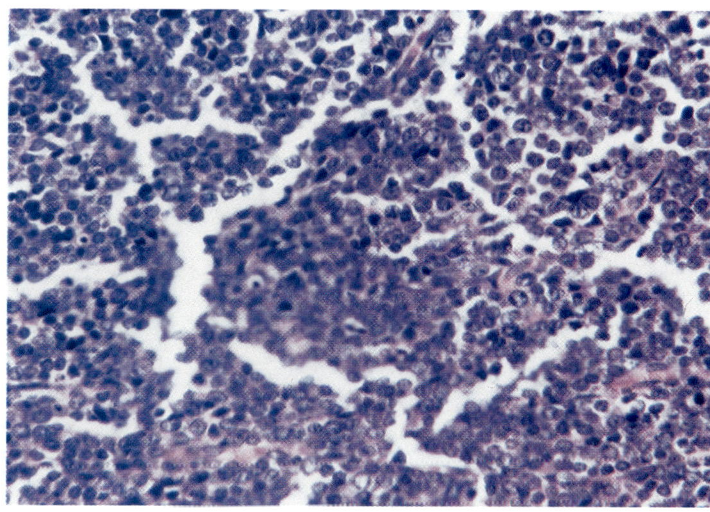

Fig. 240
This is a section of lymph node demonstrating an extensive amount of wrinkling. The fact that these wrinkles are all in the vertical or slightly oblique plane strongly suggests that this is due to the effects of a dull knife or a floatation water bath which is not warm enough. (Chapter 17)

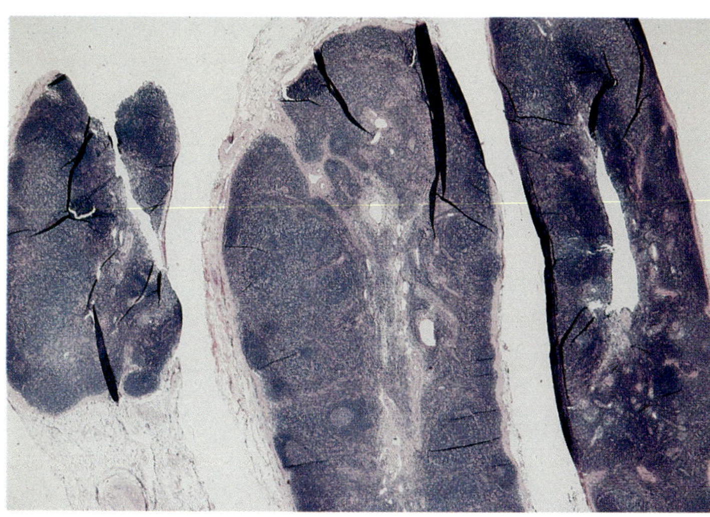

Fig. 241
This section of brain demonstrates the effects of improper processing. Generally, this can be determined by the fact that the wrinkling is in various directions which is always a strong indication of inadequate exposure of specimens to paraffin impregnation. (Chapter 17)

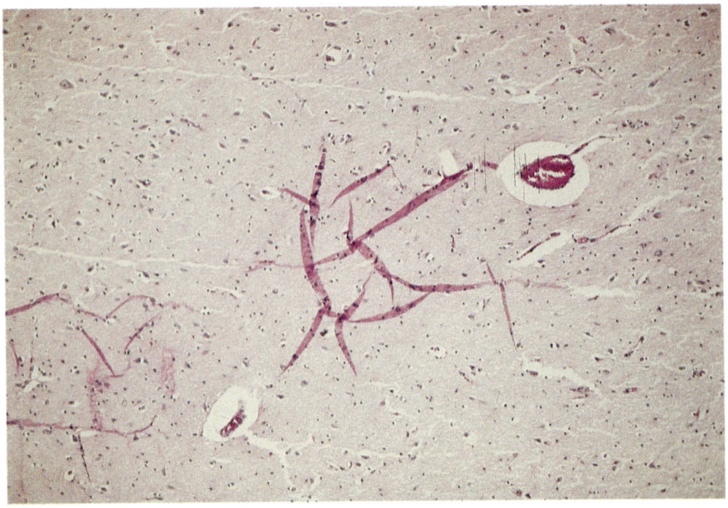

Fig. 242
This section of liver demonstrates excessive exposure to heat which can be seen on the right side. In most instances this is the result of overextended exposure to paraffin above 60°C. (Chapter 17)

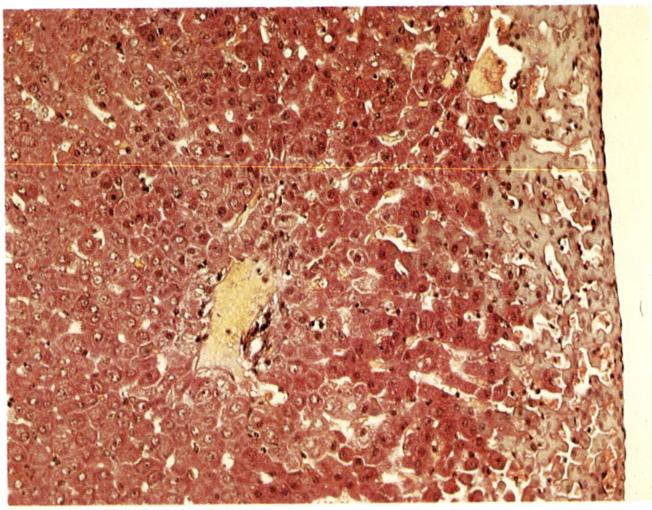

Fig. 243
This section is another example of that which is illustrated in Fig. 242 above. Note the denaturing effect of the entire section. Again this is a result of overexposure to heat during paraffin exposure of the specimen. (Chapter 17)

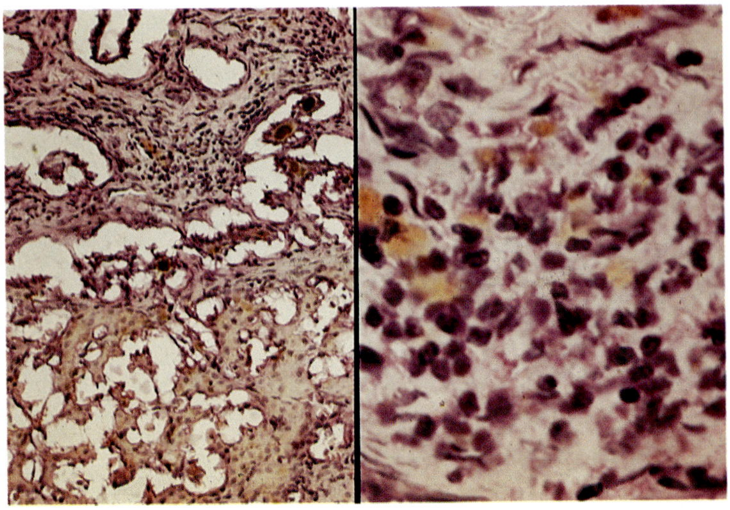

Fig. 244
This photograph shows the effects of processing on a specimen which has been underfixed. The center portion of this section of liver (illustrated in red) is improperly fixed. The cell at the bottom represents a cell which was viewed from the middle of the liver specimen which was unfixed. The cell on the upper right corner illustrates a cell which was taken from the peripheral margins of the liver which received proper fixation. (Chapter 17)

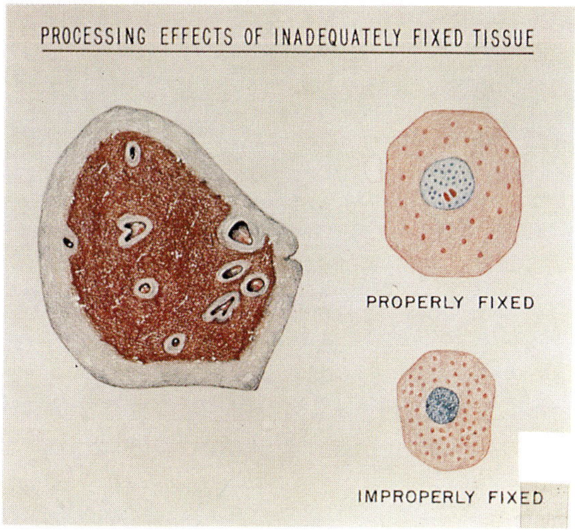

Fig. 245
This is a section of brain containing vacuoles. These vacuoles are due to long exposure of animal brain to alcohol prior to processing. In this instance the vacuoles are produced in rats or mice. Specimens from this animal should not be exposed to alcohol more than an hour from trimming to processing. A corrective measure is to leave the specimens in the formalin fixative before processing. (Chapter 17)

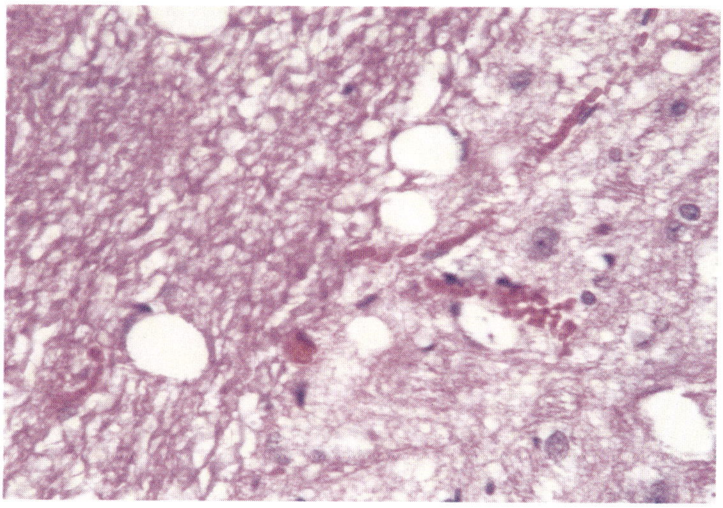

Microtomy Artifacts

Fig. 246
This section demonstrates microvibration which is the result of inadequate introduction of water (moisture) into the specimen prior to microtomy. Compare with Fig. 247. (Chapter 17)

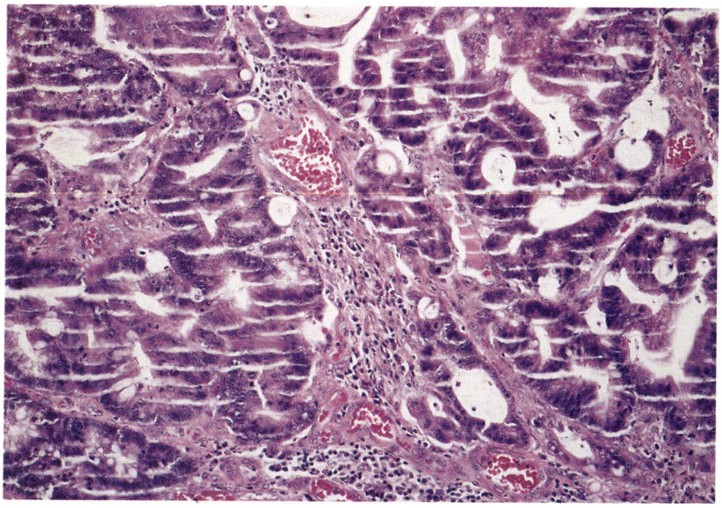

Fig. 247
This section demonstrates a similar section in which the proper amount of moisture was introduced into the paraffin block prior to microtomy. In this particular instance a piece of cotton soaked with lukewarm water was applied to the surface of the specimen for 5 seconds. Compare with Fig. 246 which did not receive this treatment. (Chapter 17)

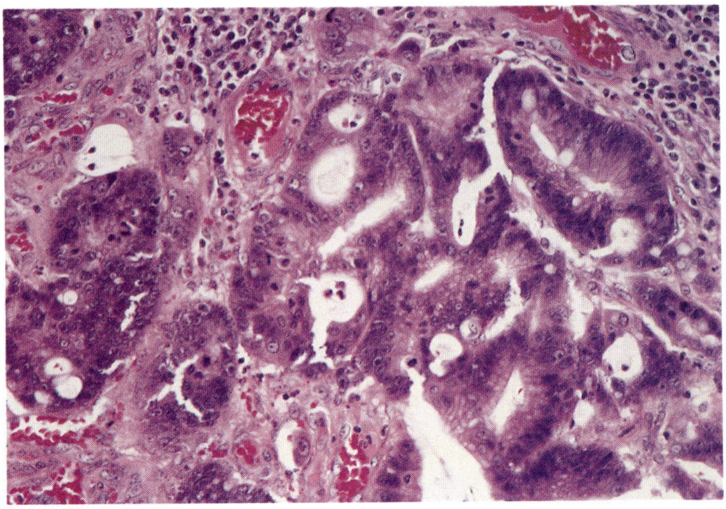

Fig. 248
This is another section illustrating microvibration throughout the section as a result of lack of reintroduction of moisture. (Chapter 17)

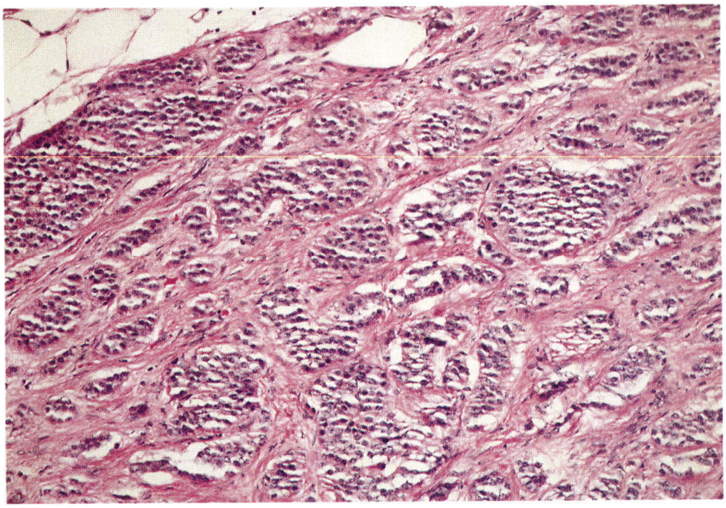

Fig. 249
This is an example of venetian blind effect. Venetian blind effect is generally characterized by thick and thin broad areas throughout the section. This is due to lack of moisture in tissue prior to sectioning. This can be compared with microvibration which can be seen in Fig. 248. (Chapter 17)

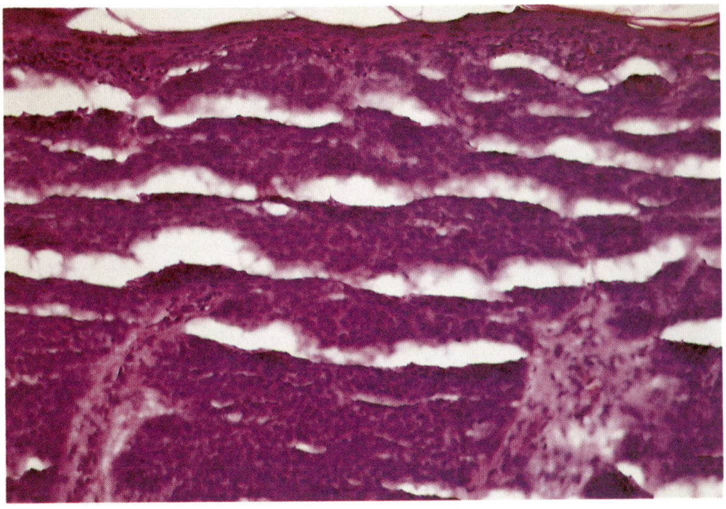

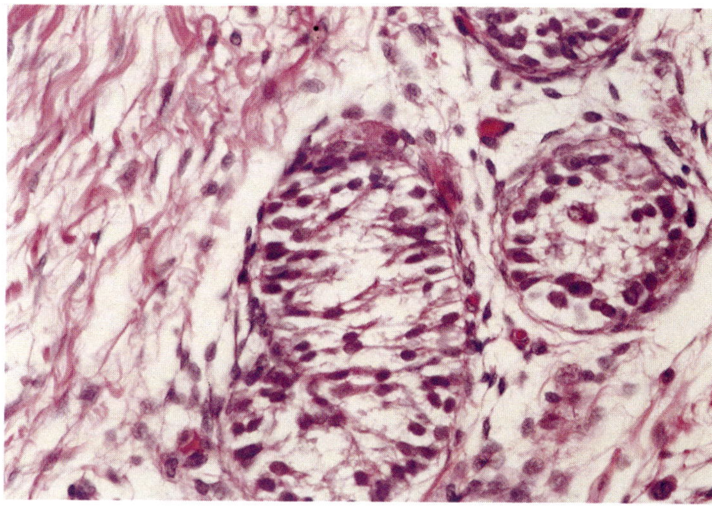

Fig. 250
This is a section of testis demonstrating microvibration throughout the sartoli cell. (Chapter 17)

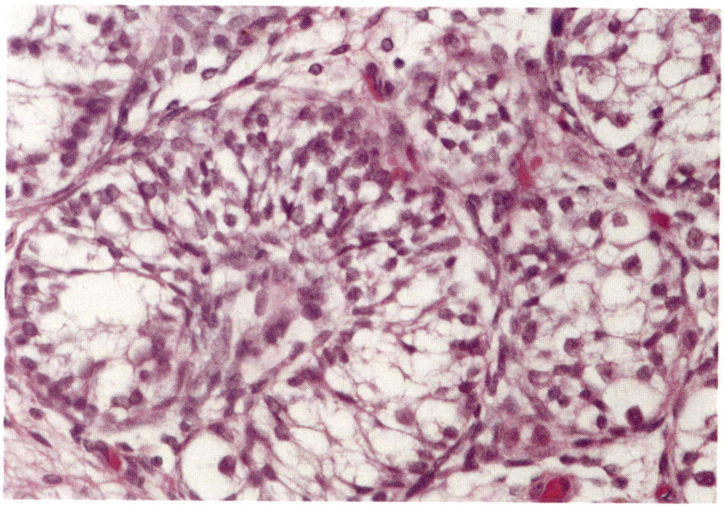

Fig. 251
This is a section identical to that seen in Fig. 250 which shows the effect of introduction of sufficient moisture prior to cutting. Note the lack of microvibration. (Chapter 17)

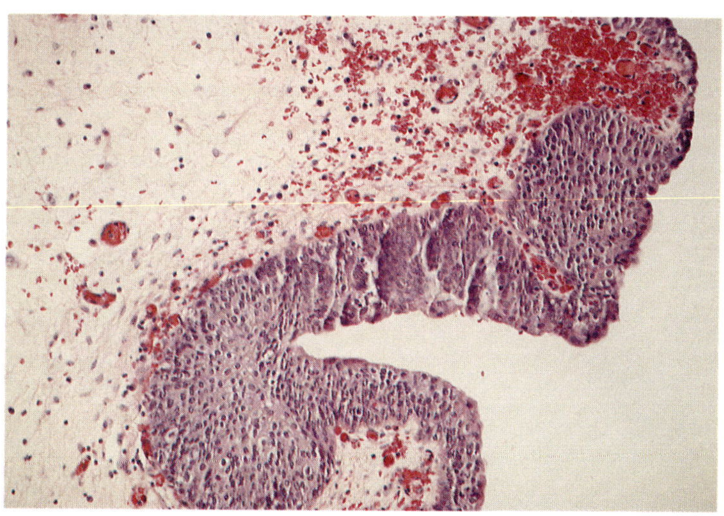

Fig. 252
This particular section shows two effects due to the lack of reintroduction of moisture into the specimen. Microvibration can be seen at the upper portion of the section while venetian blind effect can be seen in the middle of the section. In both instances this is due to lack of reintroduction of moisture prior to microtomy. (Chapter 17)

Fig. 253
This section of skin demonstrates extensive microvibration, again due to lack of introduction of moisture prior to microtomy. (Chapter 17)

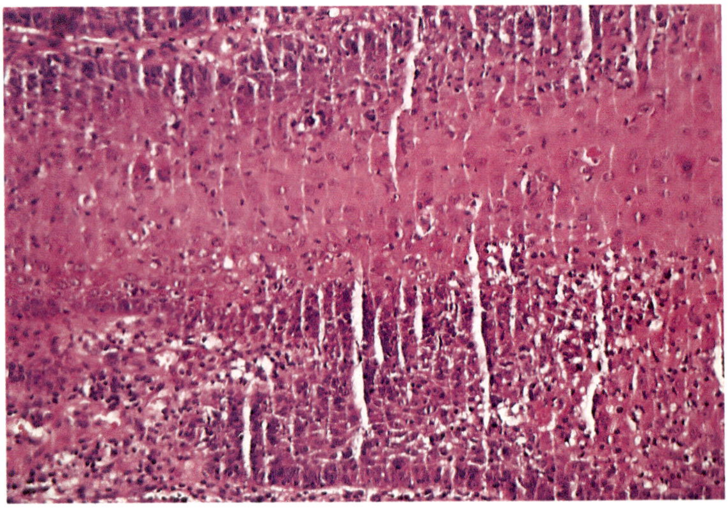

Fig. 254
This section of lymph node shows extensive microvibration on the peripheral margins of the specimen. The peripheral margins in this instance were slightly harder than the center thus creating the problem. Again this is due to the lack of reintroduction of moisture into the tissue prior to sectioning. The lines which can be seen throughout the specimen are those produced by the hard outer core surface material which was dragged by the knife as it cut through the specimen. (Chapter 17)

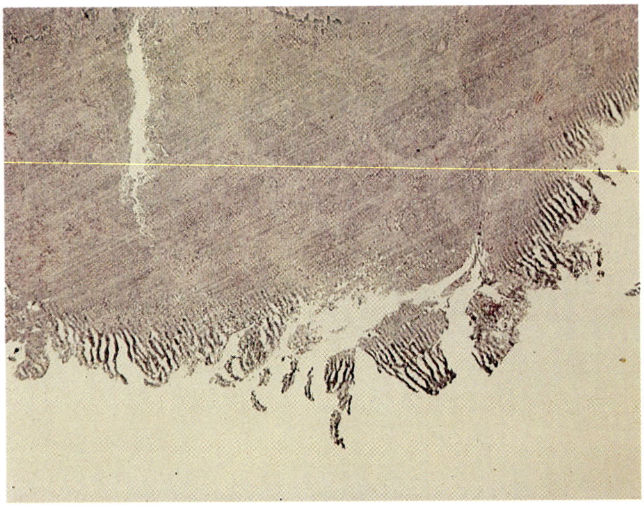

Fig. 255
This photograph illustrates the effects of microvibration at the microscopic level. Note the thick and thin areas throughout the section. Compare with Fig. 256. (Chapter 17)

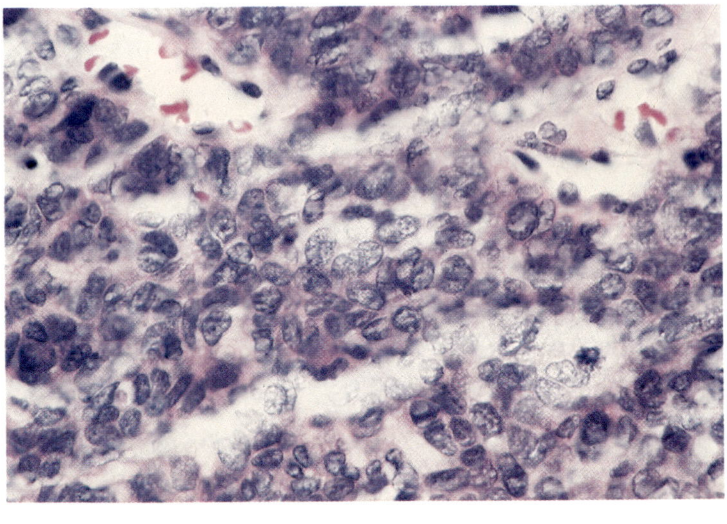

Fig. 256
This section was cut from the same specimen as Fig. 255 after reintroducing lukewarm water into the surface of the specimen by the aid of a piece of cotton. Note the difference in the cells.
(Chapter 17)

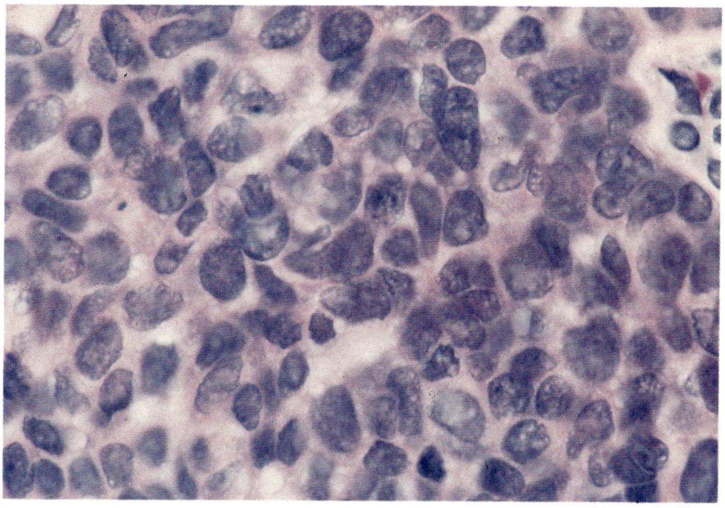

Fig. 257
This section shows blood vessel walls which have been displaced slightly by the microtome knife as it traveled through the specimen. This is another example of the effects which can be produced by the lack of moisture in a tissue specimen.
(Chapter 17)

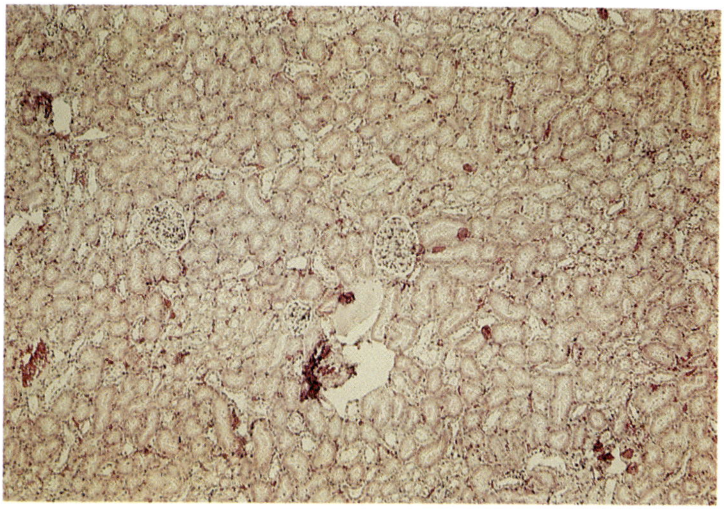

Fig. 258
This is a section of liver demonstrating a moth eaten effect throughout. These artifacts are also produced by the lack of reintroduction of moisture into the tissue prior to microtomy. (Chapter 17)

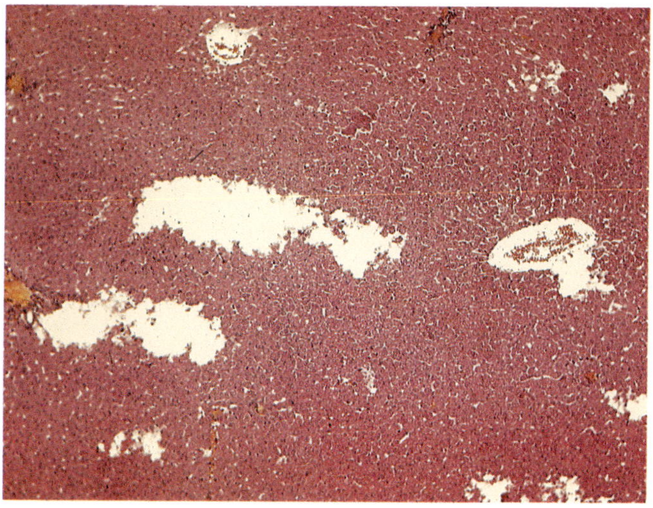

Fig. 259
This section shows leprae bacilli which has been displaced from its normal site due to the lack of reintroduction of moisture into the specimen. It should be understood that many artifacts in microtomy are due to this one factor, the reintroduction of sufficient moisture into the tissue prior to microtomy. (Chapter 17)

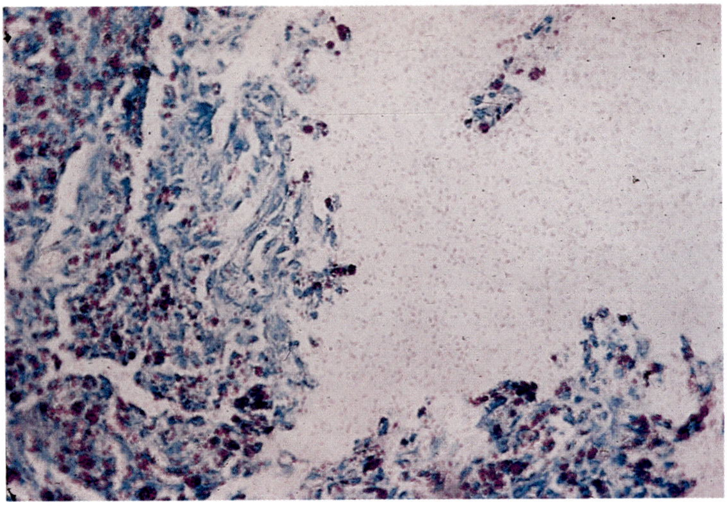

Fig. 260
This is a section demonstrating red structures throughout which represent dandruff. Dandruff can be found in almost every section produced in the histology laboratory. It is a very difficult artifact to eliminate. (Chapter 17)

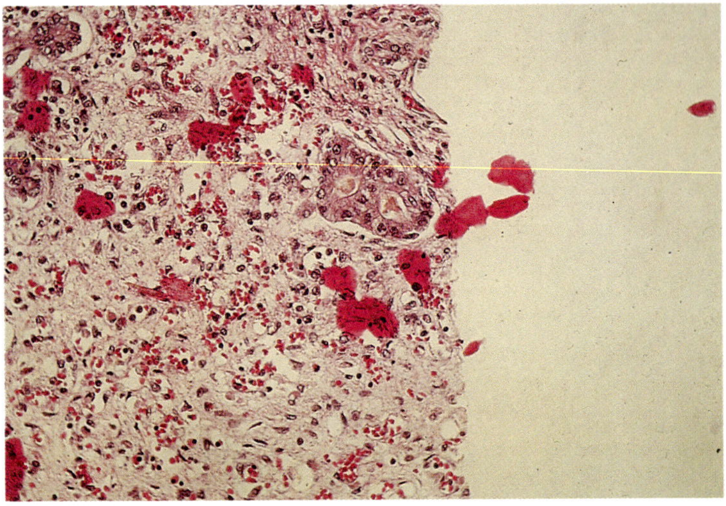

Fig. 261
This is a section of the Warthin-Starry method demonstrating the silver reaction potential of dandruff. The black structures that can be seen throughout the section are dandruff flakes. Note the silver reaction of the residual melanin in the dandruff flakes. (Chapter 17)

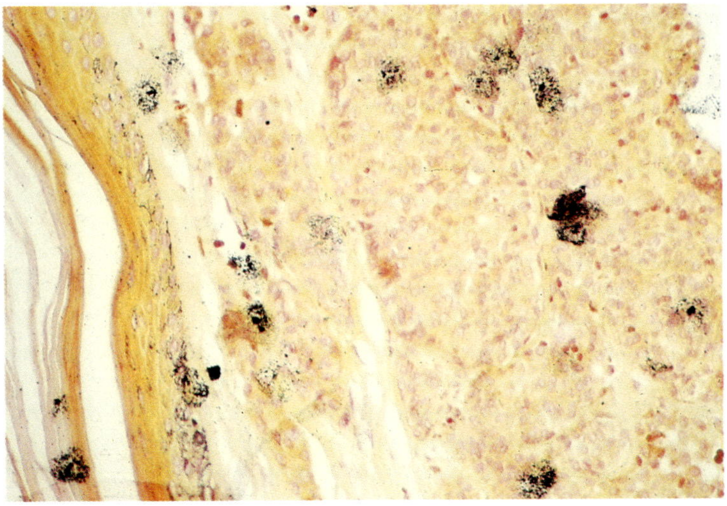

Fig. 262
This is a section demonstrating tissue cracking from freezing the paraffin specimen with a skin refrigerant. Note the clear delineated cracking throughout the section. These clear delineated cracks are consistent with cracking due to the use of these refrigerants. (Chapter 17)

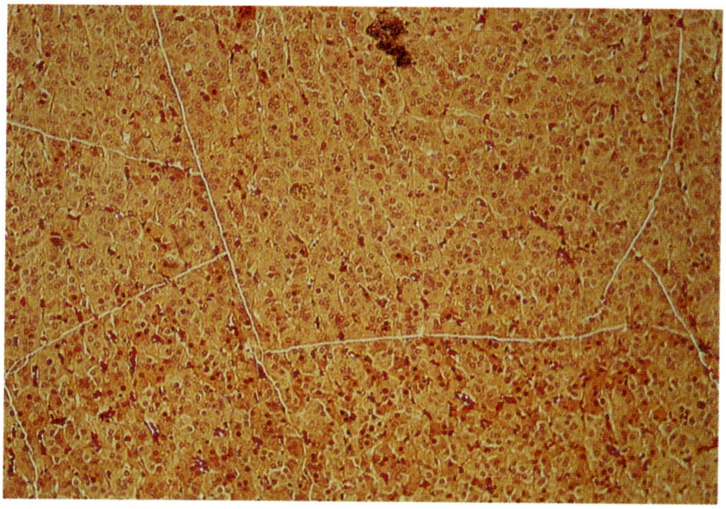

H&E Staining Artifacts

Fig. 263
This is a section of skin stained with an H&E. There are several problems with this section: (1) the section is too thick; (2) microvibrations are present throughout; and (3) the staining is too dark with both the hematoxylin and eosin. (Chapter 17)

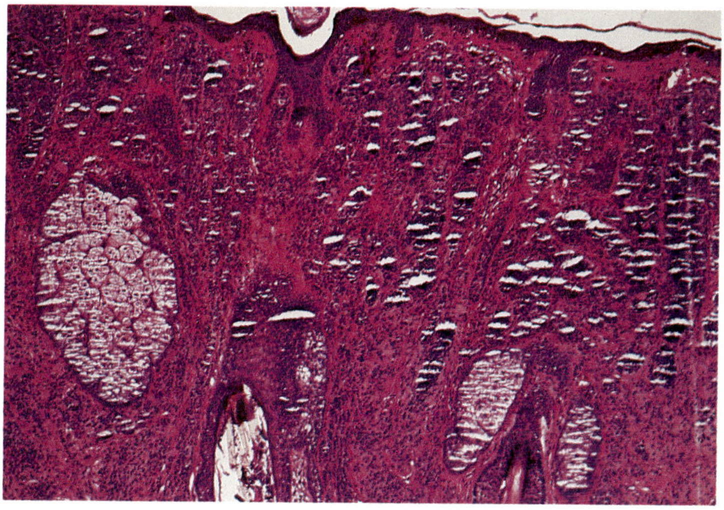

Fig. 264
This is a section of skin demonstrating what happens when overoxidized hematoxylin is used. Note the dark staining of the peripheral margins. (Chapter 17)

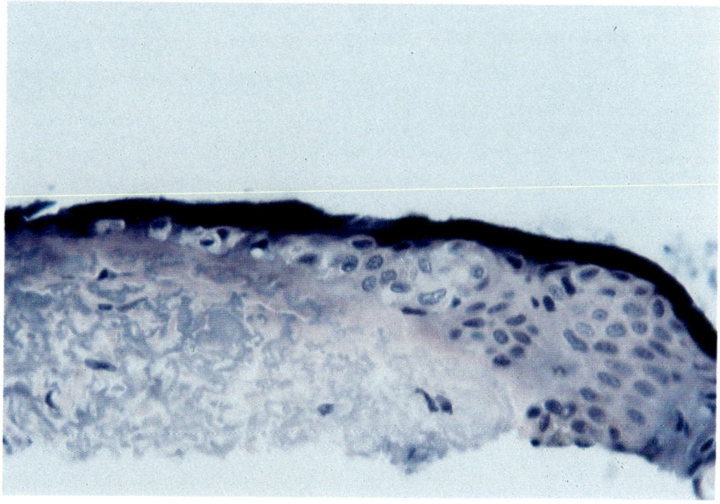

Fig. 265
This composite shows the difference between a well differentiated eosin and that which has not been differentiated well. The section on the left was not differentiated properly resulting in dark staining of cytoplasm including overshadowing of the nuclei with eosin. The section on the right demonstrates a properly differentiated eosin making the nuclei more prominent because of the lack of overshadowing of eosin. (Chapter 17)

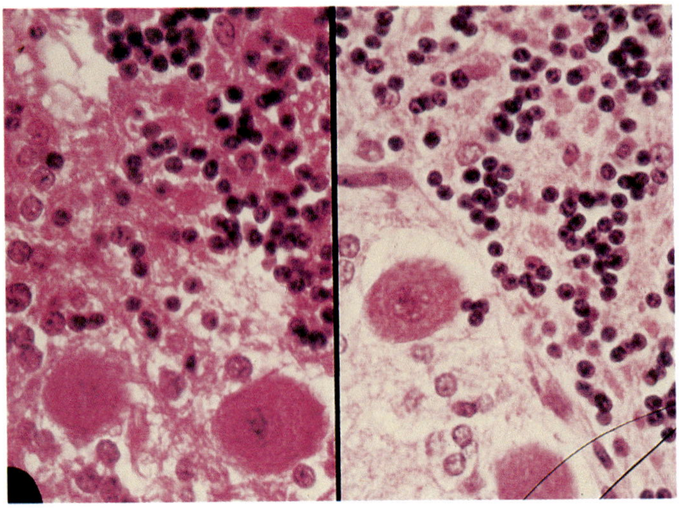

Fig. 266
This photograph shows mold contamination throughout. This was obtained from a hematoxylin solution which had not been changed for 3 weeks. It is always possible to grow mold in most solutions if they are not changed frequently. (Chapter 17)

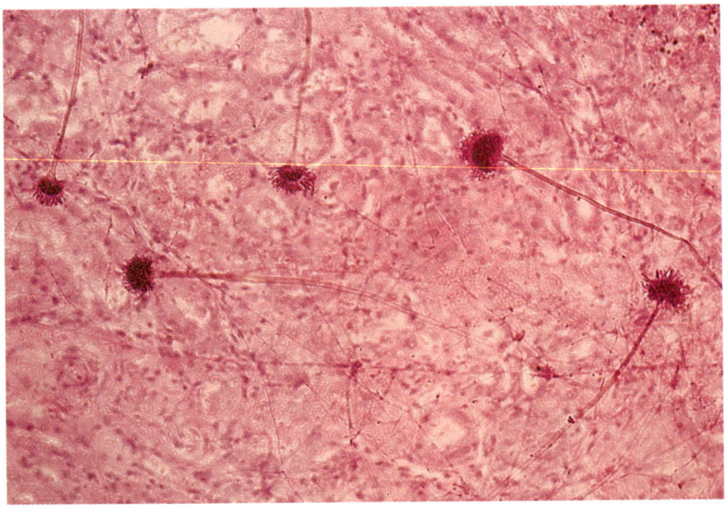

Fig. 267
This composite illustrates black staining material. This is the result of hematoxylin that has reached the point of overoxidation. In this particular instance the hematoxylin converts back to modified hematein crystals producing a precipitate throughout the section. (Chapter 17)

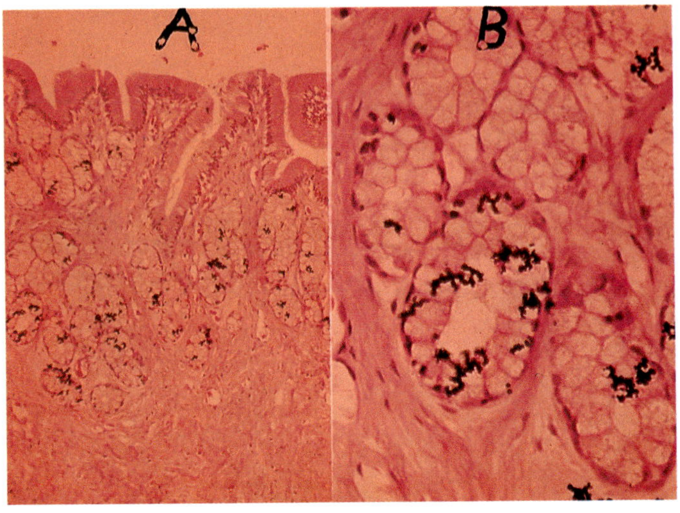

Fig. 268
This photograph demonstrates aluminum ammonium sulfate which is used as a mordant in many hematoxylins. In this particular instance, the solution was not mixed sufficiently resulting in the aluminum converting back to its original form in the staining container. It is not unusual to find fine crystals deposited on tissue sections in instances where hematoxylin is not mixed properly. (Chapter 17)

Fig. 269
This photograph demonstrates water bubbles (light circles) seen throughout the photograph. This is a case in which there was very poor dehydration and clearing of the stained section. This effect can be seen better by lowering the substage of the microscope. (Chapter 17)

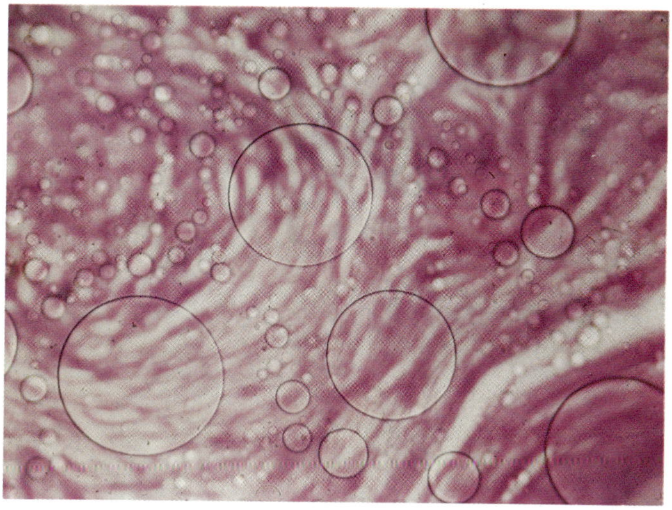

Fig. 270
This Leder stain vividly demonstrates the effects of improper removal of paraffin during the deceration process. This is a partially polarized section demonstrating residual paraffin in white. (Chapter 17)

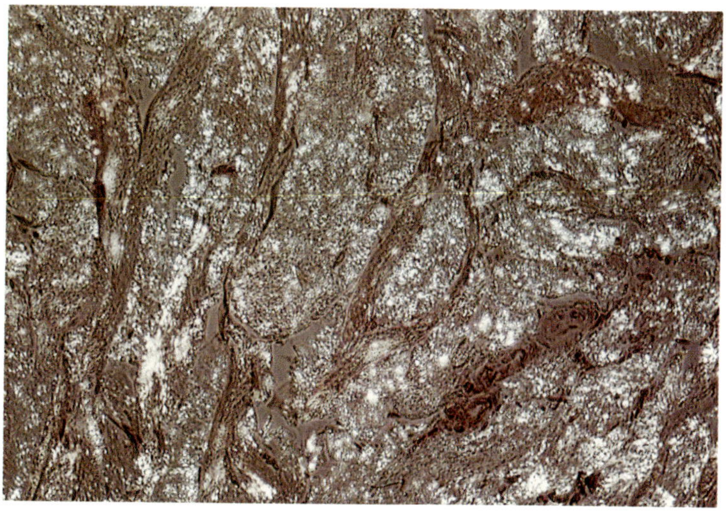

Fig. 271
This is a section stained with hematoxylin which has undergone overoxidation. Notice the fine-granular black material in the middle of the section which is staining mucoid substances. Mucinous type materials will stain frequently when hematoxylin is undergoing overoxidation. (Chapter 17)

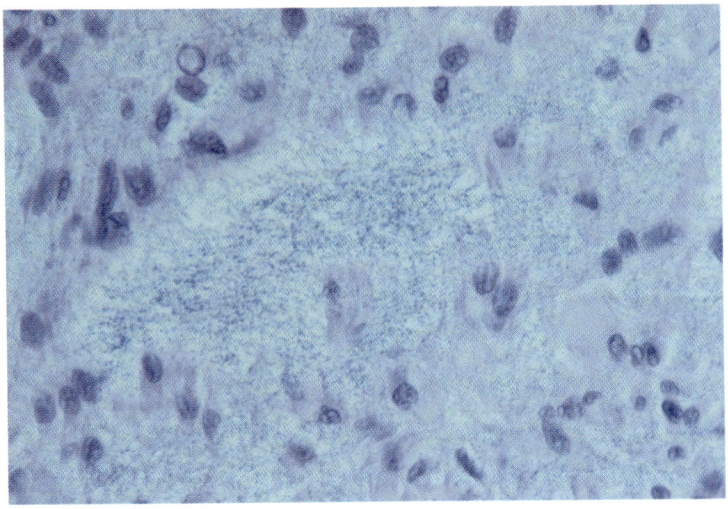

Fig. 272
This photograph shows the effect of limited exposure to 0.25% citric acid eliminating the hematoxylin from the mucous material seen in Fig. 271. (Chapter 17)

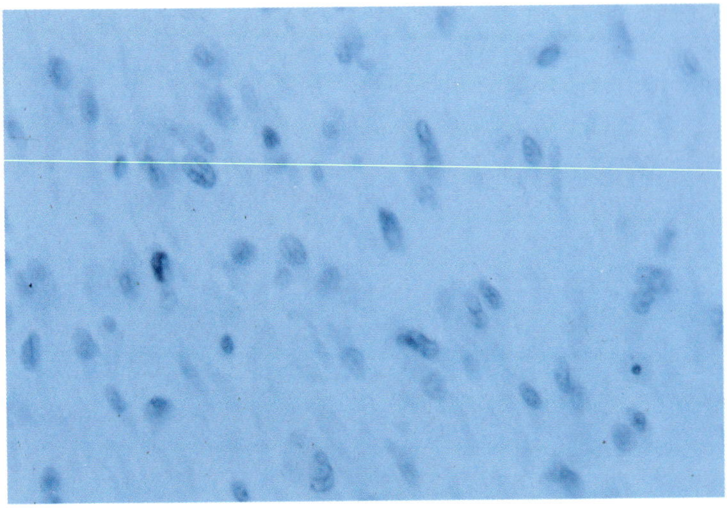

Fig. 273
This section of intestine demonstrates a similar problem such as that seen in Fig. 271; namely the staining of mucinous material such as that seen in the goblet cells and lumen of this specimen. (Chapter 17)

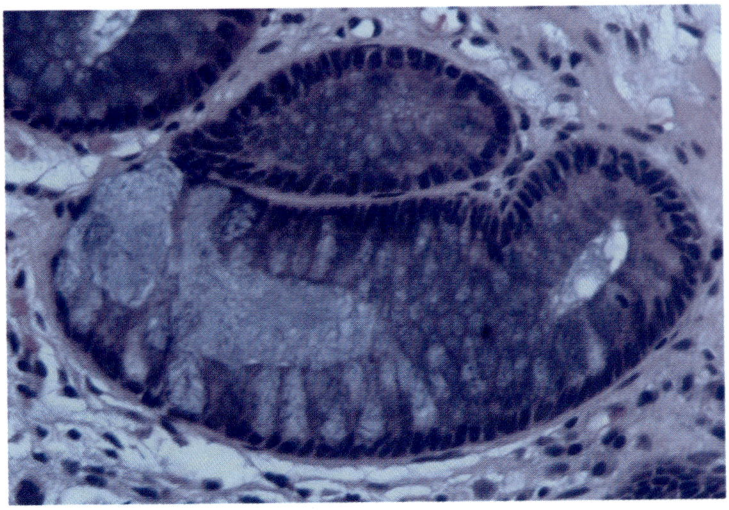

Fig. 274
This composite demonstrates the use of excessive or thick mounting medium on a stained slide. The photograph on the left demonstrates the proper amount and/or thickness of the mounting medium. The photograph on the right shows a slightly milky or cloudy appearance demonstrating an excessive amount of mounting medium. Excessive mounting medium will raise the coverslip and therefore change the refractive index. A simple test to check for thick coverglass mounting medium is to take a 10 ml pipette and drain 10 ml of mounting medium from the container being tested. Allow the upper 5 ml of the pipette to evacuate. The evacuation should take 7-9 seconds if the mounting medium is of proper consistency. (Chapter 17)

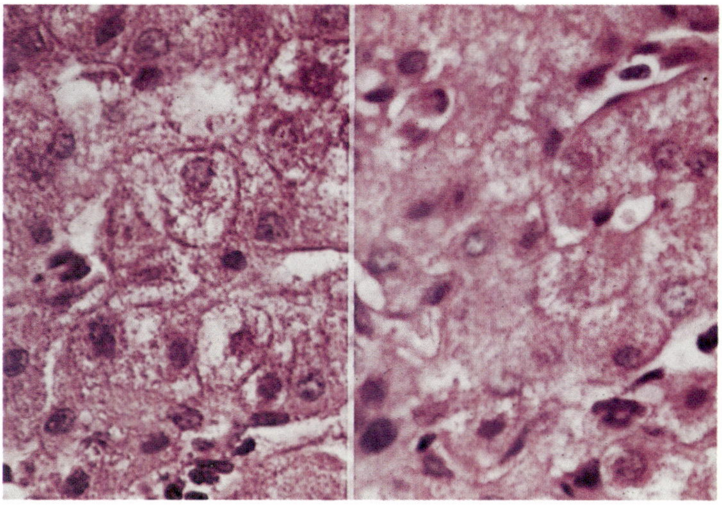

Fig. 275
Excessive amount of water on the stained slide due to lack of improper dehydration and/or clearing. This problem can be recognized by the tinctorial quality which can be seen in the photograph at the bottom. Note this section is considerably browner than the one on top. This tinctorial quality of a section often suggests an excessive amount of water in the section prior to mounting of the coverglass. (Chapter 17)

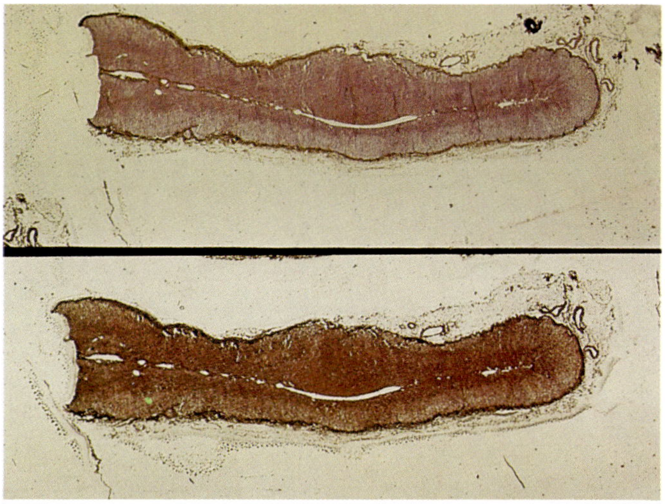

Fig. 276
This composite shows another example of a section which has been allowed to dry prior to the application of the mounting medium. The section on the left was properly coverslipped. The section on the right was allowed to dry prior to the application of the mounting medium. Notice the brown stippling throughout the section which is trapped air. (Chapter 17)

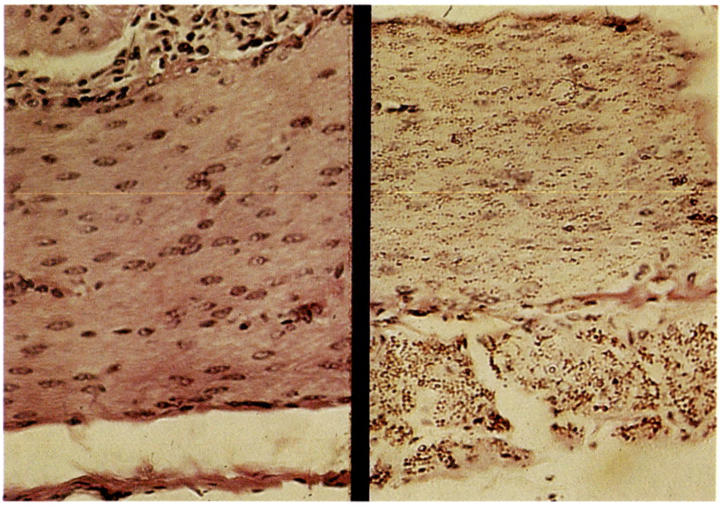

Fig. 277
This is a section similar to Fig. 276 which shows a piece of intestine in which the mounting medium was applied after the section was allowed to dry for a short period of time. Notice the air bubbles trapped in the nuclei creating a false sense of nuclear staining. In this case, one should recognize that all sections should be coverslipped while the section is saturated with the clearing agent and prior to the application of the coverglass. (Chapter 17)

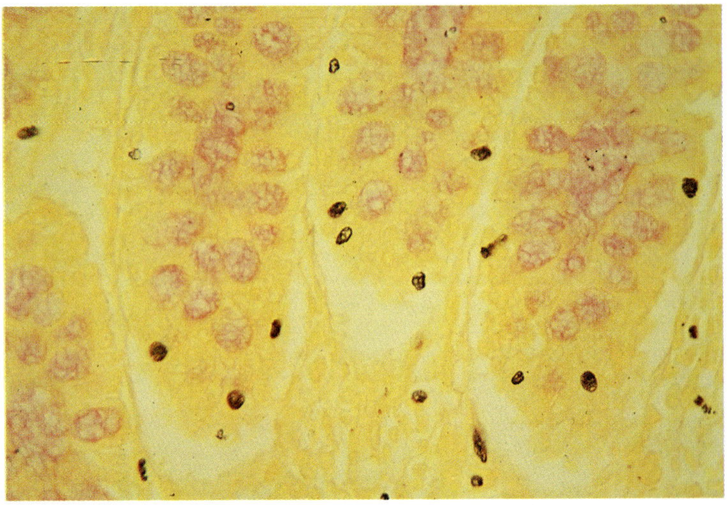

Fig. 278
This illustration demonstrates a fairly well stained H&E with a modest amount of an eosin counterstain. Note that nuclei are not overshadowed by the eosin. Also notice that the nuclei are distinct indicating good blueing of the hematoxylin. (Chapter 17)

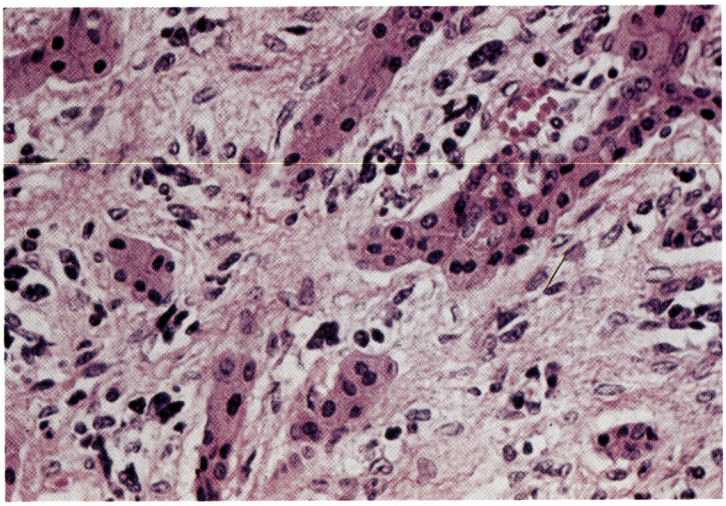

Fig. 279
This section of kidney demonstrates another H&E containing slightly more eosin for those desiring that type of an effect. Both Fig. 278 and this figure demonstrate what roughly could be considered a good H&E. See page 71 for more detail on proper H&E staining. (Chapter 17)

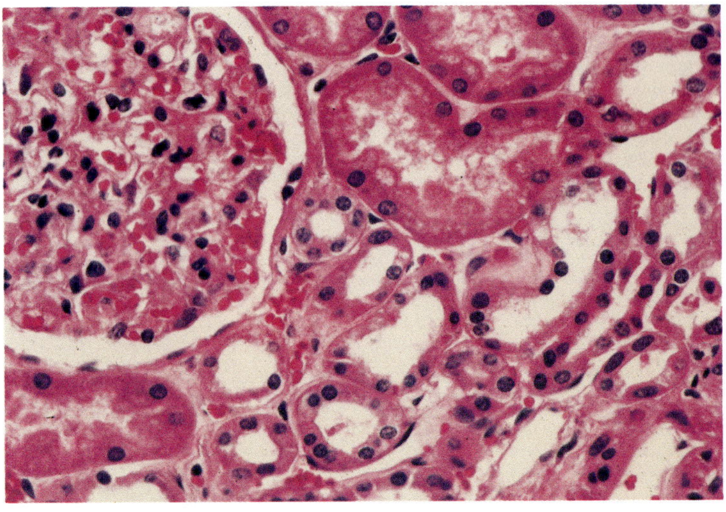

Fig. 280
This is a photograph illustrating the effects of oil red O crystal precipitate. In this particular instance, the oil red O is insufficiently dissolved in the solvent resulting in the oil red O converting back to its original crystalline form and depositing throughout the section. (Chapter 17)

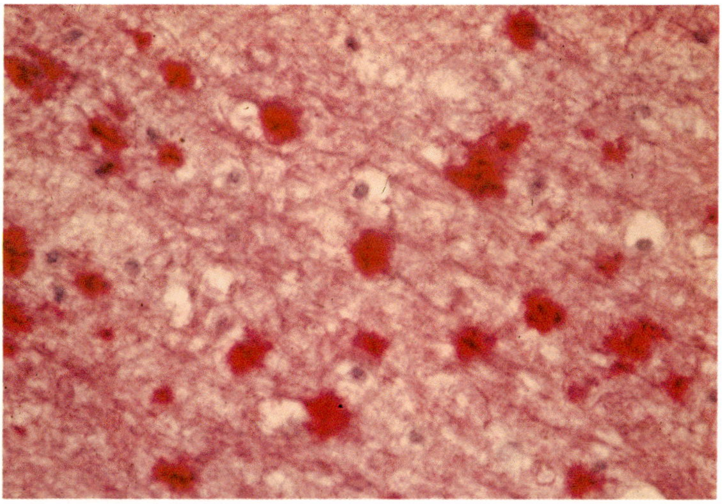

Special Stain Artifacts

The above photo included in this group

Fig. 281
This is a section of a mucin stained specimen showing a great deal of residual material stained throughout. In this instance the mucicarmine solution was very old leaving a mold type effect on the section. (Chapter 17)

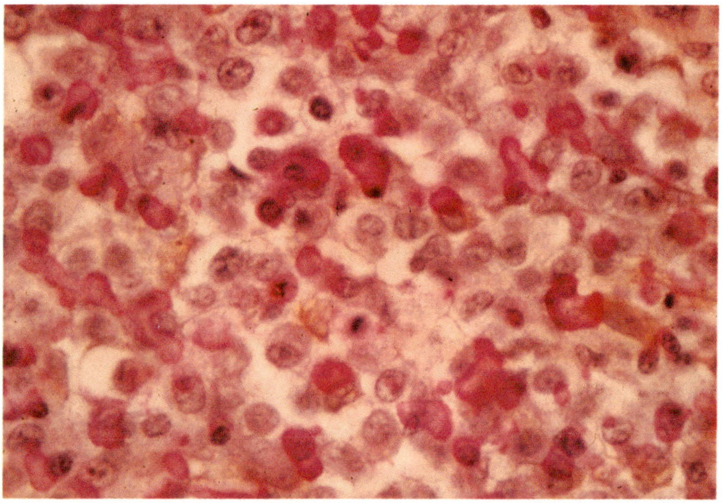

Fig. 282
This photograph illustrates picric acid crystal contamination; the result of picric acid being allowed to remain too long in the staining container. It is also important to note that when picric acid is allowed to remain in dry form for an extended period of time it could become explosive. (Chapter 17)

Fig. 283
This is a section of a silver reaction counterstained with light green. Note that there is a fungal contamination seen across the middle of the section. This is produced by a solution that has been on the shelf too long. It is possible for many solutions to produce fungal or mold type growths and therefore, one should change solutions often. (Chapter 17)

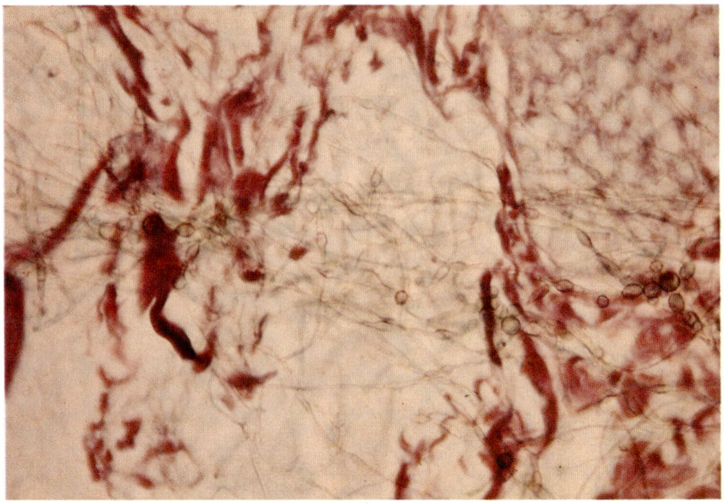

Fig. 284
A section of heart which is properly mordanted in Bouin's prior to staining. Note the well delineation between the staining of the collagen and the heart muscle. (Chapter 17)

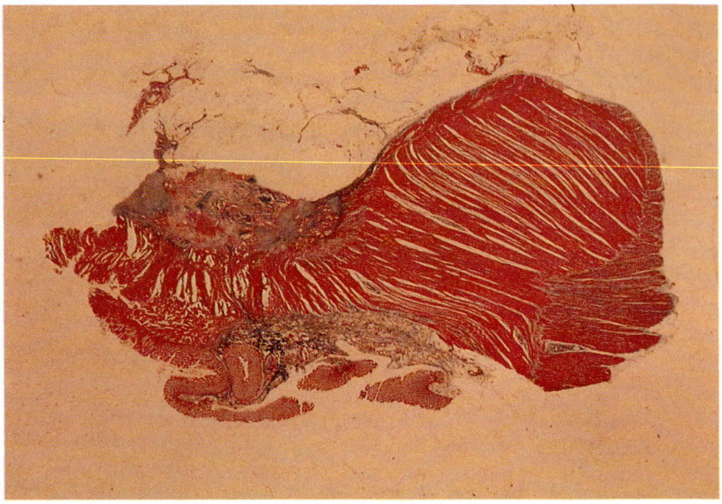

Fig. 285
A similar section which did not receive sufficient mordanting by Bouin's solution. Note that much of the heart muscle contains bluish to purplish staining due to the improper staining of the aniline blue. This section was facing the inner portion of the staining dish which resulted in lack of sufficient molecular movement. (Chapter 17)

Fig. 286
Section of the same heart muscle demonstrating complete lack of staining of muscle due to lack of proper mordanting by Bouin's. This is all related to heat convection. (Chapter 17)

Fig. 287
This composite demonstrates fungal growth due to slides being stored in highly humid conditions. This problem can also be produced if slides wrapped in polyethylene are allowed to sit in sunlight where the ultraviolet rays produce heat, which in turn causes moisture between slides. The result is the development of these fungal organisms. (Chapter 17)

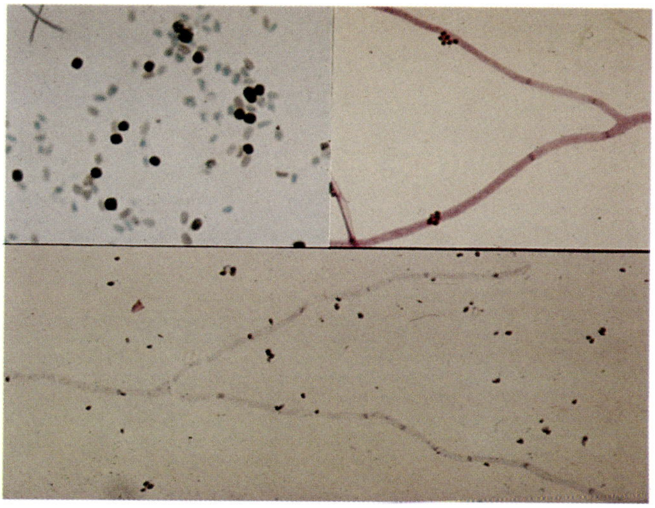

Fig. 288 See next page

Fig. 289
Gas bubble formation is evident in this photograph. (Chapter 17)

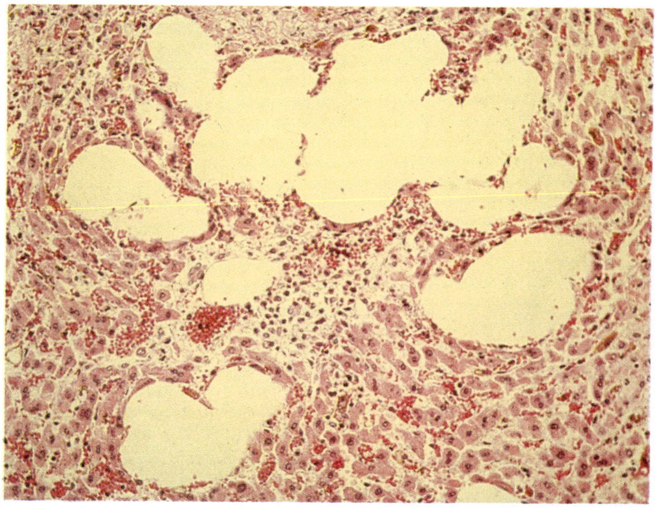

Fig. 288a
The presence of microscopic particulate matter between the tissue section and the slide results in poor section attachment. (Chapter 17)

Fig. 288b
Poor attachment of section to microscopic slide due to uneven underside surface of tissue section. This effect is produced by improper processing or venetian blind effect (thick and thin) on the section during microtomy. (Chapter 17)

Fig. 288c
Slide devoid of particulate matter and tissue section which has an even underside surface. Note the presence of an adhesive which has provided homogeneous section attachment. (Chapter 17)

Fig. 288d
This schematic demonstrates the need for the presence of water between the section and slide before slide drying, for proper adhesion of section to the slide. A vacuum effect is produced between the section and slide as water evaporates during exposure to a slide drying oven or slide warming plate. This vacuum effect produces better section adhesion to the slide which in turn prevents sections from falling off during subsequent staining. (Chapter 17)

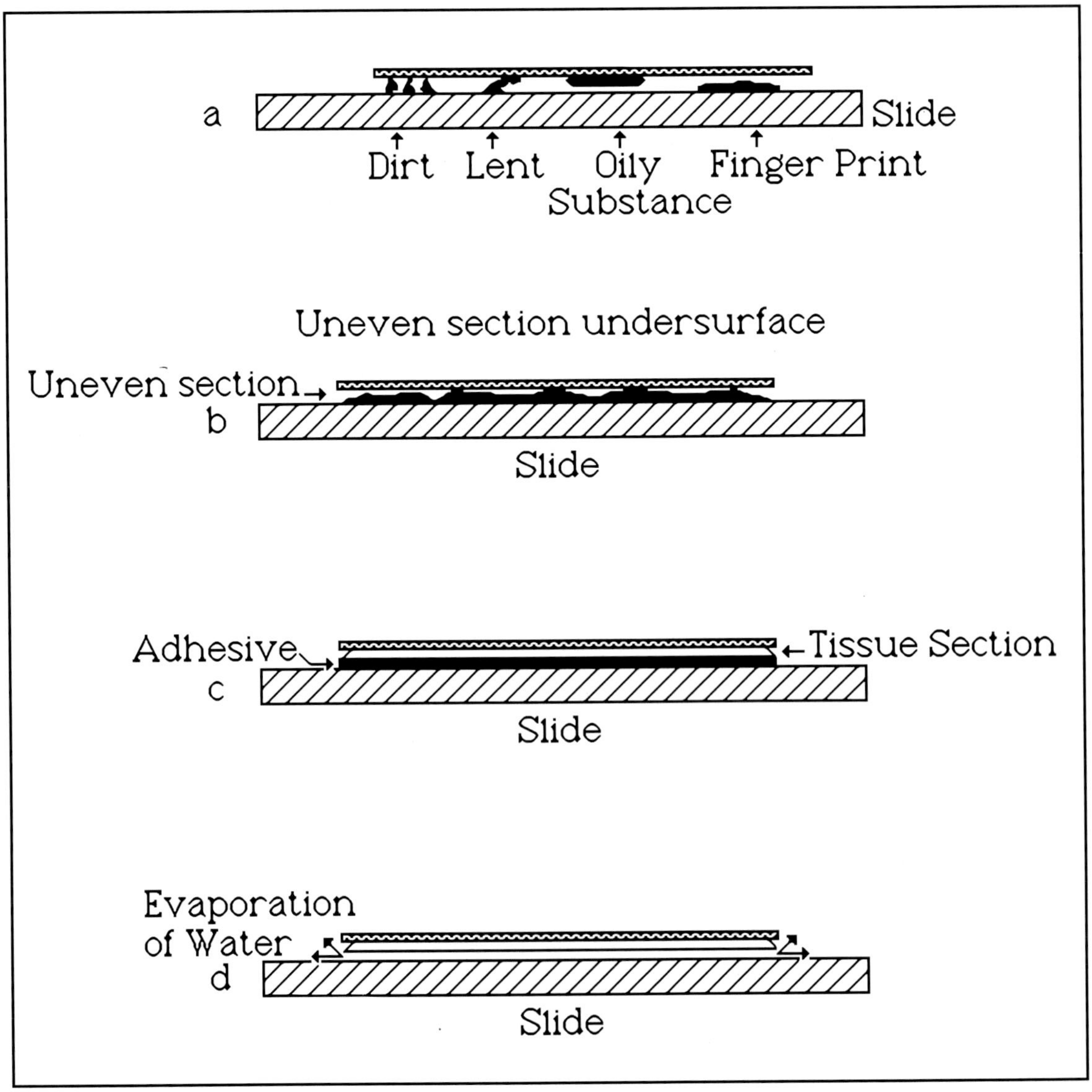

References

1 Wolman, M.: Problems of fixation in cytology, histology and histochemistry. Intern. Rev. Cytol. 4:79-102, 1955.

2 Hayat, M.A.: Fixation for electron microscopy. Academic Press, New York, 58-60, 1981.

3 Luna, L.G.: A simple test to detect presence of formalin. Histo-Logic, Vol. 20, No. 1:29, 1990.

4 Noble, L. and Davis, S.: Color coding of formalin. Histo-Logic, Vol. 20, No. 1:28, 1990.

5 Russell, W.D.: The substitution of zinc chloride for mercuric chloride in Zenker's fluid. J. Tech. Meth., Bull. Inter. Assoc. Med. Mus, 21:47-49, 1941.

6 Heaman, G.E.; Chlipala, E.; Bochenski, G.; Sabin, L. and Elfont, E.: Zinc formalin fixative for automated tissue processing. The Journal of Histotechnology, Vol. 11, No. 2:85, 1988.

7 Puchtler, H., et. al.: Carnoys fixation: practical and theoretical considerations. Histochemie, 16:361-371, 1968, published by Springer-Verlag.

8 Luna, L.G. and Cawley, M.: Some significant morphologic and chromatic alterations of tissue and staining reactions. Advances in Automatic Analysis: Technician International Congress, 1:143-152, 1970.

9 Galigher, A.E. and Kozloff, E.N.: Essentials of practical microtechniques. Second Edition, Philadelphia, Lea and Febiger, pgs. 67-132, 1971. Reproduced with permission.

10 Luna, L.G.: Evaluation of processing and fixation efficiency with the Tissue-Tek VIP System. Miles, Inc. Diagnostics Division, Elkhart, Indiana, pg. 10, 1988.

11 Ansman, T.R.: Stat endocardial biopsies. The Journal of Histotechnology, Vol. 9, No. 4:249-250, 1982.

12 Moore, J.: Vacuum processing for small biopsies. Histo-Logic, Vol. 7, No. 2:95-96, 1977.

13 Solie, B.A.: Improved technique in cell block preparations. Laboratory Medicine, Vol. 4, No. 1:40-42, 1973.

14 DeMarca, M.: Preparation of cell blocks using Bouin's fluid. Laboratory Medicine, Vol. 4, No. 2:31-32, 1973.

15 Graves, E.: Preparation of cell blocks containing no visible button. Histo-Logic, Vol. 10, No. 4:151-152, 1980.

16 Foltz, L.D.: Foam sponge for cell block preparations in paraffin sections. Laboratory Medicine, Vol. 16, No. 11:696-697, 1985.

17 Havener, V.: Processing aqueous taps and vitrectomies. Histo-Logic, Vol. 10, No. 2:143, 1980.

18 Sims, E.: Whole organ processing with the Tissue-Tek vacuum infiltration processor (VIP). The Journal of Histotechnology, Vol. 6, No. 1:29-30, 1983.

19 Luna, L.G.: Methods for reprocessing dried tissue specimens. Histo-Logic, Vol. 8, No. 2:111, 1978.

20 Willard, P.S.: Rapid procedure for rehydration of dissicated tissue. Histo-Logic, Vol. 4, No. 1:43-44, 1974.

21 Meyers, D.L.: Identification of vasectomies during processing. Histo-Logic, Vol. 4, No. 4:57, 1974.

22 Axley, D.: An aid to cleaning paraffin. Histo-Logic, Vol. 4, No. 4:192, 1974.

23 Wachtel, A.W., et al: Physical techniques in biological research. Second Edition, Academic Press, New York, 173-250, 1966.

24 Collins, E.M.: Improved paraffin sectioning with etched microtome knife edges. Stain Technology, 44:33-37, 1969. Copyright Williams and Wilkins Co.

25 Roberts, I.M.: The use of tungsten coated glass knives lowers coefficient of friction between knife and specimen. Journal of Microscopy, 103:113-117, 1974.

26 Marengo, N.P.: Increase facet angle reduces compression in cutting 4 micrometer paraffin sections. Stain Technology, 42:23-25, 1967. Copyright Williams and Wilkins Co.

27 Shapiro, S.H.: Postembedding adjuvants in routine microtomy of paraffin-processed tissues. Stain Technology, Vol. 53, No. 5:269-272, 1978. Copyright Williams and Wilkins Co.

28 Meloan, S.N. and Puchtler, H.: Harris hematoxylin, what Harris really wrote and the mechanism of hemalum stains. The Journal of Histotechnology, Vol. 10, No. 4:257-261, 1987.

29 Schenk, E.A.; Churukian, C.; Willis, C. and Stotz, E.: Staining problems with eosin y: a note from the biological stain commission. Stain Technology, Vol. 58, No. 6:377-379, 1983. Copyright Williams and Wilkins Co.

30 Bloomfield, M.M.: The chemistry and the living organism. Second Edition, John Wiley and Sons, Inc., New York, 238-250, 1980. Reprinted by permission of John Wiley and Sons, Inc.

31 Linder, J.E.; Hopps, R.M. and Johnson, N.W.: Methyl green-pyranine with hematoxylin and orange g. for the identification of inflammatory cells in tissue sections. Stain Technology, Vol. 52, No. 4:233-236, 1977. Copyright Williams and Wilkins Co.

32 Terry, B.T.: The rapid preparation of polychrome methylene blue stain for frozen sections of fresh and fixed tissue. Journal of Laboratory Clinical Medicine, Vol. 1, No. 3:159-164, 1922.

33 Brain, E.B.: The preparation of decalcified sections. Courtesy of Charles C. Thomas, Springfield, Illinois, 69-76, 1966.

34 Fenyves, M.J.: Rapid decalcification of teeth - nitric acid method modification. Laboratory Medicine, Vol. 6, No. 1:35, 1975.

35 Wakita, M.; Kobayashi, M. and Shioi, T.: Decalcification for electron microscopy with L-ascorbic acid. Stain Technology, Vol. 58, No. 6:337-341, 1983. Copyright Williams and Wilkins Co.

36 Shapiro, L.: Staining hard keratin in specimens for microscopic sections. Histo-Logic, Vol. 7, No. 2:98, 1977.

37 Luna, L.G.: Luna's method for softening keratin. Histo-Logic, Vol. 5, No. 2:65, 1975.

38 Dotson, W.A.: Additional information on methods for softening keratin. Histo-Logic, Vol 6, No. 2:81, 1976.

39 Turner, J. and Holtom, D.B.: The use of fabric softener in the reconstruction of mummified tissue prior to paraffin wax sectioning for light microscopical examination. Stain Technology, Vol. 56, No. 1:35, 1981. Copyright Williams and Wilkins Co.

40 Kuwabara, T. and Cogan, D.G.: Trypsin digestion for demonstration of the eye retinal vessels. Archives of Ophthalmology, 64:904-911, copyright 1960, American Medical Association.

41 Yoshiki, S.; Nikaido, T.; Yamazaki, T and Tachikawa, T.: Combined silver nitrate and cyanuric chloride for histological identification of asteroid and unmineralized tissue in decalcified sections of bone. The Journal of Histotechnology, Vol. 10, No. 1:17, 1987.

42 Mizgala, J.; Schmitz, J.P. and Hollinger, J.O.: A thick sectioning technique for preservation of bone-metal interfaces. The Journal of Histotechnology, Vol. 10, No. 1:31, 1987.

43 Repassy, G. and Lapis, K.: A new embedding medium for firm materials applied to the sectioning of plastic vascular prostheses. Stain Technology, Vol. 52, No. 2:118, 1977. Copyright Williams and Wilkins Co.

44 Hinshaw, J.R. and Pearse, H.E.: Histologic techniques for the differential staining of burned and normal tissue. By permission of Surgery, Gynecology and Obstetrics, Vol. 103, No. 6:726-736, 1959.

45 Trombetta, L.D. and Forbes, J.: A double infiltration technic for serial sectioning of heavily sclerotized insects. Journal of Histechnology, Vol. 1, No. 1:25-26, 1977.

46 Lopez, M.K.: Solutions to your delutions. Histo-Logic, Vol. 19, No. 2:76-79, 1989.

47 Puchtler, H.; McGowan, A.L. and Meloan, S.N.: Why do stains fade? part I history, fastness rating, light absorption and fluorochromes. The Journal of Histotechnology, Vol. 12, No. 1:57, 1989.

48 Puchtler, H.; McGowan, A.L. and Meloan, S.N.: Why do some stains fade? part I effects of dye structure, molecular orbital system, metals, substracts, moisture and chemical reactions. The Journal of Histotechnology, Vol. 12, No. 2:129, 1989.

49 Greenberg, S.R.: The comparative histologic application of phorwite fabric brighteners. The Journal of Histotechnology, Vol. 12, No. 1:43-45, 1989.

50 Bosch, P.F.: An alginate matrix double-embedding method for paraffin sectioning of minute specimens. Stain Technology, Vol. 61, No. 4:1-235, 1986. Copyright Williams and Wilkins Co.

51 Riley, D.A. and Slocum, G.R.: Contraction-free fume fixed longitudinal sections of fresh frozen muscle. Stain Technology, Vol. 63, No. 2:93, 1988. Copyright Williams and Wilkins Co.

52 Liem, R.S.B. and Van Willigen, J.D.: In toto staining and preservation of peripheral nervous tissue. Stain Technology, Vol. 63, No. 2:113, 1988. Copyright Williams and Wilkins Co.

53 Rubin, N.H. and Baur, P.S.: Simple grouping technique for paraffin embedment. Stain Technology, Vol. 58, No.3:157, 1983. Copyright Williams and Wilkins Co.

54 Swan, R.W. and Davis, H.J.: The biopsy-cucumber unit a method to improve tissue orientation. Obstetrics and Gynecology, Vol. 15, No. 5:803, 1970.

55 Luna, L. G.: Manual of histologic staining methods of the Armed Forces Institute of Pathology. McGraw-Hill, New York, NY, 1960, 1968.

55a Fite, G.L., Cambre, P.J., and Turner, M.H.: Fite's method for acid fast organisms. Arch. Path., 43:624-625, 1947.

55b Truant, J.P.: Truant's fluorescent method for acid fast organisms. Henry Ford Hospital Med. Bull., 10:287, 1962.

55c Kinyoun, J.J.: Kinyoun's method for acid fast bacteria. American Journal of Public Health, 5:867, 1915.

55d Mallory, F.B.: Ziehl-Neelsen method for acid-fast bacteria. Pathological Technique, New York, Hafner Publishing Co., 275, 1961.

55e Wade, H.W.: Wade's method for acid fast organisms. American Journal of Pathology, 28:157, 1952.

55f Brown, J.H. and Brenn, L.: Brown and Brenn method for Gram positive and Gram negative bacteria. Bull. Johns Hopkins Hospital, 48:69-73, 1931.

55g MacCallum, W.G.: MacCallum-Goodpasture method for Gram positive and Gram negative bacteria. Journal of the American Medical Association, 72:193, 1919.

55h Gridley, M.F.: Gridley's method for fungi. American Journal of Clinical Pathology, 25:975-979, 1955. Copyright Williams and Wilkins Co.

55i Grocott, R.G.: Grocott's method for fungi (GMS). American Journal of Clinical Pathology, 25:975-979, 1955. Copyright Williams and Wilkins Co.

55j Pickett, J.P., et. al: Pickett's fluorescence method for fungi. American Journal of Clinical Pathology, 34:197-202, 1960. Copyright Williams and Wilkins Co.

55k Price, D.L.: Price's giemsa method. Mil. Med., 133:363-367, 1968.

55l Lendrum, A.C.: Lendrum's method for inclusion bodies. Journal of Path. Bact., 59:399-404, 1947.

55m Bosch, R.: Bosch's method for negri bodies. Stain Technology, 41:250-251, 1966. Copyright Williams and Wilkins Co.

55n Schleifstein, J.: schleifstein's method for negri bodies. American Journal of Public Health, 27:1283-1285, 1937.

55o Luna, L.G.: Giemsa method for rickettsia.

55p Thomas, J.T.: Thomas' method for malarial parasites. Stain Technology, 28:311-312, 1953. Copyright Williams and Wilkins Co.

55q Mallory, F.B.: Maximow's method for bone marrow. Pathological Technique, New York, Hafner Publishing Co., 196, 1961.

55r Mallory, F.B.: Mallory's method for hemofuchsin. Pathological Technique, New York, Hafner Publishing Co., 136, 1961.

55s Lillie, R.D.: Lillie's method for nucleic acids. Histopathologic Technic and Practical Histochemistry, 3rd ed., McGraw-Hill Book Co., 149-150, 1965. Used by permission by McGraw-Hill, Inc.

55t Menzies, D.W.: Menzie's method for nucleic acids. Stain Technology, 38:157-160, 1963. Copyright Williams and Wilkins Co.

55u Clayden, E.C.: Diazo method for argentaffin granules. Practical Section Cutting and Staining, 3rd ed., London, W2, J. and A. Churchill, Ltd., 53, 1955.

55v Sevier, A.C. and Munger, B.L.: Sevier-Munger method for neutral tissues. J. Neuropath. Exp. Neurol., 24:130-135, 1965.

55w Glenner, G.G. and Lillie, R.D.: Glenner-Lillie method for pituitary. Stain Technology, 32:187-190, 1957. Copyright Williams and Wilkins Co.

55x Wilson, W.D. and Ezrin, E.: Wilson-Ezrin method for pituitary. American Journal of Pathology, 30:891-899, 1954.

55y Luna, L.G.: Luna's method for mast cells.

55z Mallory, F.B.: Unna's method for mast cells. Pathological Technique, New York, Hafner Publishing Co., 175, 1961.

55aa Smith, D.L.: Smith's method for juxtaglomerular granules. Stain Technology, 41:291-294, 1966. Copyright Williams and Wilkins Co.

55ab Berg, J.W.: Berg's method for spermatozoa. American Journal of Clinical Pathology, 23:513-515, 1963. Copyright Williams and Wilkins Co.

55ac DeFano, C.: DeFano's method for golgi apparatus. J. Micr. Soc., 157-161, 1920.

55ad Cain, A.J.: Cain's method for mitochondria. Quarterly J. Micr. Sci., 89:229-231, 1948.

55ae Carr, L.B., Rambo, O.N. and Feichtmeir, T.V.: Carr's method for calcium. J. Histochem. Cytochem., 9:415-417, 1961. Copyright Williams and Wilkins Co.

55af Mallory, F.B.: Von Kossa's method for calcium. Pathological Technqiue, New York, Hafner Publishing Co., 144, 1961.

55ag Pizzolato, P.: Pizzolato's method for calcium oxalate. Journal of Histochem. Cytochem., 12:333-336, 1964. Copyright Williams and Wilkins Co.

55ah Hall, M.J.: Hall's method for bilirubin. American Journal of Clinical Pathology, 34:313-316, 1960. Copyright Williams and Wilkins Co.

55ai Puchtler, H. and Sweat, F.: Puchtler-Sweat method for basement membrane. Stain Technology, 39:163-166, 1964. Copyright Williams and Wilkins Co.

55aj Jones, D.B.: Jones' method for kidney. American Journal of Pathology, 27:991-1009, 1951.

55ak Luna, L.G. and Ishak, K.: Luna-Ishak method for bile canaliculi. American Journal of Medical Technology, 33:49-466, 1967.

55al Mallory, F.B.: Van Gieson's method for collagen fibers. Pathological Technqiue, New York, Hafner Publishing Co., 144, 1961.

55am Lendrum, A.C., Fraser, D.S., Slidders, W. and Henderson, R.: Fraser-Lendrum method for fibrin. Journal of Clinical Pathology, 15:401-413, 1962.

55an Ayoub, P. and Shklar, G.: Ayoub-Shklar method for keratin and prekeratin. J. Oral Surg., 16:580-581, 1963.

55ao Mallory, F.B.: Van Gieson's method for collagen fibers. Pathological Technqiue, New York, Hafner Publishing Co., 144, 1961.

55ap Gomori, G.: Gomori's method for reticulum. American Journal of Pathology, 13:933-1002, 1937. Copyright Williams and Wilkins Co.

55aq Gomori, G.: Gomori's one step trichrome method. American Journal of Clinical Pathology, 20:661-664, 1950. Copyright Williams and Wilkins Co.

55ar Masson, P.J.: Masson's trichrome method. J. Techn. Methods, 12:75-90, 1929.

55as Mallory, F.B.: Van Gieson's method for collagen fibers. Pathological Technique, New York, Hafner Publishing Co., 144, 1961.

55at Gaffney, E.: Personal communication.

56 Rojas, M.; Darby, J. and Nochomovitz, L.: A modification of the Fite method for acid-fast organisms. The Journal of Histotechnology, Vol. 10, No. 3:181-182, 1987.

57 Garvey, W.; Fathi, A.; Bigelow, F.; Carpenter, B. and Jimenez, C.: A reliable method for demonstrating gram-positive and gram-negative bacteria. Stain Technology, Vol. 61, No. 4:251-253, 1986. Copyright Williams and Wilkins Co.

58 Taylor, R.D.: Taylor's method for gram-positive and gram-negative organisms. American Journal of Clinical Pathology, 46:472, 1966.

59 Churukian, C.J. and Schenk, E.A.: A method for demonstrating gram-positive and gram-negative bacteria. The Journal of Histotechnology, Vol. 5, No. 3:127-128, 1982.

60 Brown, R.C. and Hopps, H.C.: Staining of bacteria in tissue sections: a reliable gram stain method. American Journal of Clinical Pathology, 60:234, 1973.

61 Greer, P.A.; Chandler, F.W. and Hicklin, M.D.: Rapid demonstration of Legionella pneumophila in unembedded tissue. American Journal of Clinical Pathology, 73:788-790, 1980.

62 Roy, K.H.: Visualization of Legionella pneumophila in paraffin sections using hexamine silver. Stain Technology, Vol. 58, No. 4:215-218, 1983. Copyright Williams and Wilkins Co.

63 Mahan, C.T. and Sale, G.E.: Rapid methenamine silver stain for pneumocystis and fungi. Arch. Pathol. Laboratory Medicine, 102:351-352, Copyright 1978, American Medical Association.

64 Churukian, C.J.; Schenk, E.A. and Clark, G.: Dilute ammoniacal silver as a substitute for methenamine silver to demonstrate Pneumocystis carinii and fungi. Laboratory Medicine, Vol. 17, No. 4:87-90, 1986.

65 Swartz, J.H.; Medrek, T.F. and Robboy, S.J.: A new stain for systemic fungi in tissue. American Journal of Clinical Pathology, 57:27-29, 1972.

66 Hummer, D.A.: Cytodiagnosis of Pneumocystis carinii infection using toluidine blue O. Laboratory Medicine, Vol. 19, No. 12:821, 1988.

67 Genaw, C.: Use of cresyl echt violet for the staining of pneumocystis carinii as compared to Grocott's (GMS) and Giemsa methods. The Journal of Histotechnology, Vol. 12, No. 1:39-40, 1989.

68 Elliott, K.G.: A simple step to enhance the demonstration of Leptospira in Young's Warthin-Starry technique. Stain Technology, Vol. 63, No. 2:122-179, 1988. Copyright Williams and Wilkins Co.

69 Staples, T.C.: Methods for staining Campylobacter pyloridia. Histo-Logic, Vol. 18, No. 4:25-27, 1988.

70 Warren, J.R.: Unidentified bacilli on gastric epithelum in active chronic gastritis. Lancet, 1:1273, 1983.

71 Jones, D.M.; Lessells, A.M. and Eldridge, J.: Campylobacter like organisms on the gastric mucosa culture, histological and serological studies. Journal of Clinical Pathology, 37:1002, 1984.

72 Goodwin, C.S.; Armstrong, J.A. and Marshall, B.J.: Campylobacter pyloridis gastric and peptic ulceration. Journal of Clinical Pathology, 39:353, 1986.

73 Garvey, W.; Fathi, A. and Bigelow, F.: Modified Steiner for the demonstration of spirochetes. The Journal of Histotechnology, 8:15, 1985.

74 McMullen, L.; Walker, M.M.; Bain, L.A.; Karim, Q.N. and Baron, J.H.: Histological identification of campylobacter using Gimenez technique in gastric mucosa. Journal of Clinical Pathology, 40:464, 1987.

75 Razzaboni, F.: A Warthin-Starry silver method for the demonstration of bacillus piliformis. Histo-Logic, Vol. 21, No. 1:214-215, 1991.

76 Tomlinson, Maj. W.J. and Grocott, R.: A simple method of staining malaria protozoa and other parasites in paraffin sections.

77 Richard, D.F.; Hunter D.T. and Janis, B.: Detection of plasmodia by acridine orange stain. Technical Bulletin of the Registry of Medical Technologist, Vol. 39, No. 1:4, 1969.

78 Wilson, H.R. and Fairchild, D.G.: Modified Whipf's polychrome: a connective tissue stain with special application for demonstrating leishmania. Stain Technology, Vol. 52, No. 2:105-111, 1977. Copyright Williams and Wilkins Co.

79 Massignani, A.M. and Malferrari, R.: Massignani-Malferrari method for negri bodies. Stain Technology, 36:5-8, 1961. Copyright Williams and Wilkins Co.

80 Kass, L.; Cooper, K.L. and Choi, C.M.: Acid red 137: a new selective stain for eosinophils in smears and tissue sections. The Journal of Histotechnology, Vol. 12, No. 1:29-32, 1989.

81 Luna, L.G.: Methods for staining charcott-leyden crystals. Histo-Logic, Vol. 18, No. 5:38-40, 1988.

82 Puchtler, H.; Rosenthal, S. and Sweat, F.: Revision of the amido black stain for hemoglobin. Archives of Pathology, 78:76-78, copyright 1964, American Medical Association.

83 Guard, H.R.: A new technic for differential staining of the sex chromatin and the determination of its incidence in exfoliated vaginal epithelial cells. American Journal of Clinical Pathology, 32:145, 1959.

84 Klinger, H.P. and Ludwig, K.S.: A universal stain for the sex chromatin body. Stain Technology, 32:235, 1957. Copyright Williams and Wilkins Co.

85 Leder, L.D.: The selective enzymo-chemical demonstration of neutrophilic myploid cells and tissue mast cells in paraffin sections. Klin. Wchnschr., 42:533, 1964, published by Springer-Verlag.

86 Kass, L.: Lycramine brilliant blue JL: a new stain for human megakaryocytes. Stain Technology, Vol. 60, No. 4:233, 1985. Copyright Williams and Wilkins Co.

87 Kass, L.: Acid red 52: a new stain for cells of granulocytic origin. Stain Technology. Vol. 62, No. 2:77-84, 1987. Copyright Williams and Wilkins Co.

88 Doe, K.A.; Gryzbac, M. and Schumacher, H.R.: A new modified rapid noncarcinogenic myeloperoxidase staining technique using 4-chloro-1-napthol. Laboratory Medicine, Vol. 19, No. 6:374-375, 1988.

89 Boon, M.E.; Schleicher, A.; Wijsman-Grootendorst, A.; Lyon, H. and Kok, L.P.: Staining of the nucleolus with protein and RNA stains for automatic measurement of nucleolar size in paraffin sections. Stain Technology, Vol. 63, No. 5:289-292, 1988. Copyright Williams and Wilkins Co.

90 Kass, L.: Basic blue 41: a new panoptic stain for blood and bone marrow cells. The Journal of Histotechnology, Vol. 11, No. 1:10-14, 1988.

91 Taft, E.B.: Taft's method for nucleic acids. Stain Technology, 26:205-212, 1951. Copyright Williams and Wilkins Co.

92 Staples, T.C. and Grizzle, W.E.: A methyl green nuclear stain for agyrophil procedures. Laboratory Medicine, Vol. 17, No. 9:532-534, 1986.

93 Staples, T.C. and Grizzle, W.E.: Effect of temperature on argyrophil impregnation: development of a high temperature rapid argyrophil procedure. Stain Technology, Vol. 62, No. 1:41-49, 1987. Copyright Williams and Wilkins Co.

94 Churukian, C.J. and Schenk, E.A.: Churukian-Schenk method for argentaffin and argyophil cell granules. The Journal of Histotechnology, Vol. 2, No. 3:107, 1979.

95 Grimelius, L.A.: Silver nitrate stain for A_2 cell in human pancreatic islets. Acta. Soc. Med. Upsal., 73:243, 1968.

96 Tanaka, K.; Mori, W. and Suwa, K.: Victoria blue-nuclear fast red stain for HBS antigen detection in paraffin sections. Acta. Pathol. Jpn., 31(1):93, 1981.

97 Lehy, T. and Zeitoun, P.: Influence of fixatives on the metachromatic reactions to toluidine blue at pH 5; A_1 and A_2 cells of human endocrine pancreas. Stain Technology, Vol. 45, No. 2:63-70, 1970. Copyright Williams and Wilkins Co.

98 Kito, H. and Hosoda, S.: Triple staining for simultaneous visualization of cell types in islet of Langerhans of pancreas. Journal of Histochemistry and Cytochemistry, Vol. 25, No. 8:1019-1020, 1977.

99 Maldonado, R. and San Jose, H.: Maldonado's method for pancreatic islet cells. Stain Technology, 42:11-13, 1967. Copyright Williams and Wilkins Co.

100 Wenger, J.B.: Modified method for pancreatic beta cell granules. Histo-Logic, Vol. 17, No. 4:258-259, 1987.

101 Heath, E.H.: Heath's method for pituitary. American Journal of Veterinary Research, 26:368-373, 1965.

102 Lauren, P.A. and Sorvari, T.E.: Staining of paneth cells with Best's carmine after methylation. Stain Technology, Vol. 42, No. 6:311-315, 1967. Copyright Williams and Wilkins Co.

103 Roque, A.L.: Chromotrope aniline blue method of staining Mallory bodies of laennec's cirrhosis. Laboratory Investigation, Vol. 2, No. 15:15-21, 1953. Copyright United States and Canadian Academy of Pathology, Inc.

104 Jhung, J.W.; Edelstein, L.M.; Brookman, C. and Cariglia, N.: Congo red: a simple stain for the stratum lucidum. American Journal of Medical Technology, Vol. 39, No. 1:1-4, 1974.

105 Akita, M.; Murata, E.; Fujita, K. and Kaneko, K.: A new method using a modified Mayer's hemalum at pH 6 for demonstrating mucous neck cells. Stain Technology, Vol. 60, No. 5:261-263. Copyright Williams and Wilkins Co.

106 Gomori, G.: Gomori's method for chromaffin. American Journal of Clinical Pathology, 16:115, 1946.

107 Takaya, K.: Luxol fast blue MBS and phloxine: a stain for mitochondria. Stain Technology, Vol. 42, No. 4:207-211, 1967. Copyright Williams and Wilkins Co.

108 Dahl, L.K.: Dahl's method for calcium. Reproduced from the Journal of Experimental Medicine, 1952, 95:474-479, by copyright permission of the Rockefeller University Press.

109 Puchtler, H.: Puchtler's method for calcium. Journal of Histochemistry and Cytochemistry, Vol. 17, No. 2, 1969.

110 Fenton, R.H.; Johnson, F.B. and Zimmerman, L.E.: Fenton-Johnson-Zimmerman method for iron (microincineration). Journal of Histochemistry and Cytochemistry, 12:153, 1964.

111 Mallory, F.B.: Turnbull blue method for hemosiderin. Pathological Technique, pg. 138, 1961, published by Springer-Verlag.

112 Gomori, G.: Gomori's method for iron. American Journal of Pathology, 12:655-663, 1936.

113 Hukill, P. and Putt, A.: Hukill-Putt method for iron. Journal of Histochemistry and Cytochemistry, 10:490-494, 1962.

114 Perl, M.: Perl's method for iron. Virchow. Arch. Path. Anat., 39:42, 1867.

115 Lillie, R.D.; Geer, J.C. and Gutierrez, J.: The removal of histochemically demonstratable iron from tissue sections by brief exposure to sodium dithionite solution. Journal of Histochemistry and Cytochemistry, Vol. 11, No. 5:664, 1963.

116 Linquist, R.R.: Studies on the pathogenesis of hepatolenticular degeneration II cytochemical methods for the localization of copper. Archives of Pathology, Vol. 87, No. 4:370-379, copyright 1969, American Medical Association.

117 Uzman, L.L.: Uzman's method for copper. Laboratory Investigation, 5:299-305, 1956. Copyright United States and Canadian Academy of Pathology, Inc.

118 Roscher, A.A.: A new histochemical method for the demonstration of calcium oxalate in tissue following ethylene glycol poisoning. American Journal of Pathology, 55:99, 1971.

119 Johnson, F.B. and Pani, K.: Histochemical identification of calcium oxalate. Archives of Pathology, 74:347-351, copyright 1962, American Medical Association.

120 Warkel, R.L.; Luna, L.G. and Helwig, E.B.: A modified Warthin-Starry procedure at low pH for melanin. American Journal of Clinical Pathology, Vol. 73, No. 6:812-815, 1980.

121 Frangioni, G. and Borgioli, G.: Rapid bleach for melanin. Stain Technology, Vol. 61, No. 4:239-242, 1986. Copyright Williams and Wilkins Co.

122 Southgate, H.W.: Note for preparing mucicarmine. Journal of Pathology and Bacteriology. 30:729, 1927.

123 Sweat, F.W.; Puchtler, H. and Valentine, L.S.: Fluorescence microscopy of amyloid. Archives of Pathology, Vol. 95:37-41, copyright 1973, American Medical Association.

124 Hinds, I.L.: Oxalic acid-crystal violet staining method for demonstration of amyloid. Laboratory Medicine, Vol. 14, No. 5:295-297, 1983.

125 Ono, K.; Katsuyoma, T. and Hotchi, M.: Histochemical application of mild alkaline hydrolysis for selective elimination of o-glycosidically linked glycoproteins. Stain Technology, Vol. 58, No. 5:309, 1983. Copyright Williams and Wilkins Co.

126 Culling, C.F.A.; Reid, P.E.; Burton, J.D. and Dunn, W.L.: A histochemical method of differentiating lower gastrointestinal tract mucin from other mucins in primary or metastic tumors. Journal of Clinical Pathology, 28:656-658, 1975.

127 Churukian, C.J. and Schenk, E.A.: A modification of Kreyberg's method for demonstrating keratin and mucin. The Journal of Histotechnology, Vol. 7, No. 3:146-147, 1984.

128 Gowali, F.M.: Iron pyronin: a simple stain for sulfated mucosubstances. Laboratory Medicine, Vol. 17, No. 5:284-285, 1986.

129 Churukian, C.J. and Schenk, E.A.: Eastwood congo red method for demonstrating amyloid. The Journal of Histotechnology, Vol. 11, No. 2:105, 1988.

130 Hayama, M.; Katsuyama, T.; Nakayama, J.; Akamatsu, T. and Honda, T.: A combined stain for identifying epithelial cells of the gastric mucosa. Stain Technology, Vol. 62, No. 1:35-40, 1987. Copyright Williams and Wilkins Co.

131 Avallone, F.: Personal communication.

132 Lillie, R.D.: Lillie's method for basement membrane, muscle and arteriosclerotic lesions. American Journal of Clinical Pathology, 21:484-488, 1951.

133 Miller, P.J.: An elastin stain. Histo-Logic, Vol. 14, No. 1:227 & 229, 1984.

134 Luna, L.G.: A useful method for demonstrating bone canaliculi. Histo-Logic, Vol. 18, No. 5:46-47, 1988.

135 Barbeito and Lopez, J.: Barbeito-Lopez trichrome stain. American Journal of Clinical Pathology, Technical Section, 16:53-56, 1946.

136 Ehrenreich, T. and Espinosa, T.: Chromotrope silver methenamine stain of glomerular lesions. American Journal of Clinical Pathology, 56:448-451, 1971.

137 Clark, G.; Spooner, G.H.; Dodds, H.M. and Churukian, C.J.: A simple silver stain for fine connective tissue fibers. The Journal of Histotechnology, Vol. 8, No. 3:163-164, 1985.

138 Garvey, W.; Fathi, A.; Bigelow, F.; Carpenter, B. and Jimenez, C.: A combined elastic, fibrin and collagen stain. Stain Technology, Vol. 62, No. 6:365, 1987. Copyright Williams and Wilkins Co.

139 Musto, L.: Modified Movat pentachrome stain. The Journal of Histotechnology, Vol. 9, No. 3:173-174, 1986.

140 Lascano, E.F.; Gomez, R.M. and Berria, M.I.: Controlled silver impregnation of cultured cardiomyocytes. Stain Technology, Vol. 63, No. 4:221-227, 1988. Copyright Williams and Wilkins Co.

141 Meloan, S.N. and Puchtler, H.: On the chemistry of phosphotungstic acid-hematein: development of a rapidly ripening PTAH solution. The Journal of Histotechnology, Vol. 11, No. 3:153-157, 1988.

142 Pinkus, H.: Pinkus' acid orcein-giemsa method. Arch. Dermatol. & Syph., 49:355, copyright 1944, American Medical Association.

143 Roush, J.K.; Breur, G.J. and Wilson, J.W.: Picro sirus red staining of dental structures. Stain Technology, Vol. 63, No. 6:363-367, 1988. Copyright Williams and Wilkins Co.

144 Gordon, H. and Sweet, H.: A simple method for the silver impregnation of reticulum. American Journal of Pathology, 12:545, 1956.

145 Hinds, I.L.; Murry, S.A. and Kitynsky, S.: A rapid reliable silver impregnation method for glomeular basement membrane. The Journal of Histotechnology, Vol. 10, No. 4:237, 1987.

146 Terry, B.T.: The rapid preparation of polychrome methylene blue stains for frozen sections of fresh and fixed tissue. Journal of Laboratory Clinical Medicine, Vol. 8, No. 3:159-164, 1922.

147 Berenbaum, M.C.: The histochemistry of bound lipids. Quarterly Journal of Microscopical Science, Company of Biologists, Ltd., 99:231, 1958.

148 Luna, L.G.: Luna's modification of Bodian's method for axons and nerve endings. American Journal of Medical Technology, 30:355, 1964.

149 D'Amelio, F.E.: The Golgi-Hortega-Lorilla technique with a useful additional step for application to brain tissue after prolonged fixation. Stain Technology, Vol. 58, No. 2:79, 1983. Copyright Williams and Wilkins Co.

150 deBold, M.L.; DeBold, A.J. and Kraicer, J.: Demonstration of stellate cells of the pars intermedia of the pituitary gland using a new silver impregnation technique. Stain Technology, Vol. 59, No. 1:49-52, 1984. Copyright Williams and Wilkins Co.

151 Zimny, M.L.; Onge, M.S. and Schutte, M.: A modified gold chloride method for the demonstration of nerve endings in frozen sections. Stain Technology, Vol. 60, No. 5:305, 1985. Copyright Williams and Wilkins Co.

152 DeGirolami, U. and Zvaigzne, O.: Modification of the Achucarro-Hortega pineal stain for paraffin embedded formalin fixed tissue. Stain Technology, Vol. 48, No. 1: 48-50, 1973. Copyright Williams and Wilkins Co.

153 Garvey, W.; Fathi, A.; Bigelow, F.; Jimenez, C. and Carpenter, B.: A reliable silver impregnation technique for the nervous system. The Journal of Histotechnology, Vol. 10, No. 4:245-247, 1987.

154 Braak, H.; Braak, E.; Ohm, T. and Bohl, J.: Silver impregnation of Alzheimer's neurofibrillary changes counterstained for basophilic material and lipofuscin pigment. Stain Technology, Vol. 63, No. 4:197-200, 1988. Copyright Williams and Wilkins Co.

155 Feeback, D.L. and Brumback, R.A.: A simplified staining procedure for degenerating nerve terminals. The Journal of Histotechnology, Vol. 11, No. 2:101-103, 1988.

156 Nauta, W.J.H. and Gygax, P.A.: Nauta-Gygax method for degenerating axon terminals. Stain Technology, 29:91-95, 1954. Copyright Williams and Wilkins Co.

157 Penfield, W.: Penfield's method for oligodendroglia and microglia. American Journal of Pathology, 4:153-157, 1928.

158 Lapham, L.W.; Johnstone, M.A. and Brundjar, K.H.: Lapham's method for myelin and glial fibers. Journal of Neuropath. Exp. Neurol., 23:156-160, 1964.

159 Sevier, A.C. and Munger, B.L.: Sevier-Munger method for neural tissues. Journal of Neuropath. Exp. Neurol., 24:130-135, 1965.

160 Sirkin, P.W.: Critical defatting of frozen brain sections for optimal differentiation with cresyl violet stain. Stain Technology, Vol. 58, No. 2:121. Copyright Williams and Wilkins Co.

161 Edwards, R and Price, R.: Butvar B-98 resin as a section adhesive. Stain Technology, Vol. 57, No. 1:50, 1982. Copyright Williams and Wilkins Co.

162 Henderson, C.: Aminoalkylsilone, an inexpensive simple preparation for slide adhesion. The Journal of Histotechnology, Vol. 12, No. 2:123, 1989.

163 Smith, D.J. and Fulton, W.: Control for identification of herpes simplex virus type 1 and 2 tissue sections. Laboratory Medicine, Vol. 14, No. 10:663, 1983.

Index